WHOOPING CRANES:
BIOLOGY AND
CONSERVATION

T0335631

This is a Volume in the

Biodiversity of the World:
Conservation from Genes to Landscapes
Edited By Philip J. Nyhus

WHOOPING CRANES: BIOLOGY AND CONSERVATION

BIODIVERSITY OF THE WORLD: CONSERVATION FROM GENES TO LANDSCAPES

Series editor

PHILIP J. NYHUS
Colby College, Waterville, ME, United States

Volume editors

JOHN B. FRENCH, JR.
U.S. Geological Survey, Patuxent Wildlife Research Center, Laurel, MD, United States

SARAH J. CONVERSE
U.S. Geological Survey, Patuxent Wildlife Research Center, Laurel, MD, United States
U.S. Geological Survey, Washington Cooperative Fish and Wildlife Research Unit,
School of Environmental and Forest Sciences (SEFS) & School of Aquatic and Fishery Sciences (SAFS),
University of Washington, Seattle, WA, United States

JANE E. AUSTIN
U.S. Geological Survey, Northern Prairie Wildlife Research Center, Jamestown, ND, United States

Original section drawings by

JACK H. DELAP
Cornish College of Arts, Seattle, WA, United States

ACADEMIC PRESS
An imprint of Elsevier

Academic Press is an imprint of Elsevier
125 London Wall, London EC2Y 5AS, United Kingdom
525 B Street, Suite 1650, San Diego, CA 92101, United States
50 Hampshire Street, 5th Floor, Cambridge, MA 02139, United States
The Boulevard, Langford Lane, Kidlington, Oxford OX5 1GB, United Kingdom

Notices
Knowledge and best practice in this field are constantly changing. As new research and experience broaden our understanding, changes in research methods, professional practices, or medical treatment may become necessary.

Practitioners and researchers must always rely on their own experience and knowledge in evaluating and using any information, methods, compounds, or experiments described herein. In using such information or methods they should be mindful of their own safety and the safety of others, including parties for whom they have a professional responsibility.

To the fullest extent of the law, neither the Publisher nor the authors, contributors, or editors, assume any liability for any injury and/or damage to persons or property as a matter of products liability, negligence or otherwise, or from any use or operation of any methods, products, instructions, or ideas contained in the material herein.

Library of Congress Cataloging-in-Publication Data
A catalog record for this book is available from the Library of Congress

British Library Cataloguing-in-Publication Data
A catalogue record for this book is available from the British Library

ISBN: 978-0-12-803555-9

For information on all Academic Press publications visit our website at
https://www.elsevier.com/books-and-journals

  Working together
to grow libraries in
Book Aid developing countries
International

www.elsevier.com • www.bookaid.org

Publisher: Andre Gerharc Wolff
Acquisition Editor: Anna Valutkevich
Editorial Project Manager: Pat Gonzalez
Production Project Manager: Mohana Natarajan
Designer: Christian Bilbow

Typeset by Thomson Digital

This volume is dedicated to past and future Whooping Crane conservationists: Robert Porter Allen, whose pioneering contributions set the stage for decades of Whooping Crane research and management, and the next generation of conservationists, may they be inspired to take up the charge and finish the work of recovering this species to a self-sustaining status.

Contents

Section A
WHOOPING CRANES PAST AND PRESENT

1. Whooping Cranes Past and Present
JOHN B. FRENCH, JR., SARAH J. CONVERSE, JANE E. AUSTIN

2. Phylogenetic Taxonomy of Cranes and the Evolutionary Origin of the Whooping Crane
CAREY KRAJEWSKI

3. Revisiting the Historic Distribution and Habitats of the Whooping Crane
JANE E. AUSTIN, MATTHEW A. HAYES, JEB A. BARZEN

Section B
POPULATION AND BREEDING BIOLOGY

4. Population and Breeding Range Dynamics in the Aransas-Wood Buffalo Whooping Crane Population
SCOTT WILSON, MARK T. BIDWELL

5. Monitoring Recruitment and Abundance of the Aransas-Wood Buffalo Population of Whooping Cranes: 1950–2015
BRADLEY N. STROBEL, MATTHEW J. BUTLER

Section C
BEHAVIOR AND SOCIAL STRUCTURE

Section D
HABITAT USE

Section E
CAPTIVE BREEDING AND WHOOPING CRANE HEALTH

List of Contributors

Jane E. Austin U.S. Geological Survey, Northern Prairie Wildlife Research Center, Jamestown, ND, United States

Jeb A. Barzen International Crane Foundation, Baraboo; Private Lands Conservation LLC, Spring Green, WI, United States

Mark T. Bidwell Canadian Wildlife Service, Environment and Climate Change Canada, Prairie and Northern Wildlife Research Centre, Saskatoon, SK, Canada

Sandra R. Black Calgary Zoological Society, Calgary, AB, Canada

David A. Brandt U.S. Geological Survey, Northern Prairie Wildlife Research Center, Jamestown, ND, United States

William B. Brooks U.S. Fish and Wildlife Service, Jacksonville, FL, United States

Megan Brown Smithsonian Conservation Biology Institute, National Zoological Park, Front Royal, VA, United States

Felipe Chavez-Ramirez Gulf Coast Bird Observatory, Lake Jackson, TX, United States

Elisabeth Condon International Crane Foundation, Baraboo, WI, United States

Sarah J. Converse U.S. Geological Survey, Patuxent Wildlife Research Center, Laurel, MD; U.S. Geological Survey, Washington Cooperative Fish and Wildlife Research Unit, School of Environmental and Forest Sciences (SEFS) & School of Aquatic and Fishery Sciences (SAFS), University of Washington, Seattle, WA, United States

Tim A. Dellinger Fish and Wildlife Research Institute, Florida Fish and Wildlife Conservation Commission, Gainesville, FL, United States

Joseph W. Duff Operation Migration Inc., Port Perry, ON, Canada; Operation Migration USA, Niagara Falls, NY, United States

William F. Fagan University of Maryland, College Park, MD, United States

Megan J. Fitzpatrick Department of Zoology, University of Wisconsin-Madison, Madison, WI, United States

Lara E.A. Fondow International Crane Foundation, Baraboo, WI; Natural Resources Conservation Service, Rexburg, ID, United States

John B. French, Jr. U.S. Geological Survey, Patuxent Wildlife Research Center, Laurel, MD, United States

Andrew P. Gossens International Crane Foundation, Baraboo, WI, United States

Barry K. Hartup International Crane Foundation, Baraboo, WI, United States

Matthew A. Hayes International Crane Foundation, Baraboo, WI, United States

Matthew J. Butler U.S. Fish and Wildlife Service, Albuquerque, NM, United States

Sammy L. King U.S. Geological Survey, Louisiana Cooperative Fish and Wildlife Research Unit, Baton Rouge, LA, United States

Carey Krajewski Department of Zoology, Southern Illinois University, Carbondale, IL, United States

Anne E. Lacy International Crane Foundation, Baraboo, WI, United States

Julie Langenberg International Crane Foundation, Baraboo, WI, United States

Davin Lopez Wisconsin Department of Natural Resources, Madison, WI, United States

Luz Lumb Harte Research Institute for Gulf of Mexico Studies, Texas A&M University-Corpus Christi, Corpus Christi, TX, United States

Paul D. Mathewson Department of Zoology, University of Wisconsin-Madison, Madison, WI, United States

Clinton T. Moore U.S. Geological Survey, Georgia Cooperative Fish and Wildlife Research Unit, Warnell School of Forestry and Natural Resources, University of Georgia, Athens, GA, United States

Thomas Mueller Senckenberg Biodiversity and Climate Research Center and Goethe University Frankfurt, Frankfurt, Germany

Glenn H. Olsen U.S. Geological Survey, Patuxent Wildlife Research Center, Laurel, MD, United States

Aaron T. Pearse U.S. Geological Survey, Northern Prairie Wildlife Research Center, Jamestown, ND, United States

Warren P. Porter Department of Zoology, University of Wisconsin-Madison, Madison, WI, United States

Michael C. Runge U.S. Geological Survey, Patuxent Wildlife Research Center, Laurel, MD, United States

Will Selman Biology Department, Millsaps College, Jackson, MS, United States

Sabrina Servanty Colorado State University, Colorado Cooperative Fish and Wildlife Research Unit, Fort Collins, CO, United States

Elizabeth H. Smith International Crane Foundation, Texas Program, Fulton, TX, United States

Nucharin Songsasen Smithsonian Conservation Biology Institute, National Zoological Park, Front Royal, VA, United States

Bradley N. Strobel U.S. Fish and Wildlife Service, Necedah National Wildlife Refuge, Necedah, WI, United States

Kelly D. Swan Centre for Conservation Research, Calgary Zoological Society, Calgary, AB, Canada

Eva K. Szyszkoski International Crane Foundation, Baraboo, WI; Louisiana Department of Wildlife and Fisheries, Gueydan, LA, United States

Claire S. Teitelbaum Senckenberg Biodiversity and Climate Research Center and Goethe University Frankfurt, Frankfurt, Germany

Hillary L. Thompson International Crane Foundation, Baraboo, WI; Clemson University, Clemson, SC, United States

Richard P. Urbanek U.S. Fish and Wildlife Service, Necedah National Wildlife Refuge, Necedah, WI, United States

Phillip L. Vasseur Louisiana Department of Wildlife and Fisheries, Rockefeller Wildlife Refuge, Grand Chenier, LA, United States

Scott Wilson Wildlife Research Division, Environment and Climate Change Canada, National Wildlife Research Centre, Ottawa, ON, Canada

Sara E. Zimorski International Crane Foundation, Baraboo, WI; Louisiana Department of Wildlife and Fisheries, Gueydan, LA, United States

Foreword

George Archibald

Co-founder, International Crane Foundation Member,
Whooping Crane Recovery Team, 1990 – present
October 2016

In 1954, the year the last breeding area for Whooping Cranes was discovered in Wood Buffalo National Park, in a one-room schoolhouse in rural Nova Scotia, Canada, at the tender age of eight, I developed a life-long interest in this species after hearing a dramatization of the terrified response of a female crane when an airplane circled low over her nest. She had been discovered and she feared the museum collectors would soon arrive! I like to think that her mate comforted her, assuring her that they were secure with a nest within the boundaries of the Park, established in 1922, and through the protections of the Migratory Bird Treaty Act of 1918. Like millions of others around the world, I have followed — and eventually participated in — the drama of the conservation of North America's tallest bird, and one of its rarest and most majestic species.

The tall grass prairie stretching from Indiana to Saskatchewan apparently was the breeding stronghold for migratory Whooping Cranes, while the vast wetlands on the prairies of southwest Louisiana provided habitat throughout the year for a resident population. The number of Whooping Cranes alive when Europeans arrived on the continent will remain a mystery: they were reported to be much less common than Sandhill Cranes, possibly in the low thousands.

Wetland loss, subsistence hunting, and finally, collecting led to their severe decline. The last pair on the prairie breeding grounds of the migratory whooping cranes was shot near Luceland, Saskatchewan in 1922. Their eggs were collected for a museum. The last nesting of the resident whooping cranes in Louisiana was 1939, and the last bird observed there in the wild was 1950. But migrant birds still appeared each winter along the Gulf Coast of Texas and on nearby inland wetlands of the King Ranch. The core wintering area on the Blackjack Peninsula for those migratory Whooping Cranes was protected in 1935 as the Aransas National Wildlife Refuge. Their breeding area remained obscure until 1954 when the mystery of where the birds bred was finally resolved.

The Canadian and the US governments have provided protection for the Whooping Crane since 1918, especially within Wood Buffalo National Park and Aransas National Wildlife Refuge, and in various protected areas along the 2700 mile migration route of the cranes. However, Whooping Cranes migrate in small groups, feeding and resting in various shifting, widely-scattered, and often unknown locations that are spread over a large swath of the central plains of the North American continent; a spatial distribution providing great challenges for protection.

Starting in 1940, the National Aubudon Society became a major player in Whooping Crane conservation through the research and conservation activities of Robert Porter Allen.

His comprehensive studies of cranes on their wintering grounds in Texas defined their specialized need for wetland habitat. His work with community groups and the mass media of the day, along the migration route from Texas to southern Canada, led to widespread interest in identifying and protecting the Great White Birds. And his monumental book, *The Whooping Crane*, published in 1952 by the National Audubon Society, stands as a milestone upon which subsequent activities have been based. Very quickly, the serious plight of the Whooping Crane population, as described in Allen's book, alarmed and motivated conservation-minded persons.

To promote the welfare of the Whooping Crane, in 1961 a group of private citizens from both Canada and the United States established the Whooping Crane Conservation Association (WCCA). They urged both governments to collaborate in establishing a viable captive population of Whooping Cranes as a species bank and as a source of birds for reintroductions. Following WCCA's encouragement, in 1967 the Canadian Wildlife Service and the United States Fish and Wildlife Service collaborated in collecting hatching eggs from the nests of the wild cranes at Wood Buffalo. One egg was collected from each of several nests containing two eggs. The eggs were transported to the Patuxent Wildlife Research Center in Maryland, where they were hatched and the chicks reared. There was no noticeable detriment to the productivity of the wild cranes and soon a flourishing captive population was established. Today, there are approximately 160 cranes in captivity at several centers in both nations. Hundreds of captive-produced eggs and birds have been used in release programs in Idaho, Florida, Wisconsin, and Louisiana during a period in which the wild flock has increased from a low of about 15 cranes in 1940 to approximately 350 as of this writing in 2016.

The Endangered Species Act (ESA) of 1973 in the USA, and The Wildlife Act of 1998 in Saskatchewan, were milestones in the welfare of the Whooping Cranes. Under the ESA, direction was provided to establish a recovery team of experts from private and public sectors, to write and regularly update a Recovery Plan. In addition the ESA required protection for critical habitats of listed species. Including specialists from both Canada and the United States, the Whooping Crane Recovery Team has promoted and undertaken a variety of programs throughout North America to help the cranes. These have included establishing a network of centers willing to house the captive population, encouraging introduction and reintroduction experiments, and assuring the welfare of the original population through the protection of additional critical habitat and mitigation of a variety of threats. The longevity, variety, and indeed success of the conservation efforts for whooping cranes are impressive, and are models for conservation of other species.

As the new millennium unfolds, there remain many threats to Whooping Cranes and serious pressure on their populations. To name a few: reduction of fresh water inflow to coastal wetlands where cranes winter, spread of black mangrove north that makes unavailable the wetland habitats formerly used by wintering cranes, proliferation of wind farms along the migration corridor, and tar sands development with associated toxic pollution of wetlands near Wood Buffalo National Park and along the migration route. Experimental reintroduction of Whooping Cranes in Idaho and Florida were not successful, and the continuing reintroduction programs in Wisconsin and Louisiana have yet to produce a self-sustaining population. Hence, the work to conserve this majestic species is not done; more research leading to more conservation action is needed before the Whooping Crane's future is secure.

This volume brings together a series of papers by leading researchers and conservationists who have dedicated large portions of their lives to Whooping Cranes. The papers are written in a style for the educated lay reader, as well as for those that will carry the torch for Whooping Crane conservation in future decades. This book documents what we have learned since Robert Porter Allen published his 1952 volume, and will help lead the way forward toward a more secure future for the Whooping Crane. A secure future for the Whooping Crane would mean that the dream of that 8-year-old boy from Nova Scotia had come true.

Acknowledgments

We are grateful to Philip Nyhus, series editor for *Biodiversity of the World,* for the opportunity to contribute this volume to the series. Philip provided important guidance on the overall structure and tone of the book, and provided valuable editorial comments on the chapters. The editors at Elsevier, especially Pat Gonzalez, were patient and helpful throughout the process.

Lynda Garrett, *emeritus* librarian from the USGS Patuxent Wildlife Research Center, was a true hero of this project. Lynda was tireless in checking references, organizing manuscript reviews and correspondence, providing copy edits, and motivating us. Her dedication was impressive, and her help was indispensable. We are tremendously indebted.

The chapter manuscripts benefitted greatly from the efforts of many reviewers. We are deeply grateful of their willingness to give their time to improve this work.

Jack Delap, an accomplished wildlife illustrator and ecologist, contributed the drawings of Whooping Cranes, which add interest and beauty to this volume. The cover photo was contributed by Ted Thousand, who has produced a large number of beautiful photographs of birds in the Eastern Migratory Population, and has generously allowed us to use these photos in multiple publications over the years.

Most of all, this volume would not have been possible without the expertise and hard work of the chapter authors. Each author brought something unique as we attempted to build a picture of this complex species on a near-continental scale. We appreciate their willingness to stick with us over the long course of this project.

Crane people are passionate about cranes. This volume has benefitted from the work of countless very dedicated crane biologists and conservationists over many years. We are grateful for their contributions and acknowledge that this volume could not have been written without them.

Finally, we thank our families for their patience and support. Living with a book editor has its costs, which they shouldered gracefully.

John B. French, Jr.
Sarah J. Converse
Jane E. Austin

WHOOPING CRANES PAST AND PRESENT

Extant crane species: (Left to right) Grey Crowned Crane (*Balearica regulorum*), Black Crowned Crane (*Balearica pavonina*), Siberian Crane (*Leucogeranus leucogeranus*), Sandhill Crane (*Grus canadensis*), White-naped Crane (*Grus vipio*), Sarus Crane (*Grus antigone*), Brolga (*Grus rubicunda*), Wattled Crane (*Bugeranus carunculatus*), Blue Crane (*Anthropoides paradiseus*), Demoiselle Crane (*Anthropoides virgo*), Red-crowned Crane (*Grus japonensis*), Whooping Crane (*Grus americana*), Eurasian Crane (*Grus grus*), Hooded Crane (*Grus monacha*), Black-necked Crane (*Grus nigricollis*)

Whooping Cranes Past and Present

John B. French, Jr.*, Sarah J. Converse*,**,
Jane E. Austin†

*U.S. Geological Survey, Patuxent Wildlife Research Center, Laurel, MD, United States
**U.S. Geological Survey, Washington Cooperative Fish and Wildlife Research Unit,
School of Environmental and Forest Sciences (SEFS) & School of Aquatic and Fishery
Sciences (SAFS), University of Washington, Seattle, WA, United States
†U.S. Geological Survey, Northern Prairie Wildlife Research Center,
Jamestown, ND, United States

INTRODUCTION

The Whooping Crane (*Grus americana*) is the rarest of the 15 species in the family Gruidae (iucnredlist.org; accessed 28 October 2016). It is endemic to North America and listed as Endangered in the United States and Canada (Canadian Wildlife Service and U.S. Fish and Wildlife Service, 2005) and on the IUCN Red List (iucnredlist.org). Whooping Cranes have several traits that have contributed to their rarity. They have delayed maturity (3–5 years in the wild; Kuyt and Goossen, 1987), high adult survival, and a low annual reproductive rate. Species with this type of life history cannot sustain significant decreases in adult survival rate. Hence, the unregulated hunting of Whooping Cranes during the 19th and early 20th centuries represented

a serious threat to the species (Allen, 1952). In addition, the wetlands and surrounding prairies used by Whooping Cranes were transformed by agriculture in the 19th century (Allen, 1952), reducing habitat for this specialized, wetland-dependent crane species. The impact of these threats was rapid population decline leading to an extreme population bottleneck: all Whooping Cranes alive today are descended from 16 or fewer birds that were alive in 1941 in the Aransas-Wood Buffalo Population (AWBP; Canadian Wildlife Service and U.S. Fish and Wildlife Service, 2005), resulting in an estimated loss of 66% of all genetic material (Glenn et al., 1999).

Due perhaps partly to their rarity, but also certainly due to the aesthetic appeal of these large and graceful birds, Whooping Cranes elicit widespread and strong interest by the general

Whooping Cranes: Biology and Conservation
http://dx.doi.org/10.1016/B978-0-12-803555-9.00001-3

public. They are described as iconic, charismatic, and majestic (Bernacchi et al., 2015). Support for Whooping Cranes has resulted in a wide variety of private organizations investing in Whooping Crane conservation, in addition to the conservation measures implemented by the governments of the United States and Canada (Cannon, 1996). Importantly, Whooping Cranes have enjoyed legal protection for many decades under species protection laws in Canada and the United States.

The conservation investments in this species have paid off to an encouraging degree. Whooping Cranes have experienced resurgence from the nadir of 21 to approximately 764 individuals in winter 2016–17. Just over 430 of those individuals exist in the AWBP, which has grown steadily (at a mean rate of $r = 0.0383$ since the late 1930s; Butler et al., 2013). In addition, approximately 174 cranes are found in three reintroduced populations, although none of these are currently self-sustaining (Converse et al., Chapter 7, this volume). Also, approximately 160 birds are housed at several captive centers in the United States and Canada. The captive population has been the source for all of the existing reintroduced Whooping Cranes.

In describing Whooping Crane conservation through the present day, we have found it useful to conceive of two distinct eras: before 1950 and 1950 to present. Before 1942 the species was in decline; in that year the AWBP hit a low count of 16 birds, and there were 6 birds left in the wild Louisiana nonmigratory population, which was extirpated in 1950 and left no descendants. Since 1955 the species has generally been increasing. The focus of conservation in these two eras has been substantially different. During the first era, publicizing the plight of Whooping Cranes and halting the impacts of serious threats (habitat loss and unregulated shooting) were paramount. In the second era, efforts have shifted toward the restoration of the species through habitat protection and reintroductions. In this chapter, we trace the history of efforts to conserve this species through these two eras.

There is still more work to do. Only one self-sustaining population of the species exists, the remnant AWBP, and threats to its winter range in coastal Texas are a continuing concern (Chavez-Ramirez and Wehtje, 2012; Smith et al., Chapter 13, this volume). Reintroduction efforts have not enjoyed complete success, and the Whooping Crane research and management communities are addressing hard questions about whether reintroduction can be a viable conservation strategy. This volume anticipates a third era of Whooping Crane conservation, in which new information and management strategies are likely to change the face of Whooping Crane conservation over the next several decades. Throughout this book, we look ahead toward this coming era, by summarizing current information with a view toward preparing for the management decisions that will result in continued improvements in the outlook for Whooping Cranes.

TWO ERAS OF WHOOPING CRANE CONSERVATION

First Era (Before 1950)

The Whooping Crane historically had a wide distribution that covered diverse regions across North America, extending from the Arctic southward through the Great Plains to central Mexico, with scattered locations eastward to the central and southern Atlantic coast (Allen, 1952; Austin et al., Chapter 3, this volume). Historical breeding records extended from the boreal forest of southern Northwest Territories, Canada, through the Prairie Pothole Region, with the highest densities of breeding records in northern Iowa in the United States. These birds were migratory, and winter habitats were found mainly along the coast of the Gulf of Mexico in Texas and Louisiana. Additional records placed Whooping Cranes along the Atlantic coast and in the central highlands of northern Mexico in winter. A nonmigratory population also occurred along the Gulf coast.

Despite such wide distribution, Whooping Cranes may never have been abundant. Population estimates for 1860–70 ranged from 500 to 700 cranes (Banks, 1978) to 1300–1400 cranes (Allen, 1952, p. 83). As Euro-American settlers expanded westward, Whooping Cranes were increasingly threatened by shooting, habitat loss from wetland drainage or grassland conversion to agriculture, and general human disturbance. Shooting mortality was identified as the primary cause of the species' decline (Allen, 1952), and Whooping Cranes likely had declined precipitously by the early 20th century. The last documented breeding event in the Prairie Pothole Region occurred in Saskatchewan in the late 1920s (Hjertaas, 1994). The last reported breeding event in the nonmigratory population in Louisiana was in 1939, when just 13 cranes were found there (Drewien et al., 2001; Gomez, 1992).

The decline in Whooping Crane numbers caught the attention of professional biologists in the early 20th century. The discovery in 1921 of the wintering grounds of the small remaining migratory population on the Blackjack Peninsula of coastal Texas, later protected as part of the Aransas National Wildlife Refuge (NWR), and a growing interest in rare species in general brought focus to the plight of Whooping Cranes by ornithologists and naturalists across the continent (Dunlap, 1991). Biologists with the Provincial Museum of Natural History in Saskatchewan, National Audubon Society (NAS), and U.S. Fish and Wildlife Service (USFWS) started extensive surveys in the 1920s to learn more about the species' status and distribution. These dedicated biologists also raised awareness by giving talks, distributing leaflets and posters, and engaging with the media and state and provincial organizations (Dunlap, 1991). Publications ranging from the *Saturday Evening Post* to the *Proceedings of the International Ornithological Congress* highlighted concerns about the species' demise (Allen, 1952). Reports from the American Ornithologists' Union Committee on Bird Protection identified threats to the population in coastal Texas (Cottam et al., 1942) and called on the USFWS and NAS to "take immediate steps to learn the exact status of the species throughout its range and institute practical measures to forestall its extinction" (Allen et al., 1944).

Conservation actions to protect Whooping Cranes and their habitat during the first half of the 20th century were part of a broader interest in protecting migratory birds and their habitats, particularly for sustaining hunting. The first legal protection for Whooping Cranes was provided under the 1918 Migratory Bird Treaty Act (United States) and the Migratory Birds Convention Act (Canada), which ended unregulated hunting of all migratory birds. The treaties gave legal responsibility for the conservation of migratory birds to federal governments. In 1937, the Aransas Migratory Waterfowl Refuge (19,126 ha) was established to serve as a refuge and breeding ground for migratory waterfowl and Whooping Cranes; the area was renamed the Aransas NWR in 1939. Additional protection of the wintering area was provided in 1938 by closing the adjoining bay waters (5,236 ha) surrounding the Blackjack Peninsula to migratory bird hunting. However, wintering cranes remained threatened by oil development, activity on the new Intracoastal Waterway, and the establishment of an airport and bombing range on nearby Matagorda Island in 1942 (Cottam et al., 1942). The northern breeding grounds of Whooping Cranes remained unknown.

Second Era (1950–Present)

The period from roughly 1950 to the present has been one of accelerating interest in Whooping Crane conservation, increasing organization and institutionalization of effort, and producing significantly more information and making it available for decision making about Whooping Crane management.

The Cooperative Whooping Crane Project, a partnership between the NAS and the USFWS, was born out of the rising alarm over the decline of Whooping Cranes in the first era, and was the first concrete action of the second era of Whooping Crane conservation. The project was to be "an investigation into the status and biology of the species," and Robert Porter Allen was hired in 1946 to undertake the project (Sprunt, 1969). Allen teamed up with Robert H. Smith, a pilot-biologist with the USFWS, to help with field work on Whooping Cranes in Texas and in the Northwest Territories. Partnerships were established between state, provincial, and federal agencies and nongovernmental organizations. Allen's work in Canada was assisted by the Canadian Wildlife Service, the Royal Canadian Mounted Police, Ducks Unlimited, and the Hudson's Bay Company (Allen, 1952; Preface and Acknowledgements). Such wide-ranging partnerships characterize Whooping Crane conservation efforts to the present day. Allen's (1952) monograph, *The Whooping Crane*, comprised a summary of the extant knowledge about Whooping Cranes, and a report of the research findings stemming from his tireless field work (Kaska, 2012). The monograph has remained the definitive source for Whooping Crane biology throughout the second era.

However, even after extensive searching by airplane and on the ground, the breeding grounds of the AWBP remained unknown when Allen's (1952) monograph was published. Nests were discovered in 1954 in Wood Buffalo National Park (NP) by firefighters knowledgeable about cranes (Allen, 1956). Wood Buffalo NP is a vast area of boreal forest and muskeg (4,288,542 ha) located in northern Alberta and southern Northwest Territories of Canada, and had been established in 1922 to protect the last free-roaming herds of wood bison (*Bison bison athabascae*) in northern Canada. Thus by fortunate happenstance, protection was already provided for the breeding and summering areas of the few remaining migratory Whooping Cranes in North America.

During this era, national legislation addressing the specific needs of endangered species was passed in both Canada and the United States which enabled more aggressive conservation actions for Whooping Cranes. The Endangered Species Act in the United States, passed in 1973, established a means for official recognition of endangered species and set guidelines for their protection and recovery by requiring a recovery plan and designation of critical habitat. Whooping Cranes were immediately designated as endangered upon passage of the Endangered Species Act. Canada's Species at Risk Act (SARA) was passed in 1993, and while different in the details, provided similar protections to Whooping Cranes. Prior to 1993, protection in Canada was provided through the Committee on the Status of Endangered Wildlife in Canada (COSEWIC).

The first International Whooping Crane Recovery Team (IRT) was established in 1976, and the first recovery plan was written in 1979 by the IRT; the current recovery plan is the third revision (Canadian Wildlife Service and U.S. Fish and Wildlife Service, 2005). A population viability analysis was conducted in 1991 (Mirande et al., 1994), which led to recommended management actions and down-listing criteria found in subsequent recovery plans. Canadian documents to meet the requirements of SARA and outline habitat protections in Canada were written jointly (Johns and Stehn, 2005a, 2005b).

The IRT is the fundamental partnership for Whooping Crane conservation in North America, and is co-chaired by a Canadian Wildlife Service (CWS) employee and a USFWS employee. The membership of the team recognizes the variety of governmental and private interests (state, provincial, federal, and nongovernmental organizations) and the wide range of protection and recovery activities the team oversees (e.g., conservation of the AWBP, reintroductions, captive breeding). In the United States, the work of recovery teams is advisory to the USFWS, the bureau that has ultimate responsibility for protecting and managing Whooping Cranes. Several other important partnerships have been established in this era for Whooping Crane conservation, notably for reintroductions or research. Various USFWS refuges and the state wildlife agencies in Florida (Florida Fish and Wildlife Commission) and Louisiana (Louisiana Department of Wildlife and Fisheries) have worked closely with institutions that bred captive Whooping Cranes to undertake reintroductions of nonmigratory populations in Florida and Louisiana, and migratory populations in both the western and eastern United States.

The reintroduction of the Eastern Migratory Population (EMP) has been a particularly complex project requiring many partners and a large formal organization, The Whooping Crane Eastern Partnership (WCEP; see http://www.bringbackthecranes.org/). WCEP was established in 2000 with seven founding partners: USFWS, IRT, U.S. Geological Survey (USGS) laboratories including the Patuxent Wildlife Research Center (PWRC) and the National Wildlife Health Center, International Crane Foundation (ICF), Operation Migration, Wisconsin Department of Natural Resources, and the Natural Resources Foundation of Wisconsin). Beginning in 2009, the collaborative work became more structured and various teams were constituted (e.g., Rearing and Release, Monitoring and Management, Research and Science) to oversee aspects of the project. Another important WCEP team established in 2009 was the Communication and Outreach Team (COT). WCEP recognized from the start, and later the COT capitalized on, the opportunity for public engagement in this compelling project through education about Whooping Crane conservation. WCEP convened annual meetings for partners to identify common goals and facilitate the often substantial work of coordinating activities among the multiple partners.

A growing amount of both scientific research and public interest in Whooping Cranes supported conservation work in this second era. The Whooping Crane Conservation Association was established in 1961 by a group of biologists and others to advance the conservation of Whooping Cranes, a role that continues today (see http://whoopingcrane.com/). Publications on Whooping Cranes prior to 1971 are listed in a bibliography produced by the USFWS Office of Migratory Bird Management (Reeves, 1975); most entries are records of sightings or habitat associations from local journals. The International Crane Foundation was established in 1971 in Wisconsin and provided an important focus of work on all crane species world-wide, and especially on the two North American species, Sandhill Cranes (*Grus canadensis*) and Whooping Cranes. Technical research often found outlets in the proceedings of international meetings published by ICF (Archibald and Pasquier, 1987; Harris, 1991; Lewis, 1981). A group of crane biologists in the United States who had been organizing meetings to share their work since 1975 formalized their association in 1987 as the North American Crane Working Group (NACWG). Fourteen North American Crane Workshops have been held since 1975, each producing proceedings (see http://www.nacwg.org/publications.html). In 1996, the PWRC and ICF jointly published a volume on cranes that became a milestone in crane research (Ellis et al., 1996), concentrating mostly on studies of captive birds and captive breeding techniques. The majority of crane research in North America has focused on Sandhill Cranes, an abundant species that has provided important information relevant to Whooping Cranes (e.g., Ellis et al., 1996).

Aransas-Wood Buffalo Population – The AWBP is crucial to the conservation of Whooping Cranes and the survival of the species. The AWBP was the focus of Allen's 1952 monograph, which is still the most comprehensive account of Whooping Crane biology available. The AWBP is the only remnant of historical Whooping Crane populations. Much of what we know about the ecology and behavior of Whooping Cranes in the wild derives from study of the AWBP. Care should be taken to avoid assuming that the AWBP encompasses the historical variability of Whooping Crane biology, as the current AWBP passed through an extreme bottleneck in 1942 (Glenn et al., 1999). All extant Whooping Cranes today derive from a flock of 16 birds of the AWBP; it is unlikely that all 16 were founders as some were likely related as parent–offspring, and some likely did not leave offspring.

The iconic first member of the captive population, CANUS ("CAN" for Canada and "US" for the United States, named to honor the cooperation between the countries), was captured in Wood Buffalo NP in 1964 after injuring his wing around the time of fledging and brought to PWRC. His injured wing prevented CANUS from copulating properly, leading scientists at PWRC to develop procedures for artificial insemination, procedures now used widely for captive Whooping Cranes. CANUS was the only wild bird taken from the AWBP, and the AWBP

was also the source of all the eggs that started the captive population.

Fortunately, the AWBP has increased substantially in number since 1950 (Fig. 1.1). The 2016–17 winter estimate of the population at Aransas NWR was 431 (95% confidence interval of 371–493; see Table 1.1). Given that the reintroduced populations are not self-sustaining at this writing, the importance of the AWBP is obvious: continued persistence of the AWBP is almost certainly required for Whooping Crane persistence in the wild. However, threats to the population include the loss of winter habitat in and around Aransas NWR. Human population expansion has meant that oak (*Quercus spp.*) uplands adjacent to Aransas NWR, where Whooping Cranes often feed, are being lost to development. The estuarine habitat of Aransas itself is affected by

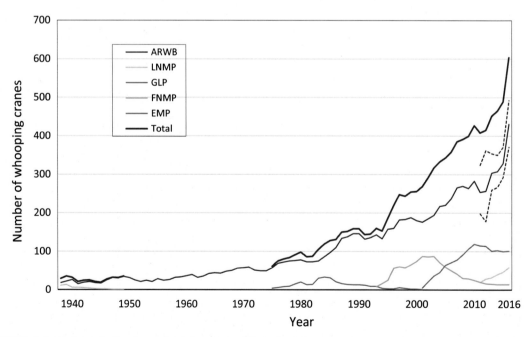

FIGURE 1.1 Estimated abundance through winter 2015–16 for Whooping Crane populations, including the Aransas-Wood Buffalo Population (AWBP), the Louisiana Nonmigratory Population (LNMP; including both the pre-1950 population prior to extirpation and the reintroduced population established in 2012), the Grays Lake Population (GLP), the Florida Nonmigratory Population (FNMP), and the Eastern Migratory Population (EMP). Beginning in 2011, 95% confidence intervals are provided for the AWBP.

TABLE 1.1   Estimated Whooping Crane Abundance in Wild and Captive Populations, Winter 2016–17. The Populations Are Assessed with Different Methods.

Population	Number
AWBP	431[a,b]
EMP	101[c]
LNMP	58[d]
FNMP	14[e]
Total in wild populations	604
Captive Whooping Cranes	160[f]
TOTAL	764

[a] *M.J. Butler and W. Harrell, Whooping Crane Survey results: winter 2016–17. (https://www.fws.gov/uploadedFiles/Region_2/NWRS/Zone_1/Aransas-Matagorda_Island_Complex/Aransas/Sections/What_We_Do/Science/Whooping_Crane_Updates_2013/WHCR_Update_Winter_2016-2017.pdf), accessed August 2017.*
[b] *The AWBP is the only population for which a statistically rigorous population estimate, including uncertainty, is available. The 95% confidence interval is 371–493.*
[c] *WCEP homepage, Project Update, 1 March 2017 (http://www.bringbackthecranes.org/), accessed August 2017.*
[d] *Friends of the Louisiana WHOOPING CRANES newsletter vol. 6, Issue 1, March 2017 (http://www.wlf.louisiana.gov/sites/default/files/pdf/simplenews/40976-friends-louisiana-whooping-crane-newsletter/whoopingcranenewsletter-march_2017.pdf), accessed August 2017.*
[e] *Personal communication: Tim Dellinger, Florida Fish and Wildlife Conservation Commission, 9 October 2017.*
[f] *S. Black and K. Swan (Chapter 16, this volume).*

human activity (see Smith et al., Chapter 13, this volume), including increased removal of water from the rivers that feed the estuaries to support human uses upstream. There are large areas of suitable habitat for an expanding AWBP in and around Wood Buffalo NP, Northwest Territories, but in both the United States and Canada most of the habitat south of the boreal forest once used for breeding is no longer available (see Austin et al., Chapter 3, this volume). Recent study of the timing and causes of mortality among AWBP has found that migration is less risky than previously thought (Pearse et al., Chapter 6, this volume), but increased development of wind farms and oil and gas operations throughout the long migration route of the AWBP (~4,000 km) suggests that this may be a topic for further investigation.

Reintroduction efforts – There is vulnerability inherent in having only a single self-sustaining population of Whooping Cranes. This vulnerability – along with the recognition that the species was much more widely distributed prior to the 20th century (e.g., Austin et al., Chapter 3, this volume) – has focused attention on reintroduction of Whooping Cranes with the goal of producing additional, separate populations. Unfortunately, reintroduction has proven to be challenging.

The first effort to establish a reintroduced population of Whooping Cranes began in 1975. In the spring of that year, Whooping Crane eggs were placed in the nests of Greater Sandhill Cranes (*Grus canadensis tabida*) at Grays Lake NWR in Idaho. The hope was that Sandhill Crane adults would successfully raise the young birds and that they would later mate with conspecifics. Eggs for the reintroduction were collected from AWBP nests and from the newly established captive breeding population at PWRC (a fertile egg was first produced at Patuxent in 1975; Ellis et al., 1992). From 1975 through 1988, 289 eggs (73 from PWRC and 216 from Wood Buffalo NP) were placed in nests at Grays Lake NWR. The effort was beset by challenges from the beginning. A total of 210 (72.7%) of the transferred eggs hatched, and only 84 (40%) of the hatched chicks fledged (Ellis et al., 1992). Only 49.4% of fledged birds survived to 1 year of age. After that time, annual survival was 0.843 (calculated from summaries presented in Garton et al., 1989) through age 10, which is low compared to the AWBP (Wilson et al., 2016; Wilson and Bidwell, Chapter 4, this volume). Most disappointing was that no Whooping Cranes in the Grays Lake Population (GLP) produced long-term pair bonds or bred with conspecifics. It is thought that cross-fostering by Sandhill Cranes led to inappropriate sexual imprinting, and indeed at least one hybrid chick was produced (male Whooping Crane × female Sandhill Crane) in the population, perhaps encouraged by a relative paucity of female Whooping Cranes (Canadian Wildlife Service and U.S. Fish and Wildlife Service, 2005). By early 2002, the last individuals in the GLP had died.

In early 1989, the USFWS and the CWS approved a reintroduction program with the goal of establishing a nonmigratory Whooping Crane population in Florida. This approval followed approximately a decade of research. Doubts about the effectiveness of release methods were an outcome of the Grays Lake experiment. Furthermore, because no birds were left that were descendants of nonmigratory individuals, it wasn't known whether released birds would migrate north (i.e., whether migratory behavior was genetically determined). Nesbitt and Carpenter (1993) determined that migratory Greater Sandhill Cranes could be either cross-fostered in nests of Florida Sandhill Cranes (*G. canadensis pratensis*) or captive-reared and soft-released (released into temporary enclosures), and in either case would not migrate north. Also, of the two groups, captive-reared birds had higher survival. Thus, the authors inferred that captive-reared Whooping Cranes could be released in Florida to establish a nonmigratory population. In 1993, the state of Florida, working with the USFWS, PWRC, and ICF, initiated releases to establish a nonmigratory population. The Florida Nonmigratory Population (FNMP) was the first to demonstrate successful reproduction of released birds in the wild (Dellinger,

Chapter 9, this volume; Folk et al., 2005; Moore et al., 2012). However, both survival and breeding success were lower than required to establish a self-sustaining population (Converse et al., Chapter 7, this volume; Moore et al., 2012). In 2004, releases were terminated, and in 2008, a modeling and decision-analytic exercise was conducted to evaluate whether additional releases should be considered (Converse et al., Chapter 7, this volume; Converse et al., 2013a, 2013b; Moore et al., 2012). In early 2009, the IRT decided that no additional releases would be attempted. The population in winter 2016–17 numbered approximately 14 individuals (Table 1.1).

In 2001, the USFWS and the CWS approved a reintroduction program with the goal of establishing a migratory population of Whooping Cranes in the eastern United States, separated from the AWBP. The unconventional start of reintroducing the EMP, where captive-reared Whooping Crane chicks were taught to migrate behind ultralight aircraft from Necedah NWR in central Wisconsin to the Gulf coast of Florida, garnered tremendous public attention (Duff, Chapter 21, this volume; Urbanek et al., 2005). The image of young Whooping Cranes flying behind an ultralight aircraft will linger in the public's imagination for years to come, although the method itself was discontinued in 2016 in favor of techniques with less direct human intervention. Beginning in 2005, direct autumn releases were initiated as a second release method, where costume-reared Whooping Crane chicks were soft-released on the summering grounds in a manner designed to encourage birds to follow conspecifics or Sandhill Cranes on migration (Urbanek et al., 2010a, 2010b). In 2011, the ultralight release program was relocated to White River Marsh State Wildlife Area (Green Lake and Marquette counties, WI) and the direct autumn release program to Horicon NWR (Dodge and Fond du Lac counties, WI) because high densities of biting black flies (Simuliidae) appeared to impede nest attendance and success in birds nesting at Necedah NWR (Juneau County, WI) (Converse et al., 2013b; Converse et al.,

Chapter 8, this volume; Urbanek et al., 2010b) (see Fig. 1.2). From 2013 to 2015, small numbers of birds reared by captive Whooping Cranes were also released at Necedah to test the parent-rearing method (Converse et al., Chapter 8, this volume). In 2016, all birds released in the EMP were parent reared. Concerns about the potential impact of costume-rearing and ultralight-led migration

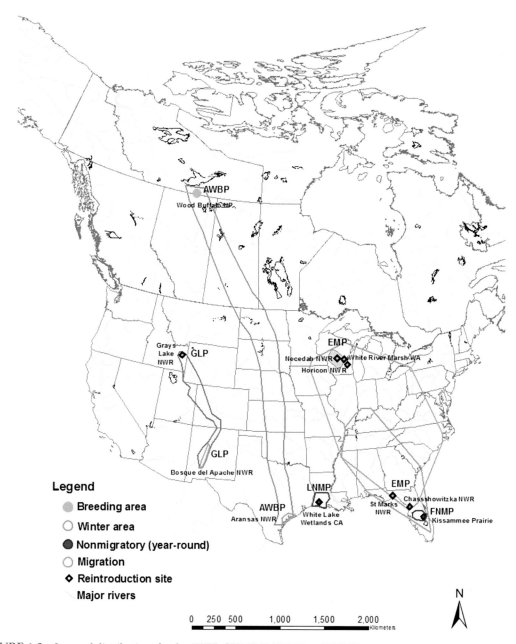

FIGURE 1.2 Seasonal distributions for the AWBP, GLP, EMP, FNMP, and LNMP.

on early learning and later reproductive behavior led to this decision. Despite challenges with reproduction, the population has come as close as any reintroduced population to success (Converse et al., Chapter 7, this volume; Servanty et al., 2014). Reproductive failure, however, must be addressed for the population to be self-sustaining, and it remains to be seen whether hatch rate and chick survival can be sufficiently increased for the population to become self-sustaining.

White Lake Wetlands Conservation Area in Louisiana was a natural place to consider for a reintroduction. Whooping Cranes persisted there in a nonmigratory population through 1950 (Canadian Wildlife Service and U.S. Fish and Wildlife Service, 2005), and White Lake is surrounded by a vast area of coastal marsh sparsely inhabited by humans. The area and habitats have changed somewhat in the 60 years since cranes were last present (King et al., Chapter 22, this volume) yet the extent of wetland habitat and remoteness of the area indicates that is was suitable for cranes. Releases of costume-reared cranes began in Louisiana in 2011 (Gomez, 2014). In 2016, the first chick was fledged by a reintroduced pair in the Louisiana Nonmigratory Population (LNMP), to the great delight of the Whooping Crane conservation community. Interestingly, the pair chose to nest in a crawfish (*Procambus spp.*) farm pond, rather than more remote marsh habitat (http://www.wlf.louisiana.gov/news/40095; accessed 8 December 2017). Future years will reveal whether this population can successfully reproduce, or whether the problems that have hampered growth of the EMP and other reintroduced populations will also limit success in Louisiana.

While reintroduction projects have claimed much of the focus and attention of the Whooping Crane conservation community, there is not yet a viable population resulting from reintroduction. The community continues to evaluate the reasons for poor demographic performance in these birds, particularly poor reproductive performance, with the goal of cracking the code: how to create a viable Whooping Crane population via reintroduction.

LOOKING AHEAD, AND THE CONTENTS OF THIS VOLUME

This volume was written in anticipation of a third era of Whooping Crane conservation. We expect the work of this next era will be based on the substantial operations experience and the large amount of scientific analysis now available on Whooping Cranes, much of which is summarized in the following chapters. Through their decision-making processes, managers will need to develop clear management objectives and efficient management plans to meet those objectives, given tight budgets and the heightened accountability that comes with tight budgets. An important effort is now underway to conduct a population viability analysis that accounts for all the populations of Whooping Cranes and is designed to inform future management. Some important decisions that should arise from management planning include decisions about the captive breeding of Whooping Cranes given anticipated future reintroduction efforts, and decisions about management actions for improving breeding outcomes in reintroduced populations.

There is a lot to be done in the coming third era. It is heartening that there are more Whooping Cranes in the wild now than at any time in perhaps a hundred years. Certainly, the chances of success are bolstered by the much larger knowledge base now available to the Whooping Crane conservation community. The aim of this volume is to provide the information needed for planning in the next era of Whooping Crane conservation.

This volume is divided into six sections. The first section describes the status, taxonomy, and distribution of the species. In Chapter 2, Krajewski (2018) discusses the phylogenetic history of the crane family, the Gruidae, including the current taxonomy of Whooping Cranes. (Note that Krajewski retains the use of *Grus canadensis* for the Sandhill Crane, the only other crane in North America, and we have followed that practice in this volume; recently, others have adopted the name *Antigone canadensis*.) In Chapter 3, Austin et al. (2018) revisit the data used by Robert Porter

Allen to define the range of Whooping Cranes in 1952, add new location records, and more fully define the ecological niche of Whooping Cranes, which should help identify areas for possible reintroduction or expansion of the current range.

The chapters in the second section, *Population and Breeding Biology*, present data on all populations established to date except for the LNMP, for which population data have not yet been systematically analyzed. The focus is on three populations: the AWBP, FNMP, and EMP. Chapters 4, 5, and 6 focus on the AWBP: Chapter 4 (Wilson and Bidwell, 2018) reviews historical demographic (abundance, survival, reproduction, and growth rate) and range information and evaluates future growth potential for the population; Chapter 5 (Strobel and Butler, 2018) reviews new methods for assessing abundance of the AWBP on and around the Aransas NWR wintering grounds; and Chapter 6 (Pearse et al., 2018) provides more detail on mortality of Whooping Cranes in the AWBP based on a satellite telemetry study. In Chapter 7, Converse Servanty, et al. (2018) consider demography of reintroduced populations, with a focus on modeling efforts for the EMP and the FNMP, noting how important population models have been for understanding these populations and guiding management actions. Chapter 8 (Converse, Strobel, and Barzen, 2018) outlines the efforts to understand reproductive failure in the EMP. Chapter 9 (Dellinger, 2018) gives an overview of the reintroduction of the FNMP, with a focus on reproduction, and interesting observations of nesting behavior.

The chapters in the third section, *Behavior and Social Structure*, provide more detail on Whooping Cranes in the EMP. We learn about the importance of the first few years of life for pair formation and subsequent breeding of reintroduced Whooping Cranes in Chapter 10 (Urbanek et al., 2018), and the influence of social interactions and experience on movements and changes in large-scale migration patterns in Chapter 11 (Mueller et al., 2018). In Chapter 12, Fitzpatrick et al. (2018) describe a mechanistic energetics model for individual cranes, parameterized with

data from the EMP, to evaluate the constraints on ecology and behavior of Whooping Cranes. All of these chapters provide insights into the conditions needed for success of reintroduced populations.

Almost every endangered species is at risk due to habitat loss, and Chapters 13 (Smith et al., 2018), 14 (Barzen et al., 2018), and 15 (Barzen 2018) in the fourth section, *Habitat Use*, provide information on various aspects of habitat selection by Whooping Cranes in the AWBP and the EMP. There are conservation implications arising from these analyses: habitat is clearly a potential limiting factor in all extant Whooping Crane populations.

Whooping Crane conservation during the second era has focused heavily on captive breeding and reintroduction. The fifth section, *Captive Breeding and Whooping Crane Health*, reviews important aspects of those efforts. Chapter 16 (Black and Swan, 2018) is an overview of the management challenges of captive Whooping Cranes among the primary institutions that hold them, and Chapter 17 (Songsasen et al., 2018) focuses on reproduction in captivity and suggests opportunities for incorporating new technologies to improve captive reproduction. The prevention of disease among Whooping Cranes kept in captivity (Olsen et al., 2018), has been a priority from the outset, and strategies to minimize risk of introducing novel diseases into the wild with birds raised in captivity has been a focus of managers. Chapter 18 (Hartup 2018b) describes important data gathered on the health of wild Whooping Cranes in the AWB, data that are quite difficult to gather.

The final section of this volume addresses *reintroduction and conservation*. The methods of preparing chicks for reintroduction and of releasing fledged chicks into wild habitats are areas of intense scrutiny because they are expensive, can be evaluated in a structured manner, and can be modified for improvement. Rearing techniques are now recognized as impacting later performance in the wild. In Chapter 20, Hartup (2018a) reviews the various goals (and expense) of methods used for rearing and release in the FNMP, the EMP, and the LNMP. In Chapter 21,

Duff (2018) describes the complicated operations needed for reintroduction of a migratory flock that involves migration training behind an ultralight aircraft – a visually compelling activity and one that has drawn the public into Whooping Crane conservation. King et al. (2018) provide an early analysis in Chapter 22 on the development of a new reintroduction project in the coastal marshes of southwest Louisiana, a vast, relatively remote area within the historical range of Whooping Cranes, but an area with well-recognized challenges for reintroduction success. One of those challenges for management is illegal shooting of Whooping Cranes. Chapter 23 examines illegal shooting and opportunities to strengthen protections in (Condon et al., 2018). In the final chapter of the volume, Chapter 24 (Converse et al., 2018c), the editors take a look forward at the challenges to be met for continued success in the growth of Whooping Crane numbers and for reduced risk of extinction.

References

Allen, A.A., Errington, P.L., Hickey, J.J., Munro, J.A., Stoner, D., 1944. Report of the A.O.U. Committee on Bird Protection for 1943. Auk 61 (4), 622–635.

Allen, R.P., 1952. The Whooping Crane. Research Report 3. National Audubon Society, New York.

Allen, R.P., 1956. A Report on the Whooping Cranes' Northern Breeding Grounds: A Supplement to Research Report No. 3, The Whooping Crane. National Audubon Society, New York, pp. 1–60.

Archibald, G.W., Pasquier, R.F. (Eds.), 1987. Proceedings of the International Crane Workshop, 1983. International Crane Foundation, Baraboo, WI, p. 595.

Austin, J.E., Hayes, M.A., Barzen, J.A., 2018. Revisiting the historic distribution and habitats of the whooping crane (Chapter 3). In: French, Jr., J.B., Converse, S.J., Austin, J.E. (Eds.), Whooping Cranes: Biology and Conservation. Biodiversity of the World: Conservation from Genes to Landscapes. Academic Press, San Diego, CA.

Banks, R. 1978. The size of early whooping crane populations. Unpublished report. U.S. Fish and Wildlife Service files. 10pp.

Barzen, J.A., 2018. Ecological implications of habitat use by reintroduced and remnant whooping crane populations (Chapter 15). In: French, Jr., J.B., Converse, S.J., Austin, J.E. (Eds.), Whooping Cranes: Biology and Conservation.

Biodiversity of the World: Conservation from Genes to Landscapes. Academic Press, San Diego, CA.

Barzen, J.A., Lacy, A.E., Thompson, H.L., Gossens, A.P., 2018. Habitat use by the reintroduced Eastern Migratory Population of whooping cranes (Chapter 14). In: French, Jr., J.B., Converse, S.J., Austin, J.E. (Eds.), Whooping Cranes: Biology and Conservation. Biodiversity of the World: Conservation from Genes to Landscapes. Academic Press, San Diego, CA.

Bernacchi, L.A., Ragland, C.J., Peterson, T.R., 2015. Engaging active stakeholders in implementation of community-based conservation: whooping crane management in Texas, USA. Wildl. Soc. Bull. 39, 564–573.

Black, S.R., Swan, K.D., 2018. Advances in conservation breeding and management of whooping cranes (Chapter 16). In: French, Jr., J.B., Converse, S.J., Austin, J.E. (Eds.), Whooping Cranes: Biology and Conservation. Biodiversity of the World: Conservation from Genes to Landscapes. Academic Press, San Diego, CA.

Butler, M.J., Harris, G., Strobel, B.N., 2013. Influence of whooping crane population dynamics on its recovery and management. Biol. Conserv. 162, 89–99.

Canadian Wildlife Service and U.S. Fish and Wildlife Service, 2005. International Recovery Plan for the Whooping Crane. Recovery of Nationally Endangered Wildlife (RENEW), Ottawa, and U.S. Fish and Wildlife Service, Albuquerque, NM.

Cannon, J.R., 1996. Whooping crane recovery: a case study in public and private cooperation in the conservation of endangered species. Conserv. Biol. 10, 813–821.

Chavez-Ramirez, F., Wehtje, W., 2012. Potential impact of climate change scenarios on whooping crane life history. Wetlands 32, 11–20.

Condon, E., Brooks, W. B., Langenberg, J., Lopez, D., 2018. Whooping crane shootings since 1967 (Chapter 23). In: French, Jr., J.B., Converse, S.J., Austin, J.E. (Eds.), Whooping Cranes: Biology and Conservation. Biodiversity of the World: Conservation from Genes to Landscapes. Academic Press, San Diego, CA.

Converse, S.J., Moore, C.T., Folk, M.J., Runge, M.C., 2013a. A matter of tradeoffs: reintroduction as a multiple objective decision. J. Wildl. Manage. 77, 1145–1156.

Converse, S.J., Royle, J.A., Adler, P.H., Urbanek, R.P., Barzen, J.A., 2013b. A hierarchical nest survival model integrating incomplete temporally varying covariates. Ecol. Evol. 3, 4439–4447.

Converse, S.J., Servanty, S., Moore, C.T., Runge, M.C., 2018a. Population dynamics of reintroduced whooping cranes (Chapter 7). In: French, Jr., J.B., Converse, S.J., Austin, J.E. (Eds.), Whooping Cranes: Biology and Conservation. Biodiversity of the World: Conservation from Genes to Landscapes. Academic Press, San Diego, CA.

Converse, S.J., Strobel, B.N., Barzen, J.A., 2018b. Reproductive failure in the Eastern Migratory Population: interaction of research and management (Chapter 8). In: French, Jr., J.B., Converse, S.J., Austin, J.E. (Eds.), Whooping Cranes: Biology and Conservation. Biodiversity of the World: Conservation from Genes to Landscapes. Academic Press, San Diego, CA.

Converse, S.J., French, Jr., J.B., Austin, J.E. 2018c. Future of Whooping Crane Conservation and Science (chapter 24). In: French, Jr., J.B., Converse, S.J., Austin, J.E. (Eds.), Whooping Cranes: Biology and Conservation. Biodiversity of the World: Conservation from Genes to Landscapes. Academic Press, San Diego, CA.

Cottam, C., Leopold, A., Finley, W., Cahalane, W.H., 1942. Report of the Committee on Bird Protection 1941. Auk 59 (2), 286–300.

Dellinger, T.A., 2018. Florida's non-migratory whooping cranes (Chapter 9). In: French, Jr., J.B., Converse, S.J., Austin, J.E. (Eds.), Whooping Cranes: Biology and Conservation. Biodiversity of the World: Conservation from Genes to Landscapes. Academic Press, San Diego, CA.

Drewien, R.C., Tautin, J., Courville, M.L., Gomez, G.M., 2001. Whooping cranes breeding at White Lake, Louisiana, 1939: observations by John J. Lynch, U.S. Bureau of Biological Survey. In: Proceedings of the North American Crane Workshop 8, 24–30.

Duff, J., 2018. The operation of an aircraft-led migration: goals, successes, challenges 2001 to 2015 (Chapter 21). In: French, Jr., J.B., Converse, S.J., Austin, J.E. (Eds.), Whooping Cranes: Biology and Conservation. Biodiversity of the World: Conservation from Genes to Landscapes. Academic Press, San Diego, CA.

Dunlap, T.R., 1991. Organization and wildlife preservation: the case of the whooping crane in North America. Soc. Stud. Sci. 21, 197–221.

Ellis, D.H., Gee, G.F., Mirande, C.M., International Crane Foundation, and Patuxent Wildlife Research Center, 1996. Cranes: Their Biology, Husbandry and Conservation. Washington, DC and Baraboo, WI: U.S. Department of the Interior, National Biological Service and International Crane Foundation, vol. 12, p. 308.

Ellis, D.H., Lewis, J.C., Gee, G.F., Smith, D.G., 1992. Population recovery of the whooping crane with emphasis on reintroduction efforts: past and future. In: Proceedings of the North American Crane Workshop 6, 142–150.

Fitzpatrick, M.J., Mathewson, P.D., Porter, W.P., 2018. Ecological energetics of whooping cranes in the Eastern Migratory Population (Chapter 12). In: French, Jr., J.B., Converse, S.J., Austin, J.E. (Eds.), Whooping Cranes: Biology and Conservation. Biodiversity of the World: Conservation from Genes to Landscapes. Academic Press, San Diego, CA.

Folk, M.J., Nesbitt, S.A., Schwikert, S.T., Schimdt, J.A., Sullivan, K.A., Miller, T.J., Baynes, S.B., Parker, J.M., 2005. Breeding biology of re-introduced non-migratory whooping cranes in Florida. In: Proceedings of the North American Crane Workshop 9, 105–109.

Garton, E.O., Drewein, R.C., Brown, W.M., Bizeau, E.G., Hayward, P.H., 1989. Survival rates and population prospects of whooping cranes at Grays Lake NWR. Unpublished Report, Fish and Wildlife Department, University of Idaho. Prepared for the U.S. Fish and Wildlife Service, Albuquerque, NM.

Glenn, T.C., Wolfgang, S., Braunm, M.J., 1999. Effects of a population bottleneck on whooping crane mitochondiral DNA variation. Conserv. Biol. 13 (5), 1097–1105.

Gomez, G., 2014. The history and reintroduction of whooping cranes at White Lake Wetlands Conservation Area, Louisiana. In: Proceedings of the North American Crane Workshop 12, 76–79.

Gomez, G.M. 1992. Whooping cranes in southwest Louisiana: history and human attitudes. In: Proceedings of the North American Crane Workshop 6, 19–23.

Harris, J. (Ed.), 1991. Proceedings of the 1987 International Crane Workshop. International Crane Foundation, Baraboo, WI, p. 456.

Hartup, B.K., 2018a. Rearing and release methods for reintroduction of captive-reared whooping cranes (Chapter 20). In: French, Jr., J.B., Converse, S.J., Austin, J.E. (Eds.), Whooping Cranes: Biology and Conservation. Biodiversity of the World: Conservation from Genes to Landscapes. Academic Press, San Diego, CA.

Hartup, B.K., 2018b. Health of Whooping Cranes in the central flyway (Chapter 18). In: French, Jr., J.B., Converse, S.J., Austin, J.E. (Eds.), Whooping Cranes: Biology and Conservation. Biodiversity of the World: Conservation from Genes to Landscapes. Academic Press, San Diego, CA.

Hjertaas, D.G., 1994. Summer and breeding records of the whooping crane in Saskatchewan. Blue Jay 52 (2), 99–115.

Johns, B.W., Stehn, T.V., 2005a. Action Plan for the Whooping Crane. Government Report, Environment Canada, Ottawa, p. 22.

Johns, B.W., Stehn, T.V., 2005b. National Recovery Strategy for the Whooping Crane (*Grus americana*). Government Report, Environment Canada, Ottawa, p. 27.

Kaska, K., 2012. The Man Who Saved the Whooping Crane: The Robert Porter Allen Story. University Press of Florida, Gainesville, p. 234.

King, S.L., Selman, W., Vasseur, P., Zimorski, S.E., 2018. Louisiana non-migratory whooping crane reintroduction (Chapter 22). In: French, Jr., J.B., Converse, S.J., Austin, J.E. (Eds.), Whooping Cranes: Biology and Conservation. Biodiversity of the World: Conservation from Genes to Landscapes. Academic Press, San Diego, CA.

Krajewski, C., 2018. Phylogenetic taxonomy of cranes and the evolutionary origin of the whooping crane

(Chapter 2). In: French, Jr., J.B. Converse, S.J., Austin, J.E. (Eds.), Whooping Cranes: Biology and Conservation. Biodiversity of the World: Conservation from Genes to Landscapes. Academic Press, San Diego, CA.

Kuyt, E., Goossen, J., 1987. Survival, age composition, sex ratio, and age at first breeding of whooping cranes in Wood Buffalo National Park, Canada. In: Lewis, J., Ziewitz, J. (Eds.), Proceedings of the 1985 Crane Workshop. Platte River Whooping Crane Habitat Maintenance Trust and USFWS, Grand Island, NE.

Lewis, J.C., 1981. Crane Research Around the World: Proceedings of the International Crane Symposium. International Crane Foundation, Baraboo, WI, p. 259.

Mirande, C., Lacy, R., Seal, U. (Eds.), 1994. Whooping Crane (*Grus americana*) – Conservation Viability Assessment Workshop Report. A publication of the Captive Breeding Specialist Group (CBSG/SSC/IUCN), Apple Valley, MN, p. 194.

Moore, C.T., Converse, S.J., Folk, M.J., Runge, M.C., Nesbitt, S.A., 2012. Evaluating release alternatives for a long-lived bird species under uncertainty about long-term demographic rates. J. Ornithol. 152, S339–S353.

Mueller, T., Teitelbaum, C.S., Fagan, W.F., Converse, S., 2018. Movement ecology of reintroduced migratory whooping cranes (Chapter 11). In: French, Jr., J.B., Converse, S.J., Austin, J.E. (Eds.), Whooping Cranes: Biology and Conservation. Biodiversity of the World: Conservation from Genes to Landscapes. Academic Press, San Diego, CA.

Olsen, G.H., Hartup, B.K., Black, S.R., 2018. Health and disease treatment in captive and reintroduced Whooping Cranes (Chapter 19). In: French, Jr., J.B., Converse, S.J., Austin, J.E. (Eds.), Whooping Cranes: Biology and Conservation. Biodiversity of the World: Conservation from Genes to Landscapes. Academic Press, San Diego, CA.

Nesbitt, S.A., Carpenter, J.W., 1993. Survival and movements of greater sandhill cranes experimentally released in Florida. J. Wildl. Manage. 57, 673–679.

Pearse, A.T., Brandt, D.A., Hartup, B.K., Bidwell, M., 2018. Mortality in Aransas-Wood Buffalo whooping cranes: timing, location, and causes (Chapter 6). In: French, Jr., J.B., Converse, S.J., Austin, J.E. (Eds.), Whooping Cranes: Biology and Conservation. Biodiversity of the World: Conservation from Genes to Landscapes. Academic Press, San Diego, CA.

Reeves, H.M., 1975. A Contribution to an Annotated Bibliography of North American Cranes, Rails, Woodcock, Snipe Doves and Pigeons. U.S. Fish and Wildlife Service, Washington, DC, p. 527.

Servanty, S., Converse, S.J., Bailey, L.L., 2014. Demography of a reintroduced population: moving toward management models for an endangered species, the Whooping Crane. Ecol. Appl. 24, 927–937.

Smith, E.H., Chávez-Ramirez, F., Lumb, L., 2018. Wintering habitat ecology, use, and availability for the Aransas-Wood Buffalo Population of whooping cranes (Chapter 13). In: French, Jr., J.B., Converse, S.J., Austin, J.E. (Eds.), Whooping Cranes: Biology and Conservation. Biodiversity of the World: Conservation from Genes to Landscapes. Academic Press, San Diego, CA.

Songsasen, N., Converse, S.J., Brown, M., 2018. Reproduction and reproductive strategies relevant to management of whooping cranes ex situ (Chapter 17). In: French, Jr., J.B., Converse, S.J., Austin, J.E. (Eds.), Whooping Cranes: Biology and Conservation. Biodiversity of the World: Conservation from Genes to Landscapes. Academic Press, San Diego, CA.

Sprunt, A., 1969. In memoriam: Robert Porter Allen. Auk 86, 26–34.

Strobel, B.N., Butler, M.J., 2018. Monitoring recruitment and abundance of the Aransas-Wood Buffalo Population of Whooping Cranes: 1950–2015 (Chapter 5). In: French, Jr., J.B., Converse, S.J., Austin, J.E. (Eds.), Whooping Cranes: Biology and Conservation. Biodiversity of the World: Conservation from Genes to Landscapes. Academic Press, San Diego, CA.

Urbanek, R.P., Fondow, L.E.A., Satyshur, C.D., Lacy, A.E., Zimorski, S.E., Wellington, M., 2005. First cohort of migratory whooping cranes reintroduced to eastern North America: the first year after release. In: Proceedings of the North American Crane Workshop 9, 213–223.

Urbanek, R.P., Fondow, L.E.A., Zimorski, S.E., 2010a. Survival, reproduction, and movements of migratory whooping cranes during the first seven years of reintroduction. In: Proceedings of the North American Crane Workshop 11, 124–132.

Urbanek, R.P., Szyszkoski, E.K., Zimorski, S.E., Fondow, L.E.A., 2018. Pairing dynamics of reintroduced migratory whooping cranes (Chapter 10). In: French, Jr., J.B., Converse, S.J., Austin, J.E. (Eds.), Whooping Cranes: Biology and Conservation. Biodiversity of the World: Conservation from Genes to Landscapes. Academic Press, San Diego, CA.

Urbanek, R.P., Zimorski, S.E., Fasoli, A.M., Szyszkowski, E.K., 2010b. Nest desertion in a reintroduced population of migratory whooping cranes. In: Proceedings of the North American Crane Workshop 11, 133–141.

Wilson, S., Bidwell, M., 2018. Population and breeding range dynamics in the Aransas-Wood Buffalo whooping crane population (Chapter 4). In: French, Jr., J.B., Converse, S.J., Austin, J.E. (Eds.), Whooping Cranes: Biology and Conservation. Biodiversity of the World: Conservation from Genes to Landscapes. Academic Press, San Diego, CA.

Wilson, S., Gil-Weir, K.C., Clark, R.G., Robertson, G.J., Bidwell, M.T., 2016. Integrated population modeling to assess demographic variation and contributions to population growth for endangered whooping cranes. Biol. Conserv. 197, 1–7.

Phylogenetic Taxonomy of Cranes and the Evolutionary Origin of the Whooping Crane

Carey Krajewski

Department of Zoology, Southern Illinois University, Carbondale, IL, United States

INTRODUCTION

Taxonomy is the science of naming and classifying organisms – in effect, identifying the units of biodiversity. In recent decades, systematists have reached a consensus that formally recognized taxa should represent not only units of biodiversity but also branches on the tree of life. Thus the names that we now apply to species, genera, families, and so on provide a common vocabulary for evolutionary, ecological, and comparative studies of organisms. The goal of conservation biology is to develop effective strategies for preserving this evolutionary heritage. However, the tree of life's branching pattern (phylogeny) is often difficult to infer, especially for closely related species like cranes. Reviewing the historical struggles of crane systematists, particularly during the recent application of powerful molecular and statistical techniques, allows us appreciate what is robust (and what is not) in our knowledge of crane relationships and the classification that reflects those relationships.

CRANE CLASSIFICATION PRIOR TO PHYLOGENETIC STUDIES

Six species of cranes were recognized by Linneaus (1758), the standard starting point for animal taxonomy. Linneaus placed the Eurasian, Sandhill, Whooping, Sarus, Demoiselle, and Black Crowned cranes in his genus *Ardea* along with other wading birds. Between Linneaus and the present, avian systematists have generally come to accept that there are 15 species of cranes (Table 2.1), with the Black-necked Crane (*Grus nigricollis*) the last of the currently recognized species to be formally described, and the Grey Crowned Crane (*Balearica regulorum*) the last described taxon to be accorded species status (Walkinshaw 1964). Subsequent to the monograph by Blythe and Tegetmeier (1881), most authors accepted placement of cranes in their own family (Gruidae). Gruidae eventually settled into the order Gruiformes following the bird classification of Sharpe (1899).

TABLE 2.1   Classification of Living Cranes
(Krajewski et al., 2010)

Order Gruiformes (Bonaparte, 1854)
  Suborder Grues (Bonaparte, 1854): gruoids, rallids, heliornithids
    Superfamily Gruoidea (Vigors, 1825): cranes, limpkins, trumpeters
      Family Gruidae (Vigors, 1825): cranes
      Subfamily Balearicinae (Brasil, 1913): crowned cranes
        Genus *Balearica* (Brisson, 1760)
          B. pavonina (Linneaus, 1758): Black Crowned Crane
          B. regulorum (Bennett, 1833): Grey Crowned Crane
      Subfamily Gruinae (Vigors, 1825): typical cranes
        Genus *Leucogeranus* (Bonaparte, 1855)
          L. leucogeranus (Pallas, 1773): Siberian Crane
        Genus *Bugeranus* (Gloger, 1841)
          B. carunculatus (Gmelin, 1789): Wattled Crane
        Genus *Anthropoides* (Vieillot, 1816)
          A. paradisea (Lichtenstein, 1793): Blue Crane
          A. virgo (Linneaus, 1758): Demoiselle Crane
        Genus *Grus* (Pallas 1766)
          Species Group Canadensis
            G. canadensis (Linneaus, 1758): Sandhill Crane
          Species Group Antigone
            G. antigone (Linneaus, 1758): Sarus Crane
            G. rubicunda (Perry, 1810): Brolga Crane
            G. vipio (Pallas 1811): White-naped Crane
          Species Group Americana
            G. japonensis (Müller, 1776): Red-crowned Crane
            G. americana (Linneaus, 1758): Whooping Crane
            G. grus (Linneaus, 1758): Eurasian Crane
            G. monacha (Temminck, 1835): Hooded Crane
            G. nigricollis (Prezhwalsky, 1876): Black-necked Crane

Genus-level taxonomy of cranes has been much more controversial, particularly in the late 19th and early 20th centuries. The landmark classification of Peters (1934), however, brought a measure of stability in the recognition of four genera (*Balearica* for crowned cranes, *Bugeranus* for the Wattled Crane, *Anthropoides* for the Blue and Demoiselle cranes, and *Grus* for the remaining species). Peters adopted a subfamily division used in the fourth "Checklist of North American Birds" (AOU, 1931), recognizing Balearicinae for crowned cranes and Gruinae for "typical" cranes. Crowned cranes are anatomically distinct from gruines in several important characteristics. For example, the furcular process of gruine species is fused to the anteroventral tip of the sternal keel; in balearicines, the furcula forms a "wishbone" as in most other birds. Johnsgard (1983) argued that many morphological traits exhibited by crowned cranes are primitive for Gruidae, with gruine species showing derived states.

Bugeranus is unique in its possession of wattles, fleshy and mostly feathered extensions of the anterior neck region. *Anthropoides* cranes have fully feathered heads, in contrast to other gruines, which have some amount of exposed skin on the crown or face. In all but one species of Peters' (1934) *Grus*, the trachea extends into – and coils within – an excavated sternal keel prior to its bifurcation into bronchi. The Siberian Crane is the exception; its trachea, like those of *Bugeranus* and *Anthropoides*, curves along an anteriorly sculpted but unexcavated keel. This distinction, as it turns out, has some phylogenetic relevance (see later).

PHYLOGENETIC SYSTEMATICS OF GRUIDAE: IDENTIFYING THE CLOSEST LIVING RELATIVES OF CRANES

Since at least Wetmore (1934) and Peters (1934), cranes have been closely associated with the New World limpkins (Aramidae) and trumpeters (Psophiidae), and somewhat less consistently with rails (Rallidae). Their affinities with other putative gruiforms, such as sun grebes (Heliornithidae), kagus (Rhynochetidae), sun bitterns (Eurypygidae), seriemas (Cariamidae), and bustards (Otitidae) have been much less clear, reflecting deep uncertainty about the monophyly of Gruiformes. Wetmore (1934) and Peters (1934) solidified the use of Gruoidea as a superfamily for cranes, limpkins, and trumpeters, and authors as recent as Olson (1985) argued that cranes and limpkins are closely related. Cracraft's (1982) cladistic analysis of anatomical traits placed cranes as sister to a diverse clade of gruiforms (including limpkins and trumpeters), but Livezey's (1998) morphocladistic analysis returned a [trumpeters, (limpkin, cranes)] clade to the exclusion of other gruiforms. DNA hybridization data (Sibley and Ahlquist, 1990) inserted heliornithids into Gruoidea as sister to limpkins, and analysis of mitochondrial 12S rDNA sequences by Houde et al. (1997) linked trumpeters with rails and heliornithids apart from a limpkin-crane clade. A phylogenetic estimate based on beta-fibrinogen intron 7 sequences (Fain and Houde, 2004) recovered a monophyletic group of "core gruiforms" consisting of cranes, limpkins, trumpeters, rails, and sun grebes, to which other putative gruiform families were more or less distantly related. Most recently, Fain et al. (2007) analyzed DNA sequence data from four mitochondrial and three nuclear gene sequences from core gruiforms and found strongly supported pairings of cranes with limpkins, and heliornithids with rallids. Although trumpeters formed the sister to this clade, statistical support for Gruoidea was only moderate. The suprafamilial classification in Table 2.1 reflects a consensus of these studies. The high-level classification of living birds, including gruiforms, is currently the subject of vigorous research using whole-genome data sets (e.g., Jarvis et al., 2014).

PHYLOGENETIC RELATIONSHIPS AMONG CRANES

Archibald (1976) meticulously documented the unison call behavior of crane species. Unison calls are ritualized displays involving vocalizations and body movements that occur between males and females of mated pairs. Archibald found that these displays are highly species-specific, but also show a nested pattern of similarity among species that allowed construction of a hierarchical classification. Unison-call traits separate *Balearica* from all other cranes, supporting the distinction between Balearicinae and Gruinae. Gruine species formed two major clusters, one comprising nine *Grus* species and the other including *Anthropoides*, *Bugeranus*, and the Siberian Crane. Archibald found the similarity between Wattled and Siberian cranes sufficiently compelling to reclassify the latter as a member of *Bugeranus*. Within *Grus*, unison calls delimited three species groups: Canadensis (Sandhill Crane); Antigone (Sarus, Brolga, and White-naped cranes); and Americana (Hooded, Eurasian, Whooping, and Red-crowned cranes). Although Archibald was not able to include the Black-necked crane in these analyses, the extent to which his taxonomy anticipated the consensus of numerous phylogenetic studies over the ensuing 40 years (Table 2.1) is striking.

Krajewski (1989) used the DNA–DNA hybridization method to estimate the phylogeny of cranes using genetic-distance analyses. Whatever the limitations of DNA hybridization data, being "inherently phenetic" is not one of them. Phenetic methods infer phylogenetic groups based on character similarity alone, as in Archibald's unison call study. In the case of DNA distances, phenetic grouping would place species showing the smallest distances from one another in close proximity on a tree. Grouping by similarity produces reliable phylogenetic estimates only when rates of character (e.g., DNA) evolution are uniform among species. Additive methods, in contrast, find the tree on which observed pairwise distances can be decomposed into branch lengths that come as close as possible to regenerating the original pairwise distances. They do not necessarily put the most similar species together, can be consistent even when evolutionary rates vary, and will be accurate when measured distances are accurate (Springer and Krajewski, 1989). It was, therefore, quite surprising that Krajewski's (1989) phylogeny of cranes matched – with a single exception – Archibald's taxonomy from unison calls. Two more different data sets (behavior patterns versus DNA hybridization distances) or methods of analysis (qualitative phenetic similarity versus additive distance) are difficult to imagine, yet their results were almost identical. The exception was the Siberian Crane, which DNA hybridization placed as sister to all other gruines rather than as sister to *Bugeranus*.

Other systematic studies of cranes were published between 1976 and 2010, including morphological data (Livezey, 1998; Wood, 1979) and molecular data (Dessauer et al., 1992; Fain, 2001; Fain et al., 2007; Ingold et al., 1987; Krajewski and Fetzner, 1994; Krajewski et al., 1999; Krajewski and King, 1996; Krajewski and Wood, 1995). None of these, however, produced significant departures from the Archibald-Krajewski consensus. Indeed, even the most recent and comprehensive molecular analysis (Krajewski et al., 2010) represents only a refinement of that consensus.

Krajewski et al. (2010) obtained the complete sequence (16.5–16.8 kilobases) of the mitochondrial genome of each crane species and, from those sequences, estimated crane phylogeny using maximum-likelihood and Bayesian methods. In addition, divergence dates were placed on the tree with a "relaxed clock" approach (i.e., one that does not assume rate uniformity) calibrated by the meager fossil record of cranes. The result is shown in Fig. 2.1. As in all modern systematic studies of cranes, the deepest separation is between balearicines and gruines, estimated at some 31–37 million years ago (Mya). Within Gruinae, the Siberian Crane diverges from the ancestor of all other species 12–14 Mya. Strong statistical support for this result led Krajewski et al. (2010) to endorse Livezey's (1998) suggestion that the Siberian Crane be placed in its own genus, *Leucogeranus*. The remaining species form four major clades corresponding to Archibald's (1976) Canadensis, Antigone, and Americana species groups of *Grus*, and *Bugeranus* + *Anthropoides*. Mitochondrial sequences provide strong support for sister-pairing of Canadensis and Antigone groups, which Krajewski et al. (2010)

referred to as the "Pacific Rim clade" reflecting the distributions of its species. Within Antigone, Sarus and Brolga cranes are sisters. Within Americana, the Red-crowned Crane diverged early (7.6–9.0 Mya) from the other four species, followed by successive speciation events leading to Whooping, Eurasian, Hooded, and Black-necked cranes during the past 4 million years.

The only point on which the mitogenomic sequences were equivocal was the position of the *Bugeranus-Anthropoides* group; specifically, it was unclear whether this group is more closely related to some members of *Grus* than to others. Because of this uncertainty, Krajewski et al. (2010) did not recommend further classificatory changes. Some authors (e.g., Del Hoyo et al. 2014), however, took the estimated tree at face value and placed all members of the Pacific Rim clade in a new genus "*Antigone*." This seems premature for the reasons described later.

The phylogeny in Fig. 2.1 and the classification based on it (Table 2.1) are unlikely to be the last word in crane systematics. Although the broad outlines of crane relationships have been consistently supported by multiple data sets and analyses for several decades, the branching order within and between species groups is not firmly established. Although we are reasonably confident that Fig. 2.1 is an accurate (if not totally precise) representation of relationships among the 15 mitochondrial genomes sampled, two important issues remain unaddressed: (1) Would a larger sample of mitogenomes (i.e., more than one from each species) show a tree that is consistent with Fig. 2.1? (2) Will sequences from the much larger nuclear genome have trees that are consistent with Fig. 2.1? DNA hybridization (Krajewski, 1989) and mitochondrial DNA (mtDNA) (Krajewski et al., 2010) analyses suggest that many internal branches of crane phylogeny are short relative to terminal ones. This implies that some cladogenic episodes happened

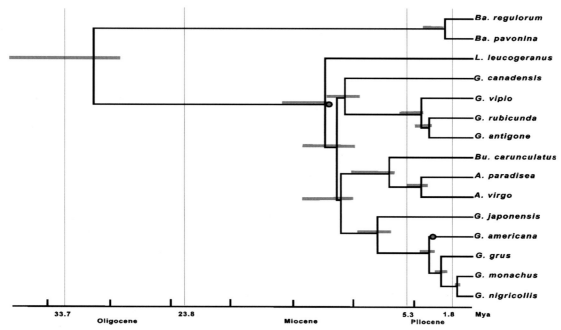

FIGURE 2.1 Time-calibrated phylogeny of cranes from mitochondrial genome DNA sequences, partitioned by codon position, with control region and gap sites deleted (redrawn from Figure 3 of Krajewski et al., 2010). Tree was obtained with BEAST software (Drummond and Rambaut, 2007) implementing Bayesian analysis with lognormal prior distribution (mean 0, standard deviation 1) on fossil calibration dates (indicated by filled circles). Tick marks on the time axis represent 5-million-year intervals. Shaded bars on nodes are 95% confidence intervals on nodal ages.

rapidly, with very little time between successive speciation events (e.g., the radiation that produced Pacific Rim, *Bugeranus-Anthropoides*, and Americana clades 10–13 Mya). Theoretical studies have shown that this is precisely the scenario under which a particular "gene tree" (i.e., mutation history of a DNA sequence) may differ from the phylogeny of species in which it is evolving (see Baum and Smith, 2012). Thus, we will not be sure of crane phylogeny until multiple complete genomes from each species have been analyzed, and maybe not even then (if the intervals between speciation events are too short). In the meantime, maintaining the stability of taxon names that have been in use for many decades facilitates communication about biodiversity – which is, after all, the primary goal of taxonomy.

PHYLOGENETIC POSITION OF THE WHOOPING CRANE

The Whooping Crane (*Grus americana*) is a member of Species Group Americana (see above). According to mitogenomic analyses, it is the sister group of a clade that includes the Eurasian, Hooded, and Black-necked cranes, from which it diverged some 3.7–3.9 Mya in the Pliocene. This recent coancestry is reflected in the low levels of mtDNA sequence and whole-genome DNA hybridization divergence (ca. 3% and 0.5%, respectively) between the Whooper and each of the Eurasian, Hooded, and Black-necked cranes (Krajewski, 1989; Krajewski et al., 2010).

The historical biogeography of cranes has never been studied, but ancestral area reconstructions would almost certainly find that the Americana group arose in Asia, with the Whooping Crane dispersing into North America during the Pliocene, presumably via the Bering land bridge that was open intermittently during the Cenozoic (Lomolino et al., 2010). This dispersal would be concomitant with the speciation event separating the Whooping Crane from the *G. grus-monacha-nigricollis* ancestor in East Asia. Consistent with this timing, fossil fragments ascribed to a Whooping Crane have been reported from the late Pliocene (ca. 3.5 Mya) of Idaho (Feduccia, 1967).

Grus canadensis (Sandhill Crane), the only other gruid with a New World distribution, is a much older lineage than the Whooping Crane, and its affinities are with the Pacific Rim clade. Mitogenomic analyses place Sandhills as sister to the Antigone group, the two sharing a common ancestor some 10–12 Mya in the Miocene. Ancestral area inferences are less obvious for the Pacific Rim species, but one plausible scenario is that Sandhills also entered North America via Beringia, though 6–8 million years before Whoopers arrived. In any case, the two New World members of the crane family are something on the order of "second cousins" in terms of phylogeny, genetic and temporal divergence, and historical biogeography.

Their very different residence times in North America may be reflected in the different genetic structures observed within Whooping and Sandhill cranes. Whoopers experienced the well-known population bottleneck in the late 1930s, losing much of their mtDNA diversity (Glenn et al., 1999; Snowbank and Krajewski, 1995). Glenn et al. (1999) found six distinct mtDNA haplotypes in museum specimens of 10 prebottleneck Whoopers, one prebottleneck and two additional haplotypes in 17 postbottleneck birds, and no obvious geographic structuring of genetic variation. In contrast, Rhymer et al. (2001) found 54 mtDNA haplotypes in a sample of 74 Sandhills, as well as a deep and relatively old (Plio-Pleistocene) phylogeographic division

between arctic-nesting Lesser Sandhills and all other subspecies (see also Glenn et al., 2002; Jones et al., 2005; Petersen et al., 2003). One wonders whether, despite its broad prebottleneck distribution in North America, the Whooping Crane with its characteristically long generation time may have retained low effective population size and genetic diversity from the Pliocene founder event that isolated it from its Asian ancestors. If so, the human-induced bottleneck of the 20th century was a genetic double-whammy to a relatively young species trying to establish itself in the New World.

SUMMARY AND OUTLOOK

For over a century, cranes have been recognized as the family Gruidae within the avian order Gruiformes; their closest relatives within the order are limpkins (Aramidae) and trumpeters (Psophiidae). Since 1934, avian taxonomists have recognized four crane genera in two subfamilies: Balearicinae (crowned cranes, two species of *Balearica*) and Gruinae (13 species in *Anthropoides*, *Bugeranus*, and *Grus*). Analyses of morphological, behavioral, and molecular data agree on the monophyly of crane subfamilies, but show much less consensus on phylogenetic relationships among gruines. In recent decades, morphological and molecular studies have shown that the Siberian Crane is sister to all other gruines, such that it is now recognized as a distinct, monotypic genus (*Leucogeranus*). DNA analyses indicate that the remaining gruines comprise four clades: *Anthropoides* + *Bugeranus*; the Sandhill Crane (*Grus canadensis*); Sarus (*G. antigone*) + Brolga (*G. rubicunda*) + White-naped (*G. vipio*) cranes; and Red-crowned (*G. japonensis*), Whooping (*G. americana*), Eurasian (*G. grus*), Hooded (*G. monacha*), and Black-necked (*G. nigricollis*) cranes. Most recently, analysis of complete mitochondrial genomes (Krajewski et al., 2010) showed strong support for Whooping Crane as sister to the Eurasian + Hooded + Black-necked clade, and Sandhill Crane as sister to the Sarus + Brolga + White-naped group. However, this study could not resolve the position of the *Anthropoides-Bugeranus* group; hence, monophyly of *Grus* (*sensu stricto*) cannot yet be established or rejected. Analyses of nuclear DNA sequences may help resolve these short branches of the crane phylogenetic tree.

References

American Ornithologists' Union (AOU), 1931. Check-List of North American Birds, fourth ed. American Ornithologists' Union, Lancaster, PA.

Archibald, G.W., 1976. Crane taxonomy as revealed by the unison call. In: Lewis, J.C., Masatomi, H. (Eds.), Crane Research Around the World. International Crane Foundation, Baraboo, WI, pp. 225–251.

Baum, D.A., Smith, S.D., 2012. Tree Thinking: An Introduction to Phylogenetic Biology. Roberts and Company Publishers, Greenwood Village, CO.

Blyth, E., Tegetmeier, W.G., 1881. The Natural History of the Cranes. Horace Cox, London.

Cracraft, J., 1982. Phylogenetic relationships and transantarctic biogeography of some gruiform birds. In: Hoffstetter, R., Buffetaut, E. (Eds.), Phylogénie et Paléobiogeographie: Livre Jubilaire en l'honneur de Robert Hoffstetter. Edution de l Université Claude Bernard, Lyon, France, pp. 393–402, Geobios Mémoire spécial no. 6.

Del Hoyo, J., Collar, N.J., Christie, D.A., Elliott, A., Fishopool, L.D.C., 2014. HBW and BirdLife International Illustrated Checklist of the Birds of the World, vol. 1, Nonpasserines. Lynx Edicions in association with BirdLife International.

Dessauer, H.C., Gee, G.F., Rogers, J.S., 1992. Allozyme evidence for crane systematics and polymorphisms within populations of Sandhill, Sarus, Siberian, and Whooping Cranes. Mol. Phylogenet. Evol. 1 (4), 279–288.

Drummond, A.J., Rambaut, A., 2007. BEAST: Bayesian evolutionary analysis by sampling trees. BMC Evol. Biol. 7, 214.

Fain, M.G. 2001. Phylogeny and evolution of cranes (Aves: Gruidae) inferred from DNA sequences of multiple genes. PhD dissertation, Southern Illinois University, Carbondale.

Fain, M.G., Houde, P., 2004. Parallel radiations in the primary clades of birds. Evolution 58 (11), 2558–2573.

Fain, M.G., Krajewski, C., Houde, P., 2007. Phylogeny of "core Gruiformes" (Aves: Grues) and resolution of the limpkin-sungrebe problem. Mol. Phylogenet. Evol. 43 (2), 515–529.

Feduccia, A., 1967. *Ciconia maltha* and *Grus americana* from the Upper Pliocene of Idaho. Wilson Bull. 79 (3), 316–318.

Glenn, T.C., Stephan, W., Braun, M.J., 1999. Effects of a population bottleneck on whooping crane mitochondrial DNA variation. Conserv. Biol. 13 (5), 1097–1107.

Glenn, T.C., Thompson, J.E., Ballard, B.M., Roberson, J.A., French, J.O., 2002. Mitochondrial DNA variation among wintering midcontinent Gulf Coast sandhill cranes. J. Wildl. Manage. 66 (2), 339–348.

Houde, P., Cooper, A., Lesli, E., Strand, A.E., Montaño, G.A., 1997. Phylogeny and evolution of 12S rDNA in Gruiformes (Aves). In: Mindell, D.P. (Ed.), Avian Molecular Evolution and Systematics. Academic Press, San Diego, CA, pp. 121–158.

Ingold, J.L., Guttman, S.I., Osborne, D.O., 1987. Biochemical systematics and evolution of the cranes (Aves: Gruidae). In: Archibald, G.W., Pasquier, R.F. (Eds.), Proceedings of the 1983 International Crane Workshop. International Crane Foundation, Baraboo, WI, pp. 575–584.

Jarvis, E.D., Mirarab, S., Aberer, A.J., et al., 2014. Whole-genome analyses resolve early branches in the tree of life of modern birds. Science 346 (6215), 1320–1331.

Johnsgard, P.A., 1983. Cranes of the World. Indiana University Press, Bloomington.

Jones, K.L., Krapu, G.L., Brandt, D.A., Ashley, M.V., 2005. Population genetic structure in migratory sandhill cranes and the role of Pleistocene glaciations. Mol. Ecol. 14 (9), 2645–2657.

Krajewski, C., 1989. Phylogenetic relationships among cranes (Gruiformes: Gruidae) based on DNA hybridization. Auk 106 (4), 603–618.

Krajewski, C., Fain, M.G., Buckley, L., King, D.G., 1999. Dynamically heterogenous partitions and phylogenetic inference: an evaluation of analytical strategies with cytochrome *b* and ND6 gene sequences in cranes. Mol. Phylogenet. Evol. 13 (2), 302–313.

Krajewski, C., Fetzner, Jr., J.W., 1994. Phylogeny of cranes (Gruiformes: Gruidae) based on cytochrome-*b* DNA sequences. Auk 111 (2), 351–365.

Krajewski, C., King, D.G., 1996. Molecular divergence and phylogeny: rates and patterns of cytochrome *b* evolution in cranes. Mol. Biol. Evol. 13 (1), 21–30.

Krajewski, C., Sipiorski, J.T., Anderson, F.E., 2010. Complete mitochondrial genome sequences and the phylogeny of cranes (Gruiformes: Gruidae). Auk 127 (2), 440–452.

Krajewski, C., Wood, T.C., 1995. Mitochondrial DNA relationships within the Sarus crane species group (Gruiformes: Gruidae). Emu 95 (2), 99–105.

Linneaus, C., 1758. Systema Naturae per Regna Tria Naturae, tenth ed. Impensis Salvii, Stockholm.

Livezey, B.C., 1998. A phylogenetic analysis of the Gruiformes (Aves) based on morphological characters, with an emphasis on the rails (Rallidae). Philos. Trans. R. Soc. Lond. B 353 (1378), 2077–2151.

Lomolino, M.V., Riddle, B.R., Whittaker, R.J., Brown, J.H., 2010. Biogeography, fourth ed. Sinauer Associates, Sunderland, MA.

Olson, S.L., 1985. The fossil record of birds. In: Farner, D.S., King, J.R., Parkes, K.C. (Eds.), Avian Biology, 8, Academic Press, New York, pp. 79–238.

Peters, J.L., 1934. Check-List of Birds of the World, vol. 2. Harvard University Press, Cambridge, MA.

Petersen, J.L., Bischof, R., Krapu, G.L., Szalanski, A.L., 2003. Genetic variation in the midcontinental population of sandhill crane, *Grus canadensis*. Biochem. Genet. 41 (1), 1–12.

Rhymer, J.M., Fain, M.G., Austin, J.E., Johnson, D.H., Krajewski, C., 2001. Mitochondrial phylogeography, subspecific taxonomy, and conservation genetics of sandhill cranes, *Grus canadensis* (Aves: Gruidae). Conserv. Genet. 2 (3), 203–218.

Sharpe, R.B., 1899. A Hand-List of the Genera and Species of Birds, vol. 1. British Museum (Natural History), London.

Sibley, C.G., Ahlquist, J.E., 1990. Phylogeny and Classification of Birds: A Study in Molecular Evolution. Yale University Press, New Haven, CT.

Snowbank, S.A., Krajewski, C., 1995. Lack of restriction-site variation in the mitochondrial-DNA control region of whooping cranes (*Grus americana*). Auk 112 (4), 1045–1049.

Springer, M.S., Krajewski, C., 1989. DNA hybridization in animal taxonomy: a critique from first principles. Q. Rev. Biol. 64 (3), 291–318.

Walkinshaw, L.H., 1964. The African crowned cranes. Wilson Bull. 76 (4), 355–377.

Wetmore, A., 1934. A systematic classification of the birds of the world, revised and amended. Smith. Misc. Coll. 89 (13), 1–11.

Wood, D.S., 1979. Phenetic relationships within the family Gruidae. Wilson Bull. 91 (3), 384–399.

Revisiting the Historic Distribution and Habitats of the Whooping Crane

Jane E. Austin*, Matthew A. Hayes**, Jeb A. Barzen**

*U.S. Geological Survey, Northern Prairie Wildlife Research Center, Jamestown, ND, United States
**International Crane Foundation, Baraboo, WI, United States

INTRODUCTION

Understanding the historic range and habitats of an endangered species can assist in conservation and reintroduction efforts for that species. Individuals reintroduced into a species' historic core range have a higher survival rate compared to individuals introduced near the periphery or outside the historic range (Falk and Olwell, 1992; Griffith et al., 1989). Individuals on the periphery of a species' range tend to occupy less favorable habitats and have lower and more variable densities than those near the core of their range (Brown, 1984; Brown et al., 1995, 1996). Such conclusions, however, presume that historic habitats have not changed since a species was extirpated from core areas – a difficult assumption for many areas, and particularly for wetland

habitat (Prince, 1997). Many endangered species persist only on the periphery of their historic range because of habitat loss or modification in their core range (Channell and Lomolino, 2000), which can bias our understanding of the species' habitat preferences. Further, habitat models based on locations where species persist necessarily emphasize local conditions rather than historical conditions (Kuemmerle et al., 2011). For example, habitat models for the European bison (*Bison bonasus*) suggested it was a woodland species, but assessment of the bison's historic range indicated it preferred mosaic-type landscapes and had a more eastern and northern distribution than previously reported (Kuemmerle et al., 2011, 2012). Hence, accurate determination of the historic range and habitat conditions for endangered species can improve our understanding of their ecology and assist in conservation and reintroduction efforts. Examining the historic range from an ecological perspective can also help identify where appropriate habitat still exists that could sustain a population.

The main source of information on the historic distribution of Whooping Cranes is a monograph, *The Whooping Crane*, a detailed report by Allen (1952). Allen collected and analyzed all available information about Whooping Cranes, including historic observations. From these observations, Allen constructed a historic Whooping Crane distribution map (Allen, 1952, p. 2) to illustrate historic breeding and wintering distribution and habitat use. Allen's maps and information have been utilized by many sources since their initial publication (e.g., Doughty, 1989; Meine and Archibald, 1996; Walkinshaw, 1973), including the International Whooping Crane Recovery Plan (Canadian Wildlife Service and U.S. Fish and Wildlife Service, 2005), and Allen's report remains a crucial source for Whooping Crane biology today. However, it has never been updated and questions remain about some areas identified by Allen. For example, when comparing recorded observations and the historical principal range map in Allen (1952, p. 2), many observations listed are not included in the map (e.g., southern Michigan), and some observations indicated on the map are not found in the table (Allen 1952, pp. 51–64). Also, some areas encompassed by the former breeding (e.g., northern Illinois) or wintering ranges that Allen described do not appear to be supported by his tabular data.

Given the influence of Allen's (1952) report in planning reintroduction efforts and the need to manage for a growing Aransas-Wood Buffalo Population (AWBP), Allen's original analyses deserved a review. In this chapter, we update Allen's work with records found subsequent to the publication of *The Whooping Crane*, and evaluate the accuracy of Allen's range maps and habitat use descriptions, taking advantage of newly discovered information, new tools such as geographic information systems (GIS), and contemporary knowledge of wetland and species ecology. We review the historic range, adjust the ecological niche for Whooping Cranes described by Allen by linking historic habitat use patterns to the life history of Whooping Cranes (most of which was developed after Allen's publication), and discuss how our analysis could help to direct current and future reintroduction efforts. Finally, we hypothesize that the ecological niche for Whooping Cranes is defined primarily by wetland productivity, which meets three primary needs in the annual cycle: rapid chick growth in spring and summer, habitat for molting birds in summer, and acquisition of large energy reserves in winter. We focus our results and discussion on breeding, summering, and wintering records; we do not consider habitat use during migration.

METHODS

All location records contained in *The Whooping Crane* (Allen 1952, pp. 51–64), as well as records published after that document was published, were entered into a tabular database (available on USGS ScienceBase; doi:10.5066/F7QZ282R). We used online search tools to find observations of Whooping Cranes in the literature published after 1952; some records were also found in early publications that are now available electronically that may not have been available to Allen. We used only observations up to 1941, the low point of 15 or 16 Whooping Cranes in the AWBP (Allen, 1952), to determine their historic distribution. Thus, our evaluation of Allen's work was constrained to the period before the intensification of agricultural practices that occurred after the 1940s, but encompassed the years of rapid settlement and cropland expansion across the Great Plains during the 1870s–1930 (Waisanen and Bliss, 2002). Records were organized by date, location (place, city, county, state, country, and latitude/longitude), type of record (nesting, observation, specimen, or captured bird), and life-history stage (breeding, summering, migrating, or wintering). We categorized observations or specimens of a nest, eggs, or flightless chick as breeding; other observations between mid-May and 30 August were categorized as summering. For observations or specimens that included a year but no date, we categorized life-history stage as unknown. We included observer identity, if known, source for each record, and, for birds that were killed or collected, current location of existing specimens. Some records lacked information on multiple aspects (e.g., observer name, specific location, season) and, without other supporting information, were included in the database but not included in analyses. Records were exported to a GIS, ArcMap10 (ESRI, Redlands, CA, USA), and sorted by life-history stage and season.

Landscape features and patterns of habitat use were derived by overlaying locations and range polygons on the World Wildlife Fund (WWF) terrestrial biomes and ecoregions (Abell et al., 2008). These data layers were selected because of their use of a comprehensive set of characteristics at the ecoregion level (wetlands, climate, topography, and vegetation). Additional information was obtained from two other ecoregion classification systems: the Canadian Ecological Framework (CEF), which closely matches the WWF ecoregion boundaries (ESWG, 1995), and the Prairie Pothole Region (PPR) in the northern Great Plains (Euliss et al., 1999). Finally, we used descriptions of habitat types in Allen (1952), landscape and wetland information from WWF global habitat types and ecoregions, and contemporary information about landscape and wetland characteristics in those locations to assess landscape and habitat features of Whooping Crane locations. Where locations were identified as a "lake" (biome = 98)

or unknown (biome = 0), we assigned the location to the nearest adjacent biome and ecoregion. We also replicated the hand-drawn map ranges from Allen (1952, pp. 2, 19) in the GIS to the best of our ability, given uncertainty about the map projection used by Allen.

RESULTS

Updating Allen's Historic Distribution

We compiled a total of 884 records from Allen (1952), dating from 1722 to 1941: 25 (2.8%) records preceded 1850, 390 (44.1%) occurred from 1850 to 1899, 221 (25.0%) from 1900 to 1919, and 201 (22.7%) from 1920 to 1941; 47 (5.3%) had no known date specified. We added 74 records to those listed in Allen (1952): 20 breeding locations (total 87), 21 summering locations (total 169), 18 migrating locations (total 367), 5 wintering locations (total 121), and 10 unknown life-history stages (total 140) (Table 3.1; Appendix). The additional records were from 28 published sources, including eight that had been published before 1952 but either may not have been available to Allen or we felt deserved to be reinterpreted. Among the most striking additions were the 13 breeding records, 14 summering records, and 7 migration records reported for Saskatchewan by Hjertaas (1994) and 4 breeding records in southern Texas (Oberholser, 1938, 1974). Also notable were six records for Wisconsin and six each for Florida and Indiana.

Verifying the location of some records was problematic. For example, in a report of his explorations of the Canadian north, Samuel Hearne indicated that "[t]his bird visits Hudson's Bay in the Spring, though not in great numbers. They are generally seen only in pairs, and then not very often… It seldom has more than two young and retires Southward early in the fall" (Hearne, 1795, pp. 271–272). Unfortunately, Hearne did not give specific locations of

TABLE 3.1 Distribution of Historic Whooping Crane Locations by (WWF) Ecoregions (Indented) within Biomes (Bold), in Order of Total Number of Records. Breeding Records Were Those That Had Observations or Specimens of a Nest, Eggs, or Flightless Chick; Other Observations between Mid-May and 30 August Were Categorized as Summering. Observation Records in September–Mid-May or Late March–Mid-May Were Classified as Migrating and Those of December–Mid-March as Wintering. Records Lacking a Date Were Classified as Unknown.

Biome	Ecoregion	Total No.	Total %	Breeding No.	Breeding %	Summering No.	Summering %	Migrating No.	Migrating %	Wintering No.	Wintering %	Unknown No.	Unknown %
Temperate Grasslands, Savannas, and Shrublands		616	69.8	61	70.9	123	72.8	341	92.9	0	0	91	65.0
	Central and Southern Mixed Grasslands	120	13.6	0	0	1	0.6	109	29.7	0	0	10	7.1
	Central Tall Grasslands	115	13.1	18	21.2	7	4.1	66	18.0	0	0	24	17.1
	Canadian Aspen Forests and Parklands	105	11.9	17	20.0	43	25.4	24	6.5	0	0	21	15.0
	Northern Mixed Grasslands	61	6.9	8	9.4	23	13.6	28	7.6	0	0	2	1.4
	Northern Short Grasslands	61	6.9	6	7.1	31	18.3	18	4.9	0	0	6	4.3
	Northern Tall Grasslands	50	5.7	12	14.1	12	7.1	22	6.0	0	0	4	2.9
	Central Forest-Grasslands Transition	40	4.5	0	0	4	2.4	27	7.4	0	0	9	6.4
	Texas Blackland Prairies	26	2.9	0	0	0	0	21	5.7	0	0	5	3.6
	Nebraska Sand Hills Mixed Grasslands	14	1.6	0	0	0	0	11	3.0	0	0	3	2.1
	Western Short Grasslands	12	1.4	0	0	1	0.6	5	1.4	0	0	6	4.3
	Flint Hills Tall Grasslands	9	1.0	0	0	0	0	8	2.2	0	0	1	0.7
	Edwards Plateau Savanna	3	0.3	0	0	1	0.6	2	0.5	0	0	0	0
Tropical and Subtropical Grasslands, Savannas, and Shrublands		120	13.6	9	10.6	7	4.1	2	0.5	87	72.5	15	10.7
	Western Gulf Coastal Grasslands	120	13.6	9	10.6	7	4.1	2	0.5	87	72.5	15	10.7
Temperate Broadleaf and Mixed Forest		52	5.9	5	5.8	10	5.9	20	5.4	4	3.3	13	9.3
	Upper Midwest Forest-Savanna Transition	18	2.1	4	4.7	5	3.0	6	1.6	0	0	3	2.1
	Southern Great Lakes Forests	9	1.0	0	0	2	1.2	4	1.1	0	0	3	2.1

Central U.S. Hardwood Forests	7	0.8	0	0	1	0.6	4	1.1	0	0	2	1.4
Mississippi Lowland Forests	7	0.8	0	0	0	0	3	0.8	2	1.7	2	1.4
Southeastern Mixed Forests	4	0.4	0	0	1	0.6	3	0.8	0	0	0	0
Northeastern Coastal Forests	2	0.2	0	0	0	0	0	0	2	1.7	0	0
East Central Texas Forests	1	0.1	1	1.2	0	0	0	0	0	0	0	0
Eastern Forest-Boreal Transition	1	0.1	0	0	0	0	0	0	0	0	1	0.7
Eastern Great Lakes Lowland Forests	1	0.1	0	0	1	0.6	0	0	0	0	0	0
Appalachian Mixed Mesophytic Forests	2	0.2	0	0	0	0	0	0	0	0	2	1.4
Deserts and Xeric Shrublands	33	3.7	0	0	1	0.6	4	1.1	17	14.2	11	7.9
Tamaulipan Mezquital	17	1.9	0	0	0	0	1	0.3	15	12.5	1	0.7
Wyoming Basin Shrub Steppe	11	1.2	0	0	1	0.6	2	0.5	0	0	8	5.7
Chihuahuan Desert	2	0.2	0	0	0	0	0	0	1	0.8	1	0.7
Great Basin Shrub Steppe	1	0.1	0	0	0	0	0	0	0	0	1	0.7
Meseta Central Matorral	1	0.1	0	0	0	0	0	0	1	0.8	0	0
Snake-Columbia Shrub Steppe	1	0.1	0	0	0	0	1	0.3	0	0	0	0
Boreal Forest/Taiga	24	2.7	9	10.6	14	8.3	0	0	0	0	1	0.7
Mid-Continental Canadian Forests	16	1.8	8	9.3	8	4.7	0	0	0	0	0	0
Muskwa-Slave Lake Forests	5	0.6	1	1.2	4	2.4	0	0	0	0	0	0
Southern Hudson Bay Taiga	2	0.2	0	0	1	0.6	0	0	0	0	1	0.7
North Canadian Shield Taiga	1	0.1	0	0	1	0.6	0	0	0	0	0	0
Temperate Coniferous Forests	21	2.4	1	1.2	5	3.0	0	0	7	5.8	8	5.7
Southeastern Conifer Forests	7	1.0	0	0	0	0	0	0	4	4.2	4	2.9
Middle Atlantic Coastal Forests	6	0.7	0	0	0	0	0	0	2	1.7	4	2.9

(Continued)

TABLE 3.1 Distribution of Historic Whooping Crane Locations by (WWF) Ecoregions (Indented) within Biomes (Bold), in Order of Total Number of Records. Breeding Records Were Those That Had Observations or Specimens of a Nest, Eggs, or Flightless Chick; Other Observations between Mid-May and 30 August Were Categorized as Summering. Observation Records in September–Mid-May or Late March–Mid-May Were Classified as Migrating and Those of December–Mid-March as Wintering. Records Lacking a Date Were Classified as Unknown. (cont.)

Biome	Ecoregion	Total No.	Total %	Breeding No.	Breeding %	Summering No.	Summering %	Migrating No.	Migrating %	Wintering No.	Wintering %	Unknown No.	Unknown %
	South Central Rockies Forests	4	0.4	1	1.2	3	1.8	0	0	0	0	0	0
	North Central Rockies Forests	1	0.1	0	0	1	0.6	0	0	0	0	0	0
	Piney Woods Forests	1	0.1	0	0	1	0.6	0	0	0	0	0	0
Tropical and Subtropical Coniferous Forests		2	0.2	0	0	0	0	0	0	2	1.7	0	0
	Trans-Mexican Volcanic Belt Pine-Oak Forests	2	0.2	0	0	0	0	0	0	2	1.7	0	0
Tundra		9	1.0	0	0	9	5.3	0	0	0	0	0	0
	Arctic Coastal Tundra	8	0.9	0	0	8	4.7	0	0	0	0	0	0
	Low Arctic Tundra	1	0.1	0	0	1	0.6	0	0	0	0	0	0
Tropical and Subtropical Dry Broadleaf Forests		4	0.4	0	0	0	0	0	0	3	2.5	1	0.7
	Baja Dry Forests	4	0.4	0	0	0	0	0	0	3	2.5	1	0.7
Total		880	100.0	85	9.7	169	19.2	367	41.7	119	13.5	140	15.9

Whooping Crane nests or chicks, so his comments could apply to the broader region he explored, which extended from the Churchill River to the Great Slave Lake, then north to the Coppermine River (Hearne 1795). Therefore, Allen (1952) centralized Hearne's findings at Churchill, Manitoba, as we did. Whooping Cranes did occur near Hudson Bay; records show a Whooping Crane adult collected in 1771 at York Factory, located south of Churchill, and others observed along the Hudson Bay coast (Houston et al., 2003). Migrating Whooping Cranes were also observed around Churchill in 1953 and 1964 (Jehl and Smith, 1970). Nelson's (1876) reference of breeding pairs in central Illinois was similarly problematic, as it provided no date, specific location, observer, or note of actual nest, eggs, or chicks. An authoritative reference of the era (Ridgway, 1889) listed Whooping Crane only as a summer resident of Illinois. We considered but excluded a nesting report in Brown County, Wisconsin, reported in Carr (1890) but not cited in Allen (1952), and two observations in Florida noted in Bent (1926) but never supported in Allen (1952) or Nesbitt's (1981) review of Whooping Cranes in Florida. We chose to exclude Nelson (1876), Carr (1890), and Bent's (1926) records from our updated map and biome and ecoregion assessments because of their vague nature, but those records are included in the Appendix. We were unable to determine specific locations for two private ranches listed in Allen (1952, p. 56), but placed them near La Barca, Jalisco (west-central Mexico), as described in Allen's text. Allen's extension of the wintering range into northern Mexico is not supported by any listed locations. However, discussions with biologists with extensive knowledge of northern Mexico suggest that including that area within the wintering range may have been accurate (R. Drewein, U.S. Fish and Wildlife Service [retired], and J. Taylor, U.S. Fish and Wildlife Service [deceased], personal communications).

Our revised set of records alters the historic distribution of Whooping Cranes from that of Allen (Fig. 3.1). Additional breeding records in Iowa and North Dakota fall within the former breeding area Allen delineated, and the breeding area in Saskatchewan now appears larger. There is no direct evidence to support Allen's extension of the breeding range into northern Illinois but Allen clearly believed northern Illinois was breeding habitat for Whooping Cranes, a contradiction that was unresolvable with data that we could evaluate. Four additional breeding records were found for Texas, possibly associated with the original nonmigratory population. New records extend the occurrence of summering birds into central Saskatchewan, southern Wisconsin, and northeastern Illinois. Some of the summer observations may well suggest breeding birds (e.g., failed breeders that lacked evidence of eggs or chicks) or they may have included birds that were not yet breeding but ranging widely from their natal areas in summer (see Barzen et al., Chapter 14, this volume). Our additional summer records in Wisconsin and Illinois provide somewhat more support for Whooping Cranes breeding in that area, but no actual records of nests, eggs, or chicks exist.

We accepted one record as breeding around the Greater Yellowstone-Teton area, coinciding with summering records for that area (Fig. 3.1). Kemsies (1930) noted that "regular reports are received of the breeding of the Whooping Crane," including observation of two young and later two adults in the southwest part of Yellowstone National Park by a reputable observer. Allen (1952) included Kemsies' record but suspected his observations to be summering subadult birds in association with Sandhill Cranes. In this particular case, we believe the breeding records from Kemsies (1930) are viable, given the dates of observations and size differential between Whooping Cranes chicks near fledging (100–120 cm; Urbanek

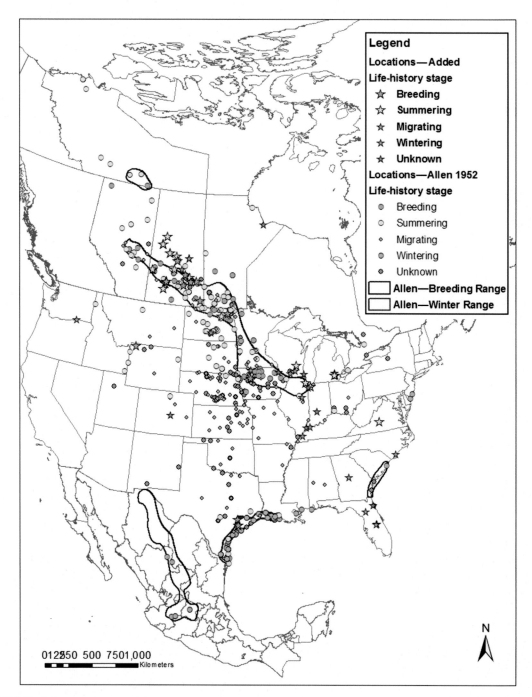

FIGURE 3.1 Distribution of historic Whooping Crane locations 1722–1941, by life-history stage. Records from Allen (1952) and additional records (this chapter), with breeding and wintering ranges delineated by Allen. Multiple records may be represented by a single symbol. Scale = 1:25,000,000.

and Lewis, 2015) and adult greater Sandhill Cranes in this region (*Grus canadensis tabida*, 130–150 cm).

Our revised distribution of wintering birds along the Gulf of Mexico is similar to that described by Allen (1952) but extended further east into Mississippi. Along the Atlantic coast we assess the range to have been much broader and to have extended from North Carolina into central Florida (Fig. 3.1). Most of the Atlantic coast records occurred before 1880, but several additional records reporting Whooping Cranes in Florida as late as 1936 suggest that a few cranes continued to winter in the state into the early 20th century.

Biome Affiliations

Historic observations of Whooping Cranes occurred across nine biomes, extending from the Tundra and Boreal Forest/Taiga of interior Canada to the Tropical/Subtropical Grasslands, Savannas, and Shrublands along the Gulf coast, and to the Desert and Xeric Shrublands of interior Mexico (Table 3.1, Fig. 3.2). Most (69.8%) of the locations, however, occurred within the Temperate Grasslands, Savannas, and Shrublands biome, which extends from the northern Great Plains to the Gulf coast. This biome also encompassed more than two-thirds of both breeding and summering records. All of the southern (nonmigratory) breeding records and two-thirds of the wintering locations occurred in the Tropical/Subtropical Grasslands, Savannas, and Shrublands biome along the Gulf coast.

Ecoregion Affiliations

Breeding and summering locations – Most breeding records were affiliated with the Central Tall Grasslands, Canadian Aspen Forest and Parkland (hereafter Aspen Parkland), and Northern Tall Grasslands (primarily in northern Iowa) ecoregions (Table 3.1, Fig. 3.3). Primary summering ecoregions also were the Aspen Parkland and Northern Mixed Grasslands, as well as adjacent Northern Short Grasslands. These ecoregions coincide with the PPR, a glacially formed region in the northern midcontinent grasslands that is largely defined by its high wetland densities (Kantrud et al., 1989; Smith et al., 1964). Clearly once an important area for Whooping Cranes, the PPR encompassed 57.6% of all breeding records, 52.1% of all summering records, and 26.7% of migrating records. Collectively, the PPR accounted for 30.1% of all 880 records. The highest density of breeding records occurred in the Central Tall Grasslands of Iowa, especially that portion of northern Iowa that corresponds to the southernmost extent of the PPR. This area was identified by Allen as likely the historic center of breeding. The Upper Midwest Forest-Savanna Transition ecoregion was used to a lesser extent; it is more mesic and found at lower latitudes than the PPR or other grasslands.

Within the Boreal Forest/Taiga biome, the majority of breeding and summering cranes were observed in the Midcontinental Canadian forests. Four summer observations and one breeding observation fall within the Muskwa-Slave Lake Forests ecoregion, which encompasses the Hay River, Slave River, Peace, and Wabasca Lowlands, Boreal Transition, and the Mid-Boreal Upland CEF ecoregions. Locations fall in areas of high wetland density, or along rivers or large lakes.

Records of nonmigratory breeding and southern summering cranes were located primarily in the Western Gulf Coast Grasslands. The last remnants of the nonmigratory population at White Lake, Louisiana, were observed in this ecoregion as late as the 1930s. The breeding location at Eagle Lake in Wharton County, Texas, appears to fall within the East Central Texas Forest ecoregion but more likely is at the edge of the Western Gulf Coast Grasslands.

Other records of breeding or summering cranes are scattered across various ecoregions

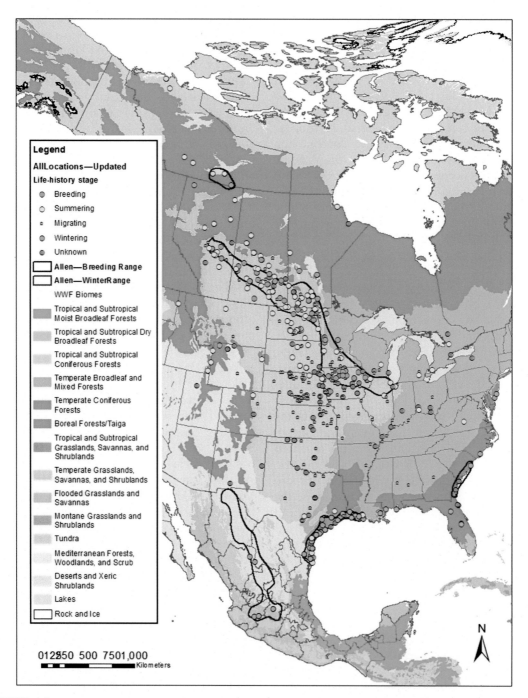

FIGURE 3.2 Distribution of historic Whooping Crane locations, 1722–1941, by life-history stage, overlain on WWF biomes. Records from Allen (1952) and additional records (this chapter). Multiple records may be represented by a single symbol. Scale = 1:28,000,000.

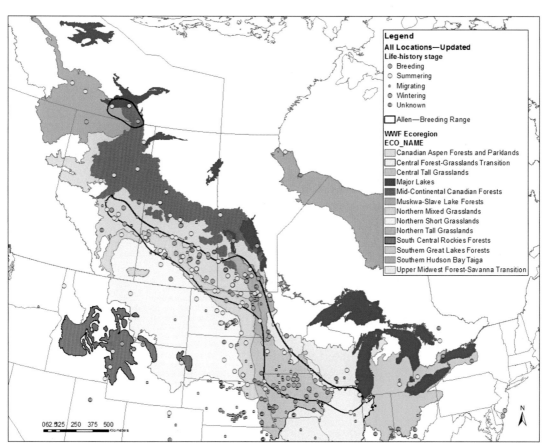

FIGURE 3.3 Locations of historic Whooping Crane records, 1722–1941, by life-history stage, in northern breeding and summering areas, overlaid on WWF ecoregions. The core breeding range was delineated by Allen (1952). Records from Allen (1952) and additional records (this chapter). Only those ecoregions having at least two breeding or summering crane records are identified in the legend. Multiple records may be represented by a single symbol. Scale = 1:15,000,000.

from the Arctic tundra and northern forest ecoregions to southern forests and grasslands. Like Allen (1952), we believe many summer records were wandering subadult or nonbreeding birds; such wandering has been documented by satellite tracking of subadult and nonbreeding cranes in the Eastern Migratory Population (Mueller et al., Chapter 11, this volume; Urbanek et al., 2014). Unlike Allen, however, we suspect some of the summering records and records of unknown life-history stage were breeding birds because we now know that, in any one year, cranes that fail at nesting still retain their nest

territory (Kuyt, 1993; Novakowski, 1966; Van Schmidt et al., 2014) and can appear in late summer as adults in small groups or as pairs.

Wintering locations – The Western Gulf Coast Grasslands also was the most important wintering ecoregion, encompassing 72.5% of the wintering records, with records dispersed along much of its length (Fig. 3.4). As noted by Allen (1952, p. 28), 20 specimens were taken in the area between 1889 and 1904. Today, the AWBP Whooping Cranes still winter in this ecoregion. Allen identified another important wintering area (12.5% of winter records)

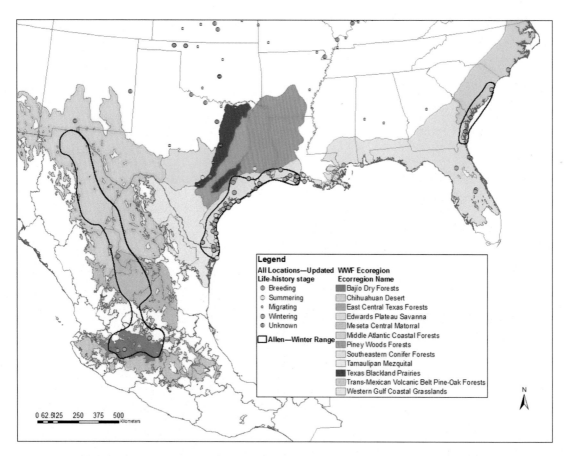

FIGURE 3.4 Distribution of historic Whooping Crane locations, 1722–1941, in southern breeding and wintering areas, with ranges delineated by Allen, overlain on WWF ecoregions. Records from Allen (1952) and additional records (this chapter). Only those ecoregions having at least two breeding, summering, or wintering crane records are identified in the legend. Multiple records may be represented by a single symbol. Scale = 1:18,000,000.

along the mouth of the Rio Grande River near Brownsville, Texas, located within the Tampaulipan/Mezquital ecoregion, just inland from the Western Gulf Coast Grasslands ecoregion. Whooping cranes were found wintering in the interior of Florida and along the Atlantic coast in association with the Southeastern Coniferous Forest and the Northern, Middle Atlantic, and Southeastern coastal forests. All coastal locations were associated with coastal wetlands or river deltas (e.g., Savannah River and Waccamaw River, South Carolina). Inland

locations in Florida were likely associated with wetlands embedded in prairie (e.g., Kissimmee prairie).

Allen (1952) listed 10 records for wintering cranes in the interior of Mexico; we found no additional records. None of the records supported Allen's extension of the wintering range into the northern interior highlands of Mexico. Whooping cranes were observed at Bolson del Mapimi, Chihuahua (Chihuahan Desert); on the Plains of Silao, Guanajuato (Trans-Mexican Volcanic Belt Pine-oak Forests); near

Lerdo, Durago (Mesata Central Matorral); and in the state of Jalisco (Bajio Dry Forests; the southernmost locations), all of which appear to be associated with large playa lake or riparian systems. The latter included three records somewhere near La Barca, possibly along the Rio Lerma or Rio Santiago, and one on marshes of Lake Chapala, the largest freshwater lake in Mexico (1,112 km^2).

DISCUSSION

Allen's (1952) description of the historic breeding and wintering distribution of Whooping Cranes appears to be largely accurate. The additional 70 locations, however, combined with GIS-based information about landscape and wetland features within ecoregions, provide new insights about the landscapes and habitats used by the species before some of the major environmental alterations of habitats occurred in the mid to late 1900s. We discuss the life-history strategies of Whooping Cranes as understood under contemporary conditions, then reexamine and redefine Allen's description of niche based on habitat features of the ecoregions and ecological processes. Here we define the species' ecological niche in terms of habitat – that is, as the noninteractive (scenopoetic) variables and environmental conditions on broad scales, relevant to understanding coarse-scale ecological and geographic properties of the species (Soberón, 2007, p. 1117), such as climate, terrain, and land cover features.

Habitat Attributes Important to Life Strategies

Historic and current patterns of Whooping Crane distribution demonstrate that breeding and wintering populations use a diversity of wetland and wet grassland habitats. The dependence of Whooping Cranes on fresh and brackish wetlands across life-history stages is well recognized (Canadian Wildlife Service and U.S. Fish and Wildlife Service, 2005; Urbanek and Lewis, 2015; Walkinshaw, 1973). The species' large body size, energetic demands, and flightlessness during wing molt (Urbanek and Lewis, 2015) necessitate a reliance on productive wetland habitat that maintains permanent water for safety from predators and ensures sufficient food resources for themselves and, during breeding, for their chicks (Wellington et al., 1996, p. 82). Foods consumed by Whooping Cranes during breeding and wintering periods tend toward higher trophic levels, such as crabs, snakes, minnows, and small mammals; energy-rich tubers, berries, corn, wheat, and acorns are also seasonally important (Allen, 1952; Pugesek et al., 2013; Urbanek and Lewis, 2015; International Crane Foundation, unpublished data). Abundance of high-energy foods on the wintering grounds is critical to the survival of over-wintering Whooping Cranes (Chavez-Ramirez, 1996; Hunt and Slack, 1989; Pugesek et al., 2013), as well as for the acquisition of sufficient fat reserves to provision energy demands to support their 4,000-km spring migration (Chavez-Ramirez, 1996; Urbanek and Lewis, 2015) and egg formation upon arrival. Territorial fidelity in breeding and winter periods confers benefits through greater familiarity with habitat quality (e.g., nest sites, food resources) and risks of predation or disturbance (Piper, 2011; Switzer, 1993).

Periodic drawdown is important to sustaining wetland productivity (Mitsch and Gosselink, 2000) and hence important to sustain long-term habitat quality for cranes. However, drawdown conditions may limit both food resources and safe feeding areas and contribute to poor chick survival (Kuyt, 1981a; Spalding et al., 2009). Drawdown conditions may be similarly detrimental to adult Whooping Cranes in summer, as adults become flightless during a synchronous remigial molt every 2–3 years (Folk et al., 2008; Lacy and McElwee, 2014). Wetlands or wetland complexes would therefore need to be large enough to maintain some water in a substantial portion of the cranes' territory so as to provide safe areas for foraging and molting habitat (i.e., shallow open water and flooded emergent vegetation) in summer, while also having the ecological processes in place to support high productivity (Mitsch and Gosselink, 2000). Areas where most wetlands dry out entirely annually or in most years would not provide suitable habitat for breeding or flightless Whooping Cranes. For deltaic or coastal wetlands, productivity is sustained by inflow of nutrients via flooding or tidal influxes but may similarly be reduced during low-water conditions (Mitsch and Gosselink, 2000).

Given these life history features, areas with extensive productive wetlands would be most likely to support crane territories on the breeding grounds in most years, provide safe feeding and molting sites in summer (molt cycles are variable, and thus unpredictable, among individual birds; Folk et al., 2008), and provide adequate food to support cranes during winter. Further, shallow flooded habitat provides for isolation of nests, chicks, flightless cranes, and roost sites (all seasons) from mammalian predators. Habitats that are visually open (i.e., little vertical vegetation obstructing view) enhance cranes' ability to detect threats and have long been recognized as an important feature of areas used by cranes throughout their life cycle (Armbruster, 1990; Howe, 1989; Timoney, 1999).

Redefining Allen's Niche Description

Allen (1952) provided the first detailed assessment of habitats used by whooping cranes. He frequently referred to their "biotic niche" – habitat features that were similar across breeding and wintering locations. Allen (1952, p. 48) described the "preferred niche, especially for nesting, as a flat or slightly rolling, open area interspersed with bulrush, cattail, and

sedge marshes and swales, covered with standing water and having the biotic characteristics found in the willow communities of the Aspen Parkland. There must be a great abundance of small animal life, including basic invertebrate forms. The entire area must be several hundred (or even several thousand) acres in size and completely isolated from human disturbance of any sort." These features have been used to inform selection of new sites for reintroduction efforts (Canadian Wildlife Service and U.S. Fish and Wildlife Service, 2005, p. 150; Cannon, 1999). Allen's general features are consistent with findings of contemporary habitat-use studies at Wood Buffalo National Park (WBNP) (Kuyt, 1981b, 1993; Olson and Olson, 2003; Timoney et al., 1997; Timoney, 1999), ANWR (Labuda and Butts, 1979), and southwestern Lousiana (Kang and King, 2014).

Based on our evaluation of the habitat features of the historic locations and contemporary knowledge of wetland ecology and Whooping Crane life history (see Barzen, Chapter 15, this volume), we identify four features that are important for nesting and wintering habitats. First, habitats used by Whooping Cranes have gentle to rolling topography with an interspersion of wetland and low meadow or prairie habitats and relatively sparse cover of trees and shrubs. Second, Whooping Cranes were commonly found in areas having high densities of wetlands or wetland complexes that provide open, shallowly flooded habitat. Primary examples of such conditions are found in the PPR, portions of the southern boreal-taiga region, inland river deltas, large playa lakes in the high plains of Mexico, the Chenier Plain of coastal Louisiana and Texas, and estuarine or deltaic systems along the Gulf Coast. Third, wetland complexes collectively provide reliable habitat conditions important for nesting, brood rearing, molt, or feeding in most years, even though hydrological variation within an individual wetland might be large. Fourth, historical locations were usually in areas that are considered highly productive due to fertile soils, hydrological pulsing, periodic inflow of nutrients, or other periodic perturbations such as fire or storms (Euliss et al., 1999; Mitsch and Gosselink, 2000; Prince, 1997). Allen (1952) touched on many of the features we examine in more detail below but lacked the depth of knowledge about ecological processes important to sustaining these conditions. In the following sections, we discuss the significance of these four features in the three primary regions used by breeding and wintering Whooping Cranes: PPR and northern grasslands, boreal forest and taiga, and southern region.

PPR and northern grasslands – As recognized by Allen (1952, pp. 24–25), the greatest number of historic nesting records occurred in the Canadian Aspen Parkland, Central Tall Grasslands, and Northern Tall Grasslands, largely corresponding to the PPR (Fig. 3.3). Compared to boreal regions, frequent nest records from the PPR and northern grasslands may be explained in part by the greater opportunity for detection by explorers and early settlers in the late 1800s and the growth of interest in ornithology in that era. However, the density of records also indicates these regions were historically important for breeding and summering Whooping Cranes.

The PPR is broadly delineated by its glacial history, with high wetland densities, high soil fertility, warm summers, and seasonal and interannual hydrological pulsing (Acton et al., 1998; Euliss et al., 1999; Mitsch and Gosselink, 2000; van der Valk, 1998). Precipitation in the PPR increases to the east and north (Millett et al., 2009); therefore, the more northern and eastern portions of the ecoregions tend to have more reliable water conditions across years and are less prone to extended periods of drought than the more westerly PPR ecoregions (Acton et al., 1998; Smith et al., 1964). This corresponds to Whooping Crane locations in the PPR, which were largely in the Aspen Parkland and the Northern Tall Grasslands (Table 3.2, Fig. 3.3). A cluster of nesting records in northern Iowa,

which Allen considered the historic center of breeding, corresponds to the southernmost extent of the PPR. Before European settlement, wetlands in that portion of Iowa were predominantly temporarily flooded to saturated wetlands (84% of wetland area; Miller et al., 2012), embedded in tallgrass prairie. By the 1870s, much of Iowa was settled, and most wetlands were drained by the 1930s (Prince, 1997; Schrader, 1955). Severe drought during the 1930s, combined with hunting and other persecution, appeared to be key factors in the extirpation of Whooping Cranes from the Saskatchewan prairies (Hjertaas, 1994) and likely elsewhere in the PPR.

Crane observations in the Northern Tall Grasslands largely fall within the area of rich, deep lake deposits of glacial Lake Agassiz, which once extended along the Red River Valley from northern South Dakota, across the northern third of Minnesota, and extending into Manitoba and western Ontario. This flat, open prairie region once had a fractured system of low intersecting ridges and shallow wetlands (Augustadt, 1955). The deep rich soils, relatively high annual precipitation, hydrological pulses of wet/dry years, and wildfires would have supported highly productive wetlands and grasslands, and periodic extensive flooding and wildfires would have severely limited woody growth. Most wetlands in this region were lost to extensive drainage efforts in the 1920s and 1930s (Prince, 1997).

Extensive wet prairies and wetlands also occurred in northern and central Illinois and northeastern Indiana (Prince, 1997; Robertson et al., 1997), which fall within the Central Forest-Grassland Transition ecoregion. Whooping cranes were known to frequent that area, largely as migrants (Butler, 1898; Woodruff, 1907). The Upper Midwest Forest-Savanna Transition supported some breeding and summering cranes, largely in locations having extensive lake, wetland, and bog complexes. The relatively few records suggest these two ecoregions were of

TABLE 3.2 Distribution of Historic Whooping Crane Breeding and Summering Locations within the PPR by WWF Ecoregion. Breeding Records Were Those That Had Observations or Specimens of a Nest, Eggs, or Flightless Chick; Other Observations between Mid-May and 30 August Were Categorized as Summering. Observation Records in September–Mid-May or Late March–Mid-May Were Classified as Migrating and Those of December–Mid-March as Wintering. Records Lacking a Date Were Classified as Unknown.

	Total		Breeding		Summering		Migrating		Wintering		Unknown	
Ecoregion	No.	%	No.	%	No.	%	No.	%	No.	%	No.	%
Canadian Aspen forests and parklands	88	32.6	12	24.5	39	44.3	17	17.3	0	0	20	57.1
Northern mixed grasslands	58	21.5	8	16.3	22	25.0	26	26.3	0	0	2	5.7
Central tall grasslands	53	19.6	13	26.5	4	4.5	31	31.6	0	0	5	14.3
Northern tall grasslands	40	14.8	10	20.4	7	7.9	19	19.4	0	0	4	11.4
Northern short grasslands	29	10.7	6	12.4	15	17.0	5	5.1	0	0	3	8.6
Upper Midwest Forest-Savanna Transition	2	0.7	0	0	1	1.1	0	0	0	0	1	2.9
Total	*270*	*100.0*	*49*	*18.1*	*88*	*32.6*	*98*	*36.3*	*0*	*0*	*35*	*13.0*

less importance when compared to Northern Tall Grasslands, and cranes in some of these far-flung locations may have been summer wanderers or birds that were unable to complete migration. Only one observation (summering) occurred at Horicon Marsh, a 12,950-ha extinct glacial lake within the Upper Midwest Forest-Savanna Transition of southeastern Wisconsin. Horicon Marsh had been manipulated by damming, drainage, and restoration during the 1800s, coinciding with much of the period when Whooping Crane records were acquired.

The westernmost cluster of Whooping Crane observations falls within the Greater Yellowstone Ecosystem, located in the juncture of Montana, Idaho, and Wyoming (Ricketts et al., 1999). This region has extensive riparian wet meadow systems and some large shallow wetlands, fed largely by spring run-off from mountain snowpack, which support diverse waterbird communities (Austin and Pyle, 2004; Cody, 1996; Debinski et al., 1999), including breeding Sandhill Cranes in the Rocky Mountain Population (Gerber et al., 2014). Allen considered the Whooping Cranes reported in southeastern Idaho and northwestern Wyoming to be summer wanderers rather than breeding birds. However, the frequency of reports of Whooping Cranes in the Greater Yellowstone Area during 1906–35 and information in Kemsies (1930) suggest there might have been a small population in the central Rocky Mountains. Moreover, the historic locations of Whooping Cranes are similar to areas used by breeding and summering Greater Sandhill Cranes (*Grus canadensis tabida*) of the Rocky Mountain Population (Drewien et al., 1999). Breeding in this region may have been constrained, however, by cool climate and short growing season. The disappearance of Whooping Cranes from the Greater Yellowstone

Area coincides with the decline of Sandhill Crane numbers as the region was settled during the early 1900s (Drewien and Bizeau, 1974), both likely the victims of shooting.

Boreal forest and taiga – Observations of Whooping Cranes in the huge expanse of the Boreal Forest/Taiga biome and its ecoregions were much scarcer. Observations are likely biased toward locations near large rivers and lakes, as these were the main travel routes of explorers and settlers who recorded biological information, and these geographic features were among the relatively few named features of the landscape. Despite imprecise locations, the features for existing records correspond to our proposed ecological niche. Broadly, the landscapes where cranes were observed have flat or rolling topography and high wetland densities – lowlands, river deltas, and shoreline wetlands on large lakes. Ten locations occurred in association with river deltas or large riparian-wetland complexes. Such areas have a high density and interspersion of wetlands, extensive shorelines, diverse wildlife, rich growth of emergent and submergent vegetation supported by river inflows, sediments, and annual spring floods, relatively warm summers, and long summer day length, which maximize the growing season for that latitude (Mitsch and Gosselink, 2000; Peters et al., 2006). Allen (1952) also often recognized the importance of marshes associated with river deltas.

Lakes used by Whooping Cranes were located in areas with extensive wetland and lake complexes. Four records were reported on two huge lakes in central Manitoba (three on Lake Winnipeg and one on Lake Winnipegosis) in the late 1800s. While vague in location, these two lakes are remnants of glacial Lake Agassiz and lie within large lowland areas of extensive shallow palustrine wetlands and peatlands.

Crane locations in and around WBNP fall at the edge of the Hay River Lowlands and Northern Alberta Uplands CEF ecoregions, in an area having a high density of small ponds interspersed with sedge (*Carex*) meadows and sparse shrubs and conifers. The area is typified by long summer day length and short growing season, but climate varies among these CEF ecoregions from cool to warm summers (ESWG, 1995). The current breeding area in WBNP is in an area with high densities of shallow wetland complexes and sparse woody cover (Novakowski, 1966; Olson and Olson, 2003; Timoney, 1999). Many wetlands currently used by breeding Whooping Cranes are associated with smaller creeks or rivers. Riparian flow, periodic drought, and fire likely support wetland productivity, although it is relatively low. During dry years, cranes tended to shift use areas closer to riparian habitat, which would provide more stable water conditions and provided foods, such as minnows (Kuyt, 1993). In this system, cranes likely need large breeding territories (averaging 4.1 km^2; Kuyt, 1993) that encompass both the more stable riparian-supported wetlands and those that dry more frequently and may be more productive.

Interestingly, the current breeding area lies just north of the Peace-Athabasca Delta, the largest freshwater inland delta in the world, recognized for high productivity, ecological diversity, and high densities of breeding waterfowl (Mitsch and Gosselink, 2000). Despite European settlements starting in the late 1700s and active research in the area over the last 50 years or more, only one Whooping Crane observation was recorded in those deltaic wetlands (Fort Chipewyan in 1898). We speculate that breeding cranes may be unable to establish territories or successfully nest in the large delta due to the magnitude or timing of flood pulses and limited availability of unflooded habitat during the early breeding season, but it doesn't explain the absence of summering birds. Traditional ecological knowledge of the aboriginal communities of the area may yield further insights into this uncertainty.

A few summering cranes were observed in the far north in the Taiga Plains of the Northwest Territories. This is an expansive area encompassing several major river deltas and other areas that have high densities of small lakes and fens, dominated by sedges and *Sphagnum* moss, interspersed with shrubby tundra (ESWG, 1995; Ricketts et al., 1999). The area has a very short growing season, long summer day length, a subarctic climate, and continuous permafrost that would support limited productivity of larger animal food resources important to Whooping Cranes.

Southern areas for wintering and nonmigratory breeding cranes – Records added to Allen's (1952) wintering range indicate a more extensive if spotty distribution along the south Atlantic and Gulf coast (e.g., North Carolina and Florida), as well as three inland sites in Florida. Most of the eastern records are coastal locations associated with river mouths (deltas) or coastal freshwater to brackish wetland systems. The majority of winter records fall within the Western Gulf Coast Grasslands ecoregion, which also sustained the only known nonmigratory breeding population at White Lake, Louisiana. This ecoregion is one of the most wetland-rich regions of the world, with high productivity sustained through tidal pulsing and nutrient-rich inflows of fresh water (Esslinger and Wilson, 2001; Mitsch and Gosselink, 2000; Smith et al., 1989). It is characterized by flat topography, high soil fertility, high average rainfall, and a long growing season. Periodic disturbance from wildfire, grazing (bison historically, domestic cattle since the early 1900s), and hurricanes, and nutrient inflow from river inflows, storm surges, and daily tidal fluctuations sustain the high productivity of the wetland and grassland systems of the region (Frost, 1995). Importantly, the inland portion of this ecoregion encompasses what was once tallgrass prairie, interspersed with numerous small wetlands and scattered oak savanna (Esslinger and Wilson, 2001), where cranes could forage and find freshwater water when food availability in coastal wetlands was low (Blankenship, 1976; Labuda and Butts, 1979). Juxtaposition of these

habitats for foraging, drinking water, and safe roost sites appears to be an important aspect of this ecoregion for wintering birds. Allen (1952, pp. 28–39) identified 10 different habitat types within this ecoregion, ranging from tallgrass prairie inland to coastal lagoons and beaches, and noted differences in use by wintering versus breeding birds (discussed later). Much of the inland portion was converted to rice production by 1900 (Gomez et al., 2003), but remaining freshwater marshes in the region sustain a high biodiversity (Esslinger and Wilson, 2001).

Eight of the 10 southern breeding records also fall within the Western Gulf Coast Grasslands. Four of the added breeding records were away from the coast in Texas, possibly in Allen's prairie swale and prairie marsh type. Five of the breeding records were in the vicinity of White Lake, an area of extensive and highly productive shallow marshes located in the Chenier Plain of western Louisiana (see King et al., Chapter 22, this volume). These marshes were described in the 1930s as 16,000 ha in area, 12–20 cm deep, and largely inaccessible until the Intracoastal Waterway was extended in 1929–30 (J. Lynch, in Drewien et al., 2001). The freshwater, maidencane (*Panicum hemitomon*) marshes such as those found at White Lake were considered the most important to breeding birds (Allen, 1952; Lynch in Drewien et al., 2001) but little used by wintering birds. White Lake today is relatively intact ecologically and is one of the largest undeveloped freshwater marshes in the southeastern United States (Gomez et al., 2003). Recent assessment of the coastal Louisiana habitats and food resources demonstrated that the freshwater marshes could support Whooping Cranes, particularly in spring and summer (Kang and King, 2014). The final one of the 10 southern breeding records was located on Boca del Rio (mouth of the Rio Grande), a once-extensive coastal delta of the Rio Grande near Bagdad Matamoros, Tamaulipas, Mexico.

The wetlands in the interior of Mexico once supported some wintering Whooping Cranes. Deltiac areas associated with large interior lakes, such as the marshes of Lake Chapala, were likely important wintering sites. River inflows would have provided nutrient inputs, periodic flooding disturbance, and reliable water conditions in most years. Other observations in the Mexican interior were in areas of semiarid to arid shrub/grasslands with small to large ephemeral lakes that were often brackish or saline (Henrickson and Johnston, 1986; Saunders and Saunders, 1981). Under good water conditions, such basins provided safe roost sites and high productivity of aquatic plant foods (Goldman, 1951). However, interannual variability in water conditions of these playa-like basins may have been too large to support consistent winter use. The Whooping Crane's wintering distribution in interior Mexico may therefore have shifted annually in response to changing water and wetland conditions and food availability, as observed for Sandhill Cranes (Drewien et al., 1996; Saunders and Saunders, 1981). Intensified land and water use have degraded many of Mexico's interior wetlands, including those at Lake Chapala (Cervantes and Abarca, 1996; Drewien et al., 1996; Limón and Lind, 1990; Perez-Arteaga et al., 2002; Saunders and Saunders, 1981).

Allen's winter distribution extended into the interior highlands of northcentral Mexico but he provided no specific records associated with that area. That delineation appears to be based on communications about Sandhill Crane migration and wintering areas described in communications to Allen from George B. Saunders, a U.S. Fish and Wildlife Service biologist who spent many years surveying waterfowl and other birds across Mexico (Allen, 1952, pp. 37–38). Saunders's descriptions of wetlands used by wintering Sandhill Cranes match those noted above for Whooping Cranes (broad, shallow lake basins, surrounded by grasslands, or extensive marshes and wet meadows in river deltas). Allen did not speculate on the source of the Whooping Cranes in interior Mexico, as he did for birds wintering in Louisiana and Texas. We speculate that a small, western group of Whooping Cranes once migrated from the Idaho-Wyoming area through Great Salt Lake, Utah (Allen, 1952, p. 17) down the Front Range of the Rocky Mountains to winter along the Rio Grande, New Mexico, and on interior Mexican wetlands, similar to the path now taken by the Rocky Mountain Population of Sandhill Cranes. Riparian and shallow lake habitats along that path (e.g., San Luis Valley, Colorado, Rio Grande in New Mexico) would have provided appropriate habitats for migrant cranes. Additional evidence suggesting a western flyway and wintering area includes the depiction of a Whooping Crane on Kiva murals in New Mexico (Hibben, 1975), observations of Whooping Cranes at Fort Thorn on the Rio Grande in southern New Mexico in 1855 (Bailey, 1928), and observations of "*G. americantus*" in New Mexico during migration in March and October (Henry, 1859). Further research into historic Spanish and Native American resources may yield additional support for this possible migration path.

Most sources used by Allen (1952) for his historic distribution map came from English, French, and American explorers and ornithologists. There are various sources that have yet to be examined that may yield new information, particularly older Spanish sources for Mexico, southwestern United States, Florida, and possibly Cuba. Tapping into the traditional ecological knowledge of aboriginal societies also may yield greater insights into distribution and the ecology of areas historically used by Whooping Cranes.

SUMMARY AND OUTLOOK

Ideally, priority for reintroduction sites should be placed within areas where Whooping Cranes historically bred and wintered. Unfortunately, this option is often not feasible due to habitat loss or degradation. Wetland and grassland habitats through much of the central portion of the Whooping Crane's historic breeding range have been dramatically altered, primarily by agriculture (Dahl and Gifford, 1996; Prince, 1997; Samson et al., 1998). Agriculture and industrial development (chiefly timber harvest and oil development; Foote and Krogman, 2006) also are expanding into the southern boreal forest (midcontinent Canadian forest ecoregion). In the boreal plain of Saskatchewan, Hobson et al. (2002) reported 73% of the forest in the historic boreal transition zone was converted to agriculture between 1966 and 1994. Human impacts on habitats in other boreal ecoregions are substantially less (Kerr and Deguise, 2004). Although northern areas unaffected by habitat change may be promising for alternative reintroduction sites, they conflict with the current recovery goal of establishing biologically separate Whooping Crane populations (Canadian Wildlife Service and U.S. Fish and Wildlife Service, 2005). Breeding Whooping Cranes from new reintroduction sites in northern areas would likely overlap with the current AWBP outside the breeding season. Further, many currently intact habitats are at risk on the main wintering grounds. Coastal wetlands continued to be threatened by development, pollution, invasive species, and erosion (Dahl and Gifford, 1996; Güneralp et al., 2013; Montagna et al., 2011).

These challenges suggest that the most promising opportunities for reintroduction or restoration may lie in the periphery of the historic range, where suitable habitats are more intact and less likely to be impacted in the near future (Channell and Lomolino, 2000). Iowa has lost 89% of its original wetlands, but immediately to the northeast, Wisconsin – most of which is in the Upper Midwest Forest-Savanna Transition ecoregion – has retained just over 50% of its wetlands (Dahl, 1990). Those key habitat features that should be present in any remaining

core or peripheral area are (1) subtle to rolling topography with an interspersion of wetland and low meadow or prairie habitats, with relatively sparse cover of trees and shrubs; (2) high densities of shallow, open wetlands or wetland complexes; (3) hydrological regimes providing reliable conditions for nesting, brood rearing, and flightless adults; and (4) high productivity due to fertile soils, hydrological pulsing, periodic inflow of nutrients, or other periodic perturbations. Some important wetland areas within the historic range but outside of ANWR and WBNP remain relatively intact and have been protected under state, provincial, or federal ownership or wetland regulations (e.g., Quill Lakes, Saskatchewan; Hails, 1997). Both traditional and more innovative approaches (e.g., conservation agreements with private landowners and industry) may be needed to protect additional reintroduction sites that have potential. Reintroduction evaluations additionally should consider potential impacts of changing climate, in particular the potential impacts on wetland hydrology and food resources. Further research into the hydrology regimes and wetland productivity of current reintroduction sites would be valuable to deepen our understanding of the ecological niche and seasonal habitat needs of cranes.

This study provides new insights into the historic distribution and habitats used by Whooping Cranes during the 1700s to mid-1900s. We believe site-specific studies of current habitat use by Whooping Cranes would benefit from greater consideration of hydrological regime and site productivity, and the ecological processes that sustain productivity at higher trophic levels (e.g., seasonal hydrological cycles, nutrient dynamics, vulnerability to long-term drought). These historic data and insights should stimulate new discussions about habitats assessments used for selection of reintroduction locations and ultimately enhance the long-term success of recovery efforts.

Acknowledgments

We would like to recognize the late Robert P. Allen for the extensive time and effort put into collecting and organizing information in his 1952 monograph about the Whooping Crane. The Canadian and U.S. Whooping Crane Recovery Teams, and all those involved, have worked diligently to protect the Whooping Crane and the habitats it utilizes. Additional historical Whooping Crane records were collected through the assistance of George Archibald, Ann Burke, Mike Putnam, and Betsy Didrickson. Dorn Moore provided guidance in building the database. Hugh Britten and Bill Mitsch provided useful comments in the earlier development of the manuscript. Rod Drewien, Felipe Chavez-Ramirez, and the late John Taylor provided useful comments concerning the distribution map. We thank Leigh Fredrickson, Sammy King, and an anonymous reviewer for their comments on later versions of this manuscript. Any use of trade, firm, or product names is for descriptive purposes only and does not imply endorsement by the U.S. Government.

References

Abell, R., Thieme, M.L., Revenga, C., Bryer, M., Kottelat, M., Bogutskaya, N., Coad, B., Mandrak, N., Contreras, S.B., Bussing, W., Stiassny, M.L.J., Skelton, P., Allen, G.R., Unmack, P., Naseka, A., Ng, R., Sindorf, N., Robertson, J., Armijo, E., Higgins, J.V., Heibel, T.J., Wikramanayake, E., Olson, D., López, H.L., Reis, R.E., Lundberg, R.J., Sabaj, M.H.P., Petry, P., 2008. Freshwater ecoregions of the world: a new map of biographic units for freshwater biodiversity conservation. BioScience 58 (5), 403–414.

Acton, D.F., Padbury, G.A., Stushnoff, C.T., 1998. Ecozones of Saskatchewan. University of Regina Press, Saskatchewan, Canada.

Allen, R.P., 1952. The Whooping Crane. Research Report No. 3. National Audubon Society, New York.

Armbruster, M.J., 1990. Characterization of Habitat Used by Whooping Cranes During Migration. US Fish Wildl. Serv. Biol. Rep. 90 (4), U.S. Fish and Wildlife Service, Washington, DC.

Augustadt, W.W., 1955. Drainage in the Red River Valley of the North. Water: Yearbook of Agriculture 1955. U.S. Department of Agriculture, Washington, DC, pp. 569–575.

Austin, J.E., Pyle, W.H., 2004. Nesting ecology of waterbirds at Grays Lake, Idaho. West. N. Am. Nat. 64 (3), 277–292.

Bailey, F.M., 1928. Birds of New Mexico. New Mexico Department of Game and Fish. Judd and Detweiler, Inc., Washington, DC.

Barzen, J.A., 2018. Ecological implications of habitat use by reintroduced and remnant whooping crane populations (Chapter 15). In: French, Jr., J.B., Converse, S.J., Austin, J.E. (Eds.), Whooping Cranes: Biology and Conservation. Biodiversity of the World: Conservation from Genes to Landscapes. Academic Press, San Diego, CA.

Barzen, J.A., Lacy, A., Thompson, H.L., Gossens, A.P., 2018. Habitat use by the reintroduced Eastern Migratory Population of Whooping Cranes (Chapter 14). In: French, Jr., J.B., Converse, S.J., Austin, J.E. (Eds.), Whooping Cranes: Biology and Conservation. Biodiversity of the World: Conservation from Genes to Landscapes. Academic Press, San Diego, CA.

Bent, A. C., 1926. Life histories of North American marsh birds. Bull. U.S. Natl. Mus. 135, Smithsonian Institution.

Blankenship, D.R., 1976. Studies of whooping cranes on the wintering grounds. In: Lewis, J.C. (Ed.), Proceedings of the 1975 International Crane Workshop. Oklahoma State University Publishing and Printing, Stillwater, pp. 197–207.

Brown, J.H., 1984. On the relationship between abundance and distribution of species. Am. Nat. 124 (2), 255–279.

Brown, J.H., Mehlman, D.W., Stevens, G.C., 1995. Spatial variation in abundance. Ecology 76 (7), 2028–2043.

Brown, J.H., Stevens, G.C., Kaufman, D.M., 1996. The geographic range: size, shape, boundaries, and internal structure. Annu. Rev. Ecol. Syst. 27, 597–623.

Butler, A. W., 1898. The Birds of Indiana. Twenty-second annual report, Indiana Department of Geology and Natural Resources, 575–1187.

Canadian Wildlife Service and U.S. Fish and Wildlife Service, 2005. International Recovery Plan for the Whooping Crane. Recovery of Nationally Endangered Wildlife (RENEW), Ottawa, and U.S. Fish and Wildlife Service, Albuquerque, NM.

Cannon, J.R., 1999. Wisconsin whooping crane breeding site assessment. Final Report to the Canadian-United States Whooping Crane Recovery Team.

Carr, C.F., 1890. A list of the birds known to nest within the boundaries of Wisconsin, with a few notes thereon. Wis. Nat. 1 (1), 77–79.

Cervantes, M., Abarca, F., 1996. Manual Para el Manejo y la Conservacion de los Humedales. Wetlands International, Guaymas.

Channell, R., Lomolino, M.V., 2000. Dynamic biogeography and conservation of endangered species. Nature 403 (6765), 84–86.

Chavez-Ramirez, F. H., 1996. Food availability, foraging ecology and energetics of whooping cranes wintering in Texas. PhD dissertation, Texas A&M University, College Station.

Cody, M.L., 1996. Bird communities in the central Rocky Mountains. In: Cody, M.L., Smallwood, J.A. (Eds.), Long-term Studies of Vertebrate Communities. Academic Press, San Diego, CA, pp. 291–342.

Dahl, T.E., 1990. Wetlands losses in the United States, 1780s to 1980s. Report to the Congress. National Wetlands Inventory, St. Petersburg, FL.

Dahl, T.E., Gifford, G.J., 1996. History of wetlands in the conterminous United States. In: Fretwell, J.D., Williams, J.S., Redman, P.J. (Compilers), National Water Summary on Wetland Resources. United States Geological Survey Water-Supply Paper 2425, Reston, VA, pp. 19–26.

Debinski, D.M., Kindscher, K., Jakubauskas, M.E., 1999. A remote sensing and GIS-based model of habitats and biodiversity in the Greater Yellowstone Ecosystem. Int. J. Remote Sens. 20 (17), 3281–3291.

Doughty, R.W., 1989. Return of the Whooping Crane. University of Texas Press, Austin.

Drewien, R.C., Bizeau, E.G., 1974. Status and distribution of greater sandhill cranes in the Rocky Mountains. J. Wildl. Manage. 38 (4), 720–742.

Drewien, R.C., Brown, W.M., Benning, D.S., 1996. Distribution and abundance of sandhill cranes in Mexico. J. Wildl. Manage. 60 (2), 270–285.

Drewien, R.C., Brown, W.M., Varley, J.D., Lockman, D.C., 1999. Seasonal movements of sandhill cranes radio-marked in Yellowstone National Park and Jackson Hole, Wyoming. J. Wildl. Manage. 63 (1), 126–136.

Drewien, R.C., Tautin, J., Courville, M., and Gomez, G.M., 2001. Whooping cranes breeding at White Lake, Louisiana, 1939: observations by John J. Lynch, U.S. Biological Survey. Proceedings of the North American Crane Workshop 8, 24–30.

Ecological Stratification Working Group (ESWG), 1995. A national ecological framework for Canada. Centre for Land and Biological Resources Research and Environment Canada. State of the Environment Directorate, Ecozone Analysis Branch, Ottawa.

Esslinger, C.G., Wilson, B.C., 2001. North American Waterfowl Management Plan, Gulf Coast Joint Venture: Chenier Plain Initiative. North American Waterfowl Management Plan, Albuquerque, NM.

Euliss, Jr., N.H., Mushet, D.M., Wrubleski, D.A., 1999. Wetlands of the Prairie Pothole Region: invertebrate species composition, ecology, and management. In: Batzer, D.P., Rader, R.B., Wissinger, S.A. (Eds.), Invertebrates in Freshwater Wetlands of North America: Ecology and Management. John Wiley & Sons, New York, pp. 471–514.

Falk, D.A., Olwell, P., 1992. Scientific and policy considerations in restoration and reintroduction of endangered species. Rhodora 94 (879), 287–315.

Folk, M., Nesbitt, S.A., Parker, J.M., Spalding, M.G., Baynes, S.B., Candelora, K.L., 2008. Feather molt of non-migratory Whooping Cranes in Florida. Proceedings of the North American Crane Workshop 10, 128–132.

Foote, L., Krogman, N., 2006. Wetlands in Canada's western boreal forest: agents of change. Forest. Chron. 82 (6), 825–833.

Frost, C.C., 1995. Presettlement fire regimes in southeastern marshes, peatlands, and swamps. In: Cerulean, S.I., Engstrom, R.T., (Eds), Fire in Wetlands: A Management Perspective. Proceedings of the 19th Tall Timbers Fire Ecology Conference. Tall Timbers Research Station, Tallahassee, FL, pp. 39–60.

Gerber, B.D., Dwyer, J.F., Nesbitt, S.A., Drewien, R.C., Littlefield, C.D., Tacha, T.C., Vohs, P.A., 2014. Sandhill Crane (*Antigone canadensis*). In: Rodewald, P.G. (Ed.), The Birds of North America. Cornell Lab of Ornithology, Ithaca, NY. Available from: https://birdsna.org/Species-Account/bna/species/sancra/introduction [5 August 2015].

Goldman, E.A., 1951. Biological investigations in Mexico. Smith. Misc. Coll. 115.

Gomez, G.M., Drewien, R.C., Courville, M.L., 2003. Historical notes on Whooping Cranes at White Lake, Louisiana: The John J. Lynch interview, 1947–1948. Proceedings of the North American Crane Workshop 9, 111–116.

Griffith, B., Scott, J.M., Carpenter, J.W., Reed, C., 1989. Translocation as a species conservation tool: status and strategy. Science 245 (4917), 477–480.

Güneralp, B., Güneralp, I., Castillo, C.R., Filippi, A.M., 2013. Land change in the Mission-Aransas coastal region, Texas: implications for coastal vulnerability and protected areas. Sustainability 5 (10), 4247–4267.

Hails, A.J. (Ed.), 1997. Wetlands, Biodiversity and the Ramsar Convention: The Role of the Convention on Wetlands in the Conservation and Wise Use of Biodiversity. Ramsar Convention Bureau, Gland, Switzerland, Available from: http://www.ramsar.org/sites/default/files/documents/library/wetlands_biodiversity_and_the_ramsar_convention.pdf [3 August 2016].

Hearne, S.A., 1795. A Journey from Prince of Wales's Fort in Hudson's Bay to the Northern Ocean. Macmillan, Toronto.

Henrickson, J., Johnston, M.C., 1986. Vegetation and community types of the Chihuahuan Desert. In: Barlow, J.C., Powell, A.M., Timmermann, B. (Eds.), Second Symposium on the Resources of the Chihuahuan Desert Region. United States and Mexico, Chihuahuan Desert Research Institute and Sul Ross State University, Alpine, TX, pp. 20–39.

Henry, T.C., 1859. Catalogue of the birds of New Mexico as compiled from notes and observations made while in that territory, during a residence of six years. Proc. Acad. Nat. Sci. 11, 104–109.

Hibben, F.C., 1975. Kiva Art of the Anasazi at Pottery Mound. KC Publications, Las Vegas, NV.

Hjertaas, D.G., 1994. Summer and breeding records of the Whooping Crane in Saskatchewan. Blue Jay 52 (2), 99–115.

Hobson, K.A., Bayne, E.M., Van Wilgenburg, S.L., 2002. Large-scale conversion of forest to agriculture in the boreal plains of Saskatchewan. Conserv. Biol. 16 (6), 1530–1541.

Houston, C.S., Ball, T., Houston, M., 2003. Eighteenth-Century Naturalists of Hudson Bay. McGill-Queen's Native and Northern Series. McGill-Queen's University Press, Montreal.

Howe, M.A., 1989. Migration of Radio-Marked Whooping Cranes from the Aransas-Wood Buffalo Population: Patterns of Habitat Use Behavior and Survival. U.S. Fish and Wildlife Technical Report 21, U.S. Fish and Wildlife Service, Washington, DC.

Hunt, H.E., Slack, R.D., 1989. Winter diets of Whooping and Sandhill cranes in south Texas. J. Wildl. Manage. 53 (4), 1150–1154.

Jehl, Jr., J.R., and Smith, B.A., 1970. Birds of the Churchill Region, Manitoba. Manitoba Museum of Man and Nature Special Publication 1.

Kang, S.-R., King, S.L., 2014. Suitability of coastal marshes as Whooping Crane (*Grus americana*) foraging habitat in southwest Louisiana, USA. Waterbirds 37 (3), 254–263.

Kantrud, H.A., Krapu, G.L., Swanson, G.A., 1989. Prairie Basin Wetlands of North Dakota: A Community Profile. US Fish Wildl. Serv. Biol. Rep. 85 (7.28).

Kemsies, E., 1930. Birds of the Yellowstone National Park, with some recent additions. Wilson Bull. 42 (3), 198–210.

Kerr, J.T., Deguise, I., 2004. Habitat loss and the limits to endangered species recovery. Ecol. Lett. 7 (12), 1163–1169.

King, S.L., Selman, W., Vasseur, P., Zimorski, S.E., 2018. Louisiana non-migratory Whooping Crane reintroduction (Chapter 22). In: French, Jr., J.B., Converse, S.J., Austin, J.E. (Eds.), Whooping Cranes: Biology and Conservation. Biodiversity of the World: Conservation from Genes to Landscapes. Academic Press, San Diego, CA.

Kuemmerle, T., Hickler, T., Olofsson, J., Schurgers, G., Radeloff, V.C., 2012. Reconstructing range dynamics and range fragmentation of European bison for the last 8000 years. Divers. Distrib. 18 (1), 47–59.

Kuemmerle, T., Radeloff, V.C., Perzanowski, K., Kozlo, P., Sipko, T., Khoyetskyy, P., Bashta, A.-T., Chikurova, E., Parnikoza, I., Baskin, L., Angelstam, P., Waller, D.M., 2011. Predicting potential European bison habitat across its former range. Ecol. Appl. 21 (3), 830–843.

Kuyt, E., 1981a. Clutch size, hatching success, and survival of Whooping Crane chicks, Wood Buffalo National Park, Canada. In: Lewis, J.C., Masatomi, H. (Eds.), Crane Research Around the World. International Crane Foundation, Baraboo, WI, pp. 126–129.

Kuyt, E., 1981b. Population status, nest site fidelity, and breeding habitat of whooping cranes. In: Lewis, J.C., Masatomi, H. (Eds.), Crane Research Around the World. International Crane Foundation, Baraboo, WI, pp. 119–125.

Kuyt, E., 1993. Whooping crane, *Grus americana*, home range and breeding range expansion in Wood Buffalo National Park 1970–1991. Can. Field Nat. 107 (1), 1–12.

Labuda, S.E., Butts, K.O., 1979. Habitat use by wintering whooping cranes on the Aransas National Wildlife Refuge. In: Lewis, J.C. (Ed.), Proceedings of the 1978 Crane Workshop. Colorado State University and National Audubon Society, Fort Morgan, pp. 151–157.

Lacy, A., McElwee, D., 2014. Observations of molt in reintroduced whooping cranes. Proceedings of the North American Crane Workshop 12, 75.

Limón, J.G., Lind, O.T., 1990. The management of Lake Chapala (Mexico): considerations after significant changes in the water regime. Lake Reservoir Manage. 6 (1), 61–70.

Meine, C.D., Archibald, G.W., 1996. The Cranes: Status Survey and Conservation Action Plan. IUCN, Gland, Switzerland, and Cambridge, England.

Miller, B.A., Crumpton, W.G., van der Valk, A.G., 2012. Wetland hydrologic class change from prior to European settlement to present on the Des Moines Lobe, Iowa. Wetl. Ecol. Manage. 20 (1), 1–8.

Millett, B.V., Johnson, W.C., Guntenspergen, G., 2009. Climate trends of the North American Prairie Pothole Region 1906–2000. Clim. Change 93 (1), 243–267.

Mitsch, W.J., Gosselink, J.G., 2000. Wetlands, third ed. John Wiley & Sons, New York.

Montagna, P.A., Brenner, J., Gibeaut, J., Morehead, S., 2011. Coastal impacts. In: Schmandt, J., North, G.R., Clarkson, J. (Eds.), The Impact of Global Warming on Texas, second ed. University of Texas Press, Austin, pp. 96–123.

Mueller, T., Teitelbaum, C.S., Fagan, W.F., Converse, S.J., 2018. Movement ecology of reintroduced migratory Whooping Cranes (Chapter 11). In: French, Jr., J.B., Converse, S.J., Austin, J.E. (Eds.), Whooping Cranes: Biology and Conservation. Biodiversity of the World: Conservation from Genes to Landscapes. Academic Press, San Diego, CA.

Nelson, E.W., 1876. Birds of northeastern Illinois. Bull. Essex Inst. 8 (9–12), 90–155.

Nesbitt, S.A., 1981. The past, present, and future of whooping cranes in Florida. In: Lewis, J.C. (Ed.), Proceedings of the 1981 Crane Workshop. National Audubon Society, Tavernier, FL, pp. 151–154.

Novakowski, N.S., 1966. Whooping crane population dynamics on the nesting grounds, Wood Buffalo National Park, Northwest Territories, Canada. Canadian Wildlife Service Report Series 1, Ottawa.

Oberholser, H.C., 1938. The bird life of Louisiana. Bulletin 28, Department of Conservation, State of Louisiana, Baton Rouge.

Oberholser, H.C., 1974. The Birds of Texas. University of Texas Press, Austin.

Olson and Olson, 2003. Final report: whooping crane potential habitat mapping project. Unpublished report presented to Parks Canada and Environment Canada by Olson and Olson Planning and Design Consultants.

Perez-Arteaga, A., Gaston, K.J., Kershaw, M., 2002. Population trends and priority conservation sites for Mexican duck *Anas diazi*. Bird Conserv. Int. 12 (1), 35–52.

Peters, D.L., Prowse, T.D., Pietroniro, A., Leconte, R., 2006. Flood hydrology of the Peace-Athabasca Delta, northern Canada. Hydrol. Process. 20 (19), 4073–4096.

Piper, W.H., 2011. Making habitat selection more "familiar": a review. Behav. Ecol. Sociobiol. 65 (7), 1329–1351.

Prince, H., 1997. Wetlands of the American Midwest: A Historical Geography of Changing Attitudes. University of Chicago Press, Chicago, IL.

Pugesek, B.H., Baldwin, M.J., Stehn, T., 2013. The relationship of blue crab abundance to winter mortality of whooping cranes. Wilson J. Ornithol. 125 (3), 658–661.

Ricketts, T.H., Dinierstein, E., Olson, D.M., Loucks, C.J., Eichbaum, W., Dalla-Sala, D., Kavanagh, K., Hedao, P., Hurley, P.T., Carney, K.M., Abell, R., Walters, S., 1999. Terrestrial Ecoregions of North America. Island Press, Washington, DC.

Ridgway, R., 1889. The Ornithology of Illinois: Part I. Descriptive Catalogue, vol. 1. H.W. Rokker, printer, Springfield, IL.

Robertson, K.R., Anderson, R.C., Schwartz, M.W., 1997. The tallgrass prairie mosaic. In: Schwartz, M.W. (Ed.), Conservation in Highly Fragmented Landscapes. Springer, New York, pp. 55–87.

Samson, F.B., Knopf, F.L., Ostlie, W.R., 1998. Grasslands. In: Mac, M.J., Opler, P.A., Puckett Haecker, C.E., Doran, P.D. (Eds.), Status and Trends of the Nation's Biological Resources, 2. U.S. Department of the Interior, Reston, VA, pp. 437–472.

Saunders, G.B., Saunders, D.C., 1981. Waterfowl and their wintering grounds in Mexico, 1937–64. Resource Publication 138, U.S. Fish and Wildlife Service, Washington, DC.

Schrader, T.A., 1955. Waterfowl and the potholes of the north central states. Water: Yearbook of Agriculture 1955. U.S. Department of Agriculture, Washington, DC., pp. 596–604.

Smith, A.G., Stoudt, J.H., Gollop, J.B., 1964. Prairie potholes and marshes. In: Linduska, J.P. (Ed.), Waterfowl Tomorrow. U.S. Fish and Wildlife Service, Washington, DC., pp. 39–50.

Smith, L.M., Pederson, R.L., Kaminski, R.M. (Eds.), 1989. Habitat Management for Migrating and Wintering Waterfowl in North America. Texas Tech University Press, Lubbock.

Soberón, J., 2007. Grinnellian and Eltonian niches and geographic distributions of species. Ecol. Lett. 10 (12), 1115–1123.

Spalding, M.G., Folk, M.J., Nesbitt, S.A., Folk, M.L., Kiltie, R., 2009. Environmental correlates of reproductive success for introduced resident whooping cranes in Florida. Waterbirds 32 (4), 538–547.

Switzer, P.V., 1993. Site fidelity in predictable and unpredictable habitats. Evol. Ecol. 7 (6), 533–555.

Timoney, K., 1999. The habitat of nesting whooping cranes. Biol. Conserv. 89 (2), 189–197.

Timoney, K., Zoltai, S.C., Goldsborough, L.G., 1997. Boreal diatom ponds: a rare wetland associated with nesting whooping cranes. Wetlands 17 (4), 539–551.

Urbanek, R.P., Lewis, J.C., 2015. Whooping crane (*Grus americana*). In: Rodewald, P.G. (Ed.), The Birds of North America Online. Cornell Lab of Ornithology, Ithaca, NY. Retrieved from the Birds of North America: https://birdsna.org/Species-Account/bna/species/whocra, doi:10.2173/bna.153 [24 January 2017].

Urbanek, R.P., Zimorski, S.E., Szyszkoski, E.K., Wellington, M.M., 2014. Ten-year status of the eastern migratory whooping crane reintroduction. Proceedings of the North American Crane Workshop 12, 33–42.

van der Valk, A., 1998. Northern Prairie Wetlands. Iowa State University Press, Ames.

Van Schmidt, N.D., Barzen, J.A., Engels, M.J., Lacy, A.E., 2014. Refining reintroduction of whooping cranes with habitat use and suitability analysis. J. Wildl. Manage. 78 (8), 1404–1414.

Waisanen, P.J., Bliss, N.B., 2002. Changes in population and agricultural land in conterminous United States counties, 1790 to 1997. Glob. Biogeochem. Cy. 16 (4), 1137.

Walkinshaw, L.H., 1973. Cranes of the World. Winchester Press, New York.

Wellington, M., Burke, A., Nicolich, J.M., O'Malley, K., 1996. Chick rearing. In: Ellis, D.H., Gee, G.F., Mirande, C.M. (Eds.), Cranes: Their Biology, Husbandry and Conservation. Department of the Interior, National Biological Service, Washington, DC, and International Crane Foundation, Baraboo, WI, pp. 77–95.

Woodruff, F.M., 1907. Order Paludicolae. In: Woodruff, F.M. (Ed.), The Birds of the Chicago Area. Chicago Academy of Sciences, Bull. Ill. Nat. Hist. Surv. 6, pp. 56–57.

APPENDIX

Historical Records of Whooping Crane Observations, Ordered by State/Province and Life-History Stage

Source	Location	City	County	State/province	Country	Latitude	Longitude	Year	Month	Day	Life stage	Observer
Allen (1952)		Killam		AB	Can	52.780	−111.850	1905				F.L. Farley
Allen (1952)	Jackfish Creek			AB	Can	54.420	−110.620	1917				Bobbie Herron
Allen (1952)	Reed Deer River			AB	Can	52.070	−114.150					Dr. George
Allen (1952)		Killam		AB	Can	52.780	−111.850	1904			Breed	F.L. Farley
Allen (1952)		Edmonton		AB	Can	53.550	−113.470	1906			Breed	Sidney Stansell
Allen (1952)		Edmonton		AB	Can	53.550	−113.470	1907			Breed	Sidney Stansell
Allen (1952)		Edmonton		AB	Can	53.550	−113.470	1908			Breed	Sidney Stansell
Allen (1952)	Whitford Lake			AB	Can	53.870	−112.250	1909			Breed	F.L. Farley
Allen (1952)		Edmonton		AB	Can	53.550	−113.470	1909			Breed	Sidney Stansell
Allen (1952)	Wood Buffalo National Park	Wainright		AB	Can	59.320	−118.460	1914			Breed	Park keeper
Allen (1952)	Battle River	Camrose		AB	Can	53.020	−112.830	1886			Migr	F.L. Farley
Allen (1952)	Battle River	Camrose		AB	Can	53.020	−112.830	1886			Migr	F.L. Farley
Allen (1952)	Buffalo Lake			AB	Can	52.770	−111.000	1906	Oct	11	Migr	C.H. Davis
Allen (1952)	Buffalo Lake			AB	Can	52.770	−111.000	1910	Oct		Migr	C.H. Davis
Allen (1952)		Islay		AB	Can	53.400	−110.550	1919	Oct	15	Migr	J. Dewey Soper
Allen (1952)	Birch Lake			AB	Can	53.750	−114.530	1922	Apr		Migr	C.E. Mills
Allen (1952)	Paddle River			AB	Can	54.080	−114.250	1933			Migr	J.P. Gillese
Allen (1952)	Fort Chipewyan			AB	Can	58.700	−111.130	1835	Jun	22	Summ	Richard King
Allen (1952)	Willow River			AB	Can	55.970	−113.920	1905	May	13	Summ	H.H. Jones
Allen (1952)	Fort McMurray			AB	Can	56.730	−111.380	1907	Oct	16	Summ	Seton and Preble
Allen (1952)	Stony Plain			AB	Can	53.530	−114.000	1909	May	21	Summ	Sidney Stansell
Allen (1952)	Swan River			AB	Can	55.430	−115.300	1913	May		Summ	F.L. Farley
Allen (1952)		Tofield		AB	Can	53.370	−112.670	1919			Summ	F.L. Farley

(Continued)

Historical Records of Whooping Crane Observations, Ordered by State/Province and Life-History Stage (*cont.*)

Source	Location	City	County	State/province	Country	Latitude	Longitude	Year	Month	Day	Life stage	Observer
Allen (1952)	Sullivan Lake			AB	Can	52.000	−112.000	1920	Sep	13	Summ	F.L. Farley
Allen (1952)		Camrose		AB	Can	53.020	−112.830	1923	Jun	4	Summ	F.L. Farley
Allen (1952)	Bittern Lake	Camrose		AB	Can	53.050	−113.080	1927	Sep	22	Summ	F.L. Farley
Allen (1952)	Willow River			AB	Can	55.970	−113.920				Summ	
Allen (1952)	Mulberry Creek	Prattville	Autauga	AL	US	32.379	−86.506	1899	Nov	30	Migr	Thomas Hook
Allen (1952)	Cypress Slough	Millwood	Hale	AL	US	32.661	−87.751				Migr	Unknown
Allen (1952)	Dauphin Island		Mobile	AL	US	30.249	−88.184				Wint	Captain Sprinkle
Allen (1952)	White River	Crockett's Bluff	Arkansas	AR	US	34.444	−91.220	1882	Nov	5	Migr	D. B. Wiler
Allen (1952)		Corning	Clay	AR	US	36.408	−90.580	1914	Apr	22	Migr	Spencer
Bailey and Niedrach (1965)		Colorado Springs	El Paso	CO	US	38.834	−104.821	1880			Migr	
Allen (1952)		Loveland	Larimer	CO	US	40.398	−105.074	1889	Apr	8	Migr	Smith
Allen (1952)		Loveland	Larimer	CO	US	40.398	−105.074	1890	Apr	16	Migr	Smith
Allen (1952)	Kit Carson Refuge		Cheyenne	CO	US	38.761	−102.789	1941	Oct	13	Migr	F. F. Poley and E. R. Kalmbach
Allen (1952)			Adams	CO	US	39.867	−104.383				Migr	Hershey and Rockwell
Allen (1952)	Fort Collins		Larimer	CO	US	40.585	−105.084	1931	Jun	20	Summ	Mrs. Clara Gordon
Allen (1952)		Lerdo		DR	Mex	25.330	−103.320	1889			Wint	Dr. Fischer
Allen (1952)	Bolson del Mapimi			DR	Mex	25.817	−103.850	1894			Wint	T. S. Van Dyke
Allen (1952)	Rivers near city	Saint Augustine	Saint Johns	FL	United States	29.895	−81.323	1722				
Allen (1952)	Rivers near city	Saint Augustine	Saint Johns	FL	US	29.895	−81.323	1723				

Hallman (1965)	Saint Mark's Pond		Saint John's	FL	US	29.989	-81.398	1927				Lloyd Crichlow
Sprunt (1954)	East of Kissimmee River and Lake Okechobee		Osceola	FL	US	28.193	-81.282					Captain Dummit
Shaffer (1940)	East of Kissimmee River		Osceola	FL	US	28.193	-81.282	1936	Jan	19	Wint	
Harmon (1954)		Micanopy	Alachua	FL	US	29.504	-82.280	1911			Wint	
Bent (1926)*		Hastings	Saint John's	FL	US	29.718	-81.508				Wint	
Bent (1926)*			Lee	FL	US	26.567	-81.883				Wint	
Allen (1952)	Altamaha	Darien	McIntosh	GA	US	31.364	-81.373	1722				A white man
Allen (1952)	Savannah River	Savannah	Chatham	GA	US	32.057	-80.923	1722				
Allen (1952)	Savannah River	Savannah	Chatham	GA	US	32.057	-80.923	1723				A white man
Allen (1952)	Altamaha	Darien	McIntosh	GA	US	31.364	-81.373	1723				
Griffin (1957)		Macon	Bibb	GA	US	32.841	-83.633	1885	Nov	12	Migr	Edward Hodgkins
Allen (1952)	Plains of Silao			GU	Mex	21.010	-101.170	1869			Wint	Alfred Duges
Allen (1952)	Plains of Silao			GU	Mex	21.010	-101.170	1891			Wint	Alfred Duges
Allen (1952)			Jefferson	IA	US	41.033	-91.950	1870				
Allen (1952)			Decatur	IA	US	40.751	-93.783	1872				Parker
Allen (1952)			Jefferson	IA	US	41.033	-91.950	1873				
Allen (1952)			Tama	IA	US	41.967	-92.557	1873				Parker
Allen (1952)	Lake Mills		Winnebago	IA	US	43.420	-93.538	1873				John Krider
Allen (1952)	Lake Mills		Winnebago	IA	US	43.420	-93.538	1879				
Allen (1952)		Cedar Rapids	Linn	IA	US	42.008	-91.644	1880				Sinclair
Allen (1952)			Poweshiek	IA	US	41.683	-92.533					Kelsey

(Continued)

Historical Records of Whooping Crane Observations, Ordered by State/Province and Life-History Stage (cont.)

Source	Location	City	County	State/province	Country	Latitude	Longitude	Year	Month	Day	Life stage	Observer
Allen (1952)			Jackson	IA	US	42.167	−90.583					
Allen (1952)			Pottawat-tamie	IA	US	41.333	−95.533					Trostler
Allen (1952)			Mills	IA	US	41.033	−95.617					Trostler
Allen (1952)		Jefferson	Greene	IA	US	42.015	−94.377					Hall
Allen (1952)			Polk	IA	US	41.628	−93.583					Johnson
Allen (1952)		Humboldt	Humboldt	IA	US	42.721	−94.215					
Allen (1952)	Eagle Lake		Hancock	IA	US	43.129	−93.734					
Allen (1952)			Sac	IA	US	42.383	−95.117	1868			Breed	S. Tiberghin
Allen (1952)			Dubuque	IA	US	42.483	−90.867	1868	Apr	25	Breed	Blackburn
Allen (1952)			Black Hawk	IA	US	42.467	−92.317	1871	May	12	Breed	J. H. Bowles
Allen (1952)			Black Hawk	IA	US	42.467	−92.317	1871	May	15	Breed	John Krider
Allen (1952)	Oakland Valley		Pottawata-mie	IA	US	41.310	−95.397	1874	May	6	Breed	E. Dickinson
Allen (1952)			Cherokee	IA	US	42.733	−95.617	1877	May	8	Breed	W. Rice
Dinsmore (1994)		Pomeroy	Calhoun	IA	US	42.551	−94.684	1877			Breed	
Allen (1952)	Lake Mills		Winnebago	IA	US	43.420	−93.538	1879	May	12	Breed	John Krider
Allen (1952)			Franklin	IA	US	42.733	−93.267	1880	May	2	Breed	W. C. Reice
Allen (1952)	Clear Lake		Cerro Gordo	IA	US	43.143	−93.386	1880	May	4	Breed	S. Howland
Allen (1952)			Kossuth	IA	US	43.200	−94.217	1881	May	14	Breed	C. M. Jones
Allen (1952)			Wright	IA	US	42.733	−93.717	1881	May	8	Breed	C. M. Jones
Allen (1952)	Clear Lake		Cerro Gordo	IA	US	43.143	−93.386	1882	May	2	Breed	
Allen (1952)	Midway		Floyd	IA	US	43.005	−91.697	1883			Breed	Farmer
Allen (1952)	Eagle Lake		Hancock	IA	US	43.129	−93.734	1883	May	4	Breed	J. W. Preston

Source	Location	City	County	State	Country	Latitude	Longitude	Year	Month	Day	Status	Observer
Allen (1952)	Eagle Lake		Hancock	IA	US	43.129	−93.734	1894	May	26	Breed	R. M. Anderson
Allen (1952)	Eagle Lake		Hancock	IA	US	43.129	−93.734	1897	Jun	5	Breed	Local farmers
Allen (1952)	Eagle Lake		Hancock	IA	US	43.129	−93.734	1897	May	5	Breed	R. M. Anderson
Allen (1952)			Decatur	IA	US	40.751	−93.783	1871	Nov	10	Migr	Parker
Allen (1952)		Waterloo	Black Hawk	IA	US	42.493	−92.192	1873	Apr		Migr	
Allen (1952)	Spirit Lake		Dickinson	IA	US	43.424	−95.110	1876	Oct		Migr	D. O. of B.
Allen (1952)	Cedar Creek		Calhoun	IA	US	42.495	−94.500	1877	Oct	30	Migr	Sandpiper
Allen (1952)	Wolf Creek Slough		Woodbury	IA	US	42.383	−96.033	1884	Apr	13	Migr	
Allen (1952)	Storm Lake		Buena Vista	IA	US	42.643	−95.202	1884	Mar	26	Migr	Bond
Allen (1952)		Iowa City	Johnson	IA	US	41.661	−91.530	1884	Mar	29	Migr	Preston
Allen (1952)		Emmetsburg	Palo Alto	IA	US	43.108	−94.674	1885	Apr	1	Migr	Cline
Allen (1952)		Sioux City	Woodbury	IA	US	42.500	−96.400	1885	Apr	5	Migr	Scougal
Allen (1952)		Emmetsburg	Palo Alto	IA	US	43.108	−94.674	1885	Mar	23	Migr	Cline
Allen (1952)		Sioux City	Woodbury	IA	US	42.500	−96.400	1885	Mar	30	Migr	Scougal
Allen (1952)		La Porte City	Black Hawk	IA	US	42.315	−92.192	1885	Mar	30	Migr	Peck
Allen (1952)	Spirit Lake		Dickinson	IA	US	43.424	−95.110	1886	Apr	7	Migr	Mosker
Allen (1952)	Storm Lake		Buena Vista	IA	US	42.643	−95.202	1886	Mar	22	Migr	Bond
Allen (1952)		Holly Springs	Woodbury	IA	US	42.271	−96.078	1887	Apr		Migr	Talbot
Allen (1952)		Sioux City	Woodbury	IA	US	42.500	−96.400	1887	Apr	3	Migr	Scougal
Allen (1952)		Holly Springs	Woodbury	IA	US	42.271	−96.078	1887	Apr	8	Migr	Talbot
Allen (1952)		Sioux City	Woodbury	IA	US	42.500	−96.400	1887	Apr	8	Migr	Scougal
Allen (1952)	Storm Lake		Buena Vista	IA	US	42.643	−95.202	1887	Mar	25	Migr	Bond
Allen (1952)	Storm Lake		Buena Vista	IA	US	42.643	−95.202	1888	Apr	3	Migr	Bond
Allen (1952)	Storm Lake		Buena Vista	IA	US	42.643	−95.202	1888	Mar	25	Migr	Bond

(Continued)

Historical Records of Whooping Crane Observations, Ordered by State/Province and Life-History Stage (cont.)

Source	Location	City	County	State/ province	Country	Latitude	Longitude	Year	Month	Day	Life stage	Observer
Allen (1952)	Storm Lake		Buena Vista	IA	US	42.643	−95.202	1888	Mar	30	Migr	Bond
Allen (1952)		Garner	Hancock	IA	US	43.104	−93.602	1889	Apr	2	Migr	Byington
Allen (1952)		Hawarden	Sioux	IA	US	42.996	−96.485	1890			Migr	Johnson
Allen (1952)		Glidden	Carroll	IA	US	42.057	−94.729	1891	Apr	3	Migr	Collett
Allen (1952)		Indianola	Warren	IA	US	41.358	−93.557	1900	Apr	1	Migr	Jeffrey
Allen (1952)		Alden	Harding	IA	US	42.520	−93.376	1900	Apr	2	Migr	Bigelow
Allen (1952)		Indianola	Warren	IA	US	41.358	−93.557	1900	Apr	2	Migr	Jeffrey
Allen (1952)		Indianola	Warren	IA	US	41.358	−93.557	1900	Mar	29	Migr	Jeffrey
Allen (1952)		Indianola	Warren	IA	US	41.358	−93.557	1900	Mar	30	Migr	Jeffrey
Allen (1952)		Indianola	Warren	IA	US	41.358	−93.557	1900	Sep	23	Migr	Jeffrey
Allen (1952)	Wall Lake		Sac	IA	US	42.271	−95.093	1904	Mar	21	Migr	J. A. Spurrell
Allen (1952)		Sioux City	Woodbury	IA	US	42.500	−96.400	1906	Apr	10	Migr	Rich
Allen (1952)		Webb	Clay	IA	US	42.949	−95.012	1911	Apr	9	Migr	Ira N. Gabrielson
Dinsmore (1994)	High Lake		Emmet	IA	US	43.304	−94.715	1922	Apr		Migr	
Allen (1952)		Sioux City	Woodbury	IA	US	42.500	−96.400	1885	May	7	Summ	Scougal
Allen (1952)		Hawarden	Sioux	IA	US	42.996	−96.485	1890			Summ	Berry
Allen (1952)		Forest City	Winnebago	IA	US	43.262	−93.633	1897	Jun	5	Summ	Anderson
Allen (1952)		Forest City	Winnebago	IA	US	43.262	−93.633	1897	May	15	Summ	Anderson
Allen (1952)		Sioux City	Woodbury	IA	US	42.500	−96.400	1909	May	5	Summ	Rich
Allen (1952)	Bear River	Montpelier	Bear Lake	ID	US	42.297	−111.309	1834	Jul	8	Summ	J. K. Townsend and Thomas Nuttall
Allen (1952)		Rathdrum	Kootenai	ID	US	47.813	−116.896	1899	Apr	8	Summ	Danby
Allen (1952)*	Central Illinois marshes	Peoria	Peoria	IL	US	40.620	−89.630	1870s			Breed	Nelson 1877

Allen (1952)	Cairo	Alexander	IL	US	36.990	−89.186	1880				Fuchs
Allen (1952)		Champaign	IL	US	40.133	−88.200	1871	Mar	27	Migr	
Bogardus (1878)		Ford	IL	US	40.583	−88.251	1877	Mar		Migr	Adam H. Bogardus
Allen (1952)	Warsaw	Hancock	IL	US	40.359	−91.434	1879	Oct	23	Migr	
Allen (1952)	Mount Carmel	Wabash	IL	US	38.411	−87.761		Mar	6	Migr	
Baird et al. (1884)	Chicago	Cook	IL	US	41.850	−87.650	1858	Jun		Summ	T. Blackney
Allen (1952)	Weston	McLean	IL	US	40.747	−88.622	1881	Apr	15	Summ	
Allen (1952)	Old Apple River	Jo Daviess	IL	US	42.350	−90.183	1891	Apr		Summ	
Woodruff (1907)	Chicago	Cook	IL	US	41.850	−87.650		Aug	6	Summ	B. T. Gault
Baczkowski (1955)		Porter	IN	US	41.473	−87.061	1905			Summ	F. Baczkowski
Butler (1898)	Bloomington	Monroe	IN	US	39.165	−86.526					Charles Dury
Butler (1898)	Lower Wabash Valley	Posey	IN	US	37.798	−88.027					Dr. Stein
Coale (1912)		Lake	IN	US	41.417	−87.365					
Allen (1952)		La Porte	IN	US	41.611	−86.723	1881	Mar	28	Migr	
Butler (1898)		Porter	IN	US	41.473	−87.061	1887	Apr	25	Summ	Trouslot
Allen (1952)	La Barca		JL	Mex	20.280	−102.570	1894				Jouy
Allen (1952)	Lago de Chapala		JL	Mex	20.190	−102.950	1890			Wint	E. W. Wilson
Allen (1952)	La Barca		JL	Mex	20.280	−102.570	1903				
Allen (1952)	La Barca		JL	Mex	20.280	−102.570	1903				
Allen (1952)	Manhattan	Riley	KS	US	39.184	−96.571	1878	Jan		Wint	D. E. Lantz
Allen (1952)		Sedgwick	KS	US	37.717	−97.450	1907	Jan		Wint	H. D. Burchell and Dr. R. Mathews
Allen (1952)	Burton	Harvey	KS	US	38.024	−97.669	1912				C. B. Heinricks

(Continued)

Historical Records of Whooping Crane Observations, Ordered by State/Province and Life-History Stage (cont.)

Source	Location	City	County	State/province	Country	Latitude	Longitude	Year	Month	Day	Life stage	Observer
Allen (1952)	Saline River		Saline	KS	US	38.858	−97.506	1867	Apr	15	Migr	A. Crocker
Allen (1952)		Burlington	Coffey	KS	US	38.194	−95.743	1882	Apr	8	Migr	Blachly
Allen (1952)			Riley	KS	US	39.184	−96.571	1884	Mar	18	Migr	D. E. Lantz
Allen (1952)			Riley	KS	US	39.184	−96.571	1884	Mar	27	Migr	Kellogg
Allen (1952)		Emporia	Lyon	KS	US	38.404	−96.181	1885	Mar	18	Migr	Kellogg
Allen (1952)		Emporia	Lyon	KS	US	38.404	−96.181	1885	Mar	21	Migr	Smith
Allen (1952)		Richmond	Franklin	KS	US	38.403	−95.254	1887	Mar	15	Migr	
Allen (1952)		Richmond	Franklin	KS	US	38.403	−95.254	1887	Apr	8	Migr	
Allen (1952)		Richmond	Franklin	KS	US	38.403	−95.254	1887	Apr	28	Migr	
Allen (1952)		Richmond	Franklin	KS	US	38.403	−95.254	1887	Mar	7	Migr	
Allen (1952)		Richmond	Franklin	KS	US	38.403	−95.254	1887	Mar		Migr	
Allen (1952)			Riley	KS	US	39.184	−96.571	1890	Mar	18	Migr	D. E. Lantz
Allen (1952)			Riley	KS	US	39.184	−96.571	1891	Apr	18	Migr	D. E. Lantz
Allen (1952)			Jackson	KS	US	39.400	−95.833	1893			Migr	J. A. Bryant
Allen (1952)			Riley	KS	US	39.184	−96.571	1896	Apr	6	Migr	
Allen (1952)		Durham	Marion	KS	US	38.486	−97.228	1903	Apr		Migr	F.L. Jaques
Allen (1952)		Baldwin	Douglas	KS	US	38.775	−95.186	1904	Mar	18	Migr	Monahan
Allen (1952)		Hays	Ellis	KS	US	38.879	−99.326	1905			Migr	
Allen (1952)		Blue Rapids	Marshall	KS	US	39.682	−96.659	1906	Mar	31	Migr	P. B. Peabody
Allen (1952)			Ford	KS	US	37.700	−99.900	1906	Oct	13	Migr	W. M. McClom
Allen (1952)			Stafford	KS	US	37.962	−98.600	1907	Oct		Migr	Hal G. Everts
Allen (1952)		Onaga	Pottawatomie	KS	US	39.489	−96.170	1907	Oct	18	Migr	Crevecoeur
Allen (1952)		Onaga	Pottawatomie	KS	US	39.489	−96.170	1907	Oct	20	Migr	Crevecoeur
Allen (1952)			Stafford	KS	US	37.962	−98.600	1919			Migr	Hal G. Everts
Allen (1952)			Stafford	KS	US	37.962	−98.600	1922			Migr	Hal G. Everts
Allen (1952)			Graham	KS	US	39.350	−99.883	1929			Migr	

Reference	Location	City	County	State	Country	Latitude	Longitude	Year	Month	Day	Season	Observer
Allen (1952)		Louisville	Jefferson	KY	US	38.254	−85.759	1810	Mar	20	Migr	J. J. Audubon and A. Wilson
Mengel (1965)		Henderson	Henderson	KY	US	37.836	−87.590	1810	Oct	28	Migr	Audubon and Wilson
Mengel (1965)	Mouth of Tennessee River		Livingston	KY	US	37.065	−88.563	1820	Nov	14	Migr	Audubon
Allen (1952)		Hickman	Fulton	KY	US	36.571	−89.186	1886	Aug	20	Migr	
Allen (1952)	Grand Chenier to Johnson's Bayou		Cameron	LA	US	29.767	−92.975	1895				Duncan Crain
Allen (1952)	Hellhole		Vermilion	LA	US	29.817	−92.300	1895				Grevillen Chote
Allen (1952)	White Lake Marsh		Cameron	LA	US	29.717	−93.111	1901				O'Neil and Nunez
Allen (1952)	White Lake Marsh		Cameron	LA	US	29.717	−93.111	1934				L. J. Merovka
Allen (1952)	Avery Island		Iberia	LA	US	29.898	−91.906					E. A. McIlhenny
Allen (1952)	White Lake Marsh	Pecan Island	Vermilion	LA	US	29.642	−92.433					Ulyese Veazey
Allen (1952)	White Lake Marsh		Vermilion	LA	US	29.858	−92.403					Duncan Crain
Allen (1952)	White Lake Marsh		Vermilion	LA	US	29.858	−92.403	1883			Breed	Mrs. Gaspard
Allen (1952)	White Lake Marsh		Cameron	LA	US	29.717	−93.111	1890			Breed	O'Neil and Nunez
Allen (1952)	White Lake Marsh		Cameron	LA	US	29.717	−93.111	1900			Breed	O'Neil and Nunez
Allen (1952)	White Lake Marsh		Cameron	LA	US	29.717	−93.111	1939	May	15	Breed	J. J. Lynch
Allen (1952)	White Lake Marsh		Cameron	LA	US	29.717	−93.111				Breed	Duncan Crain
Allen (1952)	White Lake Marsh		Cameron	LA	US	29.717	−93.111	1937	Jun	5	Summ	E. A. McIlhenny
Allen (1952)	Avery Island		Iberia	LA	US	29.898	−91.906	1937	Jun	5	Summ	E. A. McIlhenny
Allen (1952)		New Orleans	Orleans	LA	US	29.986	−89.949	1821	Apr		Wint	J. J. Audubon

(Continued)

Historical Records of Whooping Crane Observations, Ordered by State/Province and Life-History Stage (*cont.*)

Source	Location	City	County	State/province	Country	Latitude	Longitude	Year	Month	Day	Life stage	Observer
Allen (1952)	Big Bayou Constance		Cameron	LA	US	29.894	−92.744	1895			Wint	Delcambre
Allen (1952)		Delcambre	Iberia	LA	US	29.948	−91.989	1895			Wint	Delcambre
Allen (1952)	Chenier la Croix	Marsh Island	Iberia	LA	US	29.566	−91.846	1895			Wint	Delcambre
Allen (1952)	Tall Grass Prairies		Calcasieu	LA	US	30.233	−93.350	1899			Wint	Bailey
Allen (1952)	Louisiana State Refuge		Vermilion	LA	US	29.817	−92.300	1916			Wint	
Allen (1952)	Big Bayou Constance		Cameron	LA	US	29.894	−92.744	1916	Feb		Wint	S. C. Arthur
Allen (1952)	Chenier au Tigre		Vermillion	LA	US	29.817	−92.300	1916	Nov		Wint	A. M. Bailey and S. C. Arthur
Allen (1952)	Chenier au Tigre		Vermillion	LA	US	29.817	−92.300	1917			Wint	A. M. Bailey
Allen (1952)	Tall Grass Prairies		Calcasieu	LA	US	30.233	−93.350	1918			Wint	Alcie Daigle
Allen (1952)	Big Bayou Constance		Cameron	LA	US	29.894	−92.744	1928			Wint	S. C. Arthur
Allen (1952)		Pecan Island	Vermilion	LA	US	29.642	−92.433	1928	Mar	11	Wint	E. W. Nelson and Ulyese Veazey
Allen (1952)	Chenier au Tigre		Vermillion	LA	US	29.817	−92.300	1929			Wint	
Allen (1952)		Pecan Island	Vermilion	LA	US	29.642	−92.433	1929	Jan	7	Wint	Ulyese Veazey
Allen (1952)	Chenier au Tigre		Vermillion	LA	US	29.817	−92.300	1930			Wint	
Allen (1952)	Chenier au Tigre		Vermillion	LA	US	29.817	−92.300	1931			Wint	
Allen (1952)	Mulberry Island		Vermilion	LA	US	29.547	−92.363	1931			Wint	Ralph Sagrera

Reference	Location	Parish	State	Country	Latitude	Longitude	Year	Month	Day	Season	Collector
Allen (1952)	Mulberry Island	Vermilion	LA	US	29.547	−92.363	1932			Wint	Ralph Sagrera
Allen (1952)	Mulberry Island	Vermilion	LA	US	29.547	−92.363	1933			Wint	Ralph Sagrera
Allen (1952)	White Lake Marsh	Cameron	LA	US	29.717	−93.111	1933	Jan		Wint	R. B. Worthen
Allen (1952)	Chenier au Tigre	Vermillion	LA	US	29.817	−92.300	1934			Wint	Trappers
Allen (1952)	Mulberry Island	Vermilion	LA	US	29.547	−92.363	1934			Wint	Ralph Sagrera
Allen (1952)	Chenier au Tigre	Vermillion	LA	US	29.817	−92.300	1934	Dec	16	Wint	Lionel LeBlanc
Allen (1952)	Mulberry Island	Vermilion	LA	US	29.547	−92.363	1935			Wint	Ralph Sagrera
Allen (1952)	White Lake Marsh	Cameron	LA	US	29.717	−93.111	1935	Dec		Wint	Jno. Gaspard and Ovid Abshire
Allen (1952)	White Lake Marsh	Cameron	LA	US	29.717	−93.111	1935	Nov	23	Wint	Ambrose Daigre
Allen (1952)	White Lake Marsh	Cameron	LA	US	29.717	−93.111	1936			Wint	George Welch
Allen (1952)	White Lake Marsh	Cameron	LA	US	29.717	−93.111	1936	Apr		Wint	Jno. Gaspard and Ovid Abshire
Allen (1952)	White Lake Marsh	Cameron	LA	US	29.717	−93.111	1937			Wint	C. E. Gillham
Allen (1952)	White Lake Marsh	Cameron	LA	US	29.717	−93.111	1937	Jan	16	Wint	J. H. Baker and Richard Gordon
Allen (1952)	White Lake Marsh	Cameron	LA	US	29.717	−93.111	1937	Jan	27	Wint	A. Simmons, R. Gordon and Schexnayder
Allen (1952)	White Lake Marsh	Cameron	LA	US	29.717	−93.111	1939	Jan	30	Wint	J. J. Lynch
Allen (1952)	White Lake Marsh	Cameron	LA	US	29.717	−93.111	1940			Wint	J. J. Lynch

(Continued)

Historical Records of Whooping Crane Observations, Ordered by State/Province and Life-History Stage (cont.)

Source	Location	City	County	State/province	Country	Latitude	Longitude	Year	Month	Day	Life stage	Observer
Allen (1952)	Grand Chenier to Johnson's Bayou		Cameron	LA	US	29.767	−92.975	1941			Wint	A. O. U. Committee
Allen (1952)	Chenier au Tigre		Vermillion	LA	US	29.817	−92.300				Wint	
Allen (1952)	Tall Grass Prairies		Calcasieu	LA	US	30.233	−93.350				Wint	E. A. McIlhenny
Houston et al. (2003)	York Factory			MB	Can	57.000	−92.310	1771				F. Jacobs
Allen (1952)		Portage la Prairie		MB	Can	49.970	−98.300	1884				F. Cresswell
Allen (1952)		Winnipeg		MB	Can	49.880	−97.150	1890				William Hine
Allen (1952)		Carberry		MB	Can	49.870	−99.370	1890				E. E. Thompson
Allen (1952)	Shoal Lake			MB	Can	50.400	−100.620	1890				E. E. Thompson
Allen (1952)	Whitewater Lake			MB	Can	50.000	−100.360	1895				
Allen (1952)	Shoal Lake			MB	Can	50.400	−100.620	1901				Ward brothers
Allen (1952)	Whitewater Lake			MB	Can	50.000	−100.360	1904				
Allen (1952)	Shoal Lake			MB	Can	50.400	−100.620	1916				Ward Brothers
Allen (1952)	Shoal Lake			MB	Can	50.400	−100.620	1917				Ward brothers
Allen (1952)	Shoal Lake			MB	Can	50.400	−100.620	1924				F. C. Ward
Allen (1952)		Winnipeg		MB	Can	49.880	−97.150	1871	Aug		Breed	
Allen (1952)	Lake Winnipeg			MB	Can	52.130	−97.270	1877			Breed	L. D. Schultz
Allen (1952)	Lake Winnipeg			MB	Can	52.130	−97.270	1883	May	18	Breed	
Allen (1952)	Oak Point			MB	Can	50.500	−98.030	1885			Breed	A. T. Small
Allen (1952)	Lake Winnipegosis			MB	Can	52.450	−99.920	1885	May	9	Breed	
Allen (1952)	Shoal Lake			MB	Can	50.400	−100.620	1886			Breed	

Source	Location	Station	Prov	Country	Lat	Long	Year	Month	Day	Status	Observer
Allen (1952)	Lake Winnipeg		MB	Can	52.130	-97.270	1891			Breed	W. Raine
Allen (1952)	Oak Lake		MB	Can	49.700	-98.400	1891	Jun	17	Breed	Walter Raine
Allen (1952)	Oak Lake		MB	Can	49.700	-98.400	1893	May	21	Breed	Walter Raine
Allen (1952)	Oak Lake		MB	Can	49.700	-98.400	1894	May	13	Breed	Walter Raine
Allen (1952)	Oak Lake		MB	Can	49.700	-98.400	1900	May	16	Breed	
Allen (1952)	Oak Lake		MB	Can	49.700	-98.400	1900	May	18	Breed	
Allen (1952)	Morris River		MB	Can	49.350	-97.350	1906	May	30	Breed	C. P. Forge
Allen (1952)	Oak Point		MB	Can	50.500	-98.030	1885	Apr	15	Migr	A. T. Small
Allen (1952)	Shell River		MB	Can	50.970	-101.400	1890	Apr	16	Migr	E. Calcutt
Allen (1952)		Aweme	MB	Can	49.720	-99.600	1895	Apr	12	Migr	Talbot Criddle
Allen (1952)		Neepawa	MB	Can	50.230	-99.470	1896	Apr	10	Migr	Wemyss
Allen (1952)		Neepawa	MB	Can	50.230	-99.470	1896	Apr	17	Migr	Wemyss
Allen (1952)		Reaburn	MB	Can	50.080	-97.870	1897	Apr	15	Migr	Wemyss
Allen (1952)		Reaburn	MB	Can	50.080	-97.870	1897	Apr	19	Migr	Wemyss
Allen (1952)		Reaburn	MB	Can	50.080	-97.870	1900	Apr	19	Migr	Wemyss
Allen (1952)		Aweme	MB	Can	49.720	-99.600	1900	Apr	6	Migr	Talbot Criddle
Allen (1952)		Reaburn	MB	Can	50.080	-97.870	1900	Apr	9	Migr	Wemyss
Allen (1952)	Morris River		MB	Can	49.350	-97.350	1900	Oct	19	Migr	C. K. Worthen
Allen (1952)		Aweme	MB	Can	49.720	-99.600	1904	Oct	12	Migr	Talbot Criddle
Allen (1952)		Churchill	MB	Can	58.750	-94.080	1748			Summ	Isham
Allen (1952)		Carberry	MB	Can	49.870	-99.370	1882	Apr	19	Summ	E. E. Thompson
Allen (1952)	Oak Point		MB	Can	50.500	-98.030	1884	May	1	Summ	A. T. Small
Allen (1952)	Shell River		MB	Can	50.970	-101.400	1885	Apr	30	Summ	E. Calcutt
Allen (1952)	Shell River		MB	Can	50.970	-101.400	1885	May	3	Summ	E. Calcutt
Allen (1952)	Westbourne		MB	Can	50.130	-98.580	1890			Summ	C. W. Nash
Allen (1952)	Westbourne		MB	Can	50.130	-98.580	1890			Summ	C. W. Nash
Allen (1952)		Reaburn	MB	Can	50.080	-97.870	1898	May	23	Summ	Wemyss
Allen (1952)		Reaburn	MB	Can	50.080	-97.870	1898	May	27	Summ	Wemyss

(Continued)

Historical Records of Whooping Crane Observations, Ordered by State/Province and Life-History Stage (cont.)

Source	Location	City	County	State/province	Country	Latitude	Longitude	Year	Month	Day	Life stage	Observer
Allen (1952)	Whitewater Lake			MB	Can	50.000	−100.360	1905	Sep	15	Summ	A. M. Laing
Allen (1952)		Margaret		MB	Can	49.400	−99.850	1909	Apr	25	Summ	Black
Allen (1952)		Margaret		MB	Can	49.400	−99.850	1909	Apr	29	Summ	Black
Allen (1952)		Margaret		MB	Can	49.400	−99.850	1909	Aug	10	Summ	Black
Allen (1952)		Margaret		MB	Can	49.400	−99.850	1909	Jul	14	Summ	Black
Allen (1952)		Margaret		MB	Can	49.400	−99.850	1909	Oct	20	Summ	Black
Allen (1952)		Margaret		MB	Can	49.400	−99.850	1911	Aug	10	Summ	Black
Allen (1952)		Margaret		MB	Can	49.400	−99.850	1911	Aug	20	Summ	Black
Allen (1952)		Margaret		MB	Can	49.400	−99.850	1911	Sep	12	Summ	Black
Allen (1952)		Margaret		MB	Can	49.400	−99.850	1912	Apr	18	Summ	Black
Allen (1952)		Margaret		MB	Can	49.400	−99.850	1912	Apr	27	Summ	Black
Allen (1952)		Margaret		MB	Can	49.400	−99.850	1912	May	10	Summ	Black
Allen (1952)		Margaret		MB	Can	49.400	−99.850	1912	May	3	Summ	Black
Allen (1952)		Margaret		MB	Can	49.400	−99.850	1913	Apr	12	Summ	Black
Allen (1952)	Rocky Lake			MB	Can	54.150	−101.500	1936			Summ	Cree Indian
Barrows (1912)		Brighton	Livingstone	MI	US	42.529	−83.780	1882	Apr		Migr	Charles Cushing
Bailey (1881)		Ann Arbor	Washtenaw	MI	US	42.271	−83.726				Migr	
Barrows (1912)		Geddesburg	Washtenaw	MI	US	42.251	−83.850	1877	Jun	8	Summ	
Allen (1952)		Fergus Falls	Otter Trail	MN	US	46.283	−96.077	1890				Washburn
Allen (1952)		Brainerd	Morrison	MN	US	46.017	−94.300	1874	Jul		Breed	R. B. Christ
Allen (1952)		Brainerd	Morrison	MN	US	46.017	−94.300	1874	Jun	10	Breed	R. B. Christ
Allen (1952)	Elbow Lake		Grant	MN	US	46.006	−96.008	1876	May	20	Breed	J. N. Sanford
Allen (1952)	Elbow Lake		Grant	MN	US	46.006	−96.008	1876	May	21	Breed	G. B. Sennet
Allen (1952)		Herman	Grant	MN	US	45.809	−96.143	1879	Jun		Breed	T. S. Roberts
Allen (1952)	Thief Lake		Marshall	MN	US	48.483	−95.883	1889	Jun	19	Breed	E. L. Brown
Allen (1952)	Southern Minnesota			MN	US	44.000	−95.000	1864	Feb	27	Migr	Hatch

Source	Location	Place	County	State	Country	Latitude	Longitude	Year	Month	Day	Type	Observer
Allen (1952)	Heron Lake		Jackson	MN	US	43.798	−95.286	1884	Mar	30	Migr	Miller
Allen (1952)		Lanesboro	Fillmore	MN	US	43.721	−91.977	1884	Mar	31	Migr	J. C. Huoslef
Allen (1952)	Heron Lake		Jackson	MN	US	43.798	−95.286	1885	Apr	3	Migr	P. B. Peabody
Allen (1952)	Heron Lake		Jackson	MN	US	43.798	−95.286	1885	Mar	31	Migr	P. B. Peabody
Allen (1952)	Heron Lake		Jackson	MN	US	43.798	−95.286	1885	Nov	13	Migr	P. B. Peabody
Allen (1952)	Heron Lake		Jackson	MN	US	43.798	−95.286	1887	Apr	12	Migr	P. B. Peabody
Allen (1952)		North Star	Nicollet	MN	US	44.291	−94.079	1888	Apr	11	Migr	Schrooten
Allen (1952)		North Star	Nicollet	MN	US	44.291	−94.079	1888	Apr	14	Migr	Schrooten
Allen (1952)		North Star	Nicollet	MN	US	44.291	−94.079	1888	Apr	15	Migr	Schrooten
Allen (1952)	Heron Lake		Jackson	MN	US	43.798	−95.286	1888	Apr	3	Migr	P. B. Peabody
Allen (1952)	Heron Lake		Jackson	MN	US	43.798	−95.286	1888	Apr	7	Migr	P. B. Peabody
Allen (1952)		Waverly	Wright	MN	US	45.076	−93.966	1889	Apr	7	Migr	Schrooten
Allen (1952)	Heron Lake		Jackson	MN	US	43.798	−95.286	1889	Mar	20	Migr	P. B. Peabody
Allen (1952)		Waverly	Wright	MN	US	45.076	−93.966	1889	Mar	28	Migr	Schrooten
Allen (1952)		Waverly	Wright	MN	US	45.076	−93.966	1889	Mar	30	Migr	Schrooten
Allen (1952)		North Star	Nicollet	MN	US	44.291	−94.079	1890	Apr	7	Migr	Schrooten
Allen (1952)		North Star	Nicollet	MN	US	44.291	−94.079	1890	Mar	28	Migr	Schrooten
Allen (1952)		North Star	Nicollet	MN	US	44.291	−94.079	1890	Mar	30	Migr	Schrooten
Allen (1952)	Heron Lake		Jackson	MN	US	43.798	−95.286	1891	Apr	17	Migr	P. B. Peabody
Allen (1952)	Heron Lake		Jackson	MN	US	43.798	−95.286	1891	Apr	6	Migr	P. B. Peabody
Allen (1952)	Heron Lake		Jackson	MN	US	43.798	−95.286	1891	Mar	25	Migr	P. B. Peabody
Allen (1952)	Warren		Marshall	MN	US	48.197	−96.773	1892	Apr	12	Migr	B. M. Slee
Allen (1952)	Warren		Marshall	MN	US	48.197	−96.773	1892	Apr	15	Migr	C. B. Miller
Allen (1952)	Dawson		Lac Qui Parle	MN	US	44.993	−96.054	1893	Apr	10	Migr	Albert Lano
Allen (1952)	14 miles south of Madison		Lac Qui Parle	MN	US	44.809	−96.194	1895	Apr	10	Migr	G. T. Oium
Allen (1952)	Heron Lake		Jackson	MN	US	43.798	−95.286	1895	Apr	19	Migr	P. B. Peabody

(Continued)

Historical Records of Whooping Crane Observations, Ordered by State/Province and Life-History Stage (cont.)

Source	Location	City	County	State/province	Country	Latitude	Longitude	Year	Month	Day	Life stage	Observer
Allen (1952)		Saint Vincent	Kittson	MN	US	48.968	−97.225	1896	Apr	15	Migr	P. B. Peabody
Allen (1952)		Saint Vincent	Kittson	MN	US	48.968	−97.225	1896	Apr	16	Migr	P. B. Peabody
Allen (1952)			Faribault	MN	US	43.667	−93.933	1898	Mar	29	Migr	A. Hewitt
Allen (1952)		Brainerd	Morrison	MN	US	46.017	−94.300	1873			Summ	Coues
Allen (1952)		Cedar Mills	Meeker	MN	US	44.943	−94.522	1878	Sep	17	Summ	William Howling
Allen (1952)		Saint Peter	Nicollet	MN	US	44.324	−93.958	1883			Summ	
Allen (1952)		Waverly	Wright	MN	US	45.076	−93.966	1889	Apr	25	Summ	Schrooten
Allen (1952)		Madison	Lac Qui Parle	MN	US	45.010	−96.196	1894	Apr	21	Summ	Albert Lano
Allen (1952)		Hallock	Kittson	MN	US	48.774	−96.946	1899	Apr	29	Summ	P. B. Peabody
Allen (1952)		Badger	Roseau	MN	US	48.783	−96.014	1917	Apr	23	Summ	A farmer
Allen (1952)	The Grand Prairie		Dunklin	MO	US	36.108	−90.097	1864				
Allen (1952)		Saint Louis	Saint Louis	MO	US	38.627	−90.198	1884	Mar	17	Migr	
Allen (1952)		Mount Carmel	Audrain	MO	US	39.236	−91.871	1885	Mar	25	Migr	Musik
Allen (1952)		Freistatt	Lawerence	MO	US	37.018	−93.898	1886	Mar	27	Migr	Hy Nehrling
Allen (1952)		Saint Louis	Saint Louis	MO	US	38.627	−90.198	1888	Mar	25	Migr	Otto Widmann
Allen (1952)		Laclede	Linn	MO	US	39.786	−93.166	1889	Mar	20	Migr	P. L. Ong
Allen (1952)			Jackson	MO	US	39.017	−94.350	1893			Migr	J. A. Bryant
Allen (1952)		Stotesbury	Vernon	MO	US	37.972	−94.563	1894	Mar	10	Migr	T. Surber
Allen (1952)		Stotesbury	Vernon	MO	US	37.972	−94.563	1894	Mar	15	Migr	T. Surber
Allen (1952)		Stotesbury	Vernon	MO	US	37.972	−94.563	1894	Mar	9	Migr	T. Surber
Allen (1952)			Jackson	MO	US	39.017	−94.350	1904	Apr	14	Migr	Charles Dankers
Allen (1952)			Jackson	MO	US	39.017	−94.350	1906			Migr	Charles Dankers
Allen (1952)		Corning	Atchison	MO	US	40.439	−95.421	1907	Mar	23	Migr	Charles Dankers
Allen (1952)		Corning	Atchison	MO	US	40.439	−95.421	1907	Mar	26	Migr	Charles Dankers
Allen (1952)			Jackson	MO	US	39.017	−94.350	1913	Mar	27	Migr	Charles Dankers

Reference	Location	County	State	Country	Latitude	Longitude	Year	Month	Day	Season	Observer
Allen (1952)	Corning	Atchison	MO	US	40.439	−95.421	1913	Mar	27	Migr	Charles Dankers
Allen (1952)	Bay Saint Louis	Hancock	MS	US	30.309	−89.330	1902	Apr	15	Wint	
Allen (1952)	Terry	Prairie	MT	US	46.794	−105.313	1904	Oct	5	Migr	Cameron
Allen (1952)	Billings	Yellowstone	MT	US	45.783	−108.500	1918	Apr	8	Migr	Thomas
Allen (1952)	Big Sandy	Chouteau	MT	US	48.179	−110.113	1903	May	1	Summ	Coubeaux
Bailey (1881)	Wilmington	New Hanover	NC	US	34.226	−77.945	1875	Apr	22	Wint	
Allen (1952)	Big Slough	Pembina	ND	US	48.854	−97.388	1899				W. H. Williams
Allen (1952)	Mandan	Morton	ND	US	46.827	−100.889	1908				
Allen (1952)	Lakota	Nelson	ND	US	48.043	−98.336	1919				W. H. Williams
Allen (1952)	Cashel	Walsh	ND	US	48.485	−97.298	1935				H. V. Williams
Allen (1952)	Calvin	Cavalier	ND	US	48.853	−98.935					
Allen (1952)	Dawson	Kidder	ND	US	46.869	−99.751					Mershon
Allen (1952)	Dawson	Kidder	ND	US	46.869	−99.751					William B. Mershon
Allen (1952)	Ina	Rolette	ND	US	48.783	−99.814	1871	Jun	3	Breed	Delos Hatch
Allen (1952)	Larimore	Grand Forks	ND	US	47.907	−97.626	1894	May	18	Breed	Eastgate
Allen (1952)	Lakota	Nelson	ND	US	48.043	−98.336	1908			Breed	Eastgate
Hibbard (1956)	Adams	Walsh	ND	US	48.497	−97.862	1909	May	18	Breed	Hibbard
Allen (1952)	Towner	McHenry	ND	US	48.346	−100.405	1915			Breed	E. T. Judd
Allen (1952)		Rolette	ND	US	48.942	−100.066	1917			Breed	
Allen (1952)	Mouth of the Little Missouri River	Dunn	ND	US	47.597	−102.323	1805	Apr	11	Migr	Lewis and Clark
Allen (1952)	Couteau des Prairies	Burke	ND	US	48.767	−102.533	1876			Migr	C. E. McChesney
Allen (1952)	Couteau des Prairies	Burke	ND	US	48.767	−102.533	1876			Migr	C. E. McChesney

(Continued)

Historical Records of Whooping Crane Observations, Ordered by State/Province and Life-History Stage (cont.)

Source	Location	City	County	State/province	Country	Latitude	Longitude	Year	Month	Day	Life stage	Observer
Allen (1952)		Couteau des Prairies	Burke	ND	US	48.767	−102.533	1877			Migr	C. E. McChesney
Allen (1952)		Couteau des Prairies	Burke	ND	US	48.767	−102.533	1877			Migr	C. E. McChesney
Allen (1952)		Couteau des Prairies	Burke	ND	US	48.767	−102.533	1878			Migr	C. E. McChesney
Allen (1952)		Couteau des Prairies	Burke	ND	US	48.767	−102.533	1878			Migr	C. E. McChesney
Allen (1952)		Menoken	Burleigh	ND	US	46.821	−100.531	1885	Apr	12	Migr	Tyler
Allen (1952)		Menoken	Burleigh	ND	US	46.821	−100.531	1885	Apr	5	Migr	Tyler
Allen (1952)		Dawson	Kidder	ND	US	46.869	−99.751	1889	Oct	11	Migr	Louis A. Yorke
Allen (1952)		Bathgate	Pembina	ND	US	48.877	−97.476	1890	Apr	18	Migr	Bowen
Allen (1952)			Towner	ND	US	48.683	−99.200	1890	Oct	5	Migr	E. T. Judd
Allen (1952)		Larimore	Grand Forks	ND	US	47.907	−97.626	1893	Apr	8	Migr	Eastgate
Allen (1952)			Rolette	ND	US	48.942	−100.066	1894			Migr	E. T. Judd
Allen (1952)		Bathgate	Pembina	ND	US	48.877	−97.476	1894	Apr	10	Migr	Bowen
Allen (1952)		Bathgate	Pembina	ND	US	48.877	−97.476	1895	Apr	11	Migr	Bowen
Allen (1952)		Glasston	Pembina	ND	US	48.706	−97.447	1899	Apr		Migr	W. H. Williams
Allen (1952)	Devils Lake		Ramsey	ND	US	48.028	−98.931	1903	Apr	11	Migr	Bowman
Allen (1952)		Inkster	Grand Forks	ND	US	48.151	−97.644	1904	Oct	18	Migr	Colling
Allen (1952)	Lake George		Kidder	ND	US	46.733	−99.489	1910			Migr	
Allen (1952)		Hamilton	Pembina	ND	US	48.809	−97.453	1912	Apr	12	Migr	D. D. Warren
Allen (1952)	Chase Lake		Stutsman	ND	US	47.009	−99.443	1913	Oct	11	Migr	H. H. McCumber
Allen (1952)		Rolette	Rolette	ND	US	48.942	−100.066	1917	Apr	13	Migr	E. T. Judd
Allen (1952)		LeRoy	Pembina	ND	US	48.923	−97.752	1919			Migr	E. T. Judd
Allen (1952)		Steele	Kidder	ND	US	46.855	−99.916	1920	Oct	18	Migr	G. Bruening

Allen (1952)	Long Lake		Kidder	ND	US	46.709	−100.174	1921	Oct	2	Migr	C. E. Boardman
Allen (1952)		Bismark	Burleigh	ND	US	46.808	−100.783	1922	Apr	8	Migr	Russell Reid
Allen (1952)		Edinburg	Walsh	ND	US	48.497	−97.862	1923			Migr	H. B. Williams
Allen (1952)	Slough west of Edinburg		Walsh	ND	US	48.500	−97.900	1923			Migr	H. B. Williams
Allen (1952)		Grafton	Walsh	ND	US	48.412	−97.410	1923	Apr		Migr	H. B. Williams
Bent (1926)	Long Lake		Kidder	ND	US	46.709	−100.174	1923	Oct	1	Migr	C. E. Boardman
Allen (1952)		Grafton	Walsh	ND	US	48.412	−97.410	1924	Oct	15	Migr	H. V. Williams
Allen (1952)		Hazelton	Emmons	ND	US	46.485	−100.279	1933	Oct		Migr	
Allen (1952)	Fort Union		Williams	ND	US	48.303	−103.433	1833	Sep	22	Summ	Maximilian
Allen (1952)		Fort Stevenson	McLean	ND	US	47.642	−101.300	1874	Jun		Summ	Elliot Coues
Allen (1952)		Pembina	Pembina	ND	US	48.966	−97.243	1879	Jul	6	Summ	W. L. Abbott
Allen (1952)		Fort Berthold	McLean	ND	US	47.642	−101.300	1881	Sep		Summ	W. J. Hoffman
Allen (1952)			Dickey	ND	US	46.209	−98.762	1883	Aug	2	Summ	E. S. Gaylord
Allen (1952)		Mandan	Morton	ND	US	46.827	−100.889	1891	Jun		Summ	
Allen (1952)		Larimore	Grand Forks	ND	US	47.907	−97.626	1893	Apr	21	Summ	Eastgate
Allen (1952)		Bathgate	Pembina	ND	US	48.877	−97.476	1893	May	6	Summ	Bowen
Allen (1952)		Bathgate	Pembina	ND	US	48.877	−97.476	1894	Apr	17	Summ	Bowen
Allen (1952)		Argusville	Cass	ND	US	47.052	−96.934	1894	Apr	28	Summ	Edwards
Allen (1952)	Devils Lake		Ramsey	ND	US	48.028	−98.931	1903	Apr	16	Summ	Bowman
Allen (1952)		Calvin	Cavalier	ND	US	48.853	−98.935	1907	Apr	20	Summ	W. R. Ross
Allen (1952)	Chase Lake		Stutsman	ND	US	47.009	−99.443	1908	May	1	Summ	H. H. McCumber
Allen (1952)		Antler	Bottineau	ND	US	48.971	−101.282	1908	May	3	Summ	Currie
Allen (1952)		Sherwood	Renville	ND	US	48.960	−101.632	1908	Sep	19	Summ	A. J. Clark
Allen (1952)		Mandan	Morton	ND	US	46.827	−100.889	1912	May	14	Summ	J. D. Allan
Hibbard (1956)		Edmone	Ramsey	ND	US	48.233	−98.733	1912	May	2	Summ	Hibbard
Allen (1952)	Chase Lake		Stutsman	ND	US	47.009	−99.443	1913	May	7	Summ	H. H. McCumber

(Continued)

Historical Records of Whooping Crane Observations, Ordered by State/Province and Life-History Stage (*cont.*)

Source	Location	City	County	State/province	Country	Latitude	Longitude	Year	Month	Day	Life stage	Observer
Allen (1952)		Bismark	Burleigh	ND	US	46.808	−100.783	1920	Sep	16	Summ	Russell Reid
Allen (1952)		Medina	Stutsman	ND	US	46.894	−99.299	1921	Jun	24	Summ	N. A. Wood
Allen (1952)	Long Lake		Kidder	ND	US	46.709	−100.174	1922	Sep	1	Summ	
Allen (1952)	Long Lake		Kidder	ND	US	46.709	−100.174	1922	Sep	15	Summ	
Allen (1952)	Long Lake		Kidder	ND	US	46.709	−100.174	1922	Sep	8	Summ	
Allen (1952)	Long Lake		Kidder	ND	US	46.709	−100.174	1923	Sep		Summ	C. E. Boardman
Allen (1952)	Long Lake		Kidder	ND	US	46.709	−100.174	1930	Aug	13	Summ	Bernie Maurek
Allen (1952)		Mercer	McLean	ND	US	47.491	−100.711	1930	Aug	13	Summ	Bernie Maurek
Allen (1952)		Steele	Kidder	ND	US	46.855	−99.916	1931	Sep	29	Summ	T. G. Pearson
Allen (1952)		Steele	Kidder	ND	US	46.855	−99.916	1932	Sep	30	Summ	Bernie Maurek and Russell Reid
Allen (1952)		Lincoln	Lancaster	NE	US	40.833	−96.686	1900				J. S. Hunter
Allen (1952)		Omaha	Douglas	NE	US	41.251	−95.931	1918				A hunter
Allen (1952)	Red Deer Lake		Cherry	NE	US	42.565	−100.498	1918				F. G. Caldwell
Allen (1952)		Brady Island	Lincoln	NE	US	41.027	−100.468	1936	May			
Allen (1952)		Grand Island	Hall	NE	US	40.926	−98.342					F. J. Brezee
Allen (1952)		Valentine	Cherry	NE	US	42.873	−100.551					J. M. Bates
Allen (1952)			Holt	NE	US	42.450	−98.767					L. Bruner
Allen (1952)			Gage	NE	US	40.267	−96.683					F. A. Colby
Allen (1952)		West Point	Cuming	NE	US	41.842	−96.708					L. Bruner
Allen (1952)		Craig	Burt	NE	US	41.786	−96.364					L. Bruner
Allen (1952)		Omaha	Douglas	NE	US	41.251	−95.931					L. Bruner and L. Skow
Allen (1952)		Omaha	Douglas	NE	US	41.251	−95.931					
Allen (1952)			Washington	NE	US	41.456	−96.026	1820	Mar	19	Migr	Thomas Say
Allen (1952)		Alda	Hall	NE	US	40.869	−98.468	1884	Mar	24	Migr	Powell

Source	City	County	State	Country	Latitude	Longitude	Year	Month	Day	Migr	Observer
Allen (1952)	Elm Creek	Buffalo	NE	US	40.719	−99.372	1884	Nov	7	Migr	
Allen (1952)	Wood River	Hall	NE	US	40.819	−98.600	1884	Oct	24	Migr	
Allen (1952)	O'Neill	Boyd	NE	US	42.900	−98.783	1887	Apr	2	Migr	Miller
Allen (1952)	O'Neill	Boyd	NE	US	42.900	−98.783	1887	Apr	9	Migr	Miller
Allen (1952)	Gibbon	Buffalo	NE	US	40.749	−98.844	1889	Apr	10	Migr	Thatcher
Allen (1952)	Falls City	Richardson	NE	US	40.061	−95.602	1890	Mar	16	Migr	Wilson
Allen (1952)	Gibbon	Buffalo	NE	US	40.749	−98.844	1890	Mar	19	Migr	Powell
Allen (1952)	Chambers	Boyd	NE	US	42.205	−98.749	1891	Apr	11	Migr	Earl
Allen (1952)	Falls City	Richardson	NE	US	40.061	−95.602	1891	Sep	19	Migr	
Allen (1952)	Valentine	Cherry	NE	US	42.873	−100.551	1893	Oct	1	Migr	J. M. Bates
Allen (1952)	Valentine	Cherry	NE	US	42.873	−100.551	1894	Apr	1	Migr	J. M. Bates
Allen (1952)	Gresham	York	NE	US	41.028	−97.402	1896	Oct	11	Migr	Dickinson
Allen (1952)	Gothenburg	Dawson	NE	US	40.926	−100.163	1897			Migr	J. Kennedy
Allen (1952)	Long Pine	Brown	NE	US	42.534	−99.699	1897	Mar	31	Migr	Bates
Allen (1952)	Holdrege	Harlan	NE	US	40.436	−99.369	1898			Migr	Loren Bunney
Allen (1952)	Holdrege	Harlan	NE	US	40.436	−99.369	1898			Migr	Loren Bunney
Allen (1952)	Grand Island	Hall	NE	US	40.926	−98.342	1899			Migr	Fred Gunther
Allen (1952)	Neligh	Antelope	NE	US	42.129	−98.029	1899	Mar	20	Migr	Merritt Cary
Allen (1952)	Badger	Gage	NE	US	40.088	−96.595	1899	Oct	12	Migr	Colt
Allen (1952)	Neligh	Antelope	NE	US	42.129	−98.029	1899	Oct	15	Migr	Merritt Cary
Allen (1952)	Lincoln	Lancaster	NE	US	40.833	−96.686	1899	Oct	27	Migr	Wolcott
Allen (1952)	Badger	Gage	NE	US	40.088	−96.595	1900	Apr	18	Migr	Colt
Allen (1952)	Badger	Gage	NE	US	40.088	−96.595	1900	Mar	23	Migr	Colt
Allen (1952)	Kearney	Buffalo	NE	US	40.695	−99.081	1900	Oct	8	Migr	Colt
Allen (1952)	Badger	Gage	NE	US	40.088	−96.595	1901	Apr	12	Migr	
Allen (1952)	Badger	Gage	NE	US	40.088	−96.595	1901	Apr	6	Migr	Colt
Allen (1952)	Badger	Gage	NE	US	40.088	−96.595	1901	May	14	Migr	Colt
Allen (1952)	Badger	Gage	NE	US	40.088	−96.595	1901	Oct	1	Migr	Colt

(Continued)

Historical Records of Whooping Crane Observations, Ordered by State/Province and Life-History Stage (cont.)

Source	Location	City	County	State/province	Country	Latitude	Longitude	Year	Month	Day	Life stage	Observer
Allen (1952)		Badger	Gage	NE	US	40.088	−96.595	1902	Apr	18	Migr	Colt
Allen (1952)		Badger	Gage	NE	US	40.088	−96.595	1902	Apr	24	Migr	Colt
Allen (1952)		Gibbon	Buffalo	NE	US	40.749	−98.844	1902	Apr	6	Migr	Ashburn
Allen (1952)		Badger	Gage	NE	US	40.088	−96.595	1902	May	11	Migr	Colt
Allen (1952)		Lincoln	Lancaster	NE	US	40.833	−96.686	1903			Migr	L. Brunter and Myron Swenk
Allen (1952)		Badger	Gage	NE	US	40.088	−96.595	1903	Apr	30	Migr	Colt
Allen (1952)		Badger	Gage	NE	US	40.088	−96.595	1903	Apr	6	Migr	Colt
Allen (1952)		Badger	Gage	NE	US	40.088	−96.595	1903	Oct	5	Migr	Colt
Allen (1952)		Gothenburg	Dawson	NE	US	40.926	−100.163	1905	Oct		Migr	William Kennedy
Allen (1952)		Grand Island	Hall	NE	US	40.926	−98.342	1907	Oct	18	Migr	Fred Gunther
Allen (1952)		Harvard	Clay	NE	US	40.617	−98.097	1908	Mar	12	Migr	George Schupan
Allen (1952)		Atkinson	Holt	NE	US	42.533	−98.978	1909	Apr	22	Migr	
Allen (1952)	Wood Lake		Cherry	NE	US	42.639	−100.238	1912	Oct	16	Migr	H. T. Clark and Mr. Quick
Allen (1952)		Grand Island	Hall	NE	US	40.926	−98.342	1912	Oct	20	Migr	Goose hunters
Allen (1952)		Greenwood	Cass	NE	US	40.963	−94.441	1913	Mar	29	Migr	J. Armstrong
Allen (1952)		Newark	Kearney	NE	US	40.641	−98.963	1914	Apr		Migr	Ed Larson
Allen (1952)		Prosser	Adams	NE	US	40.687	−98.576	1915			Migr	L. Pitcarthley
Allen (1952)		Gothenburg	Dawson	NE	US	40.926	−100.163	1915			Migr	J. Kennedy
Allen (1952)		Ogallala	Keith	NE	US	41.123	−101.719	1915	Mar	10	Migr	J. Koehr
Allen (1952)	Post Lake		Brown	NE	US	42.433	−99.950	1915	Oct	10	Migr	
Allen (1952)		Overton	Dawson	NE	US	40.739	−99.537	1916			Migr	Link Milburn
Allen (1952)		Mindern	Kearney	NE	US	40.503	−98.950	1917			Migr	
Allen (1952)		Loup City	Sherman	NE	US	41.276	−98.966	1917	Apr		Migr	H. Jenner
Allen (1952)		Kearney	Buffalo	NE	US	40.695	−99.081	1917	Oct	3	Migr	Hunter
Allen (1952)		Omaha	Douglas	NE	US	41.251	−95.931	1918	Mar	12	Migr	

Source	Location	County	State	Country	Latitude	Longitude	Year	Month	Day	Status	Observer
Allen (1952)	Kearney	Buffalo	NE	US	40.695	−99.081	1918	Mar	15	Migr	Hunter
Allen (1952)	Kearney	Buffalo	NE	US	40.695	−99.081	1919	Mar	29	Migr	Harry Connor
Allen (1952)	Kearney	Buffalo	NE	US	40.695	−99.081	1920	Apr	2	Migr	C. A. Black
Allen (1952)	Kearney	Buffalo	NE	US	40.695	−99.081	1920	Oct	15	Migr	C. A. Black
Allen (1952)	Kearney	Buffalo	NE	US	40.695	−99.081	1920	Oct	7	Migr	C. A. Black
Allen (1952)	Kearney	Buffalo	NE	US	40.695	−99.081	1921	Apr	2	Migr	C. A. Black
Allen (1952)	Red Deer Lake	Cherry	NE	US	42.565	−100.498	1921	Oct	14	Migr	Farm boy
Allen (1952)	Kearney	Buffalo	NE	US	40.695	−99.081	1921	Oct	20	Migr	C. A. Black
Allen (1952)	Gibbon	Buffalo	NE	US	40.749	−98.844	1922	Apr	14	Migr	C. A. Black
Allen (1952)	Kearney	Buffalo	NE	US	40.695	−99.081	1922	May	1	Migr	A. R. Golay
Allen (1952)	Kearney	Buffalo	NE	US	40.695	−99.081	1922	Oct	20	Migr	C. A. Black
Allen (1952)	Red Deer Lake	Cherry	NE	US	42.565	−100.498	1922	Oct	22	Migr	A. Nooka
Allen (1952)	Inland	Clay	NE	US	40.594	−98.224	1923	Sep		Migr	C. A. Black
Allen (1952)	Kearney	Buffalo	NE	US	40.695	−99.081	1924	Apr	13	Migr	C. A. Black
Allen (1952)	Amherst	Buffalo	NE	US	40.769	−99.269	1924	Apr	2	Migr	C. A. Black
Allen (1952)	Kearney	Buffalo	NE	US	40.695	−99.081	1924	Oct	16	Migr	C. A. Black
Allen (1952)	Kearney	Buffalo	NE	US	40.695	−99.081	1925	Apr	5	Migr	L. Pitcarthley, C. A. Black, Guy Smith
Allen (1952)	Odessa	Buffalo	NE	US	40.702	−99.257	1925	Oct	12	Migr	R. Swanson and William Hicks
Allen (1952)	Kearney	Buffalo	NE	US	40.695	−99.081	1925	Oct	19	Migr	George Tracy and Oscar Blevens
Allen (1952)	Odessa	Buffalo	NE	US	40.702	−99.257	1925	Oct	25	Migr	William Hicks
Allen (1952)	Overton	Dawson	NE	US	40.739	−99.537	1925	Oct	25	Migr	J. Q. Holmes
Allen (1952)	Odessa	Buffalo	NE	US	40.702	−99.257	1926	Apr	18	Migr	A. Webert
Allen (1952)	Odessa	Buffalo	NE	US	40.702	−99.257	1926	Apr	26	Migr	Roy Knapp
Allen (1952)	Kearney	Buffalo	NE	US	40.695	−99.081	1926	Apr	4	Migr	Golay, Lilga, Black, Garvin
Allen (1952)	Lowell	Kearney	NE	US	40.648	−98.847	1926	Apr	5	Migr	J. C. Chapman
Allen (1952)	Wilcox	Kearney	NE	US	40.365	−99.170	1926	Apr	8	Migr	

(Continued)

Historical Records of Whooping Crane Observations, Ordered by State/Province and Life-History Stage (*cont.*)

Source	Location	City	County	State/province	Country	Latitude	Longitude	Year	Month	Day	Life stage	Observer
Allen (1952)		Axtel	Kearney	NE	US	40.478	−99.126	1926	Apr	8	Migr	C. A. Black
Allen (1952)		Odessa	Buffalo	NE	US	40.702	−99.257	1926	Oct	20	Migr	C. A. Black
Allen (1952)		Kearney	Buffalo	NE	US	40.695	−99.081	1927	Apr	1	Migr	A. R. Golay
Allen (1952)		Newark	Kearney	NE	US	40.641	−98.963	1927	Oct	15	Migr	Farmer
Allen (1952)		Antioch	Sheridan	NE	US	42.068	−102.582	1927	Oct	9	Migr	Keller
Allen (1952)		Wilcox	Kearney	NE	US	40.365	−99.170	1928	Apr	6	Migr	A. R. Marsteller
Allen (1952)		Gibbon	Buffalo	NE	US	40.749	−98.844	1928	Apr	9	Migr	C. A. Black
Allen (1952)		Gibbon	Buffalo	NE	US	40.749	−98.844	1928	Apr	9	Migr	B. Armitage
Allen (1952)		Newark	Kearney	NE	US	40.641	−98.963	1928	Oct	24	Migr	F. R. Kingsley
Allen (1952)		Newark	Kearney	NE	US	40.641	−98.963	1928	Sep	22	Migr	L. and H. Brown
Allen (1952)		Lowell	Kearney	NE	US	40.648	−98.847	1929	Apr	10	Migr	Charles Radborn
Allen (1952)		Kearney	Buffalo	NE	US	40.695	−99.081	1929	Apr	18	Migr	A. R. Golay
Allen (1952)		Merriman	Cherry	NE	US	42.920	−101.700	1929	Mar		Migr	Phillip Mensinger
Allen (1952)		Odessa	Buffalo	NE	US	40.702	−99.257	1929	Mar	27	Migr	B. Armitage
Allen (1952)		Kearney	Buffalo	NE	US	40.695	−99.081	1929	Oct	12	Migr	C. A. Black
Allen (1952)		Wilcox	Kearney	NE	US	40.365	−99.170	1930	Apr	19	Migr	A. R. Marsteller
Allen (1952)		Kearney	Buffalo	NE	US	40.695	−99.081	1930	Apr	2		A. R. Golay, H. Ligga, C. A. Black. Mrs. Frances Garvin
Allen (1952)		Eli	Cherry	NE	US	42.944	−101.491	1930	Oct	6	Migr	A hunter
Allen (1952)	Wood River		Hall	NE	US	40.819	−98.600	1931			Migr	Mrs. Will Burmwood
Allen (1952)		Gothenburg	Dawson	NE	US	40.926	−100.163	1931			Migr	J. P. Kennedy
Allen (1952)	Elm Creek		Buffalo	NE	US	40.719	−99.372	1931	Oct	25	Migr	Anthony Roeser
Allen (1952)		Wilcox	Kearney	NE	US	40.365	−99.170	1932	Apr	5	Migr	A. R. Marsteller
Allen (1952)		Gothenburg	Dawson	NE	US	40.926	−100.163	1933			Migr	J. P. Kennedy
Allen (1952)		Overton	Dawson	NE	US	40.739	−99.537	1933	Apr	4	Migr	B. Armitage

Reference	Locality	County	State	Country	Latitude	Longitude	Year	Month	Day	Season	Observer
Allen (1952)	Elm Creek	Buffalo	NE	US	40.719	−99.372	1933	Apr	6	Migr	Herbert Richardson
Allen (1952)	Odessa	Buffalo	NE	US	40.702	−99.257	1933	Oct	1	Migr	J. Flannery
Allen (1952)	Kearney	Buffalo	NE	US	40.695	−99.081	1934	Apr	1	Migr	C. A. Black and Procter
Allen (1952)	Lowell	Kearney	NE	US	40.648	−98.847	1934	Apr	15	Migr	Chris Zwink
Allen (1952)	Wood River	Hall	NE	US	40.819	−98.600	1934	Apr	17	Migr	G. H. Phillips
Allen (1952)	Cozad	Dawson	NE	US	40.775	−99.740	1936	Apr	1	Migr	Conober and Foley
Allen (1952)	Alda	Hall	NE	US	40.869	−98.468	1936	Apr	1	Migr	Adams
Allen (1952)	Overton	Dawson	NE	US	40.739	−99.537	1936	Mar	31	Migr	C. Lanphear
Allen (1952)	Kearney	Buffalo	NE	US	40.695	−99.081	1937	Apr	10	Migr	George Eaglestrom
Allen (1952)	Gibbon	Buffalo	NE	US	40.749	−98.844	1937	Apr	2	Migr	J. Shields
Allen (1952)	Cozad	Dawson	NE	US	40.775	−99.740	1939	Oct	22	Migr	
Allen (1952)	Gothenburg	Dawson	NE	US	40.926	−100.163	1941	Apr	19	Migr	C. Swanson
Allen (1952)	Niobrara River	Knox	NE	US	42.776	−98.047	1889	Jun	24	Summ	Baker
Allen (1952)	Wood River	Hall	NE	US	40.819	−98.600	1934	May	2	Summ	S. W. Wells
Allen (1952)	Cape May	Cape May	NJ	US	38.959	−74.928	1810			Wint	Alexander Wilson
Allen (1952)	Beesley's Point	Cape May	NJ	US	39.277	−74.637	1857			Wint	William P. Turnbull
Allen (1952)	Fort Thorn	Dona Ana	NM	US	32.312	−106.778	1853				Dr. T. C. Henry
Allen (1952)	Portales	Roosevelt	NM	US	34.186	−103.334	1938				W. G. Vinzant
Allen (1952)	Fort Resolution		NT	Can	61.170	−113.670	1864			Breed	J. Lockhart
Allen (1952)	Salt River		NT	Can	60.120	−112.230				Breed	J. Lockhart
Allen (1952)	Fort Simpson		NT	Can	61.870	−121.350	1861	Jun	10	Summ	B. R. Ross
Allen (1952)	Anderson River	Fort Anderson	NT	Can	69.700	−129.000	1862			Summ	R. R. MacFarlane
Allen (1952)	Anderson River	Fort Anderson	NT	Can	69.700	−129.000	1862			Summ	R. R. MacFarlane

(Continued)

Historical Records of Whooping Crane Observations, Ordered by State/Province and Life-History Stage (cont.)

Source	Location	City	County	State/province	Country	Latitude	Longitude	Year	Month	Day	Life stage	Observer
Allen (1952)	Anderson River	Fort Anderson		NT	Can	69.700	−129.000	1863			Summ	R. R. MacFarlane
Allen (1952)	Anderson River	Fort Anderson		NT	Can	69.700	−129.000	1863			Summ	R. R. MacFarlane
Allen (1952)	Anderson River	Fort Resolution		NT	Can	61.170	−113.670	1864			Summ	J. Lockhart
Allen (1952)	Anderson River	Fort Anderson		NT	Can	69.700	−129.000	1864			Summ	R. R. MacFarlane
Allen (1952)	Anderson River	Fort Anderson		NT	Can	69.700	−129.000	1864			Summ	R. R. MacFarlane
Allen (1952)	Anderson River	Fort Anderson		NT	Can	69.700	−129.000	1865	Jun		Summ	R. R. MacFarlane
Allen (1952)	Anderson River	Fort Anderson		NT	Can	69.700	−129.000	1865	May	25	Summ	R. R. MacFarlane
Allen (1952)	Mackenzie River Delta			NT	Can	61.500	−119.500	1900			Summ	Douglas Oniak
Allen (1952)	Hay River			NT	Can	60.870	−115.730	1908	May	12	Summ	
Allen (1952)	Hay River			NT	Can	60.870	−115.730	1923			Summ	Slavey Indian
Allen (1952)	70 miles in from Eskimo Point			NT	Can	68.750	−127.750	1930			Summ	Koonook
Allen (1952)	Cayuga Lake		Cayuga	NY	US	42.947	−76.736					
Allen (1952)		Cincinnati	Hamilton	OH	US	39.310	−84.430	1876				
Allen (1952)	Scioto River		Scioto	OH	US	38.731	−83.013	1902				
Allen (1952)		Waverly	Pike	OH	US	39.127	−82.986	1902				
Allen (1952)		Granville	Licking	OH	US	40.068	−82.520	1887	Mar	28	Migr	Tight
Allen (1952)	Little Miami River	Indian Hill Station	Hamilton	OH	US	39.078	−84.433	1895	Aug		Summ	F. B. Magill
Allen (1952)			Cleveland	OK	US	35.200	−97.300	1901				C. D. Bunker
Allen (1952)			Cleveland	OK	US	35.200	−97.300	1902				C. D. Bunker
Allen (1952)			Beaver	OK	US	36.767	−100.483	1909				

Reference	Locality	County	State	Country	Latitude	Longitude	Year	Month	Day	Season	Observer
Allen (1952)		Harper	OK	US	36.767	−99.683	1909				W. E. Lewis
Allen (1952)		Beaver	OK	US	36.767	−100.483	1910				
Allen (1952)		Harper	OK	US	36.767	−99.683	1910				
Allen (1952)		Beaver	OK	US	36.767	−100.483	1911				
Allen (1952)		Harper	OK	US	36.767	−99.683	1911				
Allen (1952)		Beaver	OK	US	36.767	−100.483	1912				
Allen (1952)		Harper	OK	US	36.767	−99.683	1912				
Allen (1952)		Cleveland	OK	US	35.200	−97.300	1832	Oct	29	Migr	Latrobe
Allen (1952)		Comanche	OK	US	34.609	−98.390	1899	Oct		Migr	W. A. Mayer
Allen (1952)		Woods	OK	US	36.767	−98.800	1907	Apr	7	Migr	A hunter
Allen (1952)	Stillwater	Payne	OK	US	36.116	−97.058	1928	Oct		Migr	William Loane
Allen (1952)	Toronto		ON	Can	43.650	−79.380	1880				Handy
Allen (1952)	Emsdale		ON	Can	45.530	−79.320	1895				
Allen (1952)	Camden		ON	Can	42.580	−82.080	1871	Sep	27	Summ	Wesley Potter
Allen (1952)	Ardtrea		ON	Can	44.680	−79.420	1885	May	9	Summ	Blair
Allen (1952)	Waccamaw River	Georgetown	SC	US	33.358	−79.257	1850			Wint	
Allen (1952)		Clay	SD	US	42.917	−96.958	1885				G. S. Agersborg
Allen (1952)		Union	SD	US	42.817	−96.694	1885				G. S. Agersborg
Allen (1952)		Yankton	SD	US	43.050	−97.383	1885				G. S. Agersborg
Allen (1952)	12 miles SW of Bowdle	Walworth	SD	US	45.355	−99.736	1888				Oliver
Allen (1952)	Sioux Falls	Minnehaha	SD	US	43.556	−96.723	1905				
Allen (1952)	Mobridge	Walworth	SD	US	45.537	−100.428					
Allen (1952)		Edmunds	SD	US	45.400	−99.200	1883	Oct	19	Migr	C. K Worthen
Allen (1952)	Huron	Bradle	SD	US	44.363	−98.214	1887	Oct	13	Migr	Cheney
Allen (1952)	Grandview	Brule	SD	US	43.644	−99.297	1888	Apr	6	Migr	Blanchard
Allen (1952)	Harrison	Douglas	SD	US	43.430	−98.527	1889	Oct	18	Migr	Colt
Allen (1952)	Harrison	Douglas	SD	US	43.430	−98.527	1889	Oct	26	Migr	Colt

(Continued)

Historical Records of Whooping Crane Observations, Ordered by State/Province and Life-History Stage (cont.)

Source	Location	City	County	State/province	Country	Latitude	Longitude	Year	Month	Day	Life stage	Observer
Allen (1952)		Harrison	Douglas	SD	US	43.430	−98.527	1890	Apr	2	Migr	Colt
Allen (1952)		Harrison	Douglas	SD	US	43.430	−98.527	1890	Mar	25	Migr	Colt
Allen (1952)		Harrison	Douglas	SD	US	43.430	−98.527	1890	Oct	15	Migr	Colt
Allen (1952)		Harrison	Douglas	SD	US	43.430	−98.527	1890	Oct	8	Migr	Colt
Allen (1952)		Harrison	Douglas	SD	US	43.430	−98.527	1891	Apr	11	Migr	Colt
Allen (1952)		Harrison	Douglas	SD	US	43.430	−98.527	1891	Apr	5	Migr	Colt
Allen (1952)		Harrison	Douglas	SD	US	43.430	−98.527	1891	Nov	1	Migr	Colt
Allen (1952)		Ipswich	Edmunds	SD	US	45.444	−99.029	1893	Apr	10	Migr	
Allen (1952)		Ipswich	Edmunds	SD	US	45.444	−99.029	1893	Apr	5	Migr	
Allen (1952)		Roswell	Miner	SD	US	44.007	−97.696	1898	Oct	4	Migr	
Allen (1952)		Harrison	Douglas	SD	US	43.430	−98.527	1899	Apr	7	Migr	
Allen (1952)	12 miles SW of Bowdle	Walworth		SD	US	45.355	−99.736	1885	Sep		Summ	Oliver
Allen (1952)		Grandview	Brule	SD	US	43.644	−99.297	1888	Apr	22	Summ	Blanchard
Allen (1952)		Grandview	Brule	SD	US	43.644	−99.297	1888	May	1	Summ	Blanchard
Allen (1952)		Harrison	Douglas	SD	US	43.430	−98.527	1889	Sep	12	Summ	Colt
Allen (1952)		Harrison	Douglas	SD	US	43.430	−98.527	1889	Sep	25	Summ	Colt
Allen (1952)		Harrison	Douglas	SD	US	43.430	−98.527	1890	Apr	29	Summ	Colt
Allen (1952)		Harrison	Douglas	SD	US	43.430	−98.527	1890	Sep	12	Summ	Colt
Allen (1952)		Harrison	Douglas	SD	US	43.430	−98.527	1890	Sep	8	Summ	Colt
Allen (1952)		Harrison	Douglas	SD	US	43.430	−98.527	1891	May	15	Summ	Colt
Allen (1952)		Harrison	Douglas	SD	US	43.430	−98.527	1891	Sep	14	Summ	Colt
Allen (1952)		Harrison	Douglas	SD	US	43.430	−98.527	1891	Sep	24	Summ	Colt
Allen (1952)		Harrison	Douglas	SD	US	43.430	−98.527	1891	Sep	8	Summ	Colt
Allen (1952)			Brown	SD	US	45.553	−98.308	1893	Apr	22	Summ	
Allen (1952)		Pierre	Hughes	SD	US	44.368	−100.351	1909	May	3	Summ	H. E. Lee

Allen (1952)	Loon Creek		SK	Can	50.970	−104.380	1858				H. Y. Hind
Allen (1952)	Indian Head		SK	Can	50.330	−103.670	1858				H. Y. Hind
Allen (1952)	Touchwood Hills		SK	Can	51.570	−104.270	1858				Hind
Allen (1952)	White Sand River		SK	Can	51.570	−101.930	1884				Christy
Allen (1952)		Prince Albert	SK	Can	53.200	−105.770	1892				Thomas McKay
Allen (1952)	Indian Head		SK	Can	50.330	−103.670	1908				C. G. Harold
Allen (1952)	Beaver Hills	Ituna	SK	Can	51.300	−103.430	1913				A farmer
Allen (1952)		Ituna	SK	Can	51.170	−103.500	1919				Indian
Allen (1952)		Balcarres	SK	Can	50.800	−103.550	1920				
Allen (1952)		Ituna	SK	Can	51.170	−103.500	1925				Indian
Hjertaas (1994)	Big Quill Lake		SK	Can	51.920	−104.370	1872	Aug	12	Breed	Framk Fleming and John Macoun
Allen (1952)	Moose Mountain		SK	Can	49.780	−102.580	1880	Jul		Breed	Macoun
Allen (1952)	North Saskatchewan River	Battleford	SK	Can	52.730	−108.320	1884			Breed	Macdonald
Allen (1952)	Twelve Mile Lake		SK	Can	49.480	−106.230	1895	Jun	6	Breed	Macoun
Hjertaas (1994)		Prince Albert	SK	Can	53.200	−105.770	1896	Jun	10	Breed	Hugh Richardson
Allen (1952)		Yorkton	SK	Can	51.220	−102.470	1900	May	16	Breed	Cowboy Brown
Hjertaas (1994)	Cussed Creek		SK	Can	51.400	−102.550	1900			Breed	William Fernie
Hjertaas (1994)	Beaver Hills	Ituna	SK	Can	51.300	−103.430	1901	May	21	Breed	Edward Arnold
Hjertaas (1994)		Wauchope	SK	Can	49.600	−101.900	1902			Breed	
Hjertaas (1994)		Demaine	SK	Can	50.900	−107.250	1907			Breed	Ernest J. Demaine and Fred Swann
Allen (1952)		Davidson	SK	Can	51.270	−105.980	1911			Breed	R. Lloyd

(Continued)

Historical Records of Whooping Crane Observations, Ordered by State/Province and Life-History Stage (*cont.*)

Source	Location	City	County	State/province	Country	Latitude	Longitude	Year	Month	Day	Life stage	Observer
Allen (1952)		Bradwell		SK	Can	51.950	−106.230	1911			Breed	L. G. Moore
Hjertaas (1994)	Shallow Lake			SK	Can	54.620	−108.300	1911			Breed	Archie Smith
Hjertaas (1994)		Southey		SK	Can	50.930	−104.500	1911	Jun		Breed	H. M. Dahl
Allen (1952)		Bradwell		SK	Can	51.950	−106.230	1912			Breed	L. G. Moore
Hjertaas (1994)	Shallow Lake			SK	Can	54.620	−108.300	1921			Breed	W. W. Smith
Allen (1952)		Baliol		SK	Can	52.020	−109.280	1922	May	19	Breed	Neil Gilmour
Allen (1952)	Kiyiu Lake			SK	Can	52.320	−109.100	1922	May	28	Breed	Fred Bradshaw
Hjertaas (1994)	Shallow Lake			SK	Can	54.620	−108.300	1928			Breed	Archibald Smith
Hjertaas (1994)	Luck Lake			SK	Can	51.080	−107.080	1929			Breed	Frank Miller
Hjertaas (1994)	Luck Lake			SK	Can	51.080	−107.080	1929	Aug		Breed	Steve West
Hjertaas (1994)	Luck Lake			SK	Can	51.080	−107.080	1929	Oct		Breed	Steve West
Allen (1952)	Indian Head			SK	Can	50.330	−103.670	1904	Oct	2	Migr	Lang
Allen (1952)		Semans		SK	Can	51.420	−104.730	1913			Migr	Thomas L. James
Allen (1952)		Forget		SK	Can	49.650	−102.870	1914	Oct		Migr	
Hjertaas (1994)		Lake Lenore		SK	Can	52.400	−104.980	1919	Oct	8	Migr	N. T. Kingsley
Allen (1952)		Liberty		SK	Can	51.130	−105.430	1927	Nov	3	Migr	
Allen (1952)		Estevan		SK	Can	49.130	−102.980	1927	Oct	29	Migr	A farmer
Hjertaas (1994)	White Fox River			SK	Can	53.530	−104.020	1927			Migr	Andrew Wytoski
Hjertaas (1994)	White Fox River			SK	Can	53.530	−104.020	1927			Migr	Andrew Wytoski
Hjertaas (1994)	White Fox River			SK	Can	53.530	−104.020	1928			Migr	Andrew Wytoski
Hjertaas (1994)	White Fox River			SK	Can	53.530	−104.020	1928			Migr	Andrew Wytoski

Source	Location	Sub-location	Prov	Country	Lat	Long	Year	Month	Day	Season	Observer
Hjertaas (1994)	White Fox River		SK	Can	53.530	−104.020	1929			Migr	Andrew Wytoski
Hjertaas (1994)	White Fox River		SK	Can	53.530	−104.020	1929			Migr	Andrew Wytoski
Allen (1952)		Isham	SK	Can	51.080	−108.570				Migr	R. H. Carruthers
Allen (1952)	Saskatchewan River		SK	Can	53.250	−105.680	1827	May	7	Summ	
Allen (1952)	Indian Head		SK	Can	50.330	−103.670	1884	Apr	28	Summ	Guernsey
Hjertaas (1994)	Rush Lake		SK	Can	50.400	−107.400	1891	Jun	13	Summ	MacDonald
Allen (1952)		Osler	SK	Can	52.370	−106.550	1893	May	1	Summ	Colt
Allen (1952)		Osler	SK	Can	52.370	−106.550	1893	May	25	Summ	Colt
Allen (1952)	Indian Head		SK	Can	50.330	−103.670	1901	Apr	20	Summ	Cates
Allen (1952)	Indian Head		SK	Can	50.330	−103.670	1904	Apr	19	Summ	Lang
Allen (1952)	Indian Head		SK	Can	50.330	−103.670	1904	Apr	24	Summ	Lang
Allen (1952)	Indian Head		SK	Can	50.330	−103.670	1904	Sep	25	Summ	Lang
Allen (1952)	Indian Head		SK	Can	50.330	−103.670	1905	Apr	26	Summ	Lang
Allen (1952)	Indian Head		SK	Can	50.330	−103.670	1905	May	1	Summ	Lang
Allen (1952)		Yellow Grass	SK	Can	49.800	−104.170	1905	Sep		Summ	H. H. Hanson
Allen (1952)		Lajord	SK	Can	50.230	−104.150	1908	May		Summ	
Allen (1952)		Resource	SK	Can	52.730	−104.530	1909			Summ	
Allen (1952)	Big Quill Lake		SK	Can	51.920	−104.370	1909			Summ	Ferry
Allen (1952)	Last Mountain Lake and vicinity		SK	Can	51.080	−105.230	1909	Apr		Summ	
Allen (1952)	Indian Head		SK	Can	50.330	−103.670	1910	Apr	27	Summ	Lang
Hjertaas (1994)	Foam Lake		SK	Can	51.720	−103.620	1910			Summ	H. C. Grose
Allen (1952)		Kerrobert	SK	Can	51.920	−109.130	1913			Summ	E. Margaret Estlin

(Continued)

Historical Records of Whooping Crane Observations, Ordered by State/Province and Life-History Stage (cont.)

Source	Location	City	County	State/province	Country	Latitude	Longitude	Year	Month	Day	Life stage	Observer
Hjertaas (1994)		Big River		SK	Can	53.830	−107.020	1916	Jun		Summ	H. Sharpe
Allen (1952)	Indian Head			SK	Can	50.330	−103.670	1920	Apr	15	Summ	Lang
Allen (1952)	Indian Head			SK	Can	50.330	−103.670	1921			Summ	J. R. Garden
Allen (1952)		Kerrobert		SK	Can	51.920	−109.130	1921			Summ	E. Margaret Estlin
Hjertaas (1994)	Luseland			SK	Can	52.080	−109.400	1921			Summ	J. V. Finley and Joe Perry
Hjertaas (1994)	Shallow Lake			SK	Can	54.620	−108.300	1922	Sep	30	Summ	Hoyes Lloyd
Allen (1952)	Indian Head			SK	Can	50.330	−103.670	1923			Summ	
Allen (1952)	Findlater			SK	Can	50.780	−105.400	1924	May	3	Summ	H. L. Felt
Hjertaas (1994)	Luck Lake			SK	Can	51.080	−107.080	1925	May		Summ	Neil Gilmour
Allen (1952)		Wiseton		SK	Can	51.320	−107.650	1926	Apr	18	Summ	Miss B. M. Dickson
Hjertaas (1994)	Luck Lake			SK	Can	51.080	−107.080	1926	Jun	10	Summ	Neil Gilmour
Hjertaas (1994)	Ladder Lake			SK	Can	53.830	−107.000	1927	Aug	10	Summ	G. H. Cartwright
Hjertaas (1994)	Luck Lake	Vonda		SK	Can	52.320	−106.100	1930	Jun	13	Summ	
Hjertaas (1994)	Luck Lake			SK	Can	51.080	−107.080	1930	May	15	Summ	Fred Bradshaw
Hjertaas (1994)	Luck Lake			SK	Can	51.080	−107.080	1930	May	27	Summ	Fred Bradshaw
Hjertaas (1994)	Emma Lake			SK	Can	53.600	−105.900	1932			Summ	John N. Hachett
Hjertaas (1994)	Luck Lake			SK	Can	51.080	−107.080	1935	Aug	12	Summ	Emil Lestin
Hjertaas (1994)	Beaver River			SK	Can	55.430	−107.750	1936	Aug		Summ	George Bauman
Allen (1952)	Boca del Rio Grande	Bagdad		TM	Mex	26.000	−97.260	1863	Jun		Breed	H. E. Dresser
Allen (1952)		Matamoros		TM	Mex	25.520	−97.300	1863	Jun		Summ	H. E. Dresser
Allen (1952)	Mouth of Brazos River		Brazoria	TX	US	28.876	−95.378	1860				
Allen (1952)	Galveston Island		Galveston	TX	US	29.222	−94.909	1860				
Allen (1952)	Port Isabel		Cameron	TX	US	26.117	−97.517	1863				H. E. Dresser

Reference	Locality	Place	County	State	Country	Latitude	Longitude	Year	Breed	Collector
Allen (1952)		San Antonio	Bexar	TX	US	29.424	−98.493	1863		H. E. Dresser
Allen (1952)	Blackjack Peninsula		Aransas	TX	US	28.124	−96.946	1885		J. A. Brundett
Allen (1952)		Waco	McLennan	TX	US	31.549	−97.146	1886		Elanoides
Allen (1952)		Waco	McLennan	TX	US	31.549	−97.146	1888		Elanoides
Allen (1952)		Waco	McLennan	TX	US	31.549	−97.146	1889		Elanoides
Allen (1952)	Worsham Ranch, 8 miles east of Henrietta	Henrietta	Clay	TX	US	33.817	−98.195	1908		R. L. More
Allen (1952)	Blackjack Peninsula		Aransas	TX	US	28.124	−96.946	1910		Thomas Webb
Allen (1952)		Dallas/Lake Worth	Tarrant	TX	US	32.791	−97.414	1920		Jno. B. Litsey
Allen (1952)		Dallas/Lake Worth	Tarrant	TX	US	32.791	−97.414	1927		Mrs. Bruce Reid
Allen (1952)		Maxwell	Caldwell	TX	US	29.881	−97.793	1928		R. W. Strandtmann
Allen (1952)		Matagorda Island	Calhoun	TX	US	28.227	−96.640	1938		J. O. Stevenson
Allen (1952)		Matagorda Island	Calhoun	TX	US	28.227	−96.640	1939		J. O. Stevenson
Allen (1952)		Matagorda Island	Calhoun	TX	US	28.227	−96.640	1940		J. O. Stevenson
Allen (1952)		Matagorda Island	Calhoun	TX	US	28.227	−96.640	1941		J. O. Stevenson
Allen (1952)	Red River	Vernon	Wilbarger	TX	US	34.215	−99.190			
Oberholser (1974)	Southern		Wharton	TX	US	29.027	−96.275	1867	Breed	J. D. Mitchell
Oberholser (1938)	Southern		Wharton	TX	US	29.027	−96.275	1869	Breed	J. D. Mitchell
Oberholser (1938)	Southern		Wharton	TX	US	29.027	−96.275	1878	Breed	J. D. Mitchell
Oberholser (1938)		Eagle Lake	Colorado	TX	US	29.564	−96.342		Breed	Oliver Davie

(Continued)

Historical Records of Whooping Crane Observations, Ordered by State/Province and Life-History Stage (cont.)

Source	Location	City	County	State/province	Country	Latitude	Longitude	Year	Month	Day	Life stage	Observer
Allen (1952)		San Antonio	Bexar	TX	US	29.424	−98.493	1845	Nov		Migr	Colonel G. A. McCall
Allen (1952)		San Angelo	Tom Green	TX	US	31.464	−100.437	1884			Migr	W. Lloyd
Allen (1952)			Williamson	TX	US	30.650	−97.600	1884	Apr		Migr	G. B. Benners
Allen (1952)			Comal	TX	US	29.817	−98.300	1884	Apr		Migr	G. B. Benners
Allen (1952)			Williamson	TX	US	30.650	−97.600	1884	Mar		Migr	G. B. Benners
Allen (1952)			Comal	TX	US	29.817	−98.300	1884	Mar		Migr	G. B. Benners
Allen (1952)			Williamson	TX	US	30.650	−97.600	1884	May		Migr	G. B. Benners
Allen (1952)		San Angelo	Tom Green	TX	US	31.464	−100.437	1885			Migr	W. Lloyd
Allen (1952)		Bonham	Fannin	TX	US	33.577	−96.178	1885	Apr	4	Migr	Peters
Allen (1952)		Bonham	Fannin	TX	US	33.577	−96.178	1885	Mar	23	Migr	Peters
Allen (1952)		Bonham	Fannin	TX	US	33.577	−96.178	1885	Mar	27	Migr	Peters
Allen (1952)		Bonham	Fannin	TX	US	33.577	−96.178	1885	Mar	30	Migr	Peters
Allen (1952)		Gainesville	Cooke	TX	US	33.626	−97.133	1885	Mar	31	Migr	Ragsdale
Allen (1952)		San Angelo	Tom Green	TX	US	31.464	−100.437	1885	Mar	5	Migr	W. Lloyd
Allen (1952)		Bonham	Fannin	TX	US	33.577	−96.178	1885	Nov	16	Migr	Peters
Allen (1952)		Bonham	Fannin	TX	US	33.577	−96.178	1885	Nov	9	Migr	Peters
Allen (1952)		Bonham	Fannin	TX	US	33.577	−96.178	1886	Apr	2	Migr	Peters
Allen (1952)		Bonham	Fannin	TX	US	33.577	−96.178	1886	Apr	9	Migr	Peters
Allen (1952)		Bonham	Fannin	TX	US	33.577	−96.178	1886	Mar	25	Migr	Peters
Allen (1952)		San Angelo	Tom Green	TX	US	31.464	−100.437	1886	Mar	5	Migr	W. Lloyd
Allen (1952)		Bonham	Fannin	TX	US	33.577	−96.178	1888	Nov	22	Migr	Peters
Allen (1952)		Bonham	Fannin	TX	US	33.577	−96.178	1888	Oct	15	Migr	Peters
Allen (1952)		Bonham	Fannin	TX	US	33.577	−96.178	1888	Oct	8	Migr	Peters
Allen (1952)		Bonham	Fannin	TX	US	33.577	−96.178	1889	Nov	18	Migr	Peters
Allen (1952)		Bonham	Fannin	TX	US	33.577	−96.178	1889	Nov	8	Migr	Peters
Allen (1952)		Brownsville	Cameron	TX	US	25.901	−97.497	1890	Apr	2	Migr	F. B. Armstrong

Source	Locality	City	County	State	Country	Latitude	Longitude	Year	Month	Day	Season	Observer
Allen (1952)		Bonham	Fannin	TX	US	33.577	-96.178	1890	Apr	2	Migr	Peters
Allen (1952)		Bonham	Fannin	TX	US	33.577	-96.178	1890	Mar	23	Migr	Peters
Allen (1952)		Waco	McLennan	TX	US	31.549	-97.146	1899	Apr		Migr	Mrs. Bruce Reid
Allen (1952)		Port Arthur	Jefferson	TX	US	29.899	-93.929	1913	Nov	2	Migr	W. H. Bauer
Allen (1952)	Mission Lake		Calhoun	TX	US	28.468	-96.809	1932	Nov		Migr	G. B. Benners
Allen (1952)			Comal	TX	US	29.817	-98.300	1884	May		Summ	W. Lloyd
Allen (1952)	Head of Padre Island		Kennedy	TX	US	26.844	-97.368	1891	Aug	20	Summ	W. Lloyd
Allen (1952)	Head of Padre Island		Kennedy	TX	US	26.844	-97.368	1891	Aug	30	Summ	W. Lloyd
Allen (1952)		Corpus Christi	Nueces	TX	US	27.800	-97.396	1891	Oct	7	Summ	W. Lloyd
Allen (1952)			Cameron	TX	US	26.418	-97.368	1900	May	6	Summ	Vernon Bailey
Allen (1952)	Tarkington Prairie	Cleveland	Liberty	TX	US	30.324	-94.965	1905	Apr	23	Summ	Gaut
Allen (1952)		Corpus Christi	Nueces	TX	US	27.800	-97.396	1845			Wint	Colonel G. A. McCall
Allen (1952)		Brownsville	Cameron	TX	US	25.901	-97.497	1877	Apr	1	Wint	George B. Sennett
Allen (1952)		Brownsville	Cameron	TX	US	25.901	-97.497	1877	Mar	20	Wint	George B. Sennett
Allen (1952)	Head of Padre Island		Kennedy	TX	US	26.844	-97.368	1878	Mar		Wint	George B. Sennett
Allen (1952)		Houston	Harris	TX	US	29.763	-95.363	1881	Dec		Wint	H. Nehrling
Allen (1952)		Houston	Harris	TX	US	29.763	-95.363	1881	Nov		Wint	H. Nehrling
Allen (1952)		Houston	Harris	TX	US	29.763	-95.363	1882	Feb		Wint	W. Lloyd
Allen (1952)		Houston	Harris	TX	US	29.763	-95.363	1882	Jan		Wint	H. Nehrling
Allen (1952)		Houston	Harris	TX	US	29.763	-95.363	1882	Mar		Wint	W. Lloyd
Allen (1952)	Matagorda Peninsula, Kanes Landing		Matagorda	TX	US	28.585	-96.018	1885	Jan	23	Wint	
Allen (1952)		Beaumont	Jefferson	TX	US	30.086	-94.102	1886	Dec	23	Wint	
Allen (1952)		Brownsville	Cameron	TX	US	25.901	-97.497	1889			Wint	Worthen

(Continued)

Historical Records of Whooping Crane Observations, Ordered by State/Province and Life-History Stage (cont.)

Source	Location	City	County	State/province	Country	Latitude	Longitude	Year	Month	Day	Life stage	Observer
Allen (1952)		Corpus Christi	Nueces	TX	US	27.800	−97.396	1891	Dec	17	Wint	W. Lloyd
Allen (1952)		Padre Island	Kennedy	TX	US	26.844	−97.368	1891	Dec	20	Wint	George B. Sennett
Allen (1952)			Hidalgo	TX	US	26.333	−98.200	1891	Feb	22	Wint	F. S. Webster
Allen (1952)		Brownsville	Cameron	TX	US	25.901	−97.497	1891	Jan	5	Wint	
Allen (1952)	Head of Padre Island		Kennedy	TX	US	26.844	−97.368	1891	Nov	1	Wint	W. Lloyd
Allen (1952)	Head of Padre Island		Kennedy	TX	US	26.844	−97.368	1891	Nov	10	Wint	W. Lloyd
Allen (1952)		Corpus Christi	Nueces	TX	US	27.800	−97.396	1891	Nov	11	Wint	W. Lloyd
Allen (1952)	Head of Padre Island		Kennedy	TX	US	26.844	−97.368	1891	Nov	12	Wint	W. Lloyd
Allen (1952)	Head of Padre Island		Kennedy	TX	US	26.844	−97.368	1891	Nov	7	Wint	W. Lloyd
Allen (1952)			Hidalgo	TX	US	26.333	−98.200	1892	Dec	14	Wint	F. S. Webster
Allen (1952)			Hidalgo	TX	US	26.333	−98.200	1892	Dec	2	Wint	
Allen (1952)			Hidalgo	TX	US	26.333	−98.200	1892	Dec	7	Wint	F. S. Webster
Allen (1952)	Matagorda Peninsula, Kanes Landing		Matagorda	TX	US	28.585	−96.018	1892	Feb	20	Wint	W. Lloyd
Allen (1952)	Matagorda Peninsula, Kanes Landing		Matagorda	TX	US	28.585	−96.018	1892	Jan	6	Wint	W. Lloyd
Allen (1952)			Hidalgo	TX	US	26.333	−98.200	1892	Nov	15	Wint	F. S. Webster
Allen (1952)			Hidalgo	TX	US	26.333	−98.200	1893	Jan	2	Wint	F. S. Webster
Allen (1952)		Houston	Harris	TX	US	29.763	−95.363	1893	Jan	4	Wint	Jason Whyte
Allen (1952)		Brownsville	Cameron	TX	US	25.901	−97.497	1894	Feb	18	Wint	

Allen (1952)	Blackjack Peninsula	Aransas	TX	US	28.124	−96.946	1895			Wint	E. Hough
Allen (1952)	Padre Island	Kennedy	TX	US	26.844	−97.368	1896	Nov	29	Wint	Worthen
Allen (1952)	Padre Island	Kennedy	TX	US	26.844	−97.368	1897	Feb	20	Wint	
Allen (1952)	Matagorda Island	Calhoun	TX	US	28.227	−96.640	1900	Apr	2	Wint	H. C. Oberholser
Allen (1952)	Matagorda Island	Calhoun	TX	US	28.227	−96.640	1900	Mar	29	Wint	H. C. Oberholser
Allen (1952)	Padre Island	Kennedy	TX	US	26.844	−97.368	1904	Feb	10	Wint	F. B. Armstrong
Allen (1952)	Corpus Christi	Nueces	TX	US	27.800	−97.396	1904	Feb	3	Wint	
Allen (1952)	Padre Island	Kennedy	TX	US	26.844	−97.368	1904	Jan	16	Wint	F. B. Armstrong
Allen (1952)	Brownsville	Cameron	TX	US	25.901	−97.497	1911	Feb	15	Wint	L. R. Cowen
Allen (1952)	Brownsville	Cameron	TX	US	25.901	−97.497	1911	Feb	22	Wint	L. R. Cowen
Allen (1952)	Brownsville	Cameron	TX	US	25.901	−97.497	1911	Feb	27	Wint	L. R. Cowen
Allen (1952)	Laguna Larga	Kleberg	TX	US	27.521	−97.398	1915			Wint	Richard Kleberg
Allen (1952)	Laguna Larga	Kleberg	TX	US	27.521	−97.398	1921	Dec	23	Wint	T. G. Pearson and Richard Kleberg
Allen (1952)	South of Baffin Bay	Kennedy	TX	US	27.233	−97.498	1923	Jan		Wint	Ludlow Griscom and Maunsell Crosby
Allen (1952)	Laguna Larga	Kleberg	TX	US	27.521	−97.398	1923	Jan	12	Wint	Ludlow Griscom and Maunsell Crosby
Allen (1952)	Brownsville	Cameron	TX	US	25.901	−97.497	1924	Mar	4	Wint	R. D. Camp
Allen (1952)	Laguna Larga	Kleberg	TX	US	27.521	−97.398	1926	Feb	12	Wint	Baker
Allen (1952)	Blackjack Peninsula	Aransas	TX	US	28.124	−96.946	1931			Wint	J. G. Fuller
Allen (1952)	63 miles north of Brownsville	Kennedy	TX	US	26.938	−97.581	1933	Feb	8	Wint	T. H. Clegg
Allen (1952)	Espiritu Santo Bay	Calhoun	TX	US	28.347	−96.526	1933	Jan	10	Wint	H. C. Oberholser

(Continued)

Historical Records of Whooping Crane Observations, Ordered by State/Province and Life-History Stage (*cont.*)

Source	Location	City	County	State/province	Country	Latitude	Longitude	Year	Month	Day	Life stage	Observer
Allen (1952)	63 miles north of Brownsville		Kennedy	TX	US	26.938	−97.581	1933	Jan	23	Wint	H. C. Oberholser
Allen (1952)		Laguna Larga	Kleberg	TX	US	27.521	−97.398	1934	Jan	5	Wint	J. J. Carroll
Allen (1952)	Dewberry Island		Calhoun	TX	US	28.385	−96.511	1934	Mar	14	Wint	J. J. Carroll
Allen (1952)	Dewberry Island		Calhoun	TX	US	28.385	−96.511	1934	Mar	28	Wint	J. J. Carroll
Allen (1952)	Blackjack Peninsula		Aransas	TX	US	28.124	−96.946	1935			Wint	G. B. Saunders
Allen (1952)		Laguna Larga	Kleberg	TX	US	27.521	−97.398	1935	Mar	23	Wint	J. J. Carroll
Allen (1952)	Blackjack Peninsula		Aransas	TX	US	28.124	−96.946	1936	Feb	29	Wint	Neil Hotchkiss
Allen (1952)	Blackjack Peninsula		Aransas	TX	US	28.124	−96.946	1936	Feb	8	Wint	Neil Hotchkiss
Allen (1952)		Laguna Larga	Kleberg	TX	US	27.521	−97.398	1936	Jan	10	Wint	J. J. Carroll
Allen (1952)	31 miles south of	Corpus Christi	Kleberg	TX	US	27.322	−97.475	1936	Jan	14	Wint	Bob Snow
Allen (1952)	East Bay		Galveston	TX	US	29.511	−94.671	1936	Jan	19	Wint	Mrs. Bruce Reid
Allen (1952)	Blackjack Peninsula		Aransas	TX	US	28.124	−96.946	1936	Jan	25	Wint	Neil Hotchkiss
Allen (1952)		Laguna Larga	Kleberg	TX	US	27.521	−97.398	1936	Mar	1	Wint	Neil Hotchkiss
Allen (1952)	Blackjack Peninsula		Aransas	TX	US	28.124	−96.946	1936	Mar	3	Wint	Neil Hotchkiss
Allen (1952)	Blackjack Peninsula		Aransas	TX	US	28.124	−96.946	1936	Mar	8	Wint	Neil Hotchkiss
Allen (1952)		Laguna Larga	Kleberg	TX	US	27.521	−97.398	1937	Feb	7	Wint	J. J. Carroll
Allen (1952)		Austwell	Refugio	TX	US	28.390	−96.842	1941	Feb		Wint	J. O. Stevenson
Allen (1952)		Austwell	Refugio	TX	US	28.390	−96.842	1941	Jan		Wint	J. O. Stevens
Allen (1952)	Great Salt Lake		Davis	UT	US	40.703	−112.387	1880			Wint	

Bailey (1881)	Lynchburg	Lynchburg	Lynchburg	VA	US	37.414	−79.143	1876	Jun	21	Summ	
Jewett et al. (1953)	Umatilla Rapids—Mouth of the Walla Walla River		Walla Walla	WA	US	46.058	−118.910	1805	Oct	19	Migr	Lewis and Clark
Kumlien and Hollister (1951)	Southwest Wisconsin	Prairie du Chien	Crawford	WI	US	43.046	−91.139	1840				P. R. Hoy
Kumlien and Hollister (1951)	Lake Michigan	Racine	Racine	WI	US	42.726	−87.783	1840				P. R. Hoy
Allen (1952)	Sugar River		Dane	WI	US	43.033	−89.660	1854				
Carr (1890)*			Brown	WI	US	44.467	−87.967				Breed	
Kumlien and Hollister (1951)			Green	WI	US	42.833	−89.600	1878	Oct		Migr	
Kumlien and Hollister (1951)	Lake Mills		Jefferson	WI	US	43.081	−88.912	1935			Migr	Douglas E. Wade
Hunt and Gluesing (1976)	Horicon Marsh		Dodge	WI	US	43.550	−88.656	1900	Apr		Summ	W. Snyder
Allen (1952)		Cody	Park	WY	US	44.526	−109.056	1906				William Richard
Allen (1952)		Cody	Park	WY	US	44.526	−109.056	1907				William Richard
Allen (1952)		Cody	Park	WY	US	44.526	−109.056	1908				William Richard
Allen (1952)		Cody	Park	WY	US	44.526	−109.056	1909				William Richard
Allen (1952)		Cody	Park	WY	US	44.526	−109.056	1910				William Richard
Allen (1952)		Cody	Park	WY	US	44.526	−109.056	1911				William Richard
Allen (1952)		Cody	Park	WY	US	44.526	−109.056	1912				William Richard
Allen (1952)	Shoshone River		Bighorn	WY	US	44.862	−108.204	1915				
Bent (1926)	Yellowstone Park	Cody	Park	WY	US	44.767	−110.233	1914	Aug	4	Breed	M. P. Skinner
Allen (1952)	Jackson Lake		Carbon	WY	US	42.437	−107.339	1906			Migr	William Richard

(Continued)

Historical Records of Whooping Crane Observations, Ordered by State/Province and Life-History Stage (*cont.*)

Source	Location	City	County	State/province	Country	Latitude	Longitude	Year	Month	Day	Life stage	Observer
Allen (1952)	Jackson Lake		Carbon	WY	US	42.437	−107.339	1912	Feb		Migr	William Richard
Allen (1952)	Bechler River		Teton	WY	US	44.148	−110.996	1930			Summ	T. G. Pearson and Ranger Bicknell
Allen (1952)		Osage	Weston	WY	US	43.987	−104.421	1934	Apr	23	Summ	J. F. Bock
Allen (1952)	Star Valley		Lincoln	WY	US	43.123	−111.027	1934	May		Summ	Ray Wolfley
Allen (1952)	Star Valley		Lincoln	WY	US	43.123	−111.027	1935	May		Summ	Theone Wolfley

Breed, breeding; *Summ*, summering; *Migr*, migrating; *Wint*, wintering.
Records marked with an *asterisk* were questionable and not included in maps or in biome and ecoregion summaries.
Added sources in addition to those listed in Allen 1952, pp. 51–64:

Baczkowski, F., 1955. Whooping Cranes in Porter County 50 years ago. Indiana Audubon Q. 33, 43–44.
Bailey, H.B., 1881. Bird notes: an index and summary. Forest and Stream I–XII, 70–71.
Bailey, A.M., Niedrach, R.J., 1965. Birds of Colorado. Denver Museum of Natural History, Denver, CO.
Baird, S.F., Brewer, T.M., Ridgway, R., 1884. The water birds of North America, 2 vols. Little, Brown, Boston, MA.
Barrows, W.B., 1912. Michigan Bird Life. Michigan Agricultural College, Lansing, 822 pp.
Bent, A.C., 1926. Life Histories of North American Marsh Birds. Smithsonian Institution, United States National Museum Bulletin 135, Washington, DC.
Bogardus, A.H., 1878. Wild geese, cranes, and swans (Chapter 12). In: Field, Cover, and Trap Shooting. Wolfe Publishing Co, Prescott, AZ.
Butler, A.W., 1898. The Birds of Indiana. Indiana Department of Geology and Natural Resources Annual Report 22, pp. 575–1187.
Carr, C.F., 1890. Wisconsin Natur. 1, 78.
Coale, H.K., 1912. The Birds of Lake County, Illinois. In: History of Lake County, pp. 353–370.
Dinsmore, J.J., 1994. A Country So Full of Game: The Story of Wildlife in Iowa. University of Iowa Press, Iowa City.
Griffin, W.W., 1957. A Whooping Crane from Macon. Oriole 22.
Hallman, R.C., 1965. Record of Whooping Crane (*Grus americana*) killed in St. Johns County, Florida. Florida Natur. 38, 23.
Harmon, W.Z., 1954. Notes on cranes. Florida Natur. 27, 22.
Hibbard, E.A., 1956. An old nesting record for the Whooping Crane in North Dakota. Wilson Bull. 68, 73–74.
Hjertaas, D.G., 1994. Summer and breeding records of the Whooping Crane in Saskatchewan. Blue Jay 52 (2), 99–115.
Houston, C. S., Ball, T., Houston, M., 2003. Eighteenth-century naturalists of Hudson Bay. McGill-Queen's Native and Northern Series, Vol. 34. McGill-Queen's University Press, Montreal.
Hunt, R.A., Gluesing, E.A., 1976. The sandhill crane in Wisconsin. In: Proceedings of the International Crane Workshop, International Crane Foundation, Baraboo, WI, pp. 19–34.
Jewett, S.G., Taylor, W.P., Shaw, W.T., Aldrich, J.W., 1953. Birds of Washington State. University of Washington Press, Seattle.
Kemsies, E., 1930. Birds of the Yellowstone National Park, with some recent additions. Wilson Bull. 42 (3), 198–210.
Kumlien, L., Hollister, N., 1951. The Birds of Wisconsin. Wisconsin Society for Ornithology, Madison.
Mengel, R.M., 1965. The Birds of Kentucky. AOU Monograph 3.
Oberholser, H.C., 1938. The Bird Life of Louisiana. Bulletin 28, Department of Conservation, State of Louisiana, Baton Rouge.
Oberholser, H.C., 1974. The Bird Life of Texas. University of Texas Press, Austin.
Shaffer, C., 1940. Whooping Crane, *Grus americana*. Redstart 7, 65.
Sprunt Jr., A., 1954. Florida Bird Life. Coward-McCann and National Audubon Society, New York, 527 pp.
Woodruff, F.M., 1907. The Birds of the Chicago Area. Cornell University Library, 264 pp.

POPULATION AND BREEDING BIOLOGY

Whooping Crane parents and offspring

Population and Breeding Range Dynamics in the Aransas-Wood Buffalo Whooping Crane Population

Scott Wilson*, Mark T. Bidwell**

*Wildlife Research Division, Environment and Climate Change Canada, National Wildlife Research Centre, Ottawa, ON, Canada
**Canadian Wildlife Service, Environment and Climate Change Canada, Prairie and Northern Wildlife Research Centre, Saskatoon, SK, Canada

INTRODUCTION

In the mid-19th century, the migratory Whooping Crane (*Grus americana*) breeding range covered the southern Canadian prairies and northern Great Plains where primary breeding habitats were extensive marshes in mixed and tall grass prairie (Allen, 1956). Although still relatively widespread, the species' abundance at the time was low, and estimated at 1,300–1,400 individuals in the 1860s (Allen, 1956). The widespread conversion of breeding habitat to agriculture combined with overhunting led to a dramatic reduction in the migratory population by the early 1900s, and by 1942 only 16 individuals

remained in what is now known as the Aransas-Wood Buffalo Population (AWBP; Urbanek and Lewis, 2015). The breeding grounds for this remnant population were initially unknown but subsequently identified in 1954 as being along the border of Alberta and the Northwest Territories in and around what is now Wood Buffalo National Park (WBNP; Allen, 1956).

The AWBP has afforded excellent opportunities to study the population ecology of an endangered species as it has begun to recover from near extinction. A detailed survey of the AWBP has been conducted annually at Aransas National Wildlife Refuge (NWR) since 1938, providing annual counts of hatching-year and after-hatching-year individuals (Butler et al., 2013; Stehn and Taylor, 2008). Most study on the dynamics of the population has been based on the information provided by this survey. Miller et al. (1974) conducted one of the initial analyses of the dynamics of the population using a stochastic exponential growth model applied to survey data from 1941 to 1973. They showed that the population had experienced exponential growth since 1941 with a pattern suggesting the early stages of a logistic growth curve. They also noted that the growth of the population during this period was due primarily to a stabilizing of the mortality rate for subadults and adults, and that the reproductive rate had declined, potentially due to density-dependent effects. Even though the population was only 53 individuals in 1973, the extinction probability was estimated to be low (0.021). In a subsequent study, Binkley and Miller (1983) observed that the annual population growth rate appeared to have increased from an average of about 2.2% from 1938 to 1956 to about 4.2% from 1956 to 1980. Recent analyses confirm an annual average population growth of approximately 4% over the period from 1938 to 2010 (Butler et al., 2013; Converse et al., Chapter 8, this volume; Wilson et al., 2016).

A characteristic feature of the AWBP is the periodicity of population growth. Binkley and Miller (1983) first showed that despite an overall increasing trend, the population underwent periods of decline at about 10-year intervals. They ascribed these fluctuations to variation in annual recruitment, the cause of which was unknown. Boyce and Miller (1985) also showed the approximate 10-year periodicity using autocorrelation and periodogram analysis on the detrended Aransas NWR survey data. In contrast to the results of Binkley and Miller (1983), their analyses were inconsistent with variation in recruitment as the cause of these fluctuations and they suggested it may be due to factors operating on mortality. Although the underlying source of the 10-year pattern was still unknown, Boyce and

Miller (1985) noted the similarity in the periodicity of the Whooping Crane time series with the approximate 10-year fluctuations of predators associated with the hare-lynx cycle (Elton and Nicholson, 1942; Smith and Davis, 1981), including Canada lynx (*Lynx canadensis*), grey wolves (*Canis lupus*), and red fox (*Vulpes vulpes*). More recent studies using the Aransas NWR survey data provided further evidence of the periodicity of Whooping Crane population growth and the possible links with predators (Boyce et al., 2005; Butler et al., 2013).

The remote nature of the AWBP breeding grounds, combined with the difficulty of capturing and marking individuals, has made it challenging to collect vital rate data on the population. Consequently, initial demographic models focused on the estimation of recruitment, survival, and age structure from the Aransas NWR survey because it was possible to identify the number of hatch-year juveniles ("brown birds" sensu; Binkley and Miller, 1980) and after-hatch-year subadults and adults (white birds) based on plumage. Binkley and Miller (1980) modeled age-specific survival by regressing the number of white birds on previous counts of brown birds. They used a relationship that identified the probability of an individual surviving for a given number of years based on a presumed maximum life span and specific form of the survivorship function. Link et al. (2003) used the survey information to construct a demographic model for estimation of age and year-specific survival. Their approach treated the population age structure as latent and the data as summaries of this latent structure. Their model revealed that through 2001 there was a declining trend in recruitment, possibly due to density dependence, and a slight increase in survival.

While most study of the population ecology of Whooping Cranes has been based on the Aransas NWR survey, the Canadian Wildlife Service has conducted aerial surveys of the nesting grounds since 1966, providing annual estimates of the number of nesting pairs and the number of fledged young. In addition, from 1977 to 1988 biologists captured and color-banded juveniles on the breeding grounds at about 60–65 days of age (Kuyt, 1979). Gil-Weir et al. (2012) used these data sets on reproduction and survival to create an age-structured demographic matrix model for the population (see also Gil-Weir, 2006). Population projections based on this model and assuming different winter carrying capacities suggested population growth may slow by about 2030, with a consequent delay in the time to reach International Recovery Plan goals (Canadian Wildlife Service and U.S. Fish and Wildlife Service, 2007).

Taken together, the studies just described revealed much important information on the population ecology of Whooping Cranes as they recovered from historic lows in the 1940s. However, other key information has been lacking, including (1) the extent of annual variability in the vital rates; (2) temporal changes in proportion of breeding age versus subadult individuals in the population; (3) the mechanism by which predators associated with the hare-lynx cycle influence the population; and (4) characteristics of range expansion as the population increased, particularly on the breeding grounds. In addition, there is continued debate on the impacts of egg collecting on the breeding grounds (Canadian Wildlife Service and U.S. Fish and Wildlife Service, 2007). Egg collection was conducted from 1967 to 1996 and was used to build the captive population that has been key to the establishment of reintroduced populations (French et al., Chapter 1, this volume). Earlier studies showed no evidence for negative effects of egg collection on recruitment success for breeding pairs (Boyce et al., 2005; Ellis and Gee, 2001) or population-level recruitment (Link et al., 2003). No recent studies have compared mean fledge rates between the period with egg collection to the more recent period after egg collection ceased

(1997–present) to evaluate whether population-level fledge rates differed during the two periods. The resumption of egg collection is being considered as a management strategy and evaluating impacts of this practice on reproduction and population growth is therefore important.

Wilson et al. (2016) developed an integrated population model (IPM) to address several of the aforementioned questions. This modeling approach combined the Aransas NWR winter count data with data used for estimation of reproduction and survival in a joint analysis. In this chapter, we review the development of the IPM and the main findings from Wilson et al. (2016), and discuss how they enhance our understanding of the population dynamics and ecology of Whooping Cranes.

Along with rapid growth in population size of the AWBP, its breeding range also expanded considerably since surveys began on the breeding grounds in 1966. Importantly, breeding range expansion has occurred both within and outside the borders of WBNP, yet to our knowledge no assessment of breeding season range dynamics has been conducted for the AWBP. A better understanding of historical and current breeding range dynamics could allow conservation planners to predict and protect areas that Whooping Cranes will rely on in the future. Specifically, knowledge of breeding range dynamics could inform definition and protection of critical habitat, which, under endangered species legislation in Canada and the United States, is defined as habitat necessary for the survival or recovery of a listed wildlife species. Therefore, in this chapter we also investigate the expansion of the AWBP breeding range during the period of rapid population increase since 1966. Our specific objectives are to describe the size, growth rate, and structure of the AWBP breeding range over time, and in relation to growth in population size, using information obtained from surveys conducted during the breeding season.

USING INTEGRATED POPULATION MODELS TO ASSESS WHOOPING CRANE POPULATION DYNAMICS

Integrated Population Models

IPMs combine information relevant for estimating abundance with information used to estimate any of the four main demographic rates (survival, reproduction, immigration, emigration) under a common analytical framework (Brooks et al., 2004; Schaub et al., 2007). Count data typically provide information about population abundance, which is the outcome of the four demographic rates and, thus, the data sets share information on the underlying processes that influence population dynamics. This information sharing allows for greater precision in the estimation of population parameters and the potential to estimate parameters for which direct data are unavailable (e.g., immigration rates; Schaub et al., 2007). The ability to estimate parameters for which direct data are unavailable makes IPMs particularly well suited for population modeling of endangered species where abundance is often well known but data on the demographic rates can be difficult to collect. Examples of their use for the conservation of rare or threatened species include studies on Golden-cheeked Warblers (*Setophaga chrysoparia*; Duarte et al., 2016), California Spotted Owls (*Strix occidentalis occidentalis*; Tempel et al., 2014), and Sierra Nevada Bighorn Sheep (*Ovis canadensis sierra*; Johnson et al., 2010). To our knowledge, IPMs have not been used with populations of any other crane species.

Data Sets Used in the IPM

Three data sets providing information on abundance, reproduction, and survival were used to construct the IPM for AWBP Whooping Cranes. Abundance data were obtained from the U.S. Fish and Wildlife Service (USFWS) aerial counts of the population on the wintering grounds (late November to late March) in and around Aransas National Wildlife Refuge (Stehn and Taylor, 2008). The information from these surveys was used to provide an annual estimate of hatch-year and after-hatch-year individuals, which were multiplied by 0.5 to estimate the number of hatch-year ($C_{f,hy,t}$) and after-hatch-year ($C_{f,ahy,t}$) females. The winter surveys also provided an estimate of the number of winter territories held by breeding-age pairs (Wt_t), and resighting information for all marked birds detected. The survey was assumed to be a census from the first year in 1938 through 2010, but the growing size of the population led to increased uncertainty on the accuracy of the number of individuals recorded (Butler et al., 2014). In 2011, the USFWS switched to line transects and abundance estimation methods, which allowed for only a combined estimate of hatch-year and after-hatch-year individuals (Butler et al., 2014).

Reproductive data were obtained during Canadian Wildlife Service (CWS) annual aerial surveys covering ~925 km^2 of Wood Buffalo National Park between 1977 and 2013. Surveys were conducted in May, June, and August, and recorded the number of active nests and fledged chicks from nesting pairs, which were used to estimate the annual fledging rate. The collection of single eggs from two-egg nests for captive rearing and reintroduction occurred in the first 20 years of this time series (1977–96; Ellis and Gee, 2001).

From 1977 to 1988, 132 Whooping Cranes were banded on the breeding grounds at 60–65 days of age (Kuyt, 1979). Of these individuals, 46 were female, 47 were male, and 39 were unknown sex. Systematic searches for banded individuals occurred at Aransas each winter. The resighting data for each color-banded individual across all years was used to create individual encounter histories for survival estimation.

IPM Development

The IPM was comprised of separate likelihoods from the three data sources on abundance, reproduction, and survival. In the abundance likelihood, the estimated size of the population in winter was constructed as a state-space model (Brooks et al., 2004), with a state process model describing the abundance of the AWBP and an observation model linking the winter counts at Aransas to the estimated abundance. The process model was female-based and contained six age classes after fledging (Fig. 4.1). Individuals were assumed to occupy the initial hatch-year (hy) age class for a 6-month period from June 1 to December 1 and then the next four age classes for successive 12-month periods from December 1 in year $t-1$ to December 1 in year t. These age classes included subadult 1 (sad1, 6–18 months), subadult 2 (sad2, 18–30 months), subadult 3 (sad3, 30–42 months), and adult 1 (ad1, 42–54 months). Females were assumed to breed for the first time at age 4 (Urbanek and Lewis, 2015) as they transitioned from the sad3 to the ad1 age class. When individuals reached 54 months (4.5 years) they entered the final adult age class (ad2).

For each age class, we specified N_t as the annual estimated abundance from the model. The first age class, the number of female hatch-year individuals reaching the wintering grounds at Aransas ($N_{f.hy,t}$), was specified as:

$$N_{f.hy,t} = \left[\left(N_{f.sad3,t-1} \cdot \phi_{ad1,t-1}^{5/12} \right) \right.$$
$$\left. + \left(\left(N_{f.ad1,t-1} + N_{f.ad2,t-1} \right) \cdot \phi_{ad2,t-1}^{5/12} \right) \right]$$
$$\cdot bp_{t-1} \cdot f_{t-1} \cdot 0.5 \cdot \phi_{hy,t-1}$$

where $N_{f.sad3,t-1}$, $N_{f.ad1,t-1}$, and $N_{f.ad2,t-1}$ are the number of sad3, ad1, and ad2 females, respectively, in the winter of the previous time step (i.e., 12 months previous), bp_{t-1} and f_{t-1} are the breeding propensity and fledging rate in the previous summer, and $\phi_{hy,t-1}$ is the 4-month survival rate for fledglings from August 1 through November 30 (when birds were assumed to arrive on the wintering grounds). We assumed that 50% of fledglings were female. To allow for the possibility of mortality during the 5-month period from December 1 of the previous year to the start of the breeding season (assumed to be May 1) we multiplied $N_{f.sad3,t-1}$, $N_{f.ad1,t-1}$ and $N_{f.ad2,t-1}$ by a

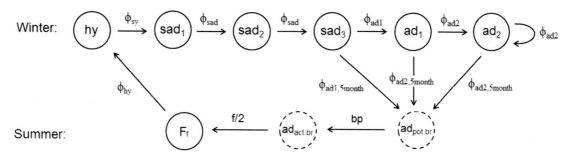

FIGURE 4.1 Life cycle diagram for the integrated population model (IPM) of the Aransas-Wood Buffalo Whooping Crane population (Wilson et al., 2016). The stages in the model are represented by nodes with demographic rates reflecting the transitions between nodes. Stages include female fledged young (F_f), hatch-year (hy), subadults (sad1, sad2, sad3), adult first breeding (ad1), adult second breeding or older (ad2). We estimate the potential (ad$_{pot.br}$) and actual number of breeding adults after adjusting for breeding propensity (ad$_{act.br}$). Demographic rates are estimated in the IPM and include breeding propensity (bp), fledge rate (f), and age-class-specific survival (ϕ). Survival estimated over a 5-month period from 1 December to 1 May ($\phi_{5\ month}$) was used to account for possible mortality of potential breeders between the start of winter (1 December) and the onset of breeding (1 May).

5-month survival estimate derived from the annual survival for each of these age classes.

The abundance of the three subadult age classes and the first adult age class were modeled as:

$$N_{j,t} \sim binomial\left(\phi_{j,t-1}, N_{j-1,t-1}\right)$$

where the number of individuals in subadult age class j at time t follows a binomial distribution with probability equal to survival (ϕ) to subadult age class j from year $t-1$ to t and N equal to the number of individuals in the preceding age class ($j-1$) at time $t-1$. The abundance of the final age class was:

$$N_{f.ad2,t} \sim binomial\left(\phi_{ad2,t-1}, N_{f.ad1,t-1}\right)$$
$$+ binomial\left(\phi_{ad2,t-1}, N_{f.ad2,t-1}\right)$$

with $\phi_{ad2,t-1}$ equal to the annual survival of ad1 and ad2 females from years $t-1$ to t and $N_{f.ad1,t-1}$ and $N_{f.ad2,t-1}$ equal to the abundance of ad1 and ad2 females, respectively, in year $t-1$. The abundance of all age classes in winter for both sexes was derived assuming a 50:50 sex ratio:

$$\left(N_{tot,t}\right) = \left(N_{f.hy,t} + N_{f.sad1,t} + N_{f.sad2,t} + N_{f.sad3,t} \right.$$
$$\left. + N_{f.ad1,t} + N_{f.ad2,t}\right) * 2$$

The total abundance was then used to estimate population change between years (λ_t) as:

$$\lambda_t = N_{tot,t} / N_{tot,t-1}$$

The observation model described the relationship between the true annual population size (N_t) and the number of observed individuals from the winter counts. A Poisson distribution was assumed for the observation model describing the number of observed hatch-year ($C_{f.hy,t}$) and after-hatch-year ($C_{f.ahy,t}$) females in year t:

$$C_{f.hy,t} \sim Poisson\left(N_{f.hy,t}\right)$$

$$C_{f.ahy,t} \sim Poisson\left(N_{f.sad1,t} + N_{f.sad2,t} \right.$$
$$\left. + N_{f.sad3,t} + N_{f.ad1,t} + N_{f.ad2,t}\right)$$

where N_t for each age class is the model-predicted abundance of females at time t. We used winter abundance data from the winters of 1976–77 to 2012–13 for the IPM.

Estimation of Model Demographic Rates

The annual number of breeding females was estimated as $J_t \sim$ binomial (bp_t, Wt_t) where one active nest (J_t) equaled one breeding female, Wt_t was the number of winter territories, and bp_t was the population-level breeding propensity. Breeding ground surveys recorded the activity of nesting pairs but were not a complete count of potential breeders, and therefore we used the winter territory count as an estimate of potential breeders. Only paired breeding-age individuals hold territories on the wintering grounds, with subadults and unpaired adults forming small flocks outside the area of occupied territories (Bishop and Blankinship, 1982). Therefore, estimates of bp_t assumed that the number of breeding territories in year t equaled the number of winter territories just prior to spring migration in the same year (Wt_t). Winter territory counts were missing for some years (1977–80, 2010–13) and in those cases we specified a prior, following methods in King et al. (2010) and Kéry and Royle (2016). Specifically, the prior incorporated a linear trend in Wt_t because of the trend in population size (see Appendix A in Wilson et al., 2016). Because some nests may have been missed, our estimates of bp_t are minimum estimates.

The number of fledglings produced each year (F_t) was assumed to follow a Poisson distribution with the expected annual mean (ρ_t) equal to the product of the fledging rate for the population (i.e., young per breeding female, f_t) and the number of active nests in year t. Age-specific survival was analyzed with a multistate framework with age classes as states (Lebreton et al., 1992, 1999). There were initially six survival rates (ϕ_{hy}, ϕ_{sy}, ϕ_{sad1}, ϕ_{sad2}, ϕ_{ad1}, ϕ_{ad2}). Note that ϕ_{sy} refers to second-year survival (6–18 months) while ϕ_{sad1} and ϕ_{sad2} refer to survival between 18 and 30 months and 30 and 42 months respectively. We assumed that $\phi_{sad1} = \phi_{sad2}$ (i.e., ϕ_{sad}, see Fig. 4.1) for subsequent analyses in the model. We assumed equal survival for males and females; sex-specific survival was not estimated because sex was known for only 70% of the marked juveniles. We also estimated the resighting probability of marked individuals (p), which was assumed to be constant across age and sex classes. Additional detail on the structure of the age-specific survival model is presented in Wilson et al. (2016).

All annual demographic rates were estimated as random variables governed by a common mean and temporal variance (i.e., annual temporal variation was accounted for). Logit link functions were used in the estimation of survival and breeding propensity, and a log link was used for fledge rate. After estimating demographic rates we examined the correlation in each rate in year $t-1$ with population growth (λ) in year t, as an indicator of how variation in a demographic rate influences variation in population growth. For each sample of the posterior distribution (see later) we had an estimate of all rates and λ for years where data were available. Each of these samples allowed us to calculate the correlation across years between a particular rate and λ. Multiple calculations of this correlation across samples in the posterior allowed us to generate a posterior distribution of the correlation coefficient; thus we calculated the mean and the 95% confidence interval (CI) of this distribution, as well as the probability that the correlation coefficient was greater than 0 (Schaub et al., 2013).

IPMs assume independence among data sets, but this will rarely be the case for endangered species where, due to small population sizes, the data on abundance and demographic rates will often include the same individuals. In our model, this would have been the case in particular for fledging rate and abundance because the former was estimated based on nests for nearly the entire population. Our survival data, by contrast, included a smaller proportion of the population with less overlap. Abadi et al. (2010) showed that violating the assumption of independence had minimal effects on the parameter estimates when there are sufficient data for each of the likelihoods. This was the case in our study for fledge rate and survival of the oldest adult age class, but we acknowledge that fewer data existed for survival estimates of the younger age classes. A useful future analysis could examine how model output is affected by differing amounts of data across all model parameters. There are currently no goodness-of-fit tests for IPMs (Schaub and Abadi, 2011). Therefore, we used a bootstrap goodness-of-fit test for the mark-recapture data in program MARK (White and Burnham, 1999) using a model that allowed survival to vary by age and time while the recapture probability was constant. This model indicated no evidence of overdispersion within the data ($\hat{c} = 1.15$).

Model Implementation

The IPM was fit in a Bayesian framework with estimation of parameters using Markov chain Monte Carlo methods. We used uninformative priors with a uniform distribution for all ϕ (0,1), p (0,1), bp (0,1), and f (0,2). The priors for the initial age class abundances were specified with a normal distribution and low precision (0.01). The means of the distribution were based on the observed winter abundance of juvenile and adult females assuming a 50:50 sex ratio (i.e., $C_{f.hy,t}$ and $C_{f.ahy,t}$) combined with information on age-specific survival in Gil-Weir et al. (2012) to estimate an approximate size for all age classes in the model in the winter of 1976–77. The model run included two chains, each with 100,000 iterations. We discarded the initial 75,000 iterations for each chain as a burn-in and thinned each chain by 10, giving us 5,000 samples from the posterior distribution for inference. Model convergence was examined visually using the parameter trace plots and $\hat{R}$ diagnostics (Gelman et al., 2004). Analyses were conducted in Open-BUGS (Lunn et al., 2000) through R version

3.2.1 (R Development Core Team, 2004) with the package R2OpenBUGS (Sturtz et al., 2005).

Assessment of Effects of Predators and Egg Collection on Fledge Rates

A modified version of the IPM allowed annual fledge rates to vary as a function of whether egg collection occurred in each year, as well as current and 2-year-previous (thus allowing for lagged effects) indicators of predator growth rate. Egg collection was included as a categorical covariate (collect$_t$) with 1s and 0s to represent the years with (1977–96) and without (1977–2013) egg collection, respectively. We used lynx fur return data for northern Alberta from Statistics Canada (http://www.statcan.gc.ca/) as an index of lynx abundance (Krebs et al., 2013). Lynx abundance reflects the timing of the hare-lynx cycle and associated fluctuations in the abundance of all boreal predators whose numbers vary cyclically with snowshoe hares (Boutin et al., 1995; Krebs et al., 2013). Fur returns vary over time (e.g., from one cycle to another) and instead of using the absolute fur return data, we estimated the log population growth rate of lynx (i.e., intrinsic rate) = $\ln(X_t/X_{t-1})$ based on the fur return data with X_t and X_{t-1} equal to the total fur returns in years t and $t-1$, respectively. This estimate provides a measure of where a particular year lies in the hare-lynx cycle. We expected that the effects of predators on Whooping Cranes might differ depending on the stage of the hare-lynx cycle because of differences in the availability of hares. Predator abundance typically lags behind snowshoe hare abundance by 1–2 years (O'Donoghue et al., 2010) with the collapse of snowshoe hare numbers occurring as predator numbers peak. Thus, a strong negative correlation between Whooping Crane fledge rates in year t and predator growth rates from year $t-1$ to t (lynx.curr) corresponds to a strong effect of predators as their abundance is increasing while hares are still abundant. In contrast, a strong negative correlation between Whooping

Crane fledge rates in year t and predator growth rates two years prior from year $t-3$ to $t-2$ (lynx. lag) corresponds to a period after hare numbers are expected to have collapsed while predators are still abundant. This latter scenario would be consistent with the alternative prey hypothesis (Angelstam et al., 1984) where predators turn to alternate prey in the community after their primary prey base has declined. The model structure for this version of the IPM specified fledge rate (f_t, the number of young per breeding female) as:

$$\log(f_t) = \mu + \beta_1{}^* \, collect_t + \beta_2{}^* \, lynx.curr_t + \beta_3{}^* \, lynx.lag_t$$

ASSESSMENT OF BREEDING RANGE DYNAMICS

In concert with rapid growth in population size since 1966, the size of the AWBP's breeding range also increased dramatically. Thus, to better understand the population ecology of the AWBP, we investigated its range dynamics during the breeding season. We assessed range dynamics using a data set containing nest locations determined during a census of nests conducted annually in all known nesting areas since 1966 by the CWS. For each year from 1967 to 2016, we estimated the annual and cumulative breeding ranges. Though breeding surveys began in 1966, we did not estimate the extent of occurrence (EOO) for 1966 because it is likely that surveys did not provide adequate coverage of the breeding range in that year (Johns et al., 2005).

We estimated annual breeding range size as the EOO$_t$ of nesting areas by computing the minimum convex polygon (MCP; Burgman and Fox, 2003) containing all nests observed in year t and the annual cumulative EOO$_{1967:t}$ as the MCP containing all nests observed from 1967 through year t. We use EOO as defined by a MCP to represent the breeding range, instead of area of occupancy (i.e., the area of habitats within the EOO actually occupied by nesting cranes), because EOO describes the total area within which crane nests may occur and is thus commonly used to describe the range of endangered species (International Union for Conservation of Nature, 2012) requiring protection or for official listing purposes (Committee on the Status of Endangered Wildlife in Canada, 2015). We caution, however, that EOO is not necessarily correlated with area or quality of breeding habitat, which are more readily assessed with area of occupancy.

We computed EOO_t, $EOO_{1967:t}$, and the proportions of these occurring outside the designated critical habitat, using standard tools in a geographic information system (Environmental Systems Research Institute, 2014). We estimated annual change in range size (ΔEOO_t) as:

$$\Delta EOO_t = EOO_t / EOO_{t-1}$$

for t = 1968–2016, and we estimated annual change in population size (λ_t^*), using the ratios of the previously described counts from Aransas National Wildlife Refuge, as:

$$\lambda_t^* = \left(C_{f,hy,t} + C_{f,ahy,t}\right) \Big/ \left(C_{f,hy,t-1} + C_{f,ahy,t-1}\right)$$

for t = 1968–2015. We calculated λ_t^* as described rather than using the estimated λ_t from the IPM to allow us to incorporate information prior to 1977 and after 2012; IPM predicted abundance and λ_t estimates were highly correlated with the Aransas survey counts and λ_t^* estimates (see later). To gain insight into the relationship between size of the population and size of the breeding range, and their change over time, we examined geometric mean values (and their standard deviations, SD) of $\overline{\Delta EOO}$ and $\overline{\lambda}^*$ and computed correlations between ΔEOO_t and λ_t^*. Additionally, to test the hypothesis that colonization by new breeding pairs (which form around age 4) is the primary cause of range expansion, we computed the correlation

between ΔEOO_t and λ_{t-4}^*. Finally, to determine how colonization of new nesting areas contributed to range expansion, we created maps of nesting areas and of the breeding range over time, and interpreted them along with summary statistics for the metrics described earlier.

POPULATION AND BREEDING RANGE DYNAMICS OF WHOOPING CRANES IN THE AWBP

Population Growth and Age Structure

Estimates of annual abundance (N_t) from the IPM correlated very closely with the observed population counts from the Aransas survey ($r = 0.99$), and estimates of λ_t (from the IPM) correlated closely with estimates of λ_t^* from the Aransas survey counts ($r = 0.93$). The counts indicated an increase from 43 individuals in the winter of 1966–67 to 329 in 2015–16 with an average $\overline{\lambda}^*$ of 4.2% per year (Fig. 4.2). We estimated, based on our IPM, $N = 72$ in the winter of 1977–78 (95% CI: 66, 90) increasing to $N = 285$ individuals by the winter of 2012–13 (CI: 252, 318). Average geometric mean $\overline{\lambda}$ estimated via the IPM (1977–2013) was 3.8% (CI: 3.2%, 4.4%), slightly lower than the geometric mean of $\overline{\lambda}^*$ (4.2%) estimated for the entire period from 1967 to 2013 as:

$$\overline{\lambda}_{1967:2013}^* = \left(\left(C_{f.hy,2013} + C_{f.ahy,2013}\right) \Big/ \left(C_{f.hy,1967} + C_{f.ahy,1967}\right)\right)^{1/(2013-1967)}$$

likely because of higher growth in the earlier period (1967–77) not covered by the IPM ($\overline{\lambda}_{1967:1977}^* = 5.2\%$; Fig. 4.2B). For comparison, the geometric mean estimate of $\overline{\lambda}_{1977:2013}^*$ was 4.0%, similar to though slightly higher than the estimate from the IPM at 3.8%. Our estimates are similar to those of Butler et al. (2013), who reported 3.8% average annual growth from 1938 to 2010 (see also Converse et al., Chapter 8, this volume). Although the population (as indexed

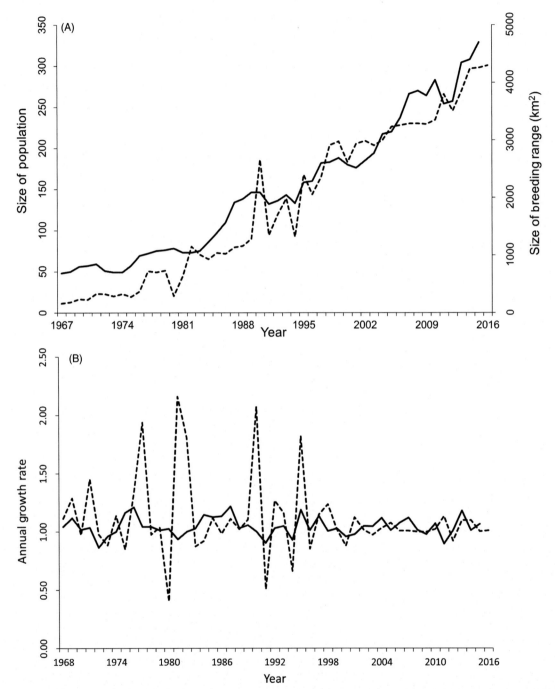

FIGURE 4.2 Estimates of (A) population size (solid line, 1966–2015) and breeding range size (dashed line, 1967–2016) and (B) annual growth rates of population size (solid line, 1968–2015) and breeding range (dashed line, 1968–2016) of the Aransas-Wood Buffalo Population of Whooping Cranes. Population size estimates are from winter surveys at Aransas NWR.

by counts) exhibited strong positive growth over the time period considered, apparent negative growth ($\lambda_t^* < 1$) occurred in 25% of the years between 1977 and 2012 and 18% of the years between 1966 and 2015 (Fig. 4.2B).

Previous analyses raised the possibility that density dependence was influencing the population prior to 2000, including possible negative density-dependent effects on recruitment (Link et al., 2003). Negative density dependence has been incorporated into recent attempts to model future population growth through a "ceiling" beyond which growth would cease, although the influence of this ceiling was judged to occur at population sizes that have yet to be reached (Gil-Weir, 2006; Miller et al., 2016). The sustained growth of the population over the full time period of this analysis suggests that density dependence is not yet operating on any of the demographic rates, although inference is limited because of missing data on subadult survival for most years and adult survival in recent years. With increasing abundance, the AWBP has expanded both its breeding range (this study) and its winter range (Butler et al., 2014), and in the latter case, individuals are now overwintering in unprotected areas outside of the Aransas refuge. The AWBP breeding range is vast and inhabited at relatively low density in comparison to the winter range. It is reasonable to expect that at some larger population size, negative density dependence will begin to influence the population, but we predict those effects will be more likely due to crowding and site-dependent mechanisms operating on the wintering grounds (Gil-Weir et al., 2012). Continued monitoring and analysis of the demographic rates will provide insight into the timing and population size at which density-dependent effects may begin to operate.

Because the Aransas survey distinguishes only juvenile "brown birds" from all subadult and adult "white birds," there has been limited information available on how the proportions of breeding-age and non-breeding-age individuals vary over time. The IPM estimates the annual abundance of all age classes via the state space model, allowing us to use those estimates to derive the estimated proportions of juveniles, subadults, and adults along with measures of uncertainty. We assumed that AWBP Whooping Cranes breed beginning in their fifth year (Urbanek and Lewis, 2015) and found that, averaged across years, the proportion of the population represented by breeding-age birds was 0.61 (CI: 0.57, 0.64), proportion of subadults was 0.26 (CI: 0.23, 0.29), and proportion of hatch-year juveniles was 0.13 (CI: 0.11, 0.13). Plots of these age distributions over time revealed annual variation with an apparent periodic pattern most evident in the adult and subadult proportions (Fig. 4.3). The proportion of adult breeding-age individuals peaked at 0.72 in 1983 and was lowest in 1988 at 0.48. Maximum and minimum proportions, respectively, for subadults were 0.36 (1989) and 0.17 (1984), and for juveniles were 0.18 (1988) and 0.06 (1982). As we discuss further next, this pattern may reflect the influence of cyclic predators on early life stages followed by relatively consistent annual survival estimates for older subadults and adults. There was no apparent long-term trend in the proportions of adult breeders, subadults, or juveniles in the population (Fig. 4.3).

Relationship between Predator Population Growth and Reproductive Output

A constrained version of the IPM was used to evaluate the influence of boreal forest predators on fledging rates in the AWBP. Lynx fur returns from northern Alberta provided an index of predator abundance associated with the hare-lynx cycle and is likely correlated with the abundance of several other species, including other potential predators of juvenile Whooping Cranes: coyote (*Canis latrans*), American mink (*Neovison vison*), red fox (*Vulpes vulpes*), and gray wolves (*Canis lupus*) (Boutin et al., 1995; Boyce et al., 2005; Bulmer, 1974). Actual predators of

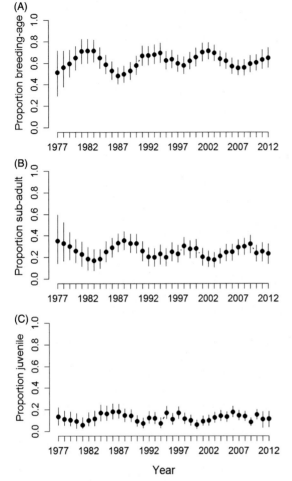

FIGURE 4.3 Estimated annual proportion of the population represented by breeding-age adults (A), non-breeding-age subadults (B), and hatch-year juveniles (C) in the Aransas-Wood Buffalo Whooping Crane population between the winters of 1977–78 and 2012–13. Bars represent the 95% confidence intervals on the estimates.

AWBP Whooping Crane eggs and chicks are poorly known because of the inaccessibility of the breeding grounds; however, predation of juveniles by gray wolves has been observed (Boyce and Miller, 1985; Kuyt et al., 1981). Coyotes and red fox were key predators of juveniles in the former Sandhill Crane cross-fostering program in the Rocky Mountain population

(Drewien et al., 1985). Bobcats (*Lynx rufus*) were a major source of predation on juvenile Whooping Cranes in the reintroduced Florida Nonmigratory Population (Nesbitt et al., 2001).

We found strong evidence for a negative relationship related to 2-year delayed, but not current, predator population growth, indicating that the strongest impacts of predators occur two years after they have reached peak population growth rates. The increase in predators following the rise in snowshoe hare abundance is due to increased predator reproductive output and immigration but typically occurs with a 1- to 2-year lag behind the increase in snowshoe hare numbers (Krebs et al., 2001, 2013; O'Donoghue et al., 2010). Our results suggest that predation pressure on juvenile Whooping Cranes does not increase as predator numbers are rising while hares are still abundant, but instead corresponds to the period after the snowshoe hare population has declined, thereby supporting the alternative prey hypothesis (Angelstam et al., 1984). When hares begin to decline, predators may seek out alternative prey following the loss of their principal prey base (Angelstam et al., 1984; Lack, 1954). Such effects have been observed for other prey species in the boreal ecosystem, including spruce grouse (*Falcipennis canadensis*), willow ptarmigan (*Lagopus lagopus*), porcupine (*Hystricomorph histricidae*), striped skunk (*Mephitis mephitis*), and sciurid mammals (Boutin et al., 1995; Keith and Cary, 1991). The effects of predator abundance on fledge rates may also interact with fluctuations in wetland depth in nesting areas by affecting the extent to which predators are able to access nest sites (Binkley and Miller, 1983; Gil-Weir, 2006; Urbanek and Lewis, 2015).

Our results add to the growing body of evidence that cyclic fluctuations in boreal forest predators play an important role in the population dynamics of Whooping Cranes (Boyce et al., 2005; Butler et al., 2013). We show that this effect occurs through an influence of predators on fledging rates, although within-year correlations between fledge rates and juvenile survival from August to December are positive ($r = 0.57$), suggesting that high predator abundance may also affect juveniles postfledging. The influence of predators was evident in the periodicity of both winter abundance (Fig. 4.2A) and the proportion of individuals in breeding and non-breeding-age classes (Fig. 4.3).

Contributions of the Demographic Rates to Population Growth

Estimates of the demographic rates indicated that (1) on average about one chick is fledged for every two breeding pairs per year (see also Converse et al., Chapter 8, this volume); (2) after-hatch-year annual survival is lowest for second-year birds, intermediate for subadults (3rd and 4th year) and first-time-breeding adults, and highest for adults at least 5 years of age (see also Gil-Weir, 2006); and (3) individuals expected to breed for the first time have similar survival rates to the subadult age classes but lower rates than adults in at least their second breeding year (Table 4.1). Both adult and hatch-year mean survival probabilities tend to be lower than those in the Eastern Migratory Population (EMP), although there is overlap in the credible intervals between the two populations. Estimates of 4-month survival for fall juveniles in the EMP were approximately 0.94–0.98, and adult annual survival varied between 0.96 and 0.98 (Servanty et al., 2014).

The survival probabilities differed markedly in their temporal variability across age groups. In general, fledge rates and survival of early life stages, such as hatch year 4-month survival, were more variable than survival for the later life stages (e.g., adult annual survival, Table 4.1). The degree to which these rates varied had strong effects on the extent to which they influenced variability in population growth between 1977 and 2013. Population growth was most strongly correlated with fledging rate ($r = 0.47$), while the correlation with adult annual survival was only $r = 0.22$ (Table 4.1). Elasticity analyses performed on long-lived species with a delayed age at maturity indicate that a proportional change in adult survival has a greater influence on population growth compared to the same proportional change in the reproductive rates (Sæther and Bakke, 2000). However, such analyses include only the influence of a change in each rate and not the extent to which the rates actually

TABLE 4.1 Estimates of the Posterior Means (Median for Fledge Rate), Temporal Variance, and Correlation with Population Growth for All Demographic Rates Estimated in the IPM. Values in Parentheses Are the 95% Credible Intervals. Hatch-Year Survival Is a 4-Month Estimate from 1 August through 30 November. For Survival Rates after Hatch Year, the Annual Survival Is Estimated from 1 December of Year t to 1 December of Year $t + 1$, and the Years Shown Represent Year t. All Rates Initially Estimated in Wilson et al. (2016).

Rate	Mean	Temporal variance	Correlation with population growth	Years
Breeding propensity (bp_t)	0.919 (0.884, 0.949)	0.006 (0.003, 0.015)	0.408 (0.146, 0.634)	1982–2009
Fledge rate (f_t)	0.515 (0.466, 0.564)	0.015 (0.005, 0.031)	0.467 (0.189, 0.678)	1977–2013
Hatch-year survival (ϕ_{hy})	0.905 (0.816, 0.976)	0.014 (0.001, 0.050)	0.294 (−0.191, 0.694)	1977–1988
Second-year survival (ϕ_{sy})	0.850 (0.754, 0.932)	0.016 (0, 0.070)	0.332 (−0.257, 0.747)	1977–1988
Subadult survival (ϕ_{sad})	0.892 (0.835, 0.952)	0.005 (0, 0.019)	0.012 (−0.516, 0.525)	1978–1990
Adult 1 survival (ϕ_{ad1})	0.895 (0.798, 0.971)	0.009 (0, 0.052)	0.104 (−0.505, 0.633)	1980–1991
Adult 2 survival (ϕ_{ad2})	0.944 (0.926, 0.962)	0.001 (0, 0.002)	0.221 (−0.220, 0.576)	1981–2005

vary. The estimate of annual temporal variance in Whooping Crane fledge rate was 15 times greater than that in adult survival, and therefore the observed fluctuations in population growth were driven to a greater extent by the more variable fledge rate. This observation was similar to that found for the Rocky Mountain Sandhill Crane (*Grus canadensis*) population where over a 23-year period, adult survival was stable with little variation, but recruitment showed high temporal variation (Gerber et al., 2015). The pattern wherein adult survival varies little over time and early life stages show greater variation and have a greater observed influence on variation in population growth has also been reported for other long-lived species such as desert tortoise (*Gopherus agassizii*; Doak et al., 1994) and orca (*Orcinus orca*; Mills et al., 1999). However, it is important to recognize that this result was observed because adult survival has low interannual variability. Substantial changes in adult survival, for instance due to an increase in anthropogenic mortality of adults, would be expected to produce a strong effect on population growth.

Effects of Egg Collecting on Population-Level Productivity

The mean population-level fledging rate was very similar during the 1977–96 period when egg collecting occurred (0.523, 95% CI: 0.470, 0.580) compared to the 1997–2013 period after egg collecting had ceased (0.518, 95% CI:0.475, 0.562). In addition, the coefficient for egg collecting effects on fledge rate was very near 0 (0.010; 95% CI: −0.126, 0.145), further suggesting no difference in mean fledging rates between the two periods. Although such an analysis would preferentially include a mix of collection and noncollection years throughout the 1977–2013 period, there was no trend in the fledge rate that might bias interpretation between the two periods. The lack of any influence of egg collecting on the AWBP is consistent with earlier studies

showing no negative effects of egg collection at the population level (Ellis and Gee, 2001; Link et al., 2003) and in fact, an increase in breeding success may have been realized for pairs that had an egg removed (Boyce et al., 2005). Whooping Cranes typically lay two eggs but rarely raise two young, with the second-hatched chick usually dying from starvation or siblicide (Bergeson et al., 2001; Urbanek and Lewis, 2015). Potential reasons for increased breeding success following egg removal include (1) sibling aggressive behavior that increases conspicuousness to predators is reduced; (2) parents are able to more capably defend a single chick against predators; and (3) elimination of the effect of predators increasing their search effort for live chicks after locating one that is dead or injured (Bergeson et al., 2001; Boyce et al., 2005). Egg collecting has benefits for alternative management efforts (e.g., head-starting programs that temporarily rear juveniles in captivity before release), as well as research aimed at increasing the productivity of captive-reared Whooping Crane populations. Our results combined with earlier studies indicate that this practice could be resumed and would be unlikely to affect the population, although any increase in egg collection should be accompanied by monitoring the responses of individual- and population-level fecundity to egg collection.

Breeding Range Dynamics

From 1967 to 2016, the breeding range increased by 26-fold, from 165 km^2 to 4,295 km^2 (Fig. 4.4). This expansion occurred via the colonization of new sites within nesting areas at the core of the breeding range (i.e., the Sass and Klewi nesting areas, where breeding cranes were first discovered in 1954) and of entirely new nesting areas on its periphery to the northwest and southeast (Fig. 4.5). Annual increases in range size tended to be small (Fig. 4.2B), likely because natal dispersal in AWBP Whooping Cranes is low, with the majority of new pairs

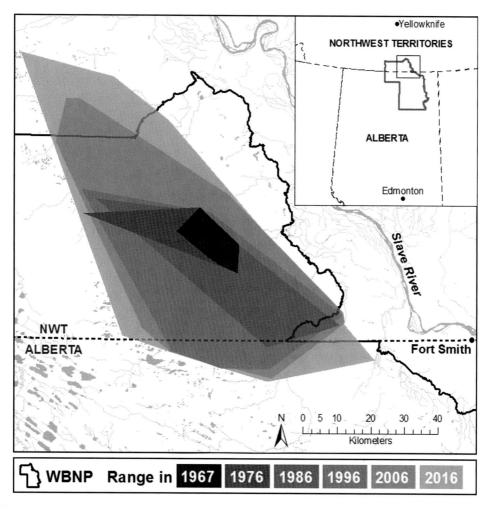

FIGURE 4.4 Map showing expansion in the breeding range of the Aransas-Wood Buffalo Whooping Crane population, 1967–2016. Shaded polygons represent annual breeding ranges in the years indicated, and correspond to minimum convex polygons derived from all nests in those years.

nesting within 20 km of their natal sites (Johns et al., 2005). Moreover, it is likely that habitat was not limiting in the core of the range, meaning new breeding pairs did not have to disperse long distances to find territories. Small decreases in breeding range size were due to established pairs apparently failing to nest in a particular year, or nesting at a site closer to the core of the range. In some years, however, the breeding range grew or shrank more dramatically. Large increases likely occurred because pioneering behavior (Johns et al., 2005) by one or a few breeding pairs resulted in colonization of new, peripheral nesting areas. With a few exceptions, new nesting areas were initially colonized close to the core, with subsequent colonization in more peripheral areas (Fig. 4.5). Outside of the core, new areas were colonized in

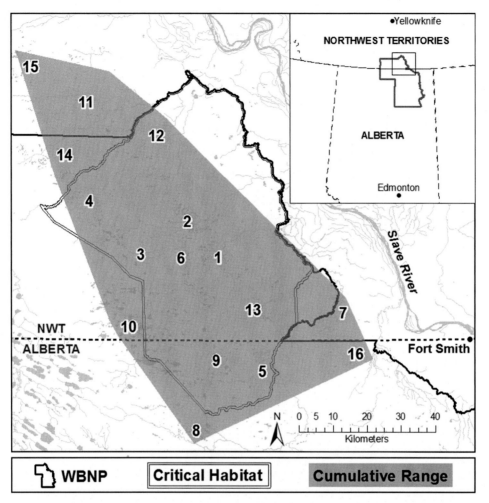

FIGURE 4.5   Map showing the cumulative breeding range of the Aransas-Wood Buffalo Whooping Crane population in relation to critical habitat in 2016. Nesting areas are numbered chronologically, according to the year they were colonized: 1 Sass, 2 Klewi, 3 Lake of the Grave, 4 Nyarling, 5 Alberta East, 6 Sass-Klewi, 7 Lobstick Creek, 8 Alberta South, 9 Alberta West, 10 Preble Creek, 11 North Boundary, 12 Bear Creek, 13 Seton Creek, 14 North Nyarling, 15 Swampy Lakes, 16 Salt Plains.

1969 (Lake of the Grave), 1971 (Nyarling), 1977 (Alberta East), 1980 (Sass-Klewi), 1982 (Lobstick Creek), 1990 (Alberta South, Alberta West), 1993 (Preble Creek), 1998 (North Boundary), 2001 (Bear Creek), 2003 (Seton Creek), 2005 (North Nyarling), 2014 (Swampy Lakes), and 2015 (Salt Plains). Once new nesting areas were colonized, they tended to be occupied continuously, although in a few cases they were unoccupied for several years before being used for nesting again; only one nesting area (Alberta South) has apparently been abandoned (Table 4.2). Large decreases in breeding range size occurred when pairs that formerly occupied a peripheral nesting area either failed to nest in a particular year or nested much closer to the core.

TABLE 4.2   Summary of Colonization History and Subsequent Use by Whooping Cranes of 16 Nesting Areas in and near Wood Buffalo National Park (WBNP), Canada. Columns Indicate the Year Nesting Was First Confirmed, and the Number and Proportion of Years with Confirmed Nesting after Colonization. Numbers in Brackets Refer to Numbered Locations as Shown in Fig. 4.5.

Nesting area	Years		
	Confirmed	Number	Proportion
Sass [01]	1966	51	1.00
Klewi [02]	1967	50	1.00
Lake of the Grave [03]	1969	27	0.56
Nyarling [04]	1971	46	1.00
Alberta East [05]	1977	39	0.98
Sass-Klewi [06]	1980	37	1.00
Lobstick Creek [07]	1982	35	1.00
Alberta South [08]	1990	1	0.04
Alberta West [09]	1990	25	0.93
Preble Creek [10]	1993	23	0.96
North Boundary [11]	1998	19	1.00
Bear Creek [12]	2001	7	0.44
Seton Creek [13]	2003	14	1.00
North Nyarling [14]	2005	12	1.00
Swampy Lakes [15]	2014	3	1.00
Salt Plains [16]	2015	2	1.00

Since 1967, there has been a strong positive relationship between the size of the population, as indexed by the winter counts ($C_{f.hy,t} + C_{f.ahy,t}$), and the size of its breeding range (EOO_t; Fig. 4.2A; $r = 0.97$, $n = 48$ years) supporting the hypothesis that ecological mechanisms such as territoriality influence range dynamics. Since 1968, the geometric mean growth rate of the range $\bar{\lambda}$ was 6.9% compared to only 4.4% for the average annual growth rate of the population (λ^*), suggesting that cranes are not limited by habitat at the scale of the breeding range because increased population growth results in a relatively larger increase in range size. In general, annual ΔEOO_t ranged from −60% to 116% and was much more variable than λ_t^*, which ranged from −14% to 22%. This difference in variability was particularly true until 1996 after which there was a dampening in ΔEOO_t (Fig. 4.2B). Variability in ΔEOO_t was driven mainly by large changes in a few years from 1967 to 1995, when geometric mean $\overline{\Delta EOO}$ averaged 10.0%, likely because cranes colonized, recolonized, or failed to occupy new nesting areas; since 1995, $\overline{\Delta EOO}$ declined and stabilized to an average of 2.8%. Taken together, these results suggest that with ongoing expansion of the population, cranes began to fill the available habitat within the range in a more uniform manner. This is consistent with the idea that range limits correspond to fundamental niche limitation (Sexton et al., 2009), that is, that the potential breeding range is partly defined by the distribution of a fixed amount of suitable breeding habitat, which, when colonized, becomes the realized range (i.e., EOO). When the population was small, large changes in EOO could occur via colonization of areas far from the core; at larger population sizes, however, there was less unoccupied habitat available for colonization far from the core within the potential breeding range, so changes in EOO were less dramatic.

Although $\bar{\lambda}^*$ and $\overline{\Delta EOO}$ have both averaged approximately 3–4% since 1996, they are not strongly correlated ($r = 0.08$ since 1996, and $r = -0.10$ since 2006). Large increases in range size may occur 3–4 years after periods of elevated recruitment, as subadults enter the breeding population at 3–4 years of age (Urbanek and Lewis, 2015), and pioneer and colonize new areas. If this is true, growth in range size at time t (ΔEOO_t) should be positively correlated with the population growth rate 4 years earlier (λ_{t-4}^*), so we tested this hypothesis for the 10-year period from 2006 to 2015 (i.e., by examining the correlation between ΔEOO_t from 2006 to 2015 and λ_t^* between 2002 and 2011) and similarly for the periods from 1976 to 2015 (40 years), 1986 to 2015 (30 years), and 1996 to 2015 (20 years). Lending some support to this hypothesis, from 2006 to 2015 the correlation of ΔEOO_t with λ_{t-4}^* ($r = 0.28$) was stronger than with λ_t^* ($r = -0.10$), although this was not the case for the other periods we considered. Thus, it is not clear if pioneering by new breeding pairs is the primary cause of range expansion. Another possibility is that established pairs may pioneer new areas after experiencing failed breeding attempts elsewhere (Clobert et al., 2012), or after pair formation following death of a mate or divorce.

When the breeding grounds of the AWBP were discovered in 1954 (Allen, 1956), the entire range fell within WBNP (Fig. 4.4). Cranes colonized the Lobstick Creek and North Boundary nesting areas outside the national park in 1982 and 1998, respectively (Fig. 4.5); since then these areas have been used continuously and expanded in area and population size. The Lobstick Creek area occurs on reserve lands of the

Salt River First Nation, and the North Boundary area is on crown land managed by the Government of the Northwest Territories. When critical habitat was identified for the AWBP under Canada's Species at Risk Act in 2007 (Environment Canada, 2007), 8 nests and 22% of the cumulative range occurred outside the area designated as containing critical habitat; this increased to 15 nests and 38% of the cumulative range in 2016 (Fig. 4.5). Cranes nesting in areas outside the critical habitat, and especially outside WBNP, could come under increasing pressure from industrial development, such as forestry, mining, or road construction. Expanding the critical habitat to include the cumulative range, within and outside WBNP, would reduce risk to cranes during the breeding season by ensuring that protections afforded under federal legislation are extended to the entire breeding population.

Whooping Cranes are listed as Endangered under Canada's Species at Risk Act and the Endangered Species Act in the United States. Recent modeling efforts suggest the AWBP will reach the population threshold for downlisting from Endangered to Threatened by 2026 (S. Wilson and M. Bidwell, unpublished analysis). In addition to population size, EOO is also used to assess status of species at risk in Canada. The AWBP is fast approaching the EOO threshold for downlisting from Endangered to Threatened, and at current growth rates could reach it before the population threshold for downlisting is reached. It is unknown if reassessment of Whooping Cranes in Canada would consider only the size of the breeding range, or if size of the wintering range would also come under consideration.

SUMMARY AND OUTLOOK

Our analyses highlight several novel features of the population and breeding range dynamics of Aransas-Wood Buffalo Whooping Cranes that not only add to our knowledge of the ecology of this species but also provide important information for conservation efforts. Some of these findings include (1) the stronger influence of early relative to later life stages on variation in population growth, which is due to much greater annual variability in the reproductive rates and survival of young birds compared to adult survival; (2) the presence of a delayed relationship between predator population growth on Whooping Crane reproductive success consistent with the alternative prey hypothesis; (3) the lack of any influence of egg collecting on population-level productivity; (4) the initial rapid breeding range expansion prior to the mid-1990s that exceeded population growth, with a more recent dampening in both the rate and variability of that expansion; and (5) the ongoing colonization, as recently as 2015, of new nesting areas within and outside WBNP and the increasing proportion of the range lying outside areas designated as critical habitat.

Several areas of research would be useful to further our knowledge on population dynamics and recovery of the AWBP. First, the connection between AWBP productivity and the hare-lynx cycle is still correlational and we do not fully understand the mechanism by which predators influence Whooping Cranes, or the key predator species involved. Detailed field study is difficult owing to the remote nature of the breeding grounds and the time period of the hare-lynx cycle. However, additional research would aid our understanding of the population dynamics of Whooping Cranes and how shifting predator–prey communities following both landscape and climate change might influence ongoing recovery efforts for this species. Second, individuals are spreading into new parts of the species range with continuing growth of the population. On both the breeding and the wintering grounds, individuals are now occupying unprotected habitats outside of Wood Buffalo National Park and Aransas NWR (Butler et al., 2014), respectively. Further study is needed to estimate the risk

from development and other threats for cranes occupying areas outside WBNP and ANWR. For example, whether the occupation of new areas affects survival or reproduction is unclear, but individual-based field projects could determine demographic rates across space in both the breeding and wintering areas. Third, more research is needed to determine which management strategies, if any, should be enacted to enhance population growth of the AWBP. While the recovery of Whooping Cranes is generally regarded as a success story, the remnant population only recently exceeded 300 individuals, with a similar total number in the captive and reintroduced populations. The AWBP is susceptible to catastrophic events, and continued research should focus on programs that would allow for the maintenance of at least current population growth. Anthropogenic mortality risks to subadults and adults, such as powerline collisions, oil spills, and accidental or illegal hunting all have the potential to be minimized (Sharp and Vogel, 1992; Stehn and Wassenich, 2008). Research could also focus on the benefits of programs to enhance reproduction and/or juvenile survival. For instance, head-starting programs could collect single eggs from two-egg nests, raise juveniles with captive birds, and then release those juveniles back into the remnant population during fall migration or on the wintering grounds. Modeling efforts could assess potential benefits and costs of these programs for the population, and decisions would need to incorporate other risks such as disturbance or the introduction of undesirable traits into the population.

Acknowledgments

The authors wish to thank many individuals who contributed to this work, including B. Johns, R. Clark, J. Conkin, S. Converse, L. Craig-Moore, K. Gil-Weir, W. Jobman, L. Scott Mills, J. Rempel, G. Robertson, T. Stehn, and M. Tacha. This research was supported by Environment and Climate Change Canada and the United States Fish and Wildlife Service.

References

Abadi, F., Gimenez, O., Arlettaz, R., Schaub, M., 2010. An assessment of integrated population models: bias, accuracy and violation of the assumption of independence. Ecology 91, 7–14.

Allen, R.P. (Ed.), 1956. A Report on the Whooping Crane's Northern Breeding Grounds: A Supplement to Research Report No. 3, The Whooping Crane. National Audubon Society, New York, pp. 1–60.

Angelstam, P., Lindstrom, E., Widen, P., 1984. Role of predation in short-term population fluctuations of some birds and mammals in Fennoscandia. Oecologia 62 (2), 199–208.

Bergeson, D.G., Johns, B.W., Holroyd, G.L., 2001. Mortality of whooping crane colts in Wood Buffalo National Park, Canada, 1997–1999. Proceedings of the North American Crane Workshop 8, 6–10.

Binkley, C.S., Miller, R.S., 1980. Survivorship of the whooping crane, Grus americana. Ecology 61 (2), 434–437.

Binkley, C.S., Miller, R.S., 1983. Population characteristics of the whooping crane, Grus americana. Can. J. Zool. 61 (12), 2768–2776.

Bishop, M.A., Blankinship, D.R., 1982. Dynamics of sub-adult flocks of whooping cranes at Aransas National Wildlife Refuge, Texas, 1978–1981. In: Lewis, J.C. (Ed.), Proceedings of the 1981 International Crane Workshop. National Audubon Society, Tavernier, FL, pp. 1–9.

Boutin, S., Krebs, C.J., Boonstra, R., Dale, M.R.T., Hannon, S.J., Martin, K., Sinclair, A.R.E., Smith, J.N.M., Turkington, R., Blower, M., Byrom, A., Doyle, F.I., Doyle, C., Hik, D., Hofer, L., Hubbs, A., Karels, T., Murray, D.L., Nams, V., O'Donoghue, M., Rohner, C., Schweiger, S., 1995. Population changes of the vertebrate community during a snowshoe hare cycle in Canada's boreal forest. Oikos 74 (1), 69–80.

Boyce, M.S., Lele, S.R., Johns, B.W., 2005. Whooping crane recruitment enhanced by egg removal. Biol. Conserv. 126 (3), 395–401.

Boyce, M.S., Miller, R.S., 1985. Ten-year periodicity in whooping crane census. Auk 102 (3), 658–660.

Brooks, S.P., King, R., Morgan, B.J.T., 2004. A Bayesian approach to combining animal abundance and demographic data. Anim. Biodivers. Conserv. 27 (1), 515–529.

Bulmer, M.G., 1974. A statistical analysis of the 10-year cycle in Canada. J. Anim. Ecol. 43 (3), 701–718.

Burgman, M.A., Fox, J.C., 2003. Bias in species range estimates from minimum convex polygons: implications for conservation and options for improved planning. Anim. Conserv. 6 (1), 19–38.

Butler, M.J., Harris, G., Strobel, B.N., 2013. Influence of whooping crane population dynamics on its recovery and management. Biol. Conserv. 162, 89–99.

Butler, M.J., Metzger, K.L., Harris, G., 2014. Whooping crane demographic responses to winter drought focus conservation strategies. Biol. Conserv. 179, 72–85.

Canadian Wildlife Service and U.S. Fish and Wildlife Service, 2007. International Recovery Plan for the Whooping Crane (Grus americana), third revision. Environment Canada, Ottawa, and U.S. Fish and Wildlife Service, Albuquerque, NM.

Clobert, J., Baguette, M., Benton, T.G., Bullock, J.M., 2012. Dispersal Ecology and Evolution. Oxford University Press, Oxford, UK, 462pp.

Committee on the Status of Endangered Wildlife in Canada, 2015. COSEWIC Assessment Process, Categories and Guidelines. Downloaded in January 2017 from http://www.cosewic.gc.ca/default.asp?lang=en&n=ED199D3B-1

Converse, S.J., Strobel, B.N., Barzen, J.A., 2018. Reproductive Failure in the Eastern Migratory Population: The Interaction of Research and Management (Chapter 8). In: French, Jr., J.B., Converse, S.J., Austin, J.E. (Eds.), Whooping Cranes: Biology and Conservation. Biodiversity of the World: Conservation from Genes to Landscapes. Academic Press, San Diego, CA.

Doak, D.F., Kareiva, P., Klapetka, B., 1994. Modeling population viability for the desert tortoise in the western Mojave Desert. Ecol. Appl. 4 (3), 446–460.

Drewien, R.C., Bouffard, S.H., Call, D.D., Wonacott, R.A., 1985. The whooping crane cross-fostering experiment: the role of animal damage control. In: Bromley, P.T. (Ed.), Proceedings 2d Eastern Wildlife Damage Control Conference, Raleigh, NC, p. 713.

Duarte, A., Weckerly, F.W., Schaub, M., Hatfield, J.S., 2016. Estimating golden-cheeked warbler immigration: implications for the spatial scale of conservation. Anim. Conserv. 19 (1), 65–74.

Ellis, D.H., Gee, G.F., 2001. Whooping crane egg management: options and consequences. Proceedings of the North American Crane Workshop 8, 17–23.

Elton, C.S., Nicholson, M., 1942. The ten year cycle in numbers of lynx in Canada. J. Anim. Ecol. 11 (2), 215–244.

Environment Canada, 2007. Recovery Strategy for the Whooping Crane (Grus americana) in Canada. Species at Risk Act Recovery Strategy Series. Environment Canada, Ottawa. 27 pp.

Environmental Systems Research Institute, 2014. ArcGIS Desktop: Release 10.3. Redlands, U.S.

French, Jr., J.B., Converse, S.J., Austin, J.E., 2018. Whooping cranes past and present (Chapter 1). In: French, Jr., J.B., Converse, S.J., Austin, J.E. (Eds.), Whooping Cranes: Biology and Conservation. Biodiversity of the World: Conservation from Genes to Landscapes. Academic Press, San Diego, CA.

Gelman, A., Carlin, J.B., Stern, H.S., Rubin, D.B., 2004. Bayesian Data Analysis. Chapman and Hall/CRC, Boca Raton, FL.

Gerber, B.D., Kendall, W.L., Hooten, M.B., Dubovsky, J.A., Drewien, R.C., 2015. Optimal population prediction of sandhill crane recruitment based on climate-mediated habitat limitations. J. Anim. Ecol. 84 (5), 1299–1310.

Gil-Weir, K.C., 2006. Whooping crane (*Grus americana*) demography and environmental factors in a population growth simulation model. PhD dissertation, Texas A&M University, College Station, TX, 159 pp.

Gil-Weir, K.C., Grant, W.E., Slack, R.D., Wang, H.H., Fujiwara, M., 2012. Demography and population trends of whooping cranes. J. Field Ornithol. 83 (1), 1–10.

International Union for Conservation of Nature, 2012. IUCN Red List Categories and Criteria, second ed. Gland, Switzerland, Version 3.1; p. 33.

Johns, B.W., Goossen, J.P., Kuyt, E., Craig-Moore, L., 2005. Philopatry and dispersal in whooping cranes. Proceedings of the North American Crane Workshop 9, 117–125.

Johnson, H.E., Scott Mills, L., Wehausen, J.D., Stephenson, T.R., 2010. Combining ground count, telemetry and mark-resight data to infer population dynamics in an endangered species. J. Appl. Ecol. 47 (5), 1083–1093.

Keith, L.B., Cary, J.R., 1991. Mustelid, squirrel and porcupine population trends during a snowshoe hare cycle. J. Mammal. 72 (2), 373–378.

Kéry, M., Royle, J.A., 2016. Applied Hierarchical Modeling in Ecology: Analysis of Distribution, Abundance and Species Richness in R and BUGS, vol. 1: Prelude and Static Models. Academic Press, London, p. 783.

King, R., Morgan, B.J.T., Gimenez, O., Brooks, S.P., 2010. Bayesian Analysis for Population Ecology. Chapman & Hall, Boca Raton, FL.

Krebs, C.J., Boutin, S., Boonstra, R., 2001. Ecosystem Dynamics of the Boreal Forest: The Kluane Project. Oxford University Press, New York.

Krebs, C.J., Kielland, K., Bryant, J., O'Donoghue, M., Doyle, F., McIntyre, C., DiFolco, D., Berg, N., Carriere, S., Boonstra, R., Boutin, S., Kenney, A.J., Reid, D.G., Bodony, K., Putera, J., Timm, H.K., Burke, T., 2013. Synchrony in the snowshoe hare (*Lepus americanus*) cycle in northwestern North America, 1970–2012. Can. J. Zool. 91 (8), 562–572.

Kuyt, E., 1979. Banding of juvenile whooping cranes on the breeding range in the Northwest Territories Canada. North Am. Bird Bander 4 (1), 24–25.

Kuyt, E., Johnson, B.E., Drewien, R.C., 1981. A wolf kills a juvenile whooping crane. Blue Jay 39 (2), 116–119.

Lack, D., 1954. The Natural Regulation of Animal Numbers. Oxford University Press, London.

Lebreton, J.-D., Almeras, T., Pradel, R., 1999. Competing events, mixtures of information and multistratum recapture models. Bird Stud. 46 (Suppl. 1), S39–S46.

Lebreton, J.-D., Burnham, K.P., Clobert, J., Anderson, D.R., 1992. Modelling survival and testing biological hypotheses using marked animals: a unified approach with case studies. Ecol. Monogr. 62 (1), 67–118.

Link, W.A., Royle, J.A., Hatfield, S., 2003. Demographic analysis from summaries of an age-structured population. Biometrics 59 (4), 778–785.

Lunn, D.J., Thomas, A., Best, N., Spiegelhalter, D., 2000. WinBUGS – a Bayesian modelling framework: concepts, structure, and extensibility. Stat. Comput. 10 (4), 325–337.

Miller, R.S., Botkin, D.B., Mendelssohn, R., 1974. The whooping crane (*Grus americana*) population of North America. Biol. Conserv. 6 (2), 106–111.

Miller, P.S., Butler, M., Converse, S., Gil-Weir, K., Selman, W., Straka, J., Traylor-Holzer, K., Wilson, S. (Eds.), 2016. Recovery Planning for the Whooping Crane – Workshop 1: Population Viability Analysis. IUCN/SSC Conservation Breeding Specialist Group, Apple Valley, MN, 34 pp.

Mills, L.S., Doak, D.F., Wisdom, M.J., 1999. Reliability of conservation actions based on elasticity analysis of matrix models. Conserv. Biol. 13 (4), 815–829.

Nesbitt, S.A., Folk, M.J., Sullivan, K.A., Schwikert, S.T., Spalding, M.J., 2001. An update on the Florida whooping crane release project through June 2000. Proceedings of the North American Crane Workshop 8, 62–72.

O'Donoghue, M., Slough, B.G., Poole, K.G., Boutin, S., Hofer, E.J., Mowat, G., Krebs, C.J., 2010. Cyclical dynamics and behaviour of Canada lynx in northern Canada. In: Macdonald, D.W., Loveridge, A.J. (Eds.), Biology and Conservation of Wild Felids. Oxford University Press, Oxford, UK, pp. 521–536.

R Development Core Team, 2004. R: A Language and Environment for Statistical Computing. R Foundation for Statistical Computing, Vienna, Austria, www.r-project.org.

Sæther, B.-E., Bakke, Ø., 2000. Avian life history variation and contribution of demographic traits to the population growth rate. Ecology 81 (3), 642–653.

Schaub, M., Abadi, F., 2011. Integrated population models: a novel analysis framework for deeper insights into population dynamics. J. Ornithol. 152 (Suppl. 1), S227–S237.

Schaub, M., Gimenez, O., Sierro, A., Arlettaz, R., 2007. Use of integrated modeling to enhance estimates of population dynamics obtained from limited data. Conserv. Biol. 21 (4), 945–955.

Schaub, M., Jakober, H., Stauber, W., 2013. Strong contribution of immigration to local population regulation: evidence from a migratory passerine. Ecology 94 (8), 1828–1838.

Servanty, S., Converse, S.J., Bailey, L.L., 2014. Demography of a reintroduced population: moving toward management models for an endangered species, the whooping crane. Ecol. Appl. 24 (5), 927–937.

Sexton, J.P., McIntyre, P.J., Angert, A.L., Rice, K.J., 2009. Evolution and ecology of species range limits. Annu. Rev. Ecol. Evol. Syst. 40, 415–436.

Sharp, D.E., Vogel, W.D., 1992. Population status, hunting regulations, hunting activity and harvests of mid-continent sandhill cranes. Proceedings of the North American Crane Workshop 6, 24–32.

Smith, C.H., Davis, J.M., 1981. A spatial analysis of wildlife's ten-year cycle. J. Biogeogr. 8 (1), 27–35.

Stehn, T.V., Taylor, T.E., 2008. Aerial census techniques for whooping cranes on the Texas coast. Proceedings of the North American Crane Workshop 10, 146–151.

Stehn, T.V., Wassenich, T., 2008. Whooping crane collisions with power lines: an issue paper. Proceedings of the North American Crane Workshop 10, 25–36.

Sturtz, S., Ligges, U., Gelman, A., 2005. R2WinBUGS: a package for running WinBUGS from R. J. Stat. Softw. 12 (3), 1–16.

Tempel, D.J., Peery, M.Z., Gutiérrez, R.J., 2014. Using integrated population models to improve conservation monitoring: California spotted owls as a case study. Ecol. Model. 289, 86–95.

Urbanek, R.P., Lewis, J.C., 2015. Whooping crane (*Grus americana*). In: Rodewald, P.G. (Ed.), The Birds of North America. Cornell Lab of Ornithology, Ithaca, NY. Retrieved from the Birds of North America: https://birdsna.org/Species-Account/bna/species/whocra/introduction doi:10.2173/bna.153 [30 May 2017].

White, G.C., Burnham, K.P., 1999. Program MARK: survival estimation from populations of marked animals. Bird Stud. 46, S120–S139.

Wilson, S., Gil-Weir, K.C., Clark, R.G., Robertson, G.J., Bidwell, M.T., 2016. Integrated population modeling to assess demographic variation and contributions to population growth for endangered whooping cranes. Biol. Conserv. 197, 1–7.

Monitoring Recruitment and Abundance of the Aransas-Wood Buffalo Population of Whooping Cranes: 1950–2015

Bradley N. Strobel*, Matthew J. Butler**

*U.S. Fish and Wildlife Service, Necedah National Wildlife Refuge, Necedah, WI,
United States
**U.S. Fish and Wildlife Service, Albuquerque, NM, United States

INTRODUCTION

When we value a natural resource, whether it's a Widgeon (*Anas americana*), a White Pine (*Pinus strobus*) or a Whooping Crane (*Grus americana*), information regarding that resource is often highly sought after. Knowledge of the status or condition of a natural resource (e.g., abundance, spatial distribution, demographics, habitat preference, etc.) provides the foundation for well-informed decisions regarding natural resource management. Often this information is collected by implementing a monitoring program. When wildlife conservation decisions are complicated by uncertainty, long-term monitoring efforts can help conservation practitioners understand how a natural resource responds to changes in its environment (e.g., food availability, climatic conditions, harvest, etc.). In addition,

long-term monitoring efforts allow natural resource managers to evaluate the efficacy of prior management decisions and incorporate that knowledge into future decisions (i.e., adaptive management; sensu; Walters, 1986; Williams et al., 2007). For these reasons, the U.S. Fish and Wildlife Service (USFWS) has conducted aerial surveys to monitor the Aransas-Wood Buffalo Whooping Crane population on the Texas coast since 1950.

Information regarding wildlife populations is valuable because it allows wildlife managers to evaluate the consequences of management decisions and avoid making suboptimal decisions (Lyons et al. 2008; Williams et al., 2002). Suboptimal decisions can be costly, and are often the result of incomplete or insufficient information (Conroy and Peterson, 2013). Decisions regarding endangered species such as Whooping

Cranes should be informed by useful data. However, the usefulness of data depends on the nature of the decision being addressed. Over the past 65 years, Whooping Crane abundance has increased, and so too has the complexity of issues surrounding the cranes' management. Naturally, as the types of conservation decisions change over time, the types of information (i.e., data) necessary to inform them may need to change.

When aerial surveys of Whooping Cranes began in 1950, the Aransas-Wood Buffalo Population (AWBP) consisted of fewer than 30 individuals that wintered on the Blackjack Peninsula of the Aransas National Wildlife Refuge, near Austwell, Texas. Since then, annual aerial surveys have documented an increase in size of the population, as well as an increase in the area occupied during winter. The USFWS has recognized impending issues for Whooping Crane conservation (e.g., coastal erosion, rising sea levels, increased human development, etc.) that were not realized when aerial survey efforts began (Metzger et al., 2014). In response to these issues, USFWS adopted technological advances to improve collection and interpretation of these data such as fixed-wing aircraft, global positioning systems (GPS), and statistical data analyses to better inform conservation and management decisions. The long-term, annual monitoring of Whooping Cranes on the Texas coast has been a valuable tool to keep up with the rapidly changing nature of Whooping Crane conservation.

Despite the 65-year history of monitoring Whooping Cranes in Texas, there are very few documents that have specifically described how and why the population has been monitored. Understanding the objectives and methods of any monitoring strategy is critical to determining the scope and limitations of the resulting data. In this chapter, we discuss the history of the objectives and methods of the aerial surveys of the AWBP of Whooping Cranes. We organize the aerial surveys into three distinct periods: 1950–94, 1995–2010, and 2011–current. Each period represents a distinct shift in objectives and/or methods. The most substantial change

to the aerial survey methods was implemented beginning in winter 2011–12 (Butler et al., 2016). Our goal for this chapter is to emphasize the contributions aerial surveys have made to our understanding of Whooping Crane ecology and conservation efforts and look ahead toward the potential future contributions of this monitoring effort.

OBJECTIVES: FORM FOLLOWS FUNCTION

Stating the objectives of a monitoring program is essential to designing a monitoring strategy that will inform conservation decisions (Conroy and Peterson, 2013; Nichols and Williams, 2006; Witmer, 2005). Monitoring can fulfil several forms of objectives, including determining the current state of a wildlife population or its habitat, evaluating the performance of past decisions, and/or improving our understanding in how the system functions (Lyons et al., 2008). Regardless of their form, objectives must be clearly articulated to determine the details of a monitoring program, such as the type of data needed, the best time to gather those data, and the amount of data necessary to inform the decision. Unfortunately, many wildlife monitoring programs lack objectives to guide them (Marsh and Trenham, 2008; Nichols and Williams, 2006). As a consequence, these programs become a "surveillance" effort to detect an ambiguously defined change in the status of the population (Nichols and Williams, 2006). They are often driven by a hope that one day practitioners might observe something interesting or useful with the data. Instead, a good monitoring program has a specific, measurable purpose. Prior to 2011, the objectives of Whooping Crane aerial surveys were not well documented. However, several sources describe the perceived value of the aerial survey data and, thus, imply its objectives (Canadian Wildlife Service and U.S. Fish and Wildlife Service, 2007; Stehn and Taylor, 2008; U.S. Fish and Wildlife Service, 1980, 1994).

In 2011, the USFWS explicitly articulated the objectives of the survey and implemented a program designed to measure the parameters of interest identified by those objectives (Butler et al., 2016).

1950–94

The importance of objectives to sampling design was not widely recognized in 1950 when the Whooping Crane aerial surveys were initiated. The entire AWBP numbered around 31 individuals (from a low of 16, less than a decade before) and very little was known about the species' ecology or population dynamics. As an example, at the inception of aerial surveys, the population's breeding grounds had not yet been discovered, and their annual arrival to the Texas coast must have carried an air of mystery and frustration for those working to protect them. At that time, any information regarding the population was potentially useful for making conservation decisions and thus warranted gathering. Therefore, it is not surprising that between 1950 and 1980 there were no formal objectives for the aerial surveys of Whooping Cranes in the AWBP. However, several sources describe the intention and perceived value of the aerial surveys to Whooping Crane conservation, the most prominent of these being the Whooping Crane recovery plan (U.S. Fish and Wildlife Service, 1980).

As a requirement of the Endangered Species Act, the first Whooping Crane recovery plan was drafted by the USFWS and the Whooping Crane Recovery Team in 1980 (U.S. Fish and Wildlife Service, 1980). The plan was intended to identify "all existing and needed research and management efforts, to move the Whooping Crane to nonendangered status" (U.S. Fish and Wildlife Service, 1980). The USFWS and the Whooping Crane Recovery Team recognized that "knowledge of crane presence and period of residence" was required to implement and evaluate most of the research and management actions identified in their plan (U.S. Fish and Wildlife Service, 1980). To obtain this knowledge, they recommended conducting aerial survey flights at weekly, or shorter, intervals to monitor wintering Whooping Crane numbers, locations, and movements (U.S. Fish and Wildlife Service, 1980). They further recommended allocating survey effort to locating cranes that were missing from their usual territories for longer than 1 week (U.S. Fish and Wildlife Service, 1980). It is evident that the information obtained from aerial surveys was relevant to many facets of Whooping Crane recovery. Specifically, the recovery plan indicates that aerial survey data were useful for identifying critical habitat of Whooping Cranes, detecting mortalities or disease outbreaks, deterring harassment and poaching by humans, and detecting chemical spills or other environmental hazards (U.S. Fish and Wildlife Service, 1980). However, the recovery plan did not describe how survey data were to be used to meet those ends or how well the survey data addressed management needs.

The Whooping Crane recovery plan was revised in 1986 and 1994. Similar to the original plan, the revisions recognized that the information obtained through aerial surveys was valuable to many facets of Whooping Crane conservation. Although specific objectives of the aerial surveys remained undefined, the second and third revisions of the recovery plan shared many of the implied objectives of the original plan. For example, aerial surveys were still used to detect immediate hazards to Whooping Cranes (i.e., harassment by humans, chemical spills), delineate Whooping Crane habitat on the wintering grounds, and detect disease and mortalities (U.S. Fish and Wildlife Service, 1994). Additional uses for the aerial survey data were identified, suggesting that the objectives for the aerial surveys were dynamic and responded to emerging research questions or conservation issues. For example, the 1994 recovery plan includes documenting pair formation, territory establishment, and the population's age structure as important outcomes of the aerial surveys (U.S. Fish and Wildlife Service, 1994), although the surveys were not particularly well designed to obtain those metrics.

1995–2010

The 2007 revision of the Whooping Crane recovery plan included detailed abundance criteria as thresholds for downlisting the species to threatened status (Canadian Wildlife Service and U.S. Fish and Wildlife Service, 2007). One of the primary intentions of the aerial surveys at that time was to track the growth of the Whooping Crane population (Stehn and Taylor, 2008), presumably in relation to the recovery criterion. In addition, aerial surveys continued to be used to obtain basic ecological information, including: extent of wintering range, habitat use, territory establishment, location of subadult use areas, and detection of unusual movements and mortalities (Stehn and Taylor, 2008). Stehn and Taylor (2008) also argued that aerial surveys could provide important surveillance of human-caused risks to Whooping Cranes, such as disturbance, harassment, or environmental hazards. However, obtaining such information was by coincidence and not the product of a careful design to ensure the survey would yield the desired information.

The aerial survey of the AWBP of Whooping Cranes has been viewed as a valuable tool for the conservation and recovery of Whooping Cranes since its inception. Yet, no objectives were clearly outlined. In the first 60 years of aerial surveys, a plethora of useful information was collected. However, in 2011, it was apparent that the aerial surveys could yield data that were more valuable for making conservation decisions, and with greater efficiency, if specific objectives were created and prioritized.

2011–Present

In 2011, the USFWS began developing a protocol that identified and prioritized five objectives for the aerial surveys of Whooping Cranes on the Texas coast (Butler et al., 2016). Similar to the emphasis of earlier survey efforts (Stehn and Taylor, 2008; U.S. Fish and Wildlife Service, 1980), the primary objective for the current aerial survey is to provide a robust means to quantify the Whooping Crane population's progress toward, or away from, the downlisting criteria identified in the recovery plan (Butler et al., 2016). Four additional objectives were identified to help meet the goals described in the recovery plan. Specifically, the current survey objectives are (Butler et al., 2016):

1. Estimate Whooping Crane abundance within the surveyed area with enough precision to detect a 10–15% annual population decline over a 3- to 4-year period.
2. Estimate the number of paired Whooping Cranes (i.e., two white-plumaged birds) and productive pairs (i.e., a pair with at least one colt) in the wintering population within the surveyed area.
3. Estimate annual recruitment rate of hatch-year Whooping Cranes into the population wintering within the surveyed area.
4. Create a spatially explicit resource use model to predict abundance of Whooping Cranes in relation to local characteristics (e.g., vegetation type, patch configuration, water quality, food availability, etc.) to use in conservation planning efforts.
5. Monitor the expansion of the Whooping Crane population into new areas by identifying and systematically searching areas of known or potential population expansion.

SURVEY AREA

Many aspects of the Whooping Crane surveys have remained unchanged since their inception. For example, the vast majority of effort put into monitoring abundance is exerted while the flock is on their Texas wintering grounds. Considering the remoteness of the flock's breeding territories, the wintering grounds seem to be the most efficient time and place to encounter and survey the population. Whooping Cranes in the Aransas-Wood Buffalo flock typically establish winter territories in coastal saltmarsh (U.S. Fish and Wildlife Service, 1980). Whooping Cranes exhibit high fidelity to their winter territories (Bonds, 2000; Stehn, 1992; Stehn and Johnson, 1987), and newly established territories are frequently established adjacent to existing territories (Stehn and Prieto, 2010). These conditions result in a high probability of detecting many of the Whooping Cranes within the population on a given survey flight. The details regarding the precise spatial boundaries of a given survey were not well documented prior to 2011. In turn, it is difficult to determine how the spatial boundaries changed between surveys and years.

1950–2010

The strong fidelity Whooping Cranes have to their winter territory makes it relative easy to predict the location of many of the birds in the AWBP (Stehn and Taylor, 2008). That is, Whooping Cranes detected in one area of the Blackjack Peninsula could often be resighted in the same area on subsequent surveys. This attribute resulted in the aerial surveys assuming a "mail route" approach where known territories were visited in succession and observed birds were tallied. Unfortunately, a sampling frame was not strictly defined and the spatial extent of the search effort used for each survey was not recorded.

In the 1950s, most of the known Whooping Crane winter territories were located on the Blackjack Peninsula with a few on Matagorda and San Jose Islands (Stehn and Johnson, 1987; Stehn and Prieto, 2010; Fig. 5.1). As the AWBP's size and winter range increased, the area surveyed for cranes also increased. It is unclear what threshold observers used to increase the spatial extent of their survey area. However, by winter 2010–11, the surveys were conducted on Blackjack Peninsula, Matagorda Island, Lamar Peninsula, San Jose Island, and Welder Flats (Fig. 5.1).

2011–Present

To improve the rigor of the Whooping Crane aerial survey, the current protocol identifies two types of survey areas called primary and secondary sampling frames (Butler et al., 2016). The regions within the primary sampling frame include those areas that recent data indicate are occupied consistently by multiple groups of Whooping Cranes each winter (Fig. 5.1) and were frequently included in earlier survey efforts. The primary sampling frame is 623 km^2 and consists of six areas. Previously collected data indicate that Whooping Cranes have occasionally used several specific areas outside of the primary sampling frame. To provide a transparent means to track the expansion of the Whooping Crane population, as well as the survey area, the protocol identified secondary sampling frames and a systematic and lower-frequency sampling strategy in these areas (Fig. 5.1). The secondary sampling frame is 685 km^2 and consists of 10 areas. The protocol also provides a criterion for promoting a secondary area into the primary sampling frame based on a minimum observed Whooping Crane density (Butler et al., 2016).

DATA COLLECTION/SURVEY TIMING

The high density of Whooping Cranes on and around Aransas NWR during winter seems to provide the most efficient opportunity to monitor the population size of the flock. However, determining the best timing and frequency of the surveys depends upon the type of data desired. For example, if the objective is to estimate the maximum abundance of the AWBP on the wintering grounds, it seems prudent to conduct surveys shortly after the population's arrival on the wintering grounds. Considering the AWBP's staggered arrival to the wintering grounds, surveys conducted at this time would capture the maximum proportion of the population within the sampling frame and preempt most population

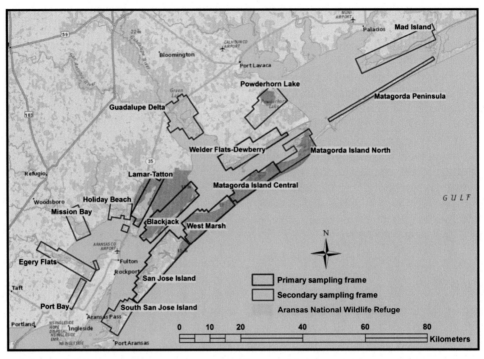

FIGURE 5.1   The primary (623 km²) and secondary (685 km²) sampling frames used for monitoring Whooping Cranes on their wintering grounds along the Texas Gulf coast, USA. Sampling frame boundaries and priority level were determined by examining the location of Whooping Crane observations during aerial surveys conducted from 1950 to 2011.

decline due to winter mortalities. Historical data indicates that this typically occurs at the end of November through the end of December (Butler et al., 2016). On the other hand, if the objective of the survey is to document arrival and departure patterns, then perhaps it makes sense to increase survey efforts near the beginning and end of the winter, as suggested in the 1994 recovery plan (U.S. Fish and Wildlife Service, 1994). However, other techniques, such as monitoring space use of radio- or GPS-marked individuals, could do this more efficiently than aerial surveys.

1950–2010

Multiple aerial surveys were conducted each winter spanning up to a 7-month period (Stehn and Taylor, 2008; Taylor et al., 2015). Within a winter, survey flights began as early as

9 October and ended as late as 24 May, but more typically ran from 22 October until 20 April (approx. 90%). Approximately 78% of survey flights were conducted between 15 November and 31 March. The frequency of flights varied from year to year (ranged from 5 to 47 per year). The 1980 recovery plan (U.S. Fish and Wildlife Service, 1980) implied that survey flights occurred at least weekly throughout the winter, and the 1994 recovery plan recommended that survey efforts decrease during midwinter (U.S. Fish and Wildlife Service, 1994).

The crew of each survey flight consisted of one pilot and at least one observer (Taylor et al., 2015). During the survey's initial 60 years, the flight pattern varied among years, pilots, survey areas, and individual surveys. Based on descriptions of Stehn and Taylor (2008), the flight paths were typically oriented parallel to the coastline with transects spaced approximately 250–800 m apart in an attempt to cover the entire survey area. Transect widths and locations were subjectively determined based on visibility and weather conditions during the survey in an effort to maximize detection of Whooping Cranes (Stehn and Taylor, 2008). Therefore, the extent and intensity of the search effort for each flight are unknown and likely inconsistent. Surveys were conducted at approximately 60 m above ground level at a speed of approximately 90 knots (Stehn and Taylor, 2008).

Upon observing a Whooping Crane group, the location was noted on paper maps by recording the number of white-plumaged (i.e., after-hatch-year; adult) and tawny-plumaged (i.e., hatch-year; juvenile) birds in each group. For example, a group of two adults and one juvenile would be recorded as "2 + 1," indicating two white-plumaged birds and one tawny-plumaged bird were present. If birds could not be confidently identified when initially detected, the pilot would alter course to allow the observer additional opportunities to confirm data. These data were typically hand-written and the approximate location of cranes was circled on a paper map. The types of paper maps varied over the 61 years of the survey (Taylor et al., 2015). Data were recorded on hand-drawn maps during winter 1950–51 until winter 1997–98. From winter 1998–99 until winter 2010–11, observations were recorded on printed digital orthrophoto quarter quadrangle (DOQQ) maps.

2011–Present

To provide a precise annual estimate of Whooping Crane abundance on the wintering grounds, survey methods were modified as follows. Intense survey efforts were allocated between 28 November and 26 December. The primary sampling frame was surveyed at least six times and the secondary sampling frame was surveyed at least twice. In addition to a Department of the Interior–certified aircraft and

pilot, the current aerial survey protocol required two observers collecting data. Each observer collected data independently from one side of the aircraft and recorded it on a digital device such as a laptop or tablet. Observers scanned the area systematically, allocating less effort as distance from the aircraft increased and attempting to ensure perfect detection rates along the transect line.

At a minimum, observers collected the size of detected Whooping Crane groups, the spatial location of Whooping Cranes, the age class of detected birds within the group (i.e., hatch-year, after-hatch-year, or unknown age class), and the track the aircraft flew. To avoid inconsistent detection rates caused by low sun angles (Stehn and Taylor, 2008; Strobel and Butler, 2014), surveys were conducted between 10:00 and 15:00 hours, or under high overcast conditions. The pilot attempted to maintain constant altitude (60 m above ground level) and airspeed (90 knots) as recommended by Stehn and Taylor (2008), while navigating along predefined transect lines spaced 1 km apart.

DATA ANALYSIS AND INTERPRETATION

1950–2010

In the absence of an explicit protocol or detailed report of the methods, it is not possible to know exactly how data were analyzed or interpreted prior to winter 2011–12. Typically, data were summarized (total number of each age class) by hand in the margins of the hard copy maps. The total raw count from all hard copy maps used during a given survey was included in annual or interim reports. The specific procedures used to resolve discrepancies, account for uncertainty, or extrapolate findings between flights or years is unknown. Taylor et al. (2015) outlined many of the weaknesses and limitations of the historical survey data and discussed their consequences. Weaknesses and limitations included lack of written protocol, unknown scale and resolution of paper maps, inconsistent determination of age class, assumed ability to identify unique individuals from the aerial survey platform, and unknown search effort.

2011–Present

Objectives 1 and 4

Several recent technological advances have allowed wildlife managers and biologists to better estimate wildlife population abundances. One such advancement, distance sampling, allows biologists to correct for the bias in population estimates that result from imperfect detection (Buckland et al., 2001, 2004; Thomas et al., 2010). The current protocol uses distance sampling to estimate the probability of detecting a Whooping Crane group and uses it to improve the rigor of population estimates (Butler et al., 2016; Strobel and Butler, 2014). More recent advances have built upon distance sampling to enable simultaneous modeling of the relationships of abundance and detectability to features of the habitat using hierarchical distance sampling (HDS; Chandler et al., 2011; Royle et al., 2004; Sillett et al., 2012; Timmer et al., 2014).

Models using HDS predict the number of groups of Whooping Cranes within discrete cells across a sampling grid while simultaneously accounting for incomplete detection of Whooping Crane groups. To do this, the models include factors that influence the ability of the observers to detect a Whooping Crane group if it were present, most importantly distance of the group from the observer and others as well (e.g., sun angle, vegetation type, average vegetation height within the grid cell, etc.). In addition, the models can incorporate factors that are presumed to affect the abundance of Whooping Cranes within grid cells (e.g., proportion of the grid cell comprised of saltmarsh vegetation,

etc.). The results are (1) an estimated number of Whooping Crane groups within the sampling frame and (2) a model of the effects of habitat conditions on Whooping Crane abundance.

To meet objective 1 of the survey, the number of individuals in the population can be estimated by multiplying the estimated number of groups by mean group size detected during the surveys. However, the mean Whooping Crane group size can be biased because some individuals in a group may be missed and larger groups are easier to detect than smaller ones (i.e., size-biased detection; Buckland et al., 2001; Strobel and Butler, 2014). Therefore, the current protocol accounts for these biases by regressing the natural log of group size against estimated detection probability to obtain an unbiased estimate of mean Whooping Crane group size (Buckland et al., 2001).

Another important improvement over the prior survey approach is the ability to calculate confidence intervals around the estimated abundance, which allows statistical comparison of abundance between years. Prior to 2011, the uncertainty surrounding Whooping Crane abundance estimates was not estimated and this hampered the ability to make inferences about ecological relationships dictating the demographics of the population.

Arguably, more important to Whooping Crane recovery than an improved population abundance estimate are the modeled relationships between habitat conditions and Whooping Crane abundance provided by HDS. The 1980 recovery plan recognized the opportunity for the aerial survey data to inform habitat management and conservation decisions but offered no indication about how to achieve it (U.S. Fish and Wildlife Service, 1980). With thoughtful construction of habitat models, HDS offers this critical missing link. Now aerial survey data can directly and transparently inform habitat conservation decisions, and predict the effects of planned management actions or the risks of unintended environmental changes (e.g., sea level rise) on Whooping Crane abundance (e.g., Metzger et al., 2014). These models, when applied to the sampling frame, also provide spatially explicit abundance estimates (i.e., spatial distribution of abundance on the wintering grounds; Fig. 5.2).

Objectives 2 and 3

The International Recovery Plan for Whooping Cranes identified the number of productive pairs as one metric on which downlisting decisions will be based (Canadian Wildlife Service and U.S. Fish and Wildlife Service, 2007). A productive pair is defined as "a pair that nests regularly and has fledged offspring" (Canadian Wildlife Service and U.S. Fish and Wildlife Service, 2007). On the surface these metrics seem straightforward; however, two conditions complicate our ability to estimate them. First,

although it is relatively easy to determine if two Whooping Cranes are spatially associated with one another, it is difficult to observe whether their spatial association in the nonbreeding season is the product of mate fidelity as the term *pair* implies. Second, some fledged offspring inevitably die prior to arriving on the wintering grounds, so estimates of the proportion of productive pairs made on the wintering ground can, at best, provide an index to the total proportion of productive pairs in the population. In any event the proportion of "pairs" and "productive pairs" in the population can be estimated as the proportion of all detected Whooping Crane groups that contained two adults and groups that contained juveniles, respectively.

Juvenile recruitment is critical for the continued growth of the AWBP (Butler et al., 2014), and therefore identifying juveniles is an important

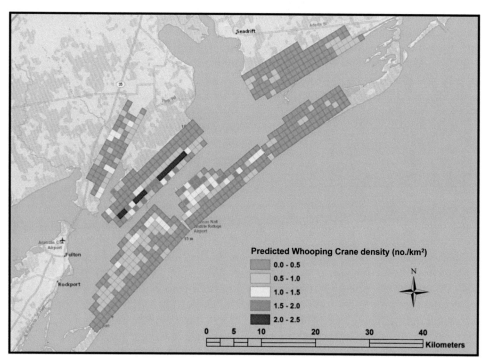

FIGURE 5.2 Distribution of Whooping Crane density in the primary sampling frame predicted using hierarchical distance sampling of aerial survey data collected during winter 2014–15, Texas, USA. Densities are predicted on a 1 km² grid oriented over the known Whooping Crane wintering range.

objective of the aerial surveys (Butler et al., 2016). An index of juvenile recruitment rate into the wintering population is estimated as the observed ratio of hatch-year to after-hatch-year birds from survey flights. As with other estimates produced under the current protocol, confidence intervals are provided to allow statistical comparison of the recruitment rates between years (Skalski et al., 2005).

Objective 5

To track the spatial expansion of the Whooping Crane wintering grounds, the current protocol prescribes a systematic approach for surveying areas that have the potential to be occupied by Whooping Cranes in the near future and are not included in the primary sampling area. These areas are known as the secondary sampling frame. When a sufficient number (as prescribed in the protocol) of Whooping Cranes have established predictable wintering territories in one of the secondary sampling areas, the area can be "promoted" into the primary sampling frame and thus be included in the annual population abundance estimate (Butler et al., 2016). The current protocol does not address how new secondary areas will be identified, but future revisions will (Butler et al., 2016).

SUMMARY AND OUTLOOK

The first 60 years of aerial surveys of the AWBP amassed more than 38,000 observations of Whooping Cranes (Taylor et al., 2015). These data have helped answer many fundamental questions regarding the ecology and conservation of this endangered species. Undoubtedly, these data have been extremely useful for solving many of the problems that challenged Whooping Crane conservation during the last century. Data collected during aerial surveys have been used to estimate population size and trends (Boyce and Miller, 1985; Butler et al., 2013; Wilson and Bidwell, Chapter 4, this volume), estimate

recruitment and survival rates (Boyce et al., 2005; Butler et al., 2014; Gil-Weir et al., 2012; Lewis et al., 1992; Link et al., 2003; Stehn and Haralson-Strobel, 2014), track the population's range expansion (Austin et al., Chapter 3, this volume; Stehn, 1992; Stehn and Prieto, 2010), and identify areas in need of conservation (Bonds, 2000; Smith et al., 2014; Smith et al., Chapter 13, this volume; Stehn and Johnson, 1987), and have been the primary means of monitoring the recovery of this endangered species (Canadian Wildlife Service and U.S. Fish and Wildlife Service, 2007). Indeed, these data have been a cornerstone of Whooping Crane recovery efforts.

The challenges to Whooping Crane conservation have changed over the past 6 decades. For example, thousands of acres of Whooping Crane habitat are being "squeezed" between rising sea levels, salt marsh subsidence, and increasing human development. The urgency and potential impact of these issues was not widely recognized 30 years ago, let alone at the inception of aerial surveys. The importance of efficient and effective conservation has never been greater. Fortunately, as the complexity of conservation issues has increased, more powerful and efficient tools have been developed to facilitate conservation decisions. Data collected under the current aerial survey protocol can be used to predict the effects of sea level rise on Whooping Crane abundance (Metzger et al., 2014), or estimate the number of Whooping Crane territories protected under different land conservation strategies. It is doubtful that these analyses would be possible with the historical data set.

We recognize that although the current aerial survey protocol has improved the capabilities of the data to answer today's conservation questions, it is by no means a panacea. We expect the challenges facing the conservation of Whooping Cranes to continue to grow in complexity in the foreseeable future. Therefore, we expect, and encourage, the protocol to be revised as the questions surrounding Whooping Crane recovery dictate. In addition, the survey's value should be measured by the usefulness of its information for making challenging decisions for the conservation of Whooping Cranes and not by the number of consecutive years of effort or the number of Whooping Cranes observed.

References

Austin, J.E., Hayes, M.A., Barzen, J.A., 2018. Revisiting the historic distribution and habitats of the whooping crane (Chapter 3). In: French, Jr., J.B., Converse, S.J., Austin, J.E. (Eds.), Whooping Cranes: Biology and Conservation. Biodiversity of the World: Conservation from Genes to Landscapes. Academic Press, San Diego, CA.

Bonds, C.J. 2000. Characterization of banded whooping crane winter territories from 1992–93 to 1996–97 using GIS and remote sensing. MS thesis, Texas A&M University, College Station.

Boyce, M.S., Lele, S.R., Johns, B.W., 2005. Whooping crane recruitment enhanced by egg removal. Biol. Conserv. 126 (3), 395–401.

Boyce, M.S., Miller, R.S., 1985. Ten-year periodicity in whooping crane census. Auk 102 (3), 658–660.

Buckland, S.T., Anderson, D.R., Burnham, K.P., Laake, J.L., Borchers, D.L., Thomas, L., 2001. Introduction to Distance Sampling: Estimating Abundance of Biological Populations. Oxford University Press, New York.

Buckland, S.T., Anderson, D.R., Burnham, K.P., Laake, J.L., Borchers, D.L., Thomas, L. (Eds.), 2004. Advanced Distance Sampling: Estimating Abundance of Biological Populations. Oxford University Press, UK.

Butler, M.J., Harris, G., Strobel, B.N., 2013. Influence of whooping crane population dynamics on its recovery and management. Biol. Conserv. 162:89–99.

Butler, M.J., Metzger, K.L., Harris, G., 2014. Whooping crane demographic responses to winter drought focus conservation strategies. Biol. Conserv. 179, 72–85.

Butler, M.J., Strobel, B.N., Eichhorn, C., 2016. Whooping Crane Winter Abundance Survey Protocol: Aransas National Wildlife Refuge. U.S. Fish and Wildlife Service, Austwell, TX, Version 1.1. Survey Identification Number: FF02RTAR00-002.

Canadian Wildlife Service and U.S. Fish and Wildlife Service, 2007. International Recovery Plan for the Whooping Crane (Grus americana), Third Revision. Environment Canada, Ottawa, and U.S. Fish and Wildlife Service, Albuquerque, NM.

Chandler, R.B., Royle, J.A., Kind, D.I., 2011. Inference about density and temporary emigration in unmarked populations. Ecology 92 (7), 1429–1435.

Conroy, M.J., Peterson, J.T., 2013. Decision Making in Natural Resource Management: A Structured, Adaptive Approach. John Wiley & Sons, Hoboken, NJ.

Gil-Weir, K.C., Grant, W.E., Slack, R.D., Wang, H.H., Fujiwara, M., 2012. Demography and population trends of whooping cranes. J. Field Ornithol. 83 (1), 1–10.

Lewis, J.C., Kuyt, E., Schwindt, K.E., Stehn, T.V., 1992. Mortality in fledged cranes of the Aransas-Wood Buffalo Population. In: Wood, D.A. (Ed.), Proceedings of the 1988 North American Crane Workshop. State of Florida Game and Fresh Water Fish Commission, Tallahassee, pp. 145–148.

Link, W.A., Royle, J.A., Hatfield, J.S., 2003. Demographic analysis from summaries of an age-structured population. Biometrics 59 (4), 778–785.

Lyons, J.E., Runge, M.C., Laskowski, H.P., Kendall, W.L., 2008. Monitoring in the context of structured decision-making and adaptive management. J. Wildl. Manage. 72 (8), 1638–1692.

Marsh, D.M., Trenham, P.C., 2008. Current trends in plant and animal population monitoring. Conserv. Biol. 22 (3), 647–655.

Metzger, K., Sesnie, S., Lehnen, S., Butler, M., Harris, G., 2014. Establishing a Landscape Conservation Strategy for Whooping Cranes in the Texas Gulf Coast. U.S. Fish and Wildlife Service, Albuquerque, NM.

Nichols, J.D., Williams, B.K., 2006. Monitoring for conservation. Trends Ecol. Evol. 21 (12), 668–673.

Royle, J.A., Dawson, D.K., Bates, S., 2004. Modeling abundance effects in distance sampling. Ecology 85 (6), 1591–1597.

Sillett, S., Chandler, R.B., Royle, J.A., Kéry, M., Morrison, S.A., 2012. Hierarchical distance sampling models to estimate population size and habitat-specific abundance of an island endemic. Ecol. Appl. 22 (7), 1997–2006.

Skalski, J.R., Ryding, K.E., Millspaugh, J.J., 2005. Wildlife Demography: Analysis of Sex, Age, and Count Data. Elsevier Academic Press, Burlington, MA.

Smith, E.H., Chávez-Ramirez, F., Lumb, L., 2018. Winter habitat ecology, use, and availability for the Aransas-Wood Buffalo Population of whooping cranes (Chapter 13). In: French, Jr., J.B., Converse, S.J., Austin, J.E. (Eds.), Whooping Cranes: Biology and Conservation. Biodiversity of the World: Conservation from Genes to Landscapes. Academic Press, San Diego, CA.

Smith, E.H., Chavez-Ramirez, F., Lumb, L., Gibeaut, J., 2014. Employing the conservation design approach on sea-level rise impacts on coastal avian habitats along the central Texas coast, final report. Gulf Coast Prairies Landscape Conservation Cooperative. http://gulfcoastprairielcc.org/science/science-projects/studying-the-effects-of-sea-level-rise-in-coastal-texas/ [9 November 2015].

Stehn, T.V. 1992. Unusual movements and behaviors of color-banded whooping cranes during winter. Proceedings of the North American Crane Workshop 6, 95–101.

Stehn, T.V., Haralson-Strobel, C., 2014. An update on mortality of fledged whooping cranes in the Aransas/Wood Buffalo Population. Proceedings of the North American Crane Workshop 12, 43–50.

Stehn, T.V., Johnson, E.F., 1987. Distribution of winter territories of whooping cranes on the Texas coast. In: Lewis, J.C. (Ed.), Proceedings of the 1985 Crane Workshop. Platte River Whooping Crane Maintenance Trust and U.S. Fish and Wildlife Service, Grand Island, NE, pp. 180–195.

Stehn, T.V., Prieto, F., 2010. Changes in winter whooping crane territories and range 1950–2006. Proceedings of the North American Crane Workshop 11, 40–56.

Stehn, T.V., Taylor, T.E., 2008. Aerial census techniques for whooping cranes on the Texas coast. Proceedings of the North American Crane Workshop 10, 146–151.

Strobel, B.N., Butler, M.J., 2014. Monitoring whooping crane abundance using aerial surveys: influences on detectability. Wildl. Soc. Bull. 38 (1), 188–195.

Taylor, L.N., Ketzler, L.P., Rousseau, D., Strobel, B.N., Metzger, K.L., Butler, M.J., 2015. Observations of Whooping Cranes during Winter Aerial Surveys: 1950–2011. Aransas National Wildlife Refuge, U.S. Fish and Wildlife Service, Austell, TX.

Thomas, L., Buckland, S.T., Rexstad, E.A., Laake, J.L., Strindberg, S., Hedley, S.L., Bishop, J.R., Marques, T.A., Burnham, K.P., 2010. Distance software: design and analysis of distance sampling surveys for estimating population size. J. Appl. Ecol. 47 (1), 5–14.

Timmer, J.M., Butler, M.J., Ballard, W.B., Boal, C.W., Whitlaw, H.A., 2014. Spatially explicit modeling of lesser prairie-chicken lek density in Texas. J. Wildl. Manage. 78 (1), 142–152.

U.S. Fish and Wildlife Service, 1980. Whooping crane recovery plan. D.L. Olsen, Team Leader. Washington, DC.

U.S. Fish and Wildlife Service, 1994. Whooping crane recovery plan, second revision. J.C. Lewis, Team Leader. Albuquerque, NM.

Walters, C.J., 1986. Adaptive Management of Renewable Resources. Blackburn Press, Caldwell, NJ.

Williams, B.K., Nichols, J.N., Conroy, M.J., 2002. Analysis and Management of Animal Populations. Academic Press, San Diego, CA.

Williams, B.K., Szaro, R.C., Shapiro, C.D., 2007. Adaptive Management: The U.S. Department of the Interior Technical Guide. Adaptive Management Working Group, U.S. Department of the Interior, Washington, DC.

Wilson, S., Bidwell, M., 2018. Population and breeding range dynamics in the Aransas-Wood Buffalo Whooping Crane Population (Chapter 4). In: French, Jr., J.B., Converse, S.J., Austin, J.E. (Eds.), Whooping Cranes: Biology and Conservation. Biodiversity of the World: Conservation from Genes to Landscapes. Academic Press, San Diego, CA.

Witmer, G.W., 2005. Wildlife population monitoring: some practical considerations. Wildl. Res. 32, 259–263.

Mortality in Aransas-Wood Buffalo Whooping Cranes: Timing, Location, and Causes

Aaron T. Pearse*, David A. Brandt*,
Barry K. Hartup**, Mark T. Bidwell[†]

*U.S. Geological Survey, Northern Prairie Wildlife Research Center,
Jamestown, ND, United States
**International Crane Foundation, Baraboo, WI, United States
[†]Canadian Wildlife Service, Environment and Climate Change Canada,
Prairie and Northern Wildlife Research Centre, Saskatoon, SK, Canada

INTRODUCTION

The Aransas-Wood Buffalo Population (AWBP) of Whooping Cranes (*Grus americana*) has experienced a population growth rate of approximately 4% for multiple decades (Butler et al., 2014a; Miller et al., 1974). Population growth for long-lived species of birds is generally highly sensitive to variation in adult mortality rates (Sæther and Bakke, 2000). A population model for endangered Red-crowned Cranes (*Grus japonensis*) in Japan conforms to this pattern, where growth rate is most sensitive to adult mortality (Masatomi et al., 2007). Earlier analyses observed that the AWBP growth rate increased in the mid-1950s and that this increase

was likely caused by reduced annual mortality rates, even while the population experienced slightly decreasing natality (Binkley and Miller, 1988; Miller et al., 1974). A more contemporary analysis of the AWBP determined that approximately 50% of variation in annual population growth could be explained by variation in annual mortality (Butler et al., 2014a). Therefore, as a vital rate, mortality is critical to the maintained growth of the AWBP.

Understanding where, when, and why animals die can be of use for setting priorities among multiple management, conservation, or reintroduction practices. The Whooping Crane recovery plan lists numerous threats that relate to mortality and includes identification of mortality factors, and reducing mortality rate specifically, as important recovery actions (Canadian Wildlife Service and U.S. Fish and Wildlife Service, 2005). Using information primarily from winter aerial surveys, Lewis et al. (1992) estimated 19% of mortality occurred during winter. Because few deaths had been documented during the breeding or summer season, Lewis et al. (1992) speculated that 60–80% of annual mortality occurred during migration and the small remainder occurred during breeding. Migration in Whooping Cranes specifically and migratory birds in general has been thought to be especially dangerous because of exposure to potential hazards encountered in unfamiliar areas (Lewis et al., 1992; Newton, 2008). Causes of mortality have been determined for a limited number of deaths and, in certain instances, causes have related to man-made structures or human activities (e.g., collision with power lines, gunshot; Stehn and Haralson-Strobel, 2014). Therefore, the notion that most mortality has occurred during migration has motivated recovery objectives and actions (Canadian Wildlife Service and U.S. Fish and Wildlife Service, 2005).

Current information on causes, timing, and location of mortality of AWBP Whooping Cranes has known biases and limitations. Specifically, few mortality events estimated from winter surveys have been confirmed with carcass recovery; thus, knowledge of where or why birds die is sparse. Furthermore, certain areas within the AWBP annual range are remote, whereas other areas are more densely populated and much more likely to yield discovery of a Whooping Crane carcass. Thus, recovered of carcasses may not be a representative sample of all deaths. The use of birds marked with satellite telemetry can provide less biased mortality information. We review what is known about patterns of mortality in Whooping Cranes, provide updates based on a sample of birds marked with transmitters, and compare our results with those presented previously. This review provides insights for management of this population and for

comparison with reintroduction efforts underway and in the future.

METHODS

During 2009–14, we captured 68 individual Whooping Cranes. The AWBP was estimated to have between 264 and 314 individuals during these years (Butler et al., 2014a, 2014b; U.S. Fish and Wildlife Service, 2015). At sites in and adjacent to Wood Buffalo National Park, we marked 31 prefledged juvenile cranes during August 2010 (9), August 2011 (12), and July and August 2012 (10). At sites along the Texas Gulf coast in the primary wintering areas, including the Aransas National Wildlife Refuge Complex, the Lamar Peninsula, and Welder Flats (Smith et al., Chapter 13, this volume), we captured 35 subadult or adult cranes during December 2009 (1), January 2011 (1), November–December 2011 (11), November 2012–January 2013 (11), and January–February 2014 (11). Finally, we captured two fledged juvenile cranes along the Texas Gulf coast, one during December 2009 and the other during January 2013. Capture teams consisted of persons with experience handling endangered cranes, including a licensed veterinarian. We captured prefledged juvenile cranes before they were capable of flight (approximately 40–60 days old) at breeding sites by locating family groups via helicopter and positioning personnel nearby for ground pursuit and hand capture (as described in Kuyt, 1979). We captured cranes in Texas using leg snares, which enclosed on the bird's lower tarsus (Folk et al., 2005). We placed the band and transmitter on the tibiotarsus of captured birds. Transmitters were platform transmitting terminals with global position system capabilities (North Star Science and Technology LLC, Baltimore, MD and Geotrak, Inc., Apex, NC) mounted on a two-piece leg band (Haggie Engraving, Crumpton, MD). Transmitters had solar panels integrated on three exposed surfaces and were expected to provide a 3–5-year life span. The transmitter and leg band weighed approximately 75 g, which was approximately 1% of body mass of adult Whooping Cranes. Survival of 651 Sandhill Cranes (*Grus canadensis*) fitted with similar-sized transmitters also mounted on leg bands had similar survival rates compared with cranes fitted only with metal U.S. Geological Survey bands, suggesting low potential for markers to negatively influence survival (Pearse et al., 2012). Transmitters were programmed to collect four or five GPS locations daily at equal time intervals and to attempt upload of location data to the Argos satellite system every 56 h (Service Argos, 2008). Capture and marking was conducted under Federal Fish and Wildlife Permit TE048806, Texas research permit SPR-1112-1042, Aransas National Wildlife Refuge special use permit, Canadian Wildlife Service Scientific Permit NWT-SCI-10-04, Parks Canada Agency Research and Collection Permit WB-2010-4998, and Northwest Territories Wildlife Research Permits WL004807, WL004821, and WL500051. Procedures were approved by the Animal Care and Use Committee at Northern Prairie Wildlife Research Center and Environment Canada's Animal Care Committee.

To summarize and compare with previously published reports, we identified four age classes of cranes: prefledged juveniles, fledged juveniles, subadults, and adults. Cranes were considered prefledged juveniles from hatching until they displayed the ability to fly by leaving their natal area. We marked prefledged juveniles exclusively at Wood Buffalo National Park. We captured all other age classes while they were on their wintering grounds. Fledged juveniles had the ability to fly and were less than 1 year old. We assigned birds to the subadult category (between 1 and 2 years of age) based on the presence of brown contour feathers. All birds with completely white body plumage were considered adults (2 years or older).

We determined mortality events primarily based on carcass and transmitter recovery.

Generally, we were able to determine death initially by lack of movement in transmitter locations, but transmitter failures introduced uncertainty in some instances (Hays et al., 2007). Thus, we included two types of mortality events in summaries (Table 6.1). We identified mortalities as confirmed with a known location only upon recovery of a carcass with transmitter or identifying bands. For two cranes only, mortality was suspected based on circumstantial evidence rather than carcass recovery. An example of such evidence was a sudden cessation of data acquisition from a transmitter along with no additional sightings of that bird, generally for years after the suspected mortality. We present summaries including and excluding those two suspected mortalities. Furthermore, we summarize with and without deaths that occurred <14 days after capture, a time period that we used to represent transmitter acclimation (Withey et al., 2001).

We assigned date of death based on interpretation of movements and motion sensor information from transmitters. We categorized mortality by season in the annual cycle of cranes: summer, spring migration, winter, and fall migration. Seasons were assigned based on time of year and migration behavioral patterns of individual cranes. Cranes were identified as wintering if they remained at a southern terminus for more than three weeks. Beginning of spring migration was identified by northerly movements from wintering areas. The summer period was defined as the northern terminus of yearly locations, and fall migration began with southerly movements from summering areas (Krapu et al., 2011; Pearse et al., 2015). We assessed the relative influence of deaths during each season using cause-specific mortality analyses. We used a nonparametric cumulative incidence function estimator to estimate mortality rates during different times of the year under a competing risks framework (Heisey and Patterson, 2006). We used this method because it accounts for multiple mortality factors and accounts for different numbers of individuals at risk throughout

TABLE 6.1 Confirmed and Suspected Deaths, 2010–15, of Whooping Cranes of the AWBP Marked with Satellite Telemetry Devices

| | Marking | | Mortality | | | | | |
Bird ID	Date	Age[a]	Type	Date	Age[a]	Season[b]	State/Province/Territory	Cause[c]
C01	01/08/11	A	Confirmed	06/12/11	A	S	NWT[d]	Unknown
B01	08/04/10	Pre-FJ	Confirmed	08/05/11	SA	S	NWT	Undetermined
B09	08/03/10	Pre-FJ	Confirmed	10/05/11	SA	S	NWT	Undetermined
C19	08/02/11	Pre-FJ	Confirmed	11/08/11	FJ	FM	Kansas	Unknown
C20	08/03/11	Pre-FJ	Confirmed	11/30/11	FJ	W	Texas	Potential bacterial infection
C18	08/02/11	Pre-FJ	Confirmed	01/06/12	FJ	W	Texas	Undetermined
C14	08/02/11	Pre-FJ	Confirmed	02/09/12	FJ	W	Texas	Undetermined
C17	08/02/11	Pre-FJ	Confirmed	06/07/12	SA	S	NWT	Undetermined
D27	08/01/12	Pre-FJ	Confirmed	08/04/12	Pre-FJ	S	NWT	Predation
D22	07/31/12	Pre-FJ	Confirmed	08/14/12	Pre-FJ	S	NWT	Undetermined
C16	08/03/11	Pre-FJ	Confirmed	08/18/12	SA	S	Alberta	Undetermined
A01	12/10/09	FJ	Confirmed	04/08/13	A	SM	South Dakota	Predation
D26	08/01/12	Pre-FJ	Confirmed	12/17/13	SA	W	Texas	Injury (see text)
D40	12/12/12	A	Confirmed	12/31/13	A	W	Texas	Undetermined
E50	02/02/14	SA	Confirmed	02/03/14	SA	W	Texas	Unknown
E54	02/03/14	SA	Confirmed	02/04/14	SA	W	Texas	Unknown
D41	01/08/13	FJ	Confirmed	03/30/15	SA	W	Texas	Undetermined
B02	08/03/10	Pre-FJ	Suspected	08/13/10	Pre-FJ	S	NWT	NA
D29	08/01/12	Pre-FJ	Suspected	11/23/12	FJ	FM	Nebraska	NA

[a] A = Adult; SA = subadult; FJ = fledged juvenile; Pre-FJ = prefledged juvenile.
[b] S = Summer; W = winter; SM = spring migration; FM = fall migration.
[c] Cause was assigned as "Undetermined" if necropsy was inconclusive as to cause of death. Cause was assigned as "Unknown" if no necropsy was attempted.
[d] NWT = Northwest Territories, Canada.

the year. We performed analyses using the *wild1* package in R (Sargeant, 2011). Finally, we estimated daily survival rates for each season by determining number of days cranes were monitored and at risk each season (Pollock et al., 1989).

Locations of mortality events were reported as latitude and longitude in the World Geodetic System, 1984 datum. We also reported locations by state, province, or territory and county or rural municipality. In instances where deaths were suspected, location of mortality refers to the last known location of the bird before cessation of data transmission; we suspected death occurred within the vicinity of the location but lacked definitive evidence.

We report cause of mortality based on assignments made from information developed from necropsies conducted by the Canadian Cooperative Wildlife Health Centre, U.S. Geological Survey Wildlife Health Laboratory, or U.S. Fish and Wildlife Service National Wildlife Forensics Laboratory. Necropsies were not available for all birds, or not practical when remains were in poor condition or incomplete (e.g., bones and feathers only). In other instances, a necropsy was carried out, but a definitive cause of death could not be assigned because of the deteriorated state of remains. We classified cause of mortality as *undetermined* when a necropsy was conducted and cause of mortality could not be diagnosed. The term *unknown* was assigned as a cause when a necropsy was not performed. Finally, we assigned *not applicable (NA)* to suspected mortalities when a carcass was not available for necropsy.

RESULTS

Among 68 Whooping Cranes marked with transmitters, we confirmed deaths of 17 by recovering remains between 12 June 2011 and 30 March 2015 using location information provided by satellite transmitters. Birds confirmed or suspected dead were marked as adults (2), subadults (2), fledged juveniles (2), and prefledged juveniles (11). At death, three birds were adults, seven subadults, four fledged juveniles, and two prefledged juveniles. Median time between estimated time of death and carcass recovery was 16 days (mean = 40 days; minimum = 1 day; maximum = 291 days).

We suspected mortality but could not confirm it in two instances. Bird B02 was marked as a prefledged juvenile and its transmitter ceased functioning 10 days after marking, suggesting a potential mortality event or transmitter malfunction. We received one additional GPS location from the transmitter on 20 February 2011 (192 days after last signal) 300 meters from its last location in Wood Buffalo National Park. To our knowledge, the bird had not been observed and reported since marking in 2010. The other suspected mortality was Bird D29, which showed unusual movements before the date of suspected death. This fledged juvenile was in its first fall migration and making typical southerly movements until reaching northern Oklahoma. After one night in Oklahoma, the bird flew north to central Kansas, where it had spent time previously. After three nights in Kansas, the bird moved farther north into south-central Nebraska. We collected six days of data in this location and then received no further information. This lengthy reverse migration of approximately 450 km was unique in our project thus far. Furthermore, the U.S. Fish and Wildlife Service received public reports of a single juvenile Whooping Crane in central Kansas at the same time and place as Bird D29 (R. Laubhan, U.S. Fish and Wildlife Service, personal communication). Because juvenile birds are typically accompanied by parents during fall and winter (Urbanek and Lewis, 2015), we suspect this juvenile may have been separated from its parents during migration. Similar to the other suspected mortality, Bird D29 has not been observed again since its suspected death in 2013.

Mortalities occurred in all seasons and over a wide time frame within summer and winter. During summer, the first mortality occurred on 7 June and the last occurred on 5 October (Table 6.1). In winter, deaths occurred between 30 November and 30 March. Both mortalities during fall migration occurred in November (8th and 23rd) and the single mortality during spring migration occurred on 8 April. We summarized mortality timing by season using various subsets of data. Including all confirmed and suspected deaths, eight mortalities occurred during summer, eight during winter, and three during migration (fall and spring combined). Estimated annual mortality related to factors occurring during winter was greatest and similar to summer (Table 6.2). Based on cause-specific rates, deaths occurring during winter accounted for 43% of annual mortality; 41% of deaths occurred during summer, and 16% during migrations. Including only events confirmed and those occurring after an acclimation period, six mortalities occurred during summer, six during winter, and two during spring and fall migrations. Cause-specific mortality rates for summer and winter were greatest and similar (Table 6.2). Mortality during winter accounted for 44% of annual mortality; 42% of deaths occurred during summer, and 14% during migrations. For fledged juvenile birds or older birds, we recorded five deaths during summer, six deaths during winter, and two deaths during migrations. Mortality during winter was greatest and accounted for 47% of overall annual mortality; 38% of deaths occurred during summer, and 15% during migration (Table 6.2).

We monitored 68 Whooping Cranes for a total of 34,948 days during winter (41%), summer (40%), and migration (19%). Daily survival during migration ($S = 0.99954$; 90% CI = 0.99910–0.99998) was slightly greater than during summer ($S = 0.99943$; 90% CI = 0.99910–0.99976) or winter ($S = 0.99944$; 90% CI = 0.99912–0.99977), yet 90% confidence intervals overlapped, suggesting differences in point estimates may have been due to chance alone.

Confirmed mortalities during migration occurred in South Dakota and Kansas, and we suspected one death in Nebraska (Fig. 6.1A). All deaths in the summer period occurred within

TABLE 6.2   Estimates of Annual Mortality (m) and Numbers of Deaths (n), 2010–15, for Whooping Cranes of the AWBP Marked with Satellite Telemetry Devices

Data subset	Season	n	m	SE	90% CI	% of overall
All	Summer	8	0.060	0.021	0.026, 0.094	41
	Winter	8	0.064	0.022	0.028, 0.100	43
	Migration	3	0.023	0.013	0.001, 0.045	16
	Overall	19	0.147	0.031	0.096, 0.199	
Restricted[a]	Summer	6	0.047	0.019	0.016, 0.079	42
	Winter	6	0.049	0.019	0.017, 0.081	44
	Migration	2	0.016	0.011	0.000, 0.035	14
	Overall	14	0.112	0.028	0.066, 0.159	
Postfledged birds only	Summer	5	0.040	0.018	0.011, 0.069	38
	Winter	6	0.049	0.020	0.017, 0.081	47
	Migration	2	0.016	0.011	0.000, 0.035	15
	Overall	13	0.106	0.028	0.060, 0.151	

[a] Includes mortality events confirmed with carcass recovery and those occurring 14 days post marking to account for potential biases from capture and marking.

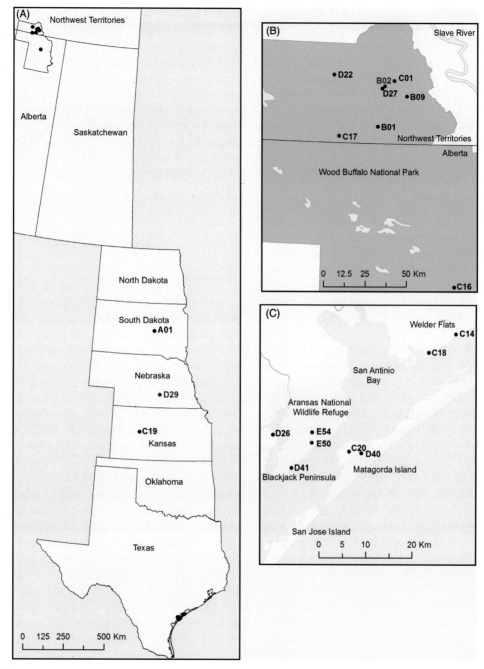

FIGURE 6.1   Locations of confirmed (black) and suspected (red) deaths of Whooping Cranes of the Aransas-Wood Buffalo Population (AWBP) marked with satellite transmitters, 2010–15 in the migration corridor (A), at Wood Buffalo National Park, Canada (B), and at the Texas Gulf Coast at and near Aransas National Wildlife Refuge (C).

Wood Buffalo National Park, and all but one within an approximately 30 km radius within the main nesting area (Johns et al., 2005; Fig. 6.1B). Finally, mortality during winter was distributed among traditional wintering locations such as the Blackjack Peninsula (three), Matagorda Island (two), and Welder Flats (two; Fig. 6.1C). Bird D26 was captured on the Lamar Peninsula and died in captivity at the San Antonio Zoo (see later).

Predation and disease were known causes of mortality for Whooping Cranes in our study (Table 6.1). For most confirmed mortalities ($n = 13$), cause of death could not be determined because of the advanced state of scavenging and/or decomposition. Cause of death was more likely determined where carcasses were recovered somewhat more quickly (median of 9 days postmortality) compared to the overall median recovery time of 16 days.

We included Bird D26 as a mortality event, although this bird was captured, removed from the remnant population, and perished in captivity. Approximately 2 months prior to capture, Bird D26 was observed at the wintering grounds near the ultimate capture site with a severed lower left leg. The cause of this injury was not known. At the request of the U.S. Fish and Wildlife Service, the bird was captured on 17 December 2013 and transported to the San Antonio Zoo for further assessment and treatment. The bird perished in captivity approximately 6 weeks after capture. We included this incident as a winter mortality because, based on the physical condition of the bird at capture and extent of the injury, our team believed it would not have survived the entirety of the winter.

We confirmed an additional mortality of a radio-marked bird that has not been included in Table 6.1. Remains of an adult Whooping Crane from our study were found without aid of transmitter location data on San Jose Island in June 2015 by a member of the public (Fig. 6.1B). The last transmission we received was within approximately 2 km of where the remains were found in early October 2014 but, at that time, we did not have sufficient information to determine whether the bird had died or the transmitter had malfunctioned. Based on location data, we could not determine timing of death. Thus, this mortality was confirmed by observation alone, similar to those reported in Lewis et al. (1992) and Stehn and Haralson-Strobel (2014) rather than via transmitter information. Because this mortality was identified solely from recovery of carcass without determining death from location data, we did not include it in summaries above.

DISCUSSION

Timing of Mortality

Previously available information on seasonal mortality of AWBP Whooping Cranes has come from three main sources. Information from the many years of winter aerial survey provided insights primarily regarding winter mortality (Strobel and Butler, Chapter 5, this volume). The proportion of mortality occurring from 1950 to 1987 during winter at Aransas NWR was estimated to be approximately 19% (Lewis et al., 1992) and 20% after an update (1950–2010; Stehn and Haralson-Strobel, 2014). Also using data from 1950–2010, Butler et al. (2014a) estimated 17% of annual mortality occurred during winter under average precipitation conditions and 43% under extreme drought conditions. Winter surveys also have been used to infer mortality at other times of the year. Based on few reported deaths during summer, researchers have speculated that 60–80% of deaths likely occurred during migration and only a small percentage of annual mortality occurred during summer (Lewis et al., 1992; Stehn and Haralson-Strobel, 2014).

A second source of information comes from a previous radio-telemetry study. Kuyt (1992) reported fates of 15 Whooping Cranes, all marked as prefledged juveniles during summers 1981–83. This work confirmed or suspected deaths of 12 cranes; 6 (50%) before fledging at Wood Buffalo National Park, 4 (33%) during winter, and 2 (17%) during migration. Finally, the third source of data on mortality comes from 48 carcasses recovered. The greatest percentage of carcasses was found in areas where birds migrate (55%). Forty percent of carcasses were found on wintering areas and only 5% in areas typically used during summer (Stehn and Haralson-Strobel, 2014). Authors noted that recoveries at Wood Buffalo National Park were likely underrepresented because of the inaccessibility of the area to people who could potentially discover carcasses.

Because all mortalities could not be confirmed and some occurred soon after capture and marking, we provided multiple estimates of seasonal mortality based on various subsets of deaths, which resulted in consistent seasonal distributions of deaths. Winter tied for or had the greatest percentage of deaths in all summarizations (43–47%). Mortalities during summer were equal to winter or ranked second (38–42%), and migration consistently had the lowest percentage (14–16%). Whooping Cranes from the AWBP generally spend approximately 2 months in migration (17%) and 5 months each at summer and winter locations (41.5%; Urbanek and Lewis, 2015). The birds we monitored were at risk for similar percentage of time each season. Furthermore, daily survival rates were comparable among seasons, implying that daily risk of mortality was relatively equal among seasons during our study.

Some similarity and numerous differences existed in timing of Whooping Crane mortality between our results and those reported previously. Similar to the past telemetry study, we documented mortality of prefledged juvenile cranes before they left the breeding area and began fall migration, which is common in other bird species as well (Bergenson et al., 2001; Grüebler et al., 2014; Johnson et al., 1992; Kuyt, 1992). In contrast with previous studies, our results were markedly different from past

information related to the AWBP and support the notion that mortality of subadult and adult Whooping Cranes during summer may have been underestimated and underappreciated. We documented more than double the number of mortality events on fledged birds than ever reported on summering grounds (two adults recovered; Stehn and Haralson-Strobel, 2014). Four subadults died in their second summer season, which corresponded with the cessation of parental attendance (Urbanek and Lewis, 2015). This timing suggested that initial independence from parents may be an especially risky time for young Whooping Cranes. Numerous deaths have been reported during summer in the Eastern Migratory Population of Whooping Cranes (53%, 9 of 17), with birds of all available age classes dying (Cole et al., 2009).

Migration has been identified as a time when 60–80% of AWBP Whooping Crane deaths occur (Lewis et al., 1992; Stehn and Haralson-Strobel, 2014). Our findings do not support these assertions and indicate that migration contributed the least proportionally to annual mortality while also representing the smallest proportion of the annual cycle. Migration posed a nearly equal rather than greater risk to Whooping Cranes as compared with other times of the year, because daily survival rates were similar among seasons. Mortality during migration has been difficult to study in birds, given their high mobility (Newton, 2008). Some studies have inferred the potential of high mortality during migration, as has been done with Whooping Cranes, when researchers were forced to combine multiple life events (e.g., breeding and migration) in survival estimates (Clausen et al., 2001; Madsen et al., 2002). High rates of mortality during migration have been observed more directly in other studies where season-specific estimates could be ascertained, providing evidence of the hazardous nature of migration (Klaassen et al., 2014; Lok et al., 2015; Oppel et al., 2015; Owen and Black, 1989; Sillett and Holmes, 2002). Yet this pattern is not universal, and numerous studies of many species have found migration of similar or less risk to birds than other times of the year [Pacific Brant (*Branta bernicla nigricans*), Ward et al., 1997; Greater Snow Goose (*Chen caerulescens atlantica*), Gauthier et al., 2001; Emperor Goose (*Chen canagica*), Hupp et al., 2008; Red Knot (*Calidris canutus*), Leyrer et al., 2013; Trumpeter Swan (*Cygnus buccinators*), Varner and Eichholz, 2012; Barn Swallow (*Hirundo rustica*), Grüebler et al., 2014; Sandhill Crane, Fronczak et al., 2015]. Relative seasonal mortality apparently varies between species and situations, suggesting our collective understanding of the risks faced by migratory birds during migration and throughout the rest of the year is incomplete and requires continued study.

Juvenile birds can experience lower survival rates than adults during their first fall migration (Owen and Black, 1989). Both mortality events during fall migration that we confirmed or suspected were juvenile birds. In Egyptian vultures (*Neophron percnopterus*), juveniles using a migration route over the Mediterranean Sea died at a much greater rate than those using an alternative overland route, and the authors suspected that lack of experienced birds in the population may have contributed to the deaths (Oppel et al., 2015). As Whooping Crane adults generally attend juveniles during their fall migration (Johns et al., 2005), risky situations where juveniles would be required to migrate alone would be rare. Accordingly, a suspected death we reported during fall migration may have been related to separation of the juvenile from its attending parents.

We found double the percentage of mortality during winter compared to previous estimates based on winter surveys. We suspect this discrepancy may be related to methodological aspects of aerial surveys, as well as winter weather and habitat conditions during our study. Determination of mortality via aerial survey of unmarked birds relies upon numerous assumptions, though Butler et al. (2014a) observed that violations would generally overestimate rather than underestimate the contribution of winter to annual mortality. Deaths occurring before many of the birds have arrived or after spring migration has begun (i.e., periods of turnover) or outside of when or where aerial surveys were conducted would potentially be missed entirely and misclassified. Such deaths that had occurred during winter but missed would be included the following year as losses occurring during migration or summer. We documented one death during winter at Aransas National Wildlife Refuge on 30 November 2011 as fall migration was ending and another death during winter on 30 March 2015, which occurred after the beginning of spring migration. If these two deaths had been incorrectly attributed to a season other than winter, the percentage of deaths during winter would have decreased to 29% (4 out of 14 confirmed and after acclimation period). It is unclear how winter surveys could consistently and reliably identify deaths that occurred during periods when birds were still arriving at or leaving the wintering grounds (Butler et al., 2014b).

The time period of our study coincided with consistent drought conditions on the wintering grounds. During our study, winter 2011–12 was classified as extreme drought, winter 2012–13 as severe drought, winter 2013–14 as moderate drought, and winter 2014–15 as mild drought (Palmer, 1965; National Climatic Data Center, 2007). Butler et al. (2014a) reported that drought conditions influenced winter mortality and extreme drought conditions could increase percentage of mortality occurring during winter up to 43%. Linkages between environmental conditions and annual or seasonal survival have been observed in other migratory birds. Kéry et al. (2006) found that pink-footed geese (*Anser brachyrhynchus*) survived at a greater rate during years when their wintering grounds were warmer and wetter. Poor habitat conditions influenced winter survival for oystercatchers (*Haematopus ostralegus*) in Europe (Duriez et al., 2012). Continued mortality monitoring of Whooping Cranes in times with less extreme drought or without drought would be useful to determine if poor environmental conditions were the primary cause of the higher incidences of winter mortality observed.

Our study was conducted over a limited time period and included few mortality events and as such may not be fully representative of AWBP mortality. Survival rate of the AWBP varies annually (Nedelman et al., 1987) as does season-specific mortality (Butler et al., 2014a). Our study was conducted in years of sustained drought conditions at wintering areas, which likely influenced results. These shortcomings will be somewhat overcome by adding more data as our project concludes, which we can use to update results and estimate season-specific mortality rates. In addition, our sample of mortality events was dominated by birds marked as juveniles and subsequently dying as young birds. Young Whooping Cranes have a greater mortality rate than older birds (Gil-Weir et al., 2012; Link et al., 2003; Nedelman et al., 1987), and they may die in different places, at different times, and from different causes. Given the high survival rate of adult Whooping Cranes, transmitters lasting greater than 5 years would be required to gather an unbiased and adequate sample of deaths from these long-lived birds.

Cause-Specific Mortality

Frequent known causes of mortality in Whooping Cranes include predation, collisions with power lines, gunshot, other trauma, and disease (Cole et al., 2009; Stehn and Haralson-Strobel, 2014). Predation and disease were the only known causes of mortality in our study. We were not able to add to existing knowledge of the causes of mortality in this population, as most recovered carcasses were degraded to a state where cause could not be determined. Power line collisions have been identified as an important cause of mortality for the AWBP and reintroduced populations of Whooping Cranes (Cole et al., 2009; Hartup et al., 2010; Stehn and

Wassenich, 2008). We did not observe direct evidence from necropsy reports or indirect evidence based on location of remains near power lines to suggest mortalities in our sample resulted from power line strikes.

Stehn and Haralson-Strobel (2014) did not assign a cause to 24% of AWBP mortalities, and Cole et al. (2009) could not determine cause of death for 35% of a sample of Eastern Migratory Population of Whooping Cranes. We classified a high rate of confirmed mortalities to unknown or undetermined causes (76%). Our protocols for defining cause relied primarily on necropsy reports and used circumstantial evidence sparingly. Many of the remains were degraded because of scavenging and decomposition; hence, cause of death could not be determined. The carcasses for which cause of death was determined may present a biased picture of mortality overall (Bumann and Stauffer, 2002; Faanes, 1987; Flint et al., 2010). Our experience underscores the difficult task of determining cause-specific mortality for a wide-ranging migratory species or whenever circumstances prevent prompt collections of fresh carcasses. If determining cause-specific mortality is a primary objective of future studies, then using different monitoring devices that transmit more frequently may be necessary to identify deaths and collect carcasses more quickly.

SUMMARY AND OUTLOOK

The AWBP has experienced positive exponential growth for decades (Butler et al., 2014a; Miller et al., 1974). Increasing annual survival through management actions would result in an increased rate of population growth, potentially allowing the population to reach recovery status more quickly (Butler et al., 2013), as long as an adequate quality and quantity of habitat will exist to support a larger future population. For a migratory bird, knowing when deaths occur in the annual cycle provides insight to effectively implement conservation actions (Klaassen et al., 2014). We provided evidence that past assessments of timing of mortality may have been based on weak assumptions, suggesting some modifications to perceptions about risks and threats. Mortality of fledged Whooping Cranes at Wood Buffalo National Park occurred at a much higher rate than had been reported previously in the AWBP. This mortality occurred in remote areas where causes, although unknown in many cases, were likely related to natural phenomena (e.g., predation), which are unlikely to be modified without intensive and costly management. Conversely, migration may be less risky than previously assumed, with birds at a similar rather than an elevated risk as compared to summer or winter. Managers should expect only modest influence on annual survival rates from efforts to reduce mortality during migration, as a low percentage of deaths occurred during this time and cranes migrate for the shortest time period annually. Finally, our results supported earlier conclusions that mortality during winter can be a large component of annual deaths during times of drought. The primary wintering grounds of the AWBP are a condensed area where the birds remain for many months each year; thus, management efforts to increase survival during winter may be more effective than those conducted at summering grounds or more feasible than in the migration corridor. Such efforts may be effective at increasing annual survival if conducted in years of high winter mortality and less so when conditions are naturally more favorable. Conservation and management activities attempting to abate winter mortality, especially those related to drought conditions, will need to be identified and tested to determine risks, efficacy, and cost effectiveness. Because we documented few deaths of adult birds, more information of their mortality, especially that of actively breeding birds, would be useful to update our results. Future work could be directed at investigating potential patterns in seasonal survival, which may be cyclic especially during summer,

as in the nature of other predator–prey relationships in northern latitudes and at times of average to good winter habitat conditions.

Acknowledgments

Logistic, administrative, and financial support was provided by the Canadian Wildlife Service, Crane Trust, U.S. Fish and Wildlife Service, Platte River Recovery Implementation Program, and U.S. Geological Survey–Platte River Priority Ecosystems Program, with additional support from the Gulf Coast Bird Observatory, International Crane Foundation, and Parks Canada. Necropsy reports were provided by the U.S. Geological Survey-National Wildlife Health Center, Canadian Cooperative Wildlife Health Centre, and U.S. Fish and Wildlife Service National Wildlife Forensics Laboratory. V.I. Shearn-Bochsler, T. Bollinger, L. Bryan, C.U. Meteyer, A.E. Ballmann, and R.A. Kagan completed necropsy exams. Any use of trade, firm, or product firm names is for descriptive purposes only and does not imply endorsement by the U.S. Government.

References

Bergenson, D.G., Johns, B.W., Holroyd, G.L., 2001. Mortality of whooping crane colts in Wood Buffalo National Park, Canada, 1997–99. Proceedings of the North American Crane Workshop 8, 6–10.

Binkley, C.S., Miller, R.S., 1988. Recovery of the whooping crane *Grus americana*. Biol. Conserv. 45 (1), 11–20.

Bumann, G.B., Stauffer, D.F., 2002. Scavenging of ruffed grouse in the Appalachians: influences and implications. Wildl. Soc. Bull. 30 (3), 853–860.

Butler, M.J., Harris, G., Strobel, B.N., 2013. Influence of whooping crane population dynamics on its recovery and management. Biol. Conserv. 162, 89–99.

Butler, M.J., Metzger, K.L., Harris, G., 2014a. Whooping crane demographic responses to winter drought focus conservation strategies. Biol. Conserv. 179, 72–85.

Butler, M.J., Strobel, B.N., Eichhorn, C., 2014b. Whooping crane winter abundance survey protocol: Aransas National Wildlife Refuge. U.S. Fish and Wildlife Service, Austwell, TX, Survey Identification Number: FF02R-TAR00-002.

Canadian Wildlife Service and U.S. Fish and Wildlife Service, 2005. International Recovery Plan for the Whooping Crane. Recovery of Nationally Endangered Wildlife (RENEW), Ottawa, and U.S. Fish and Wildlife Service, Albuquerque, NM, p. 162.

Clausen, P., Frederiksen, M., Percival, S.M., Anderson, G.Q.A., Denny, M.J.H., 2001. Seasonal and annual survival of east-Atlantic pale-bellied brant geese *Branta hrota* assessed by capture-recapture analysis. Ardea 89 (1), 101–111.

Cole, G.A., Thomas, N.J., Spalding, M., Stroud, R., Urbanek, R.P., Hartup, B.K., 2009. Postmortem evaluation of reintroduced migratory whooping cranes in eastern North America. J. Wildl. Dis. 45 (1), 29–40.

Duriez, O., Ens, B.J., Choquet, R., Pradel, R., Klaassen, M., 2012. Comparing the seasonal survival of resident and migratory oystercatchers: carry-over effects of habitat quality and weather conditions. Oikos 121 (6), 862–873.

Faanes, C.A., 1987. Bird behavior and mortality in relation to power lines in prairie habitats. U.S. Fish and Wildlife Service Technical Report 7, U.S. Department of the Interior, Washington, DC.

Flint, P.L., Lance, E.W., Sowl, K.M., Donnelly, T.F., 2010. Estimating carcass persistence and scavenging bias in a human-influenced landscape in western Alaska. J. Field Ornithol. 81 (2), 206–214.

Folk, M.J., Nesbitt, S.A., Schwikert, S.T., Schmidt, J.A., Sullivan, K.A., Miller, T.J., Baynes, S.B., Parker, J.M., 2005. Techniques employed to capture whooping cranes in central Florida. Proceedings of the North American Crane Workshop 9, 141–144.

Fronczak, D.L., Andersen, D.E., Hanna, E.E., Cooper, T.R., 2015. Annual survival rate estimate of satellite transmitter-marked eastern population of greater sandhill cranes. J. Fish Wildl. Manage. 6 (2), 464–471.

Gauthier, G., Pradel, R., Menu, S., Lebreton, J.-D., 2001. Seasonal survival of greater snow geese and effect of hunting under dependence in sighting probability. Ecology 82 (11), 3105–3119.

Gil-Weir, K., Grant, W.E., Slack, R.D., Wang, H.-H., Fujiwara, M., 2012. Demography and population trends of whooping cranes. J. Field Ornithol. 83 (1), 1–10.

Grüebler, M.U., Korner-Nievergelt, F., Naef-Daenzer, B., 2014. Equal nonbreeding period survival in adults and juveniles of a long-distant migrant bird. Ecol. Evol. 4 (6), 756–765.

Hartup, B.K., Spalding, M.G., Thomas, N.J., Cole, G.A., Kim, Y.J., 2010. Thirty years of mortality assessment in whooping crane reintroductions: patterns and implications. Proceedings of the North American Crane Workshop 11, 204.

Hays, G.C., Bradshaw, C.J.A., James, M.C., Lovell, P., Sims, D.W., 2007. Why do Argos satellite tags deployed on marine animals stop transmitting? J. Exp. Mar. Biol. Ecol. 349 (1), 52–60.

Heisey, D.M., Patterson, B.R., 2006. A review of methods to estimate cause-specific mortality in presence of competing risks. J. Wildl. Manage. 70 (6), 1544–1555.

Hupp, J.W., Schmutz, J.A., Ely, C.R., 2008. Seasonal survival of radiomarked emperor geese in western Alaska. J. Wildl. Manage. 72 (7), 1584–1595.

Johns, B.W., Goossen, J.P., Kuyt, E., Craig-Moore, L., 2005. Philopatry and dispersal in whooping cranes. Proceedings of the North American Crane Workshop 9, 117–125.

Johnson, D.H., Nichols, J.D., Schwartz, M.D., 1992. Population dynamics of breeding waterfowl. In: Batt, B.D.J., Afton, A.D., Anderson, M.G., Ankney, C.D., Johnson, D.H., Kadlec, J.A., Krapu, G.L. (Eds.), Ecology and Management of Breeding Waterfowl. University of Minnesota Press, Minneapolis, pp. 446–485.

Kéry, M., Madsen, J., Lebreton, J.-D., 2006. Survival of Svalbard pink-footed geese Anser brachyrhynchus in relation to winter climate, density and land-use. J. Anim. Ecol. 75 (5), 1172–1181.

Klaassen, R.H.G., Hake, M., Strandberg, R., Koks, B.J., Trierweiler, C., Exo, K.-M., Bairlein, F., Alerstam, T., 2014. When and where does mortality occur in migratory birds? Direct evidence from long-term satellite tracking of raptors. J. Anim. Ecol. 83 (1), 176–184.

Krapu, G.L., Brandt, D.A., Jones, K.L., Johnson, D.H., 2011. Geographic distribution of the mid-continent population of sandhill cranes and related management applications. Wildl. Monogr. 175, 1–38.

Kuyt, E., 1979. Banding of juvenile whooping cranes on the breeding range in the Northwest Territories, Canada. N. Am. Bird Bander 4 (1), 24–25.

Kuyt, E. 1992. Aerial radio-tracking of whooping cranes migrating between Wood Buffalo National Park and Aransas National Wildlife Refuge, 1981–84. Canadian Wildlife Service Occasional Paper 74.

Lewis, J.C., Kuyt, E., Schwindt, K.E., Stehn, T.V., 1992. Mortality in fledged whooping cranes of the Aransas-Wood Buffalo Population. In: Wood, D.A. (Ed.), Proceedings of the 1988 North American Crane Workshop, Florida Game and Fresh Water Fish Commission Nongame Technical Report 12, Tallahassee, FL, pp. 145–148.

Leyrer, J., Lok, T., Brugge, M., Spaans, B., Sandercock, B.K., Piersma, T., 2013. Mortality within the annual cycle: seasonal survival patterns in Afro-Siberian red knots Calidris canutus canutus. J. Ornithol. 154 (4), 933–943.

Link, W.A., Royle, J.A., Hatfield, J.S., 2003. Demographic analysis from summaries of an age-structured population. Biometrics 59 (4), 778–785.

Lok, T., Overdijk, O., Piersma, T., 2015. The cost of migration: spoonbills suffer higher mortality during trans-Saharan spring migrations only. Biol. Lett. 11, 20140944.

Madsen, J., Frederiksen, M., Ganter, B., 2002. Trends in annual and seasonal survival of pink-footed geese Anser brachyrhynchus. Ibis 144 (2), 218–226.

Masatomi, Y., Higashi, S., Masatomi, H., 2007. A simple population viability analysis of tancho (Grus japonensis) in southeastern Hokkaido. Popul. Ecol. 49 (4), 297–304.

Miller, R.S., Botkin, D.B., Mendelssohn, R., 1974. The whooping crane (Grus americana) population of North America. Biol. Conserv. 6 (2), 106–111.

National Climatic Data Center, 2007. Time Bias Corrected Divisional Temperature-Precipitation-Drought Index, TD-9640. National Climatic Data Center, Asheville,

NC. Available from: https://www.ncdc.noaa.gov/data-access/land-based-station-data/land-based-datasets [1 August 2018] README.

Nedelman, J., Thompson, J.A., Taylor, R.J., 1987. The statistical demography of whooping cranes. Ecology 68 (5), 1401–1411.

Newton, I., 2008. The Migration Ecology of Birds. Academic Press, Amsterdam.

Oppel, S., Dobrev, V., Arkumarev, V., Saravia, V., Bounas, A., Kret, E., Velevski, M., Stoychev, S., Nikolov, S.C., 2015. High juvenile mortality during migration in a declining population of a long-distance migratory raptor. Ibis 157 (3), 545–557.

Owen, M., Black, J.M., 1989. Factors affecting the survival of barnacle geese on migration from the breeding grounds. J. Anim. Ecol. 56 (2), 603–617.

Palmer, W.C., 1965. Meteorological drought. Research Paper 45, U.S. Weather Bureau, Washington, DC.

Pearse, A.T., Brandt, D.A., Harrell, W.C., Metzger, K.L., Baasch, D.M., Hefley, T.J., 2015. Whooping crane stopover site use intensity within the Great Plains. U.S. Geological Survey Open-File Report 2015–1166.

Pearse, A.T., Brandt, D.A., Krapu, G.L., Post van der Burg, M., 2012. Evaluating transmitter effects on sandhill cranes: implications for whooping crane research. In: North American Ornithological Conference. Available from: http://birdmeetings.org/naoc12/files/NAOC-V_Abstract_Book.pdf [1 August 2018].

Pollock, K.H., Winterstein, S.R., Bunck, C.M., Curtis, P.D., 1989. Survival analysis in telemetry studies: the staggered entry design. J. Wildl. Manage. 53 (1), 7–15.

Sæther, B.-E., Bakke, Ø., 2000. Avian life history variation and contribution of demographic traits to the population growth rate. Ecology 81 (3), 642–653.

Sargeant, G.A., 2011. Wild1-R Tools for Wildlife Research and Management. Available from: https://cran.r-project.org/src/contrib/Archive/wild1/ [1 October 2015].

Service Argos, 2008. User's Manual. Service Argos, Landover, MD.

Sillett, T.S., Holmes, R.T., 2002. Variation in survivorship in a migratory songbird throughout its annual cycle. J. Anim. Ecol. 71 (2), 296–308.

Smith, E.H., Chavez-Ramirez, F., Lumb, L., 2018. Wintering habitat ecology, use, and availability for the Aransas-Wood Buffalo population of whooping cranes (Chapter 13). In: French, Jr., J.B., Converse, S.J., Austin, J.E. (Eds.), Whooping Cranes: Biology and Conservation. Biodiversity of the World: Conservation from Genes to Landscapes. Academic Press, San Diego, CA.

Stehn, T.V., Haralson-Strobel, C.L., 2014. An update on mortality of fledged whooping cranes in the Aransas-Wood Buffalo Population. Proceedings of the North American Crane Workshop 12, 43–50.

Stehn, T.V., Wassenich, T., 2008. Whooping crane collisions with power lines: an issue paper. Proceedings of the North American Crane Workshop 10, 25–36.

Strobel, B.N., Butler, M.J., 2018. Monitoring recruitment and abundance of the Aransas-Wood Buffalo Population of whooping cranes: 1950–2015 (Chapter 5). In: French, Jr., J.B., Converse, S.J., Austin, J.E. (Eds.), Whooping Cranes: Biology and Conservation. Biodiversity of the World: Conservation from Genes to Landscapes. Academic Press, San Diego, CA.

Urbanek, R.P., Lewis, J.C., 2015. Whooping crane (*Grus americana*). In: Rodewald, P.G. (Ed.), The Birds of North America Online. Cornell Lab of Ornithology, Ithaca, NY. Retrieved from the Birds of North America Online: https://birdsna.org/Species-Account/bna/species/whocra/introduction [1 August 2018].

U.S. Fish and Wildlife Service, 2015. Whooping Crane Survey Results: Winter 2014–2015. Available from: https://www.fws.gov/uploadedFiles/Region_2/NWRS/Zone_1/Aransas-Matagorda_Island_Complex/Aransas/Sections/What_We_Do/Science/Whooping_Crane_Updates_2013/WHCR_Update_Winter_2014-2015.pdf [1 August 2018].

Varner, D.M., Eichholz, M.W., 2012. Annual and seasonal survival of trumpeter swans in the Upper Midwest. J. Wildl. Manage. 76 (1), 129–135.

Ward, D.H., Rexstad, E.A., Sedinger, J.S., Lindberg, M.S., Dawe, N.K., 1997. Seasonal and annual survival of adult Pacific brant. J. Wildl. Manage. 61 (3), 773–781.

Withey, J.C., Bloxtom, T.D., Marzluff, J.M., 2001. Effects of tagging and location error in wildlife radiotelemetry studies. In: Millspaugh, J.J., Marzluff, J.M. (Eds.), Radio Tracking and Animal Populations. Academic Press, San Diego, CA, pp. 43–75.

Population Dynamics of Reintroduced Whooping Cranes

Sarah J. Converse*,**, Sabrina Servanty†,
Clinton T. Moore‡, Michael C. Runge*

*U.S. Geological Survey, Patuxent Wildlife Research Center,
Laurel, MD, United States
**U.S. Geological Survey, Washington Cooperative Fish and Wildlife Research
Unit, School of Environmental and Forest Sciences (SEFS) & School of Aquatic
and Fishery Sciences (SAFS), University of Washington, Seattle, WA, United States
†Colorado State University, Colorado Cooperative Fish and Wildlife Research Unit,
Fort Collins, CO, United States
‡U.S. Geological Survey, Georgia Cooperative Fish and Wildlife Research Unit,
Warnell School of Forestry and Natural Resources, University of Georgia,
Athens, GA, United States

INTRODUCTION

Reintroduction has been a major component of recovery efforts for Whooping Cranes (*Grus americana*; French et al., Chapter 1, this volume). Demography has been an important focus of monitoring and research for the reintroduced populations for at least two reasons. First, demographic information is often key for making decisions about how to undertake reintroduction efforts. Different management options – release methods, release locations, habitat management,

and others – will vary in their effect on demographic performance. Thus, monitoring the effects of management actions on demographic outcomes can help to provide information necessary to guide management decisions. Of course, demographic rates and the relationships between management actions and those rates are often uncertain. Developing approaches to manage Whooping Cranes in the face of uncertainty is critical to the success of reintroductions (Armstrong et al., 2007; Converse et al., 2013a; Nichols and Armstrong, 2012).

Second, knowledge about demography is useful for evaluating the success of a reintroduction effort. The primary goal of reintroduction efforts for Whooping Cranes is the establishment of self-sustaining populations (Canadian Wildlife Service and U.S. Fish and Wildlife Service, 2005). Self-sustaining populations can generally be defined as those populations with a mean growth rate ≥ 1.0, or alternatively with an adequately high probability of persistence, over some acceptably long time period. Projecting population growth and persistence requires knowledge of demographic rates. Furthermore, the demographic rates themselves can tell us something about success, assuming we know what demographic rates are associated with self-sustaining populations.

Therefore, comparison of demographics between the remnant Aransas-Wood Buffalo Population (AWBP) and reintroduced populations is valuable. Wilson et al. (2016) estimated a growth rate for the AWBP of 1.04 (95% CI: 1.035, 1.046) between 1978 and 2013 (see also Wilson et al., Chapter 4, this volume). Thus, the AWBP is self-sustaining, and the demographic rates in that population represent reasonable management targets for reintroduced Whooping Crane populations. Estimated annual survival in the AWBP between 1978 and 2013 varied across age classes from 0.850 (95% CI: 0.754, 0.932) for the youngest age class (6–18 months) to 0.944 (95% CI: 0.926, 0.962) for experienced adult breeders (Wilson et al., 2016). The fledging rate in the population was estimated at 0.515 (95% CI: 0.466, 0.564) with a 0.905 (95% CI: 0.816, 0.976) survival rate for fledged chicks to 1 December (Wilson et al., 2016). Similarly, Converse et al. (Chapter 8, this volume) estimated that, on average, 0.464 young arrived at the wintering ground per nesting pair between 1968 and 2010.

No reintroduced population has yet demonstrated both survival and reproductive rates on par with the AWBP. However, it is notable that, starting with the Grays Lake Population (GLP) reintroduction through the Eastern Migratory Population (EMP) reintroduction,

there has been general improvement in the demographic performance of reintroduced Whooping Crane populations. In the GLP, eggs placed in Sandhill Crane (*Grus canadensis*) nests hatched and fledged at a rate of only 0.29 (Ellis et al., 1992). Survival of the fledged birds to breeding age was poor, and the few birds that reached breeding age failed completely to nest or even to form pairs. In the Florida Nonmigratory Population (FNMP), despite improvements compared to the GLP, observed survival and reproductive rates were lower than those in the AWBP (Moore et al., 2012). In the EMP, by contrast, survival has actually been on par with that in the AWBP, providing some hope of success (Servanty et al., 2014). But the reproductive rate in the EMP has been substantially lower than that in the AWBP (Converse et al., Chapter 8, this volume). Little demographic information is yet available for the Louisiana Nonmigratory Population (LNMP), though some preliminary information is reported in King et al. (Chapter 22, this volume).

In this chapter, we review demographic modeling of the FNMP and the EMP. The models we present have provided insight into the reasons why the populations have failed to become self-sustaining, and – more importantly – have supported decision making for ongoing management of the reintroduction efforts. In the case of both the FNMP and the EMP, the process of building the models to support decision making generally followed steps laid out by Converse and Armstrong (2016): structuring and planning models, obtaining parameter estimates from data, filling information gaps using expert judgment, constructing models, and using models to evaluate action alternatives. We review each of those steps for the two populations and then present both results and a discussion of the decision-analytic processes designed to deal with the challenges of decision making in these cases: multiple conflicting objectives and uncertainty about how reintroduced Whooping Crane populations function.

THE FLORIDA NONMIGRATORY POPULATION

Releases of Whooping Cranes were started in 1993 to establish the FNMP (French et al., Chapter 1, this volume; Dellinger, Chapter 9, this volume). The project used a soft-release technique that was novel for this species (Folk et al., 2010). Chicks (<1 year old) were shipped during winter to central Florida, grouped into mixed-sex cohorts of 4–14 birds, affixed with wing brails and radio transmitters, and maintained in an open-topped pen. After a 2-week period of acclimation, brails were removed and birds were allowed to escape the pen. The pen and automatic feeder were maintained for several weeks following the release to allow birds to return for food and roosting refuge.

Between 1993 and 2004, 289 birds were released in one to seven release cohorts per year from sites in the central Florida counties of Lake, Polk, and Osceola (Folk et al., 2005). Most (93%) had been hatched and reared at either of two captive breeding centers (USGS Patuxent Wildlife Research Center and the International Crane Foundation), and most (76%) were costume-reared; the rest were parent-reared.

The release program and the associated monitoring of bird survival and reproductive activity were conducted by biologists with the Florida Fish and Wildlife Conservation Commission (FWC) and the University of Florida. Very early in the program, biologists became concerned about higher-than-expected levels of severe injury and mortality due to predation or collisions with power lines. By altering release strategies (e.g., constructing the release pen over water to help birds develop affinity for habitat that provided safety from predators) and introducing technical modifications (e.g., installing power line markers and redesigning radio transmitters), program managers were able to mitigate these hazards to a degree. However, high mortality continued to plague the founding population over the course of the release effort (Folk et al., 2010).

Released birds initiated reproductive behaviors at about the same age as observed in the AWBP, and birds carried reproductive activities through the incubation stage (pair bonding, territory defense, nest building, egg laying, and incubation) with no indication of anomaly (Folk et al., 2005). Nesting effort was consistent, though somewhat low (as compared to the AWBP) throughout the years of the reintroduction effort (Spalding et al., 2010). However, rates of egg fertility, hatching, and fledging were extremely low (Spalding et al., 2010). A coincident severe drought undoubtedly affected recruitment through reduction of habitat and alteration of parental behavior (Folk et al., 2005); however, Spalding et al. (2010) attributed poor recruitment also to innate physiological and behavioral causes in parents.

FWC suspended releases in 2004 due to continuing concern about high mortality and low recruitment. The next few years were dedicated to assessment of the effectiveness of release methods and the likelihood of establishing a viable nonmigratory population. A demographic model of the population was developed to predict probability of population establishment in light of deep uncertainties about the biological mechanisms underlying poor demographic performance. The resulting model was central to the effort to frame and analyze the decision about whether and in what form to resume releases. Details of the population modeling effort and the decision analysis, summarized here, were previously reported in Moore et al. (2012) and Converse et al. (2013b).

Modeling Methods

Structuring and Planning Models

A stage-structured model was used to derive estimates of demographic parameters and also provided the basis for a simulation model used to project the population into the future, ultimately for the purpose of population viability analysis (PVA; Moore et al., 2012). The continued

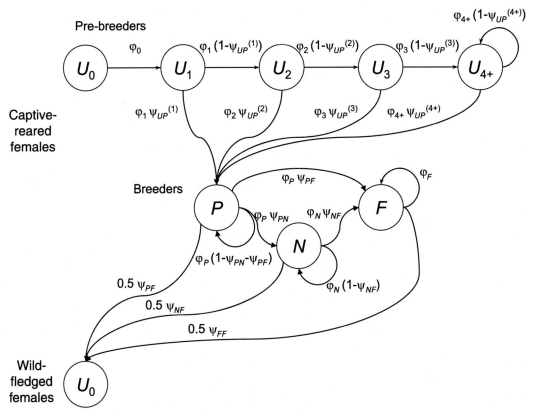

FIGURE 7.1   Female-based model for projecting dynamics of the Florida Nonmigratory Whooping Crane Population through time. Captive-reared birds released (U_0) into the population survive annually into successive age classes of unpaired birds (U_1, U_2, U_3, U_4), and unpaired birds may themselves survive and become paired (P). Paired birds may then survive and produce no young, nestling(s) that do not fledge, or fledgling(s). Likewise, birds that have ever produced only nestlings (N) may become fledgling producers (F). All fledglings produced by any class of breeder become the 0 age class of the wild-fledged segment. Symbols along arrows represent mean survival and transition rates displayed in Table 7.1. Subscript pairs ij represent transitions from class i to class j. Pathways within the wild-fledged segment (not shown) are identical to the captive-reared segment; however, rates of transition among classes may differ between the two segments according to the hypothesized model. *Source: Reproduced from Moore, C.T., Converse, S.J., Folk, M.J., Runge, M.C., Nesbitt, S.A., 2012. Evaluating release alternatives for a long-lived bird species under uncertainty about long-term demographic rates. J. Ornithol., 152 (Suppl. 2), S339–S353.*

tracking of birds following release, and the careful monitoring of reproductive activity, permitted the building of a stage-structured model (Fig. 7.1) with parameters that governed how birds of both sexes advanced through five age classes (0, 1, 2, 3, 4+ years of age) and how females advanced through four reproductive classes. These classes were nonbreeder, historical pair-bond member with no previous history

of nestling production, historical nestling producer with no previous history of fledgling production, and historical fledgling producer (Moore et al., 2012).

The PVA model tracked the fates of individual captive-reared female birds and their wild-fledged female offspring. However, parameters in the demographic model were estimated solely from data gathered on the captive-reared population segment. Because captive-reared birds produced so few fledglings, only scant information was available regarding survival of wild-fledged birds, and none was available regarding their productivity. Thus, substantial uncertainty existed as to how closely wild-fledged birds resembled the captive-reared founders in demographic performance. In other populations of reintroduced birds, there is evidence for better performance in the wild-fledged offspring than in the founders (Buner and Schaub, 2008; Heath et al., 2008; Roche et al., 2008).

The PVA model was designed to account for both aleatory uncertainty (stochastic and demographic variability) and epistemic uncertainty (parameter estimation uncertainty). Specifically, the sampling distributions resulting from Bayesian estimation were used to represent parametric uncertainty; thus the PVA model is an example of Bayesian PVA (Goodman, 2002; Kéry and Schaub, 2012; Servanty et al., 2014; Wade, 2002). In addition, multiple versions of the PVA model were built to correspond with different hypotheses about performance of wild-fledged offspring compared to captive-reared birds (described later); thus another form of epistemic uncertainty, model uncertainty, was also accounted for.

Obtaining Parameter Estimates from Data

Separate statistical models were built to estimate annual survival probability and breeding class transitions using data collected through 2007. Sixteen birds released in early October ($n = 8$), in their second year of age ($n = 5$), or via a different release method ($n = 3$) were excluded from analysis.

For survival analysis, years were divided into four seasons (i.e., 3-month periods). The number of days that a bird survived into a quarterly period was modeled as a binomial outcome with daily probability of survival expressed as a function of a fixed effect (sex) and random effects due to age/breeding class, time (60 periods),

release cohort (40 cohorts), and the individual bird, along with interactions allowing both the cohort random effect and the individual random effect to vary with the first three age classes (Moore et al., 2012). The interactions accounted for potentially reduced importance of cohort and individual effects as a bird survived through a post-release acclimation period.

Breeding class transitions were modeled in four separate conditional submodels (Moore et al., 2012). The first of these modeled the transition of females that had no history of pair bonding into the paired class. Transition probabilities were assumed to be age-specific, and they were assumed to depend on time period and individual bird random effects. The second submodel estimated the transition rates into the nestling-producer and fledgling-producer breeding classes for females that had previously paired but had not yet produced nestlings, and this model contained an individual bird random effect. Because no released bird ever produced two fledglings from a single clutch, the latter of these transition probabilities is the fledgling productivity rate of a female with a history of pair bonding but no history of nestling production. The third submodel estimated the probability of entry into the fledgling-producer breeding class given that the female had previously produced nestlings but never a fledgling; this transition rate also represents the fledgling productivity rate of this breeding class. The fourth submodel estimated the probability that a historic fledgling producer once again produces a fledgling.

Posterior distributions of all parameters in the survival models and breeding class transition submodels were derived using Markov chain Monte Carlo sampling in WinBUGS (Gilks et al., 1994), starting from uninformative priors (Moore et al., 2012). The samples from the posterior distributions provided inference on survival and breeding class transitions, and they were later accessed in the population viability modeling stage to introduce parametric uncertainty.

Filling Information Gaps Using Expert Judgment

To address uncertainty regarding demographic performance of the wild-fledged segment of the FNMP, competing PVA models were constructed around three alternative hypotheses regarding differences in demographic performance between first-generation (released) birds and second- and later-generation birds. These were:

1. Birds in the wild-fledged segment share the same demographic parameters as the captive-reared founders (FNMP_Model1);
2. Birds in the wild-fledged segment share the same demographic parameters as birds in the AWBP (FNMP_Model2); and
3. Birds in the wild-fledged segment share the same survival parameters as birds in the AWBP, and if a threshold population size is met, they also share the same productivity rate as birds in the AWBP; otherwise, they share the same productivity rate as captive-reared founders (FNMP_Model3). FNMP_Model3 was broken into eight submodels, given eight alternative settings of a threshold population size, B_T = 5, 10, 15, 20, 25, 30, 40, or 50 breeding-age (≥ 2 years old) females.

FNMP_Model1 and FNMP_Model2 represent weak and strong demographic performance, respectively, while FNMP_Model3 represents an intermediate condition. To parameterize FNMP_Model2 and FNMP_Model3, demographic rate estimates derived from Link et al. (2003) for the AWBP population were applied to wild-fledged birds in place of estimates derived from analysis of data on captive-reared birds. Allee effects (Courchamp et al., 1999; Stephens and Sutherland, 1999) as represented in FNMP_Model3 occur when demographic performance is reduced in very small populations and can be due to a number of mechanisms, for example, difficulties that individuals experience in finding a mate. FNMP_Model3 does not specify a particular mechanism; it only specifies that some mechanism exists that reduces demographic performance below a threshold population size of females ≥ 2 years old, B_T.

Given the uncertainty about which of the multiple models best reflected population processes, which could not be resolved with available data (Moore et al., 2012), scientific experts participated in an elicitation exercise that resulted in consensus relative belief weights placed on each of the 10 model variations (FNMP_Model1, FNMP_Model2, and eight settings of B_T in FNMP_Model3). About one-third of the belief weight was placed on FNMP_Model1 (0.34) and a small fraction was placed on FNMP_Model2 (0.06). The remaining weight of 0.6 was distributed among the eight variations of FNMP_Model3, with the plurality of this weight (0.16) placed on B_T = 20 breeding-age females (Converse et al., 2013b). Based on these weights, model-averaged predictions were made regarding population establishment.

Constructing Models

The PVA simulation models were executed within WinBUGS alongside the estimation procedure. Each model was simulated for each release schedule (described later) over a 131-year time frame, 100 years beyond the last release year of any schedule. Each simulation run yielded a population trend over the last 20 years of the time frame, and the proportion of positive trends was taken as the posterior probability of population increase (Moore et al., 2012).

Using Models to Evaluate Action Alternatives

The PVA models were used to predict viability of the population as a consequence of alternative schedules of bird release that FWC and its partners considered (Moore et al., 2012). For this purpose, 29 proposed release schedules were considered, each designed to accommodate specific objectives regarding use of captive bird stock for releases. One alternative was to continue the status quo (i.e., no further releases). Another 24 were combinations of varying number of releases per year (one, two, or three cohorts, with four females per cohort), varying number of years of release (5, 10, 15, or 20 years), and either of two options to delay start of releases (no delay, or 10 years from present). Four alternatives called for releases of birds in alternate years, in combinations of one or two cohorts per year over 10 or 20 years of releases.

Demographic Modeling Results

Parameter Estimates

Annual survival estimates for adult females in the FNMP ranged from a low of 0.805 for pre-breeders to 0.936 for females that had successfully fledged a chick (Table 7.1). Breeding state transitions were notably low. Fledgling production rates were 0.153 by birds that had previously been paired but hadn't produced a nestling, 0.002 for birds that had previously produced a nestling but hadn't produced a fledgling, and 0.363 for birds that had previously produced a fledgling. It was notable that birds that had previously produced a nestling but not a fledgling had the lowest chance of breeding success, suggesting that failure to successfully rear a nestling to fledging age was a good predictor for future failure. This suggests that individual quality, rather than experience, may be a determinant of successful breeding.

Evaluating Management Effects on Persistence

Under either FNMP_Model1 or FNMP_Model2, population persistence was practically insensitive to any release schedule (Moore et al., 2012). Under FNMP_Model1, every option resulted in almost certain extirpation of the population (probability of persistence ≤0.015), whereas under FNMP_Model2, every option – even that of no further releases – resulted in high probability of persistence (range 0.900–0.951). Greatest sensitivity of the outcome was observed when $B_T = 20$ breeding-age females

in FNMP_Model3: estimated persistence probability was 0.004 for the no-release scenario and was 0.756 for the most intensive release scenario (20 years of continuous release, three cohorts per year, immediate start).

Decision Analysis and Outcome

Although the modeling work was informative about the effect of each release strategy on expected population persistence (over the full range of biological uncertainty), the lack of information about other decision objectives (such as costs, public relations, etc.) held by FWC and its partners, and about the importance of those objectives relative to population recovery, prevented a full assessment of whether and how to reinstitute releases of birds into the FNMP.

Therefore, in 2008, biologists and managers from FWC convened a formal structured decision making (SDM) process. The process was designed to address FWC's decision of whether it should request that the International Whooping Crane Recovery Team (hereafter, Recovery Team) devote additional captive-reared production to continue the reintroduction effort in Florida, or whether the reintroduction effort should instead be formally terminated. Scientists and stakeholders from the University of Florida, the U.S. Fish and Wildlife Service, the U.S. Geological Survey, and the International Crane Foundation were invited to participate in the process.

The participants identified six objectives to be taken into account in the decision: population establishment, costs to FWC, costs to partners, providing captive-reared stock for alternative reintroduction projects, public relations, and information gain. Thus, the decision process needed to account for multiple objectives (Converse et al., 2013b; Keeney and Raiffa, 1976). The previously described models were used to make predictions about population establishment probability under the various release alternatives.

TABLE 7.1 Estimates of Age and Breeding-Class Specific Mean Annual Female Survival Rate (ϕ_i) and Mean Breeding Transition Rate (ψ_{ij}) Derived from 273 Captive-Reared Whooping Cranes Reintroduced in Florida, United States, 1993–2004

Age/breeding class	Model symbol	Mean	Credible interval[a]	
			Lower	Upper
ANNUAL FEMALE SURVIVAL RATE				
Means for never-paired birds				
Post-release to Age 1	ϕ_0	0.664	0.0	1.0
Age 1	ϕ_1	0.672	0.0	0.999
Age 2	ϕ_2	0.766	0.469	0.938
Age 3	ϕ_3	0.821	0.556	0.955
Age 4+	ϕ_{4+}	0.805	0.512	0.948
Means for female breeding classes[b]				
Class P	ϕ_P	0.816	0.540	0.952
Class N	ϕ_N	0.785	0.454	0.950
Class F	ϕ_F	0.936	0.703	1.000
ANNUAL FEMALE BREEDING TRANSITION RATE				
To paired class from unpaired class				
Pr(Class P | Age 1 Unpaired)	$\psi_{UP}^{(1)}$	0.257	0.005	0.788
Pr(Class P | Age 2 Unpaired)	$\psi_{UP}^{(2)}$	0.243	0.008	0.805
Pr(Class P | Age 3 Unpaired)	$\psi_{UP}^{(2)}$	0.393	0.034	0.966
Pr(Class P | Age 4+ Unpaired)	$\psi_{UP}^{(4+)}$	0.297	0.024	0.962
To nestling class from paired class				
Pr(Class N | Class P)	ψ_{PN}	0.153	0.001	0.743
To fledgling class from paired class				
Pr(Class F | Class P)	ψ_{PF}	0.052	0.0	0.232
To fledgling class from nestling class				
Pr(Class F | Class N)	ψ_{NF}	0.002	0.0	0.028
Fledging probability for class F females				
Pr(fledge chick | Class F)	ψ_{FF}	0.363	0.121	0.651

[a]*Posterior probability of true parameter value lying within indicated bounds = 0.95.*
[b]*P = Paired or historically paired, no history of nestling production; N = nestling-producer, no history of fledgling production; F = fledgling-producer.*

The two cost objectives (costs to FWC and costs to partners) were relatively straightforward to compute for each release alternative (Converse et al., 2013b). Generally, they consisted of per-bird or per-cohort costs of rearing, shipping, and release; base costs related to monitoring birds post-release; and fixed costs for restart activities associated with any strategy calling for a time delay in releases.

Because of finite productivity and rearing capacity at Whooping Crane breeding facilities, any birds provided to the FNMP meant fewer birds available to other reintroduction efforts, such as the EMP. Therefore, any release strategy that minimized the provision of birds to Florida overall or that postponed their delivery for any length of time was valued in terms of this objective (Converse et al., 2013b).

The public relations objective was assigned a 0 or 1 score, with any action that provided for the release of birds receiving a score of 1 for this attribute, and 0 otherwise (Converse et al., 2013b). Actions that received the higher score were valued over those that did not, relative to this objective.

The information objective reflected that, relative to this decision, information was valued for its own sake, rather than as a component of an adaptive management program (Williams et al., 2007) because the information was seen as valuable to Whooping Crane recovery overall, that is, contributing to the knowledge base in a way that would benefit other reintroduction efforts. The information objective was measured by how well each release alternative could generate data that reliably indicated that FNMP_Model1 was the underlying model, if in fact that was the case (Converse et al., 2013b; Moore et al., 2012). This was measured by generating data under FNMP_Model1, and by simulating decision making under that model and under alternative models. Any release strategy that quickly identified FNMP_Model1 as the true model (and could thus most rapidly provide reliable information about the inefficacy of releases for establishing a population) was of most value. FNMP_Model1 was used as the reference model in this context because FNMP_Model1 produced a worst-case scenario, whereby the poor demographic performance of the reintroduced population did not improve in later generations. Understanding this, if it was the case, could be critical to other reintroductions of Whooping Cranes; in particular it would argue for caution

in devoting resources to reintroduction efforts, especially if early performance of reintroduced populations was poor.

In this decision, objectives were scaled differently and were in conflict with one another. Generally, actions that were of high value on the population establishment, information, and public relations objectives were of less value on the cost and alternative reintroduction projects objectives. Therefore, a means of expressing all objectives in a common currency and accounting for the relative importance of the objectives was required (Converse et al., 2013b). Thus, the 29 release actions described in the previous section were assessed using the Simple Multi-Attribute Rating Technique (Converse et al., 2013b; Edwards, 1977). The swing weighting technique was used (Goodwin and Wright, 2004) to assess stakeholders' relative emphasis (W_j) on achieving each objective $j = 1, \ldots, 6$. After standardizing the range of scores on each objective to a 0–1 scale, such that each action i generates a score on each objective $j, 0 \leq N_{i,j} \leq 1$, a final score F_i for each action can be obtained by weighting and then summing over the objectives:

$$F_i = \sum_j W_j \times N_{i,j}$$

The action producing the maximum weighted mean attribute (F_i) – which represents the best compromise alternative across the set considered – was that of 10 consecutive years of releases, three cohorts per year, commencing immediately. Note that this was not the action that most benefited the population reintroduction objective (20 years of releases, three cohorts per year, starting immediately) even though that objective received the most weight (0.39) through the swing-weighting exercise. This was because the stakeholders also placed relatively high value on reducing costs and maintaining a reserve of captive stock for other reintroduction efforts (0.42 combined weight), objectives best satisfied by reducing or delaying releases.

The decision process described specifically addressed the needs of FWC, and while the results of this analysis informed FWC about the optimal decision from its perspective, ultimately the decision about whether to release more birds to the FNMP was in the hands of the Recovery Team (Converse et al., 2013b). In late 2008, the Recovery Team recommended that no further releases be conducted to support the FNMP reintroduction. Four concerns were discussed in the final recommendation (Converse et al., 2013b). First, the Recovery Team noted that "even the most optimistic assumptions ... provided no more than a 0.41 probability of achieving a selfsustaining population." This suggested that the Recovery Team didn't strongly value an ability to increase that probability from 0.09 (under no more releases) to 0.41, given the costs of doing so (financial and otherwise). Second, the Recovery Team expressed concern that 0.41 wasn't in fact achievable because "24 Whooping Crane chicks per year are not available for continued releases in Florida" presumably because they were committed to ongoing releases in the EMP, and thus viewed this as a constraint on the available alternatives. Third, the Recovery Team cited the periodic drought in Florida and its effect on reproduction in the wild. Fourth, they cited the concern that crane habitat in Florida was expected to decline under pressure from development. These last two issues suggested that the Recovery Team expected the conditions resulting in the observed demographic estimates to deteriorate going forward, such that the demographic rates in the model would not be representative of future rates.

The Recovery Team's decision to recommend against further releases, despite the relatively low rank of that alternative in this analysis, underscores the degree to which addressing the needs of a specific decision maker is critical in a decision-analytic framework. In this case, there were two linked decisions: first, FWC had to decide if they wanted to continue with releases,

and contingent on that, the Recovery Team had to decide if they would support continuation of the project. As of early 2016, the FNMP had declined to approximately 11 birds (Dellinger, Chapter 9, this volume).

THE EASTERN MIGRATORY POPULATION

The reintroduction of the EMP was initiated using captive-bred and costume-reared chicks released with the ultralight-led (UL) aircraft method (UL releases) starting in 2001 and the direct autumn release (DAR) method starting in 2005 (French et al., Chapter 1, this volume). Training and releases at the initial release site – Necedah National Wildlife Refuge (NWR) – continued using these methods until 2011. In 2011, poor reproduction and evidence that it was related to harassment by biting black flies (Simuliidae; Converse et al., Chapter 8, this volume) prompted relocation of releases to eastern Wisconsin, with UL releases on White River Marsh State Wildlife Area (SWA) and DAR releases on Horicon NWR. A new reintroduction method, autumn releases of parent-reared birds in territories of failed breeding pairs (PR releases), was initiated at Necedah NWR in 2013.

The major challenge for this population has been poor reproduction. Through 2015, from 161 nesting pair-years only nine chicks survived to 1 December, constituting a reproductive success rate of 0.056. Several different hypotheses have been posed and tested for poor reproductive success (Converse et al., Chapter 8, this volume; Runge et al., 2011). However, the uncertainty about the causes of poor reproduction has been a major factor complicating management decisions. Population models have been key to decision making in light of this uncertainty. Specifically, during a 2012–13 SDM process, population models were used to support decisions about whether and how to execute further releases to the population (Converse et al., 2014). The SDM process included an objective of maximizing probability of persistence for the EMP. Thus, it was necessary to build population models to support the SDM process (further details provided in Converse et al., 2014 and Servanty et al., 2014). These models needed to predict persistence as a function of various management alternatives, and account for fundamental uncertainties about the causes of poor reproduction in the population.

Modeling Methods

Structuring and Planning Models

The Eastern Migratory Population Simulation Model (EMPSim) was initially constructed as a population viability model, and was expanded into four submodels to represent competing hypotheses for causes of reproductive failure. The initial model (which we refer to as EMPSim_Baseline) was built to model the population based only on observed demographic rates. The four submodels representing alternative hypotheses (EMPSim_Habitat, EMPSim_RearingRelease, EMPSim_CaptiveSelection, and EMPSim_Experience, described later) that were used to project the effects of novel management actions also included parameter estimates elicited from experts to account for the possibility of different demographic rates under different management approaches.

EMPSim is a female-only, pre-breeding census, state-based matrix model (Caswell, 2001). The basic structure of EMPSim_Baseline is represented in Fig. 7.2, and is composed of age- and stage-structured models nested within various compartments, including UL and DAR birds, as well as wild-hatched birds. In the construction of the submodels, additional compartments were added to represent novel management actions (e.g., releases using a new method, or in a new location).

EMPSim was constructed to account for both aleatory and epistemic uncertainty. Specifically,

demographic and temporal uncertainty are represented via, respectively, binomial trials for demographic outcomes and inclusion of temporal effects estimated for demographic parameters. Epistemic uncertainty is represented through parametric uncertainty and also through model uncertainty (via the competing submodels). Similar to the PVA model for the FNMP, EMPSim is an example of Bayesian PVA. In EMPSim, samples from Markov chains are used to represent sampling variance in the demographic parameters.

Obtaining Parameter Estimates from Data

The demographic estimates used to parameterize EMPSim_Baseline and parts of the four hypothesis-specific submodels were obtained using a multistate mark-recapture model (Arnason, 1973; Hestbeck et al., 1991; Schwarz et al., 1993) implemented in a state-space framework (Kéry and Schaub, 2012). An earlier state-space formulation of a multistate model for this population (Converse et al., 2012) suggested a basic model structure including unpaired, paired, and nesting states.

The demographic estimation work underlying EMPSim is described in Servanty et al. (2014). Their model is based on a seasonal (i.e., three-month) time step, with data included through 31 March 2010. The Servanty et al. (2014) model expanded the state structure earlier described by Converse et al. (2012) to include captive, unpaired, first-time paired, established paired (in second and subsequent seasons that the bird is paired), first-time nester, established nester (in second and subsequent years that the bird nests), and lapsed nester (a bird that previously nested but is not nesting in a given year). Once birds entered a nesting state, subsequent transitions to a different state were allowed to occur only once per year, in April. The estimation model included effects of time, release type (UL or DAR), sex, and age in various combinations on survival and transition probabilities (see Servanty et al., 2014 for details).

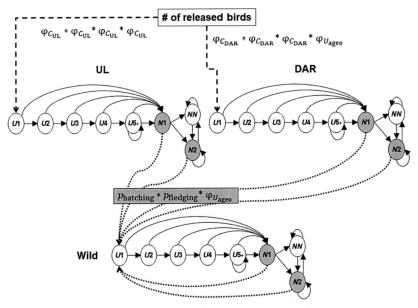

FIGURE 7.2   Basic model structure for the Eastern Migratory Population Simulation Model (EMPSim) used to predict probability of persistence in the Eastern Migratory Population (EMP) of Whooping Cranes. Captive-reared and released individuals enter either the ultralight-led (UL) release or direct autumn release (DAR) compartment depending on their release method. UL birds must survive for four 3-month seasons in the captive stage (with probability $\varphi_{C_{UL}}^{4}$), and DAR birds must survive for three 3-month seasons in the captive stage and one 3-month season in the age 0 stage (with probability $\varphi_{C_{DAR}}^{3} * \varphi_{U_{age0}}$) before passing into the prebreeder stages (dashed line). Prebreeder stages include age-specific stages for ages 1, 2, 3, 4, and 5+ years of age (stages $U1$, $U2$, $U3$, $U4$, $U5+$, respectively). Finally, birds may pass into breeder stages, including first-time nesters ($N1$), experienced nesters ($N2$), and lapsed nesters (NN). Solid arrows indicate stochastic transitions of individuals among life stages. Production of offspring (indicated by dotted lines) requires successful hatching and fledging of a chick, and survival of that chick through one 3-month season in the age 0 stage (with probability $P_{hatching} * P_{fledging} * \varphi_{U_{age0}}$). Successful production of offspring results in an entry of a bird into the Wild compartment.

The estimation model used to parameterize the EMPSim submodels did not include the first-time paired and established paired states because data to determine what birds were in pairs were increasingly difficult to obtain in later years (with less intensive monitoring of the population and more birds forming pairs). Instead, all birds that would have previously been included in the unpaired, first-time paired, or established paired states were included in a single prebreeder state. Otherwise, this estimation model was largely equivalent to the model described in Servanty et al. (2014) but included data through 31 December 2012.

Filling Information Gaps Using Expert Judgment

Given uncertainty about the cause of nest failure, EMPSim was expanded into four distinct submodels. These submodels were built to represent four competing hypotheses about the reasons for reproductive failure in the population. The hypotheses represented by the four submodels are:

1. EMPSim_Habitat. Reproductive success is largely influenced and limited by habitat. This may include winter habitat quality (which could affect condition of birds at

initiation of breeding), energy and nutrient availability in the breeding habitat, black fly density, predators, or other unspecified drivers.

2. EMPSim_RearingRelease. Captive rearing and release experiences influence early learning, which influences later reproductive success.

3. EMPSim_CaptiveSelection. Reproductive success is largely limited by captive selection in the captive breeding population. Captive selection occurs when selection in captivity favors traits that are non-adaptive post-release.

4. EMPSim_Experience. Reproductive success is largely limited by lack of breeding experience in EMP birds.

The challenge of the reintroduction planning effort was uncertainty about not just the causes of reproductive failure, but also the expected effectiveness of management alternatives to combat it. By 2012, it was clear that, at the observed demographic rates, the population could not be self-sustaining. Thus, the SDM process was focused on the possibility of changing management in some way to bring about a change in demographic performance. In this case, existing data on the population is not applicable to making predictions about performance. This is the challenge of managing in such a situation: what will be the effect of truly novel management actions?

With that in mind, an expert-elicitation process was used to develop expert judgment on the effect of novel management actions, contingent on the various hypotheses for reproductive failure (Ayyub, 2001; Burgman et al., 2011a; Martin et al., 2011; Speirs-Bridge et al., 2010). A team of experts (hereafter, Population Study Team) was established to develop predictions for demographic parameters in situations that were unobserved in the empirical data set. For example, predictions were made, under the habitat hypothesis, about how the population

would respond to releases in novel locations, which would have novel habitat conditions. Experts were asked to give predictions for survival, nesting probability, and fledging success for each of the unobserved scenarios. The experts also provided judgment on the likelihood of the four hypotheses. This information was used to identify the optimal alternative under uncertainty about the cause of nest failure via model averaging. The information provided by the Population Study Team was summarized, and average responses across experts were used for model weights, while the full range of within- and between-expert uncertainty was used to account for parametric uncertainty.

Constructing Models

EMPSim_Baseline and the hypothesis-specific submodels were constructed in R (R Development Core Team, 2012). The measurement criterion originally identified for the persistence objective was the probability that ≥ 1 female would remain 75 years after releases terminated. This criterion was identified during an elicitation exercise with the process participants. Alternative metrics derived from EMPSim submodels were also calculated and discussed, including probability of quasi-persistence (with a threshold of 25 females), mean and median female abundance estimates, and population growth rate. Each of these alternative metrics was highly correlated (>0.8) with the original criterion of probability of persistence after 75 years.

Using Models to Evaluate Action Alternatives

The components of management alternatives identified within the SDM process included: number of birds released (5 or 10 females per year), number of years of releases (15, 25, or 35 years), summer release site, rearing and release method, and egg source. The final list of strategies was based on all possible combinations of these components and ranges. Five potential release locations (which would ultimately be expected to host breeding concentrations of cranes, given high natal fidelity of the species; Barzen et al., Chapter 14, this volume) were identified, with four of them located in Wisconsin – Necedah NWR, White River Marsh SWA, Horicon NWR, Crex Meadows Wildlife Area – and the fifth in Pope County, Minnesota. Two additional rearing and release methods were included in addition to those previously discussed (UL, DAR, and PR) based on discussions with the Population Study Team: chick adoption (an alternative involving costume-rearing and summer release of chicks in the vicinity of adult pairs) and early parent-rearing (parent-rearing with release of chicks in summer rather than fall, again in the vicinity of adult pairs). The egg source alternative component included eggs from captive birds or eggs collected from the AWBP. In addition, one alternative allowing for no further releases was considered. This resulted in 251 distinct alternative release strategies to be analyzed. However, releases of 35 birds per year were considered only under alternatives involving release of chicks hatched from eggs collected at AWBP, because it was thought that 35 eggs would not be available from the captive population alone. For purposes of this chapter, we omit management actions involving releases of 35 birds per year to aid in interpretation of results. Thus we consider a total of 201 management actions.

Replicate simulations (n = 40,000) were run for each unique combination of management action and hypothesis. Some of the combinations did not produce unique results and thus did not need to be run separately. For example, under the habitat hypothesis, all different release techniques or egg sources would result in the same prediction because release technique is not a significant factor under the habitat hypothesis. Thus, under the habitat hypothesis, only a single run was necessary for any alternatives that did not differ by release site. Similarly, there were nonunique combinations under other

hypotheses. Model-averaged predictions were calculated for each management action contingent on the average weight (prior belief) for each hypothesis.

Demographic Modeling Results

Parameter Estimates

Servanty et al. (2014) found that seasonal (3-month) survival varied little by sex but did vary by release method during the captive stage, with higher survival for UL birds (0.960; 95% CI: 0.926, 0.984) than for DAR birds (0.887; 0.785, 0.955). This corresponded to a first-year survival probability of 0.852 (0.737, 0.936) for UL birds, which spent their entire first year in the captive stage. For DAR birds, first-year survival was calculated based on two seasonal periods in the captive state and two seasonal periods in the unpaired state (during which a sex effect was included) corresponding to annual probabilities of 0.721 (0.557, 0.847) for female DAR birds and 0.727 (0.560, 0.856) for male DAR birds. Annual survival in the various post-release stages was similar to that observed in the AWBP population, ranging from 0.835 (0.697, 0.935) to 0.998 (0.985, 1). Birds formed social pairs at a relatively young age with highest probability of pairing in 2-year-old birds (0.076; 0.034, 0.134). By the age of 5 years, most birds were paired. A bird in a social pair had a nearly 0.50 probability of nesting the following breeding season and, once a bird had nested, the probability of nesting in subsequent years was high (>0.9).

Population Viability Analysis

The EMPSim_Baseline model indicated that with no more releases the population had a probability of persistence in 75 years of 0.6236. This baseline model included the observed reproductive rate through December 2012 of 0.057 (five fledged chicks produced in 88 pair-years). By comparison, the AWBP has been growing at an annual rate of 1.04 (Wilson et al., 2016). If the reproductive rate in the EMP were increased to a level equivalent to that in the AWBP (0.464 chicks arriving on the wintering grounds per breeding pair; Converse et al., Chapter 8, this volume), then probability of persistence in 75 years was 0.846 with no more releases, and the mean expected growth rate was 1.011 annually. The reproductive rate at which the mean expected growth rate was 1.00 over the final 10 years was 0.35.

Evaluating Management Effects on Persistence

In all cases, the most intensive release schedules resulted in the highest probability of persistence, with the rank based on the total number of birds released, that is, in order: 10 years of 25 birds per year (250 total birds), 10 years of 15 birds per year (150 total), 5 years of 25 birds per year (125 total), and 5 years of 15 birds per year (75 total). Otherwise, factors over which the predicted probability of persistence varied depended on the specific hypothesis. It is important to recognize, however, that predictions for novel management actions (either new locations, new methods of releases, or releases of AWBP-sourced birds) were based on expert judgment, not empirical information. Predictions would be expected to change as new information influenced experts (see "Summary and Outlook"). Under the habitat hypothesis, the various release sites were ranked, from best to worst: Horicon NWR; White River Marsh SWA; Crex Meadows Wildlife Area; Pope County, Minnesota; and Necedah NWR. Under the rearing and release hypothesis, the various release methods ranked, from best to worst: DAR, UL, PR or early parent rearing, and chick adoption (under a given release intensity, PR and early parent rearing were nearly indistinguishable, indicative of how similarly experts judged that these two release methods would perform). Strategies involving release of birds hatched from AWBP eggs always ranked higher than strategies involving release of birds from captive-produced eggs under the captive selection hypothesis. Under the experience hypothesis, the only factor that

caused variation in predicted persistence was the intensity of releases, as described earlier.

Top-ranked predictions were averaged over the various models, with average elicited model weights of 0.522 (habitat hypothesis), 0.180 (rearing and release hypothesis), 0.172 (experience hypothesis), and 0.126 (captive selection hypothesis). The top-ranked management action, based only on probability of persistence, involved the most intensive releases, at Horicon NWR, using the DAR method and AWBP as the egg source (Table 7.2). The predicted probability of persistence 75 years onward was 0.743. Not surprisingly, the lowest-ranked alternative – in terms of probability of persistence – was cessation of releases, with a predicted probability of persistence of 0.608.

Decision Analysis and Outcome

Similar to the FNMP decision process, the results for the persistence objective were combined with predicted results for other objectives to solve the multiple-objective decision problem. These additional objectives included minimizing costs (Converse et al., 2014), minimizing numbers of birds released, and minimizing the risk of interaction between the AWBP and the EMP (measured by the distance, in km, from the western-most release site to the AWBP migration corridor). These objectives were weighted by process participants using a swing-weighing technique (Goodwin and Wright, 2004), indicating their relative importance; weights were then averaged across the individual participants. The objective with the highest importance was the persistence objective (average weight = 0.43), followed by costs (0.28), number of birds released (0.23), and risk of interaction with the AWBP (0.06). The Simple Multi-Attribute Rating Technique (Edwards, 1977; Goodwin and Wright, 2004) was used to identify optimal management actions (Table 7.3). The overall top-ranked alternative was 10 years of releases, 15 chicks per year, released at Horicon NWR

TABLE 7.2   The Top 10 and Bottom 5 Release Strategy Alternatives (out of a Total of 201 Considered) for Whooping Cranes in the EMP, Ranked by Predicted Probability of Persistence. The Predicted Probability of Persistence – P(Persist) – Is Model-Averaged over the EMP Simulation Submodels 1 through 4, Which Are Based, Respectively, on the Habitat Hypothesis (Model Weight = 0.522), Rearing and Release Hypothesis (0.180), Experience Hypothesis (0.172), and Captive Selection Hypothesis (0.126).

Alternative					P(Persist)	Rank
Years	Chicks/year	Location[a]	Method[b]	Egg source[c]		
10	25	HNWR	DAR	AWBP	0.743	1
10	25	WRSWA	DAR	AWBP	0.740	2
10	25	HNWR	UL	AWBP	0.738	3
10	25	WRSWA	UL	AWBP	0.736	4
10	25	CMWA	DAR	AWBP	0.736	5
10	25	HNWR	DAR	Captive	0.734	6
10	25	HNWR	ePR	AWBP	0.733	7
10	25	HNWR	PR	AWBP	0.732	8
10	25	WRSWA	DAR	Captive	0.732	9
10	25	HNWR	CA	AWBP	0.732	10
5	15	NNWR	UL	Captive	0.648	197
5	15	NNWR	ePR	Captive	0.647	198
5	15	NNWR	PR	Captive	0.646	199
5	15	NNWR	CA	Captive	0.646	200
0	0	–	–	Captive	0.608	201

[a]Release locations include Horicon National Wildlife Refuge (HNWR), White River Marsh State Wildlife Management Area (WRSWA), Crex Meadows Wildlife Area (CMWA), and Necedah National Wildlife Refuge (NNWR).
[b]Release methods include direct autumn release (DAR), ultralight-led release (UL), early parent rearing (ePR), parent-rearing (PR), and chick adoption (CA).
[c]Egg sources include the Aransas-Wood Buffalo Population (AWBP), and the captive population (Captive).

using the DAR method and using AWBP-sourced chicks. The overall lowest-ranked alternative involved releasing 25 birds per year for 10 years, in Pope County, Minnesota, using the UL method and a captive egg source. Generally, UL releases performed worst because of the cost involved. However, during discussions that occurred toward the end of the SDM process it was recognized that the characterization of costs did not account for the fact that funds for a given release method (e.g., the substantial funds raised by the nonprofit partner organization Operation Migration to support UL releases) were not always directly transferrable to other release methods.

In late 2013, the Whooping Crane Eastern Partnership (WCEP) Guidance Team took the results of these analyses and made a decision to continue with the ongoing release strategy (releases of a combination of UL and DAR birds at Horicon NWR and White River Marsh SWA). Also in 2013, formal testing of the PR method was initiated (four birds were released in 2013, four in 2014, and three in 2015). Generally, parametric uncertainty resulting from uncertainty of experts about this method drove its moderately poor performance in the modeling results, rather than a conviction that it would not work.

Since the completion of the SDM process in 2013, it appears that – in general – expert belief in

TABLE 7.3 The Top 10 and Bottom 5 Release Strategy Alternatives (out of a Total of 201 Considered) for Whooping Cranes in the EMP, Ranked by a Simple Multiattribute Rating Technique (SMART) Analysis. Scores for Each of Four Objectives Are Included in a Weighted Score for the Strategy Alternative, and Ranks Are Based on This Weighted Score. Objectives Are: Maximize Probability of Persistence, Denoted as P(Persist); Minimize Cost; Minimize Number of Birds Released; and Minimize Risk of Contact between the EMP and the AWBP, Measured by Distance in km from the Western-most Site of Release to the AWBP Migration Corridor.

Years	Chicks/year	Location[a]	Method[b]	Egg source[c]	P(Persist)	Cost ($ million)	Number released	Distance (km)	Weighted score	Rank (SMART)
10	15	HNWR	DAR	AWBP	0.721	2.64	150	612	0.703	1
5	25	HNWR	DAR	AWBP	0.707	2.22	125	612	0.700	2
10	15	WRSWA	DAR	AWBP	0.718	2.64	150	612	0.697	3
5	15	HNWR	DAR	AWBP	0.683	1.32	75	612	0.696	4
5	15	WRSWA	DAR	AWBP	0.683	1.32	75	612	0.696	5
5	25	WRSWA	DAR	AWBP	0.706	2.22	125	612	0.695	6
10	15	HNWR	PR	AWBP	0.713	2.37	150	612	0.690	7
10	15	HNWR	ePR	AWBP	0.713	2.37	150	612	0.689	8
5	15	HNWR	ePR	AWBP	0.680	1.19	75	612	0.689	9
5	15	WRSWA	ePR	AWBP	0.680	1.19	75	612	0.689	10
0	0	—	—	—	0.608	0	0	612	0.569	141
10	25	CMWA	UL	Captive	0.723	8.81	250	389	0.410	197
10	25	NNWR	UL	AWBP	0.710	8.75	250	612	0.390	198
10	25	NNWR	UL	Captive	0.701	8.75	250	612	0.377	199
10	25	POPE	UL	AWBP	0.720	8.81	250	200	0.359	200
10	25	POPE	UL	Captive	0.712	8.81	250	200	0.346	201

[a]Release locations include Horicon National Wildlife Refuge (HNWR), White River Marsh State Wildlife Management Area (WRSWA), Crex Meadows Wildlife Area (CMWA), Pope County MN (POPE), and Necedah National Wildlife Refuge (NNWR).

[b]Release methods include direct autumn release (DAR), ultralight-led (UL), early parent rearing (ePR), and parent rearing (PR).

[c]Egg sources include the AWBP and the captive population (Captive).

the various hypotheses has swung more toward bird-centric, rather than habitat-centric, hypotheses (i.e., greater weight on rearing and release and captive selection-type hypotheses rather than the habitat hypothesis; Converse et al., Chapter 8, this volume). In 2016, WCEP made the decision to terminate any further releases using the UL method, citing the need to use release techniques that better mimic the natural history of the bird (thus suggesting greater support for the rearing and release hypothesis). No direct evidence is yet available to assist in the process of updating model weights empirically, because as of 2017 there had not yet been any nesting in the eastern Wisconsin release sites (Horicon NWR and White River Marsh SWA) with the exception of a hybrid Whooping Crane-Sandhill Crane pair (A. Lacy, International Crane Foundation, personal communication). The oldest birds released in the eastern Wisconsin release sites were just 5 years old in 2016. No PR birds have yet nested; the oldest of these is just 2 years old.

To examine survival in PR birds, we ran a two-age model using data for PR birds from 1 April 2013 through 30 June 2016. Birds were assumed to join the population (i.e., hatch and age) on 1 April, but only birds that survived to be transported to Wisconsin in September of their first year were included in the analysis. Survival was estimated seasonally, with seasons including spring (1 April to 30 June), summer (1 July to 30 September), autumn (1 October to 31 December), and winter (1 January to 31 March). Annual survival (ϕ_A) was calculated based on seasonal survival (ϕ_S) as $\phi_A = \phi_S^4$. Detection probability was assumed to be 1 (these birds all wore satellite transmitters), only one live state was specified (therefore, only survival was estimated), and the only effect that was included in the survival model was differentiation by age (e.g., <1 year or >1 year of age).

Results indicate that survival for PR releases may be slightly lower in the first year compared to estimates made by Servanty et al. (2014) for DAR releases (Table 7.4), although, again,

uncertainty in the PR survival estimates is substantial. However, Servanty et al. (2014) did not include an age effect in the unpaired state, but survival in the first year for DAR birds included two seasons in the captive state and two seasons in the unpaired state, where survival in the unpaired state may have been inflated by older birds in that state. Survival of unpaired birds >1 year of age was not substantially different between PR and DAR birds, again, despite the fact that the PR birds in the unpaired state are on average younger than those unpaired DAR birds included in the Servanty et al. (2014) analysis. Further information on survival in PR birds will be valuable, but not until these birds begin nesting will empirical information be available relevant to evaluating the rearing and release hypothesis.

SUMMARY AND OUTLOOK

Reintroduction has been a focus of Whooping Crane recovery efforts for more than 40 years: at least one reintroduced population (GLP, FNMP, EMP, and/or LNMP) has received releases in all years since 1975, with the exception of a few years between the termination of the GLP reintroduction and the start of the FNMP reintroduction. No reintroduced population, however, is yet self-sustaining. In the next several years, we may begin to see signs of success in the LNMP or in the EMP under revised management strategies. But if the populations do not show signs of success in Louisiana or eastern Wisconsin, the uncertainty about causes of poor demographic performance in these populations – and more importantly, what can be done about it – will continue to pose a challenge for managers.

Survival in the EMP has been on par with that in the AWBP, and while survival in the FNMP was slightly below that in the AWBP, overall it appears that reproduction has been the major factor holding back successful reintroduction of Whooping Cranes, at least in the two reintroductions discussed in depth here. The hypotheses

TABLE 7.4 Annual Survival Pprobabilities for Whooping Cranes Released Using Three Different Release Methods in the EMP. Survival Rates Computed for Birds <1 Year of Age, and Unpaired Birds >1 Year of Age. Estimates for Parent-Reared (PR) Birds, Computed Here, Are Compared to Previous Estimates for Ultralight-Led (UL) and direct autumn release (DAR) Birds, as Reported by Servanty et al. (2014). In Some Cases, Those Estimates Were Separate for Female (♀) and Male (♂) Birds.

Release method and age	Sex	Mean	Credible interval[a]	
			Lower	Upper
PR birds <1 year[b]	–	0.637	0.352	0.878
DAR birds <1 year[c]	♀	0.721	0.557	0.847
	♂	0.727	0.560	0.856
UL birds <1 year[c]	–	0.852	0.737	0.936
Unpaired PR birds >1 year[b]	–	0.857	0.541	0.996
Unpaired DAR birds >1 year	♀	0.835	0.697	0.935
	♂	0.850	0.695	0.950
Unpaired UL birds >1 year	♀	0.926	0.847	0.976
	♂	0.936	0.879	0.976

[a]Posterior probability of true parameter value lying within indicated bounds = 0.95.
[b]Estimated based on data from 1 April 2013 through 30 June 2016.
[c]Servanty et al. (2014) did not include an age effect, but did distinguish between captive and unpaired birds. All captive birds were <1 year old, while survival for unpaired birds was calculated across all ages. For UL birds, first-year survival included all four seasons in the captive state, whereas first-year survival for DAR birds included a mix of two seasons in the captive state and two seasons in the unpaired state. Because a sex effect was included for birds in the unpaired state, DAR birds <1 year old are differentiated by sex.

for poor demographic performance in the reintroduced populations, as represented by competing models described here, can be categorized based on whether poor demographic performance would be predicted to affect just the first generation of birds released (FNMP_Model2, FNMP_Model3, EMPSim_Rearing Release) or would be predicted to affect multiple generations (FNMP_Model1, EMPSim_Habitat). The implications of these competing concepts are critical: challenges that have multigenerational effects are more likely to be related to characteristics of the release site rather than characteristics of the rearing and release method, and are perhaps a bigger impediment to success.

There are two exceptions to the categorization of hypotheses into first- or multiple-generation mechanisms. The first exception is the EMPSim_CaptiveSelection hypothesis, which postulates that poor demographic performance has a genetic basis due to selection for traits in captivity that are non-adaptive post-release, and thus performance may or may not improve with generations post-release (depending on population viability, the remaining genetic variability, and the strength of the selection pressure). The second exception is the EMPSim_Experience hypothesis, which postulates that birds simply were not old enough to display their breeding potential. Given that the birds in the EMP are now as old as 15 years, the EMPSim_Experience hypothesis has low credibility.

Thus, if we put aside experience as a factor, we are left with the idea that if reintroduction is to be successful, reintroduction efforts must either (1) improve the methods with which birds are raised and released, (2) improve the choice of habitats in which birds are released, or (3) choose birds with more favorable genetic characteristics to release. Given that we do not now know which of these general types of approaches we should take in order to improve probability of persistence in reintroduced Whooping Crane populations, it is important to consider how to proceed under uncertainty. A variety of decision-analytic approaches are designed to optimize management under uncertainty. In the case of both the FNMP and EMP decision processes described here, methods were used to predict model-averaged population outcomes, such that predictions of competing models, and the assessed belief in those models, were integrated into the analysis.

The analyses for both the FNMP and the EMP relied on expert judgment. Rigorous use of expert judgment is increasingly recognized as an important part of decision making for natural resource management (e.g., Burgman et al., 2011b; Converse et al., Chapter 8, this volume; Martin et al., 2011). While experts are subject to a variety of cognitive biases (Burgman, 2004; Burgman et al., 2011b; McBride et al., 2012), an increasing body of research has resulted in better knowledge about how elicitation methods can help to overcome biases (Burgman et al., 2011b; Martin et al., 2011; Speirs-Bridge et al., 2010). When decisions must be made and empirical information is not available, expert judgment is often the best substitute. Formalizing its use through problem decomposition and structured elicitation will continue to be important for improving Whooping Crane management decisions. However, it cannot take the place of empirical information gathered through monitoring and research. Expert judgment can perhaps be seen most usefully as providing a starting point to inform management while the empirical basis for management is strengthened.

The most thorough approach to management under uncertainty is adaptive management. Adaptive management is a specific type of decision analysis in which management proceeds under uncertainty while integrating learning to reduce uncertainty over time (McCarthy et al., 2012; Runge, 2011; Williams et al., 2007; Walters, 1986). In active, as opposed to passive, adaptive management, management actions may be chosen with consideration not just for their expected effectiveness in the short term, but also with consideration for their potential to contribute to learning (Williams et al., 2007). Development of a formal active adaptive management framework has the potential to improve long-term outcomes by accelerating learning. When considering implementation of an active adaptive management framework, it is critical to consider the time period over which learning can be applied. Specifically, there must be an adequately long period in which the benefits of learning can be realized (Canessa et al., 2016). In Whooping Cranes, learning about demographic performance of released individuals is particularly slow because of delayed sexual maturity. This emphasizes the importance of long-term planning for Whooping Crane conservation as a whole. Having an idea about the overall strategy for Whooping Crane conservation, and the degree to which a long-term commitment to reintroduction exists in the Whooping Crane conservation community, will help reintroduction programs plan for management under uncertainty, and determine to what degree it will be important to invest in learning. The Recovery Team's current process for recovery planning (W. Harrell, U.S. Fish and Wildlife Service, personal communication) will help to shape future plans for reintroduction of Whooping Cranes. This process will also be shaped by uncertainty about whether successful reintroduction of Whooping Cranes is within our grasp.

Acknowledgments

The authors wish to thank many collaborators who have contributed to this work, including M. Folk, P. Heglund, W. Link, and S. Nesbitt. We also thank G. Olsen, who assisted with compilation of data for the analysis of survival in parent-reared birds. We thank the Florida Fish and Wildlife Conservation Commission and the Whooping Crane Eastern Partnership for funding, data, and other support. Finally, we thank the many stakeholders and experts who participated in the decision processes described here. Reviews by J. Austin, J. French, E. Grant, and one anonymous reviewer greatly improved the manuscript. Any use of trade, firm, or product names is for descriptive purposes only and does not imply endorsement by the U.S. Government.

References

Armstrong, D.P., Castro, I., Griffiths, R., 2007. Using adaptive management to determine requirements of re-introduced populations: the case of the New Zealand hihi. J. Appl. Ecol. 44 (5), 953–962.

Arnason, A.N., 1973. The estimation of population size, migration rates and survival in a stratified population. Res. Popul. Ecol. 15 (1), 1–8.

Ayyub, B.M., 2001. Elicitation of Expert Opinions for Uncertainty and Risks. CRC Press, Boca Raton, FL.

Barzen, J.A., Lacy, A., Thompson, H.L., Gossens, A.P., 2018. Habitat use by the reintroduced Eastern Migratory Population of Whooping Cranes (Chapter 14). In: French, Jr., J.B., Converse, S.J., Austin, J.E. (Eds.), Whooping Cranes: Biology and Conservation. Biodiversity of the World: Conservation from Genes to Landscapes. Academic Press, San Diego, CA.

Buner, F., Schaub, M., 2008. How do different releasing techniques affect the survival of reintroduced grey partridges *Perdix perdix*. Wildl. Biol. 14 (1), 26–35.

Burgman, M., 2004. Expert frailties in conservation risk assessment and listing decisions. In: Hutchings, P., Lunney, D., Dickman, C. (Eds.), Threatened Species Legislation: Is It Just an Act? Royal Zoological Society of New South Wales, Mosman, NSW, Australia, pp. 20–29.

Burgman, M., Carr, A., Godden, L., Gregory, R., McBride, M., Flander, L., Maguire, L., 2011a. Redefining expertise and improving ecological judgment. Conserv. Lett. 4 (2), 81–87.

Burgman, M.A., McBride, M.F., Ashton, R., Speirs-Bridge, A., Flander, L., Wintle, B., Fidler, F., Rumpff, L., Twardy, C., 2011b. Expert status and performance. PLoS ONE 6 (7), e22998.

Canadian Wildlife Service and U.S. Fish and Wildlife Service, 2005. International Recovery Plan for the Whooping Crane. Recovery of Nationally Endangered Wildlife (RENEW), Ottawa, and U.S. Fish and Wildlife Service, Albuquerque, NM.

Canessa, S., Guillera-Arroita, G., Lahoz-Monfort, J., Southwell, D.M., Armstrong, D.P., Chadès, I., Lacy, R.C., Converse, S.J., 2016. Adaptive management for improving species conservation across the captive-wild spectrum. Biol. Conserv. 199, 123–131.

Caswell, H., 2001. Matrix Population Models: Construction, Analysis and Interpretation. Sinauer Associates, Sunderland, MA.

Converse, S.J., Armstrong, D.P., 2016. Demographic modeling for reintroduction decision-making. In: Jachowski, D.S., Millspaugh, J.J., Angermeier, P.L., Slotow, R. (Eds.), Reintroduction of Fish and Wildlife Populations. University of California Press, Oakland, CA, pp. 123–146.

Converse, S.J., Moore, C.T., Armstrong, D.P., 2013a. Demographics of reintroduced populations: estimation, modeling, and decision analysis. J. Wildl. Manage. 77 (6), 1081–1093.

Converse, S.J., Moore, C.T., Folk, M.J., Runge, M.C., 2013b. A matter of tradeoffs: reintroduction as a multiple objective decision. J. Wildl. Manage. 77 (6), 1145–1156.

Converse, S.J., Royle, J.A., Urbanek, R.P., 2012. Bayesian analysis of multi-state data with individual covariates for estimating genetic effects on demography. J. Ornithol. 152 (Suppl. 2), S561–S572.

Converse, S.J., Runge, M.C., Heglund, P.J., Servanty, S., 2014. Decision Analysis for Reintroduction Planning: The Whooping Crane Eastern Partnership 5-year Planning Process Final Report. USGS Patuxent Wildlife Research Center, Laurel, MD.

Converse, S.J., Strobel, B.N., Barzen, J.A., 2018. Reproductive failure in the Eastern Migratory Population: the interaction of research and management (Chapter 8). In: French, Jr., J.B., Converse, S.J., Austin, J.E. (Eds.), Whooping Cranes: Biology and Conservation. Biodiversity of the World: Conservation from Genes to Landscapes. Academic Press, San Diego, CA.

Courchamp, F., Clutton-Brock, T., Grenfell, B., 1999. Inverse density dependence and the Allee effect. Trends Ecol. Evol. 14 (10), 405–410.

Dellinger, T.A., 2018. Florida's nonmigratory whooping cranes (Chapter 9). In: French, Jr., J.B., Converse, S.J., Austin, J.E. (Eds.), Whooping Cranes: Biology and Conservation. Biodiversity of the World: Conservation from Genes to Landscapes. Academic Press, San Diego, CA.

Edwards, W., 1977. How to use multiattribute utility measurement for social decision making. IEEE Trans. Syst. Man Cybern. 7 (5), 326–340.

Ellis, D.H., Lewis, J.C., Gee, G.F., Smith, D.G., 1992. Population recovery of the whooping crane with emphasis on reintroduction efforts: past and future. Proceedings of the North American Crane Workshop, 6, 142–150.

Folk, M.J., Nesbitt, S.A., Schwikert, S.T., Schmidt, J.A., Sullivan, K.A., Miller, T.J., Baynes, S.B., Parker, J.M., 2005. Breeding biology of re-introduced non-migratory whooping cranes in Florida. Proceedings of the North American Crane Workshop 9, 105–109.

Folk, M.J., Rodgers, Jr., J.A., Dellinger, T.A., Nesbitt, S.A., Parker, J.M., Spalding, M.G., Baynes, S.B., Chappell, M.K., Schwikert, S.T., 2010. Status of non-migratory whooping cranes in Florida. Proceedings of the North American Crane Workshop 11, 118–123.

French, Jr., J.B., Converse, S.J., Austin, J.E., 2018. Whooping cranes, past and present (Chapter 1). In: French, Jr., J.B., Converse, S.J., Austin, J.E. (Eds.), Whooping Cranes: Biology and Conservation. Biodiversity of the World: Conservation from Genes to Landscapes. Academic Press, San Diego, CA.

Gilks, W.R., Thomas, A., Spiegelhalter, D.J., 1994. A language and program for complex Bayesian modelling. J. R. Stat. Soc. Ser. D, Statistician 43 (1), 169–177.

Goodman, D., 2002. Predictive Bayesian population viability analysis: a logic for listing criteria, delisting criteria, and recovery plans. In: Beissinger, S.R., McCullough, D.R. (Eds.), Population Viability Analysis. University of Chicago Press, Chicago, IL, pp. 447–469.

Goodwin, P., Wright, G., 2004. Decision Analysis for Management Judgment. John Wiley & Sons, London.

Heath, S.R., Kershner, E.L., Cooper, D.M., Lynn, S., Turner, J.M., Warnock, N., Farabaugh, S., Brock, K., Garcelon, D.K., 2008. Rodent control and food supplementation increase productivity of endangered San Clemente loggerhead shrikes (*Lanius ludovicianus mearnsi*). Biol. Conserv. 141 (10), 2506–2515.

Hestbeck, J.B., Nichols, J.D., Malecki, R.A., 1991. Estimates of movement and site fidelity using mark-resight data of wintering Canada geese. Ecology 72 (2), 523–533.

Keeney, R.L., Raiffa, H., 1976. Decisions with Multiple Objectives: Preferences and Value Tradeoffs. John Wiley & Sons, New York.

Kéry, M., Schaub, M., 2012. Bayesian Population Analysis Using WinBUGS: A Hierarchical Perspective. Academic Press, Waltham, MA.

King, S.L, Selman, W., Vasseur, P., Zimorski, S.E., 2018. Louisiana nonmigratory whooping crane reintroduction (Chapter 22). In: French, Jr., J.B., Converse, S.J., Austin, J.E. (Eds.), Whooping Cranes: Biology and Conservation. Biodiversity of the World: Conservation from Genes to Landscapes. Academic Press, San Diego, CA.

Link, W.A., Royle, J.A., Hatfield, J.S., 2003. Demographic analysis from summaries of an age-structured population. Biometrics 59 (4), 778–785.

Martin, T.G., Burgman, M.A., Fidler, F., Kuhnert, P.M., Low-Choy, S., McBride, M.F., Mengersen, K., 2011. Eliciting expert knowledge in conservation science. Conserv. Biol. 26 (1), 29–38.

McBride, M.F., Fidler, F., Burgman, M.A., 2012. Evaluating the accuracy and calibration of expert predictions under uncertainty: predicting the outcomes of ecological research. Divers. Distrib. 18 (8), 782–794.

McCarthy, M.A., Armstrong, D.P., Runge, M.C., 2012. Adaptive management of reintroduction. In: Ewen, J.G., Armstrong, D.P., Parker, K.A., Seddon, P.J. (Eds.), Reintroduction Biology: Integrating Science and Management. Wiley-Blackwell, Oxford, UK, pp. 256–289.

Moore, C.T., Converse, S.J., Folk, M.J., Runge, M.C., Nesbitt, S.A., 2012. Evaluating release alternatives for a long-lived bird species under uncertainty about long-term demographic rates. J. Ornithol. 152 (Suppl. 2), S339–S353.

Nichols, J.D., Armstrong, D.P., 2012. Monitoring for reintroductions. In: Ewen, J.G., Armstrong, D.P., Parker, K.A., Seddon, P.J. (Eds.), Reintroduction Biology: Integrating Science and Management. Wiley-Blackwell, Oxford, UK, pp. 223–259.

R Development Core Team, 2012. R: A Language and Environment for Statistical Computing. R Foundation for Statistical Computing, Vienna, Austria.

Roche, E.A., Cuthbert, F.J., Arnold, T.W., 2008. Relative fitness of wild and captive-reared piping plovers: does egg salvage contribute to recovery of the endangered Great Lakes population? Biol. Conserv. 141 (12), 3079–3088.

Runge, M.C., 2011. An introduction to adaptive management for threatened and endangered species. J. Fish Wildl. Manag. 2 (2), 220–233.

Runge, M.C., Converse, S.J., Lyons, J.E., 2011. Which uncertainty? Using expert elicitation and expected value of information to design an adaptive program. Biol. Conserv. 144 (4), 1214–1223.

Schwarz, C.J., Schweigert, J.F., Arnason, A.N., 1993. Estimating migration rates using tag-recovery data. Biometrics 49 (1), 177–193.

Servanty, S., Converse, S.J., Bailey, L.L., 2014. Demography of a reintroduced population: moving toward management models for an endangered species, the whooping crane. Ecol. Appl. 24 (5), 927–937.

Spalding, M.G., Folk, M.J., Nesbitt, S.A., Kiltie, R., 2010. Reproductive health and performance of the Florida flock of introduced whooping cranes. Proceedings of the North American Crane Workshop 11, 142–155.

Speirs-Bridge, A., Fidler, F., McBride, M., Flander, L., Cumming, G., Burgman, M., 2010. Reducing overconfidence in the interval judgments of experts. Risk Anal. 30 (3), 512–523.

Stephens, P.A., Sutherland, W.J., 1999. Consequences of the Allee effect for behaviour, ecology, and conservation. Trends Ecol. Evol. 14 (10), 401–405.

Wade, P.R., 2002. Bayesian population viability analysis. In: Beissinger, S.R., McCullough, D.R. (Eds.), Population Viability Analysis. University of Chicago Press, Chicago, IL, pp. 213–238.

Walters, C., 1986. Adaptive Management of Renewable Resources. Macmillan, New York.

Williams, B.K., Szaro, R.C., Shapiro, C.D., 2007. Adaptive Management: The U.S. Department of the Interior Technical Guide. Adaptive Management Working Group, U.S. Department of the Interior, Washington, DC.

Wilson, S., Gil-Weir, K.C., Clark, R.G., Robertson, G.J., Bidwell, M.T., 2016. Integrated population modeling to assess demographic variation and contributions to population growth for endangered whooping cranes. Biol. Conserv. 197, 1–7.

Reproductive Failure in the Eastern Migratory Population: The Interaction of Research and Management

Sarah J. Converse*,**, Bradley N. Strobel†, Jeb A. Barzen‡

*U.S. Geological Survey, Patuxent Wildlife Research Center, Laurel, MD, United States
**U.S. Geological Survey, Washington Cooperative Fish and Wildlife Research Unit, School of Environmental and Forest Sciences (SEFS) & School of Aquatic and Fishery Sciences (SAFS), University of Washington, Seattle, WA, United States
†U.S. Fish and Wildlife Service, Necedah National Wildlife Refuge, Necedah, WI, United States
‡International Crane Foundation, Baraboo, WI, United States

INTRODUCTION

In 2001, the Whooping Crane Eastern Partnership (WCEP) began releasing captive-bred Whooping Crane (*Grus americana*) chicks with the goal of establishing a self-sustaining, migratory population in the eastern United States. The effort to establish the Eastern Migratory Population (EMP) is the third attempt to reintroduce a population of Whooping Cranes in the United States. The unsuccessful reintroduction efforts at Grays Lake, Idaho, and in central Florida were indicative of the challenges in reintroducing this species (Converse et al., Chapter 7, this volume; Converse et al., 2013a; Ellis et al., 1992; Moore et al., 2012).

The EMP reintroduction was initiated with ultralight aircraft-led migrations of costume-reared birds, during their first autumn, between Necedah National Wildlife Refuge (NWR) in central Wisconsin and the Gulf coast of Florida (Urbanek et al., 2005). Releases to the EMP continued with this method through 2015, and additional release methods were developed in the intervening years. In 2005, direct autumn releases were initiated, whereby costume-reared birds were released in Wisconsin in their first autumn to follow older birds south on migration (Servanty et al., 2014; Urbanek et al., 2010a). In 2013, autumn release of birds reared by captive breeding pairs (known as parent rearing) was initiated. Parent-reared birds have been released individually in the territories of failed breeding pairs, with the objective of establishing a social bond between the released chick and the adults.

While the EMP has shown some of the most promising demographic performance to date, poor reproduction has stymied efforts to establish a self-sustaining population (Converse et al., Chapter 7, this volume; Converse et al., 2012; Servanty et al., 2014). The remnant Aransas-Wood Buffalo Population (AWBP) serves as a useful reference for understanding the scope of the challenge in the EMP. In the AWBP, 1,511 nesting pair-years were recorded on the breeding grounds from 1968 to 2010, and a total of 701 young-of-the-year were recorded at Aransas NWR in the following winters (Canadian Wildlife Service and U.S. Fish and Wildlife Service, 2005), for an average reproductive rate of 0.464 chicks surviving to winter per nesting pair. The AWBP has grown steadily during that time, at an annual population growth rate of approximately 4.3%, calculated as:

$$\left(\frac{N_{t+x}}{N_t}\right)^{1/x} \tag{8.1}$$

where N is the annual winter counts from Aransas NWR (Butler et al., 2013; Canadian Wildlife Service and U.S. Fish and Wildlife Service, 2005), $t = 1968$ (winter of 1968–69), and $x = 42$ years.

When the reproductive rate of 0.464 chicks per nesting pair is substituted into a population model for the EMP (EMPSim; Converse et al., Chapter 7, this volume; Servanty et al., 2014) and population growth rate is calculated as in Eq. (8.1) from the mean population sizes at

the beginning and end of a 30-year simulation, the result is annual growth of 3.1%, suggesting that with a reproductive rate equal to that in the AWBP, the EMP could achieve similarly robust growth. However, over the period 2005–15, during which the EMP yielded 161 nesting pair-years, just nine chicks have survived to 1 December of their first year (Table 8.1; all reproductive data for the EMP arise from intensive monitoring carried out by member organizations of WCEP). This constitutes a reproductive rate of 0.056, only approximately 12% as large as the AWBP rate, and underscores the severity of the problem that reproductive failure poses to the EMP reintroduction effort.

The problem of reproductive failure in this population can be separated into two components: poor nest success and poor chick survival. Poor nest success has been recognized as a problem for at least seven years (Runge et al., 2011; Urbanek et al., 2010b) but the problem of poor chick survival has only been recognized recently when there have been adequate numbers of chicks for evaluation. Between 2005 and 2008, 22 nesting pair-years produced only two hatched chicks (from one successful nest; Table 8.1). In 2008, Necedah NWR biologist Richard Urbanek hypothesized that nest failure in the population was due to harassment of birds by blood-feeding black flies (Simuliidae; Converse et al., 2013b; Urbanek et al., 2010b). Since that time, evidence has accumulated to support this hypothesis (e.g., Converse et al., 2013b; King et al., 2013). Meanwhile, the problem of nest failure has persisted; through 2013, only 27 chicks were hatched in 109 nesting pair-years (Table 8.1).

Nest success can be increased via treatment of black fly populations with the bacterial larvicide *Bacillus thuriengensis israelensis* (*Bti*) or by forcing failure of first nests with egg removal to encourage renesting after black fly activity has subsided (Table 8.1). And in some years, nest success can be high without *Bti* treatment or forced renesting because of low black fly abundance during nesting that is driven, presumably, by weather. For

TABLE 8.1   Summary Statistics on Nesting and Chick Production from 2005 to 2015 in the Eastern Migratory Population of Whooping Cranes. Data Were Collected via Monitoring Activities of Member Organizations of the Whooping Crane Eastern Partnership.

Year	Nesting pairs	Nests produced	Nests incubated ≥30 days[a]	Nests producing ≥1 chick	Chicks produced	Chicks surviving to 1 December
2005	2	2	0	0	0	0
2006	5	6	1	1	2	1
2007	4	5	0	0	0	0
2008	11	11	0	0	0	0
2009	12	17	2	2	2	0
2010	12	17	7	5	7	2
2011	20	22	6	4	4	0
2012	22	29	14	8	9	2
2013	21	23	3	2	3	1
2014 Treated[b]	4	6[c]	0	0	0	0
2014 Nontreated	21	22	9	8	13	1
2015 Treated[b]	8	16[c]	8	8	11	1
2015 Nontreated	19	21	12	8	13	1

[a] Incubation time in Whooping Cranes is approximately 30 days. The difference between nests incubated ≥30 days and nests producing ≥1 chick is nests in which the eggs were infertile or dead.

[b] Pairs subjected to forced renesting treatment, in which eggs from first nests are removed to encourage renesting after black fly activity has subsided.

[c] For pairs treated with forced renesting, the first nests fail by definition, including four nests in 2014 (two pairs renested) and eight nests in 2015 (eight pairs renested).

example, pairs not exposed to forced renesting in 2014 or 2015 had relatively good nest success (13 chicks from 21 pairs = 0.619 and 13 chicks from 19 pairs = 0.684, respectively; Table 8.1); black flies were largely absent or did not occur in high numbers in those years (WCEP, unpublished data). However, even in years when rates of nest success were high, chick mortality resulted in overall poor reproductive success. For example, of the 24 chicks hatched in 2015, only 2 survived to 1 December (Table 8.1). In total, through December 2015, only 9 of 64 chicks hatched in the EMP have survived until 1 December of their first year.

Since 2008, there has been an active research effort to understand the factors causing poor reproduction in this population. However, managers of the reintroduction cannot wait for uncertainty to be fully resolved to make decisions.

Thus, the management of this population, and specifically management to address poor reproduction, has been a process of integrated learning and decision making. Here, we review efforts toward understanding and managing the challenge of poor reproduction in the EMP. We also present a set of future research priorities, derived using an updated value of information analysis (Canessa et al., 2015; Runge et al., 2011; Williams et al., 2011) completed in 2015. While there is good evidence that a proximal cause of poor nest success is harassment by black flies, this may not be the only contributing factor; we present alternative hypotheses and approaches for management given those hypotheses. We also present hypotheses for factors contributing to poor chick survival, a phenomenon that is less understood than poor nest success. While much has been learned, there is still substantial uncertainty about how best to manage this reintroduction effort to improve reproductive success. It is clear, however, that the solution will not be simple, and that the challenges to reproduction in this population may have important implications for all efforts to reintroduce Whooping Cranes.

VALUE OF INFORMATION WORKSHOP: 2009

By 2009, it was becoming apparent that nesting failure was a potentially serious limiting factor for the EMP; only 1 chick had fledged in 22 nesting pair-years (Table 8.1). In March of that year, Necedah NWR hosted a structured decision making (SDM; Gregory et al., 2012; Martin et al., 2009) workshop, including members of WCEP and other experts, with the goal of identifying an approach for addressing the issue; see Runge et al. (2011) for complete details. During the workshop, the PrOACT decision-analytic process was used to guide development of a decision framework (Problem, Objectives, Alternatives, Consequences, and Tradeoffs; Hammond et al., 1999). The team began by articulating the decision problem: Which management strategy should be undertaken for the benefit of the EMP, in the face of uncertainty about the cause of reproductive failure? Next, the team identified four objectives (number of breeding pairs, reproductive success, adult survival, and body condition), which they weighted by importance. Each objective was directly associated with the higher-level objective of establishing a self-sustaining population, and each could potentially be impacted by management actions taken to address the problem of reproductive failure.

Next, to characterize uncertainty about possible management outcomes and to spur thinking about management alternatives, eight hypotheses for nest failure were identified from a longer brainstormed list. These hypotheses were:

(1) breeding birds were too young and inexperienced; (2) black flies were reducing nest success; (3) the social conditioning brought about by the rearing and release procedure was leading to poor nest success; (4) birds were nutrient-limited because of lack of energetic resources on the breeding grounds; (5) birds were nutrient-limited because of lack of energetic resources during winter or migration; (6) birds were nutrient-limited because of lack of energetic resources on both the breeding grounds and the wintering or migration grounds; (7) the act of salvaging eggs from apparently abandoned nests was the cause of nest failure; and (8) human disturbance was causing nest failure. Belief weights placed on the hypotheses were calculated by averaging individual belief weights across the $n = 8$ experts. The weights were, respectively: 0.094, 0.291, 0.119, 0.228, 0.059, 0.066, 0.044, and 0.100. Thus, the black fly hypothesis was assigned the greatest credibility by experts.

Following development of hypotheses, a set of potential alternatives was selected to address the problem of poor reproduction. Workshop participants were asked to develop management alternatives conditional on each of the hypotheses, in turn, being the best explanation for poor reproduction. Finally, the experts were asked to predict the consequences (using an expert elicitation procedure) of the alternatives in terms of each of the four objectives.

These data allowed the analysts to perform a value of information analysis (Runge et al., 2011). Value of information is a concept arising from decision theory, wherein information is evaluated in terms of its ability to improve management outcomes. Formally, the value of information is the expected value of the optimal action after new information is collected minus the expected value of the optimal action before new information is collected; see Runge et al. (2011) for details. Most interesting for our purposes, one of the quantities calculated was the Expected Value of Partial Information (EVPXI), which represents the expected improvement in management performance from reducing a subset of the uncertainty: in this case, either confirming or refuting a single hypothesis. Of the eight hypotheses considered, the one with the greatest EVPXI was the black fly hypothesis. Considering only the reproductive success attribute, the EVPXI for the black fly hypothesis represented a 12.5% improvement in expected management performance (from 0.185 young departing for southward migration per breeding pair, expected under uncertainty, to 0.208 young expected if the black fly hypothesis was either confirmed or eliminated). This finding indicated that the greatest improvement in reproductive success was expected to accrue from resolving the black fly hypothesis, compared to any other hypothesis. However, the expected improvement was relatively modest given the severity of the problem of reproductive failure.

In subsequent years, the egg salvage and disturbance hypotheses have been eliminated as major causes of nest failure (though they may have been contributing factors in some years), based on widespread nest failure in years with strict no-salvage policies, and based on the observation that the accessibility of nests does not influence nest success (WCEP, unpublished data). The first three hypotheses developed in 2009 (age and experience, black flies, social conditioning) continue to be considered today, as do versions of the nutrient limitation hypothesis (see later section, Value of Information Workshop: 2015).

STUDYING BLACK FLY EFFECTS ON NEST SUCCESS: 2009–13

Stemming from the findings of the 2009 workshop, a nest success study was undertaken to address the black fly hypothesis. The study design consisted of three elements: (1) a spatially distributed monitoring network for black flies consisting of carbon dioxide traps, (2) monitoring of nest success, and (3) treatment of black fly larval habitat in rivers and streams within 10 km of the nesting area using *Bti*. Pilot treatment with *Bti* was conducted at one treatment site in 2010, and more extensive treatment was undertaken (at up to 17 sites) in 2011 and 2012. Nontreatment years in 2009 and 2013 provided temporal controls. Because of the tightly clumped spatial distribution of Whooping Crane nests, spatial controls could not be clearly denoted. That is, we could not isolate nests that were exposed to black flies but were not affected by *Bti* treatment.

Detailed analysis of the data from all years of the study will include an examination of the evidence for multiple competing hypotheses about the causes of nest failure. The primary mode of data analysis is daily nest survival modeling, which allows for modeling the temporal pattern of nest failures (Converse et al., 2013b; Dinsmore et al., 2002). Analysis of a subset of the data, from 2009 to 2010, revealed evidence that the abundance of the black fly species *Simulium annulus* has a negative influence on daily nest survival (Converse et al., 2013b). Preliminary results based on the years 2009–13 also indicate a negative influence of *Simulium annulus* on daily nest survival (S.J. Converse, unpublished data).

It is also illuminating to undertake a coarse examination of the experimental data. For the years 2009–13, we fit a Poisson regression model of chicks hatched per pair as a function of whether a treatment occurred in a given year. We fit the model in the Bayesian analysis software JAGS (Plummer, 2003) using the package jagsUI (Kellner, 2015) in the R environment (R Development Core Team, 2012). We used standard noninformative priors for the intercept and treatment effect. The estimated treatment effect (on the natural log scale) was 0.966 (95% Bayesian confidence interval: 0.023, 2.067). We can predict, based on this simple model, that a pair in a year without *Bti* treatment would hatch 0.152 (0.049, 0.310) chicks while a pair in a year with *Bti* treatment would hatch 0.370 (0.226, 0.549) chicks. Interestingly, if we fit this same model but include the years 2014 and 2015 (while excluding the pairs subjected to the forced renesting manipulation in those years), the results look substantially different. The estimated regression coefficient is −0.147 (−0.724, 0.413) and the predicted number of chicks hatched is 0.425 (0.289, 0.587) in nontreatment years (2009, 2013–15) compared to 0.370 (0.226, 0.549) in treatment years (2010–12). That is, inclusion of these later nontreatment years resulted in higher predicted numbers of chicks hatching in nontreatment compared to treatment years. This apparent contradiction with earlier results is due to annual variation (presumably in black fly abundance) swamping the effect of the treatment.

The reproductive success experiment yielded four major findings. First, black flies are contributing to poor nest success in the EMP. Second, their reduction via *Bti* treatment may increase nest success. Third, annual variation swamps the effect of *Bti*: in our simple analysis, including years with naturally low black fly numbers (i.e., 2014 and 2015) actually resulted in more predicted chicks hatched per pair in nontreatment years than in treatment years (this also demonstrates the challenge that a lack of spatial control poses for inference). Fourth, *Bti* treatment, as applied during this experiment, does not appear capable of producing a reproductive rate of approximately 0.464 chicks per pair surviving to winter; in the 3 years it was implemented, only 0.370 chicks were hatched per pair. This is particularly true because chick survival to winter has been extremely low. Overall, while much was learned from the 2009–13 experiment, uncertainty remained about the causes of, and management actions necessary to address, poor reproduction in this population.

Treatment with *Bti* during the 2009–13 experiment was implemented to facilitate learning about the relationship between black flies and reproduction, rather than as a management strategy per se. Operational implementation of *Bti* treatments does not have widespread support within WCEP partner organizations (S.J.

Converse, personal observation). Therefore, rather than attempting to suppress black flies at the original release site, in 2011 WCEP initiated a release program at a new site in eastern Wisconsin, including Horicon NWR (using direct autumn release) and White River Marsh State Wildlife Area (using ultralight release). Black fly surveys at Horicon NWR and three other sites in eastern Wisconsin during 2010 resulted in total counts of *Simulium spp.* in carbon dioxide traps that were two to four orders of magnitude lower, depending on the site, compared to Necedah NWR (P.H. Adler, Clemson University, unpublished data). As of the 2015 breeding season, the oldest birds released at those sites were 4 years old. However, with the exception of one pair producing an early-failed nest in 2014, there had been no nesting activity in the area of the new release sites, nor had territorial pairs established in the area.

With releases of birds in new areas, and given the high natal philopatry of Whooping Cranes (Barzen et al., Chapter 14, this volume; Johns et al., 2005; Mueller et al., Chapter 11, this volume), the variety of environments used by nesting birds is likely to expand substantially. This will facilitate learning about habitat requirements (Austin et al., Chapter 3, this volume; Barzen et al., Chapter 14, this volume; Van Schmidt et al., 2014) that could guide selection of future release locations. However, if the major cause of poor reproduction is not a characteristic of habitat but is inherent to the birds themselves – such as effects of captive rearing on early learning – relocation of releases will not address the problem.

IDENTIFYING AN OPTIMAL REINTRODUCTION STRATEGY: 2012–13

In 2012–13, WCEP underwent an SDM process to evaluate different reintroduction strategies for the EMP, and identify an optimal strategy (Converse et al., Chapter 7, this volume).

A focus of the SDM process was recognizing and accounting for uncertainty about the cause of reproductive failure in the population. A total of four hypotheses were considered for poor reproductive success: (1) reproductive success is largely influenced and limited by habitat; possible drivers included black flies, predators, or other aspects of habitat quality; (2) captive rearing and release experiences influence early learning, which influences later reproductive success; (3) reproductive success is largely limited by captive selection, that is, genetic changes due to selective pressures in the captive breeding population; and (4) reproductive success is largely limited by lack of breeding experience in EMP birds. Belief weights – averaged across $n = 4$ experts – for the four hypotheses were, respectively: 0.522, 0.180, 0.126, and 0.172. Three of the four hypotheses were considered at the 2009 workshop or earlier. The captive selection hypothesis, however, represented a novel explanation for poor reproductive success in the EMP (King et al., 2013); that is, the captive environment has exerted strong selective pressures, resulting in captive birds that are less genetically fit for reintroduction. The first hypothesis (i.e., habitat) represented a combination of multiple environmental factors. In the context of this decision, isolating the factors was less important than considering whether failure was environmentally driven (thus implying the importance of relocating the reintroduction site).

During the SDM process, specific release strategy components evaluated included: (1) number of birds released per year, (2) number of years of releases, (3) release locations, (4) rearing and release methods, and (5) whether eggs should be collected from the AWBP for captive rearing and release. This last release strategy component – whether to rear and release birds from AWBP-collected eggs – was developed in light of the captive selection hypothesis.

Each of the four hypotheses was captured in a structurally distinct population projection model. The four models were constructed using a combination of expert judgment and the empirical base model EMPSim (Converse et al., Chapter 7, this volume; Servanty et al., 2014). The models were used to project population growth and abundance under various reintroduction strategies. Based on predictions calculated as a weighted average over the four hypotheses, and considering other objectives such as the desire to reduce both costs and the use of captive-reared birds, the strategy of relocating releases to eastern Wisconsin (i.e., Horicon NWR and/or White River Marsh State Wildlife Area) was affirmed as optimal (Converse et al., Chapter 7, this volume). Parent-reared releases were also identified as potentially valuable, though uncertain, and experimental releases of parent-reared birds began in 2013 to test this technique. Finally, the use of AWBP eggs as propagules for release was identified as optimal in the decision framework, but testing the associated hypothesis, captive selection, has not yet been initiated.

VALUE OF INFORMATION WORKSHOP: 2015

By early March 2015, much had been learned, and multiple management actions had been taken, to address the problem of reintroduction failure. Most notably, releases of captive-reared birds had been relocated to eastern Wisconsin. In addition, parent rearing had been implemented on a small scale (four chicks in 2013, three in 2014 on Necedah NWR). The long-term research (2009–13) was complete, and had yielded useful information about the relationship between black flies and nesting success. But it was also clear that important uncertainties remained, restricting the ability of managers to select optimal management actions going forward.

Therefore, in March 2015 the WCEP Research and Science Team organized a meeting to reevaluate the current state of knowledge and reassess the challenges limiting growth of the EMP. This meeting assembled experts in endangered species reintroductions, wetland and waterbird management, captive breeding, and Whooping Crane ecology from inside and outside of WCEP. The goal of the meeting was to revise and reprioritize the hypotheses about the causes of reproductive failure in this population.

Methods: 2015 Workshop

During the March 2015 workshop, the first day was devoted to presentation of research results so that all participants started the discussions with adequate background information. The second and third days were devoted to eliciting hypotheses, actions, and hypothesis weights from experts. Workshop participants were first asked to develop a list of hypotheses relevant to the problem of poor reproductive success in the EMP. An open brainstorming model was used, in which all ideas were recorded for further consideration. Next, participants were divided into three groups, and each group was assigned a set of the brainstormed hypotheses. Within groups, the proposed hypotheses were refined and described in greater detail. Each hypothesis was identified as relevant to nest success, chick survival, or both.

Next, experts provided an initial weighting of hypotheses. Each person was given 100 points for each set of hypotheses (nest success and chick survival) and asked to distribute those points among the hypotheses as a measure of their credibility. Finally, for the list of top-ranked hypotheses (participants opted to retain for further analysis all hypotheses that garnered ≥5% weight, averaged across participants), breakout groups identified management actions that could be taken conditional on the various hypotheses being true. At the completion of the March 2015 workshop, a list of weighted hypotheses for poor nest success and poor chick survival, along with a set of potential management actions targeting each of the top-ranked hypotheses, had been developed.

In August 2015, a subset of experts (n = 8) from the participants in the March meeting were selected for an in-depth elicitation process that was designed to develop predictions necessary to conduct a value of information analysis, thereby updating the analysis conducted by Runge et al. (2011) after the 2009 workshop (see earlier section, Value of Information Workshop: 2009). A modified Delphi elicitation process was used (Burgman et al., 2011; Martin et al., 2011). In this process, experts complete responses individually, then view anonymized responses of the group during a discussion session that allows for resolution of semantic uncertainties and introduction of additional relevant information, and finally complete a second round of individual responses. At the beginning of the process, authors SJC and BNS provided participants with the set of hypotheses (Tables 8.2 and 8.3) and actions (Table 8.4) under consideration, developed at the March 2015 workshop. The experts first met with a facilitator (SJC) for an online meeting to discuss the elicitation task and to review the hypotheses and actions. In the five days following that meeting, the experts worked independently to complete the elicitation exercise. Any clarification questions were sent directly to the facilitator, who anonymized the question, provided an answer, and sent the answer to the group. The experts' responses were then collated and presented at a second online meeting one week after the first meeting. During the second meeting, the results were discussed and clarification was provided on questions that had arisen during the first round of elicitation. The experts then had an additional four days to complete a second round of elicitation in which they could revise their answers.

The experts provided their judgment of the effect of each of the actions with respect to four attributes: nest success (the probability that a nest produces at least one chick) at 3 years after implementation of a given management action, nest success at 10 years after implementation (Appendix A), chick survival (the probability

TABLE 8.2   Hypotheses to Explain Poor Nest Success in the Eastern Migratory Population of Whooping Cranes. Each of the Listed Hypotheses Received an Average of ≥5% Weight by Experts at a March 2015 Workshop, and Was Included in Subsequent Elicitation Exercises to Predict Nest Success under Various Management Actions, Conditional on Each Hypothesis. Certain Hypotheses Are Analogous to Hypotheses Posed for Chick Survival (see Table 8.3).

Category	Hypothesis
Environmental	1. Black flies are an acute stress for Whooping Cranes that cause nest abandonment when/where black fly abundances are high.
Environmental	2. Low food availability and/or quality in the breeding area, combined with black fly stress, causes nest abandonment (see Chick Hypothesis #2).
Postrelease Experience	3. Lack of postrelease experience with nesting, due to age or opportunity, reduces nest success (see Chick Hypothesis #4).
Genetic	4. Genetic structure of the current captive population affects nest success postrelease. Mechanisms could include founder effects, inbreeding depression, genetic drift, and/or captive selection (see Chick Hypothesis #5).
Captive Rearing	5. Current costume-rearing methods do not impart information necessary for postrelease nesting success, via inadequate learning experiences regarding nest defense, incubation constancy, resource selection, or behavioral plasticity (see Chick Hypothesis #6).
Captive Rearing	6. The social structure (i.e., groups) of costume-reared cohorts is detrimental to the development of normal behaviors due to muting of natural behavioral variation (see Chick Hypothesis #7).

TABLE 8.3   Hypotheses to Explain Poor Chick Survival in the Eastern Migratory Population of Whooping Cranes. Each of the Listed Hypotheses Received an Average of ≥5% Weight by Experts at a March 2015 Workshop, and Was Included in Subsequent Elicitation Exercises to Predict Chick Survival under Various Management Actions, Conditional on Each Hypothesis. Certain Hypotheses Are Analogous to Hypotheses Posed for Nest Success (see Table 8.2).

Category	Hypothesis
Environmental	1. Predator effects are high, resulting in poor fledging success for Whooping Cranes.
Environmental	2. Low food availability and/or quality in the breeding area, combined with black fly stress, causes chick mortality (see Nest Hypothesis #2).
Environmental	3. Low food availability (limited abundance or distribution) and/or poor nutrition profile of food during the chick-rearing period limits reproductive success. Mechanisms could include poor soil productivity, impoundment management, or invasive species limiting the quality and/or availability of food resources.
Postrelease Experience	4. Lack of postrelease experience with chick rearing, due to age or opportunity, reduces chick survival (see Nest Hypothesis #3).
Genetic	5. Genetic structure of the current captive population affects chick survival postrelease. Mechanisms could include founder effects, inbreeding depression, genetic drift, and/or captive selection (see Nest Hypothesis #4).
Captive Rearing	6. Current costume-rearing methods do not impart information necessary for postrelease chick survival, via inadequate learning experiences regarding nest defense, incubation constancy, resource selection, or behavioral plasticity (see Nest Hypothesis #5).
Captive Rearing	7. The social structure (i.e., groups) of costume-reared cohorts is detrimental to the development of normal behaviors due to muting of natural behavioral variation (see Nest Hypothesis #6).

TABLE 8.4   Management Actions Proposed to Address the Challenge of Poor Reproduction (Either Poor Nest Success, Poor Chick Survival, or Both) in the Eastern Migratory Population of Whooping Cranes. The Hypotheses That Actions Were Developed Specifically to Address Are Coded to Hypotheses in Tables 8.2 and 8.3 (N = Nesting Success, Table 8.2; C = Chick Survival, Table 8.3).

Action	Hypotheses addressed
Do nothing.	N3, C4
Instill predator defense behaviors in adults: implement postrelease training.	C1
Instill predator defense behaviors in young: implement prerelease training.	C1
Time slow drawdowns with chick rearing to provide habitat far from predator cover.	C1
Implement *Bti* treatment.	N1
Implement forced renesting (removal of first nests to encourage renesting).	N1
Conduct habitat management in the Yellow River to reduce black fly populations at the source: management of impoundment releases, or improving water and stream quality to increase habitat quality for black fly larval predators like stone flies.	N1
Conduct wetland management: alter hydrology, take out top competitors, etc.	N2, C2, C3
Conduct wet meadow restoration, that is, ditch plugging.	N2, C2, C3
Provide dummy egg and then return nest's eggs at the end of incubation.	N3, C4
Bring eggs from Wood Buffalo National Park into the captive population as breeders or release directly.	N4, C5
Rear and release parent-reared chicks in mid-September.	N5, C6
Put fertile eggs from captivity in every nest that are further along in incubation.	N5, C6
Chick adoption: put very young chicks from captive facilities with failed breeding pairs (chicks released at 1–2 weeks old).	N5, C6
Costume rear 1–2 birds at a time (by a keeper), do not socialize birds in groups, and release in these numbers.	N6, C7

that a chick survives until 1 October) at 3 years after implementation, and chick survival at 10 years after implementation (Appendix B). Each of the responses was given conditional on the hypotheses for nest success or chick survival. The experts also provided their relative belief in the hypotheses. As before, each expert was given 100 points to distribute among the hypotheses for nest success and another 100 points to distribute among the hypotheses for chick survival.

Experts were asked to focus their analysis on the situation at Necedah NWR, where there was a clear problem with reproduction. While nesting occurs outside of Necedah NWR, it is not yet substantive enough to understand whether poor reproduction is a serious limitation in other areas. Experts were asked to consider two different time periods because resolving uncertainties at different time scales has implications for management of the population. First, with continuing reproductive failure there is a question about how long the reintroduction effort should be sustained: evidence of the value of continuing release efforts may be needed in the short term if the program is to continue. Also, some uncertainties are unlikely to be resolved in the short term: namely, uncertainties that require the release of additional birds to resolve (e.g., captive-rearing hypothesis, genetic structure hypothesis). By providing multiple time scales, decisions makers can be provided with an understanding of the time required to both learn and benefit from such learning.

We calculated Expected Value of Perfect Information (EVPI; the increase in expected management outcomes if uncertainty could be fully resolved) for each response variable using the mean responses across experts and the mean hypothesis weights; a more involved analysis including uncertain judgments elicited using the methods of Speirs-Bridge et al. (2010) is planned but is beyond the scope of this chapter. We also conducted an EVPXI analysis for each response variable by calculating the improvement in management outcome that can be expected from resolving (either confirming or refuting) each hypothesis. We direct readers to Runge et al. (2011) for a description of the methods for calculating EVPI and EVPXI and more detail about the concept of value of information.

Results: 2015 Workshop

At the 3-year time scale, for nest success, the optimal action under uncertainty was forced renesting, with an expected value of 0.267 (Table 8.5). With resolution of uncertainty, the expected value was 0.312, for an EVPI = 0.046, or 17.2% management improvement. The EVPXI calculations (Table 8.6) suggested that the single most valuable nest success hypothesis to resolve (either confirm or refute) in the short term was the black

fly hypothesis (followed by the genetic structure and costume-rearing hypotheses). Managing for chick survival at the 3-year time scale, the optimal action under uncertainty was predator training for young birds, with an expected value of 0.286 (Table 8.5). With resolution of uncertainty, the expected value was 0.309, for an EVPI of 0.023, a 7.9% improvement. The most valuable chick survival hypothesis to resolve was the predator hypothesis (followed closely by the lack of experience and genetic structure hypotheses), based on the EVPXI calculations (Table 8.6).

At the 10-year time scale, for managing nest success under certainty, the optimal action was forced renesting (the same as at the 3-year time scale), with an expected value of 0.282 (Table 8.5). Expected value under certainty was 0.368, for an EVPI of 0.086, or 30.7% improvement. Over this longer time scale the most valuable hypothesis to resolve was the genetic structure hypothesis (followed by the black fly hypothesis; Table 8.6). For chick survival, the optimal action was parent rearing (rather than predator training as at the 3-year time scale), with an expected value of 0.319 (Table 8.5). With resolution of uncertainty, the expected value was 0.351, for an EVPI of 0.033 representing a 10.2% improvement in management performance. For chick survival, as for nest success,

TABLE 8.5   Expected Value of Perfect Information (EVPI) Analysis, for Four Whooping Crane Reproductive Response Variables at USFWS Necedah National Wildlife Refuge (Nest Success 3 Years after Implementation of Management Actions, Nest Success 10 Years after Implementation, Chick Survival 3 Years after Implementation, and Chick Survival 10 Years after Implementation) Based on Information Elicited from Experts (n = 8). Information Presented Includes the Optimal Action to Take Given Uncertainty and the Expected Value of That Action, the Expected Value with Uncertainty Resolved, and the EVPI, Calculated as the Difference between the Expected Value with Uncertainty Resolved and the Expected Value under Uncertainty. Proportion EVPI Is the EVPI Divided by the Expected Value under Uncertainty.

Response variable	Optimal action under uncertainty	Expected value of optimal action under uncertainty	Expected value with uncertainty resolved	EVPI	Proportion EVPI
Nest: 3 years	Forced renesting	0.267	0.312	0.046	0.172
Nest: 10 years	Forced renesting	0.282	0.368	0.086	0.307
Chick: 3 years	Predator training	0.286	0.309	0.023	0.079
Chick: 10 years	Parent rearing	0.319	0.351	0.033	0.102

TABLE 8.6   Expected Value of Partial Information (EVPXI) and Proportional Improvement over Managing under Uncertainty (Prop EVPXI), for Four Whooping Crane Reproductive Response Variables at USFWS Necedah National Wildlife Refuge (Nest Success 3 Years after Implementation of Management Actions, Nest Success 10 Years after Implementation, Chick Survival 3 Years after Implementation, and Chick Survival 10 Years after Implementation) Based on Information Elicited from Experts ($n = 8$). Where Dashes Appear, the Given Hypothesis Was Not Proposed as a Mechanism to Explain the Given Response Variable.

Response variable	Value	Hypotheses							
		Predators	Black flies	Low food and black flies	Low food	Lack of experience	Genetic structure	Costume rearing	Group rearing
Nest: 3 years	EVPXI	–	0.286	0.267	–	0.274	0.282	0.282	0.274
	Prop EVPXI	–	0.071	0.000	–	0.028	0.056	0.058	0.029
Nest: 10 years	EVPXI	–	0.315	0.282	–	0.291	0.318	0.309	0.296
	Prop EVPXI	–	0.119	0.000	–	0.032	0.129	0.096	0.050
Chick: 3 years	EVPXI	0.297	–	0.288	0.290	0.293	0.293	0.287	0.288
	Prop EVPXI	0.036	–	0.008	0.015	0.022	0.024	0.004	0.006
Chick: 10 years	EVPXI	0.328	–	0.320	0.323	0.319	0.333	0.319	0.323
	Prop EVPXI	0.028	–	0.004	0.012	0.000	0.044	0.000	0.013

the genetic structure hypothesis was the most valuable to resolve (followed by the predator hypothesis; Table 8.6).

Discussion: 2015 Workshop

There are inevitable limitations of an expert-judgment-driven value of information analysis. Expert judgments can be biased due to framing effects, linguistic uncertainty, motivational bias, expert overconfidence, and other cognitive biases of experts (Burgman, 2004; Burgman et al., 2011; McBride et al., 2012). However, carefully designed elicitation processes can reduce the effects of these biases (Burgman et al., 2011; Martin et al., 2011; Speirs-Bridge et al., 2010), and resulting judgments can be a valuable guide for moving forward in situations where uncertainty is substantial and empirical data are sparse. In addition, expert judgment can be usefully thought of as providing a starting point that can be improved through further empirical evaluation. Following the 2009 value of information workshop with the 2009–13 study of black fly effects on nest success is an example.

Our expert-judgment-driven EVPXI analysis suggested that, in the short term, a better understanding of environmental impacts would be most valuable; that is, resolving either the black fly (nest success) or predator (chick survival) hypotheses, though the expected management gains in the short term were relatively modest (e.g., only a 7% expected improvement in nest success for resolving the black fly hypothesis; Table 8.6). This was because our ability to see the benefits of new rearing and release strategies (i.e., producing a "better bird") is limited in the short term due to the time it takes birds to reach maturity. However, in the longer term, expert judgment suggested that addressing hypotheses about characteristics of the birds themselves would be more valuable (though the environmental hypotheses still had relatively high EVPXI at this time scale). In particular, we found

that the hypothesis with the greatest EVPXI at the 10-year time horizon was the genetic structure hypothesis, for both nest success and chick survival. While the experts recognized that a variety of mechanisms may cause nonoptimal genetic shifts in the released population, the mechanism that has been most seriously considered to date is captive selection (King et al., 2013). Captive selection has been documented in a variety of vertebrate taxa, though empirical evidence in birds is sparse (Araki et al., 2007; Christie et al., 2012; Frankham, 2007; Lacy et al., 2013; Lynch and O'Hely, 2001).

In the meantime, when contemplating management actions under uncertainty, in the short term forced renesting was optimal for managing to improve nest success, and predator training of young birds was optimal to improve chick survival. In the long term, the optimal actions were forced renesting for nest success and parent rearing for chick survival. It is worth noting that the optimal actions to take under uncertainty are not necessarily the actions that would be optimal under the hypothesis with the greatest EVPXI. For example, at the 10-year time scale, the genetic structure hypothesis was identified as best to resolve for both nest success and chick survival, though the optimal actions under uncertainty at that time scale (i.e., forced renesting and parent rearing, respectively) are not the actions that are optimal under the genetic structure hypothesis (see Appendices A and B).

Finally, it is important to emphasize that our 2015 value of information analysis was circumscribed by our focus on the situation at Necedah NWR. It is likely that information gained by monitoring birds both at Necedah NWR and in eastern Wisconsin will ultimately contribute to management of birds in both locations. However, we chose to focus our analysis on Necedah NWR because we do not yet know whether birds released in eastern Wisconsin will be affected by poor nest success and chick survival.

SUMMARY AND OUTLOOK

Over the period from 2009 to present, WCEP has implemented an iterative process of investigating the causes of poor reproductive success, acting in response to that learning, and reprioritizing research questions based on previous results and management needs. Here we have highlighted major milestones in that process: the 2009 value of information workshop, the 2009–13 study of black fly effects on nest success, the 2012–13 SDM process for identifying an optimal reintroduction strategy, and the 2015 value of information workshop. The process has not been limited to these milestones, however. Another important example is the ongoing evaluation of forced renesting, first implemented in 2014 as a research and management tool. The close interaction between research and management has enabled advances in our understanding while focusing research directly on the needs of managers.

It is clear from the perspective provided here that multiple hypotheses related to environmental factors (black flies, predators, or nutrient limitation) or to bird-specific factors (captive rearing or genetic structure) have persisted for as long as 7 years. That is, they have not yet been confirmed or eliminated from further consideration even after many years of research. One challenge is that some of the proposed hypotheses represent mechanisms that may interact. For example, there is strong evidence that black flies reduce nest success for EMP birds in the area of Necedah NWR. Similarly for predators, while evidence is still sparse, it seems likely that many chicks are taken by predators (B.N. Strobel, unpublished data). However, genetic or learned traits of the birds themselves may still be important, via their effect on poor nest attendance or poor chick care. The underlying cause may be a trait of the birds, but effects may manifest themselves through responses to environmental factors.

While uncertainty does remain, there is great potential to learn important lessons in the coming years. As birds begin to nest in eastern Wisconsin, we have the potential to further differentiate the weight of evidence in favor of environmental hypotheses from that in favor of hypotheses focused on characteristics of the birds themselves. Similarly, as parent-reared birds begin to reach breeding age, there is potential to more thoroughly evaluate hypotheses about the effects of rearing and release strategies on the later reproductive success of released birds. One other major category of hypotheses – those relating to genetic structure – could perhaps best be addressed through rearing and release of birds hatched from AWBP-collected eggs. Moving forward with a proposal for such a management action will require cooperation across international boundaries and oversight by the International Whooping Crane Recovery Team.

The value of learning about the EMP extends beyond this single population. If the characteristics of the birds themselves, either learned or inherited, are the underlying cause of poor reproduction, then efforts to reintroduce Whooping Cranes elsewhere using current captive breeding and costume rearing techniques are likely to fail. With two failed reintroductions to date, and two reintroductions ongoing, the importance of identifying any bird-specific characteristics that are reducing reproductive rates in Whooping Cranes is substantial. Indeed, the implications of such a finding would reach beyond reintroductions to the captive population itself, calling into question the usefulness of its current role in recovery of the species. More structured consideration of hypotheses for poor reproductive success in reintroduced populations, including value of information analyses, would likely be beneficial if carried out at the scale of the entire recovery program, that is, considering reintroduction as just one facet of Whooping Crane recovery.

Acknowledgments

We thank the following individuals for participation in the 2015 workshop: *A. Lacy, *A. Pearse, *B. Brooks, B. Hartup, B. Tarr, C. Bowden, C. Sadowski, D. Gawlik, *D. Lopez, G. Archibald, G. Olsen, J. Duff, J. French, J. Howard, J. Langenberg, M. Mace, *M. McPhee, M. Wellington, N. Lloyd, P. Fasbender, P. Miller, P. Nyhus, *S. Hereford, S. King, *S. Matteson, S. Warner, and W. Harrell (*indicates a workshop participant that also participated in the postworkshop elicitation). We thank the many other collaborators who have contributed substantially to this work, especially P. Adler, A. Gossens, E. Gray, P. Heglund, J. Lyons, M. Runge, S. Servanty, and R. Urbanek. Comments by J. Austin, J. French, C. Hauser, and A. Royle improved an earlier draft of this chapter. Any use of trade, firm, or product names is for descriptive purposes only and does not imply endorsement by the U.S. Government.

References

Araki, H., Cooper, B., Blouin, M.S., 2007. Genetic effects of captive breeding cause a rapid, cumulative fitness decline in the wild. Science 318 (5847), 100–103.

Austin, J.E., Hayes, M.A., Barzen, J.A., 2018. Revisiting the historic distribution and habitats of the whooping cranes (Chapter 3). In: French, Jr., J.B., Converse, S.J., Austin, J.E. (Eds.), Whooping Cranes: Biology and Conservation. Biodiversity of the World: Conservation from Genes to Landscapes. Academic Press, San Diego, CA.

Barzen, J.A., Lacy, A., Thompson, H.L., Gossens, A.P., 2018. Habitat use by the reintroduced Eastern Migratory Population of whooping cranes (Chapter 14). In: French, Jr., J.B., Converse, S.J., Austin, J.E. (Eds.), Whooping Cranes: Biology and Conservation. Biodiversity of the World: Conservation from Genes to Landscapes. Academic Press, San Diego, CA.

Burgman, M., 2004. Expert frailties in conservation risk assessment and listing decisions. In: Hutchings, P., Lunney, D., Dickman, C. (Eds.), Threatened Species Legislation: Is It Just an Act? Royal Zoological Society of New South Wales, Mosman, NSW, Australia, pp. 20–29.

Burgman, M.A., McBride, M.F., Ashton, R., Speirs-Bridge, A., Flander, L., Wintle, B., Fidler, F., Rumpff, L., Twardy, C., 2011. Expert status and performance. PLoS ONE 6 (7), e22998.

Butler, M.J., Harris, G., Strobel, B.N., 2013. Influence of whooping crane population dynamics on its recovery and management. Biol. Conserv. 162, 89–99.

Canadian Wildlife Service and U.S. Fish and Wildlife Service, 2005. International Recovery Plan for the Whooping Crane. Recovery of Nationally Endangered Wildlife (RENEW), Ottawa, and U.S. Fish and Wildlife Service, Albuquerque, NM.

Canessa, S., Guillera-Arroita, G., Lahoz-Monfort, J.J., Southwell, D.M., Armstrong, D.P., Chadès, I., Lacy, R.C., Converse, S.J., 2015. When do we need more data? A primer on calculating the value of information for applied ecologists. Methods Ecol. Evol. 6 (10), 1219–1228.

Christie, M.R., Marine, M.L., French, R.A., Blouin, M.S., 2012. Genetic adaptation to captivity can occur in a single generation. Proc. Natl. Acad. Sci. 109 (1), 238–242.

Converse, S.J., Moore, C.T., Folk, M.J., Runge, M.C., 2013a. A matter of tradeoffs: reintroduction as a multiple objective decision. J. Wildl. Manage. 77 (6), 1145–1156.

Converse, S.J., Royle, J.A., Adler, P.H., Urbanek, R.P., Barzen, J.A., 2013b. A hierarchical nest survival model integrating incomplete temporally varying covariates. Ecol. Evol. 3 (13), 4439–4447.

Converse, S.J., Royle, J.A., Urbanek, R.P., 2012. Bayesian analysis of multi-state data with individual covariates for estimating genetic effects on demography. J. Ornithol. 152 (Suppl. 2), S561–S572.

Converse, S.J., Servanty, S., Moore, C.T., Runge, M.C., 2018. Population dynamics of reintroduced whooping cranes (Chapter 7). In: French, Jr., J.B., Converse, S.J., Austin, J.E. (Eds.), Whooping Cranes: Biology and Conservation. Biodiversity of the World: Conservation from Genes to Landscapes. Academic Press, San Diego, CA.

Dinsmore, S.J., White, G.C., Knopf, F.L., 2002. Advanced techniques for modeling avian nest survival. Ecology 83 (12), 3476–3488.

Ellis, D.H., Lewis, J.C., Gee, G.F., Smith, D.G. 1992. Population recovery of the whooping crane with emphasis on reintroduction efforts: past and future. Proceedings of the North American Crane Workshop 6, 142–150.

Frankham, R., 2007. Genetic adaptation to captivity in species conservation programs. Mol. Ecol. 17 (1), 325–333.

Gregory, R., Failing, L., Harstone, M., Long, G., McDaniels, T., Ohlson, D., 2012. Structured Decision Making: A Practical Guide to Environmental Management Choices. Wiley-Blackwell, Oxford, UK.

Hammond, J.S., Keeney, R.L., Raiffa, H., 1999. Smart Choices: A Practical Guide to Making Better Life Decisions. Broadway Books, New York.

Johns, B., Goossen, J.P., Kuyt, E. Craig-Moore, L., 2005. Philopatry and dispersal in whooping cranes. Proceedings of the North American Crane Workshop 9, 117–126.

Kellner, K., 2015. jagsUI: a wrapper around 'rjags' to streamline JAGS analyses. R Package Version 1.3.7. Available from: https://CRAN.R-project.org/package=jagsUI.

King, R.S., Trutwin, J.J., Hunter, T.S., Varner, D.M., 2013. Effects of environmental stressors on nest success of introduced birds. J. Wildl. Manage. 77 (4), 842–854.

Lacy, R.C., Alaks, G., Walsh, A., 2013. Evolution of Peromyscus leucopus mice in response to a captive environment. PLoS One 8 (8), e72452.

Lynch, M., O'Hely, M., 2001. Captive breeding and the genetic fitness of natural populations. Conserv. Genet. 2 (4), 363–378.

Martin, T.G., Burgman, M.A., Fidler, F., Kuhnert, P.M., Low-Choy, S., McBride, M.F., Mengersen, K., 2011. Eliciting expert knowledge in conservation science. Conserv. Biol. 26 (1), 29–38.

Martin, J., Runge, M.C., Nichols, J.D., Lubow, B.C., Kendall, W.L., 2009. Structured decision making as a conceptual framework to identify thresholds for conservation and management. Ecol. Appl. 19 (5), 1079–1090.

McBride, M.F., Fidler, F., Burgman, M.A., 2012. Evaluating the accuracy and calibration of expert predictions under uncertainty: predicting the outcomes of ecological research. Divers. Distrib. 18 (8), 782–794.

Moore, C.T., Converse, S.J., Folk, M.J., Runge, M.C., Nesbitt, S.A., 2012. Evaluating release alternatives for a long-lived bird species under uncertainty about long-term demographic rates. J. Ornithol. 152 (Suppl. 2), S339–S353.

Mueller, T., Teitelbaum, C.S., Fagan, W.F., Converse, S.J., 2018. Movement ecology of reintroduced migratory whooping cranes (Chapter 11). In: French, Jr., J.B., Converse, S.J., Austin, J.E. (Eds.), Whooping Cranes: Biology and Conservation. Biodiversity of the World: Conservation from Genes to Landscapes. Academic Press, San Diego, CA.

Plummer, M., 2003. JAGS: a program for analysis of Bayesian graphical models using Gibbs sampling. In: Proceedings of the 3rd International Workshop on Distributed Statistical Computing (DSC 2003), Vienna, Austria.

R Development Core Team, 2012. R: A Language and Environment for Statistical Computing. R Foundation for Statistical Computing, Vienna, Austria.

Runge, M.C., Converse, S.J., Lyons, J.E., 2011. Which uncertainty? Using expert elicitation and expected value of information to design an adaptive program. Biol. Conserv. 144 (4), 1214–1223.

Servanty, S., Converse, S.J., Bailey, L.L., 2014. Demography of a reintroduced population: moving toward management models for an endangered species, the whooping crane. Ecol. Appl. 24 (5), 927–937.

Speirs-Bridge, A., Fidler, F., McBride, M., Flander, L., Cumming, G., Burgman, M., 2010. Reducing overconfidence in the interval judgments of experts. Risk Anal. 30 (3), 512–523.

Urbanek, R.P., Fondow, L.E.A., Satyshur, C.D., Lacy, A.E., Zimorski, S.E., Wellington, M., 2005. First cohort of migratory whooping cranes reintroduced to eastern North America: the first year after release. Proceedings of the North American Crane Workshop 9, 213–223.

Urbanek, R.P., Fondow, L.E.A., Zimorski, S.E., 2010a. Survival, reproduction, and movements of migratory whooping cranes during the first seven years of reintroduction. Proceedings of the North American Crane Workshop 11, 124–132.

Urbanek, R.P., Zimorski, S.E., Fasoli, A.M., Szyszkoski, E.K., 2010b. Nest desertion in a reintroduced population of migratory whooping cranes. Proceedings of the North American Crane Workshop 11, 133–141.

Van Schmidt, N.D., Barzen, J.A., Engels, M.J., Lacy, A.E., 2014. Refining reintroduction of whooping cranes with habitat use and suitability analysis. J. Wildl. Manage. 78 (8), 1404–1414.

Williams, B.K., Eaton, M.J., Breininger, D.R., 2011. Adaptive resource management and the value of information. Ecol. Model. 222 (18), 3429–3436.

APPENDIX A

Mean of the Experts' (n = 8) Most Likely Predictions for the Response of Nest Success (the Probability That a Nest Successfully Hatches at least One Chick) 3 Years After Implementation of a Management Action (Described in Greater Detail in Table 8.4; Action Numbers in Table 8.4 Are Same as Action Numbers Here) and 10 Years After Implementation, Conditional on Hypotheses for the Cause of Nest Failure (Described in Greater Detail in Table 8.2). The Optimal Action Under Each Hypothesis (the Largest Value Within a Column) Is Highlighted in Bold. Also Presented Are the Mean Weights, Across Experts, Placed on the Hypotheses (Wts)

		Hypotheses										
	BF		BF + Food		Experience		Genetics		Costume		Group	
Wts	0.331		0.089		0.160		0.162		0.165		0.092	
Actions	3 year	10 year	3 year	10 year	3 year	10 year	3 year	10 year	3 year	10 year	3 year	10 year
1. Do nothing	0.159	0.158	0.159	0.145	0.226	0.282	0.159	0.150	0.159	0.164	0.159	0.164
2. Predator train – Ad	0.165	0.164	0.178	0.176	0.231	0.282	0.184	0.181	0.201	0.201	0.189	0.189
3. Predator train – Yng	0.165	0.164	0.178	0.176	0.231	0.295	0.178	0.175	0.201	0.214	0.182	0.195
4. Slow drawdown	0.178	0.176	0.201	0.195	0.220	0.270	0.171	0.169	0.184	0.182	0.171	0.170
5. Bti treatment	0.376	0.376	0.304	0.316	0.201	0.245	0.165	0.162	0.159	0.151	0.171	0.164
6. Forced re-nesting	**0.398**	**0.422**	**0.348**	**0.366**	0.214	0.258	0.165	0.159	0.165	0.161	0.171	0.168
7. BF habitat Mgmt	0.300	0.281	0.275	0.269	0.195	0.245	0.159	0.156	0.159	0.151	0.171	0.164
8. Hydrology Mgmt	0.176	0.182	0.238	0.229	0.195	0.245	0.159	0.156	0.159	0.151	0.171	0.164
9. Meadow restoration	0.170	0.176	0.201	0.201	0.195	0.245	0.159	0.156	0.159	0.151	0.171	0.164
10. Dummy eggs	0.214	0.214	0.201	0.214	**0.260**	0.295	0.228	0.225	0.252	0.232	0.171	0.236
11. WBNP eggs	0.190	0.189	0.184	0.182	0.208	0.258	**0.258**	**0.382**	0.178	0.182	0.184	0.180
12. Parent-rearing	0.165	0.170	0.165	0.169	0.251	**0.314**	0.184	0.186	**0.259**	**0.325**	0.244	0.281
13. Captive eggs in nests	0.204	0.204	0.185	0.185	0.232	0.281	0.180	0.195	0.190	0.251	0.174	0.189
14. Chick adoption	0.189	0.188	0.178	0.188	0.258	0.295	0.190	0.192	0.234	0.256	0.176	0.189
15. Small group rearing	0.184	0.182	0.184	0.182	0.231	0.294	0.190	0.186	0.196	0.175	**0.256**	**0.319**

APPENDIX B

Mean of the Experts' (n = 8) Most Likely Predictions for the Response of Chick Survival (the Probability That a Chick Survives until 1 October) 3 Years After Implementation of a Management Action (Described in Greater Detail in Table 8.4; Action Numbers in Table 8.4 Are Same as Action Numbers Here) and 10 Years After Implementation, Conditional on Hypotheses for the Cause of Poor Chick Survival (Described in Greater Detail in Table 8.3) The Optimal Action Under Each Hypothesis (Largest Value Within a Column) is Highlighted in Bold. Also Presented Are the Mean Weights, Across Experts, Placed on the Hypotheses (Wts)

	Predators		BF + Food		Food		Experience		Genetics		Costume		Group	
Wts	0.325		0.026		0.051		0.201		0.081		0.180		0.135	
Actions	3yr	10yr	3yr	10yr	3yr	10yr	3yr	10yr	3yr	10yr	3yr	10yr	3yr	10yr
1. Do nothing	0.184	0.178	0.196	0.190	0.196	0.190	0.216	0.260	0.196	0.178	0.209	0.202	0.221	0.209
2. Predator train – Ad	0.316	0.329	0.196	0.190	0.196	0.190	0.279	0.335	0.221	0.209	0.265	0.265	0.252	0.240
3. Predator train – Yng	**0.341**	**0.366**	0.196	0.190	0.196	0.190	0.272	0.322	0.228	0.209	0.279	0.301	0.271	0.259
4. Slow drawdown	0.300	0.312	0.251	0.245	0.264	0.258	0.260	0.304	0.209	0.190	0.221	0.215	0.221	0.209
5. *Bti* treatment	0.196	0.190	**0.279**	**0.281**	0.196	0.190	0.204	0.248	0.215	0.196	0.209	0.202	0.209	0.196
6. Forced renesting	0.240	0.234	0.275	0.277	0.216	0.210	0.216	0.260	0.209	0.190	0.215	0.209	0.215	0.202
7. BF habitat Mgmt	0.209	0.215	0.271	0.268	0.219	0.212	0.216	0.260	0.209	0.190	0.215	0.209	0.209	0.196
8. Hydrology Mgmt	0.254	0.248	0.260	0.272	**0.279**	**0.285**	0.216	0.254	0.209	0.190	0.234	0.228	0.221	0.209
9. Meadow restoration	0.209	0.215	0.234	0.252	0.271	0.277	0.216	0.254	0.209	0.190	0.221	0.215	0.209	0.196
10. Dummy eggs	0.259	0.228	0.228	0.221	0.196	0.190	0.275	0.329	0.221	0.202	0.221	0.215	0.209	0.196
11. WBNP eggs	0.260	0.258	0.215	0.209	0.209	0.202	0.241	0.279	**0.314**	**0.389**	0.221	0.215	0.221	0.209
12. Parent-rearing	0.291	0.339	0.221	0.228	0.215	0.209	**0.304**	**0.366**	0.221	0.215	**0.285**	**0.329**	**0.285**	0.308
13. Captive eggs in nests	0.252	0.221	0.209	0.202	0.196	0.190	0.252	0.291	0.221	0.215	0.234	0.261	0.234	0.228
14. Chick adoption	0.209	0.229	0.215	0.209	0.209	0.209	0.246	0.279	0.221	0.215	0.235	0.285	0.235	0.264
15. Small group rearing	0.209	0.202	0.221	0.215	0.209	0.202	0.254	0.298	0.221	0.202	0.234	0.228	0.283	**0.339**

Florida's Nonmigratory Whooping Cranes

Tim A. Dellinger

Fish and Wildlife Research Institute, Florida Fish and Wildlife Conservation
Commission, Gainesville, FL, United States

INTRODUCTION

Background

The Whooping Crane Recovery Plan lists well-defined measurable criteria to be met for downlisting the species from endangered to threatened status (Canadian Wildlife Service and U.S. Fish and Wildlife Service, 2007). One criterion is to establish two self-sustaining populations in the wild. In the 1970s, the first reintroduction was underway at Grays Lake National Wildlife Refuge, in Idaho (Canadian Wildlife Service and U.S. Fish and Wildlife Service, 2007) and conservationists were contemplating the next Whooping Crane reintroduction site, possibly for a nonmigratory population. Florida and Louisiana were both proposed at this time (Canadian Wildlife Service and U.S. Fish and Wildlife Service, 2007); a nonmigratory Whooping Crane population survived in Louisiana until the 1940s (Allen, 1952). In Florida, fossilized Whooping Cranes have been found and sightings were sporadically reported

in the 1930s (Nesbitt, 1982). In 1983, Florida was included on the official short list of eastern U.S. reintroduction sites (Canadian Wildlife Service and U.S. Fish and Wildlife Service, 2007) and 5 years later Florida was recommended as the next Whooping Crane reintroduction site. Florida was selected over Louisiana due to Louisiana's proximity to the Texas wintering sites of the Aransas-Wood Buffalo Population (AWBP). The U.S. Whooping Crane Recovery Team (USWCRT) wanted reintroductions to be separate from other Whooping Crane populations. Also at the time, a Louisiana reintroduction could result in conflicts with that state's widespread hunting of waterfowl. The USWCRT cited the amount of available wetland habitat, land use practices, and sparse human population at the proposed central Florida release site location as reasons for choosing Florida. They described the release sites as having habitats similar to areas used by the nonmigratory Whooping Cranes of the 1930s and 1940s in Louisiana (Canadian Wildlife Service and U.S. Fish and Wildlife Service, 2007).

Florida researchers began wrestling with numerous uncertainties and operational challenges of an attempt to establish a nonmigratory flock of Whooping Cranes in Florida, including whether eggs or young from migratory birds could be used to establish a nonmigratory population. To address this concern, Florida Fish and Wildlife Conservation Commission (FWC) researchers replaced eggs in nests of nonmigratory Florida Sandhill Cranes (*Grus canadensis pratensis*) with eggs of migratory Greater Sandhill Cranes (*G. canadensis tabida*) to assess the degree that migration is innate (Nesbitt and Carpenter, 1993). Five of 34 exchanged eggs survived to fledging during the 1986 and 1987 breeding seasons. Additionally, in 1986 and 1987, 9- to 10-month-old, captive-reared Greater Sandhill Cranes were released in groups within the Florida Sandhill Crane's range; 15 of the 27 released cranes survived the first year. The young birds were kept in a large pen for 4–6 weeks, which allowed the captive-raised cranes to slowly transition to the wild. All surviving fostered chicks and released juveniles remained in the Florida Sandhill Crane range and did not migrate, providing evidence that migration is a learned behavior. At the end of this project five surviving Greater Sandhill Cranes were recaptured and taken into captivity, and others (four fostered chicks and eight released birds) were not captured and left in the wild.

The cross-fostering method was used in the first reintroduction of Whooping Cranes at Grays Lake National Wildlife Refuge, in Idaho (Drewien and Bizeau, 1978), whereby Whooping Crane eggs were placed in nests of Greater Sandhill Cranes. Initially, the USWCRT thought the Florida Nonmigratory Population (FNMP) could be established by the same method (Nesbitt et al., 1997). Specifically, Whooping Crane eggs would be placed in the nests of nonmigratory Florida Sandhill Cranes. In the 1980s, researchers (Bishop, 1988) identified Florida Sandhill Crane populations on public lands whose nesting areas could serve as sites for Whooping Crane reintroductions. However, by the time federal and state agencies had selected Florida as the second reintroduction location, the International Whooping Crane Recovery Team (IWCRT) had recommended that cross-fostering of Whooping Cranes by Sandhill Cranes be discontinued because it appeared to negatively affect later pair formation (Canadian Wildlife Service and U.S. Fish and Wildlife Service, 2007). At Grays Lake, cross-fostered Whooping Cranes did not develop long-term pair bonds with conspecifics despite great efforts to encourage pair formation (Canadian Wildlife Service and U.S. Fish and Wildlife Service, 2007). Consequently, a soft release of captive-raised Whooping Cranes was selected as the means of trying to establish the nonmigratory population.

The Kissimmee Prairie, a ~500,000 ha area of freshwater marsh and open grasslands in Osceola and Polk counties, was chosen as the initial release area due to its relative isolation from human population centers. The vast area is comprised of public landholdings (federal, state, and local conservation areas) and large, sparsely populated private ranches (Bishop, 1988). The habitat on the Kissimmee Prairie and other release locations was comparable to that used by Florida Sandhill Cranes (Nesbitt and Williams, 1990). Shallow freshwater wetlands were used for nesting, roosting, and foraging. Cranes loafed and foraged on upland habitat, typically active cattle ranches, sod farms, and dry prairie. Nesbitt and Williams (1990) reported that oaks provided seasonally abundant acorn mast. Areas of ecotone between lake or marsh edges and upland habitat also provided a variety of food items.

Releases

From November 1992 through April 1993, chicks of 5–8 months old were flown in cohorts of 5–14 birds from two breeding centers, the U.S. Geological Survey Patuxent Wildlife Research Center in Laurel, Maryland, and the International Crane Foundation in Baraboo, Wisconsin, to central Florida. The release method was gradual, and termed "soft release." Whooping Crane open-topped pens 0.03–0.36 ha in area were built at the edge of marshes, with half of the pen in shallow water and half in pastureland or open upland habitat (Nesbitt et al., 1997). Typically, large pens were permanent, while smaller pens were portable and could be erected at sites with the most suitable habitat and water levels in a given release season; both allowed a soft release. Young cranes were checked for health and fitted with a USFWS aluminum band, color bands, and a radio transmitter before they were released into a pen (Nesbitt et al., 1997). The birds were kept in the pens for 2–4 weeks for acclimation (Nesbitt et al., 1997). While in the soft-release pens, young cranes' wings were brailed to prevent flight and they were monitored daily to ensure they were healthy. Release involved capturing the chicks at night, removing brails, and releasing the birds back into the open-topped pens, allowing them to fly out of the pens when they were inclined to do so: most birds departed the pens within 84 h after being captured for brail removal (Nesbitt et al., 1997). The first birds left the pen in February 1993 – the most recent previous observation of a nonmigratory Whooping Crane was 50 years earlier, in Louisiana (Allen, 1952).

Subsequent releases took place in fall and spring through 2005. A total of 289 birds were released in cohorts of 2–14 individuals (Fig. 9.1) at six release sites (Fig. 9.2). FNMP Whooping Cranes were monitored daily the first 4–6 months after they were released, and then 2–3 times per week after that (Folk et al., 2005). Cranes were observed more frequently during the breeding season, while molting, or when an individual appeared injured. Researchers located cranes through radio tracking, noting the general location, behavior, and whether other FNMP birds were in the area. FNMP birds that could not be located by ground searches were found via fixed-wing aircraft (Nesbitt et al., 1997). Data on the precise location and habitat use was not collected uniformly during the 19 years FWC monitored the FNMP. As a result, habitat selection and home ranges could not be formally analyzed.

SURVIVAL

High mortality during the initial release years was attributed to inexperienced birds roosting on dry ground rather than in shallow water and using heavily vegetated areas instead of areas of open, low vegetation (Nesbitt et al., 1997).

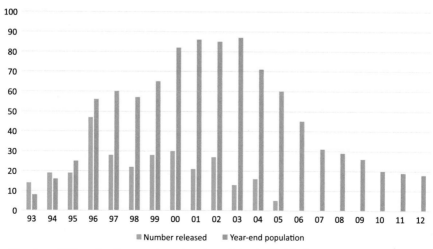

FIGURE 9.1   Number of Whooping Cranes released and surviving in Florida, by year.

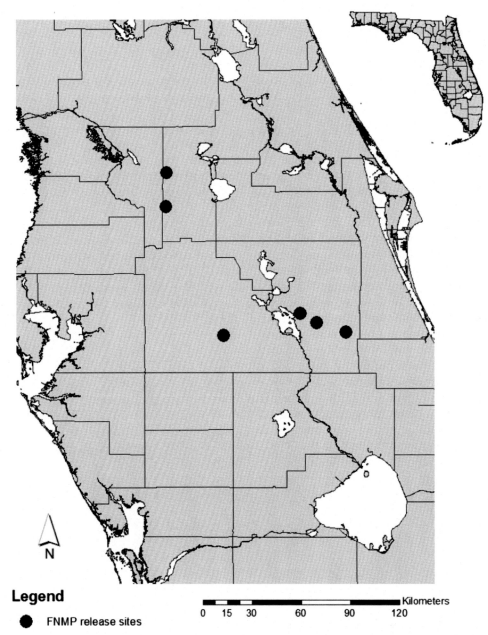

Legend

● FNMP release sites

					Kilometers
0	15	30	60	90	120

FIGURE 9.2   Release sites of Whooping Cranes in Lake, Osceola, and Polk counties in Florida, 1993–2005.

Both behaviors presumably increased the likelihood of predation (Nesbitt et al., 1997). Nesbitt et al. (1997) reported 31 of the 52 Whooping Cranes released between 1993 and 1995 were killed by bobcats. Roosting in water is a normal behavior seen in the remnant AWBP. The chicks of the AWBP are hatched in and spend the first few months in proximity to wetlands, but young captive cranes raised in pens without ponds continued to roost on dry land even when released into a landscape containing hundreds of marshes.

Modification of the rearing methods and the management of the release sites increased postrelease survival. Specifically, ponds were added to rearing pens and all young cranes learned to roost in the ponds at night (Gee et al., 2001). During the following 4 years, survival was greater among young cranes reared in pens with ponds than those raised without ponds. Rearing facility staff also cleared dense vegetation around pens to provide an unobstructed view of the surrounding area to acclimate young birds to open areas (Nesbitt et al., 1997). In addition, large areas of dense vegetation near release sites were mowed or burned, and bobcats were trapped and translocated (Nesbitt et al., 1997).

A challenge created by the release method in the early years of the reintroduction effort was the possibility of zinc toxicosis. Birds often ingested small pieces of metal debris in rearing and release pens that may have compromised their health (Spalding et al., 1997). Of the nine birds killed by predators within 6 months of the initial 1993 release, six had metal objects in their stomachs, discovered in prerelease X-rays or during necropsy, and elevated zinc levels in their blood. Researchers noted lethargy and abnormal posture among two Whooping Cranes 6 and 8 weeks after the birds arrived in Florida. Blood samples collected before release were re-examined after the birds left the pens and in these two individuals revealed anemia and elevated white blood cell counts (Spalding et al., 1997). These cranes were captured and the metal fragments surgically removed, but both were killed by predators within 4 months. Many species of birds, especially those that consume seeds or grain, ingest small stones or grit, which are stored as gastroliths in part of the stomach and aid in grinding food (Proctor and Lynch, 1993). The metal fragments may have been consumed

to serve as gastroliths. Most metal fragments were determined to have come from debris from the chain-link fencing used to build the pens. Given these findings, technicians made several passes with metal detectors in the release pens, removed metal debris, and provided appropriately sized stones for use as gastroliths. Also, preshipment protocols were modified to include X-rays and surgery when necessary, to minimize the risk of releasing cranes with metal in their stomachs (Spalding et al., 1997; Nesbitt et al., 2001). These efforts did reduce zinc exposure, but mortality rates remained high throughout the 20-year period FWC monitored the FNMP.

The average annual mortality rate was 24.1%, with the highest rate found in the first year birds were released (57.9%), and lowest rate (0.0%) in the final year of the program when 18 birds remained, 3 wild-hatched females and 5% of the captive released Whooping Cranes. The Eastern Migratory Population (EMP) annual mortality rate was 7.1% during 2001–6 (Urbanek et al., 2010). However, this population experienced an annual mortality rate similar to the FNMP, 26.7%, during 2006–7, which overlapped a period of severe drought in both wintering and breeding areas (Urbanek et al., 2010). For the FNMP, predation accounted for 131 or 46% of mortalities for 287 cranes released between 1992 and 2004. Bobcats (*Lynx rufus*) were responsible for 73% of the predation mortalities and American alligators (*Alligator mississippiensis*) 8%. Among all years, the average number of days between Whooping Crane release and mortality by bobcat was 209.8 days ($n = 95$; range: 0–3,433 days; SE = 42.04) and 539.5 days ($n = 11$; range: 2–3,397 days; SE = 326.79) by alligators. Other causes of mortality noted in the FNMP include power line, fence, or vehicle collisions, monofilament entanglement, or other man-made hazards, which together resulted in 44 deaths (Miller et al., 2010). Lightning strike ($n = 4$), gunshot ($n = 3$), choking on food items ($n = 2$), stepped on by a cow ($n = 1$), and bee stings ($n = 1$), also occurred, albeit rarely. Many

FNMP cranes, often with nonfunctional trans-mitters, simply disappeared over the years and could not be found.

Survival between the sexes was different and the FNMP became sex-skewed over time. The ratio of males to females was near 1:1 during the first several years of the reintroduction effort, but by 2007 the population was skewed to a female bias (1:1.27; Spalding et al., 2010). Annual mortality during 1999–2007 was 13% for both males and females 3–9 years old (Spalding et al., 2010). Mortality was lower for females 10–14 years old (1.9%), but no males lived >10 years during the 1999–2007. Folk et al. (2013) reported that males tended to lead when the pair flew or walked and so were the first to encounter a predator or power line. When FWC stopped monitoring the FNMP in 2012, 18 birds (7 male and 11 female) remained, including 7 pairs. The average age of males was 11.57 years (range: 9–13 years), and the average age of females was 11.90 years (range: 3–19 years).

REPRODUCTION

Territories

FNMP release sites were in areas inhabited by Florida Sandhill Cranes. These areas consist of pastureland or native prairie (uplands), and shallow freshwater marshes (wetlands) and are among the most productive Florida Sandhill Crane habitat. The uplands were maintained through grazing, burning, or mechanical manipulation (mowing or roller-chopping). The wetlands are shallow and support emergent vegetation. Many Whooping Cranes established territories in the vicinity of the release sites or on adjacent ranches. Generally, territories are used by a bird or pair and defended against other cranes, Whooping or Sandhill, often fiercely while nesting or if a chick is present (Allen, 1952; Kuyt, 1993). Whooping Cranes will use the same territory year after year (Kuyt, 1993), when

sufficient water is available (Folk et al., 2008). As of 2016, the cranes with the longest pair bond in the FNMP also have occupied a territory for the longest time, since August 2003.

Size and proximity of FNMP territories to other Whooping or Sandhill Cranes territories varied greatly most likely based on habitat quality (Folk et al., 2005). In the FNMP, Whooping Crane territories ranged in area from approximately 0.40 km^2 to roughly 2 km^2 (Folk et al., 2005). Territories in other populations ranged larger, for example, AWBP territories varied in area from 1.1 to 18.9 km^2 (Kuyt, 1993) and the EMP territory size ranged from 1.84 to 4.04 km^2 (Barzen, Chapter 15, this volume).

Nonmigratory Whooping Crane pairs in Florida have exhibited substantial variability in territorial behavior. Some pairs defend their territories from all other cranes and even from herons, egrets (*Ardea alba*, *Egretta* spp.), ibises (*Eudocimus albus*, *Plegadis falcinellus*), and storks (*Mycteria americana*), while other pairs are tolerant of other wading bird species nearby (Fig. 9.3). For example, for several consecutive years a pair of Whooping Cranes was observed nesting in a large marsh (0.53 km^2) during aerial surveys that also supported 4 or 5 Sandhill Crane nests. Resources did not appear to be limited given the number of nesting cranes reported and a lack of agonistic behavior observed between the cranes during those years. The marsh is connected to a large lake that provides a source of water even in drought years and helps to maintain emergent vegetation. Both shallow water and emergent vegetation are important components of appropriate nesting habitat for cranes (Folk et al., 2005). Furthermore, the marsh is surrounded by open areas of pastures that the cranes use for foraging and loafing. Elsewhere, on cattle ranches and sod farms with numerous small (<0.05 km^2) freshwater marshes, Whooping and Sandhill Cranes were recorded nesting in adjacent marshes with nests <0.35 km apart. Territorial behavior, however, varies among individuals and we assume site

FIGURE 9.3 Wading birds foraging near active Whooping Crane nests in Florida, Great Egret (*Ardea alba*) (A); and White Ibis (*Eudocimus albus*) (B); note Whooping Cranes at nests in the background. Photo taken with cameras placed approximately 8 m from nest platform.

conditions and available resources can influence agonistic behavior. For example, a pair of Whooping Cranes and a pair of Sandhill Cranes have overlapping territories that comprise two small marshes, a pond, pastureland, and a residential area. The pairs have frequent altercations, with the Whooping Cranes often chasing the Sandhill Cranes to the outskirts of their territory. During the 2012 breeding season, intense aggression was observed between the male Whooping Crane and the Sandhill Crane pair during the time the Sandhill Cranes had a 15-day-old chick. The male Whooping Crane attacked the Sandhill Crane family as they foraged in a pasture, and in the end the chick was dead; this event may have been related to territorial behavior. The chick was not eaten and the carcass was recovered the following day.

Pair Bonds

As in most birds, reproduction does not occur without a pair bond, and conversely, without reproduction, the pair bond does not last (Nesbitt and Tacha, 1997). In Florida, once birds were released they would often associate with another crane (i.e., a few days to many months), then move on and associate with a different individual, an activity termed "consort pairing." This behavior is similar to that in Florida Sandhill Cranes, which also form consort pairing as subadults before establishing a true pair bond and proceeding to reproduce (Nesbitt and Wenner, 1987). As the FNMP Whooping Cranes matured, they formed pair bonds characterized by unison calling by the pair, defending of a common territory, copulating, platform building, egg laying, incubating, and brood rearing. The first pairing in the Florida population occurred between an 18-month-old male and a 20-month-old female 10 months after the first cohort was released. The longest pair bond in the population was formed in August 2003 between a 3-year-old male and a 10-year-old female; these individuals remain together as of this writing, in September 2016. Although they have never fledged a chick, they have nested three times, most recently in 2011.

Most Whooping Cranes released into the FNMP died before forming a pair bond. Among those that survived, a total of 52 cranes (27 males and 25 females) formed at least one pair bond.

Incomplete records prevent a comprehensive review of all pair bonds, but among 43 FNMP, the mean age of birds that first formed a pair bond was 4.63 years (male: $n = 20$, = $x^- = 4.54$ years, SE = 0.38, range = 2.39–8.71 years; female: $n = 23$, = $x^- = 4.71$ years, SE = 0.47, range = 2.38–10.28 years). These are similar to values reported by Urbanek and Lewis (2015) for the AWBP (2–3 years) but slightly older than those for the Eastern Migratory Population (EMP) (2.43 years for males, 1.74 for females). The greatest number of paired Whooping Cranes was observed during the 2004 and 2006 breeding seasons. The greatest number of pairs nesting was also recorded in those years (Fig. 9.4).

All of the wild-hatched cranes fledged from the FNMP near the end of reintroduction or after reintroductions ended when only a small number of birds remained in the population. Most birds were already in an established pair; consequently, wild fledglings often formed associations with FNMP individuals of the same sex or with Florida Sandhill Cranes. Of the 11 Whooping Cranes that fledged from FNMP nests, only two (one male and one female) formed a true pair bond. The mean age for the two individuals

that formed a pair bond was 5.57 years (male: 5.48 years; female: 5.65 years) and both formed pair bonds with captive raised cranes. These two wild hatched cranes were siblings from different years. The wild-hatched male associated with his female sibling for 36 months but they did not breed; the two separated and he paired and nested with his mother. The female sibling subsequently formed a pair bond with another old female and built a nest and laid eggs. Researchers observed the female–female pair for behavioral evidence that the younger crane might have been mistakenly identified as female (from a blood test) but her behavior was not characteristic of male behavior; she was clearly a female. This was the first-ever documented female–female Whooping Crane pair. A small population and a female sex-skewed population may explain this pairing, and female–female pairing has been observed in other birds [e.g., gulls and terns (Nisbet and Hatch, 1999), shearwaters (Bried et al., 2009), petrels (Lorentsen et al., 2000), and albatrosses (Young et al., 2008)]. The other wild-hatched females in the FNMP formed 12-month bonds with multiple males, but neither from this population ever bred. Clearly, this is reproductive

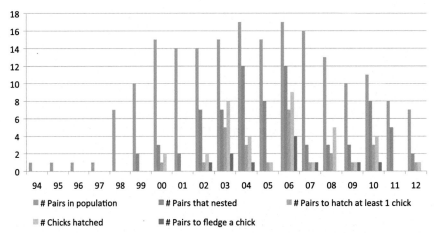

FIGURE 9.4   Pairs of Whooping Cranes and the number that nested, hatched, and fledged chicks in Florida, 1994–2012. A pair is defined as two individuals that unison call, defend a common territory, nest-build, and raise young; nesting is defined as a pair of Whooping Cranes that build a platform, lay eggs, and incubate the eggs to hatching; fledged chicks are individuals that survive beyond the age of their first flight.

behavior that could not sustain the FNMP of Whooping Cranes.

Nesting and Nest Defense

The majority of nest sites were found in freshwater marshes of moderate size (mean = 1.17 km^2 in area, range 0.005–8.27 km^2). Mean water depth at 10 nests during the 2011–13 breeding seasons was 29.11 cm (SE = 2.60), similar to that observed for AWBP nests (25 cm; Kuyt, 1981). The nest platform rose and fell with the water level and was constructed of the surrounding emergent vegetation. Common vegetation in the marshes consisted of pickerelweed (*Pontederia cordata*), maidencane (*Panicum hemitomon*), and sagittaria (*Sagittaria* spp.). Whooping Cranes occasionally nested in emergent vegetation along lake and pond edges when marshes were dry during extreme drought. These nests were susceptible to human disturbance and destruction from airboats used in these areas.

Both sexes helped build the platform by pulling emergent vegetation and placing it in a large pile. Nest construction typically took 3–5 days, but varied with the size of the nest, water depth, plant material used, and density of the vegetation. The nesting birds continued to add to the platform through the incubation period. Whooping Cranes build several platforms in the nest marsh before egg laying, similar to Florida Sandhill Cranes (Folk et al., 2005). Some new Whooping Crane pairs were observed building nest platforms the breeding season before that in which they produced their first clutch (Folk et al., 2005). Folk et al. (2005) documented a Whooping Crane pair that built a number of platforms that they used primarily for nocturnal roosting, brooding, and loafing for their chicks. Posthatching platforms were built in new locations as water levels receded as the marsh dried.

Researchers visiting nests were met with a variety of nest defense behaviors. Most adults would give the distraction display of drooped wings, feigning injury and drawing attention away from the nest. Others would remain near the nest, calling, foot-stamping, spreading wings, and at times displaying aggression toward the researchers. Eleven nests from five pairs were approached by researchers during the 2011–13 breeding seasons, and the average distance the incubating cranes flushed from the nest due to the approaching personnel was 139.24 m (range: 21.03–362.11 m, SE = 32.94). Three of these nests were abandoned immediately after just one visit by the researcher or installation of data-collection equipment near the nest.

Surveillance by nest cameras revealed some surprising behavior. Whooping Cranes defended their nests from raccoons by assuming preattack postures (see Urbanek and Lewis, 2015) and attempting to stab the intruder with their beaks. At one suburban nest, domestic dogs approaching the nest were typically engaged by one of the pair running at the dog and diverting its attention from the nest. A female Whooping Crane on a nest in a marsh within an active cattle pasture continued to incubate as a grazing cow proceeded to step on her, and killed her. Through video surveillance, cranes were observed standing with spread wings and giving alarm calls as Bald Eagles (*Haliaeetus leucocephalus*) flew over the nest area (Fig. 9.5). At one nest a Bald

FIGURE 9.5 Spread-wing display of Florida Whooping Crane pair in response to a Bald Eagle flying over the nest. Note the erect body feathers, the upright posture, and the bills open as they vocalize.

Eagle pair made numerous attacks and eventually took one of two hatchlings. When an eagle attempted to capture the other chick, the parents attacked the eagle and injured it to the extent that it required rehabilitative care.

Laying and Incubation

As with other Whooping Crane populations, egg laying in the FNMP typically occurred in the morning (Urbanek and Lewis, 2015). Incubation began immediately for single-egg nests or, otherwise, after the second egg was laid. Among nests whose contents were confirmed, the average clutch size was 1.57 ($n = 69$), with a maximum of 2. Thirty-nine unhatched eggs were collected and the average egg size was 102.7 × 61.7 mm (Spalding et al., 2010), similar to AWBP egg size, 98.4 × 62.4 mm (Bent, 1926), and EMP eggs, 103.5 × 63.1 (C. Gitter, International Crane Foundation, Baraboo, WI, personal communication).

A total of 90 clutches was laid by 37 separately identified pairs from 1999 through 2012. Most of these pairs were made up of captive-reared/soft-released cranes (27 males and 25 females) and only two wild-hatched individuals formed pairs (one male and one female). Thirty-two percent of the documented clutches were laid by only two females, one of which laid 19 clutches and the other 10 clutches.

The mean age of males and females at the first clutch was similar, at 5.64 years ($n = 27$, range: 3.08–9.77 years, SE = 0.31) and 5.84 years ($n = 25$, range: 2.82–12.82 years, SE = 0.52), respectively. This is slightly older than ages reported for other populations (EMP: 3.85 years for males and 3.92 years for females (Urbanek et al., 2010), AWBP: 5.0 years for both sexes (Kuyt and Goossen, 1987). A combination of factors may have influenced the delay in laying in the FNMP, such as periods of drought when birds would not nest, a female-biased sex ratio, high mortality of mates, limited encounters with potential mates due to the small population size and wide distribution, and possible but unknown behavioral issues that resulted from having been raised in captivity (Converse et al., Chapter 7, this volume).

None of the 11 wild-hatched chicks produced young: only 2 formed pairs. The mother–son pair (described earlier) nested in 2010 (Fig. 9.6); the female layed one egg that did not hatch. This pair nested only once, because the female was killed by a predator a few months after the breeding season and the male did not pair again before he died the following year at age 8 years. Incestuous pairs are rare in birds but have been noted in some species, including Common Loons (*Gavia immer*) (Piper et al., 2001), and in this case the parent–offspring pairing happened when only a few males remained in a small population made up primarily of females. The FNMP female–female pair described previously had one nest with a single egg, but the egg was taken by a predator. This pair did not nest again.

Incubation period in the FNMP was 27–32 days based on hatching information from 18 nests. The average laying date for the first clutch was 21 March ($n = 75$, range: 27 January–24 May), with 66 of 75 (88%) first attempts from 37 pairs occurring during February–April. Six pairs renested after the first clutch failed to hatch or was depredated; three of these six pairs renested after the first nest failed in multiple years. One pair renested twice in two separate years, when the first and second nests of the season failed. Generally, a pair would begin building a new platform in the same marsh as the failed nest within 2–3 weeks of the failure date. The mean laying date for the second nest attempt was 15 April ($n = 11$, range: 11 February–24 May). The pair that renested twice had laying dates for the third nests of 3 May and 27 May.

Dellinger (unpublished data) documented incubation behavior of captive-reared Whooping Cranes at successful and unsuccessful nests with video surveillance. Cranes generally exchanged nest attendance duty about every 2 h, with males and females averaging almost equal incubation times. Cranes incubated eggs in roughly 30-min

FIGURE 9.6 A mother–son Whooping Crane pair that nested in Osceola County, Florida, in 2010. The purple blossoms are pickerelweed (*Pontederia cordata*). Photo by M. Folk, Florida Fish and Wildlife Conservation Commission.

bouts of continuous sitting before standing to preen, get a drink of water, or change position, and the average incubation bout time decreased as the incubation period progressed through the season. Typically, the attending adult interrupted incubation for only seconds to a couple of minutes. The adult remained on the nest in most of these occurrences, but would sometimes step off the platform, leaving eggs completely exposed to the elements, for only a very short time (<3 min). This incubation behavior was consistent in the FNMP, and contrasts with inconsistent incubation and nest desertion observed in the EMP. In that population, widespread, often synchronous nest desertion occurs with the emergence of black flies (*Simulium* spp.) (Urbanek et al., 2010). Incubation bout time, time off the platform, and time between exchanges were similar for successful and failed nests in the FNMP (Dellinger, unpublished data); in addition, time not incubating averaged longer at failed nests and, interestingly, time spent turning eggs was typically longer for failed nests than for successful nests.

Dellinger (unpublished data) surveillance data and data from a nest predator study also documented nocturnal incubation behavior. Video footage showed pairs continuing to exchange incubation duties through the night. This behavior had never been described for Whooping Cranes. Time off the platform and time between incubation exchanges at night were similar to daytime incubation.

Hatching and Fledging Information

Over 13 years between 1999 and 2012 the FNMP hatched 37 eggs in the wild, approximately 26% of eggs laid. Eleven females hatched eggs at a mean age of 6.81 years (range: 3.82–10.04; SE = 3.16) at the time of hatching their first brood. The average age of males at the hatching of their first brood was 6.16 years ($n = 13$; range: 4.01–8.99; SE = 3.46). Twenty-three of 72 first nests (32%) hatched at least one egg, 3 of 10 (30%) second nests hatched, but neither of the third nesting attempts hatched. The 2006 breeding season was the most productive, as indicated by the total number of pairs to hatch at least one chick ($n = 7$), the total number of chicks hatched ($n = 9$), and the total number of chicks to fledge ($n = 4$).

Eleven of the 37 hatchlings (30%) between 1999 and 2012 survived to fledging. Mean survival age for unfledged chicks was 17.19 days ($n = 26$; range: 0–77 days, SE = 4.51). Fifteen of 26 chicks (58%) were not detected after 9 days post hatching, but aside from the instance of Bald Eagle depredation we observed no predation events.

Mean fledging age was 83.60 days ($n = 10$, range: 77–99 days, SE = 2.63). Chicks were observed preparing for first flights by wing stretching, flapping, and jumping. Parents would coax chicks to take their first flights. Typically, adult males would assume a preflight posture, indicating to the female and chick that a flight was forthcoming. Often adults would fly, circling nearby and calling to the chick as it ran and flapped. FNMP fledged chicks remained with their parents until the next breeding season. When nest building commenced, the year-old chick was treated as an intruder and chased from the pair's territory.

Overall FNMP were not productive due to a combination of factors. For example, reproduction issues, such as delayed reproduction, infertility, and pairing with Sandhill Cranes were problematic. Spalding et al. (2010) found 65% of the breeding-age birds through 2007 ($n = 122$), were delayed in the breeding cycle, that is, 6 years of age or older, or simply unproductive. Additionally, postmortem evaluation of Whooping Cranes found that 12% of females and 6% of males had poorly developed reproductive tracts. Evidence indicated that fertility of nesting females increased with age. Fertility of first-time-nesting females was estimated at 66%, which improved to 88% for birds older than 10 years (Spalding et al., 2010). Poor survival of breeding-age birds, especially males, toward the end of the reintroduction effort also hindered productivity of the FNMP.

Ending FNMP Reintroduction

During 2008, reintroduction partners from Florida Fish and Wildlife Conservation Commission, the U.S. Fish and Wildlife Service, the International Crane Foundation, Patuxent Wildlife Research Center, and the University of Florida came together for a structured decision-making process organized by Patuxent Wildlife Research Center scientists. The goal of the process was to develop management recommendations for the future of the FNMP. In support of this effort, Patuxent scientists developed a suite of population models that represented competing hypotheses about population function (Converse et al., 2013; Moore et al., 2008) and used the models to evaluate a variety of future release scenarios. The resulting population projections were included in the structured decision-making process along with evaluation of other objectives to recommend a release strategy (Converse et al., 2013; Moore et al., 2008). Model-averaged predictions resulted in a 20% chance or less of achieving a self-sustaining population under most release strategies, and even the most aggressive release strategy resulted in only a 41% chance of reaching that goal (Converse et al., 2013; Moore et al., 2008). The results of the process were reported to the IWCRT, which used the report and other factors to recommend that, although the flock should continue to be studied, no further releases of birds should be considered (Converse et al., 2013).

Factors leading to the decision to discontinue releases included primarily problems with survival and poor productivity, both complicated by drought, and also habitat loss, scarcity of release birds, and cost (Florida Fish and Wildlife Conservation Commission, 2008). After the first reintroduced birds nested in 1999, severe droughts occurred off and on, and many years were apparently not wet enough to provide adequate nesting conditions. Overall, periodic droughts led to high annual variability in Whooping Crane productivity

(Dellinger et al., 2013; Spalding et al., 2009), making it difficult for wild-hatched Whooping Cranes to breed or develop into a self-sustaining flock.

In addition, Florida's human population is projected to increase by 3.6–9.2 million between 2013 and 2040 (Smith and Rayer, 2014), and the predicted loss of suitable crane habitat is a major concern. FNMP cranes use the same habitats as Florida Sandhill Cranes; between 1974 and 2003 Florida Sandhill Cranes experienced an estimated population reduction of 36% as inferred from measures of habitat loss and degradation (Nesbitt and Hatchitt, 2008). Conserving habitat for either crane species will be challenging as nearly 90% of the Florida Sandhill Crane's preferred habitat is privately owned (Nesbitt and Hatchitt, 2008). Overall, the inability of the FNMP to become self-sustaining, together with the projected loss of crane habitat in Florida, combined to persuade the IWCRT that reintroduction of the FNMP would not succeed.

SUMMARY AND OUTLOOK

In working with the FNMP, researchers overcame several obstacles and pioneered techniques that are now used in other Whooping Crane reintroduction efforts. For example, research on new soft-release methods with migratory Greater Sandhill Cranes strongly suggested that a nonmigratory Whooping Crane population was possible in central Florida using migratory Whooping Crane offspring (Folk et al., 2010). Modifications of rearing procedures and release protocols increased survival because they required examining birds for ingested metal before they were released, habituating young birds to roosting in water, and preventing young birds from becoming acclimated to densely vegetated areas (Gee et al., 2001; Nesbitt et al., 1997; Spalding et al., 1997). The development of a pen system that could be easily moved to optimal release habitat also improved crane survival (Nesbitt et al., 2001). Researchers developed

several techniques for safely handling and recapturing Whooping Cranes (Folk et al., 2005) and caused no major injuries during >800 handling events (Miller et al., 2010).

Collision with power lines was the largest known cause of mortality known for fledged Aransas-Wood Buffalo Whooping Cranes (Stehn and Wassenich, 2008) and was a significant source of injury to and death of Florida cranes. In Florida, Miller et al. (2010) found that, from 1992 through 2007, 39 cranes, or 13% of the cranes released into the FNMP, collided with a power line and 57% of the collisions were fatal. Florida researchers worked with a utility company to mark a set of power lines at which several collisions had occurred (the lines coincided with the transition between upland and marsh habitat) with bird diverters (reflectors on swivels; Fig. 9.7), and only one strike occurred after the lines had been marked. Injury to legs fitted with transmitters or bands occurred in 50% of power line collisions. Biologists developed a modified transmitter with a tapered leading edge that could more easily glide over obstacles and thereby reduce injuries (Fig. 9.8; Folk et al., 2010).

Ending the Whooping Crane program was a difficult decision, but it was necessary because many of the factors limiting population growth were outside the direct control of wildlife managers. Nevertheless, the knowledge gained while working on the species in Florida should prove useful with future releases elsewhere. Of the remaining cranes, approximately 15 birds, five pairs and five wild-hatched individuals, continue to survive in central Florida. Nesting continues with some pairs, and one pair has successfully raised at least one chick in 2014, 2015, and 2016. The prospects of these and other wild-hatched/wild-reared individuals finding a nonpaired FNMP mate are very low. Among the wild-hatched group, three of the five share the same parents and those individuals whose sex has been determined are all female; the sex of twins hatched in April 2016 is currently

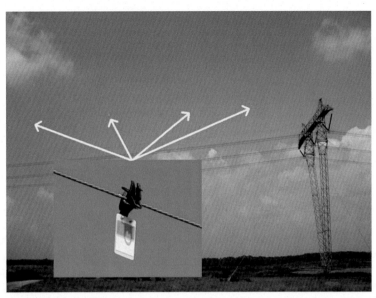

FIGURE 9.7   Bird diverters, inset, installed on power lines to deter Whooping Crane collisions. Photo by M. Folk, Florida Fish and Wildlife Conservation Commission.

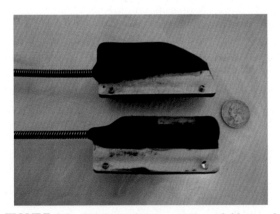

FIGURE 9.8   Old-design leg transmitter with blunt end (bottom) and new, tapered-leg transmitter (top) designed to glide over objects and minimize injury and death of Florida nonmigratory Whooping Cranes. Transmitters are mounted with the antenna extended distally down the leg such that the end toward the quarter encounters objects first.

unknown. Given the low chance of productivity of these FNMP cranes and the low number of captive-reared individuals available annually to the reintroduction effort, researchers in Florida, Louisiana, and the USFWS attempted to translocate one wild-hatched 8-month-old FNMP chick from the FNMP to the Louisiana Nonmigratory Population (LNMP) in 2015. Researchers hope that adding the genetics and life experiences of a wild-reared individual to a population of captive-reared birds could help the overall LNMP. Given there are so few Whooping Cranes remaining, we should develop techniques that will allow translocations of FNMP birds whenever possible to be with their own kind. The FNMP cranes are beloved by Floridians, many of whom agree we should attempt translocations for the benefit of the individual cranes and more importantly the species.

Acknowledgments

Editorial assistance and comments on the manuscript were provided by William Brooks, Bland Crowder, Andrew Cox, Rachel L. Dellinger, Pat Gonzalez, and Amy Schwarzer. Marty Folk provided helpful insight on the Florida nonmigratory population. Robin Grunwald and Ryan Druyor provided research and geographic information system (GIS) assistance, respectively.

References

Allen, R.P., 1952. The Whooping Crane. National Audubon Society Research Report 3, 246 pp.

Barzen, J.A., 2018. Ecological implications of habitat use by reintroduced and remnant whooping crane populations (Chapter 15). In: French, Jr., J.B., Converse, S.J., Austin, J.E. (Eds.), Whooping Cranes: Biology and Conservation. Biodiversity of the World: Conservation from Genes to Landscapes. Academic Press, San Diego, CA.

Bent, A.C., 1926. Order Paludicolae, cranes, rails, etc., Family Megalornithidae, cranes, *Megalirnis americanus* (Linnaeus) whooping crane. Life Histories of North American Marsh Birds, 135, United States National Museum Bulletin, pp. 219–231.

Bishop, M.A., 1988. Factors affecting productivity and habitat use of Florida sandhill cranes (*Grus canadensis pratensis*): an evaluation of three areas in central Florida for a nonmmigratory population of whooping cranes (*Grus americana*). PhD dissertation, University of Florida, Gainesville.

Bried, J., Dubois, M.P., Jouventin, P., 2009. The first case of female–female pairing in a burrow-nesting seabird. Waterbirds 32 (4), 590–596.

Canadian Wildlife Service and U.S. Fish and Wildlife Service, 2007. International Recovery Plan for the Whooping Crane. Recovery of Nationally Endangered Wildlife (RENEW), Ottawa, and U.S. Fish and Wildlife Service, Albuquerque, NM, p. 162.

Converse, S.J., Moore, C.T., Folk, M.J., Runge, M.C., 2013. A matter of tradeoffs: reintroduction as a multiple objective decision. J. Wildl. Manage. 77 (6), 1145–1156.

Converse, S.J., Servanty, S., Moore, C.T., Runge, M.C., 2018. Population dynamics of reintroduced whooping cranes (Chapter 7). In: French, Jr., J.B., Converse, S.J., Austin, J.E. (Eds.), Whooping Cranes: Biology and Conservation. Biodiversity of the World: Conservation from Genes to Landscapes. Academic Press, San Diego, CA.

Dellinger, T.A., Folk, M.J., Spalding, M.G., 2013. Copulatory behavior of non-migratory whooping cranes in Florida. Wilson J. Ornithol. 125 (1), 128–133.

Drewien, R.C., Bizeau, E.G., 1978. Cross-fostering whooping cranes to sandhill crane foster parents. In: Temple, S.A. (Ed.), Endangered Birds: Management Techniques for Preserving Threatened Species. University of Wisconsin Press, Madison, pp. 201–222.

Florida Fish and Wildlife Conservation Commission, 2008. FWC to Stop Releasing Non-Migratory Whooping Cranes. Available from: http://myfwc.com/media/697891/PressRelease_Nov2008_WhoopingCranes.pdf.

Folk, M.J., Dellinger, T.A., Leone, E.H., 2013. Is male-biased collision mortality of whooping cranes (*Grus americana*) in Florida associated with flock behavior? Waterbirds 36 (2), 214–219.

Folk, M.J., Nesbitt, S.A., Parker, J.M., Spalding, M.G., Baynes, S.B., Candelora, K.L., 2008. Current status of nonmigratory whooping cranes in Florida. Proceedings of the North American Crane Workshop 10, 7–12.

Folk, M.J., Nesbitt, S.A., Schwikert, S.T., Schmidt, J.A., Sullivan, K.A., Miller, T.J., Barnes, S.B., Parker, J.M., 2005. Breeding biology of reintroduced non-migratory Whooping Cranes in Florida. Proceedings of the North American Crane Workshop 9, 105–109.

Folk, M.J., Rodgers, J.A., Dellinger, T.A., Nesbitt, S.A., Parker, J.M., Spalding, M.G., Baynes, S.B., Chappell, M.K., Schwikert, S.T., 2010. Status of non-migratory whooping cranes in Florida. Proceedings of the North American Crane Workshop 11, 118–123.

Gee, G.F., Nicholich, J.M., Nesbitt, S.A., Hatfield, J.P., Ellis, D.H., Olsen, G.H., 2001. Water conditioning and whooping crane survival after release in Florida. Proceedings of the North American Crane Workshop 8, 160–165.

Kuyt, E., 1981. Population status, nest site fidelity, and breeding habitat of whooping cranes. In: Lewis, J.C., Masatomi, H. (Eds.), Crane Research around the World. International Crane Foundation, Baraboo, WI, pp. 119–125.

Kuyt, E., 1993. Whooping crane, *Grus americana*, home range and breeding range expansion in Wood Buffalo National Park, 1970–1991. Can. Field Nat. 107 (1), 1–12.

Kuyt, E., Goossen, J.P., 1987. Survival, age composition, sex ratio, and age at first breeding of whooping cranes in Wood Buffalo National Park, Canada. In: Lewis, J.C. (Ed.), Proceedings of the 1985 Crane Workshop. Platte River Whooping Crane Habitat Maintenance Trust, Grand Island, NE, pp. 203–244.

Lorentsen, S.H., Amundsen, T., Anthonisen, K., Lifjeld, J.T., 2000. Molecular evidence for extrapair paternity and female–female pairs in Antarctic petrels. Auk 117 (4), 1042–1047.

Miller, J.L., Spalding, M.G., Folk, M.J., 2010. Leg problems and power line interactions in the Florida resident flock of whooping cranes. Proceedings of the North American Crane Workshop 11, 156–165.

Moore, C.T., Converse, S.J., Folk, M.J., Boughton, R., Brooks, W.B., French, J.B., O'Meara, T.E., Putnam, M., Rodgers, J.A., Spalding, M.G., 2008. Releases of whooping cranes to the Florida nonmigratory flock: a structured decision-making approach. Fish and Wildlife Research Institute, Florida Fish and Wildlife Conservation Commission, St. Petersburg, FL, In-house Report IHR2008-009.

Nesbitt, S.A., 1982. The past, present, and future of the whooping crane in Florida. In: Lewis, J.C. (Ed.), Proceedings of the 1981 International Crane Workshop. National Audubon Society, Tavernier, FL, pp. 151–154.

Nesbitt, S.A., Carpenter, J.W., 1993. Survival and movements of greater sandhill cranes experimentally released in Florida. J. Wildl. Manage. 57 (4), 673–679.

Nesbitt, S.A., Folk, M.J., Spalding, M.G., Schmidt, J.A., Schwikert, S.T., Nicolich, J.M., Wellington, M., Lewis, J.C., Logan, T.H., 1997. An experimental release of whooping cranes in Florida: the first three years. Proceedings of the North American Crane Workshop 7, 79–85.

Nesbitt, S.A., Folk, M.J., Sullivan, K.A., Schwikert, S.T., Spalding, M.G., 2001. An update of the Florida whooping crane release project through June 2000. Proceedings of the North American Crane Workshop 8, 62–72.

Nesbitt, S.A., Hatchitt, J.L., 2008. Trends in habitat and population of Florida sandhill cranes. Proceedings of the North American Crane Workshop 10, 40–42.

Nesbitt, S.A., Tacha, T.C., 1997. "Monogamy and productivity in Sandhill Cranes." Proceedings of the North American Crane Workshop 7, 10–13.

Nesbitt, S.A., Wenner, A.S., 1987. Pair formation and mate fidelity in sandhill cranes. In: Lewis, J.C. (Ed.), Proceedings of the 1985 Crane Workshop. U.S. Fish and Wildlife Service, Grand Island, NE, pp. 117–122.

Nesbitt, S.A., Williams, K.S., 1990. Home range and habitat use of Florida sandhill cranes. J. Wildl. Manage 54 (1), 92–96.

Nisbet, I.C., Hatch, J.J., 1999. Consequences of a female-biased sex-ratio in a socially monogamous bird: female-female pairs in the roseate tern *Sterna dougallii*. Ibis 141 (2), 307–320.

Piper, W.H., Tischler, K.B., Dolsen, A., 2001. Mother-son pair formation in common loons. Wilson Bull. 113 (4), 438–441.

Proctor, N.S., Lynch, P.J., 1993. Manual of Ornithology: Avian Structure and Function. Yale University Press, New Haven, CT, p. 340.

Smith, S.K., Rayer, S., 2014. Projections of Florida Population by County, 2015–2040, with Estimates for 2013. University of Florida Bureau of Economic and Business Research vol. 47, Bulletin 168, Gainesville, pp. 1–8.

Spalding, M.G., Folk, M.J., Nesbitt, S.A., Folk, M.L., Kiltie, R., 2009. Environmental correlates of reproductive success for introduced resident whooping cranes in Florida. Waterbirds 32 (4), 538–547.

Spalding, M.G., Folk, M.J., Nesbitt, S.A., Kiltie R., 2010. Reproductive health and performance of the Florida flock of introduced whooping cranes. Proceedings of the North American Crane Workshop 11, 142–155.

Spalding, M.G., Nesbitt, S.A., Folk, M.J., McDowell, L.R., Sepulveda, M.S., 1997. Metal consumption by whooping cranes and possible zinc toxicosis. Proceedings of the North American Crane Workshop 7, 237–242.

Stehn, T.V., Wassenich, T., 2008. Whooping crane collisions with power lines: an issue paper. Proceedings of the North American Crane Workshop 10, 25–36.

Urbanek, R.P., Fondow, L.E.A., Zimorski, S.E., 2010. Survival, reproduction, and movements of migratory whooping cranes during the first seven years of reintroduction. Proceedings of the North American Crane Workshop 11, 124–132.

Urbanek, R.P., Lewis, J.C., 2015. Whooping crane (*Grus americana*). In: Rodewald, P.G. (Ed.), The Birds of North America Online. Cornell Lab of Ornithology, Ithaca, NY. Retrieved from the Birds of North America Online: http://bna.birds.cornell.edu/bna/species/153 [15 March 2016].

Young, L.C., Zaun, B.J., VanderWerf, E.A., 2008. Successful same-sex pairing in Laysan Albatross. Biol. Lett. 4 (4), 323–325.

BEHAVIOR AND SOCIAL STRUCTURE

Whooping Crane pair engaged in breeding behavioral display

Pairing Dynamics of Reintroduced Migratory Whooping Cranes

Richard P. Urbanek*, Eva K. Szyszkoski**,†,
Sara E. Zimorski**,†, Lara E.A. Fondow**,‡

*U.S. Fish and Wildlife Service, Necedah National Wildlife Refuge, Necedah, WI, United States
**International Crane Foundation, Baraboo, WI, United States
†Louisiana Department of Wildlife and Fisheries, Gueydan, LA, United States
‡Natural Resources Conservation Service, Rexburg, ID, United States

INTRODUCTION

Pairing behavior is a critical component of population social structure and life history of Whooping Cranes (*Grus americana*). This pairing behavior influences survival, reproduction, and resulting viability of crane populations. Initial steps in reproduction are social interactions leading to pairing and ultimately the formation of breeding pairs. Here we review pairing behavior in cranes (family Gruidae) and summarize a complete record of timing, location, and stages in the pairing process during the first 12 years of reintroduction of the Eastern Migratory Population (EMP) of Whooping Cranes.

Whooping Cranes are long-lived and may survive 25 or more years in the wild (Urbanek and Lewis, 2015). They are perennially monogamous, and pairs typically begin egg production at age 3 or 4 years. They hatch a maximum of one clutch of two eggs per year and seldom

fledge more than one young. Juveniles in migratory populations learn the migration route to a wintering area by following their parents, which care for them for 10–11 months. Yearlings newly separated from their parents join other Whooping Cranes and soon begin the multistep process of pair formation.

Most studies of pairing behavior in cranes have focused on the North American Sandhill Crane (*Grus canadensis*). Nonmigratory Florida Sandhill Cranes (*G. canadensis pratensis*) were studied in north-central Florida (Nesbitt, 1989, 1992; Nesbitt and Tacha, 1987; Nesbitt and Wenner, 1987; Nesbitt et al., 2001). Breeding Greater Sandhill Cranes (*G. canadensis tabida*) were studied near Briggsville, Marquette County, Wisconsin (Hayes, 2005; Hayes et al., 2006; Hayes and Barzen, 2015), where nesting density of cranes exceeded 5 pairs/km² (Barzen et al., 2016; Lacy and Su, 2008). That value is much higher than earlier reported densities of Greater Sandhill Cranes across their range (0.1–2.0 pairs/km²; Hoffman, 1983; Urbanek and Bookhout, 1992b), and higher density might contribute to differences in some pairing behaviors. Studies of pairing and social structure in Sandhill Crane populations have been based on samples, and as such they were limited by sampling methods, as well as by occurrence of many birds that were unmarked and of unknown age and history.

Sandhill Cranes have been noted to form several temporary subadult pairs before forming a breeding pair (Nesbitt and Wenner, 1987). Adult cranes generally form long-term pair bonds during their long lives (Walkinshaw, 1949); however, dissolution of pairs (not due to death of a member) has been found to be common in Florida nonmigratory Sandhill Cranes (Nesbitt, 1989; Nesbitt et al., 2001) and migratory Sandhill Cranes in the area of high density near Briggsville, Wisconsin (Hayes and Barzen, 2015). Pair fidelity in other subpopulations of Sandhill Cranes remains largely unquantified.

Quantification of pairing dynamics at the population level requires study throughout the entire range and annual cycle. Study of pairing in the remnant Aransas-Wood Buffalo Population (AWBP) has been limited because the breeding grounds are not accessible and because the sex, age, and individual identity of many birds are often unknown, even on the wintering grounds where they can be observed. However, the ages of some pairs producing eggs have been determined (Johns et al., 2005; Kuyt and Goossen, 1987). Available data collected from wintering areas suggest that some pairing may occur there and some birds may pair from within groups (Bishop, 1984; Bishop and Blankinship, 1981; Stehn, 1992, 1997).

Other Whooping Crane populations with pairing information resulted from attempted reintroductions (French et al., Chapter 1, this volume). The first reintroduction, at Grays Lake National Wildlife Refuge (NWR; Drewien and Bizeau, 1977), Idaho, failed to produce any conspecific breeding pairs (Ellis et al., 1992). Failure was likely due to improper imprinting resulting from cross-fostering of Whooping Cranes by Sandhill Cranes and exacerbated by poor survival and dispersal of females.

The second reintroduction, of the Florida Nonmigratory Population (FNMP) in central Florida, used costume-reared and parent-reared juveniles produced in captivity. These birds were properly sexually imprinted, formed breeding pairs, and produced eggs. However, overall survival and reproduction were low and not capable of supporting a self-sustaining population (Folk et al., 2005; Moore et al., 2008, 2012; Spalding et al., 2010).

In the EMP, formed in the third reintroduction attempt, rate of pairing was high, forming breeding pairs that were stable and continued nesting in subsequent years, and survival was as high as or higher than in the AWBP (Servanty et al., 2014). A relatively complete data set, obtained through intensive monitoring throughout the range of the EMP, has made detailed study of pairing in this population uniquely possible. Most other studies of pairing in cranes have been at a more general level directed only toward breeding pairs, from small samples, or based on anecdotal observations. Among studies, different measurements and definitions of terms make direct comparison of results difficult. The study presented in this chapter provides a complete summary and analysis of the pairing that occurred in the EMP during the period 2001–August 2013.

In addition, the period of social changes that occurs between the separation of juveniles from their parents or release groups through successive stages toward formation of breeding pairs has not been the focus of previous studies. In the EMP reintroduction, young birds were found to form various groupings, and some joined Sandhill Cranes. In addition to groups of three or more (which are beyond the scope of this chapter), many of these younger birds formed pairs, which are included in analysis in this chapter. Early pairings may be any combination of sexes (male–male, female–female, or male–female). These pairs form a distinct category and are important in facilitating predator avoidance and foraging, even though they are not reproductive. Early male–female pairings may but usually did not result in breeding pairs.

Objectives and Approach

A successful reintroduction of an endangered species must meet three requirements: (1) a reintroduction area with adequate suitable habitat; (2) effective rearing, release, and management methods; and (3) animals that are capable of survival and reproduction in the wild. Reproduction must compensate for mortality, and the prerequisite for reproduction in monogamous species is formation of breeding pairs.

We examine pairing by Whooping Cranes released in the core reintroduction area in central Wisconsin from HY (hatch-year) 2001 to HY2010 releases and their wild-hatched progeny through 31 August 2013 (Table 10.1). Our objectives were to (1) categorize the different types of pairs contributing to the social structure of the EMP; (2) document the formation and longevity of the different types of pairs as related to age, group status, time of year, and geographic location; (3) identify factors inhibiting pair formation or causing pair dissolution; and (4) evaluate the adequacy of breeding pair formation to potential establishment of the population.

Our study approach was to follow each released individual in the population and document its progression through pairing processes that could lead ultimately to successful reproduction. Summary data for birds released by two different methods, ultralight aircraft–led (UL) and direct autumn release (DAR) (see "Methods"), are included separately, but analyses include only combined data because of the much smaller number of observations available for DAR.

TABLE 10.1 Numbers of Whooping Cranes Alive on 31 August 2013/Number Released or Produced per Hatch Year in the EMP Reintroduced in Central Wisconsin. Rearing and Release Methods Include Ultralight-Led (UL) and Direct Autumn Release (DAR). Remaining Birds (Wild-Hatched) Were Hatched and Fledged in the Wild by Captive-Reared and Released Parents.

	Hatch year											
	2001	2002	2003	2004	2005	2006	2007	2008	2009	2010	2012	Total
UL												
Males	1/4[a]	4/6	3/11	5/10	3/11	0/1	2/9[a]	4/10	8/11	2/4	–	32/77
Females	0/3	1/10	3/5	1/3	2/8	–	4/7	1/4	8/9	2/6	–	22/55
Total	1/7	5/16	6/16	6/13	5/19	0/1	6/16	5/14	16/20	4/10	–	54/132
DAR												
Males				0/1[b]	0/1	1/3	1/3	0/3[b]	2/2	3/7	–	7/20
Females				–	1/3	0/1	1/7	1/4	5/7	2/4	–	10/26
Total				0/1	1/4	1/4	2/10	1/7	7/9	5/11	–	17/46
Wild-hatched												
Total					–	1/1	–	–	–	2/2	2/2	5/5
Grand total	1/7	5/16	6/16	6/14	6/23	2/6	8/26	6/21	23/29	11/23	2/2	76/183

[a] One 2-year-old and one 10-year-old male were transferred to permanent captivity after inadequate avoidance of humans.
[b] One male with dysfunctional flight feathers in 2004 and one male with aggression problems in 2008. These two individuals were originally reared in UL cohorts but were unsuitable for inclusion in the guided migration. They were therefore released that autumn on Necedah NWR. Neither survived to 1 year of age.

METHODS

Whooping Crane chicks hatched from eggs produced by captive propagation centers or salvaged from abandoned nests and were costume-reared in isolation from human sight and sound (Horwich, 1989; Urbanek and Bookhout, 1992a) according to either UL or DAR protocols (Duff et al., 2001; Urbanek et al., 2010a). UL juveniles were led by aircraft (Duff et al., 2001; Lishman et al., 1997) to Florida winter areas (Urbanek et al., 2010b), and DAR juveniles were released in autumn on Necedah NWR to migrate with older Whooping Cranes or Sandhill Cranes (Urbanek et al., 2014b).

Definitions

There is no standard terminology for study of pairing and related behaviors in cranes. Definitions of terms used in this study are as follows:

Pair: A pair consists of any two birds that consistently associated, usually exclusively, for 20 or more days. The temporal aspect was required to exclude relatively insignificant associations of very short duration. We define four types of pairs. Two of these are nonbreeding pairs: (1) same sex nonbreeding and (2) opposite sex nonbreeding. The remaining two pair types are breeding pairs. A breeding pair is any male and female for which copulating, nest building, or holding a territory for at least one annual cycle was documented and includes two breeding pair types: (3) breeding pairs that did not produce eggs and (4) breeding pairs that produced eggs. The nonbreeding pairs were usually young birds 1–3 years old, and the breeding pairs were usually older (3 or more years old), but there was variability and overlap in the age distributions across pair types. For the purposes of our analysis, a formerly but not currently paired bird was classified as belonging to the highest-ranked pair type (1–4 above) achieved while paired, for example, before loss of its mate. This categorization was not year-dependent.

Pair member: A pair member refers to an individual within a pair bond. This distinction was sometimes necessary in our analysis because a pair could include members from different categories (e.g., UL and DAR). Pairs were the observational unit in analyses only when specifically identified as such.

Dispersal: Dispersal refers to movement to a summer location outside the core reintroduction area (refer to description in "Study Areas"). Within a given year, for a bird that died before summer, the last occupied spring location was used in analysis.

Study Areas

The core reintroduction area (Fig. 10.1) was the candidate site in central Wisconsin selected for the reintroduction (Cannon, 1999) plus surrounding area containing crane habitat and suitable for local movements (Urbanek et al., 2005). The latter included the area within 50 km of the centroid of UL training and DAR rearing sites (44°04′N, 90°10′W) on Necedah NWR (Fig. 10.1) in Juneau and adjacent counties (Urbanek et al., 2010b, 2014b).

West of the Wisconsin River, the area included abundant wetlands in watersheds of the Yellow and Lemonweir Rivers in the former Great Central Wisconsin Swamp (U.S. Fish and Wildlife Service, 2004). This area comprised the largest complex of shallow, emergent wetlands in Wisconsin. Wetlands were much more limited east of the Wisconsin River and occurred mainly in the Roche-A-Cri Creek watershed. All UL training sites (2001–10) and DAR rearing and release sites (2005–10) were on Necedah NWR. UL juveniles trained to follow UL were led on their first autumn migration to a salt marsh release site on Chassahowitzka NWR, Citrus County (28°44′N, 82°39′W), on the central Gulf coast of Florida, during each year. Some juveniles hatched in 2008–10 were also led by aircraft to a second winter release site on St. Marks NWR (30°06′N, 84°17′W), Wakulla County, in the eastern Florida panhandle. Starting in 2011, the release program

FIGURE 10.1   (A) Distribution of primary summering areas of Whooping Cranes released or wild-fledged in 2001–12 in the core reintroduction area of central Wisconsin, Eastern Migratory Population, 2002–13. (B) Core reintroduction area of the Eastern Migratory Population (EMP) of Whooping Cranes in indicated counties of central Wisconsin, 2001–13.

was moved to eastern Wisconsin. While those birds are not subjects of this study, some paired with birds from central Wisconsin and therefore contributed to pairs including the latter birds (i.e., number of subject individuals in pairings is less than, not equal to, twice the number of pairs). The reintroduced Whooping Cranes migrated or wintered, for the most part, along a relatively direct route between Wisconsin and Florida (Mueller et al., Chapter 11, this volume; Urbanek et al., 2014a, 2014b).

Monitoring

Cranes were monitored on breeding and wintering areas and on the migration route. Monitoring effort was not quantified but was year-long, and daily monitoring was typically done for most birds in the core reintroduction area during the breeding season and at other sites of concentration, especially in the early years of the project. Birds at distant locations or locations with little use by the population were monitored less frequently, but there were usually no opportunities for new pairings at these sites. Monitoring in winter and migration areas was done at least once a week for most birds. All birds were individually marked with color-coded leg bands and carried VHF transmitters. Some also carried platform transmitter terminals (PTTs; Urbanek et al., 2010b). Most tracking was from vehicles with a through-the-roof directional yagi antenna system. Tracking from fixed-wing aircraft supplemented ground tracking, particularly during migration and breeding seasons. PTT data were used to locate birds for VHF tracking or to monitor birds at distant locations not readily accessible by ground crew (Urbanek et al., 2010b). VHF tracking facilitated visual observation, and because of accessibility provided by the extensive road network in the eastern United States, these observations could typically be made from a vehicle 200–400 m from the birds without disturbing them. The resulting data set provided a nearly complete record

of pairing in this reintroduced population during the study period. Results presented include numbers of birds in the various pair types, as well as temporal and geographical attributes of pair members. These generally represent complete counts from the extant population at any point in time.

Although mutual association between birds made identification of pairs straightforward for most birds during most of their lifetimes, determination of dates of pair formation and dissolution was complicated because of noncontinuous monitoring and limits of visual observation of behavior. Therefore, we used the following rules to identify pairs and their temporal attributes. When daily observations were not available to provide exact date of pair formation or dissolution, the mean date within the range of possible dates (usually only a few days) was used. Pair formation date was the first date that birds were observed as a discrete pair; birds for which pairing could not be confirmed because they were in a group (i.e., association of more than two birds in proximity and moving as a unit) were not identified as a pair until they separated from the group. Date of dissolution was the first date that individuals were observed apart and not reassociating within 3 days. Birds that joined a group as a pair and left the group later as a pair were considered to be a pair during the interim time within the group. Pair members temporarily apart for less than 3 days were considered paired during the separation. For long-term pair bonds spanning two or more breeding seasons, members were also considered as paired during any temporary separation period of less than 2 months if no pairing with other birds occurred during the interim. For birds that died overnight, the date of the following morning was used as the date of dissolution of the pair. For a pair that joined a larger group and for which behavior was not consistently observable within the group, if the birds did not leave as a pair, the date of entry into the group was used as date of dissolution of the pair. We increased accuracy of pair formation or dissolution dates by accounting for travel time for birds recorded at different locations on the same date; for example, birds more than 1,500 km apart must have separated at least 3 days earlier. Pairs that moved in and out of larger groups were considered pairs if they spent at least one half of their time outside of a group. For pairs with members that died but whose carcasses were not promptly recovered, date of pair dissolution was the date following the last day that the pair was observed together alive.

A chi-square test (PROC FREQ; SAS Institute, 2013) was used to evaluate difference in proportion of males and females that never paired. The generalized linear mixed model procedure in SAS (PROC GLIMMIX; SAS Institute, 2013) was used to assess differences between males and females with respect to Y (dispersal, membership in nonbreeding pairs of same sex), and differences between nonbreeding and breeding pairs with respect to Y (pair formation within groups), accommodating the repeated assessments on individual birds among years. We considered Y to have a binary distribution with a logit link function and used the default assumptions for the variance-covariance structure in GLIMMIX. We used the nonparametric procedure in SAS (PROC NPARWAY1; SAS Institute, 2013) when a Shapiro–Wilk test indicated nonnormal data distribution to conduct a Wilcoxon rank-sum two-sample test with continuity correction to evaluate differences between males and females in age of pairs of same sex, age of formation of breeding pairs, and mean re-pairing time. Confidence intervals (95%), when not supplied by the aforementioned procedures, were calculated by using the binom.confint function (exact method) in R (R Core Team, 2014).

RESULTS

Pairs formed by Whooping Cranes in the EMP consisted of four types (see "Definitions"): members of the same sex, opposite sexes but no observed breeding behavior, breeding (i.e., exhibiting behaviors of a breeding pair) but no egg production, and breeding with eggs produced (Table 10.2). The nonbreeding categories contained mainly subadults in an early social stage preceding formation of adult breeding pairs.

TABLE 10.2   Number of Pair Members in the Whooping Crane EMP Reintroduced in Central Wisconsin, 2002–13. A Member Is One of the Individuals in a Pair. Rearing and Release Methods Include UL and DAR. Also Shown (in Parentheses) Is Number of Pair Members Occurring in Pairs That Formed within a Group (More Than Two Associating Individuals) Rather Than from a New Encounter.

Sex	Method	None[a]	Nonbreeding pairs		Breeding pairs	
			Same sex[b]	Male–female	No eggs[c]	Egg-producing
Males	UL	15	53 (19)	60 (20)	19 (2)	39 (10)
	DAR	8	12 (3)	8 (3)	2 (0)	3 (1)
	Wild-hatched	0	1 (0)	0	0	0
	Total	23	66 (22)	68 (23)	21 (2)	42 (11)
Females	UL	7	14 (2)	47 (18)	13 (2)	31 (8)
	DAR	12	3 (2)	15 (4)	5 (0)	10 (3)
	Wild-hatched	0	0	8 (1)	2 (0)	1 (0)
	Total	19	17 (4)	70 (23)	20 (2)	42 (11)

[a] Number of released birds that never paired.
[b] Members of each pair were either both males or both females.
[c] Pair was observed copulating, nest-building, or defending a breeding territory through at least one annual cycle, but no eggs were produced.

Sex Ratio in the Central Wisconsin Core

The population in the core reintroduction area contained more males than females throughout the study period (Fig. 10.2). Overall, including six birds released in eastern Wisconsin in 2011–12 but which moved to the central Wisconsin core, the by-summer sex ratio was 1.36 males per female (345/253 bird-summers). Two primary factors contributed to this skewed sex ratio. Number of birds released in the core (plus five wild-hatched birds fledged through 2012) consisted of 98 males and 85 females (Table 10.1). Mean dispersal, calculated as number of all bird-summers outside the core as a proportion of total number of bird-summers, was lower for males (13.4% [53/396 bird-summers], 95% CI = 8.9–19.7%) than for females (24.9% [85/334 bird-summers], 95% CI = 17.5–34.2%) ($F_{1,158}$ = 5.48, P = 0.0205). If egg-laying pairs, of which all but one pair nested in the core, are excluded, mean dispersal remains lower for males (17.3%

[51/286 bird-summers], 95% CI = 12.0–24.3%) than for females [34.2% (83/225 bird-summers), 95% CI = 25.4–44.1%]) ($F_{1,158}$ = 8.91, P = 0.0033).

Nonbreeding Individuals and Pairs

Of the 183 birds released or produced in the population through 2012 (Table 10.1), 42 (23.0%, 95% CI = 17.1–29.7%) never paired (Table 10.2). Proportions of birds that never paired were similar for both sexes (males 23/98 = 23.5%, 95% CI = 15.1–31.9%; females 19/85 = 22.4%, 95% CI = 13.5–31.2%) (χ^2_1 = 0.03, P = 0.8578). The primary factor contributing to lack of pairing was mortality of young birds; 88.1% (37/42, 95% CI = 74.4–96.0%) of birds that never paired died at less than 3 years of age. For females, the secondary contributing factor was dispersal from the core reintroduction area; 36.8% (7/19, 95% CI = 16.3–61.6%) of females that never paired summered outside the core reintroduction area at ages 2 or more years (Table 10.3). With few

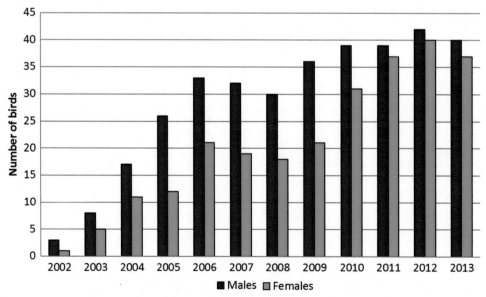

FIGURE 10.2   Numbers of male and female Whooping Cranes (released or wild-fledged in 2001–12) in the core reintroduction area in central Wisconsin, EMP, during summers 2002–13. In 2012, also included are two HY2011 females from the reintroduction in eastern Wisconsin that summered in central Wisconsin. In 2013, also included are two HY2011 male and two HY2011 females from the reintroduction in eastern Wisconsin that summered in central Wisconsin.

TABLE 10.3   Factors Contributing to Lack of Pairing by Whooping Cranes in the EMP Reintroduced in Central Wisconsin, 2001–13. More Than One Factor May Have Contributed to Lack of Pairing during the Lifetime of the Bird. Rearing and Release Methods Include UL and DAR.

Contributing factor(s)	Males			Females		
	UL	DAR	Total	UL	DAR	Total
Died <3 years	11	7	18	4	8	12
Dispersal[a]	1	–	1	1	2	3
Dispersal and died <3 years	–	–	–	2	2	4
Insufficient females[b]	1	–	1	–	–	–
Insufficient females and died <3 years	2	1	3	–	–	–
Total	15	8	23	7	12	19

[a] Dispersal outside of core reintroduction area (Fig. 10.1).
[b] Fewer females than males in core reintroduction area during breeding season.

exceptions, nonpaired females outside the core were usually associated with Sandhill Cranes (R.P. Urbanek and E.K. Szyszkoski, unpublished data). Summer dispersal pattern of these females was to different, widely separated locations that were 50–520 km distant from the center of the core reintroduction area and did not contain males. For males, the secondary factor contributing to lack of pairing, affecting 17.4% (4/23, 95% CI = 5.0–38.8%) of males, was insufficient numbers of females (i.e., fewer females than males) in the core reintroduction area (Fig. 10.1), both because of dispersal of females and because fewer females were released, especially in the earlier years of the reintroduction (Tables 10.1 and 10.3).

Transitory social associations between birds of mainly pre-reproductive age were common in the formative stages of this newly reintroduced population. Of all pair memberships recorded during the study period, 63.9% (221/346; 95% CI = 58.6–68.9%) were in nonbreeding pairs (Table 10.2). These pairs could be comprised of any combination of sexes. Males, due perhaps to their greater representation in the skewed sex ratio, had substantially more memberships in pairs of the same sex than did females (Table 10.2). Proportion of pair memberships in the same sex category was greater for males (33.2% [66/199], 95% CI = 26.5–40.7%) than for females (11.3% [17/151], 95% CI = 7.0–17.8%) ($F_{1,139}$ = 19.02, P < 0.0001). Memberships in pairs of the same sex were usually of short duration for both males and females (mean = 0.42 years, 95% CI = 0.34–0.51%, n = 83). Male members of same-sex pairs contained slightly older birds than female members of same-sex pairs at formation (mean age: males = 1.83 years, 95% CI = 1.58–2.08%; females = 1.40 years, 95% CI = 1.12–1.68%; W = 577.5, z = −1.53, P = 0.1249) and at dissolution (mean age: males = 2.25 years, 95% CI = 2.02–2.48%; females = 1.83 years, 95% CI = 1.52–2.15%; W = 544.5, z = −1.91, P = 0.0565).

Young nonbreeding male–female pairs were potentially capable of eventual reproduction; however, like birds that never paired, death at <3 years of age and dispersal were factors that precluded attainment of breeding status (Table 10.4); 48.9% (22/45; 95% CI = 33.7–64.2%) of paired, nonbreeding birds died at <3 years of age, and 23.5% (4/17; 95% CI = 3.4–43.7%) of nonbreeding females that were paired at some time dispersed outside of the core reintroduction area as single unpaired individuals. There were insufficient females to pair with 46.4% (13/28; 95% CI = 27.5–66.1%) of males in the

TABLE 10.4   Factors Contributing to Failure of Whooping Cranes That Were or Had Previously Been Members of Nonbreeding Pairs to Become Members of Breeding Pairs in the EMP Reintroduced in Central Wisconsin, 2001–13. More Than One Factor May Have Contributed to Failure to Become Members of Breeding Pairs, and Different Factors May Have Contributed through Time. Rearing and Release Methods Include UL and DAR.

Contributing factor(s)	Males			Females		
	UL	DAR	Total	UL	DAR	Total
Died after pairing but before breeding	1	–	1	–	–	–
Died <3 years	8	3	11	8	1	9
Insufficient females[a]	4	3	7	–	–	–
Insufficient females plus additional factor[b]	5	1	6	–	–	–
Dispersal[c]	–	1	1	4	–	4
Aggressive[d]; imprinted on costume	–	–	–	1	–	1
Male of pair died; then dispersal[c]	–	–	–	1	–	1
Captured and returned to captivity	1	–	1	–	–	–
Paired with younger male; died <3 years	–	–	–	1	–	1
Total	19	8	28[e]	15	1	17[e]

[a] *Fewer females than males in core reintroduction area.*
[b] *After insufficient females[a] limited earlier opportunity for pairing, additional factors after female availability increased included death after pairing but before breeding (four UL males), died <3 years (one UL male), and female of pair injured (one DAR male).*
[c] *Dispersal outside of core reintroduction area (Fig. 10.1).*
[d] *Lone female defended territory at captive chick-rearing site.*
[e] *Two yearling wild-hatched birds (one male, one female) are included only in totals.*

nonbreeding male–female pair category (i.e., this category included all nonbreeding males that were paired at any time, including those that had lost mates and were no longer paired; re-pairing of the latter males was then limited by lack of available females). More than one factor inhibited breeding by some birds (Table 10.4), for example, some males did not initially pair because females were unavailable, but then after a pairing opportunity occurred, the male died before breeding activity could occur.

Pair Formation and Group Effects

Nonbreeding pairs formed within groups (70/225 = 31.1%, 95% CI = 25.3–37.5%) more often than breeding pairs formed within groups (26/125 = 20.8%, 95% CI = 14.5–28.9%) ($F_{1,95}$ = 4.24, P = 0.0422). However, most pairs (all types) formed from new encounters

(248/346 = 71.7%, 95% CI = 66.6–76.4%) rather than from within groups (Table 10.2). Of 13 pairs in which both members were from the same release cohort, only one breeding pair formed from a continued association of two individuals within the juvenile release cohort, and that pair was separated for 2 months during their yearling spring and later classified as a breeding pair only because they met the criterion of occupying a territory. The other 12 pairs formed from reassociations long after the original release cohort had disbanded.

The first release cohort (HY2001, UL) consisted of only five individuals when they began their first spring migration from the winter release site in Florida. This group provided results for a small group of birds reintroduced into an area that initially contained no other members of their species. Although they all survived to adult age, only two of these individuals (both males) ever formed

Whooping Crane breeding pairs, low relative to birds released in later years. Of the three males, one associated primarily with Sandhill Cranes and at 4 years of age was observed on territory copulating and nest building with a female Sandhill Crane before he died later that spring. No other males released in the core reintroduction area during the course of the study paired with Sandhill Cranes. Of the two females in the first release cohort, one established a summer area in eastern Wisconsin (outside the core area) and associated primarily with Sandhill Cranes. Although she returned to the central Wisconsin core each spring as an adult, she never formed a breeding pair, even when pursued by a male at the site that she frequented. The other female socialized with other Whooping Cranes on the wintering grounds and temporarily paired with several as a subadult. As an adult, however, she established a territory at one of the chick-rearing sites on Necedah NWR and became the only female in the core area to remain unpaired as an adult. She was also the only individual in the population that remained strongly attracted to costumed humans (costumes are used during isolation rearing of captive-hatched chicks) as an adult and performed precopulatory displays to them.

Breeding Pairs

Mean initial ages of birds in male–female pair types (nonbreeding, breeding with no eggs, breeding with eggs) progressing toward reproduction were from 2.43 to 3.06 to 4.63 years, respectively, for males and from 1.74 to 2.30 to 3.68 years, respectively, for females (Table 10.5). The unequal sex ratio in the core area (Fig. 10.2) and resulting insufficient number of females available to pair with males may have resulted in females (mean = 2.30 years; 95% CI = 2.08–2.53%) becoming members of breeding pairs earlier than males (mean = 3.06 years; 95% CI = 2.73–3.40%; $W = 2,794.5$, $z = 3.7716$, $P = 0.0002$).

Causes of dissolution of breeding pairs ($n = 35$) were similar to those factors responsible for lack of pairing by some individuals (Table 10.3) and dissolution of nonbreeding pairs (Table 10.4). The primary cause of breeding pair dissolution was mortality (death of female, $n = 15$; death of male, $n = 9$). A significant, often related cause was the female of a pair being stolen by another male ($n = 8$), typically just after the original mate of the widowed male had died. This process consisted of the widowed male harassing a neighboring

TABLE 10.5 Mean Age (Years) of Bird at Formation of First Male–Female Pair, First Breeding Pair, and First Egg Production for Whooping Cranes in the EMP Reintroduced in Central Wisconsin, 2001–13. Values Include Only Birds That Formed Breeding Pairs. Rearing and Release Methods Include UL and DAR.

Sex	Method	First male–female pair[a]		First breeding pair			First egg production[b]		
		Mean	SE	Mean	SE	n	Mean	SE	n
Male	UL	2.52	0.16	3.10	0.18	43	4.63	0.29	32
	DAR	1.40	0.35	2.61	0.50	4	4.67	1.20	3
	All	2.43	0.16	3.06	0.17	47	4.63	0.28	35
Female	UL	1.84	0.13	2.29	0.14	33	3.71	0.15	28
	DAR	1.61	0.25	2.37	0.24	13	3.67	0.24	9
	Wild-hatched	1.30	0.32	2.16	0.36	3	3.00	–	1
	All	1.74	0.11	2.30	0.11	49	3.68	0.13	38

[a] The same individuals in the breeding pair column; that is, this includes their membership in earlier male–female pairs that never became breeding pairs. It does not include individuals that never became members of breeding pairs.
[b] Age of egg production is approximate (year of first egg production – hatch year of parent).

TABLE 10.6   Mean Re-Pairing Time (Days) for Members of Breeding Pairs[a] after Pair Dissolution for Whooping Cranes in the EMP Reintroduced in Central Wisconsin, 2001–13

	Males			Females		
Cause of dissolution	No. re-pairing	Time to re-pairing	No. not re-pairing	No. re-pairing	Time to re-pairing	No. not re-pairing
Death of mate	15	236[b]	2	8	47.5[c]	2
Female stolen by another male	8	102	1	8	0.6	0
Male injured	1	225	0	1	0	0
Pair member switch	2	0	1	2	0	1
Total	36		4	19		3

[a] Breeding pairs = breeding pairs that did not produce eggs plus egg-producing pairs.
[b] Range for males = 6.5–1,514 days.
[c] Range for females = 2.5–68 days.

pair until the pair separated and the female joined the widowed male. In spring 2013, a chain reaction of mate stealing occurred involving three pairs. Pair member switching resulted in one new pair forming from members of two former pairs ($n = 2$ pairs dissolved). The only instance of a female stealing a male was of a 4-year-old female translocated (Zimorski and Urbanek, 2010) from an area in New York, far outside of the reintroduction flyway, to the core, where she replaced a yearling female paired with the 3-year-old male on the following day. The remaining breeding pair dissolution ($n = 1$) involved a female pairing with a different male after her mate was injured.

Overall, females were present in smaller numbers in the core and quickly paired with surplus or widowed males. Therefore, although highly variable (Table 10.6), mean re-pairing time after loss of a mate was generally greater for males (mean = 182.3 days, 95% CI = 27.3–337.3%, $n = 22$) than females (mean = 18.1 days, 95% CI = 2.5–33.8%, $n = 16$; $W = 201.5$, $z = -3.28$, $P = 0.0010$).

Six instances of temporary interruption (17–48 days) of long-term pair bonds (i.e., pair bonds that lasted through at least two breeding seasons) were recorded. These included a female that temporarily left the core area, a female temporarily (unsuccessfully) stolen by a neighbor male whose mate had died, two temporary pairings of females with neighbor males, and a young male temporarily paired with another female while the original female joined several groups. The remaining and most unusual interruption involved a female that paired with a male from the reintroduced FNMP in Florida during the winter. Her original (EMP) mate migrated alone back to Necedah NWR and began pairing there with another female. The EMP female that had remained behind in Florida later migrated to the core with the male from the FNMP. This latter association quickly dissolved, and the female rejoined her original long-term mate on Necedah NWR. The male from the FNMP then left Wisconsin and returned to Florida.

No clear effect of nest success (i.e., hatching eggs) on longevity of breeding pairs was evident. Of 42 egg-producing pairs that formed in the population during 2005–13, 16 ended because of mortality of a member, and 31.3% ($n = 5$, 95% CI = 11.0–58.7%) of these had previously hatched eggs. Of four pairs that dissolved due to some reason other than mortality, 25% ($n = 1$, 95% CI = 0.6–80.6%) had previously hatched eggs.

Of the remaining 22 egg-producing pairs that did not separate prior to the end of our study, 40.9% ($n = 9$, 95% CI = 20.7–63.6%) had previously hatched eggs.

Geographic Location and Seasonal Pattern of Pair Formation and Dissolution

A minimum of 87.1% (54/62; 95% CI = 76.1–94.3%) of all breeding pairs formed in the core reintroduction area in central Wisconsin (Table 10.7) rather than outside the core, demonstrating the importance of this area to pair formation. Most breeding pairs formed at or just after return from spring migration with peak in April, during which 43.5% (27/62; 95% CI = 31.0–56.7%; Fig. 10.3A) of pairs formed. The pattern for nonbreeding pairs was similar, though pairing frequency was more erratic and prolonged, extending later into the spring (Fig. 10.3A).

Dissolutions of breeding pairs were most often the result of mortality, and no strong seasonal relationship was evident (Fig. 10.3B). Nonbreeder dissolutions peaked at or just after return from spring migration; 18.3% of

nonbreeder pair dissolutions occurred in April (20/109; 95% CI = 11.6–26.9%; Fig. 10.3B). For nonbreeders, there was also a secondary peak in both pair formation and dissolution just after completion of autumn migration (Fig. 10.3).

DISCUSSION

Age and Frequency of Pair Formation and Dissolution

Comparison of pairing dynamics among Whooping Crane populations is difficult because of limited data for other populations or differently defined categories of pairs used in other studies (Converse et al., Chapter 7, this volume; Folk et al., 2005; Moore et al., 2008, 2012). With that caveat, the rate of formation of breeding pairs in the reintroduced EMP appeared no less than that in the remnant AWBP. Kuyt and Goossen (1987) reported age of first egg production by 14 color-banded cranes was 5.0 (range 3–7) years. Survival in the EMP was generally greater than in the reintroduced FNMP but similar to that of the remnant AWBP (Moore et al., 2008; Servanty et al., 2014).

TABLE 10.7   Geographic Locations of Pair Formation for Whooping Cranes in the EMP Reintroduced in Central Wisconsin, 2001–13

Location[a]	Nonbreeding		Breeding		
	Same sex	Male–female	No eggs produced	Egg-producing	Total
Core[b]	28	51	19	35	133
Core or other north[c]	1			1	2
Other north[c]	2	2		2	6
Core or migration	2			2	4
Migration	3	6	1		10
Winter	7	11		1	19
Unknown				1	1
Total	43	70	20	42	175

[a] More than one location indicates that exact location of pair formation was unknown but formation occurred in one of the indicated locations.
[b] Central Wisconsin reintroduction area (Fig. 10.1).
[c] Other north location refers to areas used between completion of spring migration and beginning of the following fall migration.

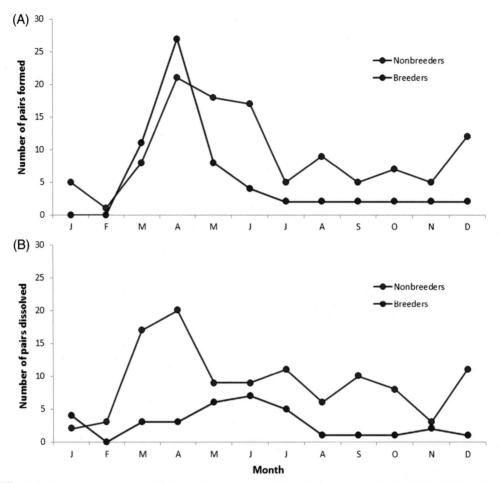

FIGURE 10.3 Seasonal occurrence of (A) pair formations and (B) pair dissolutions in the EMP of Whooping Cranes reintroduced in central Wisconsin, 2002–13.

However, mortality of young birds was the most common cause of failure to pair and of dissolution of pairs in the EMP.

Pair formation in the EMP exceeded that of the attempted FNMP reintroduction, mainly because of high mortality in the latter population (Spalding et al., 2010). However, because methods of defining pairs and determining their longevity in the latter population were different or not completely described, comparisons are limited. Spalding et al. (2010) found that nonmigratory Whooping Cranes in Florida appeared to pair later than in the AWBP or EMP, similar to Sandhill Cranes in nonmigratory and migratory populations (Nesbitt, 1992). There was a high pair separation rate in the FNMP and much member switching, even during chick rearing. For Whooping Cranes that survived through their first potential breeding season (3 years of age) during 1995–2007, Spalding et al. (2010) reported that pairs lasted an average of 2.03 years ($n = 77$), including nonlaying pairs (1.48 years), egg-producing pairs (2.69 years), pairs hatching a chick (3.17 years), and pairs fledging chicks (2.60 years). Small population size and widespread distribution limited encounters with potential mates, further limiting pairing opportunities. All Whooping Crane pairs in Florida that fledged chicks did so on their first attempt, an indication that reproductive capability may be mainly innate rather than learned (Spalding et al., 2010).

Cranes have a long-term monogamous mating system. However, pair fidelity has been correlated with breeding success in nonmigratory Florida Sandhill Cranes (Nesbitt, 1989; Nesbitt and Wenner, 1987) but not necessarily for pairs together for more than 3 years (Nesbitt and Tacha, 1987). Preliminary assessments suggest no apparent effect of reproductive success on pair fidelity, as measured by hatching of eggs, in Whooping Cranes in the EMP. However, the effect, as measured by production of fledged young, could not be evaluated because of population-wide reproductive problems resulting in insufficient successful pairs for evaluation (Converse et al., Chapter 8, this volume). Hayes (2005) also found no relationship between pair fidelity and breeding success in a dense population of Sandhill Cranes in Wisconsin. Pair fidelity in that population was most related to limited territory availability (Hayes and Barzen, 2015). The main cause of pair dissolution in the EMP was mortality of a pair member, and some pairs were disrupted because of mortality of the female of a neighboring pair followed by subsequent mate stealing by the male. Occurrence of mate stealing behavior has also been noted in Sandhill Cranes (Hayes, 2005).

Location and Season of Pairing

Most pair formation in the EMP, especially of breeding pairs, occurred in the core reintroduction area just after arrival from spring migration. Nonbreeding pairs followed this pattern but with a more extended period of pairing related to the later return from migration and spring wandering of young birds (Urbanek

et al., 2014b). No strong seasonal relationship in dissolutions was evident because dissolutions, especially of breeding pairs, were most often the result of mortality, and mortalities were evenly distributed throughout the annual cycle (Urbanek et al., 2014b).

The importance to pair formation of the concentration of Whooping Cranes in the core reintroduction area is evident in the EMP. Unlike at Aransas NWR, where members of the AWBP are concentrated during the winter and may sometimes pair at that location (Bishop, 1984; Stehn, 1997), Whooping Cranes in the EMP wintered at widely dispersed locations. This distribution reduced opportunities for pairing in winter in the EMP during the study period (Urbanek et al., 2014a). Limited data on the wintering grounds and, more importantly, almost no data on pair formation on the breeding area prevent any detailed analysis of pairing in the AWBP.

Importance of a Core Area to Pairing in a Reintroduced Population

Central Wisconsin was selected as the reintroduction site because it contained the largest complex of potentially suitable wetland habitat in Wisconsin (Cannon, 1999). Whooping Crane recovery goals of 25–30 productive pairs in this population (Canadian Wildlife Service and U.S. Fish and Wildlife Service, 2005) could be easily accommodated. During 2014–16 the 17,683-ha Necedah NWR alone supported 18–20 egg-producing pairs with additional nesting habitat available (Urbanek, 2015). Necedah NWR has remained the primary summer area of the EMP. During 2002–11, 42.4–80.0% of the population summered annually on Necedah NWR; the grand total on the refuge during that period was 54.7% (311/569 bird-summers; R.P. Urbanek, unpublished data).

The importance of this core area to pairing and eventual reproduction is accentuated by the fact that a minimum of 87% of pairs formed there. In the EMP the higher density of Whooping Cranes in the core was much more conducive to pairing and reproduction than were the lower densities outside the core. The latter may illustrate Allee effects (Converse et al., 2013a; Converse et al., Chapter 7, this volume; Moore et al., 2012), which may be expected in small, establishing populations (Courchamp et al., 1999; McCarthy, 1997). Inadequate numbers reduce opportunities for pairing and result in other social dysfunctions that could undermine growth of a reintroduced population. Without establishment of a critical mass, Whooping Cranes, especially females, may associate primarily with Sandhill Cranes. In this study one male from the small initial cohort, which included only five birds, paired with a Sandhill Crane and several females dispersed into areas without Whooping Cranes and became members of Sandhill Crane flocks. Dispersal and association of female Whooping Cranes with Sandhill Cranes were factors contributing to failure of the attempted reintroduction in the Rocky Mountains (Drewien and Bizeau, 1977; Ellis et al., 1992), although a probable primary cause was imprinting on the Sandhill Crane foster parent species. A study to force-associate captured cross-fostered Whooping Cranes with captive-reared juveniles was also unsuccessful (Drewien et al., 1997). As of the end of this study of the EMP, pairing and reproduction of female Whooping Cranes with Sandhill Cranes had not been recorded, even though some female Whooping Cranes had integrated into Sandhill Crane flocks. The few documented interspecific pairings (Grays Lake, FNMP, EMP) have included only male Whooping Cranes and may have resulted from male opportunism and female mate selection. Hypothetically, only a small proportion of properly imprinted female Sandhill Cranes would potentially pair with a Whooping Crane, but because of much larger numbers, Sandhill Crane populations might always be a source of such females.

In cranes, like most species of birds other than Anseriformes, males tend to home and females tend to disperse unless they are paired with males (Greenwood, 1980). This pattern was evident in the EMP with strong philopatry demonstrated by the males (Mueller et al., Chapter 11, this volume; Urbanek et al., 2010a, 2014b). Female dispersal appeared to limit opportunity for pairing. This factor was exacerbated by the smaller number of females released. The smaller number of females in the core was not due to mortality (Servanty et al., 2014; Urbanek et al., 2014b).

Pairing within Groups

Some associations within groups may have facilitated pairing, but study of group dynamics was beyond the scope of this chapter. Nesbitt and Carpenter (1993) suggested that birds reared in the same release cohort would not pair. In the EMP this hypothesis had some support because no egg-producing pairs formed from members that continuously associated, beginning as juveniles, in the same release group. However, in general, the hypothesis was not supported because 13 pairs formed later from members of the same release group after a period of separation.

Reintroduction Techniques

Early thought based on minimal empirical data (Nesbitt, 1979) assumed that cranes reared by humans would not be suitable reintroduction candidates because of sexual imprinting on humans or inability to adapt to natural environments. These views have been largely dispelled by subsequent experimental work with Sandhill Cranes (Horwich, 1989; Nagendran et al., 1996; Urbanek and Bookhout, 1992a), development of costume rearing, which can produce large numbers of birds conditioned for life in the wild, and results of the EMP reintroduction of Whooping Cranes (Urbanek et al., 2010a, 2014a, 2014b). The costume-rearing techniques were successful in producing a population of Whooping Cranes capable of surviving, migrating, pairing, and establishing territories on which to nest.

Because of the smaller number of released DAR birds, later application of the DAR method, and free pairing among individuals in the population, direct comparison of pairing between DAR and UL birds is not possible; both techniques resulted in breeding pairs. Costume-reared birds successfully produced pairs through the social stages culminating in breeding pairs that produced eggs.

Although pairing was not a problem in this population, the reintroduction has been hampered by two different reproductive problems: (1) nest failure and (2) prefledging chick mortality. Chronic nest failure was related to attack of incubating birds by black flies (*Simulium annulus* and *johannseni*) and resulting nest desertion (Converse et al., 2013b; Converse et al., Chapter 8, this volume; Urbanek et al., 2010c, 2017). Parental care of chicks (feeding, brooding, attentiveness) by costume-reared parents appeared sufficient and typical of behaviors of wild-hatched cranes (R.P. Urbanek and E.K. Szyszkoski, unpublished data). On Necedah NWR, environmental conditions related to the interactions of water levels, vegetation, and predators appear more likely factors responsible for chick mortality (Urbanek, 2015). Other hypotheses imply deficient ability of costume-reared adults to rear chicks because the former were not reared by crane parents (Converse et al., Chapter 8, this volume). Recent releases of parent-reared juveniles from captive propagation facilities into the EMP have begun to explore this issue (Olsen and Converse, 2017). However, chicks reared in a pen by captive parents may not have any greater ability to avoid predation in the wild than do costume-reared chicks. No significant differences were found in survival or reproduction between parent-reared and costume-reared cranes in the supplemented Mississippi Sandhill Crane (*G. c. pulla*) population (Ellis et al., 2000; Hereford et al., 2014) or FNMP of Whooping Cranes (Spalding et al., 2010). In addition, costume rearing has major advantages over parent rearing

in ability to produce more birds for release and in manageability of birds after release. The latter could be especially important in a migratory reintroduction (Nagendran et al., 1996; Urbanek and Bookhout, 1992a). Because other aspects of costume rearing, including production of breeding pairs, have been successful, it is important to continue to retain this technique in a viable reintroduction strategy unless it is proven incompatible for fully successful reproduction.

SUMMARY AND OUTLOOK

We reviewed pairing dynamics of Whooping Cranes with detailed attention to the EMP, including types of pairs, when and where pairs formed, why they dissolved, and the social progression toward pairs producing eggs. Pair formation was concentrated at the end of spring migration, and at least 87% of breeding pairs formed in the core reintroduction area in central Wisconsin. We recommend that size of the initial release cohort in a reintroduction be at least 10 birds, that additional females be released to compensate for any lost to dispersal, and that a core area with habitat capable of sustaining at least one half of the minimum final population goal be part of reintroduction planning.

Costume rearing produced release candidates that successfully paired and resulted in adequate numbers of breeding pairs that would ultimately be needed for success of the reintroduction. Both UL and DAR methods were effective in producing pairs (UL slightly more so because of higher survival, but the UL method is also more expensive because of training chicks to follow aircraft, the aircraft-led migration, and continued care through the first winter). However, because of reproduction problems not related to pairing dynamics, resulting production of young by costume-reared parents has been too low to completely evaluate this technique. The two unsolved problems consist of nest desertion caused by black fly attacks on incubating birds and high prefledging chick mortality. The first and perhaps the second problem appear environmentally caused and, if so, will require environmental management to solve. Changing rearing and reintroduction techniques, of which other aspects have already been proven effective, will not solve these remaining problems and may create additional, unnecessary problems.

As a result of this study, pairing behavior in the EMP is now well documented and better understood. The significant information gap that needs to be filled to make the EMP reintroduction successful is identification of the exact causes of chick mortality. If costume rearing is found to significantly contribute to mortality, then appropriate modifications in techniques will need to be developed if the reintroduction is to proceed. However, if not, costume rearing, which remains the most effective method to produce adequate numbers of high-quality release candidates, should continue if addition of birds to the population is needed. The future of the EMP would depend on solving the environmental problems hindering reproduction.

Acknowledgments

We thank all partners of the Whooping Crane Eastern Partnership (WCEP) and the many additional organizations and individuals involved in this effort. We especially thank interns of the International Crane Foundation and U.S. Fish and Wildlife Service who tracked released birds or assisted in rearing of DAR chicks. We thank Operation Migration and Patuxent Wildlife Research Center for rearing, training, and leading migration of UL chicks; Necedah, Chassahowitzka, and St. Marks NWRs for support and facilities; Wisconsin Natural Resources Foundation for financial support; and Windway Capital Corporation, LightHawk, and Wisconsin Department of Natural Resources for aircraft support. We appreciate statistical assistance by Wesley Newton, Sarah Converse, and Jane Austin, U.S. Geological Survey. We especially thank John Christian (retired), U.S. Fish and Wildlife Service, for his commitment to recovery of the Whooping Crane and for making this reintroduction effort a reality. The findings and conclusions in this article are those of the authors and do not necessarily represent the views of the U.S. Fish and Wildlife Service.

References

Barzen, J.A., Su, L., Lacy, A.E., Gossens, A.P., Moore, D.M, 2016. High nest density of sandhill cranes in central Wisconsin. Proceedings of the North American Crane Workshop 13, 13–24.

Bishop, M.A., 1984. The Dynamics of Subadult Flocks of Whooping Cranes Wintering in Texas, 1978–1979 through 1982–1983. MS thesis, Texas A&M University, College Station.

Bishop, M.A., Blankinship, D.R., 1981. Dynamics of subadults flocks of whooping cranes at Aransas National Wildlife Refuge, Texas, 1978–1981. In: Lewis, J.C. (Ed.), Proceedings of the 1981 International Crane Workshop. National Audubon Society, Tavernier, FL, pp. 180–189.

Canadian Wildlife Service and U.S. Fish and Wildlife Service, 2005. International Recovery Plan for the Whooping Crane. Recovery of Nationally Endangered Wildlife (RENEW), Ottawa, ON, and U.S. Fish and Wildlife Service, Albuquerque, NM.

Cannon, J.R., 1999. Wisconsin whooping crane breeding site assessment. Final Report to Canadian-United States Whooping Crane Recovery Team. Front Royal, VA.

Converse, S.J., Moore, C.T., Folk, M.J., Runge, M.C., 2013a. A matter of tradeoffs: reintroduction as a multiple objective decision. J. Wildl. Manage. 77 (6), 1145–1156.

Converse, S.J., Royle, J.A., Adler, P.H., Urbanek, R.P., Barzen, J.A., 2013b. A hierarchical nest survival model integrating incomplete temporally varying covariates. Ecol. Evol. 3 (13), 4439–4447.

Converse, S.J., Servanty, S., Moore, C.T., Runge, M.C., 2018. Population dynamics of reintroduced whooping cranes (Chapter 7). In: French, Jr., J.B., Converse, S.J., Austin, J.E. (Eds.), Whooping Cranes: Biology and Conservation. Biodiversity of the World: Conservation from Genes to Landscapes. Academic Press, San Diego, CA.

Converse, S.J., Strobel, B.N., Barzen, J.A., 2018. Reproductive failure in the Eastern Migratory Population: interaction of research and management (Chapter 8). In: French, Jr., J.B., Converse, S.J., Austin, J.E. (Eds.), Whooping Cranes: Biology and Conservation. Biodiversity of the World: Conservation from Genes to Landscapes. Academic Press, San Diego, CA.

Courchamp, F., Clutton-Brock, T., Grenfell, B., 1999. Inverse density dependence and the Allee effect. Trends Ecol. Evol. 14 (1), 405–410.

Drewien, R.C., Bizeau, E.G., 1977. Cross-fostering whooping cranes to sandhill crane foster parents. In: Temple, S.A. (Ed.), Endangered Birds: Management Techniques for Preserving Threatened Species. University of Wisconsin Press, Madison, pp. 201–222.

Drewien, R.C., Munroe, W.L., Clegg, K.R., Brown, W.M., 1997. Use of cross-fostered whooping cranes as guide birds. Proceedings of the North American Crane Workshop 7, 86–95.

Duff, J.W., Lishman, W.A., Clark, D.A., Gee, G.F., Ellis, D.H., 2001. Results of the first ultralight-led sandhill crane migration in eastern North America. Proceedings of the North American Crane Workshop 8, 109–114.

Ellis, D.H., Gee, G.F., Hereford, S.G., Olsen, G.H., Chisolm, T.D., Nicolich, J.M., Sullivan, K.A., Thomas, N.J., Nagendran, M., Hatfield, J.S., 2000. Post-release survival of hand-reared and parent-reared Mississippi sandhill cranes. Condor 102 (1), 104–112.

Ellis, D.H., Lewis, J.C., Gee, G.F., Smith, D.G., 1992. Population recovery of the whooping crane with emphasis on reintroduction efforts: past and future. Proceedings of the North American Crane Workshop 6, 142–150.

Folk, M.J., Nesbitt, S.A., Schwikert, S.T., Schmidt, J.A., Sullivan, K.A., Miller, T.J., Baynes, S.B., Parker, J.M., 2005. Breeding biology of re-introduced non-migratory whooping cranes in Florida. Proceedings of the North American Crane Workshop 9, 117–125.

French, Jr., J.B., Converse, S.J., Austin, J.E., 2018. Whooping cranes past and present (Chapter 1). In: French, Jr., J.B., Converse, S.J., Austin, J.E. (Eds.), Whooping Cranes: Biology and Conservation. Biodiversity of the World: Conservation from Genes to Landscapes. Academic Press, San Diego, CA.

Greenwood, P.J., 1980. Mating systems, philopatry and dispersal in birds and mammals. Anim. Behav. 28 (4), 1140–1162.

Hayes, M.A., 2005. Divorce and Extra-Pair Paternity as Alternative Mating Strategies in Monogamous Sandhill Cranes. Thesis, University of South Dakota, Vermillion.

Hayes, M.A., Barzen, J.A., 2015. Territory availability best explains fidelity in sandhill cranes (Chapter 2). In: Hayes, M.A. (Ed.), Dispersal and Population Genetic Structure in Two Flyways of Sandhill Cranes. Dissertation, University of Wisconsin, Madison, WI, pp. 11–56.

Hayes, M.A., Britten, H.B., Barzen, J.A., 2006. Extra-pair fertilizations in sandhill cranes revealed using microsatellite DNA markers. Condor 108 (4), 970–976.

Hereford, S., Grazia, T., Billodeaux, L., 2014. Effect of rearing technique on age of first reproduction of released Mississippi sandhill cranes. Proceedings of the North American Crane Workshop 12, 92 (Abstract).

Hoffman, R.H., 1983. Changes in the wetlands selected by an increasing sandhill crane population. Jack-Pine Warbler 61 (2), 51–60.

Horwich, R.H., 1989. Use of surrogate parental models and age periods in a successful release of hand-reared sandhill cranes. Zoo Biol. 8 (4), 379–390.

Johns, B.W., Goossen, J.P., Kuyt, E., Craig-Moore, L., 2005. Philopatry and dispersal in whooping cranes. Proceedings of the North American Crane Workshop 9, 117–125.

Kuyt, E., Goossen, J.P., 1987. Survival, age composition, sex ratio, and age at first breeding of whooping cranes in Wood Buffalo Park, Canada. In: Lewis, J.C. (Ed.), Proceedings of the 1985 Crane Workshop. Platte River

Whooping Crane Habitat Maintenance Trust, Grand Island, NE, pp. 230–244.

Lacy, A.E., Su, L., 2008. Habitat characteristics influencing sandhill crane nest site selection. Proceedings of the North American Crane Workshop 10, 165 (Abstract).

Lishman, W.A., Teets, T.L., Duff, J.W., Sladen, W.J.L., Shire, G.G., Goolsby, K.M., Bezner Kerr, W.A., Urbanek, R.P., 1997. A reintroduction technique for migratory birds: leading Canada geese and isolation-reared sandhill cranes with ultralight aircraft. Proceedings of the North American Crane Workshop 7, 96–104.

McCarthy, M.A., 1997. The Allee effect, finding mates and theoretical models. Ecol. Model. 103 (1), 99–102.

Moore, C.T., Converse, S.J., Folk, M., Boughton, R., Brooks, B., French, J.B., O'Meara, T.E., Putnam, M., Rodgers, J., Spalding, M., 2008. Releases of whooping cranes to the Florida nonmigratory flock: a structured decision-making approach. Report to the International Whooping Crane Recovery Team. Florida Fish and Wildlife Research Institute Report IHR2008-009.

Moore, C.T., Converse, S.J., Folk, M.J., Runge, M.C., Nesbitt, S.A., 2012. Evaluating release alternatives for a longlived bird species under uncertainty about long-term demographic rates. J. Ornithol. 152 (Suppl. 2), S339–S353.

Mueller, T., Teitelbaum, C.S., Fagan, W.F., Converse, S.J., 2018. Movement ecology of reintroduced migratory whooping cranes (Chapter 11). In: French, Jr., J.B., Converse, S.J., Austin, J.E. (Eds.), Whooping Cranes: Biology and Conservation. Biodiversity of the World: Conservation from Genes to Landscapes. Academic Press, San Diego, CA.

Nagendran, M., Urbanek, R.P., Ellis, D.H., 1996. Special techniques, part D: reintroduction techniques. In: Ellis, D.H., Gee, G.F., Mirande, C. (Eds.), Cranes: Their Biology, Husbandry, and Conservation. National Biological Service; International Crane Foundation, Washington, DC; Baraboo, WI, pp. 231–240.

Nesbitt, S.A., 1979. Notes on suitability of captive-reared sandhill cranes for release into the wild. In: Lewis, J.C. (Ed.), Proceedings of the 1978 Crane Workshop. Colorado State University Printing Service, Fort Collins, pp. 85–88.

Nesbitt, S.A., 1989. The significance of mate loss in Florida sandhill cranes. Wilson Bull. 101 (4), 648–651.

Nesbitt, S.A., 1992. First reproductive success and individual productivity in sandhill cranes. J. Wildl. Manage. 56 (3), 573–577.

Nesbitt, S.A., Carpenter, J.W., 1993. Survival and movements of greater sandhill cranes experimentally released in Florida. J. Wildl. Manage. 57 (4), 673–679.

Nesbitt, S.A., Folk, M.J., Schwikert, S.T., Schmidt, J.A., 2001. Aspects of reproduction and pair bonds in Florida sandhill cranes. Proceedings of the North American Crane Workshop 8, 31–35.

Nesbitt, S.A., Tacha, T.C., 1987. Monogamy and productivity in sandhill cranes. Proceedings of the North American Crane Workshop 7, 10–13.

Nesbitt, S.A., Wenner, A.S., 1987. Pair formation and mate fidelity in sandhill cranes. In: Lewis, J.C. (Ed.), Proceedings of the 1985 Crane Workshop. Platte River Whooping Crane Habitat Maintenance Trust, Grand Island, NE, pp. 117–122.

Olsen, G.H., Converse, S.J., 2017. Releasing parent-reared whooping cranes in Wisconsin: a pilot study 2013–2015. Abstract of paper presented at North American Crane Workshop 14, 12–15 January 2017, Chattanooga, TN.

R Core Team, 2014. R: A Language and Environment for Statistical Computing. R Foundation for Statistical Computing, Vienna, Austria.

SAS Institute, 2013. SAS Version 9.4. SAS Institute, Cary, NC.

Servanty, S., Converse, S.J., Bailey, L.L., 2014. Demography of a reintroduced population: moving toward management models for an endangered species, the whooping crane. Ecol. Appl. 24 (5), 927–937.

Spalding, M.G., Folk, M.J., Nesbitt, S.A., Kiltie, R., 2010. Reproductive health and performance of the Florida flock of introduced whooping cranes. Proceedings of the North American Crane Workshop 11, 142–155.

Stehn, T.V., 1992. Re-pairing of whooping cranes at Aransas National Wildlife Refuge. In: Wood, D.A. (Ed.), Proceedings of the 1988 North American Crane Workshop, Florida Game and Fresh Water Fish Commission Nongame Wildlife Program Technical Report 12, pp. 185–187.

Stehn, T.V., 1997. Pair formation by color-marked whooping cranes on the wintering grounds. Proceedings of the North American Crane Workshop 7, 24–28.

U.S. Fish and Wildlife Service, 2004. Necedah National Wildlife Refuge comprehensive conservation plan and environmental assessment. U.S. Fish and Wildlife Service, Ft. Snelling, MN.

Urbanek, R.P. 2015. Research and management to increase whooping crane chick survival on Necedah National Wildlife Refuge. U.S. Fish and Wildlife Service, Necedah, WI. Available from: http://ecos.fws.gov/ServCat/DownloadFile/55027?/Reference+54492. [3 March 2018].

Urbanek, R.P., Adler, P.H., Zimorski, S.E., Gray, E.W., Szyszkoski, E.K., 2017. The importance of black fly monitoring to understanding nest desertion by whooping cranes in the Eastern Migratory Population. Abstract of paper presented at North American Crane Workshop 14, 12–15 January 2017, Chattanooga, TN.

Urbanek, R.P., Bookhout, T.A., 1992a. Development of an isolation-rearing/gentle release procedure for reintroducing migratory cranes. Proceedings of the North American Crane Workshop 6, 120–130.

Urbanek, R.P., Bookhout, T.A., 1992b. Nesting of greater sandhill cranes on Seney National Wildlife Refuge. In: Wood, D.A. (Ed.), Proceedings of the 1988 North American Crane Workshop, Florida Game and Fresh Water Fish Commission Nongame Wildlife Program Technical Report 12, pp. 161–172.

Urbanek, R.P., Duff, J.W., Swengel, S.R., Fondow, L.E.A., 2005. Reintroduction techniques: post-release performance of sandhill cranes (1) released into wild flocks and (2) led on migration by ultralight aircraft. Proceedings of the North American Crane Workshop 9, 203–211.

Urbanek, R.P., Fondow, L.E.A., Zimorski, S.E., 2010a. Survival, reproduction, and movements of migratory whooping cranes during the first seven years of reintroduction. Proceedings of the North American Crane Workshop 11, 124–132.

Urbanek, R.P., Fondow, L.E.A., Zimorski, S.E., Wellington, M.M., Nipper, M.A., 2010b. Winter release and management of reintroduced migratory whooping cranes *Grus americana*. Bird Conserv. Int. 20 (1), 43–54.

Urbanek, R.P., Lewis, J.C., 2015. Whooping crane (*Grus americana*). In: Rodewald, P.G. (Ed.), The Birds of North America,. Cornell Lab of Ornithology, Ithaca, NY, Issue 153. Available from: http://bna.birds.cornell.edu/bna/species/153. [8 September 2016].

Urbanek, R.P., Szyszkoski, E.K., Zimorski, S.E., 2014a. Winter distribution dynamics and implications to a reintroduced population of migratory whooping cranes. J. Fish Wildl. Manage. 5 (2), 340–362.

Urbanek, R.P., Zimorski, S.E., Fasoli, A.M., Szyszkoski, E.K., 2010c. Nest desertion in a reintroduced population of migratory whooping cranes. Proceedings of the North American Crane Workshop 11, 133–141.

Urbanek, R.P., Zimorski, S.E., Szyszkoski, E.K., Wellington, M.M., 2014b. Ten-year status of the eastern migratory whooping crane reintroduction. Proceedings of the North American Crane Workshop 12, 33–42.

Walkinshaw, L.H., 1949. The Sandhill Cranes. Cranbrook Institute of Science Bulletin 29, Bloomfield Hills, MI.

Zimorski, S.E., Urbanek, R.P., 2010. The role of retrieval and translocation in a reintroduced population of migratory whooping cranes. Proceedings of the North American Crane Workshop 11, 216 (Abstract).

11

Movement Ecology of Reintroduced Migratory Whooping Cranes

Claire S. Teitelbaum*, Sarah J. Converse**,†, William F. Fagan‡, Thomas Mueller*

*Senckenberg Biodiversity and Climate Research Center and Goethe University Frankfurt, Frankfurt, Germany
**U.S. Geological Survey, Patuxent Wildlife Research Center, Laurel, MD, United States
†U.S. Geological Survey, Washington Cooperative Fish and Wildlife Research Unit, School of Environmental and Forest Sciences (SEFS) & School of Aquatic and Fishery Sciences (SAFS), University of Washington, Seattle, WA, United States
‡University of Maryland, College Park, MD, United States

INTRODUCTION

Migration is common in cranes; of the 15 species of cranes worldwide, 10 are migratory or have populations that migrate (Del Hoyo and Collar, 2014; Meine and Archibald, 1996). The migration of Whooping Cranes is well known in North America, though the species has both migratory and nonmigratory populations. Whooping Cranes were probably abundant and widely distributed in North America during the Pleistocene, but numbers declined prior to European settlement and the species has remained rare ever since (Allen, 1952). During the

19th and early 20th centuries, Whooping Cranes bred on the prairies of central North America, ranging from present-day Iowa northwest into present-day Alberta, and into the boreal forest, and wintered on the coasts of the Gulf of Mexico and the Atlantic Ocean, from present-day Texas and Mexico to the Carolinas (Allen, 1952; Austin et al., Chapter 3, this volume). However, by the early 20th century, it is likely that no Whooping Cranes migrated in eastern North America, though a small resident population lived in southwestern Louisiana until 1950. The remnant migratory population that persists today is the Aransas-Wood Buffalo Population (AWBP),

Whooping Cranes: Biology and Conservation
http://dx.doi.org/10.1016/B978-0-12-803555-9.00011-6

which breeds at Wood Buffalo National Park in northeastern Alberta and southern Northwest Territories and winters at Aransas National Wildlife Refuge (NWR) on the Gulf coast of Texas. The second migratory population is the reintroduced Eastern Migratory Population (EMP). This population breeds in central Wisconsin and migrates to the southeastern United States, ranging from the Gulf coast of Florida northwest into Georgia, Alabama, Tennessee, and, more recently, Indiana. More details about both populations and reintroduction methods for the EMP are described elsewhere (French et al., Chapter 1, this volume).

For migratory animals, migration behavior is one of the main determinants of year-round fitness and, ultimately, population viability (Lack, 1968). Seasonal variability in the environment is the fundamental driver of migration across species, where animals are able to experience optimal conditions year-round by using multiple locations. Moving between seasonally varying environments may allow animals to access more resources, minimize competition, or avoid unfavorable weather conditions (Alerstam et al., 2003). Migratory Whooping Cranes track favorable seasonal environments by summering in northern North America during periods where long days, available nesting habitat, and abundant foraging resources provide greater energy availability and presumably enhance breeding capacity. Whooping Cranes winter in the southern United States, where temperatures and food resources are similarly amenable throughout that time of the year (Chavez-Ramirez and Wehtje, 2012).

Even within a single area, different species, populations, and individuals can exhibit varying degrees of residency or migration, where some individuals migrate and others do so conditionally or not at all; further, substantial variation among migrating individuals can exist even within a single population (Dingle and Drake, 2007; Lack, 1968). Whooping Cranes exhibit vast differences in their movement behavior both between populations (e.g., the AWBP migrates farther than the EMP, and populations in Florida and Louisiana are nonmigratory) and within a single population (e.g., individuals arrive on their breeding grounds at different times) (Gomez, 1991; Wright et al., 2014). As in other migratory species, migration behavior in Whooping Cranes is determined by a combination of internal and external factors, including genetics, social cues, and the environment. Different aspects of migratory behavior – for instance, timing, routes, and orientation – appear to be controlled by different combinations of processes, in particular (1) genetic inheritance, (2) individual learning, and (3) social learning (Mueller and Fagan, 2008). Better understanding of the mechanisms driving migratory behavior can guide conservation attempts, particularly by highlighting which behaviors are plastic and which are fixed, and by understanding the importance of migratory behavior to species persistence (Wilcove and Wikelski, 2008).

Genetic inheritance can be particularly important in determining the direction of migration in birds, where individuals that are more closely related have more similar migratory orientation than those that are more distantly related (Berthold, 1988; Berthold et al., 1992). In species ranging from songbirds to large social migrants, juveniles maintain a similar migratory direction on their first fall migration even in the absence of adults (Chernetsov, 2004; Pulido and Berthold, 2010), indicating that orientation depends on innate genetic mechanisms rather than social or familial cues (Åkesson and Hedenström, 2007). However, this genetic determination does not mean that migratory direction is inflexible on the population level, as rapid changes in the direction and destinations of migratory birds are possible (Berthold et al., 1992). Whooping Cranes suffered a severe bottleneck in the early 20th century, meaning that individuals are more closely related than in many other migratory species (Glenn et al., 1999; Johns et al., 2005). Thus, Whooping Cranes could potentially have relatively little genetic variation and genetic adaptability in their migratory behavior, and genetically determined migratory patterns may have been lost, compared to historic populations of Whooping Cranes or other species.

In addition to these genetic mechanisms, in long-lived species like Whooping Cranes, lifetime experiences and social cues can form an integral part of the control of migration behavior. An individual's migratory behavior may change over the course of its lifetime depending on conditions it has experienced in the past. For instance, young Honey Buzzards (*Pernis apivorus*) take different migration routes and migrate more slowly than adults, possibly because adults are able to use more complex navigation strategies to choose an optimal route (Hake et al., 2003). A similar phenomenon accounts for raptors' ability to compensate for wind drift as they migrate, though the different physiology and habitat requirements of adults compared to juveniles and subadults may combine with learning to produce these different patterns (Sergio et al., 2014; Thorup et al., 2003). Thus, differences between individuals in a single population may be related to the number of years of experience or the conditions that have been present during that individual's lifetime. In Whooping Cranes, the ability to learn may be important for individuals to be able to find appropriate stopover wetland habitat during drought years, or learning which areas present a higher risk for disturbance or predation.

Finally, in social species, individuals may modify their behavior based not only on their own experiences but also on the experiences of conspecifics. The effects of social interactions apply to a number of aspects of migration, including migratory tendency (Palacín et al., 2011), performance (Mueller et al., 2013), and seasonal site selection (Betts et al., 2008), in both Whooping Cranes (Mueller et al., 2013) and other social bird species. Whooping Cranes gather in small flocks on the wintering grounds and during migration (Stehn, 1997) and associate with Sandhill Cranes (*Antigone canadensis*) during both migration and overwintering (Urbanek et al., 2014), pointing to the importance of both intra- and interspecific social interactions throughout the migratory cycle.

Understanding the mechanisms by which individuals and populations learn and maintain their migration patterns is critical for successfully reintroducing migratory species. Reintroduction programs aim to produce self-sustaining populations; in order to reach this goal, they must mimic natural conditions to the extent necessary for effective survival and reproduction. Some species will naturally migrate with no training (e.g., some passerines will migrate even after captive breeding and no contact with conspecifics; Pulido and Berthold, 2010). In contrast, others, such as cranes, will require more extensive training because of the importance of parental care and social learning (Urbanek et al., 2005a). Successful reintroduction will thus require a balance between mimicking natural conditions using artificial methods and allowing captive-bred individuals to naturally adapt to their new environment.

In Whooping Cranes, reintroduction of the EMP relies on both innate and external mechanisms of learning, where some birds have initially been trained to migrate following an ultralight airplane but after this training perform all movements, such as their first return (spring) migration, with no guidance (Duff, Chapter 21, this volume). Additionally, those birds released using the direct-autumn release method rely on associating with older birds or Sandhill Cranes for their first fall migration (French et al., Chapter 1, this volume). Two particularly important aspects of migratory behavior are migration routes and timing of migration. Unlike in other groups, such as songbirds, where juveniles migrate separately from adults (Åkesson and Hedenström, 2007; Thorup et al., 2007), cranes hatched in the wild perform their first migration with their parents (Alonso et al., 1984; Drewien et al., 1999; Johns et al., 2005). Thus, the initial migration – whether the timing, route, or both – is unlikely to be innate but instead is determined by a migratory culture. In contrast, migration of other reintroduced migratory bird species, such as released Little Bustards (*Tetrax tetrax*), seems to depend on genetics, rather than social context (Villers et al., 2010).

The reintroduction of the EMP provides a unique opportunity to study the mechanisms of migration in Whooping Cranes. Because each individual in the population has been tracked since its release or birth, we can analyze changes in migration behavior on both population and individual levels. From the beginning of the reintroduction in 2001 through July 2014, very high frequency (VHF) telemetry resulted in over 100,000 unique observations of 275 individual Whooping Cranes and Global Positioning System (GPS) telemetry added over 10,000 locations for 11 individuals. In this chapter, we use these tracking data to provide new analyses that describe the spatial and temporal patterns of migration in the EMP. Specifically, we examine the locations of seasonal ranges, timing of arrival and departure at those ranges, and migration routes used by the population over time. Identifying seasonal ranges and migration routes is important for protection of this endangered species, particularly habitat protection. We further describe how these characteristics of migration have changed over time, which highlights where flexibility is needed for management of Whooping Cranes. Finally, we use the specific example of the directness of migration routes to illustrate changes in the population's migration patterns over time (as analyzed and presented by Mueller et al., 2013). This example further highlights the plasticity of migration patterns in Whooping Cranes and the importance of social groups and population structure to the establishment and maintenance of migration patterns.

METHODS

Data Set

The data used for analyses presented here were collected and provided by the Whooping Crane Eastern Partnership, a public-private partnership dedicated to the reintroduction of the EMP. Released birds are fitted with VHF transmitters and unique leg bands, which allows them to be located and identified throughout their lifetimes. The transmitters are used to locate individuals and groups, and the identities of individuals are then confirmed using the leg bands; only in rare cases are transmitters non-functional or are birds identified via telemetry but not confirmed visually using the leg bands. In addition to these VHF-based data, GPS telemetry has added over 10,000 locations for 11 individuals since 2010. This long-term monitoring (LTM) data has resulted in over 110,000 unique observations of 275 individual Whooping Cranes between the beginning of reintroduction in 2001 and the ending of the period included in these analyses, July 2014. Unless noted otherwise, all analyses were performed using both VHF and GPS data. In addition, this data set provides demographic information on individuals (e.g., sex, release year), which we used to characterize movement behavior of different groups. Throughout this chapter, we define juveniles as birds in their first year before returning to the breeding grounds for the first time, subadults as those in the next year (first summer to second spring migration), and all others as adults.

Analysis of LTM Data

Seasonal Site Definitions

Because winter and summer ranges, migration routes, and migratory timing differ by individual, we use each individual's movement track to define different portions of its annual movements. For the purposes of this chapter, we define an individual's wintering range as all locations south of the Wisconsin–Illinois border where it spent at least 15 days between December and May (as in Teitelbaum et al., 2016). We define the center of an individual's summer range as the centroid (median latitude and longitude) of all observations of that individual in June–September; an individual is considered on its summer range any time it is observed within 50 km of this centroid. Accordingly, fall migration begins when a bird moves at least 50 km south of its summer centroid and ends at the first observation of the individual on its wintering grounds; spring migration follows the same criteria, but for northward movements.

Land Cover Data Sets

To analyze the protected status of land in seasonal and migration areas, we use the World Database on Protected Areas, which includes federal, state, and local parks and protected areas (IUCN and UNEP-WCMC, 2015). We use land-cover data from the 2014 CropScape data set, which provides land cover over the contiguous United States at 30-m resolution, with a focus on agricultural land (National Agricultural Statistics Service, 2015). We extracted data from each of these layers to Whooping Crane relocation points. In order to assess whether 2014 crop data was an appropriate proxy for land cover over a longer time period, we also used CropScape data from 2008 (the first year for which national data is available) and confirmed that our results did not change substantially over this period. When analyzing data extracted these layers, we used average metrics across seasons (e.g., protected land used during spring migration) to account for the variable resolution of the VHF data, which may have resulted in misassignment of the land cover of a small number of individual points.

Arrival and Departure Dates

When analyzing the timing of the movements of the EMP, it is important to keep in mind that, because locations are obtained through a combination of GPS telemetry, VHF telemetry, and direct observations, available data depend on when and where sampling was concentrated. In particular, the arrival and departure times at the wintering grounds are not precise because monitoring is focused more intensively on the breeding areas. Thus, for the purpose of

this chapter, we focus primarily on the timing related to the breeding grounds, though we use departure date from the wintering grounds to calculate a very rough upper estimate of the duration of migration. The time estimates given here were obtained in two ways: the maximum possible migration duration is the number of days elapsed between the first observation on the breeding grounds and the last observation on the wintering grounds (or vice versa), and the minimum possible duration is the amount of time elapsed between the first and last observations during migration (for the subset of individuals that were observed multiple times during a single migration season).

Analysis of Social Learning of Migration

The last section of this chapter analyzes social learning of migration patterns. These analyses were previously published in Mueller et al. (2013) and use the same LTM data as for the other analyses in the chapter, but using data that extend through 2009 instead of through 2014. For more details on the exact methods and data used, see Mueller et al. (2013).

EMP WHOOPING CRANE MOVEMENTS AND MIGRATION

An Overview of Whooping Crane Space Use

In summer, almost all Whooping Cranes in the EMP reside in Wisconsin, with a few individuals found in Michigan, Minnesota, Illinois, and Iowa (Fig. 11.1). The population is more concentrated and displays less variation over time during summer than during winter, showing very little change in distribution over the June–September period within any single year and a consistent summer distribution across years. In contrast, the winter locations of the population are highly variable both within and

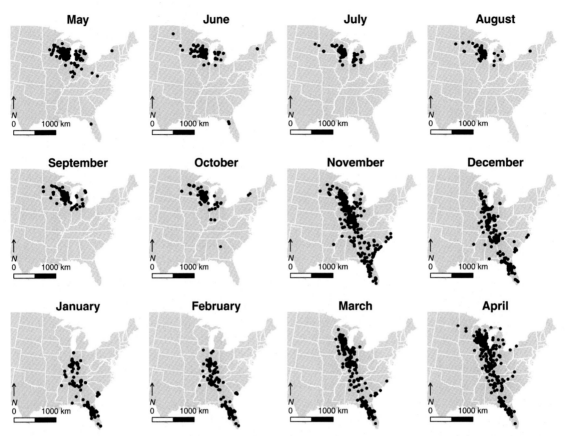

FIGURE 11.1   Monthly locations of EMP Whooping Cranes, 2001–14. Each point represents an observation of an individual. Almost all individuals arrive on their summering grounds by the beginning of May and remain there until October or November. In January and February, most birds are on their wintering grounds. Individuals are first tracked on their wintering grounds, so locations do not include juveniles on their first fall migration.

between years; January locations of the population range almost 1,500 km in latitude, stretching from central Florida in the south to southern Indiana and Illinois in the north (Fig. 11.1).

Individuals in the EMP show relatively uniform timing of migratory behavior, with fall migration beginning in early November and spring migration beginning in March. Except in rare cases of late migrants (usually subadults), all birds reach their breeding grounds by the beginning of May (Fig. 11.1; for detailed analyses see sections "Seasonal Sites" and "Temporal Aspects of Whooping Crane Migration"). Due to the wide spread in wintering locations, it is

difficult to discern from mapped snapshots of population distribution when birds arrive on their wintering grounds, but more detailed analyses (see section "Winter") show that most birds arrive on their wintering grounds by mid-December. Ultralight-guided migration takes longer and makes more stopovers than subsequent independent migrations. On their first training migration with the ultralight aircraft, Whooping Cranes begin their fall migration in October and arrive between November and January (Urbanek et al., 2005b); this timing of arrival on the wintering grounds is consistent in subsequent years during unguided migrations, though the median departure date of ultralight training is earlier (ultralight departs Wisconsin in the first week of October; median departure date of independent migrations is November 10; see section "Temporal Aspects of Whooping Crane Migration").

Seasonal Sites

Summer

Whooping Cranes were initially reintroduced at Necedah NWR, with recent ultralight releases moved to White River Marsh State Wildlife Area (SWA) (2011–present) and direct autumn releases (where juveniles are released with no ultralight training) at Necedah NWR (2005–10) and Horicon NWR (2011–present); all three protected areas are in the same region of central Wisconsin (Fig. 11.2). In the years following their release and migration, Whooping Cranes generally return to these protected areas and surrounding areas (Fig. 11.2). Breeding site fidelity is common in Whooping Cranes; Johns et al. (2005) found that the dispersal distance in the AWBP was less than 20 km for at least 76% of first-time breeders. This pattern appears to be replicated in the reintroduced EMP, where birds center on their release areas during the summer. In June–September, the mean distance of an individual to the protected area (either Necedah NWR, Horicon NWR, or White River Marsh SWA) where it was released is 36 km after age 1 ($n = 619$, SE = 4 km), with the majority of birds remaining within the same protected area where they were released.

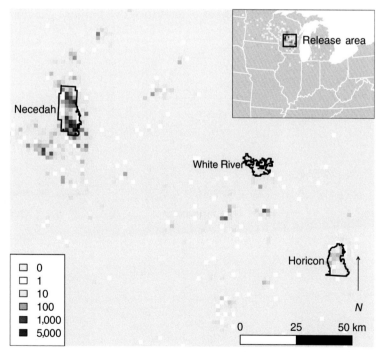

FIGURE 11.2 Three release locations (Necedah National Wildlife Refuge (NWR), Horicon NWR, and White River Marsh State Wildlife Area) and observations of EMP Whooping Cranes in Wisconsin during summer (May–September), 2001–14. Note that not all release sites were used in all years. Colors represent the number of Whooping Crane observations near the release sites in 1×1 km cells. Areas in red have the highest frequency of observation; gray areas have no crane observations.

This concentration of birds inside their release areas may be controlled not only by natal site fidelity, but also by specific habitat requirements for breeding. The breeding grounds of both migratory populations of Whooping Cranes as well as nonmigratory populations are generally characterized by wetland habitat, with proximity to open water, such as lakes and marshes (Chavez-Ramirez and Wehtje, 2012; U.S. Fish and Wildlife Service, 2001; Urbanek and Lewis, 2015). The protected areas used as release areas contain shallow marsh, oak forest, and open savanna habitats (Urbanek et al., 2005b). Outside these areas, the summering grounds of the EMP include a number of human-dominated habitats, including pasture and cropland. The individuals that are observed outside these protected areas in summertime are generally subadult nonbreeders (Wisconsin Department of Natural Resources, 2006), indicating that though birds of all ages use wetlands for foraging and roosting, breeding birds are more heavily reliant on these habitats.

Though the majority of Whooping Cranes spend the summer months near release sites in central Wisconsin, each year some individuals are observed hundreds of kilometers from the core breeding area. These occurrences are particularly common in the earlier months of the breeding season (Fig. 11.3). The spread in summer locations is closely tied to the age of birds, where younger individuals are significantly farther from the core breeding area than are older birds (Fig. 11.4). This behavior is called spring wandering and is a phenomenon observed not only in the Whooping Cranes of the EMP (Allen, 1952; Urbanek et al., 2005b) but also in the AWBP and reintroduced Sandhill Cranes (Johns, 2010; Littlefield et al., 1994), which also tend to show similar behavior during their first

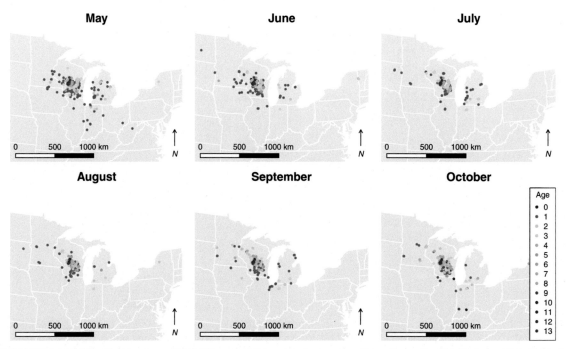

FIGURE 11.3   Monthly locations of EMP Whooping Cranes during summer (May–September) by bird age, 2001–14. Each point represents an observation of an individual. Older birds are largely concentrated in the center of the range, while younger birds exhibit a wider spread in summer locations.

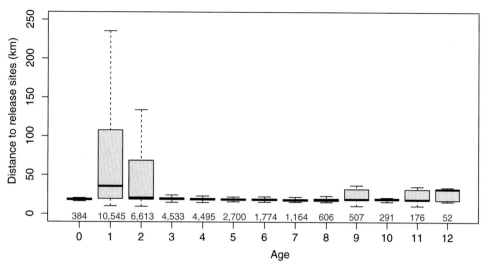

FIGURE 11.4 Distance to the centroid of summering area (May–September) by age. The box plots show that birds of pre-breeding age (<3 years) are much farther from the centroid than those of breeding age. Birds of age 0 are those that have not yet migrated (i.e., wild-born and direct autumn release birds in their year of birth). Sample sizes are shown above ages; boxes show median and interquartile range, and whiskers show range out to 1.5 times the interquartile range. Outliers are not shown.

summer (Urbanek et al., 2005a, 2005b). Not much is known about the purpose or origin of this wandering behavior, except that it presumably familiarizes birds with the areas surrounding their summering grounds (Urbanek et al., 2005b) and may have been evolutionarily favored because it facilitates selection of optimal unoccupied habitats or mate finding, as for natal dispersal in other species (Bowler and Benton, 2005). However, wandering differs from natal or breeding dispersal because most or all Whooping Cranes in the EMP breed very close to their release or natal grounds. In contrast, wandering behavior (i.e., birds more than 50 km from the release sites) all but disappears in birds older than 2 years old (Figs. 11.3 and 11.4), which also corresponds to the age at which Whooping Cranes can begin to breed (Urbanek and Lewis, 2015).

Almost all breeding sites are within Necedah NWR (through 2013, 78% of all attempted nests, *n* = 132, were within the boundaries of the refuge), indicating that the wandering behavior has

no relationship to breeding ground dispersal. The remaining 22% of nests outside the refuge were, on average, 22 km away from Necedah NWR, substantially closer than subadult wandering distances. Nevertheless, it is possible that breeding dispersal distances will increase as the number of nests in the EMP grows and the pressure to breed outside of Necedah increases because of habitat limitation or other factors favoring dispersal. In addition, through 2013 all breeding birds had been released at Necedah, but as release locations have shifted to Horicon and White River Marsh, breeding locations are also likely to shift toward these new release sites.

Summer wandering of subadults has conservation implications, especially considering the protected status of summering areas. Birds spend more time in protected areas during summer than in other seasons (Fig. 11.5), largely because they concentrate in their release areas, especially at Necedah NWR, where more birds have been released than at the other two sites

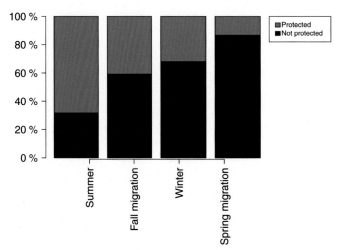

FIGURE 11.5   Protected status of diurnal sites used by EMP Whooping Cranes during migration and seasonal residency, 2001–14. Protected areas encompass all levels of protection, including national, state, and local parks and wildlife areas.

(Fig. 11.2). When subadult Whooping Cranes wander, they leave these protected areas, which could potentially affect their sources of food and exposure to power lines and other hazards, and could increase disturbance and mortality risk through interactions with humans. In some cases, subadult Whooping Cranes have wandered as far as South Dakota or New York, areas outside the intended area of reintroduction (Urbanek and Fondow, 2008), which has resulted in translocation back to Wisconsin (Zimorski and Urbanek, 2010). The conservation and management implications of spring wandering behavior require further study. In particular, in order to assess the social and fitness effects of spring wandering, it will be necessary to identify habitat selection, areas of possible conflict with humans, and intra- and interspecific social interactions during wandering bouts. The potential social implications (e.g., mate finding) of spring wandering could be important. Comparing these aspects of wandering behavior in the EMP to that in the AWBP will help identify what differences exist between the two populations, which will help inform what measures may be necessary to promote both natural movement behavior

and population persistence in the reintroduced population.

Winter

In contrast to the EMP's summer distribution, where all birds except subadults concentrate in a relatively small area, the population exhibits a much broader distribution during the winter months (Fig. 11.1) (Teitelbaum et al., 2016). The spread in distribution appeared rapidly in the years since the initial reintroduction. The population concentrated largely in Florida until the winter of 2007–08 (with a few exceptions in Tennessee and the Carolinas; Fig. 11.6). Since 2007, however, the wintering locations of the population have shifted rapidly northward (Teitelbaum et al., 2016; Urbanek et al., 2014); this northward shift was associated with a general spread in the distribution of the EMP during the winter, as indicated by increasing distance of individuals from the annual centroid of the winter distribution over time (Fig. 11.7). By the winter of 2013–14, less than 10% of the population wintered as far south as Florida ($n = 100$). This decreasing migration distance appears to be a consequence of a combination of changes in land use and climate between the times of extirpation

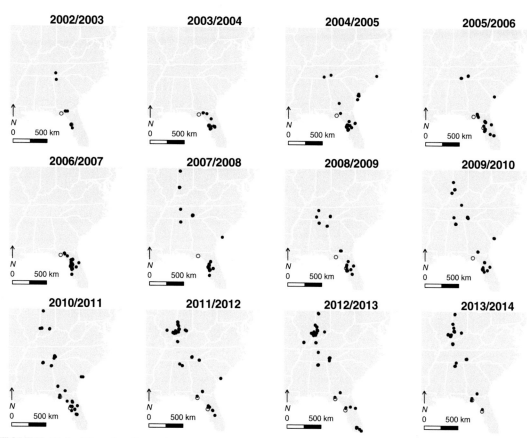

FIGURE 11.6 Wintering (December–May) locations of EMP Whooping Cranes by year (2002/2003–2013/2014). Each point represents an observation of an individual. As the beginning of the reintroduction program, there has been an increase in the spread of wintering sites and an overall northward shift. Open circles indicate pen locations at Chassahowitzka NWR and St. Marks NWR (more southern and northern, respectively).

of Whooping Cranes in eastern North America and reintroduction; new overwintering sites are in areas with high crop cover and where climate has warmed substantially in the last century, the combination of which may increase the population's ability to withstand now-milder northern winters (Teitelbaum et al., 2016).

Contrary to the common conception of migration, where individuals move between two small, discrete seasonal ranges, some Whooping Cranes may use multiple wintering areas during a single year (Fig. 11.8). Though most birds winter at a single location, 25 of 155 individuals (16%) were documented using multiple

wintering sites in a single year over the first 14 years of the reintroduction effort. On average, these sites were 233 km apart (SE = 68 km), a substantial portion of the maximum migratory distance of 1,800 km. Though the phenomenon is not well understood, several other bird species have been documented to move significant distances during a single winter (Keller et al., 2009; Oppel et al., 2008). In general, winter movements of other species are a response to a depletion of food resources at the initial site (Keller et al., 2009) or decreasing temperatures over the course of a single winter (Cox and Afton, 2000; Sauter et al., 2010). The northward

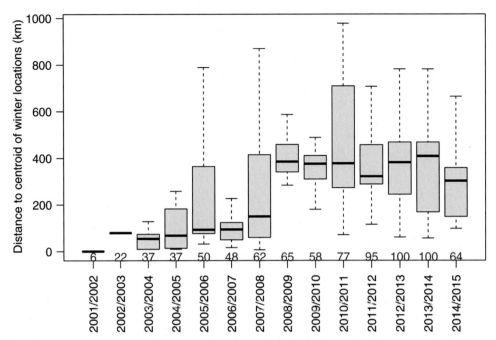

FIGURE 11.7   Spread in wintering site locations (December–May) of EMP Whooping Cranes over time. 2001/2002–2014–2015, as represented by distance from annual centroid of the wintering locations (i.e., the centroid varied from year to year). As birds increasingly used wintering sites farther north (Fig. 11.6), the distance between wintering birds also increased. This trend appears to have leveled out slightly in recent years. Boxes and whiskers are as shown in Fig. 11.4; outliers are not shown.

shift in overwintering sites in the EMP is associated with long-term landscape changes, but it is not clear how short-term events like weather and precipitation influence wintering site selection and use of multiple sites. Future research into these questions could provide predictive information about where Whooping Cranes are likely to overwinter, and how their populations will be affected by future changes in climate.

Temporal Aspects of Whooping Crane Migration

Across years, the median duration of the fall migration in the EMP lies between 17 days (calculated from locations during migration) to 31 days (calculated from the last observation on the summering grounds to the first observation on the wintering grounds), where

the median date where independently migrating (i.e., >1-year-old) birds are last observed on the breeding grounds is November 11 and median date of first observation on the wintering grounds is November 29. Similarly, the median duration of the spring migration is between 10 days and 27 days, where the median date birds are last observed on their wintering grounds is March 12 and they are first seen on their breeding grounds on April 5. Whooping Cranes are typically observed on their breeding grounds for 214 days (median, 7.1 months). The duration of the spring migration shows no detectable trend over time (Fig. 11.9), even though migration distances have decreased.

Older birds also arrive earlier on their breeding grounds than younger birds (Fig. 11.10). For instance, the average arrival date of 2-year-old birds is 18 days later than for 10-year-old

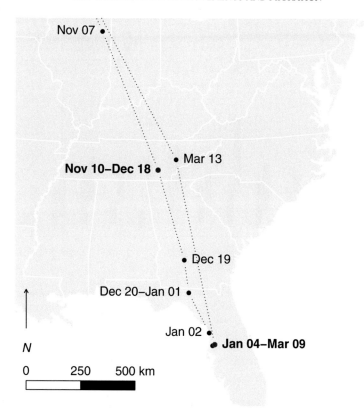

FIGURE 11.8   Example of an individual Whooping Crane (5_02) that used multiple wintering sites in 1 year (2004/2005), 2 years after this individual's initial ultralight-led migration. The first wintering site used, in southern Tennessee, was at the State of Tennessee's Hiwassee Refuge, a common wintering site for Sandhill Cranes. The second wintering site is 47 km south of Chassahowitzka NWR, Florida, which was the initial winter training site for this individual. Sites shown in black were used for shorter periods (1–10 days) in between use of these wintering sites.

birds; median arrival dates for 2-year-old birds is April 13 (SE = 2 days), as opposed to March 26 (SE = 2 days) for 10-year-old birds. The pattern of earlier arrival with age is particularly pronounced in the first 3 years of age, which is noteworthy because Whooping Cranes rarely breed before age 3. This trend in advancement of arrival date as birds approach breeding age may support theories that breeding birds are under more pressure to arrive earlier on their summering grounds than are juveniles (Gienapp and Bregnballe, 2012; Marques et al., 2010) or, alternatively, that by the time they have reached adulthood Whooping Cranes have increased the

efficiency and speed of their northward migration. However, it is important to note that, with the current data set, we cannot clearly determine whether early arrival has a direct impact on breeding success in the population.

Migration Routes

Migration is a period of high mortality in many bird populations (Lok et al., 2015). For Whooping Cranes in the AWBP, too, there is some evidence that the migration period may be a time of high mortality (Lewis et al., 1992), though Whooping Cranes may differ from

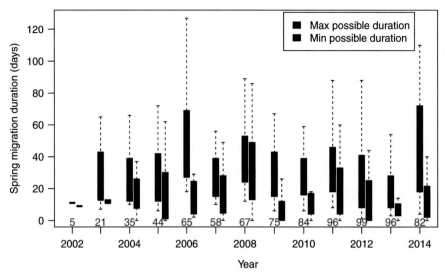

FIGURE 11.9   Duration of spring migration by EMP Whooping Cranes, 2002–2014. The maximum possible duration of migration is the amount of time between the last observation on the wintering grounds and the first observation on the breeding grounds; the minimum possible duration is the amount of time between the first and last observations during migration (see "Methods"). Boxes and whiskers are as in Fig. 11.4. Sample sizes are shown above years; outliers are not shown.

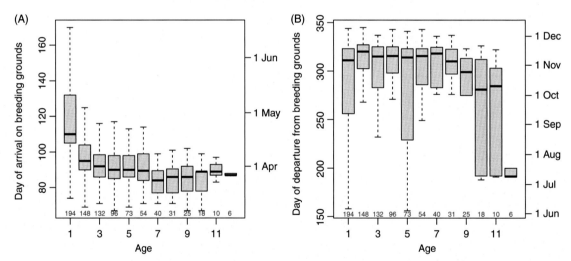

FIGURE 11.10   Ordinal day of (A) arrival at breeding grounds and (B) departure from the breeding grounds as it relates to the age of a migrating Whooping Crane. Subadults arrive at the breeding grounds later than adults (A). However, no clear trend is visible in departure dates from the breeding grounds beyond age 1 (B). Birds older than 8 years may depart earlier, but more data from older birds is needed to confirm that trend; these data will become available as the population continues to age. Boxes and whiskers are as in Fig. 11.4; outliers are not shown, and sample sizes are shown above years.

raptors (Klaassen et al., 2014) or passerines, in that more mortality occurs on summer and winter ranges (Pearse et al., Chapter 6, this volume). Nevertheless, the mortality rate during migration is at least as high as in other seasons in the AWBP (Pearse et al., Chapter 6, this volume). A primary cause of known mortalities during migration in the AWBP is collision with power lines, and reducing this mortality is an active area of research (Belaire et al., 2014). Protecting land used as stopover sites should also increase long-term population survival, since it provides more predictable stopover habitat (Belaire et al., 2014). For the EMP, the period of highest mortality is in winter and during breeding (Urbanek et al., 2014), but the primary cause of mortality during migration in the EMP is associated with interaction with humans (i.e., gunshots) (Condon et al., Chapter 23, this volume; Urbanek et al., 2014). A primary challenge for decreasing the migration mortality of the EMP will be reducing the frequency of these events, whether through education or increased protection of commonly used stopover sites.

Ultralight training used different routes over time, which could produce different migration routes among separate cohorts of reintroduced Whooping Cranes, further translating to different risks and resources for these individuals. For instance, the termination point of the training has changed over time, where birds were led to Chassahowitzka NWR on the central Gulf coast of Florida from 2001 to 2008, to Wheeler NWR in Alabama in 2011, and to St. Marks NWR on the Florida panhandle in all other years. Even when juveniles were led to the same site, each year featured slightly different stopover sites and timing during the migration training period, and the conditions of migration (e.g., number of stopovers, constraints of weather on the aircraft) did not mimic natural migration conditions. The variability in adult migration routes and wintering sites may be partially attributable to these different routes and patterns during the first migration.

In addition to the different routes introduced during ultralight training, migration routes vary between individuals within the population and for a single individual between years. However, because there is relatively little resighting data during migration periods (e.g., 2.5% of resightings are during spring migration, whereas approximately 9% of days each year are spent on spring migration), it is not possible to analyze flight paths and stopover site use for every individual in the population using the VHF-based data set. Data for most of the population

is based on VHF tracking combined with visual observations and therefore does not provide frequent tracking of routes. For 7 of the 11 GPS-tracked birds in the population, however, we do have long-term regular tracking data with several relocations a day. These data show substantial within- and between-year variability in routes (Figs. 11.8 and 11.11).

As they move between their seasonal sites, Whooping Cranes must stop along their migration route to rest and refuel (Belaire et al., 2014; Chavez-Ramirez and Wehtje, 2012). Compared to other seasons and to the breeding season in particular, a large portion of stopover sites are in agricultural areas. Between 20% and 30% of the observations of the EMP during both fall and spring migration are in areas that were either corn or soybean fields in 2014, with an additional 10–15% that were grass or pasture in 2014, based on analysis of migration data from 827 crane-years (Fig. 11.12). These locations in croplands may reflect habitat availability, since a large portion of the migratory flyway is used for agriculture, but may also reflect intentional selection, particularly for food resources provided by these crops, though there is little evidence that soybeans provide a significant source of food for Whooping Cranes (Shields and Benham, 1969). Because VHF locations are predominantly diurnal, these locations in agricultural fields probably represent foraging (rather than roosting) habitat; this use has previously been established for the AWBP (Austin and Richert, 2005). Future analyses of habitat selection on stopover sites may be important to better understand habitat and resource requirements during migration. The EMP also travels through largely unprotected land during migration (Fig. 11.5), as is true for the AWBP (Chavez-Ramirez and Wehtje, 2012). In the EMP's flyway, protected area availability is relatively low compared to that farther west in the AWBP's flyway, but Whooping Cranes in the EMP appear to select stopover sites in protected areas during migration, since the proportion of stopover sites within protected areas is significantly higher than in the regions surrounding these sites ($p < 0.001$, Wilcoxon test compared to

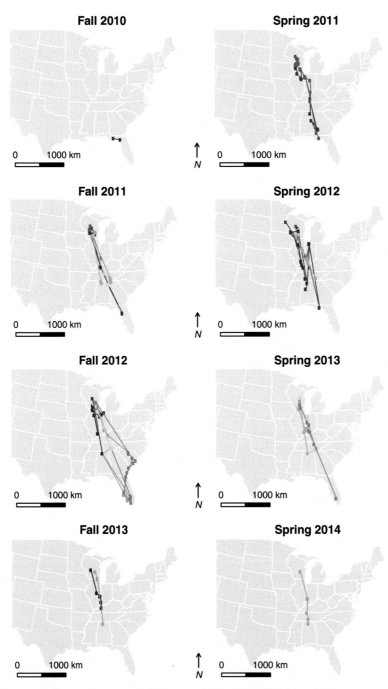

FIGURE 11.11   Migration routes of seven EMP Whooping Cranes tracked using GPS transmitters, fall 2010–spring 2014. Each point represents a GPS location of an individual. Colors represent individuals (i.e., identical colors in different years represent the same individual). Considerable variation in migration routes exists both within and between years.

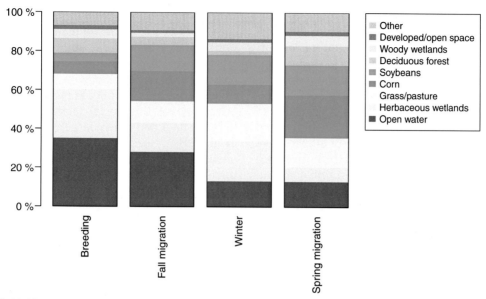

FIGURE 11.12   Land cover of areas used by EMP Whooping Cranes during different seasons, 2001–14. Land cover data are from the 2014 CropScape data set; the same analyses using 2008 CropScape data are similar (see "Methods").

a 50-km buffer zone around each stopover site). The EMP has been documented at 84 unique protected areas during migration, including in public parks and on U.S. federal land. Thus, as with the AWBP, particular attention must be paid to selection of stopover sites by the EMP in order to minimize risks during migration. Because different social groups may select for different habitat (e.g., single cranes and nonfamily groups use riverine roosts, whereas family groups do not; Austin and Richert, 2005), understanding the social structure of the EMP will also be important for making predictions about available and used stopover habitat.

SOCIAL LEARNING OF MIGRATORY PERFORMANCE IN WHOOPING CRANES

The energetic efficiency of migration, also known as migratory performance, plays a large role in determining an individual's body condition upon arrival at the breeding grounds (Åkesson and Hedenström, 2007). Migratory performance can be measured in a number of ways, including using migration speed (Åkesson and Hedenström, 2007; Sergio et al., 2014), time spent migrating (Sergio et al., 2014), and stopover duration (Sergio et al., 2014), any of which can be used to reflect one or more components of the amount of energy spent on migration. The efficiency of a migration route, measured by how much an individual deviates from a straight-line path between its wintering and summering grounds, is one further measure of migratory performance (Mueller et al., 2013). Migrating closer to a straight line presumably uses less energy, potentially improving body condition at arrival and increasing the ability to breed successfully. However, it is important to note that migration efficiency ignores the quality of stopover sites and is only one of many influences on fitness; the optimal flight path would strike a balance between stopover site quality and migration path efficiency.

In addition to reflecting possible implications for body condition and breeding success, understanding the variability in migratory performance within the EMP can help to understand how birds learn and alter their migratory behavior over time. As with other aspects of migratory behavior, these changes could be related to environment, genetic characteristics, individual learning, and social transmission of information. Mueller et al. (2013) investigated the importance of each of these drivers in determining migratory performance in the EMP. The following section summarizes these results. Social learning is the strongest determinant of migratory performance in the EMP, with increases in migratory performance over time being transmitted from older to younger individuals in the population (Mueller et al., 2013) (Fig. 11.13). Whooping Cranes migrate in small groups, and migratory performance increased about 5.5% for each year of age of the oldest bird in the group. Individual age was not a significant predictor of migratory performance, meaning that young cranes are not controlled by their physical abilities; instead, the migratory performance of young birds appears to be limited by their knowledge or experience, which they can then gain by traveling in groups with older birds. On average, 1-year-old birds in groups with older birds deviated by 34% less from a straight-line path than 1-year-old birds that traveled only with other new migrants. Further, this social learning does not extend just for migratory performance, but is also associated with the northward shift in overwintering locations, where groups with older birds were the first to use northern overwintering sites (Teitelbaum et al., 2016). These substantial changes with different migratory groups highlight the importance of social structure and cultural learning in Whooping Cranes.

The importance of social learning in Whooping Cranes has potentially important management implications. The use of ultralight training was based on knowledge that juvenile cranes learn to migrate by following conspecific adults, usually their parents (Urbanek et al., 2010), and were thus unlikely to migrate on their own. However, unlike individual birds' ability to learn and adapt over time, human-led reintroduction changes only under management recommendations. Direct autumn release methods later took advantage of this social learning component once the flock had grown large enough (Urbanek et al., 2014). Social learning indicates that by allowing juvenile cranes to migrate with other members of the flock, these young birds may be able to more quickly adopt the new migration patterns established by older individuals.

SUMMARY AND OUTLOOK

Analysis of the movement patterns of the EMP reveals substantial variability in both spatial and temporal aspects of migration both among individuals and for the same individual between years. In the summer, subadults wander far from their release sites, but nesting birds concentrate in and near Necedah NWR. This age-related behavior does not change from year to year or in a clear pattern that reflects variability in habitat conditions (e.g., precipitation); in contrast, winter site use has changed dramatically over time. Since the reintroduction began, wintering sites of the EMP have become more dispersed and shifted north. Unfortunately, it is difficult to ascertain specific migration routes of individuals across time, but analysis of flight patterns of groups indicates that the age of a migration group – a proxy for its migratory experience – results in better migratory performance. Overall, migratory performance should increase as the population ages in the years since reintroduction began.

The variability in the movement patterns of the EMP is important to consider when planning for the conservation of Whooping Cranes. Management plans should account for both the flexibility and the large spatial scale of crane movements over time. For instance, though

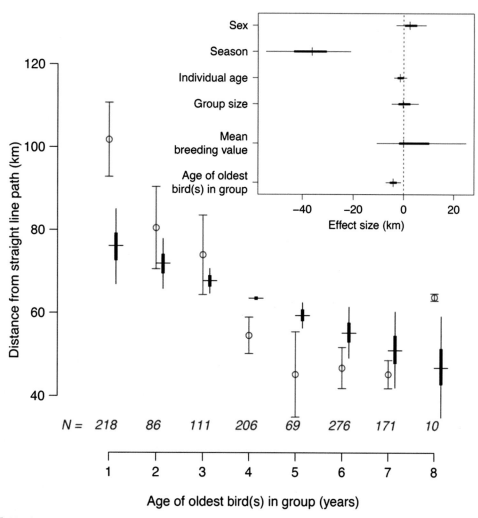

FIGURE 11.13   Migratory performance of EMP Whooping Cranes, measured as deviation of locations from a straight-line path, as it relates to the age of the oldest individual(s) in each flight group. Open circles show original data (mean and 95% confidence intervals) and solid bars and lines show model predictions (posterior modes, quartiles, and 95% highest posterior density intervals). The model incorporated migratory season, bird sex, and genetic relatedness, as well as age. Sample sizes are shown above years. *Source: This figure is reproduced with permission from Mueller, T., O'Hara, R., Converse, S., 2013. Social learning of migratory performance. Science 341 (6149), 999–1002.*

taking advantage of protected areas as wintering sites during ultralight training initially ensured some level of protection for the population, the shift in the EMP's wintering sites over time may require new management strategies. The EMP now winters in areas with different land cover types, including on private land.

Similarly, changing migration patterns have led to the use of new stopover sites over time, and all of these stopovers sites are not on protected lands. And finally, spring wandering behavior in juvenile cranes involves movements of individuals through large areas that may change as the population expands. Because population and individual behavior are not fixed, and occur at broad spatial scales, we cannot assume that populations will use protected areas only; rather, it will likely be important for the conservation of the population to work with owners of private lands across multiple states to ensure that birds are safe and their energetic needs are met on those lands. This is particularly important for migratory species like Whooping Cranes, because they inevitably travel through unprotected land even if they inhabit protected areas seasonally.

The plasticity of migration behavior is one crucial point of future research on Whooping Crane movement. Selection of optimal conservation strategies relies on being able to predict the future species distribution, especially as it changes in response to climate and land use change (Teitelbaum et al., 2016). Changes in climate and land use have been historically significant in the range of the EMP and are likely to continue into the future. In order to better conserve Whooping Cranes, it is necessary not only to predict which areas will be used as future breeding, wintering, and stopover sites, but also to determine how management agencies can best conserve habitat for the EMP when the population moves in unexpected ways.

Currently, the main challenge to persistence of the EMP is breeding and rearing success (Converse et al., Chapter 8, this volume). Not only does variation in movement behaviors have the potential to influence variation in reproductive outcomes (e.g., by influencing pairing, breeding site selection, and body condition), but lessons from migration behavior may apply to other behaviors as well. A key goal of future research should be to link movement behaviors to year-round body condition, fitness, and reproductive success.

Acknowledgments

TM and CST were supported by the Robert Bosch Foundation. We thank the Whooping Crane Eastern Partnership for collecting and providing data on EMP locations. Reviewers J. Austin, J. French, R. Urbanek, and N. Bunnefeld provided helpful comments on an earlier draft of this chapter. Thomas Mueller and Claire S. Teitelbaum contributed equally to this work. Thomas Mueller and Claire S. Teitelbaum contributed equally to this work.

References

Åkesson, S., Hedenström, A., 2007. How migrants get there: migratory performance and orientation. BioScience 57 (2), 123–133.

Alerstam, T., Hedenström, A., Åkesson, S., 2003. Long-distance migration: evolution and determinants. Oikos 103 (2), 247–260.

Allen, R.P., 1952. The Whooping Crane. Research Report 3. National Audobon Society, New York.

Alonso, J.C., Veiga, J.P., Alonso, J.A., 1984. Familienauflösung und Abzug aus dem Winterquartier beim Kranich Grus grus. J. Ornithol. 125 (1), 69–74.

Austin, J.E., Hayes, M.A., Barzen, J.A., 2018. Revisiting the historic distribution and habitats of the whooping crane (Chapter 3). In: French, Jr., J.B., Converse, S.J., Austin, J.E. (Eds.), Whooping Cranes: Biology and Conservation. Biodiversity of the World: Conservation from Genes to Landscapes. Academic Press, San Diego, CA.

Austin, J. E., A. L. Richert. 2005. Patterns of habitat use by whooping cranes during migration: summary from 1977–1999 site evaluation data. Proceedings of the North American Crane Workshop 9, 79–104.

Belaire, J.A., Kreakie, B.J., Keitt, T., Minor, E., 2014. Predicting and mapping potential whooping crane stopover habitat to guide site selection for wind energy projects. Conserv. Biol. 28 (2), 541–550.

Berthold, P., 1988. Evolutionary aspects of migratory behavior in European warblers. J. Evol. Biol. 1 (3), 195–209.

Berthold, P., Helbig, A.J., Mohr, G., Querner, U., 1992. Rapid microevolution of migratory behaviour in a wild bird species. Nature 360, 668–670.

Betts, M.G., Hadley, A.S., Rodenhouse, N., Nocera, J.J., 2008. Social information trumps vegetation structure in breeding-site selection by a migrant songbird. Proc. R. Soc. B 275 (1648), 2257–2263.

Bowler, D.E., Benton, T.G., 2005. Causes and consequences of animal dispersal strategies: relating individual behaviour to spatial dynamics. Biol. Rev. Camb. Philos. Soc. 80 (2), 205–225.

Chavez-Ramirez, F., Wehtje, W., 2012. Potential impact of climate change scenarios on whooping crane life history. Wetlands 32 (1), 11–20.

Chernetsov, N., 2004. Migratory orientation of first-year white storks (Ciconia ciconia): inherited information and social interactions. J. Exp. Biol. 207, 937–943.

Condon, E., Brooks, W.B., Langenberg, J., Lopez, D., 2018. Whooping crane shootings since 1967 (Chapter 23). In: French, Jr., J.B., Converse, S.J., Austin, J.E. (Eds.),

Whooping Cranes: Biology and Conservation. Biodiversity of the World: Conservation from Genes to Landscapes. Academic Press, San Diego, CA.

Converse, S.J., Strobel, B.N., Barzen, J.A. Reproductive failure in the eastern migratory population: the interaction of research and management (Chapter 8). In: French, Jr., J.B., Converse, S.J., Austin, J.E. (Eds.), Whooping Cranes: Biology and Conservation. Biodiversity of the World: Conservation from Genes to Landscapes. Academic Press, San Diego, CA.

Cox, Jr., R.R., Afton, A.D., 2000. Predictable interregional movements by female northern pintails during winter. Waterbirds 23 (2), 258–269.

Del Hoyo, J., Collar, N.J., 2014. HBW and BirdLife International Illustrated Checklist of the Birds of the World. Lynx Edicions, Barcelona.

Dingle, H., Drake, V.A., 2007. What is migration? BioScience 57 (2), 113–121.

Drewien, R.C., Brown, W.M., Varley, J.D., Lockman, D.C., 1999. Seasonal movements of sandhill cranes radiomarked in Yellowstone National Park and Jackson Hole, Wyoming. J. Wildl. Manage. 63 (1), 126–136.

Duff, J.W., 2018. The operation of an aircraft-led migration: goals, successes, challenges 2001 to 2015 (Chapter 21). In: French, Jr., J.B., Converse, S.J., Austin, J.E. (Eds.), Whooping Cranes: Biology and Conservation. Biodiversity of the World: Conservation from Genes to Landscapes. Academic Press, San Diego, CA.

French, Jr., J.B., Converse, S.J., Austin, J.E., 2018. Whooping cranes past and present (Chapter 1). In: French, Jr., J.B., Converse, S.J., Austin, J.E. (Eds.), Whooping Cranes: Biology and Conservation. Biodiversity of the World: Conservation from Genes to Landscapes. Academic Press, San Diego, CA.

Gienapp, P., Bregnballe, T., 2012. Fitness consequences of timing of migration and breeding in cormorants. PLoS ONE 7 (9), e46165.

Glenn, T.C., Stephan, W., Braun, M.J., 1999. Effects of a population bottleneck on whooping crane mitochondrial DNA variation. Conserv. Biol. 13 (5), 1097–1107.

Gomez, G. M. 1991. Whooping cranes in southwest Louisiana: history and human attitudes. Proceedings of the North American Crane Workshop 6, 19–23.

Hake, M., Kjellen, N., Alerstam, T., 2003. Age-dependent migration strategy in honey buzzards *Pernis apivorus* tracked by satellite. Oikos 103 (2), 385–396.

IUCN and UNEP-WCMC, 2015. The World Database on Protected Areas (WDPA) [WWW Document]. Protected Planet. Available from: www.protectedplanet.net.

Johns, B.W., 2010. Aerial survey techniques for breeding whooping cranes. Proceedings of the North American Crane Workshop 11, 83–88.

Johns, B.W., Goossen, J.P., Kuyt, E., Craig-Moore, L. 2005. Philopatry and dispersal in whooping cranes. Proceedings of the North American Crane Workshop 9, 117–125.

Keller, I., Korner-Nievergelt, F., Jenni, L., 2009. Within-winter movements: a common phenomenon in the common pochard *Aythya ferina*. J. Ornithol. 150 (2), 483–494.

Klaassen, R.H., Hake, M., Strandberg, R., Koks, B.J., Trierweiler, C., Exo, K.M., Bairlein, F., Alerstam, T., 2014. When and where does mortality occur in migratory birds? Direct evidence from long-term satellite tracking of raptors. J. Anim. Ecol. 83 (1), 176–184.

Lack, D., 1968. Bird migration and natural selection. Oikos 19 (1), 1–9.

Lewis, J.C., Schwindt, K.E., Stehn, T.V., 1992. Mortality in fledged whooping cranes of the Aransas/Wood Buffalo population. Proceedings of the 1988 North American Crane Workshop, Florida Game and Fresh Water Fish Commission Nongame Wildlife Program Technical Report, Tallahassee, pp. 145–148.

Littlefield, C.D., Stern, M.A., Schlorff, R.A., 1994. Summer distribution, status, and trends of greater sandhill crane populations in Oregon and California. Northwest. Nat. 75 (1), 1–10.

Lok, T., Overdijk, O., Piersma, T., 2015. The cost of migration: spoonbills suffer higher mortality during trans-Saharan spring migrations only. Biol. Lett. 11 (1), 1–4.

Marques, P.A.M., Sowter, D., Jorge, P.E., 2010. Gulls can change their migratory behavior during lifetime. Oikos 119 (6), 946–951.

Meine, C.D., Archibald, G.W., 1996. The Cranes: Status Survey and Conservation Action Plan. IUCN, Gland and Cambridge.

Mueller, T., Fagan, W., 2008. Search and navigation in dynamic environments – from individual behaviors to population distributions. Oikos 117 (5), 654–664.

Mueller, T., O'Hara, R., Converse, S., 2013. Social learning of migratory performance. Science 341 (6149), 999–1002.

National Agricultural Statistics Service, 2015. CropScape – Cropland Data Layer [WWW Document]. Available from: http://nassgeodata.gmu.edu/CropScape/ [27 January 2015].

Oppel, S., Powell, A.N., Dickson, D.L., 2008. Timing and distance of king elder migration and winter movements. Condor 110 (2), 296–305.

Palacín, C., Alonso, J.C., Alonso, J.A., Magaña, M., Martín, C.A., 2011. Cultural transmission and flexibility of partial migration patterns in a long-lived bird, the great bustard *Otis tarda*. J. Avian Biol. 42 (4), 301–308.

Pearse, A.T., Brandt, D.A., Hartup, B.K., Bidwell, M.T., 2018. Mortality in Aransas-Wood Buffalo whooping cranes: timing, location, and causes (Chapter 6). In: French, Jr., J.B., Converse, S.J., Austin, J.E. (Eds.), Whooping Cranes: Biology and Conservation. Biodiversity of the World: Conservation from Genes to Landscapes. Academic Press, San Diego, CA.

Pulido, F., Berthold, P., 2010. Current selection for lower migratory activity will drive the evolution of residency in a migratory bird population. Proc. Natl. Acad. Sci. USA 107 (16), 7341–7346.

Sauter, A., Korner-Nievergelt, F., Jenni, L., 2010. Evidence of climate change effects on within-winter movements of European mallards *Anas platyrhynchos*. Ibis 152 (3), 600–609.

Sergio, F., Tanferna, A., De Stephanis, R., Jiménez, L.L., Blas, J., Tavecchia, G., Preatoni, D., Hiraldo, F., 2014. Individual improvements and selective mortality shape lifelong migratory performance. Nature 515 (7527), 410–413.

Shields, R.H., Benham, E.L., 1969. Farm crops as food supplements for whooping cranes. J. Wildl. Manage. 33 (4), 811–817.

Stehn, T.V., 1997. Pair formation by color-marked whooping cranes on the wintering grounds. Proceedings of the North American Crane Workshop 7, 24–28.

Teitelbaum, C.S., Converse, S.J., Fagan, W.F., O'Hara, R.B., Böhning-Gaese, K., Lacy, A.E., Mueller, T., 2016. Experience drives innovation of new migration patterns of whooping cranes in response to global change. Nat. Commun. 7, 12793.

Thorup, K., Alerstam, T., Hake, M., Kjellén, N., 2003. Bird orientation: compensation for wind drift in migrating raptors is age dependent. Proc. R. Soc. B 270 (Suppl. 1), S8–S11.

Thorup, K., Bisson, I.-A., Bowlin, M.S., Holland, R.A., Wingfield, J.C., Ramenofsky, M., Wikelski, M., 2007. Evidence for a navigational map stretching across the continental U.S. in a migratory songbird. Proc. Natl. Acad. Sci. USA 104 (46), 18115–18119.

U.S. Fish and Wildlife Service, 2001. Endangered and threatened wildlife and plants: establishment of a nonessential experimental population of whooping cranes in the eastern United States. Fed. Reg. 66 (123), 33903–33917.

Urbanek, R.P., Duff, W.J., Swengel, S.R., Fondow, L.E. 2005a. Reintroduction techniques: post-release performance of sandhill cranes (1) released into wild flocks and (2) led on migration by ultralight aircraft. Proceedings of the North American Crane Workshop 9, 203–211.

Urbanek, R.P., Fondow, L.E.A., Satyshur, C.D., Lacy, A.E., Zimorski, S.E., Wellington, M., 2005b. First cohort of migratory whooping cranes reintroduced to eastern North America: the first year after release. Proceedings of the North American Crane Workshop 9, 213–224.

Urbanek, R.P., Fondow, L.E.A., Zimorski, S.E., Wellington, M.A., Nipper, M.A., 2010. Winter release and management of reintroduced migratory whooping cranes *Grus americana*. Bird Conserv. Int. 20 (1), 43–54.

Urbanek, R.P., Fondow, L.E., 2008. Suvival, movements, social structure and reproductive behavior during development of a population of reintroduced, migratory whooping cranes. Proceedings of the North American Crane Workshop 10, 155.

Urbanek, R.P., Lewis, J.C., 2015. Whooping crane (*Grus americana*). In: Rodewald, P.G. (Ed.), The Birds of North America. Cornell Lab of Ornithology, Ithaca, NY. Retrieved from the Birds of North America: https://birdsna. org/ Species-Account/bna/species/whocra on 11 November 2016.

Urbanek, R.P., Szyszkoski, E.K., Zimorski, S.E., 2014. Winter distribution dynamics and implications to a reintroduced population of migratory whooping cranes. J. Fish Wildl. Manage. 5 (2), 340–362.

Villers, A., Millon, A., Jiguet, F., Lett, J.M., Attie, C., Morales, M.B., Bretagnolle, V., 2010. Migration of wild and captive-bred little bustards *Tetrax tetrax*: releasing birds from Spain threatens attempts to conserve declining French populations. Ibis 152 (2), 254–261.

Wilcove, D.S., Wikelski, M., 2008. Going, going, gone: is animal migration disappearing? PLoS Biol. 6 (7), e188.

Wisconsin Department of Natural Resources, 2006. Wisconsin Whooping Crane Management Plan. Wisconsin Department of Natural Resources, Madison.

Wright, G., Harner, M., Chambers, J., 2014. Unusual wintering distribution and migratory behavior of the whooping crane (*Grus americana*) in 2011–2012. Wilson J. Ornithol. 126 (1), 115–120.

Zimorski, S. E., Urbanek, R.P. 2010. The role of retrieval and translocation in a reintroduced population of migratory whooping cranes. Proceedings of the North American Crane Workshop 11, 216.

Ecological Energetics of Whooping Cranes in the Eastern Migratory Population

Megan J. Fitzpatrick, Paul D. Mathewson, Warren P. Porter

Department of Zoology, University of Wisconsin-Madison, Madison, WI, United States

INTRODUCTION

This chapter focuses on the ecological energetics of Whooping Cranes in the Eastern Migratory Population (EMP) (French et al., Chapter 1, this volume). Energetics refers to the energy costs of the various behavioral and physiological aspects of survival and reproduction (Tomlinson et al., 2014) and the ways that animals meet those costs. Ecological energetics refers to the study of energetics within the context of environmental constraints (Tomlinson et al., 2014). In this chapter, we review ecological energetics in EMP Whooping Cranes and provide a new analysis of energy requirements of cranes in the EMP across their wintering range.

REVIEW: ECOLOGICAL ENERGETICS OF WHOOPING CRANES

The Relevance of Ecological Energetics to Conservation of Whooping Cranes

Energy is fundamental to animal survival and reproduction. Information about ecological energetics of endangered species is consequently of use in conservation decisions (Cooke and O'Connor, 2010; Cooke et al., 2013; Stevenson, 2006; Tracy et al., 2006; Wikelski and Cooke, 2006) and is of particular use in managing Whooping Cranes. In contrast to the remnant Aransas–Wood Buffalo (AWBP) (French et al., Chapter 1, this volume), the EMP is not yet self-sustaining due to low reproductive success of reintroduced cranes (Converse et al., Chapter 7, this volume; Converse et al., Chapter 8, this volume; Servanty et al., 2014). Nest abandonment caused by parasitic black flies is an important (though not the only) cause of low reproductive rates in the EMP (Converse et al., Chapter 8, this volume; Converse et al., 2013; King et al., 2013; Urbanek et al., 2010). The Whooping Crane Eastern Partnership's (WCEP's) reintroduction techniques are consequently in flux, with the goal of improving reproductive rates.

Given that energy is of fundamental importance in successful reproduction, an understanding of EMP Whooping Crane energetics is important to choosing reintroduction locations that are energetically ideal (sufficient amounts of preferred food resources for successful reproduction given local weather and disturbance conditions) as WCEP shifts reintroduction locations to areas with lower black fly populations than Necedah National Wildlife Refuge (NWR). Simultaneous changes in chick-rearing strategies (shifting from costume rearing to parent rearing of chicks) are likely to lead to behavioral changes in reintroduced cranes. As costume rearing involves demonstrating foraging behaviors to chicks, changing behaviors may include foraging strategies and preferred food resources, with implications for appropriate choice of reintroduction locations.

Further, there is evidence that energy stress (i.e., insufficient fat reserves combined with insufficient local food availability to meet energy requirements) in the spring could contribute to low reproductive rates in birds from the EMP. Energy stress could make parents more likely to abandon nests when faced with black flies. Van Schmidt et al. (2014) found that many EMP Whooping Cranes left their wetland territories to forage in crop fields, especially cornfields, after nest abandonment. In at least two pairs of Whooping Cranes, the off-nest bird has even been observed to regularly leave the nesting territory during incubation to forage in cornfields up to 10 km away from the territory

(Fitzpatrick, 2016), forsaking normal nest guarding duties and potentially disrupting nest exchange rhythms. This could indicate that food resources within wetland territories are insufficient to meet energy needs given fat reserve levels. We acknowledge that other hypotheses could explain this behavior. However, pending completion of testing of the energy stress hypothesis, or improvement of reproductive rates to levels capable of sustaining the population, we do not discount the possibility of energy stress contributing to nest abandonment. Furthermore, while EMP eggs are of normal size and have high fertility rates, changes in these factors are not necessarily expected in a bird species with a small (two-egg) clutch and a long incubation period, and therefore egg characteristics do not by themselves refute the energy stress hypothesis.

Finally, given the existence of a successfully reproducing population, the AWBP, differences in energetics between these populations could be used as an avenue to identify additional hypotheses about functioning of the EMP, additional potential threats to the EMP, and potential ways to address these threats.

Scope

We focus on the EMP in this review, but we provide some information about the remnant AWBP and captive cranes for comparison. We further focus on adult cranes, rather than growing chicks. We will start with a discussion of what is known about EMP Whooping Crane energy requirements, including how EMP energy expenditure compares to that of the AWBP and which components of daily energy expenditure are the largest. We then review available information about how Whooping Cranes meet those requirements.

Energy Requirements of Whooping Cranes

For an adult bird, basic energy costs of survival and reproduction include costs of basal metabolic processes (i.e., basal metabolic rate,

BMR), thermoregulation, locomotion and other physical activities, digestion and postdigestive food processing (specific dynamic effect), and production. Costs of production include somatic costs (e.g., fat storage, feather synthesis following molt) and reproductive costs (e.g., production of gametes, production of egg yolk and albumin). Combinations of these costs change spatially and temporally, leading to varying energy expenditure levels in different locations and throughout the year.

Energy expenditure has never been measured in wild Whooping Cranes, but it has been measured in captive Whooping Cranes. Total daily energy expenditure (the sum of the individual daily costs, e.g., BMR, physical activity) has been measured in two studies (Fitzpatrick et al., 2015; Nelson, 1995). Values in these studies suggest that captive cranes do not have unusually high or low energy requirements compared to other crane species. More specifically, Nelson (1995) measured energy expenditure of captive cranes (n = 5 cranes) using indigestible markers in food, and Fitzpatrick et al. (2015) measured energy expenditure using doubly labeled water (n = 2 cranes). Nelson (1995) found values of 1680–1835 kJ/day for Whooping Cranes of 4.3–4.85 kg. These did not appear unusual when plotted by weight alongside daily energy expenditure values for captive cranes of other species. Fitzpatrick (2016) found values of 1811 and 2381 for cranes of 5.05 and 6.15 kg. These values were larger than those found by Nelson (1995) as expected based on the larger body weights of cranes studied in Fitzpatrick et al. (2015). Mass-specific metabolic rates of cranes in Fitzpatrick et al. (2015) (359 and 387 kJ/kg) were similar to those in Nelson (1995) (range 346–478 kJ/kg).

In Whooping Cranes, costs of the individual components of daily energy expenditure have never been measured, even in captive individuals. Authors attempting to estimate the energy expenditure of wild cranes in the AWBP at the wintering grounds (Chavez-Ramirez, 1996; Nelson, 1995) and breeding grounds (Fitzpatrick, 2016), and in the EMP (Fitzpatrick, 2016) have

estimated component costs from values measured in other species (Chavez-Ramirez, 1996; Nelson, 1995) and via mechanistic modeling of animal heat balance (Fitzpatrick, 2016).

Energy expenditure has been estimated for EMP Whooping Cranes during spring. Fitzpatrick (2016) used a mechanistic model of microclimate and animal heat/energy balance incorporating time-energy budgets of wild cranes to simulate energy expenditure of individual EMP cranes breeding in Wisconsin during 2010, 2011, and 2013. A time budget refers to the proportions of time that an animal spends on various behaviors, for example, foraging, preening, and resting. A time-energy budget includes these proportions with estimates of the energy costs of these behaviors per unit time, and uses those estimates to calculate total energy spent on behaviors throughout a day. When tested with captive cranes, the model predicted average daily energy expenditure within 7% of values measured using double-labeled water (Fitzpatrick et al., 2015). For wild cranes in spring, values were similar to or slightly larger than those measured in captive cranes. During the period between arrival and egg-laying, simulated energy expenditures of individual females (modeled body weight of 5.05 kg) averaged 2009 kJ/day (range 1754–2299 kJ/day). Males (modeled body weight of 6.15 kg) averaged 2219 kJ/day (range 1946–2465 kJ/day). During the incubation period, modeled females averaged 1890 kJ/day (range 1844–1953 kJ/day), and males averaged 2297 kJ/day (range 2208–2343 kJ/day). During the first week postnest termination (nest abandonment by parents or removal of infertile eggs by managers), females averaged 2168 kJ/day (range 1903–2325 kJ/day), and males averaged 2470 kJ/day (range 2235–2703 kJ/day). The higher values for males were expected based on larger body size.

The largest components of energy expenditure were BMR and physical activity, which each comprised 40–60% of total daily energy expenditure. BMR was estimated via an allometric equation to be 1028.2 kJ/day for a 5.05 kg female crane and 1183.7 kJ/day for a 6.15 kg male crane. For females producing eggs, production costs were the next-highest source of energy expenditure. Energy needed to form a two-egg clutch was estimated at 2495.3 kJ, distributed over 14 days of yolk/egg production, with a peak energy investment of 373 kJ/day (18.6% of average daily energy expenditure during the prenesting period). Thermoregulatory costs were small, accounting for 1–2.5% of the daily energy budget on average during the prenesting period and declining through the breeding season. Energy costs of digestion and postdigestive processing (specific dynamic effect) were not considered separately because estimates of activity energy were based on values measured in fed birds.

Simulations incorporating all known spring arrival dates for the EMP and AWBP suggest that thermoregulation costs are similar for the two populations during the spring breeding season (Fitzpatrick, 2016). However, it is unknown how the longer day lengths and differences in climate in Wood Buffalo National Park affect AWBP time-energy budgets in comparison to those of the EMP. More behavioral information about the AWBP is necessary for comparison of overall energy expenditures.

EMP energy expenditures are quite similar to energy expenditures of wintering cranes in the AWBP, which Chavez-Ramirez (1996) estimated at 2245–2776 kJ/day for a 7 kg crane.

How EMP Whooping Cranes Meet Energy Requirements

Studies of EMP diet at the breeding grounds using high-resolution videos and photography (Barzen et al., 2018; Fitzpatrick, 2016) have found that cranes eat a wide variety of food items, including molluscs (e.g., snails, clams), fish, crustaceans (crayfish), aquatic insect larvae, amphibians (frogs), reptiles (snakes), berries, and sedge and cattail rhizomes in wetlands. EMP cranes also obtain food from agricultural fields, including cranberries, earthworms, and waste corn (Barzen et al., 2018; Fitzpatrick, 2016).

Fitzpatrick (2016) attempted to quantify which types of foods were most important to individual pairs of cranes during the breeding season. Prior to nesting, snails and fish made up the majority of energy consumed by one pair on the Necedah NWR. However, incomplete data from other pairs and periods indicate that variation is likely within the EMP. Rhizomes appeared more important to some pairs, and pairs with territories closely adjacent to farms foraged in agricultural fields throughout the prenesting and incubation periods. More study is needed to quantify how diet varies among years, among pairs with access to different habitats and food types on their territories, and throughout the spring.

Later in the season, following nest abandonment, Fitzpatrick (2016) found that multiple pairs spent the entire day in corn stubble fields, the behavioral pattern previously observed by Van Schmidt et al. (2014). For the one pair of cranes in which diet could be closely observed, foods consumed were earthworms and corn. Cranes returned to wetland territories only to roost during this period, and thus were likely not relying on wetland food items.

Aside from the postnest abandonment period in which EMP cranes appeared to rely entirely on earthworms and waste corn, the general types of foods consumed by EMP cranes are similar to those documented for AWBP cranes at their coastal wintering grounds at Aransas NWR, Texas. In Texas, Whooping Cranes are opportunistic in their food choices (Greer, 2010) with diets including crustaceans (e.g., crabs), molluscs (e.g., clams), berries, and insects, along with occasional larger organisms (e.g., fish, snakes) and acorns from upland areas (Chavez-Ramirez, 1996; Greer, 2010; also see Smith et al., Chapter 13, this volume).

Few studies of AWBP diet at the breeding grounds are available for comparison with breeding EMP cranes, and none have directly observed what adult cranes eat. One major difference between EMP and AWBP diets is that there are no agricultural fields available in the region of crane nesting areas in Wood Buffalo National Park.

It is relevant to note that food in agricultural fields may contain residues of herbicides that have been found to change the gut microbiome, inducing "leaky" gut, which causes chronic inflammation and oxidative stress that alters immune function in vertebrates (Abdollahi et al., 2004; Carman et al., 2013; Mezzomo et al., 2013). This is of particular concern for pesticides that may be transferred into eggs because of their lipid solubility and/or the nonionic solvents and surfactants in the formulations, potentially affecting the immune system in developing animals and causing long-term damage to the neurological, cardiovascular, endocrine, and reproductive systems (Dietert and Dietert, 2007). Energetics can be heavily influenced by immune status, which affects disease susceptibility (Demas, 2004). In addition, Warner (2013) found low levels of organochlorines, organophosphates, and many inorganic toxins (heavy metals and minerals) in EMP Whooping Crane eggs and liver tissue. Many of these chemicals are endocrine disruptors and can induce oxidative stress. While levels in cranes and eggs were below adverse effect thresholds available in the literature (Warner, 2013), endocrine-disrupting chemicals may have harmful effects on wildlife at doses lower than those typically tested in toxicology studies (Vandenberg et al., 2012; Welshons et al., 2003).

Low-dose effects are often not predicted by traditional toxicology studies searching for the lowest observable effect level because such studies have typically assumed that effects of chemicals are monotonic (dose-response curves do not change sign) in order to predict effects at low doses, when responses may in fact increase at low doses for a variety of reasons (e.g., tissue-specific receptors and cofactors, varying receptor selectivity at different chemical doses, endocrine feedback loops) (Vandenberg et al., 2012). For example, DDT has been found to impact animal neurobehavior below the Environmental Protection Agency's (EPA's) low-dose

cutoff of 0.05 µg/g (Vandenberg et al., 2012), and levels in Whooping Cranes eggs were as high as 0.06 µg/g in both eggs and livers (Warner, 2013). For these reasons, we urge continued consideration of the role of pesticides in Whooping Crane health and reproductive behavior.

Aside from the general information that cranes at Wood Buffalo National Park primarily consume food from wetlands, prior studies suggest that this population may rely heavily on small food items. This is similar to the finding by Fitzpatrick (2016) that at least one pair of cranes in Wisconsin relied heavily on snails for energy intake. Novakowski (1966) noted that a family of cranes at Wood Buffalo typically appeared to feed on benthic organisms and sampled bottom material from locations where Whooping Cranes had recently fed. Based on organisms found in the samples, Novakowski (1966) proposed that cranes fed largely on insect larvae and amphipods. The majority of organisms in the sediment were molluscs (snails and pill clams), but Novakowski (1966) discounted them as potential prey items due to their high ash content. It is possible that larger snail species are available in Wisconsin wetlands than in Wood Buffalo. Bergeson et al. (2001) found that Whooping Crane chicks in Wood Buffalo were primarily provisioned with dragonfly larvae, although chick diets may not reflect adult diets.

Diet of the EMP at wintering grounds has not been directly studied, although studies of habitat use indicate that the cranes make heavy use of cornfields for foraging (Mueller et al., Chapter 11, this volume; Teitelbaum et al. 2016; Thompson, 2018) in Indiana, Tennessee, and Alabama. This is quite different from the diet observed in AWBP cranes at Aransas NWR, which is composed primarily of foods captured in salt marshes.

Little information is available on when EMP Whooping Cranes store fat reserves via the food items described, and how/when they use fat reserves. Fat storage and body weight changes have never directly been measured in wild Whooping Cranes. Previous research on both the AWBP and the EMP centers on the potential for use of fat stores in reproduction. The only study of fat storage/use by EMP Whooping Cranes was conducted via comparing estimated energy intake to estimated energy expenditure in individual EMP cranes (Fitzpatrick, 2016). Results indicated that EMP cranes store fat upon arrival at the breeding grounds, prior to laying eggs. Fitzpatrick (2016) suggested that storing fat reserves upon arrival may be a strategy to prepare for maintaining incubation constancy, as some waterfowl species are known to do. However, sufficient energy intake data were acquired for only a single pair of cranes in a single year. Fitzpatrick (2016) noted that there may be variation in fat storage strategies among cranes in the EMP (see later).

Fat storage at the breeding grounds has not been studied in AWBP cranes, so it is unknown whether they store fat reserves at the breeding grounds as EMP cranes (at least some pairs in some years) do. However, like many arctic- and subarctic-breeding birds (Drent, 2006; Drent and Daan, 1980; Meijer and Drent, 1999), AWBP Whooping Cranes are likely to rely on fat reserves acquired at wintering grounds to help fuel their metabolic requirements for self-maintenance and reproduction. Whooping Cranes in the AWBP often arrive at Wood Buffalo National Park when water is frozen, invertebrate prey is scarce, and small vertebrates are dormant (Chavez-Ramirez, 1996). Hence, it is likely that Whooping Cranes are unable to gather much energy and must use stored energy to produce eggs and possibly to support their own energy needs through part of the incubation period. Further evidence for this hypothesis includes the findings that AWBP cranes store significant amounts of fat in winters when food is abundant (Chavez-Ramirez, 1996) and that some aspects of reproductive success at Wood Buffalo are correlated with conditions at ANWR during the prior winter (Chavez-Ramirez, 1996; Gil de Weir, 2006). This information implies that winter fat storage plays a role in reproductive success.

Although Fitzpatrick (2016) found that a pair of EMP Whooping Cranes stored fat following arrival at the breeding grounds, fat reserves stored at wintering grounds may be relevant to reproductive success of EMP cranes, as they are likely to be for AWBP cranes. First, relying on fat reserves acquired at wintering grounds may be a more typical strategy among Whooping Cranes as a species, and the pair observed by Fitzpatrick (2016) may have been unable to gather the necessary energy reserves prior to arrival at the breeding grounds. Individuals within species known to acquire fat stores prior to arrival on the breeding grounds may delay breeding to acquire fat stores at the breeding grounds if they arrive with low endogenous reserves (Drent, 2006). The pair observed by Fitzpatrick (2016) nested 8 days later than the average date for pairs in 2013, despite being an experienced breeding pair. They may have arrived with low endogenous reserves and delayed breeding as a consequence. While cranes may be able to gain sufficient energy to breed after arrival, a delay in breeding may impact reproductive success in other ways, for example, changing breeding phenology in relation to black fly outbreaks. Given that the observation was of just one pair in Fitzpatrick (2016), even if that pair's behavior upon arrival at the breeding grounds was a response to unusually low fat reserves, it is unknown whether there is variation in pair response to low endogenous reserves. Some pairs may choose to begin breeding regardless and consequently encounter energy stress later during the incubation period.

Second, it is unknown how the ability to acquire fat reserves at breeding grounds varies among years in Wisconsin. In years with inclement weather (frozen water or cold conditions making invertebrates scarce longer into the spring), fat reserves acquired at wintering grounds may be essential. Fitzpatrick (2016) noted that the focal pair of cranes in that study was not observed until 10 days after their arrival, and that a period of low food availability may have been missed even in a year where they were eventually able to acquire fat reserves.

Finally, if it is the typical strategy for Whooping Cranes to acquire fat reserves at the breeding grounds prior to nesting, similar to some other waterfowl species, it still may be advantageous to arrive with a certain base level of fat reserves. The more fat reserves must be replenished, the longer birds must delay breeding.

Regarding fat storage at other times of year, the limited available data suggest that the period following nest abandonment may be another important annual period of fat storage for EMP Whooping Cranes. The same pair of cranes that acquired fat reserves upon arrival at the breeding grounds (Fitzpatrick, 2016) acquired fat reserves on a diet of earthworms and waste corn in agricultural fields following nest abandonment. As mentioned earlier, use of agricultural fields with abundant (earthworms) and energy-dense (waste corn) food resources may be a strategy to replenish fat reserves for many pairs following nest abandonment (Van Schmidt et al., 2014). How caring for a chick would influence the ability of a successfully breeding pair to acquire fat reserves at this time of year is not known.

Review Summary

In summary, energy costs for Whooping Cranes do not appear particularly high compared to other crane species. Overall, the highest energy costs for EMP cranes during the breeding season appear to be costs of BMR and physical activity. Costs of thermoregulation are small. Daily energy expenditures appear similar between the EMP and AWBP populations in spring, although time budget information from the AWBP is necessary to confirm this result. Energy expenditures of EMP cranes in spring are also similar to those of AWBP cranes in winter. Whooping Cranes in Wisconsin meet their energy costs on a diet that includes a variety of foods and consume animals from the same broad taxa consumed by wintering AWBP cranes. Quantitatively, snails and fish are particularly important to at least some pairs in Wisconsin during the prenesting period, but there is likely variation among pairs using different habitat types available on or near their territories. EMP cranes differ from AWBP cranes in that they obtain foods from agricultural fields during spring.

Data from a single pair of breeding cranes in the EMP indicates that EMP cranes restock fat reserves upon arrival at the breeding grounds in spring, as some waterfowl species do. Such replenishment of fat reserves following spring migration could be useful in supporting incubation constancy. However, information from more pairs is necessary to determine how this behavior varies within the population. In light of evidence that AWBP cranes use fat reserves acquired at wintering grounds in reproduction, it is possible that this pair delayed breeding in response to low fat reserve levels. Comparable information about fat storage at the breeding grounds is not available for the AWBP.

ENERGETICS OF WHOOPING CRANES ON THE WINTERING GROUNDS

Here we report analyses focused on the ecological energetics of wintering cranes in the EMP, with implications for conservation. One aspect of EMP ecology that differs markedly from the AWBP is its wintering distribution. While the AWBP winters in a relatively small area around Aransas NWR on the Gulf coast of Texas, Whooping Cranes in the EMP have developed a much wider wintering range extending farther north, with cranes wintering across a range from Florida to Indiana (Mueller et al., Chapter 11, this volume; Teitelbaum et al., 2016; Urbanek et al., 2014). The expansion of the EMP wintering range from release sites in Chassahowitzka NWR and St. Mark's NWR, Florida, is likely due to a combination of factors: lack of suitable habitat in the area immediately surrounding release sites, recent drought and habitat loss in the southeastern United States, a propensity to shortstop (shorten the migratory path and winter closer to the breeding grounds) in more northern regions when mild winters make habitat and food (particularly corn in agricultural fields) available, and a tendency to reuse shortstop locations in future years (Teitelbaum et al., 2016; Urbanek et al., 2014).

Large birds like Whooping Cranes have reduced lower critical temperatures and reduced heat loss per unit body mass compared to smaller birds. However, given the extent to which EMP Whooping Cranes have expanded their wintering range, it is possible that energy costs of thermoregulation are heightened for cranes wintering in more northern locations. Choice of wintering location may therefore influence reproductive success (Runge et al., 2011; Urbanek et al., 2014), particularly if cranes rely on fat (energy) reserves acquired at wintering grounds during the breeding season (as described in "Review: Ecological Energetics of Whooping Cranes"). If cranes experience substantially heightened energy costs of thermoregulation at northern wintering locations, they may be unable to acquire the same level of fat reserves that Whooping Cranes in Florida are able to acquire, leading to reduced fat availability following spring migration. Energy stress (insufficient fat reserves combined with insufficient ability to acquire local foods on the breeding grounds) may infuence cranes' decisions to abandon nests in response to black flies. If energy stress contributes to spring nest abandonment or if low fat reserves have negative effects on reproductive success by affecting timing of breeding, an understanding of how choice of wintering location affects ability to store fat reserves will be important to mitigating low reproductive rates.

Urbanek et al. (2014) did not find any meaningful relationships between wintering region and subsequent breeding success of EMP cranes in an analysis of data from the beginning of the reintroduction through winter 2011–12. However, they pointed out that large numbers of black-fly-induced abandonment events may mask effects of different wintering locations. In addition, Urbanek et al. (2014) suggested that, once cranes begin using more northern wintering locations, they may be reluctant to move farther south when colder winters make these areas marginally suitable. Large numbers of cranes first began wintering in Indiana during the relatively warm winter of 2011–12 (Urbanek et al., 2014), and at least some cranes continued wintering there during subsequent colder winters (e.g., 2013–14).

Given the potential for fat storage to impact Whooping Crane reproduction, we have investigated the effects of wintering ground choice on Whooping Crane energy expenditure. We hypothesized that energy requirements to meet thermoregulatory costs would be substantially heightened at more northern wintering grounds compared to the southern-most wintering areas in Florida due to differing microclimate conditions, even for a large bird like the Whooping Crane. Our goal was to quantify the scope for differences between wintering locations and identify potential impacts on Whooping Cranes for future study.

To evaluate the scope for differences in energy requirements of Whooping Cranes across the wintering range, we used a mechanistic microclimate model coupled to a mechanistic model of animal heat and mass balance (Niche Mapper™). This approach was used to simulate the Whooping Crane energy expenditure across the EMP's wintering range. We compared the energy costs of overwintering at some commonly used wintering grounds at varying latitudes: Goose Pond Fish and Wildlife Area, Indiana; Muscatatuck NWR, Indiana; Hiwassee Wildlife Refuge, Tennessee; Wheeler NWR, Alabama; St. Mark's NWR, Florida; and Chassahowitzka NWR, Florida. We also considered potential effects of differences in energy expenditure on time-energy budgets and fat storage.

Methods

Introduction to the Model

The mechanistic model that we used to simulate crane energy expenditure is called Niche Mapper. At its core, Niche Mapper calculates the heat (energy) that an animal must expend to maintain its core temperature and user-input activity levels. Niche Mapper consists of a microclimate model coupled to a model of animal heat and mass balance (Fig. 12.1). Output from the microclimate model serves as input to the animal model. Niche Mapper's ability to accurately estimate Whooping Crane energy expenditure has been validated against doubly labeled water measurements of captive Whooping Crane energy expenditure (Fitzpatrick et al., 2015).

Microclimate Model Description and Parameterization

The microclimate model (Fuentes and Porter, 2013; Kearney et al., 2014a; Porter et al., 1973) uses standard meteorological measurements (e.g., air temperatures at 2 m height, wind speeds at 2 m height), topographic parameters, vegetation properties, and soil properties to simulate environmental conditions at an animal's height above ground (i.e., its microclimate). These microclimate parameters are calculated on an hourly basis for each day and location being modeled. The component heat and mass transport equations are described elsewhere (Fuentes and Porter, 2013; Kearney et al., 2014a; Porter et al., 1973). We used the model version "NicheMapR" (Kearney et al., 2014a), which runs in an R environment (R Core Team, 2014).

In this study, we simulated microclimate conditions and associated wintering crane energy expenditures across a large portion of the eastern United States (latitude 48.3°N to 24.5°N, longitude 79°W to 92°W) encompassing Whooping Crane wintering areas. We simulated average monthly climate conditions and associated crane energy expenditures for December, January, and February of the winters 2011–12 and 2013–14. These months were chosen because EMP cranes typically begin fall migration in mid-November to early December and spring migration in early March. These years were the warmest and coldest (respectively) winters during the period of the EMP reintroduction (2001 to the present).

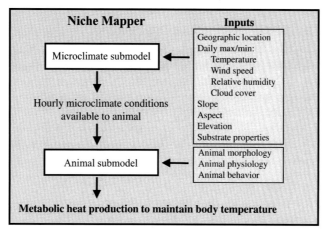

FIGURE 12.1 Conceptual diagrams of Niche Mapper submodels, inputs, and outputs used to model energy requirements of Whooping Cranes in the Eastern Migratory Population across various wintering areas and weather conditions.

We simulated energy expenditure during individual years rather than under average climate conditions for the period of the reintroduction because EMP cranes do not show high fidelity to wintering sites (Urbanek et al., 2014) and may change locations from year to year. We wanted to simulate energy expenditures using combinations of climate conditions and locations that cranes have experienced. Further details regarding parameterization of the microclimate model are discussed in Appendix 1.

Endotherm Model Description and Parameterization

For each hour of the day, the endotherm model solves a heat balance equation to find the metabolic rate that an animal must maintain to sustain a constant core body temperature in the hour's microclimate conditions, given its morphology and behavior (e.g., posture and user-input activity levels). Between the animal and environment, heat exchange through convection, conduction, solar radiation, long-wave infrared radiation, respiratory evaporation, and cutaneous evaporation are simulated. Output parameters from the microclimate model are imported into the animal heat and mass balance model to serve as variables in the heat exchange equations.

Animal size and shape also play a role in heat exchange equations. Animals with complex body shapes, such as Whooping Cranes, are modeled as a series of connected simple shapes. Whooping Cranes were modeled with an ellipsoidal torso, cylindrical legs, cylindrical neck, truncated cone-shaped head, and truncated cone-shaped beak.

Heat exchanges are modeled through three layers of each body part. Metabolic heat production is simulated as distributed heat production throughout the flesh of each body part. Whooping Cranes were also modeled with a thin layer of insulating subcutaneous fat on the torso, through which heat could pass via conduction. Finally, the model simulates simultaneous air and feather conduction and infrared radiation through the feather layer. The heat exchange equations and their derivations are described in Mathewson and Porter (2013) and Mathewson (2013).

Our goal in this analysis was to simulate an "average-sized" Whooping Crane across varying environmental conditions. Because the morphometric measurements and feather properties required to simulate individual Whooping Crane body parts were not available in the literature, we averaged measurements acquired from two captive Whooping Cranes (one male, one female) in Fitzpatrick et al. (2015). Morphological measurements are described in Table 12.A1.

Other parameters to simulate Whooping Cranes were measured or estimated from literature values (Table 12.A2). Whooping Cranes were simulated as using counter-current heat exchange between blood vessels as a heat-conservation method in the legs and beak (Table 12.A2), giving them a reduced average core temperature in the beak and legs (Fitzpatrick et al., 2015).

We simulated Whooping Cranes as inactive (not engaging in physical activities that require energy and produce heat) during the day and night, rather than engaging in activity during the day. We did this because the wintering range of Whooping Cranes spans latitudes with differing day lengths. Cranes at different latitudes may consequently be active for different numbers of hours. We wanted to exclusively simulate differences in energy expenditure due to differing energy costs of thermoregulation, rather than differences due to differing numbers of active hours with associated differing levels of daily physical activity.

In the model, energy costs of thermoregulation are simulated based on estimated BMR of the animal (Table 12.A2). If the initial simulated metabolic rate for the hour falls outside a small target range around BMR ($\pm 5\%$), behavioral and physiological thermoregulatory mechanisms engage. The range around BMR is necessary to prevent the model from entering infinite loops, and is not intended to have biological significance. The metabolic rate necessary to maintain core temperature is then recalculated accounting for new behavioral or physiological parameter values. This process continues until the calculated metabolic rate lies within the target range or the scope of thermoregulatory accommodation is exhausted. If thermoregulatory mechanisms are exhausted, the animal incurs energy costs for thermoregulation. To reduce heat loss in cold conditions, simulated Whooping Cranes in this study were allowed to undergo ptiloerection, vasoconstriction, and slight decreases in core temperature (Table 12.A2).

Simulation run time was inflated by the large number of geographic locations (2,923,661) over which Whooping Crane energy expenditure was simulated. To reduce simulation run time, thermoregulatory mechanisms for cooling the animal in warm conditions (e.g., vasodilation, shunting of blood away from counter-current exchange vessels in the legs to increase heat loss, panting) were not used. In the same vein, thermal conductivity of the flesh was automatically set to the minimum value (simulating vasoconstriction at all times) and core temperature was set to the minimum value. We then tested a cranes' ability to avoid overheating in the location with the highest daily maximum temperature. This was a location in southern Florida under February 2014 temperatures: 14.85 °C daily minimum to 29.08 °C daily maximum. Thermoregulation options allowed were panting, vasodilation, and shunting of blood to non-counter-current vessels in the legs and beak. Under these conditions, the simulated crane was able to maintain BMR without overheating. Given that the simulated crane could withstand the highest midday temperature in the simulated area, we assumed that cranes could avoid overheating in the entire simulated area.

Postprocessing of Model Output

Output from the model consisted of average simulated energy expenditure (kJ/day) for each simulated month. Estimated BMR (1105 kJ/day) was subtracted from output values to convert them into energy costs of thermoregulation above BMR. To obtain total energy expendi-

tures per winter for comparison among wintering locations, total energy expenditure for the period of December, January, and February was calculated. To create a raster containing spatially explicit estimates of total kJ/winter, each month's values (kJ/day) were multiplied by the number of days in that month to obtain kJ/month values. Monthly values were summed to obtain total kJ/winter and divided by 90 days to obtain average energy requirements on a daily basis (kJ/day) for each winter.

Shapefiles of federal and state management areas in which cranes were known to winter in 2011–12 and 2013–14 were overlaid on rasters in ArcGIS (ESRI, Redlands, CA), and values of energy expenditure within each area were averaged. Wintering areas for which averages were calculated were Goose Pond Fish and Wildlife Area, Indiana; Muscatatuck NWR, Indiana; Hiwassee Wildlife Refuge, Tennessee; Wheeler NWR, Alabama; St. Mark's NWR, Florida; and Chassahowitzka NWR, Florida. For reference, theoretical costs of overwintering at Necedah NWR were also calculated. A geographic information system (GIS) layer containing polygon boundaries of NWRs (FWS Interest Simplified layer of the USFWS National Cadastral Data data set) was obtained online from the U.S. Fish and Wildlife Service (http://www.fws.gov/gis/data/national/index.html). A layer containing Hiwassee Wildlife Refuge (Hiwassee WR) boundaries was obtained from the Tennessee Wildlife Resources Agency (http://www.state.tn.us/twra/gis/downloaddata.html). Goose Pond Fish and Wildlife Area (Goose Pond FWA) boundaries were downloaded from Indiana University Bloomington (Managed Lands IDNR layer at http://maps.indiana.edu/ layerGallery.html?category = ManagedLands).

Results

General Patterns across the Study Area

For each month, energy expenditure generally declined from north to south (Fig. 12.2;

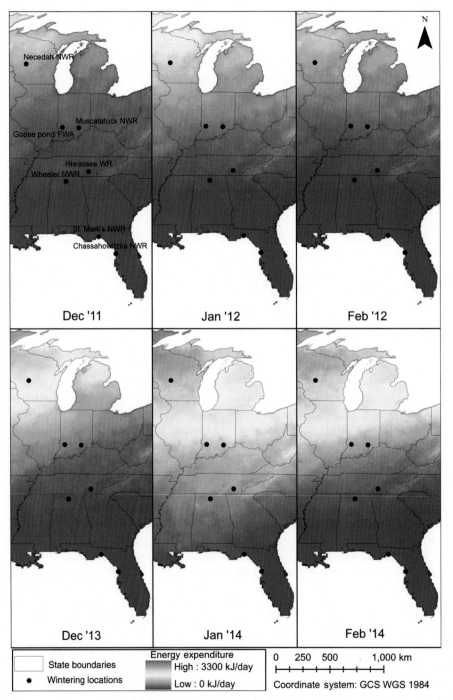

FIGURE 12.2   Simulated monthly Whooping Crane energy expenditures above basal metabolic rate (BMR) across a portion of the eastern United States under average climate conditions of December, January, and February of winter 2011–12 (mild winter) and winter 2013–14 (cold winter). Commonly used Whooping Crane wintering areas and the core breeding area of Necedah NWR are shown. Energy expenditures are shown in kJ/day above BMR (1105 kJ/day), that is, a value of 0 kJ/day indicates that the energy expenditure of simulated cranes was the BMR.

simulated monthly average Whooping Crane energy expenditures, kJ/day above BMR of 1105 kJ/day, across the area examined). Total winter (December–February) energy expenditure (kJ/winter above BMR) across the area examined also generally declined from north to south (Fig. 12.3). None of the Whooping Crane wintering locations analyzed appeared unusual in simulated energy expenditures compared to their immediate surroundings.

Comparing Wintering Locations

Simulated energy expenditures above BMR (hereafter "energy expenditures") for several wintering areas indicated that, within each year, energy expenditures in northern locations were larger than those in more southern locations (Table 12.1). Energy expenditures in wintering areas relatively near to each other were similar. Within each month, wintering areas in Muscatatuck NWR and Goose Pond FWA, Indiana, differed by a maximum of 136 kJ/day (or maximum 7% of total daily energy expenditure). St. Mark's NWR and Chassahowitzka NWR, Florida, differed by a maximum of 35 kJ/day (maximum 3% of total daily energy expenditure). Simulated energy expenditures at Hiwassee WR were consistently higher than those

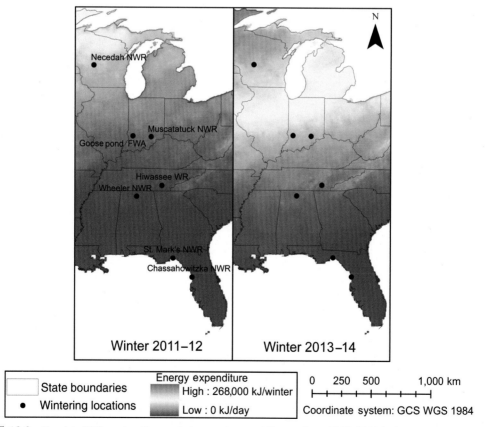

FIGURE 12.3 Simulated Whooping Crane winter energy expenditures above BMR (1105 kJ/winter) across a portion of the eastern United States under climate conditions of winter 2011–12 (mild winter) and winter 2013–14 (cold winter). Winter was defined as December, January, and February of each year. Commonly used Whooping Crane wintering areas and the core breeding area of Necedah NWR are shown. Energy expenditures are shown in kJ/day above BMR (1105 kJ/day or 99,450 kJ/winter), that is, a value of 0 kJ/day indicates that the energy expenditure of simulated cranes was the BMR.

TABLE 12.1   Mean (± SD) Simulated Whooping Crane Energy Expenditure above Basal Metabolic Rate (BMR) (1105 kJ/day) for Wintering Areas [Including National Wildlife Refuges (NWR), Fish and Wildlife Areas (FWA), and Wildlife Refuges (WRs) Used by the Eastern Migratory Population. Energy Expenditures Were Simulated under Temperature Conditions of a Mild Winter (2011–12) and a Cold Winter (2013–14). Average Monthly Values, Average Winter (Dec.–Feb.) Values, and Total Winter Values Are Shown. For Average Winter Values, Total Energy Expenditure as a Multiple of BMR Is Shown in Parentheses.

Location	Dec. (kJ/day)	Jan. (kJ/day)	Feb. (kJ/day)	Avg. winter (kJ/day)	Total winter (kJ/winter)
Winter 2011–12					
Necedah NWR, WI	900 ± 26	1289 ± 20	847 ± 19	1017 ± 21 (1.9 × BMR)	91,617 ± 1978
Goose Pond FWA, IN	222 ± 8	479 ± 9	301 ± 9	335 ± 9 (1.3 × BMR)	30,667 ± 662
Muscatatuck NWR, IN	236 ± 6	459 ± 7	325 ± 8	340 ± 7 (1.3 × BMR)	30,219 ± 823
Hiwassee WR, TN	80 ± 2	134 ± 4	72 ± 3	96 ± 3 (1.1 × BMR)	8716 ± 322
Wheeler NWR, AL	67 ± 3	65 ± 2	58 ± 3	63 ± 2 (1.1 × BMR)	5750 ± 240
St. Mark's NWR, FL	3 ± 3	6 ± 5	0 ± 0	3 ± 3 (1 × BMR)	313 ± 276
Chassahowitzka NWR, FL	0 ± 0	0 ± 0	0 ± 0	0 ± 0 (1 × BMR)	0 ± 0
Winter 2013–14					
Necedah NWR, WI	1797 ± 28	2339 ± 24	2213 ± 29	2113 ± 27 (2.9 × BMR)	190,226 ± 2438
Goose Pond FWA, IN	643 ± 11	1158 ± 21	982 ± 23	926 ± 18 (1.8 × BMR)	83,412 ± 1664
Muscatatuck NWR, IN	511 ± 9	1162 ± 11	846 ± 10	840 ± 10 (1.8 × BMR)	75,623 ± 945
Hiwassee WR, TN	151 ± 2	798 ± 7	214 ± 4	393 ± 4 (1.4 × BMR)	35,456 ± 381
Wheeler NWR, AL	105 ± 7	636 ± 17	181 ± 7	312 ± 10 (1.3 × BMR)	28,114 ± 930
St. Mark's NWR, FL	0 ± 0	76 ± 14	0 ± 0	26 ± 4 (1 × BMR)	2370 ± 445
Chassahowitzka NWR, FL	0 ± 0	41 ± 8	0 ± 0	14 ± 2 (1 × BMR)	1282 ± 264

at Wheeler NWR, but only by a maximum of 162 kJ/day (9% total daily energy expenditure). Thus, total winter energy expenditures were highest at Necedah NWR (where birds do not winter), followed by the two Indiana locations, Hiwassee WR, Wheeler NWR, and last the two Florida locations.

Differences in energy expenditure between northern and southern locations were larger in the cold winter of 2013–14 than in the mild winter of 2011–12. For example, the difference in average daily energy expenditure for the winter between Goose Pond FWA and Chassahowitzka NWR in 2013–14 (912 kJ/day) was more than double that of winter 2011–12 (335 kJ/day).

Average costs of wintering in southern Indiana (Goose Pool FWA or Muscatatuck NWR) instead of Florida (Chassahowitzka NWR or St. Mark's NWR) ranged from 332 kJ/day in a warm winter to 913 kJ/day in a cold winter. Energy savings could be obtained by migrating further south. Energy costs of thermoregulation in Tennessee instead of Florida (93 and 380 kJ/day) were 72% and 58% smaller than Indiana values. Costs declined slightly more for Alabama (60 and 298 kJ/day, or 82% and 67% smaller than Indiana costs). It is interesting to note that large thermoregulatory energy savings can be obtained even by migrating from Necedah NWR to southern Indiana (1014–2099 kJ/day).

Discussion

Energy Requirements across the Wintering Range

Our results confirm that there are heightened thermoregulation energy costs to wintering in more northern locations. These thermoregulatory costs are highest in southern Indiana (winter averages 335–926 kJ/day), and moderate in Tennessee and Alabama (winter averages 63–393 kJ/day). Thermoregulatory costs in Florida are very low (winter averages 0–26 kJ/day).

The heightened energy costs of thermoregulation at more northern wintering locations must correspond to heightened energy intake (i.e., food requirements) or increased physiological or behavioral measures for conserving energy. Depending on the availability and types of prey in different locations, cranes in more northern locations may need to increase the amount of daily time spent foraging. This could reduce time available for other activities, including vigilance, maintenance (e.g., preening), or social behaviors. Shorter day lengths at more northern locations limit total activity time for diurnal animals and may further constrain time available for nonforaging behaviors.

There is not enough information available about the winter foraging behavior, diet, or food availability of cranes in the EMP to estimate the degree to which additional foraging time impacts time-energy budgets. We estimated how additional energy costs of wintering in northern areas, compared to Florida, relate to additional food requirements for some potential Whooping Crane prey items (Table 12.2; Table 12.A3). Literature-derived values used to estimate the energy content of potential food items are also provided in Table 12.A3. Based on these estimates, the amounts of additional food items that Whooping Cranes must consume to offset additional thermoregulatory costs in Indiana are relatively large (87–338 g/day) for food types other than corn. The amount of waste corn that Whooping Cranes would have to consume (26–71 g/day) is much smaller (20–30% of the mass of noncorn foods) because corn is a particularly energy-dense food item. This may explain why Whooping Cranes in Indiana forage almost exclusively for waste corn in harvested fields (Thompson, 2018), as reported by Urbanek et al. (2014). Urbanek et al. (2014) pointed out that this food source may provide an incentive for cranes to shortstop there for the winter. Given the amounts of additional noncorn food items required to offset energy costs in Indiana, our results suggest that waste corn may be crucial to allowing Whooping Cranes to survive the winter so far north. Although heightened wintering energy costs in Tennessee and Alabama are smaller than in Indiana, Whooping Cranes at Hiwassee WR and Wheeler NWR also regularly forage in cornfields, particularly fields on the refuges that are planted and maintained for waterfowl (Thompson, 2018).

It is difficult to assess the importance of corn to Whooping Crane energetics in these areas without data regarding foraging rates and energy costs of foraging associated with noncorn food items. However, assuming that the availability of waste corn is similar between all locations, it is likely that cranes in Indiana must spend more time foraging than cranes at Hiwassee WR and Wheeler NWR, making the time budgets of cranes different in these two locations. This hypothesis could be tested in the future.

It is possible that cranes wintering in Indiana experience limited ability to store fat reserves, especially during colder winters. As illustrated in this analysis, energy costs can vary by large amounts (more than 100%) between winters. In a warm winter like 2011–12, energy costs at Goose Pond FWA and Muscatatuck NWR were similar to those at Hiwassee WR and Wheeler NWR. It was under conditions of a mild winter (winter 2006–07) that Whooping Cranes first began to overwinter in Indiana, and it was during the mild winter of 2011–12 that a large fraction of the population began wintering there (Urbanek et al., 2014). However, at least some Whooping

TABLE 12.2   Estimated Amounts of Food Items That a Whooping Crane Must Consume to Offset Increases in Energy Costs of Thermoregulation for Wintering in Northern Regions, Compared to Florida. The Difference in Simulated Energy Costs of Thermoregulation between Each Wintering Area and Florida Is Reported in the Second Column. Values Are Differences in Average Daily Energy Expenditure for the Dec.–Feb. Period of Each Winter. Ranges Correspond to Values Simulated under the Temperature Conditions of Two Winters: 2011–12 (Mild Winter) and 2013–14 (Cold Winter). The Amounts of Food with a Metabolizable Energy Content Equal to That Difference Are Listed in Subsequent Columns for Several Potential Food Types.

Wintering location	Difference from Florida[a] (kJ/day)	Corn (shelled)		Aquatic insects		Sago pondweed tubers (*Potamogeton pectinatus*)		Crayfish (*Oronectes* spp.)	
		g/day	Kernels/day	g/day	Insects/day	g/day	Tubers/day	g/day	Crayfish/day
Indiana[b]	332–913	26–71	74–203	87–240	44–120	123–338	123–338	123–338	5–14
Hiwassee WR[c]	93–380	7–29	21–84	25–100	12–50	35–141	35–141	35–141	1–6
Wheeler NWR[c]	60–298	5–23	13–66	16–78	8–39	22–110	22–110	22–110	1–5

[a] *Energy expenditures were calculated for Chassahowitzka NWR and St. Mark's NWR, Florida. Calculated energy expenditures for the two locations were similar. The larger value of the two was used to calculate the minimum value for the range in the table, and the smaller of the two was used to calculate the maximum value for the range.*

[b] *Energy expenditures were calculated for Goose Pond Fish and Wildlife Area and Muscatatuck NWR, Indiana. Calculated energy expenditures for the two locations were similar. The smaller value of the two was used to calculate the minimum value for the range in the table, and the larger of the two was used to calculate the maximum value for the range.*

[c] *WR = Wildlife Refuge, NWR = National Wildlife Refuge.*

Cranes winter in Indiana in colder winters, as in 2013–14. The heightened energy cost of thermoregulation in our simulated cold winter conditions suggests that wintering in Indiana could lead to energy stress or inability to store sufficient fat reserves for migration and breeding. If 52.4 kJ metabolizable energy must be consumed to store 1 g of fat, assuming 75% production efficiency (Krapu et al., 1985) and a fat energy density of 39.3 kJ/g (Schmidt-Nielson, 1997), then food consumed to meet additional thermoregulatory costs of wintering in Indiana compared to Florida could otherwise have gone to storage of 0.6–1.6 kg of body fat. Costs of overwintering at Hiwassee WR compared to Florida correspond to storage of 0.2–0.7 kg body fat, and those for Wheeler NWR correspond to 0.1–0.5 g body fat. To put this in perspective, maximum fat content in Sandhill Cranes (*Antigone canadensis*) at spring staging areas along the Platte River averaged about 35% of average lean body mass (Krapu

et al., 1985). For the simulated Whooping Crane composed of 5.0 kg lean body mass (see body mass and fat mass in (Table 12.A2), this percentage fat content would correspond to 1.8 kg body fat. Thus, if Whooping Cranes store fat at levels comparable to Sandhill Cranes, these values of potentially lost fat reserves correspond to increasingly larger percentages of required body fat storage as birds winter further north (up to 89%).

However, we note that birds wintering farther north experience reduced energy costs of spring migration. With the increase in shortstopping behavior over time, the average migration distance for the population has decreased by over half since the beginning of the reintroduction (Teitelbaum et al., 2016). Reduced spring migration distance and associated demands on fat reserves for birds that have wintered in northern locations could compensate for reduced levels of fat reserves at the beginning of migration. The trade-off between energy costs of thermoregulation and migration distance could be considered in more detail with more information about Whooping Crane flight (see "Summary and Outlook").

Model Sensitivity

Sensitivity analysis of the Niche Mapper model to various parameters for Whooping Cranes was conducted previously (Fitzpatrick et al., 2015). Because parameters used to simulate a Whooping Crane in this study were the same or were averages of values used in Fitzpatrick et al. (2015), those sensitivity analyses are pertinent to this study.

Among the parameters evaluated, the model was most sensitive to morphometry (length and width of body parts), solar reflectivity, changing core temperatures, and changes in the modeled minimum difference between crane leg temperature and environmental temperature (i.e., efficiency of counter-current heat exchange in legs).

There is certainly variation in morphometry among members of the population. It is expected that larger cranes would incur lower thermoregulatory costs at more northern locations, given their smaller ratio of surface area to volume compared to smaller cranes. However, because simulated cranes in the sensitivity analyses were not allowed to engage in any thermoregulatory responses aside from changing leg temperatures in the sensitivity analyses, we would expect smaller changes in energy expenditure with allometry in real wintering cranes. Given the large thermoregulatory costs simulated for Whooping Cranes in Indiana in cold winters, we do not expect this to have an impact on our conclusions.

Other noted parameters are not expected to vary much among individual cranes, although solar reflectivity has not been measured in many Whooping Cranes. We further note that our simulations in both studies were restricted to adult Whooping Cranes, and that simulations of juveniles should incorporate reflectivity measurements of brown feathers. There is no information available regarding how crane leg-air temperature differences vary among individuals, although Fitzpatrick et al. (2015) observed no major differences between the two cranes measured in that study. Further details about sensitivity analyses are available in Fitzpatrick et al. (2015).

SUMMARY AND OUTLOOK

Wintering Whooping Cranes

Given the extent to which energy costs of thermoregulation are heightened in northern wintering locations, particularly in cold years, cranes' ability to store fat reserves may be limited by choice of wintering grounds. Because cranes are continuing to use these northern areas in cold winters, continued monitoring of breeding success in relation to wintering area is warranted. This may be particularly useful if cranes begin breeding in areas with fewer black flies, which induce widespread nest desertion that could have masked effects of wintering area in previous analyses (Urbanek et al., 2014).

To simplify our analysis and examine energy costs of thermoregulation alone, we simulated cranes as resting throughout the day. We recommend carrying out simulations incorporating diurnal activity. This could decrease differences in thermoregulatory costs between northern locations and Florida because activity produces metabolic heat that can contribute to the maintenance of core body temperature in cold conditions. In this case, Whooping Cranes in more northern locations may even produce more activity-generated heat than those in more southern locations because the activity of foraging should be more frequent at higher latitudes. However, diurnal activity would not affect thermoregulatory costs at night when costs are greatest due to cooler air and sky temperatures and lack of solar radiation. In addition, impacts of activity would be lessened at more northern latitudes due to shorter day lengths. Time-energy budgets and corresponding activity energy expenditures are not available for wintering cranes in the EMP, but future analyses could examine effects of activity on thermoregulatory costs at a range of possible activity levels.

The potential for reduced spring migration distance to offset the need to store as much fat during the winter is an important factor that should also be considered. There are likely to be large differences in energy costs of migration between cranes migrating to Wisconsin from Florida versus Indiana. To estimate flight costs for Whooping Cranes, data pertaining to time spent in soaring versus flapping flight and some morphological parameters of individual cranes (e.g., wing surface area, body mass) will be necessary.

Our study results also point to the possible importance of energy-rich corn to Whooping Cranes wintering in Indiana and other northern locations. Given that such cranes are relying primarily on corn on private land (H. Thompson, International Crane Foundation, personal communication), rather than on managed property as Whooping Cranes in Hiwasssee WR and Wheeler NWR are doing, monitoring of corn availability and potential land use change near Whooping Crane roost sites is warranted (Mueller et al., Chapter 11, this volume).

As Urbanek et al. (2014) have discussed, understanding of the EMP wintering distribution will be particularly important to future reintroduction efforts because all migratory Whooping Crane populations except the AWBP were extirpated before they could be studied. No information about wintering ecology in inland habitats is available except from the EMP. Monitoring of this population will inform managers about how Whooping Cranes are likely to behave in potential future reintroductions and how those behaviors may impact their reproductive success via winter energetics. Studies of ecological energetics like the current one can also help place the behaviors of the EMP in context. For example, prior to the widespread availability of corn in agricultural fields, Whooping Cranes

may not have been able to winter as far north as southern Indiana (Teitelbaum et al., 2016).

Ecological Energetics throughout the Annual Cycle

We have described several ways in which information about ecological energetics is relevant to current Whooping Crane conservation challenges. First, information about crane energy requirements and how cranes meet their energy requirements could be used for choosing reintroduction locations that are energetically ideal (sufficient amounts of preferred food resources for successful reproduction given local weather and disturbance conditions) as WCEP shifts reintroduction locations to areas with lower black fly populations than Necedah NWR and changes rearing and release strategies (Converse et al., Chapter 8, this volume). Second, energy stress may act as a secondary factor influencing Whooping Cranes' decisions to abandon their eggs.

We do not have sufficient information about the ecological energetics of EMP Whooping Cranes to make recommendations pertaining to reintroduction locations or chick-rearing strategies. However, recent studies (Barzen et al., 2018; Fitzpatrick, 2016) have developed methods that can be used to obtain this information. In particular, we would recommend continued work in (1) quantifying diet of breeding cranes to determine which food types are most important to them from energy and nutrient intake perspectives and (2) assessing fat storage and use by comparing energy intake to energy expenditure. As noted earlier, this information is available from only one pair in one year, but variation is likely to exist within the EMP.

In addition, Fitzpatrick (2016) was unable to obtain information about energy intake during the incubation period of cranes due to limited sample size of pairs and logistical problems in the field (the observed pair of cranes nested too close to the researchers' observation blind and therefore the researchers did not use the blind to avoid disturbing the pair). However, gaining information about the energy intake and expenditure of cranes during incubation would allow for assessment of the hypothesis that energy stress contributes to nest abandonment.

Acknowledgments

We thank the University of Wisconsin–Madison Zoology Department for providing graduate student summer support and the Wisconsin Alumni Research Foundation (WARF) for a dissertation completion fellowship. We would like to acknowledge Jeremiah Yahn for preparing elevation data for the analysis. We acknowledge Jeb Barzen for comments and advice on various portions of the chapter. We would like to acknowledge the Whooping Crane Tracking Partnership (including the Canadian Wildlife Service, Crane Trust, U.S. Fish and Wildlife Service, the Platte River Recovery Implementation Program, and U.S. Geological Survey, with support from the Gulf Coast Bird Observatory, International Crane Foundation, and Parks Canada) for providing estimated arrival dates used to calculate the day lengths relevant to the AWBP. We acknowledge the Whooping Crane Eastern Partnership for estimated arrival and egg-laying dates used to calculate day lengths relevant to the EMP.

References

Abdollahi, M., Ranjbar, A., Shadnia, S., Nikfar, S., Rezaiee, A., 2004. Pesticides and oxidative stress: a review. Med. Sci. Monit. 10, RA141–RA147.

Anderson, M.G., Low, J.B., 1976. Use of sago pondweed by waterfowl on the Delta Marsh, Manitoba. J. Wildl. Manage. 40, 233–242.

Baldassarre, G.A., Whyte, R.J., Quinlan, E.E., Bolen, E.G., 1983. Dynamics and quality of waste corn available to postbreeding waterfowl in Texas. Wildl. Soc. Bull. 11, 25–31.

Barzen, J., Thousand, T., Welch, J., Fitzpatrick, M., Tran, T., 2018. Determining the diet of whooping cranes (*Grus americana*) through field measurements. Waterbirds 41, 22–34.

Bell, G., 1990. Birds and mammals on an insect diet: a primer on diet composition analysis in relation to ecological energetics. Stud. Avian Biol. 13, 416–422.

Bergeson, D.G., Bradley, M., Holroyd, G.L. 2001. Food items and feeding rates for wild whooping crane colts in Wood Buffalo National Park. Proceedings of the North American Crane Workshop 8, 36-39.

Brodzik, M., Armstrong, R., 2013. Northern Hemisphere EASE-GRID 2.0 Weekly Snow Cover and Sea Ice Extent. NASA National Snow and Ice Data Center Distributed Archive Center, Boulder, CO, Version, July 2011–June 2014.

Campbell, G.S., Norman, J.M., 1998. An Introduction to Environmental Biophysics. Springer-Verlag, New York.

Carman, J.A., Vlieger, H.R., Ver Steeg, L.J., Sneller, V.E., Robinson, G.W., Clinch-Jones, C.A., Haynes, J.I., Edwards, J.W., 2013. A long-term toxicology study on pigs fed a combined genetically modified (GM) soy and GM maize diet. J. Org. Syst. 8 (1), 38–54.

Chavez-Ramirez, F., 1996. Food availability, foraging ecology, and energetics of whooping cranes wintering in Texas. PhD dissertation, Texas A&M University, College Station.

Cho, B., 1969. Advanced Heat Transfer. University of Illinois Press, Urbana.

Converse, S.J., Royle, J.A., Adler, P.H., Urbanek, R.P., Barzen, J.A., 2013. A hierarchical nest survival model integrating incomplete temporally varying covariates. Ecol. Evol. 3 (13), 4439–4447.

Converse, S.J., Servanty, S., Moore, C.T., Runge, M.C., 2018. Population dynamics of reintroduced whooping cranes (Chapter 7). In: French, Jr., J.B., Converse, S.J., Austin, J.E. (Eds.), Whooping Cranes: Biology and Conservation. Biodiversity of the World: Conservation from Genes to Landscapes. Academic Press, San Diego, CA.

Converse, S.J., Strobel, B.N., Barzen, J.A., 2018. Reproductive failure in the Eastern Migratory Population: the interaction of research and management (Chapter 8). In: French, Jr., J.B., Converse, S.J., Austin, J.E. (Eds.), Whooping Cranes: Biology and Conservation. Biodiversity of the World: Conservation from Genes to Landscapes. Academic Press, San Diego, CA.

Cooke, S.J., O'Connor, C.M., 2010. Making conservation physiology relevant to policy makers and conservation practitioners. Conserv. Lett. 3 (3), 159–166.

Cooke, S.J., Sack, L., Franklin, C.E., Farrell, A.P., Beardall, J., Wikelski, M., Chown, S.L., 2013. What is conservation physiology? Perspectives on an increasingly integrated and essential science. Conserv. Physiol. 1 (1), cot001.

Demas, G.E., 2004. The energetics of immunity: a neuroendocrine link between energy balance and immune function. Horm. Behav. 45 (3), 173–180.

Dierenfeld, E.S., McGraw, K.J., Fritsche, K., Briggler, J.T., Ettling, J., 2009. Nutrient composition of whole crayfish (*Orconectes* and *Procambarus* species) consumed by hellbender (*Cryptobranchus alleganiensis*). Herpetol. Rev. 40, 324–330.

Dietert, R.R., Dietert, J.M., 2007. Early-life immune insult and developmental immunotoxicity (dit)-associated diseases: potential of herbal- and fungal-derived medicinals. Curr. Med. Chem. 14, 1075–1085.

Drent, R.H., 2006. The timing of birds' breeding seasons: the Perrins hypothesis revisited especially for migrants. Ardea 94 (3), 305–322.

Drent, R., Daan, S., 1980. The prudent parent: energetic adjustments in avian breeding. Ardea 68, 225–252.

Fitzpatrick, M.J. 2016. Mechanistic models and tests of whooping crane energetics and behavior locally and at landscape scales: implications for food requirements, migration conservation strategies and other bird species. PhD dissertation, University of Wisconsin–Madison.

Fitzpatrick, M.J., Mathewson, P.D., Porter, W.P.P., 2015. Validation of a mechanistic model for non-invasive study of ecological energetics in an endangered wading bird with counter-current heat exchange in its legs. PLoS ONE 10 (8), e0136677.

French, Jr., J.B., Converse, S.J., Austin, J.E., 2018. Whooping cranes past and present (Chapter 1). In: French, Jr., J.B., Converse, S.J., Austin, J.E. (Eds.), Whooping Cranes: Biology and Conservation. Biodiversity of the World: Conservation from Genes to Landscapes. Academic Press, San Diego, CA.

Fuentes, M., Porter, W., 2013. Using a microclimate model to evaluate impacts of climate change on sea turtles. Ecol. Model. 251 (4), 150–157.

Gil de Weir, K. 2006. Whooping crane (Grus americana) demography and environmental factors in a population growth simulation model. PhD dissertation, Texas A&M University, College Station.

Greer, D. 2010. Blue crab population ecology and use by foraging whooping cranes on the Texas Gulf coast. PhD dissertation, Texas A&M University, College Station.

Greenlee, K.J., Nebeker, C., Harrison, J.F., 2007. Body size-independent safety margins for gas exchange across grasshopper species. J. Exp. Biol. 210, 1288.

Hainsworth, F.R., 1981. Animal Physiology: Adaptations in Function. Addison-Wesley, Reading, MA.

Hashemi, A.M., Herbert, S.J., Putnam, D.H., 2005. Yield response of corn to crowding stress. Agron. J. 97, 839–846.

Kantrud, H.A., 1990. Sago pondweed (Potamogeton pectinatus L.): A literature review. U.S. Fish and Wildlife Resource Publication 176.

Karasov, W., 1990. Digestion in birds: chemical and physiological determinants and ecological implications. Stud. Avian Biol. 13, 1–4.

Kearney, M.R., Isaac, A.P., Porter, W.P., 2014a. Microclim: global estimates of hourly microclimate based on long-term monthly climate averages. Sci. Data 1, 140016.

Kearney, M.R., Shamakhy, A., Tingley, R., Karoly, D.J., Hoffmann, A.A., Briggs, P.R., Porter, W.P., 2014b. Microclimate modelling at macro scales: a test of a general microclimate model integrated with gridded continental-scale soil and weather data. Methods Ecol. Evol. 5 (3), 273–286.

King, R.S., Trutwin, J.J., Hunter, T.S., Varner, D.M., 2013. Effects of environmental stressors on nest success of introduced birds. J. Wildl. Manage. 77 (4), 842–854.

Krapu, G.L., Iverson, G.C., Reinecke, K.J., Boise, C.M., 1985. Fat deposition and usage by arctic-nesting sandhill cranes during spring. Auk 102 (2), 362–368.

Lafleur, P.M., 2008. Connecting atmosphere and wetland: energy and water vapour exchange. Geogr. Comp. 2 (4), 1027–1057.

Lafleur, P.M., McCaughey, J.H., Joiner, D.W., Bartlett, P.A., Jelinski, D.E., 1997. Seasonal trends in energy, water, and carbon dioxide fluxes at a northern boreal wetland. J. Geophys. Res. Atmos. (1984–2012) 102 (D24), 29009–29020.

Lafleur, P., Rouse, W.R., Hardill, S.G., 1987. Components of the surface radiation balance of subarctic wetland terrain units during the snow-free season. Arct. Alp. Res. 19 (1), 53–63.

Mathewson, P.D. 2013. Using biophysical models to protect wildlife in a changing climate: a polar bear case study. MS thesis, University of Wisconsin–Madison.

Mathewson, P.D., Porter, W.P., 2013. Simulating polar bear energetics during a seasonal fast using a mechanistic model. PLoS ONE 8 (9), e72863.

McNab, B.K., 2009. Ecological factors affect the level and scaling of avian BMR. Comp. Biochem. Physiol A: Mol. Integr. Physiol. 152, 22–45.

Meijer, T., Drent, R., 1999. Re-examination of the capital and income dichotomy in breeding birds. Ibis 141 (3), 399–414.

Mezzomo, B.P., Miranda-Vilela, A.L., Freire, I.D.S., Barbosa, L.C.P., Portilho, F.A., Lacava, Z.G.M., Grisolia, C.K., 2013. Hematotoxicity of Bacillus thuringiensis as spore-crystal strains cry1aa, cry1ab, cry1ac or cry2aa in Swiss albino mice. J. Hematol. Thrombo. Dis., http://dx.doi.org/10.4172/jhtd.1000104.

Mueller, T., Teitelbaum, C.S., Fagan, W.F., Converse, S.J., 2018. Movement ecology of reintroduced migratory whooping cranes (Chapter 11). In: French, Jr., J.B., Converse, S.J., Austin, J.E. (Eds.), Whooping Cranes: Biology and Conservation. Biodiversity of the World: Conservation from Genes to Landscapes. Academic Press, San Diego, CA.

Nelson, J.T., 1995. Nutritional Quality and Digestibility of Foods Eaten by Whooping Cranes on Their Texas Wintering Grounds. MS thesis, Texas A&M University, College Station.

Novakowski, N. 1966. Whooping crane population dynamics on the nesting grounds, Wood Buffalo National Park, Northwest Territories, Canada. Canadian Wildlife Service Report Series 1, pp. 1–20. Canadian Wildlife Service, Ottawa.

Olsen, G.H., Carpenter, J.W., Langenb, J.A., 1996. Medicine and surgery. In: Ellis, D.H., Gee, G.F., Mirande, C.M. (Eds.), Cranes: Their Biology, Husbandry, and Conservation. Department of the Interior, National Biological

Service, Washington, DC, International Crane Foundation, Baraboo, WI.

Porter, W., Mitchell, J.W., Beckman, W.A., DeWitt, C., 1973. Behavioral implications of mechanistic ecology: thermal and behavioral modeling of desert ectotherms and their micro-environment. Oecologia 13 (1), 1–54.

Porter, W.P., Munger, J., Stewart, W., Budaraju, S., Jaeger, J., 1994. Endotherm energetics: from a scalable individual-based model to ecological applications. Aust. J. Zool. 42, 125–162.

Porter, W.P., Vakharia, N., Klousie, W.D., Duffy, D., 2006. Po'ouli landscape bioinformatics models predict energetics, behavior, diets, and distribution on Maui. Integr. Comp. Biol. 46, 1143–1158.

Prinzinger, R., Pressmar, A., Schleucher, E., 1991. Body temperature in birds. Comp. Biochem. Physiol. A: Physiol. 99, 499–506.

PRISM Climate Group, O.S.U. Available from: http://prism.oregonstate.edu.

R Core Team, 2014. R: A Language and Environment for Statistical Computing. R Foundation for Statistical Computing, Vienna, Austria.

Runge, M.C., Converse, S.J., Lyons, J.E., 2011. Which uncertainty? Using expert elicitation and expected value of information to design an adaptive program. Biol. Conserv. 144 (4), 1214–1223.

Schmidt-Nielson, K., 1997. Animal Physiology: Adaptation and Environment, fifth ed. Cambridge University Press, New York.

Servanty, S., Converse, S.J., Bailey, L.L., 2014. Demography of a reintroduced population: moving toward management models for an endangered species, the whooping crane. Ecol. Appl. 24 (5), 927–937.

Sidhu, R.S., Hammond, B.G., Fuchs, R.L., Mutz, J.N., Holden, L.R., George, B., Olson, T., 2000. Glyphosate-tolerant corn: the composition and feeding value of grain from glyphosate-tolerant corn is equivalent to that of conventional corn (Zea mays L.). J. Agric. Food Chem. 48, 2305–2312.

Smith, E.H., Chavez-Ramirez, F., Lumb, L., 2018. Winter habitat ecology, use, and availability for the Aransas-Wood Buffalo Population of Whooping Cranes (Chapter 13). In: French, Jr., J.B., Converse, S.J., Austin, J.E. (Eds.), Whooping Cranes: Biology and Conservation. Biodiversity of the World: Conservation From Genes to Landscapes. Academic Press, San Diego, CA.

Stevenson, R.D., 2006. Ecophysiology and conservation: the contribution of energetics—introduction to the symposium. Integr. Comp. Biol. 46 (6), 1088–1092.

Swengel, S., 1992. Sexual size dimorphism and size indices of six species of captive cranes at the international crane foundation. In: Stahlecker, D. (Ed.), Proceedings of the North American Crane Workshop. North American Crane Working Group, Regina, Saskatchewan.

Teitelbaum, C.S., Converse, S.J., Fagan, W.F., Böhning-Gaese, K., O'Hara, R.B., Lacy, A.E., Mueller, T., 2016. Experience drives innovation of new migration patterns of whooping cranes in response to global change. Nat. Commun. 7, 12793.

Thompson, H.L., 2018. Characteristics of Whooping Crane Home Ranges during the Nonbreeding Season in the Eastern Migratory Population. MS thesis, Clemson University, Clemson, South Carolina.

Thompson, H., International Crane Foundation, Baraboo, WI, personal communication.

Tomlinson, S., Arnall, S.G., Munn, A., Bradshaw, S.D., Maloney, S.K., Dixon, K.W., Didham, R.K., 2014. Applications and implications of ecological energetics. Trends Ecol. Evol. 29 (5), 280–290.

Tracy, C.R., Nussear, K.E., Esque, T.C., Dean-Bradley, K., Tracy, C.R., DeFalco, L.A., Castle, K.T., Zimmerman, L.C., Espinoza, R.E., Barber, A.M., 2006. The importance of physiological ecology in conservation biology. Integr. Comp. Biol. 46 (6), 1191–1205.

Tucker, V.A., 1975. The energetic cost of moving about: walking and running are extremely inefficient forms of locomotion; much greater efficiency is achieved by birds, fish—and bicyclists. Am. Sci. 63 (4), 413–419.

Urbanek, R.P., Szyszkoski, E.K., Zimorski, S.E., 2014. Winter distribution dynamics and implications to a reintroduced population of migratory whooping cranes. J. Fish Wildl. Manage. 5 (2), 340–362.

Urbanek, R.P., Zimorski, S.E., Fasoli, A.M., Szyszkoski, E.K. 2010. Nest desertion in a reintroduced population of migratory whooping cranes. Proceedings of the North American Crane Workshop 11, 133–141.

Vandenberg, L.N., Colborn, T., Hayes, T.B., Heindel, J.J., Jacobs, Jr., D.R., Lee, D.-H., Shioda, T., Soto, A.M., vom Saal, F.S., Welshons, W.V., 2012. Hormones and endocrine-disrupting chemicals: low-dose effects and nonmonotonic dose responses. Endocr. Rev. 33, 378–455.

Van Schmidt, N.D., Barzen, J.A., Engels, M.J., Lacy, A.E., 2014. Refining reintroduction of whooping cranes with habitat use and suitability analysis. J. Wildl. Manage. 78 (8), 1404–1414.

Warner, S.E., 2013. Contaminants Screening in Whooping Crane Tissue – Eastern Migratory Population. U.S. Fish and Wildlife Service, Madison, WI. White paper, unpublished.

Welshons, W.V., Thayer, K.A., Judy, B.M., Taylor, J.A., Curran, E.M., Vom Saal, F.S., 2003. Large effects from small exposures I. Mechanisms for endocrine-disrupting chemicals with estrogenic activity. Environ. Health Perspect. 111, 994.

Wikelski, M., Cooke, S.J., 2006. Conservation physiology. Trends Ecol. Evol. 21 (1), 38–46.

APPENDIX 1: DETAILS OF NICHE MAPPER™ MODEL PARAMETERIZATION FOR ANALYSIS OF WINTERING WHOOPING CRANE ENERGETIC REQUIREMENTS

Microclimate Model

We simulated microclimate conditions on a monthly basis, using average climate conditions for each month, for winters 2011–12 and 2013–14. These were the warmest and coldest (respectively) winters on record during the period of the EMP reintroduction (2001 to the present). These years were selected as the warmest and coldest winters using the NOAA/NCDC Climate Division Data Mapping and Analysis Web Tool. We generated maps ranking the average temperature of each U.S. climate division (a geographic region of the United States defined by NOAA) for each winter (December through March) into percentiles (10% intervals) compared to all December–March periods for the climate division from 1895 to 2013–14. We chose winters 2011–12 and 2013–14 based on the map values for the region extending from Wisconsin to the southeastern United States. We simulated energy expenditure during individual years rather than under average climate conditions for the period of the reintroduction for these analyses because EMP cranes do not show high fidelity to wintering sites (Urbanek et al., 2014) and may change locations from year to year. We wanted to simulate energy expenditures using combinations of climate conditions and locations that real cranes have been documented to experience.

The microclimate model was parameterized for a landscape scale simulation across a large portion of the eastern United States (latitude 48.3°N to 24.5°N, longitude 79°W to 92°W) encompassing Whooping Crane wintering areas. For landscape-scale simulations, microclimate conditions were computed and exported to the endotherm model on a point-by-point basis across the area of interest. Elevation across the simulated area was obtained from Shuttle Research Topography Mission (SRTM) grids at 90-min resolution and resampled to 30-s resolution. Slope and aspect of the ground surface at each point were calculated from elevation data using ArcGIS. For each SRTM data location, we extracted average daily minimum and average daily maximum temperatures for each month of interest from Parameter-elevation Relationships on Independent Slopes Model (PRISM) grids (PRISM Climate Group, http://prism.oregonstate.edu) at 2.5-min resolution (data set AN81m). Spatial wind and cloud cover data were not available on a monthly basis for the time period and region of interest. Thus, we used monthly average values from Microclim data sets at 10-min resolution (Kearney et al., 2014a), which derived from 1961–90 data. Presence or absence of snow on the ground at each point was extracted from Northern Hemisphere EASE-Grid 2.0 Weekly Snow Cover and Sea Ice Extent grids (Brodzik and Armstrong, 2013). EASE-Grids were converted to a WGS 1984 geographic projection to match PRISM and SRTM data using ArcGIS. The weekly value for the week incorporating the middle day of each month was used as a proxy for the average condition (presence or absence of snow cover) for that month. R was used to execute all extractions from gridded data sets.

Substrate parameters were set as follows. Ground surface albedo was set to 0.75 for locations with snow cover (Campbell and Norman, 1998) and 0.15 for locations without snow cover, the middle value for sedge marsh in Lafleur et al. (1987) and within the range of values described in Lafleur (2008) and Lafleur et al. (1997). Soil wetness was set to 100% for locations with snow cover and 5% for locations without snow cover.

Soil properties, including a 5-cm organic cap to the soil, were simulated using generic values from Kearney et al. (2014a) and Kearney et al. (2014b) with an assumption of 33% clay content (a moderate value). Site-specific soil properties were not incorporated because Kearney et al. (2014b) demonstrated that generic soil properties were sufficient to simulate soil temperature profiles compared to measured values, as long as an organic cap was included.

Model Postprocessing

Energy expenditures above BMR were then converted into raster format at 30-s resolution using R.

APPENDIX 2: MODEL INPUT PARAMETERS FOR ANALYSIS OF WINTERING WHOOPING CRANE ENERGETIC REQUIREMENTS

TABLE 12.A1 Morphometric Input Parameter Values Used to Simulate Energy Expenditures of a Wintering Whooping Crane. Most Values Are the Average of Values Obtained from Two Captive Whooping Cranes (Fitzpatrick et al., 2015).

Parameter	Female	Source
Feather element diameter (μm)	18.75	Measured on plastic-embedded ostrich skin with feathers
Feather element density (cm^{-2})	14,400	Measured on plastic-embedded ostrich skin with feathers[a]
Length of feathers on neck and head, d/v (mm)	23.5/25.0	Average of head and neck values measured on study animals in Fitzpatrick et al. (2015)[b,c]
Head feather depth, d/v^d (mm)	10.0/8.25	Estimated from measurements of study animals in Fitzpatrick et al. (2015)[b]
Neck feather depth, d/v^d (mm)	10.0/9.5	Measured on study animals in Fitzpatrick et al. (2015)[b]
Torso feather length, d/v^d (mm)	60.5/57.7	Measured on study animals in Fitzpatrick et al. (2015)[b]
Torso feather depth, d/v^d (mm)	9.5/15.0[e]	Measured on study animals in Fitzpatrick et al. (2015)[b]
Head shape	Truncated cone	Approximation
Head length (cm)	10.5	Measured from photographs of study animals in Fitzpatrick et al. (2015).[f]
Head diameter, proximal (cm)	6.9	Measured from photographs of study animals in Fitzpatrick et al. (2015).[f] Average of vertical and horizontal diameters.
Head diameter, distal (cm)	3.5	Measured from photographs of study animals in Fitzpatrick et al. (2015).[f] Average of vertical diameter and horizontal diameter.
Beak shape	Truncated cone	Approximation
Beak length (cm)	12.4	Measured from photographs of study animals in Fitzpatrick et al. (2015).[f] From side view of head.
Beak diameter, proximal	3.5	Measured from photographs of study animals in Fitzpatrick et al. (2015).[f] Average of vertical and horizontal diameters.
Beak diameter, distal	0.5	Estimate
Neck shape	Cylinder	Approximation
Neck length (cm)	30.6	Measured from photographs of study animals in Fitzpatrick et al. (2015).[f]
Neck diameter (cm)	8.3	Measured from photographs of study animals in Fitzpatrick et al. (2015).[f] Average value of middle and two ends of neck.
Torso shape	Ellipsoid	Approximation
Torso length (cm)	56.6	Estimated, based on measurements[g]

TABLE 12.A1   Morphometric Input Parameter Values Used to Simulate Energy Expenditures of a Wintering Whooping Crane. Most Values Are the Average of Values Obtained from Two Captive Whooping Cranes (Fitzpatrick et al., 2015). (*Cont.*)

Parameter	Female	Source
Torso diameter, vertical (cm)	22.8	Estimated, based on measurements[g]
Torso diameter, horizontal (cm)	19.8	Estimated, based on measurements[g]
Leg shape	Ellipsoidal cylinder	Approximation
Leg length	36.7	Measured from photographs of study animals.[f] Length of unfeathered leg.
Leg diameter, front-back (cm)	1.7	Measured from photographs of study animals.[f] Average of diameter at middle of upper leg (top of unfeathered leg to tibiotarsal joint) and lower leg (tibiotarsal joint to foot).
Leg diameter, side-side (cm)	1.3	Measured from photographs of study animals.[f] Average of diameter at middle of upper leg (top of unfeathered leg to tibiotarsal joint) and lower leg (tibiotarsal joint to foot).

[a] *Density of fur or feathers had little effect on animal heat loss over a wide range of values in Porter et al. (1994).*

[b] *Measurements on study animals were made using tape measures and rulers.*

[c] *Due to missing value in Fitzpatrick et al. (2015), the female's ventral feather depth was used instead of an average value.*

[d] *Dorsal/ventral.*

[e] *Due to accidental reversal of the dorsal and ventral measurements for one crane, actual average values are 14.0 mm for the dorsal side and 3.0 mm for the ventral side. Because one half of the head feather depth was artificially increased and the other half artificially decreased, this is unlikely to make a significant difference in output energy expenditures for the whole animal.*

[f] *The known heights of the cranes' leg bands were used to scale measurements of the lengths of the cranes' unfeathered upper legs (tibiotarsal joint to bottom of feather line). Because bands were not visible in all photographs, the lengths of unfeathered upper legs were used to scale all other measurements. Measurements shown are an average of measurements from at least three photographs of each bird. To minimize effects of foreshortening, still photographs were taken in which the crane's upper leg was in a vertical position and the crane's front/back or side was perpendicular to the camera.*

[g] *The torso was modeled as an ellipsoid incorporating the volume of the wings and the feathered (top) portion of the legs. The wings were included because cranes' wings are almost always folded against the body (except when flying, which is rare for captive cranes). The feathered part of the upper legs was included because feathers have a significant impact on heat balance, making this part of the leg more similar to the torso than the unfeathered portion of the legs. Fitzpatrick et al. (2015) describe the methods of estimating ellipsoid size.*

TABLE 12.A2   Endotherm Model Input Parameters Used to Simulate Energy Expenditures of a Wintering Whooping Crane. Literature Sources and Estimates Are the Same as Those in Fitzpatrick et al. (2015).

Parameter	Value	Source
Mass (kg)	5.60	Average weight of two Whooping Cranes in Fitzpatrick et al. (2015)
Fat mass (% body mass)	10	Estimate based on values in Krapu et al. (1985)[a] and Swengel (1992)[a]
Body parts with subcutaneous fat	Torso	Assumption that the majority of subcutaneous fat is stored in torso
Animal density (kg/m^3)	633.3	Unpublished lab data from newly dead birds
Basal metabolic rate (W)	12.8	Allometric equation for non-passerine birds: Equation 3 in McNab (2009)[b]
Proportion of energy powering physical activity that is released as heat	0.8	General estimate for animals in Tucker (1975)
Core temp (°C)	37.7	Estimated minimal value based on Whooping Crane core temperature of 40.7°C (Olsen et al., 1996) and a decrease of 1–3°C in body temperature being characteristic of most birds (Prinzinger et al., 1991)[c]
Solar reflectivity of feathers	0.62	Fitzpatrick et al. (2015)[d]
Solar reflectivity of legs and beak	0.33	Estimate in Fitzpatrick et al. (2015)
Percent of skin acting as free water surface	0.2	Based on values used in Porter et al. (2006)[e]
Thermal conductivity of flesh	0.4	Minimal value from Cho (1969)[f]
Maximum O$_2$ extraction efficiency (%)	31	General value for birds in Hainsworth (1981)
Configuration factor for infrared radiation: proportion of animal facing the sky	0.5	Estimate based on crane morphology
Configuration factor for infrared radiation: proportion of animal facing the ground	0.3	Estimate based on crane morphology
Minimum difference between leg core temperature and air temperature achieved via counter-current heat exchange mechanisms in legs and beak (°C)	1.0	Based on measured values in legs and model validation in Fitzpatrick et al. (2015)
Minimum bound on leg and beak temperatures (°C). Below this the difference between appendage and air temperature increases.	3.0	Based on measured values in legs and model validation in Fitzpatrick et al. (2015)

[a] Fat masses of Whooping Cranes are not available from the literature. As described in Fitzpatrick et al. (2015), this an estimate of a moderate amount of fat based on historical weights of the study animals, information on annual weight changes in other northern-nesting crane species in captivity (Swengel, 1992), and fat content of wild Sandhill Cranes (Krapu et al., 1985).

[b] Body mass shown in this table was used.

[c] Core temperature was set to an estimated minimal value to maximize heat conservation in these simulations (see main text).

[d] Reflectivities of molted Whooping Crane feathers were measured using an ASD portable spectroreflectometer (spectral range = 350–2500 nm).

[e] Values for the Hawaiian Amakihi (Hemignathus virens) and the Hawaiian Anianiau (Hemignathus parvus) in Porter et al. (2006)[g]. In this study, modeled water loss values were within ±2SE (standard error) of water loss values measured in metabolic chambers, except for one temperature point.

[f] Flesh thermal conductivity was set to a minimum value (achieved via vasoconstriction) to maximize heat conservation in these simulations (see main text).

APPENDIX 3: ENERGY AVAILABLE IN WHOOPING CRANE FOOD ITEMS

TABLE 12.A3 Estimated Metabolizable Energy (ME) per Gram and per Item for Several Potential Whooping Crane Food Items, Along with the Associated Values Used to Estimate Them: Dry Mass, Protein Content, Lipid Content, Carbohydrate Content, Metabolizable Energy Coefficient (MEC), and Wet Mass. Values of 17.6 kJ/g for Carbohydrates, 39.3 kJ/g for Lipids, and 17.8 kJ/g of Protein (Schmidt-Nielson, 1997) Were Used to Calculate Energy Content from Composition Information in the Literature.

| Species | Species composition | | | | | | | |
	Dry mass (% wet mass)	Protein (% dry mass)	Lipid (% dry mass)	Carbohydrate (% dry mass)	MEC	Wet mass (g)	ME (kJ/item)	ME (kJ/g wet mass)
Corn (shelled)	84.7[a]	10.3[b]	3.8[b]	84.5[b]	0.84[c]	0.35[d]	4.5	12.9
Aquatic insect	27.5[e]	59.5[e]	15.5[e]	7.2[e]	0.77[f]	2.0[g]	7.6	3.8
Sago pondweed (*Potamogeton pectinatus*) tuber	30.0[h]	13.1[i]	1[i]	74.9[i]	0.56[j]	1.0[k]	2.7	2.7
Crayfish (*Oronectes* spp.)	32.73[l]	38.10[l]	3.76[l]	14.55[l]	0.77[f]	23.9[l]	65.2	2.7

ME, Metabolizable energy; MEC, metabolizable energy coefficient.

[a] Values from Baldassarre et al. (1983).
[b] Average of values for wild-type corn from Sidhu et al. (2000).
[c] Average value for six non-passerine species in Karasov (1990).
[d] Calculated using percent moisture and estimated median dry mass from Hashemi et al. (2005).
[e] Average value for insects from Bell (1990).
[f] Value for birds consuming arthropods (Karasov, 1990).
[g] Halved 4.0 g estimate for large spp. of grasshoppers from Greer (2010). This estimate was based on data in Greenlee et al. (2007).
[h] Based on values in Kantrud (1990).
[i] Average values from Anderson and Low (1976).
[j] Average value for birds eating bulbs and rhizomes (Karasov, 1990).
[k] Intermediate weight from (Kantrud, 1990).
[l] Average of values for wild-caught Oronectes spp. in: Dierenfeld et al. (2009). Average nitrogen-free extract across species (100% dry mass: 38.10% protein, 3.76% lipid, 47.16% ash, 14.55% crude fiber) was less than 0. Thus, the crude fiber value only was used for carbohydrates.

HABITAT USE

Pair of foraging Whooping Cranes

Winter Habitat Ecology, Use, and Availability for the Aransas-Wood Buffalo Population of Whooping Cranes

Elizabeth H. Smith*, Felipe Chavez-Ramirez**,
Luz Lumb†

*International Crane Foundation, Texas Program, Fulton, TX, United States
**Gulf Coast Bird Observatory, Lake Jackson, TX, United States
†Harte Research Institute for Gulf of Mexico Studies, Texas A&M University-Corpus
Christi, Corpus Christi, TX, United States

INTRODUCTION

Whooping Cranes in the Aransas-Wood Buffalo Population (AWBP) are a wetland-dependent species that inhabit freshwater marshes in the boreal forests of Canada for nesting and feeding, then migrate over 4,000 km to the Texas coast for winter where adult pairs defend territories and subadults inhabit undefended, peripheral areas of coastal salt marsh complexes (Bishop and Blankinship, 1982; Stehn and Johnson, 1987; Stehn and Prieto, 2010). They are primarily carnivorous in the wintering range, feeding principally on crustaceans, clams, snails (order Decapoda), and other estuarine animals as well as the fruits of Carolina wolfberry or desert-thorn (*Lycium carolinianum*) (Chavez-Ramirez, 1996; Hunt and Slack, 1989; Westwood and Chavez-Ramirez, 2005). The life cycles and availability of most items in Whooping Crane diets are significantly influenced by temperature, freshwater inflows, and salinity levels in this coastal estuarine environment (Gunter, 1950; Hedgpeth, 1950; Montagna and Palmer, 2012; Wozniak et al., 2012).

We describe abiotic and biotic processes that affect this ecosystem's functioning and health by generally following a conceptual ecological model developed for the wintering range of the AWBP (Fig. 13.1) (Chavez-Ramirez and

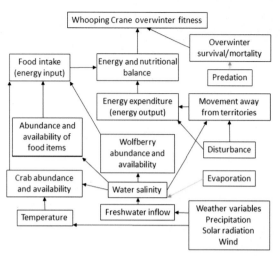

FIGURE 13.1 Conceptual ecological models of Whooping Crane life history traits and external factors impacting those traits in wintering range along the Texas coast. *Source: Adapted from Chavez-Ramirez, F., Wehtje, W., 2012. Potential impact of climate change scenarios on whooping crane life history. Wetlands 32 (1), 11–20.*

Wehtje, 2012). We also evaluate direct effects of temperature, precipitation, freshwater inflows, and water salinity, and their indirect effects on habitat diversity and food resource availability, that can guide conservation and management efforts for this recovering population. We then address habitat availability (distribution and quantity) under present conditions and those predicted with climate change impacts from sea-level rise (SLR) into the next century. Key components of conservation needs are provided to guide continuing efforts to promote the recovery of this iconic species.

INTERRELATIONS OF GEOGRAPHY, CLIMATE, AND HYDROLOGY OF TEXAS ESTUARIES

The wintering range of the AWBP is located in the northwest quadrant of the Gulf of Mexico along the central coast of Texas in the United States (Fig. 13.2), an area variously classified as within the Gulf of Mexico Biogeographic Region (Bender et al., 2005), Gulf Prairies and Marshes Ecological System (NatureServe, 2009), and Western Gulf Coastal Grasslands Ecoregion (Abell et al., 2008; U.S. EPA, 2013). The coastal region exhibits a gently sloping elevation from 46 m above sea level to the bays and Gulf shoreline, bisected by meandering coastal rivers and creeks. The remnant migratory population of Whooping Cranes was reduced to 14 adults and two yearlings by 1941–42 and they wintered only within the newly formed boundaries of the Aransas National Wildlife Refuge Complex (ANWRC) on the Blackjack Peninsula (19,126 ha) along the Guadalupe-San Antonio Estuary (ANWRC, 2010). Expansion of the ANWRC over time has increased the refuge size to 46,915 ha. The refuge complex now encompasses more upland habitats (65%) than estuarine systems (35%); the latter are primarily salt marsh, followed by tidal flats and Gulf beach habitat types. The current Whooping Crane winter range is now centered within and adjacent to ANWRC as their population has increased and the birds have inhabited new areas in adjacent estuarine areas.

Climate ranges from temperate to subtropical north to south and subhumid to semiarid east to west within the AWBP wintering range, regionally called the Texas Coastal Bend (Fulbright et al., 1990). Annual average temperatures range from 8.3–8.9°C in winter to 33.3–35.6°C in summer months; minimum temperatures below 0°C that last longer than a few days are infrequent along the coastal zone (Montagna, 2011b; Tunnell, 2002). Rainfall generally decreases southward within this region of the coast, averaging from 91.4 cm to 77.4 cm (Montagna et al., 2007). Tropical storms that make landfall along the central Texas coast can also contribute a significant amount of rainfall (Conner et al., 1989). High evaporation and evapotranspiration rates often result in a precipitation deficit during years with lower rainfall totals. Alternating wet/dry

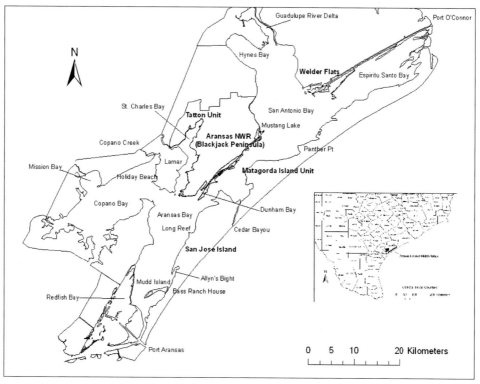

FIGURE 13.2 Wintering area of Aransas-Wood Buffalo Whooping Cranes. *Source: Modified from Stehn, T.V., Prieto, F., 2010. Changes in winter whooping crane territories and range 1950–2006. Proceedings of the North American Crane Workshop 11, 40–56).*

cycles along the Texas coast are influenced by broad-scale climate patterns originating in the Pacific and Atlantic oceans (Tolan, 2007). El Niño and La Niña events are largely explained by Pacific sea surface temperature phenomena and affect freshwater inflows into the coastal estuaries at the basin scale, resulting in decreasing and increasing estuarine salinities, respectively.

Estuarine systems are defined by the mixing of freshwater flowing from coastal basins into semienclosed bays with ocean waters flowing into and out of the bays through Gulf passes driven by regional tidal cycles (Longley et al., 1994). Tidal ranges in the northwestern Gulf of Mexico are defined as microtidal (open coast 0.6 m, bays <0.3 m); therefore, wind regimes are the predominant drivers

for water-level changes in the bay systems (Gibeaut et al., 2003). The Texas Coastal Bend experiences persistent southeast winds from March through September and periodic northnortheast winds from October through March (Behrens and Watson, 1973; Brown et al., 1976). These strong wind patterns produce two seasonal high tides (generally around October and March) and two seasonal low tides (generally around July and January) that drive inundation and exposure cycles in salt marsh/flats habitats as well as the ingress and egress of estuarine organisms (Longley et al., 1994).

Within the current AWBP wintering range (Fig. 13.2), most of the seawater from the Gulf of Mexico flows through the Matagorda Ship Channel and Pass Cavallo at the northern end of the Matagorda Island into the San Antonio Bay system, although minor flows provide connectivity and passage of aquatic organisms at Cedar Bayou at the south end of the island. Additional connectivity with the Gulf of Mexico occurs through the Aransas Pass Ship Channel in the Aransas Bay System at the southern extent of the wintering range. The estuarine intertidal environments form a fringe along Espiritu Santo, San Antonio, Ayres, Mesquite, and Aransas bays that separate Matagorda and San Jose islands from Seadrift-Port O'Connor Ridge, and Blackjack and Live Oak peninsulas (ANWRC, 2010).

The coastal hydrologic system in the primary wintering range of the AWBP is termed the Guadalupe Estuary, named after the river providing the predominant flows (Tolan, 2007). The drainage basin is small (26,330 km^2) relative to the seven estuarine systems along the Texas coast, ranking fifth, and bay area (551 km^2) ranking fourth. The Guadalupe Estuary's central geographic location along the Texas coast results in average freshwater annual inflow of 2,664 10^6 m^3/year. However, annual amounts vary widely among years, and have largely been explained by temporal signals of the El Niño-Southern Oscillation that affect freshwater inflows and estuarine-wide salinities.

Long-term average salinities for the entire estuarine system are about 16 practical salinity units (psu); however, annual variability is quite high (Tolan, 2007). For example, annual bay salinities for 1982–2004 averaged 17.7 ± 11.1 (SD) psu and monthly averages were 0.00–45.0 psu. Salinities levels within the marsh complex are affected by their level of connectivity to bay waters, in situ evaporation, and localized rainfall as well as freshwater inflows from the Guadalupe-San Antonio Basin when seasonal tides are high. All these factors affect porewater salinities of marsh soils that have a predominant influence on marsh plant composition, growth, and productivity, as well as relative abundance of estuarine prey items (Westwood and Chavez-Ramirez, 2005).

COASTAL ENVIRONMENTS IN THE AWBP WINTER RANGE

Along the Texas coast, the Gulf shoreline is bisected by a series of seven bay systems, which increases the linear extent of fringing coastal environments (Armstrong, 1987). Many Texas coastal lagoons are located leeward of barrier landforms and bordered by mainland shorelines; in the case of the central Texas coast, relict barrier landforms from previous geologic times form a series of peninsulas (Price, 1933). These ancient beach ridges display a ridge-and-swale topography that supports coastal prairies and freshwater swales similar to the present Gulf barrier islands. The shallow coastal marshes and flats adjacent to these landforms and bordering the estuarine lagoons are interspersed with brackish and salt ponds that are variously inundated during regular tidal cycles as well as flooded for longer periods of time during seasonal high tides primarily caused by persistent winds from southerly to easterly direction (Gunter, 1950).

Regularly flooded salt marshes are defined by White et al. (2002) along the Texas Gulf coast as being composed of smooth cordgrass (*Spartina alterniflora*), saltwort (*Batis maritima*), seashore saltgrass (*Distichlis spicata*), glasswort (*Salicornia* spp.), shoregrass (*Distichlis littoralis*), annual seepweed (*Suaeda linearis*), and shoreline seapurselane (*Sesuvium portulacastrum*). Irregularly flooded salt marshes are composed of sea oxeye daisy (*Borrichia frutescens*), saltmeadow cordgrass (*Spartina patens*), Gulf cordgrass (*Spartina spartinae*), seashore saltgrass, marsh fimbry (*Fimbrystylis spadicea*), aster (*Aster* spp.), and Carolina wolfberry (*Lycium carolinianum*), as well as many other species, including black mangrove (*Avicennia germinans*). Tidal flats are included within the estuarine intertidal unconsolidated shore classification (along with beaches and algal flats). The term *wind-tidal flats* refers to the microtidal range in the Gulf of Mexico and the tidal inundation/exposure process driven by predominant wind regimes (White et al., 2002).

The development of these vegetation assemblages can be influenced by distance from bay tidal flows, level of hydrologic connectivity, and topography (Fig. 13.3) (Gibeaut et al., 2003). The microtopography that influences vegetation diversity within the coastal marsh complex is difficult to measure, and general elevation data at the scale used to generate topographic maps do not adequately provide the detailed information for these low-lying, minor relief landscapes. Gibeaut et al. (2003) measured the relationship between elevation and vegetation classes for a specific site on Matagorda Island (Fig. 13.3), which is useful for evaluating ANWRC salt-marsh heterogeneity. Although vegetation types and elevation gradients follow an expected sequence from an upland-to-bay environment, the overlap of vegetation classes across the elevational gradient indicates that other factors are affecting the distribution of vegetation patterns across the elevational surface.

The influence of predominant wind regimes at the intra-annual scale, coupled with wet-drought cycle at the multiyear scale, results in more heterogeneity in the coastal vegetation assemblage than found along coastlines with higher tidal amplitude. This overlapping vegetation assemblage in the wintering range of the AWBP also creates more challenges when mapping vegetation classes in the wintering range area or using vegetation classes as habitat designations for a particular species.

EVALUATING WINTERING HABITAT USE BY AWBP WHOOPING CRANES

The term *habitat* can be described as the collection of resources and conditions necessary for occupancy by a species; therefore, the term by definition is species specific. We define wintering habitat as a geographic location where Whooping Cranes are consistently observed over an extended period (Johnson, 1980). A certain habitat is defined as selected when an individual or species demonstrates disproportionate use for one resource or habitat type over others (Block and Brennan, 1993; Johnson, 1980). A hierarchical approach to discerning habitat use at different landscape scales can be applied at the macro-, meso-, and microscale levels (Block and Brennan, 1993). The hierarchical level is dependent on the spatial scale used to address

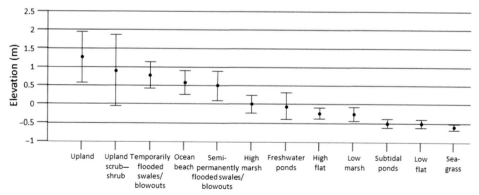

FIGURE 13.3   Elevation (mean, SD) above mean sea level of each habitat class in southern Matagorda Island in the Aransas National Wildlife Refuge Complex, Texas. *Source: Modified from Gibeaut, J.C., White, W.A., Smyth, R.C., Andrews, J.R., Tremblay, T.R., Gutierrez, R., Hepner, T.L., Neuenschwander, A., 2003. Topographic variation of barrier island subenvironments and associated habitats. In: Davis, R.A., Sallenger, A., Howd, P. (Eds.), Coastal Sediments '03: Crossing Disciplinary Boundaries: Proceedings of the 5th International Symposium on Coastal Engineering and Science of Coastal Sediment Processes, Clearwater Beach, FL. CD-ROM.*

habitat selection, as well as level of detail (spatial resolution) available in spatial data sets. Early work described habitat use patterns by Whooping Cranes at very broad and general levels (macrohabitat scale) as salt marsh or salt flats (Allen, 1952; Labuda and Butts, 1979; Stevenson and Griffith, 1946). Other macrohabitat types were later described, such as bays and uplands (Bishop, 1984; Chavez-Ramirez et al., 1996; Hunt and Slack, 1989).

Salt marsh is not a homogeneous land cover type; rather, it is a heterogeneous environment composed of a variety of smaller discrete habitat types, or patches, at finer scales (meso- and microscale) (Chavez-Ramirez, 1996). Stevenson and Griffith (1946) first recognized subdivisions within salt marsh at the mesoscale by describing brackish ponds, tidal lagoons or sloughs, shallow water bays, and inlet ponds as distinct habitats within the salt marsh environment. Later, Allen (1952) recognized several open-water categories within the salt marsh: permanent, semipermanent, and ephemeral ponds and lakes. In the mid-1990s, Chavez-Ramirez (1996) quantified use of salt marshes by wintering Whooping Cranes, including both temporal and spatial detail at the macro- and mesohabitat scales. His analysis of aerial surveys conducted over two winters (1992–93, 1993–94) showed that Whooping Cranes used salt marsh macrohabitats predominantly (87% and 86%, respectively) compared to upland (8.8% and 11%) and bay habitats (3.9% and 3.1%).

The predominant area used by AWBP Whooping Cranes within their wintering range lies within the Western Gulf Coastal Grasslands ecoregion (Bender et al., 2005). However, information about the vegetation classes that represent the habitat required by cranes at a quantifiable spatial scale was lacking. A collaborative project was designed with the objective to develop a decision support tool using methods from landscape conservation design that could be used to identify current and future wintering habitat availability that would support the recovering population (Smith et al., 2014). They constructed a Composite Habitat Type Dataset (CHTD) using a geographic information system by merging four vegetation class databases that encompassed the upland, wetland, and submerged classes for an area defined as current and potential wintering range in the central Texas coast. After standardizing the merged classes using crosswalk tables, the resulting data set was hierarchically arranged by macro-, meso-, and microhabitat, primarily using the Cowardin classification system used in National Wetland Inventory (NWI) mapping (FGDC, 2013).

This decision support tool was first used to determine which habitats are selected by wintering Whooping Cranes using the historic winter survey data collected during aerial surveys from 2004 to 2010 (Taylor et al., 2015). A total of 6,492 points representing 15,112 crane observations (points may designate >1 individual) were spatially linked to various habitat types within the CHTD using overlay analysis. Relative frequency values were ranked by grouping mesohabitat use levels as follows: High Use – Estuarine Vegetated Marsh (50.4%); Low Use – Estuarine Open Water (<0.3 m deep Mean Low Water [MLW]) (16.9%), Estuarine Vegetated Seagrass (<0.3 m deep MLW) (13.0%), and Estuarine Unvegetated Flats (8.2%), as well as Incidental Use – Upland Grassland (5.7%), Freshwater Wetland (2.8%), and Upland Shrub (1.5%). The remaining 1.5% were comprised of six mesohabitats.

In wildlife management and conservation, we should seek to better understand habitat use patterns not only as numerical value but also from a functional perspective. It is important to know not only how often, but how and why a crane uses a particularly habitat type. For example, a freshwater pond may be used <3% of the time by the population. From a numerical standpoint, this low value could be considered of little significance; however, biologically it is essential that this habitat is available in the wintering landscape if the crane requires fresh drinking water every day during drought periods. It is equally important for us to understand how the crane uses each habitat type as we seek to develop science-based conservation plans.

TIME-ACTIVITY ALLOCATIONS OF WINTERING AWBP WHOOPING CRANES

The ways in which an animal allocates resources (including time and space) will ultimately influence individual survival and reproductive success. The apportionment of time and energy may be affected by many biotic factors (i.e., body size, species physiology, food availability, competition, and predation) (Wiens, 1989) and abiotic factors (i.e., weather). In temperate species, the nonbreeding season is a period during which food is the resource of primary importance, and search and acquisition of food plays a prominent role in determining a bird's use of space (habitat use patterns) and time (time-activity budgets) (Hutto, 1985). Acquisition of food may be complicated by differences in seasonal availability of food items in different habitat patches, as well as by risks of predation associated with different habitat types. Differences in resource type and abundance between locations and differential predation risks at different sites may result in differences in time-activity budgets between individuals occupying those sites.

Over two winter seasons in the mid-1990s, Chavez-Ramirez (1996) conducted extensive focal surveys of crane behavior by habitat type on Blackjack Peninsula within ANWRC. Overall throughout both winter seasons, cranes spent more time engaged in feeding and locomotion, followed by alert, rest, maintenance, and interaction behaviors (Table 13.1). Behaviors related to searching for and procuring food averaged about two-thirds of the time-activity budget. In Chavez-Ramirez (1996)'s comparisons of behavior among habitat types in both winters, the frequency of use remained generally the same, with the notable exception of the upland habi-

tat (Table 13.1). In both years, cranes in upland habitat spent more time in locomotion and alert behaviors and less time feeding than in salt marsh habitats. Some documented sources that elicit alert behavior included natural causes from other animals, which included other Whooping Cranes and potential predators, and anthropogenic causes, such as aircraft, boats, and barges. In the case of the upland habitat, natural causes and aircraft were the most likely scenarios to elicit an alert behavior.

Feeding and location in most habitats averaged 68–76% in both winters, indicating that sufficient time was provided for foraging activities in salt marsh habitats and bay shorelines (Chavez-Ramirez, 1996). Although differences were detected between years, environmental conditions that may have impacted food resource abundance or availability were not severe enough to increase the proportion of foraging time at the expense of other activities. Alternately, if a crane spends 40% of its time in alert behavior due to disturbance and/or potential predator threat, the time available for foraging will be less than 60%, given the need for other activities (e.g., resting, maintenance). Alert behavior levels were below this value both years, even within the upland habitat, indicating that disturbance was minimal. These data serve as a baseline from which to compare future time-activity budget data, particularly in years when more variable environmental conditions occur or in areas where different levels of disturbance pressure may be present.

DIET OF WINTERING AWBP WHOOPING CRANES

The periodic wetting and drying of the marsh and pond substrates from tidal regimes, in concert with substantial variations in water and soil salinities from wet/drought cycles, limits the number of vertebrate and invertebrate species that can tolerate these extremes (Gunter, 1950;

TABLE 13.1   Time-Activity Budgets of Whooping Cranes during the 1992–93 and 1993–94 Winters, in Different Habitats on Blackjack Peninsula within Aransas National Wildlife Refuge, Texas. Values Are Means (Standard Deviation) of Percent Time Spent Showing Different Behavior; N Is Number of 30-min Observation Sessions.

Location/habitat	Behavior (%)						
	Feeding	Locomotion	Alert	Rest	Maintenance	Interaction	N
1992–93							
All habitats	35.9 (20.3)	32.2 (16.1)	20.7 (14.8)	7.6 (9.9)	3.6 (7.8)	1.2 (1.9)	84
Salt marsh vegetation	37.6 (13.9)	35.7 (11.9)	22.2 (13.7)	3.8 (3.2)	0.1 (0.3)	0.5 (1.1)	16
Salt marsh open water	44.9 (17.4)	31.7 (15.3)	18.4 (14.8)	4.5 (9.7)	0.4 (0.9)	1.6 (2.1)	34
Bay	47.8 (14.5)	28.9 (23.4)	7.5 (6.7)	6.6 (8.3)	11.5 (13.3)	0	14
Upland	10.8 (7.9)	32.6 (14.8)	32.5 (10.4)	16.7 (9.8)	6.3 (7.9)	1.9 (2.4)	20
1993–94							
All habitats	36.1 (26.9)	33.9 (23.5)	11.3 (11.7)	9.8 (22.8)	7.6 (15.1)	1 (4.4)	64
Salt marsh vegetation	34.6 (24.3)	41.8 (19.3)	18.3 (9.4)	3.3 (2.5)	1.3 (2.8)	0.9 (1.8)	12
Salt marsh open water	40.6 (28.5)	30.2 (26.5)	6.8 (8.1)	14.2 (27.7)	7.2 (16.3)	1.2 (5.5)	32
Bay	29.8 (23.5)	38.7 (12.4)	14.5 (19.1)	1.5 (1.6)	14.5 (19.1)	1 (0.9)	12
Upland	12.5 (9.8)	41.3 (16.7)	29 (12.7)	1 (0.8)	13.5 (9.8)	0	8

From: Chavez-Ramirez, F., 1996. Food Availability, Foraging Ecology and Energetics of Whooping Cranes Wintering in Texas. PhD dissertation, Texas A&M University, College Station, p. 104.

Hedgpeth, 1950). However, those species that can subsist within these ranges, or are capable of migrating and retreating under favorable to unfavorable conditions, may be quite abundant most years. In early studies, Whooping Cranes have been documented foraging on a variety of estuarine and upland food of up to a dozen species in the 1990s, as determined by fecal and stomach sample analyses as well as focal observations (Allen, 1952, 1954; Blankinship, 1976). It is apparent that Whooping Cranes can be opportunistic feeders, concentrating their feeding strategies based upon food availability within a winter season and among years.

Even in the earliest studies, blue crab (*Callinectes sapidus*) was identified as the most important food item based on total fecal volume (42%), with acorn (*Quercus virginiana*) second and stout razor clam or stout tagelus (*Tagelus plebeius*) third in importance (Allen, 1952). Blue crabs have been consistently identified as the primary prey choice for overwintering Whooping Cranes (Allen, 1952; Chavez-Ramirez, 1996; Hunt and Slack, 1989; Nelson et al., 1996; Stevenson and Griffith, 1946; Westwood and Chavez-Ramirez, 2005). Whooping Cranes have been documented eating as many as 80 blue crabs per day (Chavez-Ramirez, 1996).

In a multisite study during 1993–95, the major food items identified in Whooping Crane fecal samples included blue crab, fruit of the Carolina wolfberry, and plicate horn snail (*Cerithidea pliculosa*), with stout razor clam to a lesser extent; a total of 17 food items were identified, which increased the overall food item list to 26 (Westwood and Chavez-Ramirez, 2005).

Long-term fisheries and environmental data indicate that blue crab abundance in the San Antonio Bay system is affected by temperature, year, dissolved oxygen, and salinity (Montagna, 2011a). The crab's preferred salinities were about 10–25 psu, and their abundance was reduced below and above this range. When salinities increased across all salinity gradients within the bay by 10 psu, the probability of capturing crabs along the bay margins was greatly reduced. Salinities within inundated coastal ponds within the estuarine marsh can differ from the adjacent bay system and is related to degree of connectivity among ponds and bay (Pugesek et al., 2013). In a study conducted from 1997 to 2005, overall mean salinities of 6–38 psu within the ponds did not significantly affect crab abundance; however, those salinities were within the preferred range for crabs and the study did not coincide with drought, flooding, or storm surge conditions (Pugesek et al., 2008). In studies related to Whooping Cranes and blue crabs, crab availability and crab size were positively correlated (Greer, 2010; Pugesek et al., 2008, 2013). Crab availability is dependent on their presence in shallow, partially isolated coastal ponds, sloughs, and tidal creeks with appropriate water conditions and seasonal timing (Pugesek et al., 2008, 2013). Hedgpeth (1950) ranked the blue crab as the dominant invertebrate of salt flats, noting their ability to migrate into the flats with incoming tides from adjacent bays.

Carolina wolfberry is also an important component in the wintering Whooping Crane diet, providing short-term and often abundant fruits produced on the native shrub (Stutzenbaker, 1999).

This species is typically located at the higher elevations of the coastal marsh (Lichvar et al., 2014) and is often found with other high-marsh species such as saltwort or turtleweed (*Batis maritima*) and glasswort (*Salicornia* spp.) (Rasser et al., 2013). Carolina wolfberry exhibits a bimodal pattern of leaf production in ANWRC, peaking in the early spring and again in early fall prior to flowering and fruiting in late fall to early winter (Butzler and Davis, 2006). The highest proportion of wolfberry fruit in the Whooping Crane diet occurs during the fruiting period, contributing 21–52% of crane energy intake (Chavez-Ramirez, 1996; Westwood and Chavez-Ramirez, 2005).

The presence of plicate horn snails in fecal samples and from observations of Whooping Cranes indicates that they will consume them, but their nutritional benefits are unknown (Chavez-Ramirez, 1996; Westwood and Chavez-Ramirez, 2005). Cranes may actually be consuming empty shells as a source of grit or providing some micronutrient need, such as calcium supplement, as has been documented for other bird species. Searches for live snails were unsuccessful in the mid-1990s; however, they have been located sporadically in the high salt marsh and ponds in recent years (E. Smith, personal communication).

Dwarf surf clams (*Mulinia lateralis*) were documented as being consumed by wintering Whooping Cranes in bay habitats (Chavez-Ramirez, 1996). This mollusk is often dominant in benthic communities in Texas estuaries, including the San Antonio Bay system (Armstrong, 1987; Montagna and Kalke, 1992), inhabiting subtidal mud bottoms and seagrass meadows, and it is a primary food resource for black drum (*Pogonias cromis*). Adult dwarf surf clams have a fairly wide salinity tolerance and are typically most abundant in low to variable salinities in the San Antonio Bay system (Parker, 1959), but tolerance is lower for early development stages. Embryonic development progresses best at 20–30 psu and 17–27.5°C but declines at salinities and temperatures outside these ranges (Calabrese, 1969); larval development declines at higher temperatures, although that stage has a wider salinity tolerance (10–35 psu) at 7.5–27.5°C.

Razor clams (*Tagelus plebeius*) have been documented as prey of Whooping Cranes in tidal flats (Allen, 1952; Chavez-Ramirez, 1996), preferring stable sediments that are >2% silt and clay (Holland and Dean, 1977). These substrates support a burrow structure that allows vertical movement by the clams for feeding at the surface and fast retreat for protection from potential predators. Razor clams are a favorite prey item of American Oystercatchers (*Haematopus palliatus*) (Lomovasky et al., 2005), indicating they would also be easily accessible to Whooping Cranes. This mollusk species can be negatively impacted by the protozoan oyster pathogen *Perkinsus marinus* under certain environmental conditions (Dungan et al., 2002), which would affect its relative abundance as a prey item.

Although Whooping Cranes wintering in and around ANWRC primarily use saltmarsh habitats, their use of adjacent upland has been well documented for decades (Allen, 1952, 1954; Blankenship and Reeves, 1970; Chavez-Ramirez, 1996; Hunt and Slack, 1989; Stevenson and Griffith, 1946). Cranes often respond immediately following a management burn on nearby uplands by exploring alternate food resources on the open landscape (Chavez-Ramirez et al., 1996). Whooping Cranes may also increase their use of uplands during drought periods. Allen (1952) believed their primary focus was feeding on acorns. However, detailed evaluations of Whooping Crane fecal samples during winters of 1992–93 and 1994–95 indicated that acorns comprised only a minor proportion of the diet regardless of the amount of acorn production in a given year (Chavez-Ramirez et al., 1996; Stehn, 1995). Orthopteran insect remains were abundant in fecal samples, which suggests a higher importance as an upland food item (Westwood and Chavez-Ramirez, 2005). Other prey items such as snakes, lizards, and insects are also opportunistically consumed

while foraging in burn areas. Additional studies are necessary to fully understand the use and dietary importance of uplands by AWBP Whooping Cranes during the winter.

Nelson et al. (1996) evaluated the metabolizable energy (ME) and digestibility of five main food items documented as being eaten in the wintering range using captive-reared Whooping Cranes. The apparent ME was highest for *Rangia* clam (75.0%), acorn (43.2%), wolfberry (44.8%), and blue crab (34.1%). The digestibility coefficients for protein were higher in animal foods (blue crab, 75.2; *Rangia* clam, 69.4) than in plant foods (wolfberry, 53.4; acorn, 48.9), although total lipid digestibility was highest for acorn (87.2) and lowest for wolfberry (60.0). When total energy was evaluated for the four main food items plus stout razor clam (*Tagelus plebeius*), gross energy on a dry-weight basis was 2–5 times higher for acorn and wolfberry than for blue crab and stout razor clam. However, crude protein was 2–3 times higher for blue crab (315 g/kg) than for wolfberry (104 g/kg^{-1}) and stout razor clam (100 g/kg^{-1}). ME was highest for wolfberry (9.6 kJ/g^{-1}) and total lipid nutrient availability (80 kJ/g^{-1}).

In general, the foods examined could be separated into two categories: (1) high energy–low protein provided by wolfberry and acorns, and (2) low energy–high protein provided by blue crab and stout razor clam. Size differences (wolfberry fruit as compared to blue crab), as well as relative search time (plant versus animal) required for acquiring these foods, are highly variable (Nelson et al., 1996). In addition, these food items are not equally available throughout the typical winter season (e.g., plant fruiting phenology, animal response to changing water levels) and may be less available when environmental conditions are suboptimal. Lower food availability may affect the timing of spring migration, lower reproductive success, and even contribute to higher mortality, but these hypotheses have not been evaluated. All of these factors are integrated and must be considered in the evaluation, conservation, and management of habitat quantity, quality, and diversity in the AWBP wintering range.

SPATIAL HABITAT REQUIREMENTS OF WINTERING AWBP WHOOPING CRANES

The spatial extent and configuration of habitat patches become important because different habitat patches of the salt marsh, bay, and upland are used for different food resources when they are available (Chavez-Ramirez, 1996). The size of the area that a pair of Whooping Cranes defends needs to be large enough to encompass high habitat diversity, thus ensuring adequate food availability throughout the winter season (Stehn and Prieto, 2010). In addition, proximity to upland habitats is essential to provide additional food and fresh water during drought periods. The low relief along the Texas coast that naturally provides the geomorphic and hydrologic complexity to support a diversity of coastal habitats is exhibited within the current wintering range (Fig. 13.4A). Based on each food prey item's ecological requirements, we developed a habitat–prey matrix specific for wintering Whooping Cranes (Fig. 13.4B), and use that matrix to link to the spatial habitat heterogeneity of Blackjack Peninsula within the ANWRC. At a landscape level, the integrative components of all habitat types provide these key food and freshwater resources both within a defined crane territory, as well as in adjacent habitats that are used less frequently yet may be essential.

The amount of habitat that is needed to support a family of cranes, or a group of subadults, has been the subject of much research and is considered essential in guiding conservation efforts. Using aerial survey data collected over a 57-year period, Stehn and Prieto (2010) analyzed and mapped changes in distribution and size of winter territories. Territory boundaries were delineated by observing exclusive use of a certain area by wintering adult crane pairs or family groups (Stehn and Johnson, 1987; Stehn and Prieto, 2010). These territories were mapped in selected years (Fig. 13.5) and spatially illustrate the progressive expansion by the AWBP in their wintering range over 6 decades.

For many years, the east shoreline of the Blackjack Peninsula within the ANWRC supported all winter crane territories (Table 13.2). Average territory size decreased on Blackjack Peninsula and San Jose and Matagorda islands during 1950–2006, although the territories were larger on the islands. In all areas where all available habitats are being used, it appears that minimum size of territories (180 ha) is being approached, particularly on Blackjack Peninsula where that single area of the wintering AWBP was located in the 1950s–60s. The larger territory sizes on barrier islands versus peninsulas may be related to two factors: all habitats have not been occupied to date, leading to larger territory size, or more area is necessary to compensate for lower habitat diversity and/or habitat quality (Stehn and Prieto, 2010). Neither of these potential explanations has been evaluated; however, using territory size for a certain number of Whooping Crane pairs, as well as proximal habitat for subadult use, is the metric currently employed to guide conservation in the wintering area.

One of the main criteria for downlisting the Whooping Crane from endangered to threatened status involves several scenarios, which may involve the potential augmentation of the AWBP with one or two self-sustaining reintroduced populations (CWS and FWS, 2007). If, however, establishment of self-sustaining reintroduced populations is not successful, "the AWBP must be selfsustaining and remain above 1000 individuals (i.e., 250 productive pairs) for downlisting to occur" (CWS and FWS, 2007, p. 14). To determine if sufficient habitat is available within and around the ANWRC to support 250 pairs/families during the winter period, Stehn and Prieto (2010) used the average territory size of 172 ha.

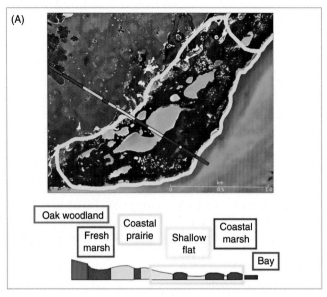

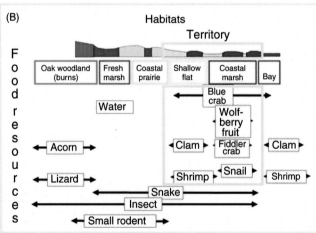

FIGURE 13.4   (A) Plan view of spatial position of the upland prairie complex and estuarine coastal habitat complex, including a representative crane territory within Aransas National Wildlife Refuge, Texas; (B) habitat types and associated food resources consumed by the wintering Aransas-Wood Buffalo Whooping Cranes; yellow box denotes those resources within cranes' winter territory.

Their analysis determined that, within a 111-km radius around the center point of ANWRC, insufficient habitat was available to support this recovery option (43,000 ha). In a similar assessment, Smith et al. (2014) quantified High Use habitat of wintering Whooping Cranes as primarily estuarine vegetated salt marsh. An extensive

delta to the south (Nueces Bay) was used in the analysis of Stehn and Prieto (2010) but was not included in the analyses of Smith et al. (2014). The latter study identified only ~37,000 ha of High Use habitat in the Matagorda, San Antonio, and Aransas bay systems (Fig. 13.6). These results indicate that under present conditions,

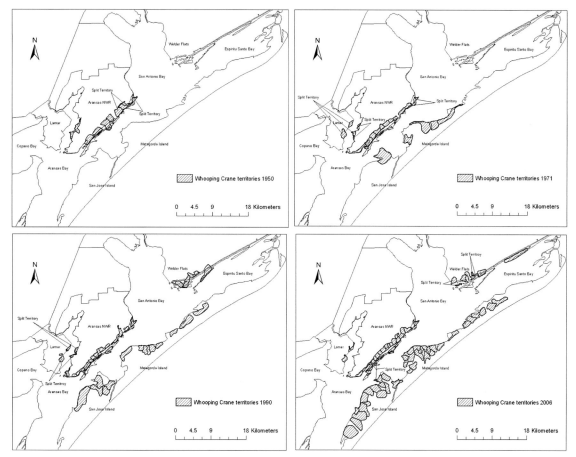

FIGURE 13.5 Winter territories of the Aransas-Wood Buffalo Whooping Cranes for 1950, 1971, 1990, and 2006 illustrating increase in territory numbers and distribution as the population increased. *Source: Adapted from Stehn, T.V., Prieto, F., 2010. Changes in winter whooping crane territories and range 1950–2006. Proceedings of the North American Crane Workshop, 11, 40–56.*

the AWBP will need to expand beyond these contiguous coastal systems and seek additional estuarine marsh mesohabitats to satisfy the target spatial requirements.

In both studies, only estuarine salt marsh habitat was considered in the target spatial extent as habitat needed to support 1,000 individuals and 250 nesting pairs (Smith et al., 2014; Stehn and Prieto, 2010). However, the latter study took the analyses further by quantifying additional habitats that were used by these wintering cranes outside the territory boundaries, including

shallow open water and shallow seagrass, which were categorized as Low Use, and palustrine marsh and coastal prairie, categorized as Incidental Use (see Fig. 13.6). Using this approach, Low Use habitat may be used only during low tide cycles for foraging on alternate prey or Incidental Use habitat used to satisfy freshwater drinking needs during drought periods or foraging in coastal prairie following upland burns. When all habitats that are variously required at some point during the wintering period are considered, it is apparent that more than just the salt

TABLE 13.2   Average Size (ha) of Territories (Excluding End Territories) within the Wintering Area of Aransas-Wood Buffalo Whooping Crane Population for Selected Years from 1950 to 2006

Landform	1950	1961	1971	1979	1985	1990	1995	2000	2006
All areas[a]	313	182	177	158[b]	236	172	214	183	196
Aransas NWR	313	182	160	156	179	108	122	101	108
West St. Charles	–	–	–	–	–	–	–	125	110
San Jose	–	–	–	–	439	349	307	339	304
Matagorda	–	–	347	222	294	221	271	212	204
Welder Flats	–	–	–	113[c]	–	265	213	139	159

[a] Excludes West St. Charles Bay from analysis.
[b] Data excluded in estimating minimum territory size since 10 of the 12 territories were in the ANWRC, where territories are smaller than in all other parts of the crane range.
[c] Data excluded from analysis since only one territory.
Source: Adapted from Stehn, T.V., Prieto, F., 2010. Changes in winter whooping crane territories and range 1950–2006. Proceedings of the North American Crane Workshop, 11, 40–56.

marsh habitat is necessary for Whooping Crane use and that these other habitats must be quantified and conserved to achieve recovery downlisting criteria.

ANTHROPOGENIC FACTORS AFFECTING HABITAT AVAILABILITY IN THE AWBP WINTERING RANGE

Habitat loss has been listed as a primary factor in the reduction of the distribution of Whooping Cranes throughout North America (CWS and FWS, 2007). In the current wintering range of the AWBP, habitat loss in coastal wetlands has been widely documented along the Texas Coast (Dahl, 2000, 2011). We discuss three anthropogenic factors that may affect the recovery of the AWBP: the direct impact of urbanization and subsequent habitat loss in rural areas along the coast, and indirect impacts of habitat degradation and human disturbance.

Urbanization of rural environments continues at a steady pace throughout Texas. Of the six major urban centers that comprise two-thirds of the human population in Texas, only two are geographically located along the coast, Houston-Sugarland-Baytown and Lower Rio Grande area (Hitchcock, 2011). The AWBP wintering range is located between these two areas, in an area having relatively low development and the greatest area under protection (Fig. 13.7A) (Smith et al., 2014). A spatial compilation of protected areas indicated that 18 tracts of land are owned (11) or managed as conservation easements (7) by two federal agencies, one state, and four nongovernmental organizations for a total of 65,000 ha (160,000 ac), including about 37,000 ha (91,000 ac) identified as potential High Use habitat for Whooping Cranes. Within the protected areas, about 9,700 ha (25,000 ac) of potential High Use habitat has been conserved as of 2016 (Fig. 13.7B). However, two of the remaining six urban areas, San Antonio and Austin-Round Rock, are located at the upper areas of the coastal basins of the present and potential future wintering range and have much less area under protection. A majority of the other privately owned salt marsh complexes are located

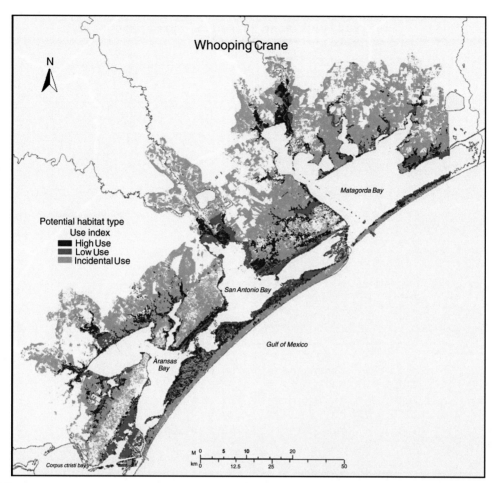

FIGURE 13.6 Potential habitat available for wintering Whooping Cranes delineated by use categories: high (primarily estuarine vegetated marsh and flats), low (primarily shallow seagrass and unvegetated bottom), and incidental (primarily coastal prairie, palustrine marsh). *Source: From Smith, E.H., Chavez-Ramirez, F.C., Lumb, L., Gibeaut, J., 2014. Employing the Conservation Design Approach on Sea-Level Rise Impacts on Coastal Avian Habitats along the Central Texas Coast. Gulf Coast Prairies Landscape Cooperative Program, 131 pp.*

in the coastal river deltas, with narrow fringes of salt marsh along the bay shorelines and back-island bay shorelines in Matagorda Bay (Smith et al., 2014). The primary concern from unregulated development is currently focused on two factors: lack of zoning ordinance control outside municipalities at the county level, and ability of landowners to receive permits to develop in flood-prone areas and receive flood insurance. Therefore, future trends of coastal development,

as well as development within the basins should be considered when evaluating the impact of urbanization on the AWBP recovery. In addition, indirect impacts of shoreline development can occur even where a shallow water environment is maintained. In a study within the Chesapeake Bay area, sites where bulkhead development had occurred had much lower mollusk diversity and abundance, and predator (primarily crabs) density was also lower compared to areas where

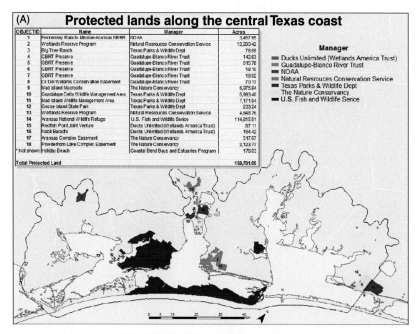

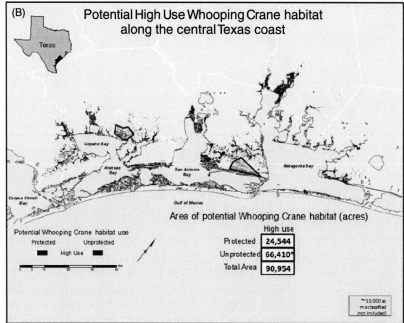

FIGURE 13.7   (A) Habitat conservation areas with associated land managers through 2014 within the current and projected wintering range of the Aransas-Wood Buffalo Whooping Crane; (B) unprotected and protected potential High Use areas through 2014. *Source: From Smith, E.H., Chavez-Ramirez, F.C., Lumb, L., Gibeaut, J., 2014. Employing the Conservation Design Approach on Sea-Level Rise Impacts on Coastal Avian Habitats along the Central Texas Coast. Gulf Coast Prairies Landscape Cooperative Program, 131 pp.*

salt marsh and adjacent shallow seagrass beds were intact (Seitz et al., 2005). As the Whooping Crane population continues to expand, development is becoming an increasing concern, particularly for Lamar Peninsula and Seadrift-Port O'Connor Ridge areas, and interactions with humans are increasing.

Various types of potential disturbance, termed human stimulus, were evaluated along the Blackjack Peninsula adjacent to the Gulf Intracoastal Waterway (GIWW) (Lafever, 2006). About one-half of the stimuli were grouped in the motorboat class, followed by barges (18%). Other classes were minor in occurrence, including other types of vessels, motor vehicles, and humans. Stimuli frequency averaged 3.83/h, varied among the five sites studied, and was significantly higher in one site that was located closest to the GIWW. Stimuli intensity, measured by a combination of speed, proximity, and auditory volume, was low in all sites, although one GIWW site had substantially more high-intensity stimuli/hour. In this site, Whooping Crane movement increased during the periods of higher intensity compared to periods of lower levels. In addition, cranes spent less time foraging during the high-intensity stimuli than before or after. This location is situated in a narrow reach of the GIWW that was excavated through the salt marsh complex in the 1940s. The Whooping Crane pair using this area is often close to the GIWW since a narrow embayment, Dunham Bay, is located behind the marsh complex as well. All boat traffic in the GIWW is ranked as higher intensity due to proximity, and the tour boats also stop longer to observe the cranes.

Lafever (2006) evaluated boat traffic to determine if current levels affected Whooping Crane behavior and use of the habitat types; however, they detected no significant differences among the five sites. At what level and frequency of disturbance these human activities become biologically significant is not known. Early reports estimated that cranes flushed at the approach of a boat <275–365 m (Stevenson and Griffith, 1946),

while a later study found cranes flushed when airboats and tour boats were <1,000 m away (Lewis and Slack, 1992). Airboat traffic can pose a more significant disturbance because of their high noise levels and their ability to traverse shallow water and across inundated marsh vegetation. Airboats have been reported to flush cranes off their territories (Mabie et al., 1989) and displace them from 15 min to several hours (Bishop, 1984).

Waterfowl hunting is not allowed in the salt marshes of Blackjack Peninsula within the Proclamation Boundary (ANWR, 2012), and access within the marsh is illegal without a special use permit. However, hunting is allowed in the salt marsh areas of Matagorda Island and shorelines of Lamar Peninsula. During aerial surveys to evaluate potential spatial interactions between cranes and hunters, Thompson and George (1987) detected crane displacement due to hunter presence and airboat activity and cranes appeared to maintain a 301–400 m distance from hunter locations, with the exception of Lamar Peninsula, where over one-half of the cranes observed were <200 m from hunters. Six of the nine cases of Whooping Crane shootings in the AWBP involving 10 Whooping Cranes have occurred in Texas since 1967 and were confirmed as being hunter-related (Condon et al., Chapter 23, this volume); four individual birds were shot in the wintering grounds, while two were killed during migration through Texas.

It is possible that the cranes that maintain winter territories in areas of higher human use may become habituated to these recurring disturbances (Mabie et al., 1989). Alternatively, they may be tolerating indirect contact to remain in an area because better habitat elsewhere is very limited (Gill et al., 2001), such as Blackjack Peninsula, or they may be reluctant to leave when food resources are limited (Lafever, 2006). It is unclear whether cranes would be more tolerant when they are also stressed by low food resources, or when few options are available elsewhere. More studies are needed to understand which of these interpretations apply to wintering Whooping Cranes to guide conservation and management actions (Butler et al., 2013, 2014).

CLIMATE CHANGE FACTORS AFFECTING AWBP RECOVERY

One primary area of concern for the long-term conservation of the AWBP Whooping Crane is the protection of winter habitat within its current range, as well as the identification and protection of future habitat areas that would support the potential growth of the population and expansion of their wintering range (CWS and FWS, 2007). In recent evaluations of potential impacts under different climate change scenarios, SLR was identified as one of the primary concerns for future habitat availability on the winter range along the Texas coast (Chavez-Ramirez and Wehtje, 2012; Harris and Mirande, 2013). Long-term changes in sea level and, to a lesser extent, freshwater inflows affect the distribution, extent, and composition of habitat types (Montagna et al., 2011) known to be important to Whooping Cranes.

Increasing minimum temperatures with concomitant fewer freezing events has accelerated the conversion of salt marsh to black mangrove habitats (Osland et al., 2013; Sherrod and McMillan, 1985). The structural characteristics of estuarine shrub physically limit access to food resources by wintering Whooping Cranes (Chavez-Ramirez and Wehtje, 2012). Changes in precipitation patterns and concomitant effects on drought severity and frequency from reduced freshwater inflows will also have major impacts on coastal habitat diversity and quality (IPCC, 2014), with potential impacts on the AWBP Whooping Cranes in the wintering range (Stehn and Prieto, 2010). These climate change factors are not mutually exclusive and can be evaluated using a biocomplexity framework approach that addresses both temporal and spatial scales of impact (Day et al., 2008).

Historic and Future Changes in Sea-Level Rise

Global SLR is an ongoing phenomenon, and evidence is widely available to demonstrate that sea levels are changing throughout the world, with higher rates since the 1850s than those during the previous 2,000 years (IPCC, 2014). Conservative estimates are given as at least 1-m SLR within this century; some models predict even higher levels in some areas of the world and those impacts are not uniformly distributed. Local, state, and federal governments have become concerned with the potential effects that predicted sea levels will have on their communities and built coastal landscapes, but less concern has been focused on the potential effects that changes in sea level will have on natural communities and particularly the animal species that depend on them (Montagna et al., 2007). Under natural conditions in gently sloping environments, coastal habitats would migrate inland and occupy areas of low-lying uplands in response to SLR. However, the continued conversion of natural habitats along the bay shorelines to urban development (Tremblay and Calnan, 2011) will constrain opportunities for habitat migration.

Sea-level data have been continuously documented in the current wintering area, and data collected at the tidal gauge at Rockport, Texas, which include the effects of land subsidence, show that sea level has risen by an average 4.6 mm/year since 1948 (Fig. 13.8) and by 3.44 mm/year at South Padre Island since 1958 (Zervas, 2001). The average rate of SLR since the 1950s has changed over time, with a high of 1.7 cm/yr from the mid-1960s to mid-1970s (Montagna et al., 2011; Snay et al., 2007). Relative SLR is defined within these studies as "the relative rise in water level with respect to a datum at the land surface, whether it is caused by a rise in mean water level or subsidence of the land surface" (White et al., 2002). The Texas coast experiences relative sea-level changes more from a factor of subsidence than eustatic

SLR. Subsidence can be attributed to both the withdrawal of groundwater, oil, and gas and to a lesser extent compaction of underlying soils. Subsidence rates along the Texas coast are among the highest in the world (Anderson, 2007; Anderson et al., 2010). Combining the rate of local land subsidence with IPCC climate models, the projected relative SLR at Rockport from 2000 to 2100 is estimated at between 0.46 and 0.87 m (Montagna et al., 2007).

Recent sea-level change that has occurred over the last 60 years within the current wintering range of the Whooping Crane provides a context to understand the trends and geographic locations of potential habitat availability. We first present these historic data of habitat conversions from relative SLR within the central Texas coast. These results also lay the foundation to guide evaluations of habitat shifts and extent that we should expect to encounter in the coming decades. We then follow with habitat projections under different sea-level projections.

The status and trends of wetland habitats in the three bay systems within the AWBP wintering range have been analyzed: Matagorda Bay (Tremblay and Calnan, 2010; White et al., 2002), San Antonio Bay (except for ANWRC) (Tremblay and Calnan, 2011; White et al., 2002), and Aransas Bay (Tremblay et al., 2008; White et al., 2006) (Fig. 13.9). In each study, similar methods were used to compile wetland trends results and are generally comparable. Spatial data of the current distribution of wetlands within bay system studies were constructed using color infrared photographs from 2002 to 2009 and compared to historic imagery from the 1950s and 1979. These studies did not evaluate the historic status and trends of wetland habitats on the Blackjack Peninsula, where the core of the population was documented when systematic winter surveys were initiated in 1950 (see Fig. 13.5). Our interest in the status and trends for other areas were focused on two key coastal habitats, estuarine marsh defined as High Use habitat and tidal flats as Low Use used by wintering Whooping Cranes (Smith et al., 2014) in relation to an average increase in sea level of ~0.3 m from 1950 to 2008 (see Fig. 13.8).

Overall, results were similar among regions, with most geographic areas exhibiting a relative increase in estuarine marshes and decrease in tidal flats (Fig. 13.9). However, some interesting patterns are evident by landform when correlating these changes to the spatial patterns of

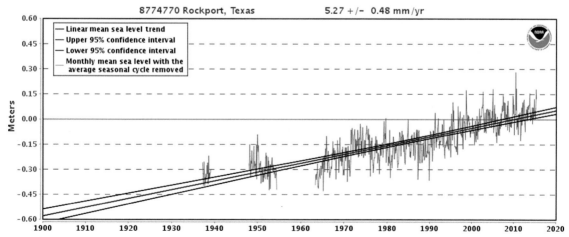

FIGURE 13.8 Long-term relative sea level change for NOAA Rockport, Texas, gauge 8774770 (https://tidesandcurrents.noaa.gov/sltrends/).

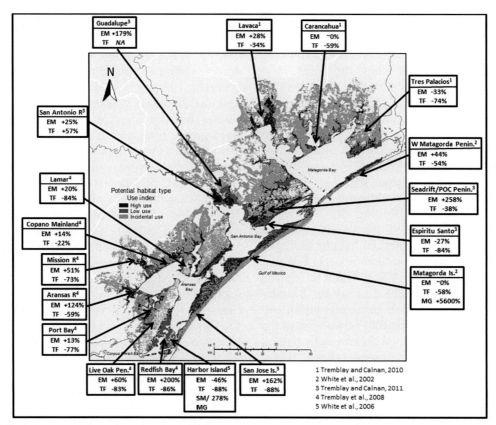

FIGURE 13.9 Changes in estuarine marsh (EM) and tidal flats (TF) from 1950s to 2000s throughout the potential distribution of wintering Whooping Cranes. *Source: Modified from Smith, E.H., Chavez-Ramirez, F.C., Lumb, L., Gibeaut, J., 2014. Employing the Conservation Design Approach on Sea-Level Rise Impacts on Coastal Avian Habitats along the Central Texas Coast. Gulf Coast Prairies Landscape Cooperative Program, 131 pp.*

AWBP population expansion. A crane pair established territories on Matagorda Island by 1958 and San Jose Island in 1969, and territories gradually expand to the north and south along these islands, respectively (Stehn and Prieto, 2010). Matagorda Island has maintained the same relative amount of estuarine marshes since the 1950s (White et al., 2002) while this same habitat has exhibited a 162% increase on San Jose Island (White et al., 2006). The concomitant decrease of tidal flats has been primarily attributed to relative SLR on these landforms where urban development is virtually nonexistent. Estuarine marshes on the West Matagorda Peninsula have

also increased at the expense of tidal flats (White et al., 2002), although the sequential expansion of wintering Whooping Cranes has not utilized this area yet.

Whooping Cranes first established territories on Lamar Peninsula in 1971 (Stehn and Prieto, 2010), although estuarine marsh on the former area has increased only 20% since 1950 (tidal flats have decreased 84%) (Tremblay et al., 2008). Few (2–4) territories have been maintained on Lamar and adjacent habitats over time; although some protected lands are located within this landform, development has been occurring on unprotected lands, which may be limiting habitat use (Stehn and Prieto, 2010). Territories on Seadrift/Port O'Connor Peninsula were established in 1973 and have increased in numbers over time (see Fig. 13.5). Some of the largest increases (258%) in estuarine marsh have been documented along the southwest shoreline of this area, with about one-third loss of tidal flats (Fig. 13.9) (Tremblay and Calnan, 2011). Substantial habitat protection has been achieved within this coastal habitat segment (Smith et al., 2014). However, the peninsula and adjacent mainland are surrounded on three sides by coastal bays, and urban development is likely (Moya et al., 2012) in areas where Whooping Cranes would be expected to expand.

River deltas have also exhibited modest to substantial relative increases of estuarine marsh, ranging from 25% to 50% in Guadalupe/San Antonio, Lavaca, and Mission deltas (Tremblay and Calnan, 2010, 2011; Tremblay et al., 2008) (Fig. 13.9). Northern portions of the Guadalupe Delta show the highest relative increases (179%) followed by the Aransas delta (124%) (Tremblay and Calnan, 2011; Tremblay et al., 2008). Interestingly, wintering Whooping Cranes have not established territories in the delta habitats. Little development has occurred in these areas; however, diminished freshwater inflows in the Lavaca and Guadalupe-San Antonio deltas during dry and drought years may affect habitat quality (Stehn and Prieto, 2010). Mission and Aransas rivers have minor freshwater diversions; however, their watersheds are small and droughts may have pronounced effects on habitat quality. Several minor drainage deltas are located in the current and potential wintering range and may afford limited habitat for subadult groups. These areas did not exhibit substantial increases in estuarine marsh (0–33%) but had much larger (22–84%) losses in tidal flat extent (Fig. 13.9).

Projected habitat changes related to future SLR have been evaluated using the Sea Level Affecting Marshes Model (SLAMM), which is a dynamic model that takes into account the dominant process in wetland change due to SLR, including inundation, erosion, overwash, saturation, and accretion. In the first use of the SLAMM within the ANWRC boundaries, as well as contextual areas (which modeled geographic areas surrounding the refuge boundaries), Clough and Larson (2010) evaluated the current availability of parameters needed for modeling the area of interest. Many of the input parameters were defined from National Oceanic and Atmospheric Administration (NOAA) tide stations within the wintering area, as well as from four sites in the northern Gulf of Mexico (Callaway et al., 1997).

In the ANWRC model (Clough and Larson, 2010), the NOAA tide station 8774770 was located at the southern extent of the study area at Rockport, Texas, with the estimate of historic (1940–present) SLR set at 5.16 mm/year. This rate represents global SLR as well as more localized subsidence rates. Accretion rates were set at 0.4 mm/year for estuarine marsh; horizontal erosion rates for marsh, swamp, and tidal flat classes were set at 0 mm/year due to lack of regional information; and elevation data were used from Lidar Digital Elevation Models from 2006. Overall, habitat changes modeled both within the ANWRC and the contextual area showed a decrease in estuarine habitats at 1- and 1.5-m SLR by 2100, and an increase at 2-m SLR. The authors cautioned that changes in intertidal flat may not be well represented and that barrier island change did not take into account aperiodic storm overwash events and associated erosional and depositional effects.

Smith et al. (2014) used the SLAMM approach developed for the ANWRC contextual area (Clough and Larson, 2010) for predicting effects of SLR on current and potential Whooping Crane habitat. They evaluated several considerations to determine if both High Use and Low Use habitats could be evaluated using the SLAMM. In the CHTD model, three land-cover data sets were combined for best available habitat coverage. Relative frequency of mesohabitat use by wintering AWBP were: High Use – Estuarine Vegetated Marsh (50.4%); Low Use – Estuarine Open Water (<0.3 m deep MLW) (16.9%), Estuarine Vegetated Seagrass (<0.3 m deep MLW) (13.0%), and Estuarine Unvegetated Flats (8.2%); and Incidental Use – Upland Grassland (5.7%), Freshwater Wetland (2.8%), Upland Shrub (1.5%); the remaining 1.5% were comprised of six mesohabitats (Smith et al., 2014).

Clough and Larson (2010) used NWI data in the SLAMM, which was the same database used in the CHTD model in Smith et al. (2014). However, only Estuarine Vegetated Marsh (combining regularly and irregularly flooded) was used in the CHTD, as Estuarine Vegetated Seagrass is not comprehensively mapped in NWI; therefore, the NOAA Benthic Atlas database was used for this habitat. In the SLAMM, tidal flat extent increased more than any other estuarine habitat (Clough and Larson, 2010), which was counter to actual historic changes discussed previously in regional status and trends reports. Efforts to explain these modeling results within this particular coastal landscape were further complicated by tidal flat being grouped with estuarine beach, accretion rates being set to 0 mm/year (whereas estuarine emergent marsh setting was 0.4 mm/year), and the relative positioning of tidal flats often occurring seaward of seagrass habitat (Clough and Larson, 2010). Therefore, Smith et al. (2014) only included High Use – Estuarine Emergent Marsh, which combined irregularly and regularly flooded habitat types, as the key factor in assessing areal habitat changes within the SLAMM contextual area. The SLAMM study covers a smaller area than the geographic extent modeled for potential available habitat for Whooping Cranes; however, the SLAMM coverage encompassed all of the current AWBP wintering range. Smith et al. (2014) generated habitat changes using the SLAMM under four SLR scenarios, IPCC A1B Mean (0.39-m SLR at 2100), A1B Maximum (0.69-m SLR at 2100), 1-m SLR, and 2-m SLR, targeting two time frames (2075 and 2100).

The IPCC scenario A1B Mean is the most conservative SLR scenario used in the SLAMM. This scenario assumes a climate change situation resulting from a balanced mix of energy from renewable and fossil-fuel sources, rapid economic and efficient technological growth, and human population growth that peaks around 2050 and then declines. The A1B Mean scenario is set to predict habitat changes from 0.39 m SLR by 2100. Under initial conditions, the key High Use habitat encompasses the bay side of Matagorda and San Jose islands, along the eastern shoreline of Blackjack Peninsula, and localized extents in the Guadalupe and Mission deltas (Fig. 13.10) and cover about 20,390 ha. Under the A1B Mean scenario, potential High Use habitat for Whooping Cranes will decrease over 12% by 2075, with the most obvious losses occurring along the bay side of the barrier islands in 2075. By 2100, the SLAMM projects a total decrease of about 23% of High Use habitat. Highest losses will continue to be geographically focused along barrier island marshes, and extensive marsh habitat is predicted as being lost in the southwest corner of Seadrift-Port O'Connor Ridge in the Welder Flats area (Fig. 13.11).

The IPCC scenario A1B Maximum is set to predict changes from 0.69 m SLR by 2100 and follows the same assumptions about energy use, economic and technological growth, and human population growth. Under this scenario, potential area of High Use habitat will decrease by about one-third in 2075. In 2075, much of the High Use habitat will be lost along the bay side of the barrier islands, as well as key areas along the eastern shorelines of the peninsulas and in Mission Delta. By 2100, potential High Use habitat will decrease by 52% from initial conditions. However, no specific geographic area can be identified where High Use habitat will be created (Fig. 13.12). The Guadalupe Delta will continue to maintain an extensive amount of fragmented habitats and Welder Flats will maintain High Use habitat on the inland side.

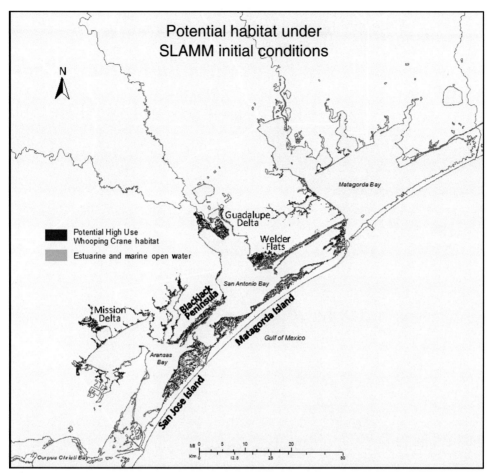

FIGURE 13.10 Current distribution of potential High Use habitat for Whooping Cranes within the SLAMM contextual area under initial (current) conditions. *Source: Modified from Smith, E.H., Chavez-Ramirez, F.C., Lumb, L., Gibeaut, J., 2014. Employing the Conservation Design Approach on Sea-Level Rise Impacts on Coastal Avian Habitats along the Central Texas Coast. Gulf Coast Prairies Landscape Cooperative Program, 131 pp.*

Under the 1-m SLR scenario, about 50% of potential High Use Whooping Crane habitat will be lost by 2075. All areas will lose extensive High Use marshes, although the Guadalupe Delta will continue to maintain relatively more continuous patches of High Use habitat. From initial conditions to 2100, potential High Use habitat for Whooping Cranes will decrease by about 54% (Fig. 13.13). Most of the High Use marsh areas are projected to be inundated with minor new marsh development in other areas,

with the exception of Guadalupe Delta, Welder Flats, and western margins of Lamar Peninsula as marsh complex migrates inland.

The 2-m SLR is the most extreme scenario used by SLAMM and is considered by some authors to be the upper limit of plausible SLR scenarios due to glaciological conditions (Pfeffer et al., 2008; Vermeer and Rahmstorf, 2009). With 2-m SLR, potential High Use habitat for Whooping Cranes will decrease by about 52% by 2075. The areas of habitat loss along the bay side

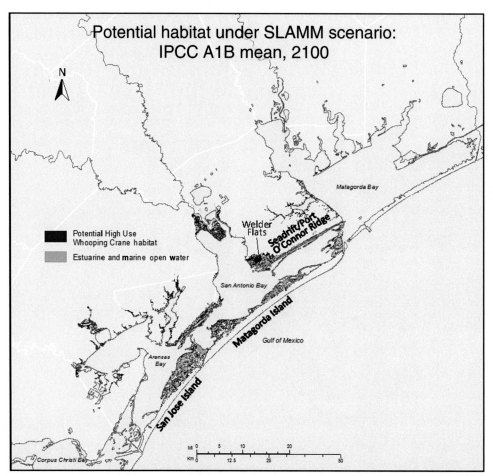

FIGURE 13.11　Predicted distribution of potential High Use habitat for Whooping Cranes within the SLAMM contextual area for IPCC A1B Mean by 2100. *Source: Modified from Smith, E.H., Chavez-Ramirez, F.C., Lumb, L., Gibeaut, J., 2014. Employing the Conservation Design Approach on Sea-Level Rise Impacts on Coastal Avian Habitats along the Central Texas Coast. Gulf Coast Prairies Landscape Cooperative Program, 131 pp.*

of the barrier islands and eastern shoreline of peninsulas will be compensated for in the Guadalupe Delta and, to a lesser extent, on Welder Flats on Seadrift-Port O'Connor Peninsula and Copano Bay side of Lamar Peninsula. By 2100, area of potential High Use habitat for Whooping Cranes will increase from predicted levels for 2075 by 5,992 ha and overall loss from initial conditions will rebound to about 15,380 ha. Areas that will exhibit the most areal extent increases are located throughout the bay shoreline of the

barrier islands, along the eastern shoreline of the peninsulas, as well as back bay areas in Powderhorn Lake, Lamar Peninsula, and Cape Valero area, and inland portions of Welder Flats and the Guadalupe and Mission river deltas (Fig. 13.14). The 2-m SLR scenario at 2100 is the only scenario where High Use habitat will potentially recover at some level toward initial conditions levels. Within the estuaries evaluated, it seems unlikely that there will be sufficient High Use habitat to meet projected habitat needs for 250 pairs, and

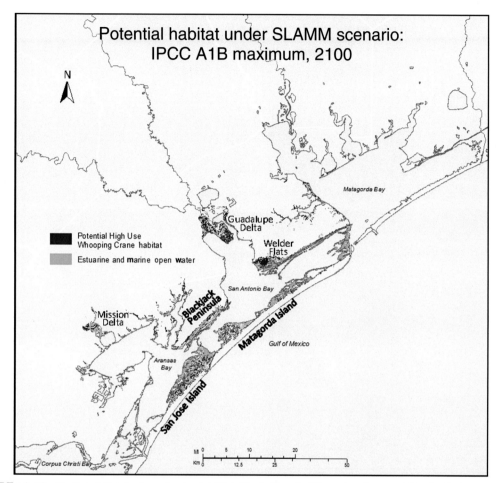

FIGURE 13.12   Predicted distribution of potential High Use habitat for Whooping Cranes within the SLAMM contextual area for IPCC A1B Max by 2100. *Source: Modified from Smith, E.H., Chavez-Ramirez, F.C., Lumb, L., Gibeaut, J., 2014. Employing the Conservation Design Approach on Sea-Level Rise Impacts on Coastal Avian Habitats along the Central Texas Coast. Gulf Coast Prairies Landscape Cooperative Program, 131 pp.*

the wintering cranes will need to expand to additional estuaries along the coast.

Historic Changes in Black Mangrove Expansion from Increasing Temperatures

Black mangroves typically occur worldwide in tropical latitudes and occur at a slightly higher elevation within the predominant mangrove community whereas the salt marsh community in tropical latitudes occurs in lesser

extent (Chapman, 1975). Black mangroves are well known as a productive marine community and provide many of the same benefits as salt marshes, including food web support, nursery habitat, and shoreline erosion control. Expansion of black mangroves within the Gulf of Mexico region has been documented for several decades, although freezing events were believed to limit the northern distribution (Sherrod and McMillan, 1985). While the correlation between rising minimum temperatures and establishment

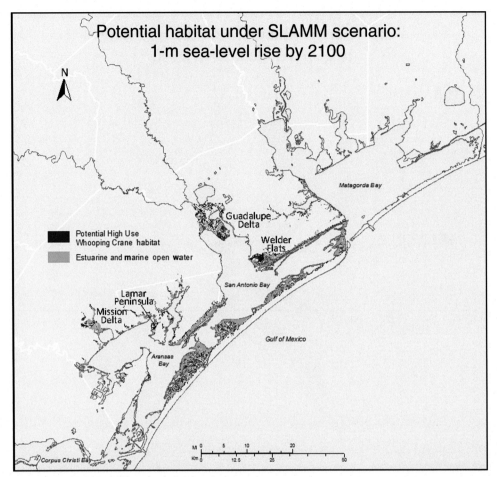

FIGURE 13.13   Predicted distribution of potential High Use habitat for Whooping Cranes within the SLAMM contextual area for a 1-m sea-level rise by 2100. *Source: Modified from Smith, E.H., Chavez-Ramirez, F.C., Lumb, L., Gibeaut, J., 2014. Employing the Conservation Design Approach on Sea-Level Rise Impacts on Coastal Avian Habitats along the Central Texas Coast. Gulf Coast Prairies Landscape Cooperative Program, 131 pp.*

of black mangroves appear to drive the distributional expansion (Osland et al., 2013), other factors of global change have also been attributed as well (CO_2, precipitation, SLR, nutrients; Saintilan et al., 2014). Certainly, this species can be used as an indicator of climate change in marine coastal environments at the temperate extent of its range (Sherrod and McMillan, 1981) and may have important ecological consequences of both functional and structural importance, including effects on biotic communities, ecosystem stability, and biogeochemical processes (Osland et al., 2013).

Concerns about the establishment of black mangroves and concomitant loss of salt marshes along the Texas coast and northern Gulf of Mexico are relevant within the current wintering range of the Whooping Crane (Chavez-Ramirez and Wehtje, 2012; Stehn and Prieto, 2010). While no studies have been initiated to quantitatively evaluate habitat use or avoidance of areas by cranes where black mangrove is predominant,

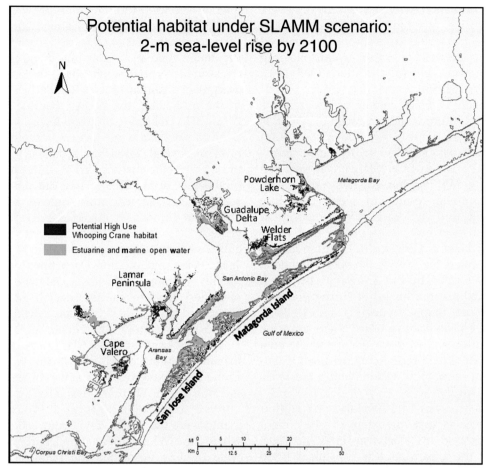

FIGURE 13.14 Predicted distribution of potential High Use habitat for Whooping Crane within the SLAMM contextual area for a 2-m sea-level rise by 2100. *Source: Modified from Smith, E.H., Chavez-Ramirez, F.C., Lumb, L., Gibeaut, J., 2014. Employing the Conservation Design Approach on Sea-Level Rise Impacts on Coastal Avian Habitats along the Central Texas Coast. Gulf Coast Prairies Landscape Cooperative Program, 131 pp.*

the physical structure of black mangrove limits any movement through or use of this habitat by Whooping Cranes. Therefore, at minimum, the extent of black mangroves throughout the wintering range reduces the amount of habitat available for crane use.

Along the northwestern Gulf shoreline, early accounts acknowledged the presence of black mangrove in Texas bays (Armstrong, 1987). The persistence of black mangroves was first quantitatively documented in the southernmost

tip of Texas, on South Padre Island, in the 1970s (Everitt et al., 2010). During a 26-year period (1966–2002), black mangrove extent had increased 477%, from 0.06 ha to 0.34 ha in that study area. Black mangroves appear to establish and concentrate around natural and channelized Gulf passes in Texas, from the southernmost area to Galveston Island and Bolivar Peninsula (Sherrod and McMillan, 1981). Recent studies document black mangrove establishment to the northern extent of the Gulf of Mexico as early as

2000 in Louisiana (Twilley et al., 2001). Severe freezes during this time did damage above-ground stems; however, resprouting occurred throughout the patches within a year following the freeze event (Lonard and Judd, 1985, 1991; Sherrod et al., 1986).

Expansion of black mangroves within the current and future wintering range of AWBP is of serious concern, primarily from a standpoint of losing estuarine salt marsh to estuarine scrub–shrub habitat. As early as 1980s, it was expected that as the AWBP increased, range expansion would continue southward along San Jose Island and westward to adjacent unoccupied salt marsh–flats complexes (Stehn and Johnson, 1987; Stehn and Prieto, 2010). Harbor Island, located south of the ANWR, is a flood-tidal delta originally formed by the Aransas Pass, which was located just northward of current location of the Aransas Ship Channel (Fig. 13.8A). Black mangroves have been recorded there since the 1930s (Britton and Morton, 1989), and the area has been described recently as the most extensive estuarine tropical wetland along the Texas coast (Pulich, 2007).

Spatial analysis of habitats on Harbor Island revealed that the extent of black mangrove increased 131% between 1930 and 2004; however, the expansion was not linear (1930 – 235.54 ha; 1979 – 548.29 ha; 1995 – 372.56 ha; 2004 – 545.58 ha) (Montagna et al., 2011). The 47.2% decrease between 1979 and 1995 was attributed to four years of significant freezes (minimum air temperature <0°C for >10 consecutive days/year) in 1982, 1983, 1985, and 1989. The subsequent recovery is indicative that black mangroves are an established component of the AWBP current wintering range; however, recovery from intermittent freeze events appears to be a slow process.

In years mapped by the NWI along the Texas coast (1950s, 1970s, 1990s, 2000s), mangroves either were not discernible at the mapping scale used, or were grouped into the estuarine intertidal scrub–shrub class with other salt-tolerant species (Moulton et al., 1997; Tremblay et al., 2008; Tremblay and Calnan, 2010, 2011; White et al., 2002, 2006). Class designations (with vegetation type and water modifiers) included Estuarine Intertidal Scrub–Shrub (broad-leaved evergreen) Irregularly and Regularly Flooded. In the mid-2000s, the NOAA Benthic Habitat Atlas (BHA) mapping program focused on estuarine habitats that were underrepresented in the NWI in several Texas bay systems, including those within the current AWBP wintering range (Finkbeiner et al., 2009). They did not have a standardized nomenclature among mapping efforts but did identify mangroves at both Habitat and Class levels in two different mapping efforts. A separate program developed by the Texas Parks and Wildlife Department to create the Texas Ecological Systems Database (TESD) identified mangroves as Coastal: Mangrove Shrubland and Coastal: Salt and Brackish High Tidal Shrub Wetland (Elliott et al., 2014).

The development of the CHTD was designed to map potential wintering habitat for Whooping Cranes, and mangrove habitat was quantified by merging the NWI, BHA, and TESD databases into one data set (Smith et al., 2014). For black mangroves, the TESD data were used except where the NWI data overlapped, and NWI data were used except when the BHA data overlapped. This approach resulted in a Mesohabitat Estuarine Vegetated Shrub, which encompassed 2,636 ha of the current wintering range and Harbor Island to the south. Considering that black mangrove extent on Harbor Island in 2004 from digitized estimates was reported as 545 ha (Montagna et al., 2011), extensive coverage of mangroves occurs within the AWBP range.

Locations of wintering Whooping Cranes were taken from the 2004 to 2010 wintering surveys and overlaid on the CHTD to determine habitat site selection. Only 14 of 15,112 individual locations of Whooping Cranes (0.09%) were linked to either Mangrove or Estuarine Intertidal Scrub–Shrub Irregularly Flooded microhabitat types that covered 1,734 ha (Smith et al., 2014).

Based on the Habitat Use Index developed for Whooping Cranes, the mesohabitat Estuarine Scrub–Shrub was not defined as suitable habitat for cranes in the wintering range. Essentially, the increase of black mangroves extent over the past few decades has eliminated >2,600 ha from the area of potential range expansion. The projected habitat availability within the Aransas-Copano, San Antonio, and Matagorda bay systems is only ~37,000 ha (Smith et al., 2014), and the recovery goal is for least 50,585 ha (CWS and FWS, 2007). Therefore, the cranes will need to move northeastward along the Texas coast to find sufficient habitat. The expansion of black mangrove continues along this coastal corridor.

A recent study addressed the question of whether black mangroves are displacing salt marsh habitat at a local and regional scale (Armitage et al., 2015). Their work encompassed the last 20 years and determined that a 24% net loss of salt marsh occurred along the Texas coast. Only 6% of these losses were attributable to replacement by black mangrove, and SLR appeared to be the main driver of salt marsh losses. However, that study also showed that black mangrove expansion is occurring primarily around Gulf passes, which are important areas for potential wintering range expansion (Fig. 13.14). One such area is located on the northern end of Matagorda Island, where increases of black mangrove have been documented near Pass Cavallo (Armitage et al., 2015).

Future predictions concerning the expansion of black mangroves are difficult when integrating potential effects of SLR (Montagna et al., 2011). As water levels increase, the potential for erosion from wave energy also increases. In other areas of similar latitudes in Florida, concerns have been raised on the substantial loss of mangrove communities with a 50-cm SLR (Ning et al., 2003). Thus, it is difficult to determine if black mangrove establishment will continue to occur in future climate change scenarios. These issues point to more immediate need for monitoring black mangrove establishment in the intertidal areas adjacent to Cedar Bayou within the core area of Whooping Crane wintering range (see Fig. 13.5, Whooping Crane Winter Territories 2006).

Future Changes in Precipitation and Freshwater Inflows

Long-term changes in precipitation patterns have been documented globally, with average precipitation increasing over the midlatitude continents in the Northern Hemisphere (IPCC, 2014). Increases in ocean salinities have also been observed in areas where evaporation exceeds precipitation. With temperatures rising in the future, this phenomenon is expected to continue. In some areas, the frequency and intensity of precipitation events have increased and are expected to increase in the future. These shifts affect vegetation and animal distributions, as well as seasonal activities and migration patterns of animals.

Drought in some regions may also be longer and more intense, although lower confidence in these predictions is linked to variability observed at decadal scales (IPCC, 2014). It is expected that climate change will continue to affect renewable surface water and groundwater resources in dry, subtropical regions as a result future drought intensification. In the current wintering range, meteorological droughts that persist for more than one season occur regularly. During 1942–2009, the central Texas coast estuaries experienced droughts over half the time (Nueces 75%, Guadalupe 58%, Lavaca 62%), with about two-thirds of the droughts lasting <3 years (Montagna and Palmer, 2012). Megadroughts (those that last longer than a decade) have been much rarer; the drought of record for this same period occurred in the 1950s and continued for 12 years in the Guadalupe Estuary. These drought cycles have serious impacts on food availability in the estuarine system and have been linked to increased winter mortality of Whooping Cranes (Chavez-Ramirez and Wehtje, 2012). An increase in megadroughts under future climate conditions could set back recovery projections.

The Guadalupe and Lavaca estuaries have similar freshwater inflow rates; however, the size of the Guadalupe Estuary is smaller and experiences more limited water exchange with the Gulf of Mexico (Montagna and Li, 1996). Therefore, the system has a lower salinity range and supports the highest abundance of phytoplankton, benthic organisms, and suspension feeders. Several species of mollusks and crustaceans can thrive in these conditions. Montagna and Li (1996) predicted that under a lower freshwater inflow scenario, these estuarine communities will change dominance patterns toward a deposit-feeding community. These changes would potentially impact key prey items required by wintering Whooping Cranes.

Another concern related to changing precipitation patterns is the linkage between decreased precipitation and habitat conversion. Along coastal habitats along the northern Gulf of Mexico, shifts from estuarine marsh to black mangrove have been attributed to decrease in freshwater inflows as a result of drought (McKee et al., 2004). Future increases in drought frequency and intensity may trigger new or expanded establishment of mangroves in secondary bays and deltas farther inland in the estuarine system, further decreasing the extent of preferred habitat for Whooping Cranes. When considered with other climate change factors, such as SLR and expected decreases in extent of estuarine marsh, the focus on conserving all current and potentially future areas of crane habitats within the Gulf Coast Prairies region becomes even more essential.

A Biocomplexity Framework for Addressing Climate Change

We address the primary factors of climate that are perceived as having the most impact on the coastal environment within the present wintering range of the AWBP. These factors are not mutually exclusive, are occurring at different scales, and are further affected by anthropogenic influences. One reasonable approach to synthesizing these impacts has been brought forward by using a biocomplexity framework to address climate change factors on ecogeomorphology of this coastal system (Day et al., 2008).

The difference in scales at which the climate change is occurring and the ability of the ecological system to respond (usually longer) is of particular significance and can be manifested both functionally and structurally. The inclusion of human impacts at all scales can further exacerbate that system response. Day et al. (2008) identified five impacts related to climate change to frame within the context of biocomplexity (Michener et al., 2001): accelerated SLR, increased temperature, changes in rainfall distribution and freshwater inputs, frequency and intensity of storms, and the interaction of human activities. Day et al. (2008) further developed the concept by linking the physical framework of a fluvial sedimentary system with ecological response within coastal systems through an ecogeomorphological approach. We can use this approach in evaluating the combined effects of the climate change variables discussed earlier to provide recommendations to guide management and conservation of the wintering range of the AWBP.

The resiliency of the estuarine marsh ecosystems along the central Texas coast are dependent upon the combined effects of sediment delivery from riverine discharge that results in elevation change (accretion or erosion), salinity and nutrient gradients (higher or lower) from freshwater inflow pulses that drive vegetation dominance and estuarine faunal productivity, and increases in temperature that shift estuarine marsh to estuarine tropical shrubs. Historic and future trends indicate that within the wintering range these combined effects are negatively affecting the spatial extent of estuarine marsh complexes important to Whooping Cranes. The five impacts identified by Day et al. (2008) are further affected by increasing human population density within the Guadalupe-San Antonio coastal basins, which will require greater freshwater diversion for human uses. The measurable increase of black mangrove due to increasing temperatures is resulting in conversion of estuarine marsh; concomitant conversion of marsh to open water will continue with increasing rates in relative SLR. While it is difficult to separate the degree of influence that these impacts have on the potential recovery of the AWBP, conservation strategies should use an inclusive approach to minimize impacts that we can control or manage that will mitigate those climate change factors we cannot.

SUMMARY AND OUTLOOK

The current wintering range and habitat for the remnant AWBP is concentrated within the Gulf Coastal Plains, primarily in and around the ANWRC and located within the Guadalupe Estuary. The natural landscape exhibits low elevation relief grading from coastal prairies and freshwater marshes into tidal flat, estuarine marsh, and seagrass habitats. Ecological diversity and productivity in this coastal system, important to Whooping Crane health and survival, is driven by freshwater inflows from the Guadalupe-San Antonio coastal basin.

Recent studies and analyses of a long-term survey data set from 1950 to 2010 have provided valuable insights about the winter range, habitat use, and territories of Whooping Cranes. As the population has increased to over 400 individuals, cranes have expanded into peripheral areas around the ANWRC. Spatial analyses identified crane habitats of High Use (i.e., estuarine vegetated marsh) or Low Use (e.g., estuarine open water and estuarine vegetated seagrass); those analyses in turn inform models to assess habitat availability and carrying capacity into the future. More detailed information linking daily movements to habitat types will be forthcoming

from an ongoing telemetry study, which should also identify key roosting habitats and short-term responses to management activities such as controlled burns in the uplands. As crane distribution and numbers expand, continuation of such research will be valuable to understand how cranes are adapting and using new areas.

The same historic survey data set also provided spatial information to define winter territories. While the size and habitat composition varies by landform and availability of habitat, at least 172 ha of estuarine marsh, shallow open water, and tidal flat habitat are necessary to support a family of cranes. One of the recovery options to downlist the species from endangered to threatened is to have 1,000 individuals and 250 nesting pairs in the AWBP, which will require an estimated ~51,000 ha to meet that goal. Estimates through 2006 indicate that less than 43,000 ha of estuarine marsh habitat is currently available within 111 km radius from ANWRC to support that goal. Models that included only two adjacent estuaries around Guadalupe Estuary resulted in less than 39,000 ha of estuarine habitat available. Two new studies have been initiated that encompass a much larger extent of the Texas coast; results from these studies will be valuable in determining where additional habitat is located, as well as the current protection status. Preliminary results from one of the new studies indicate that sufficient habitat is currently available to satisfy downlisting recovery needs within 200 km northeast along the Texas coast.

Projecting habitat needs for the AWBP over time has relied on estimating how much habitat each adult crane pair requires for a winter territory. Those projections used historic data that was collected throughout the winter seasons of 1950–2011; the surveys throughout the winter also provided key data to track winter range expansion as the AWBP population increased. In 2011, survey protocols to estimate the winter population of the AWBP were changed and focused on one multiday period in December. Discussions are ongoing to augment the protocol and include more aerial surveys throughout the winter to better understand ongoing crane habitat use and social group interactions as the population increases. These data will be useful for identifying future habitat targets for conservation and management.

The extent and diversity of habitat types across this landscape provide the diverse food resources to the Whooping Cranes during their overwintering period and are critical to their health and survival, and preparation for spring migration. The AWBP cranes feed on a variety of estuarine and upland food items, although the blue crab is an important food item when available in the shallow, open water coastal ponds. Wolfberry fruit is readily consumed when seasonally abundant within the salt marsh. A few mollusk species are variably consumed in salt marsh ponds and shallow bay margins; however, little is known about their ecology and response to changes in freshwater inflows, salinities, or temperature. Alternate habitats, such as freshwater ponds and grasslands, can be utilized in the uplands adjacent to coastal marsh complexes. Time-activity studies in the 1990s provided a good baseline for assessing the ecological importance of the habitats within this coastal environment. As ecological conditions change and wintering distribution of cranes shifts, future studies should incorporate time-activity budgets to better understand habitat use under varying environmental conditions, habitat alteration, or areas where disturbance activities may be present.

The health and extent of these coastal habitats within the Guadalupe Estuary, as well as adjacent Texas coastal estuaries are critical to sustaining and growing the AWBP as the population expands the wintering range. The ecological system can be resilient in regard to natural perturbations and climatic regimes, but the compounded stressors of reduced freshwater inflows and unchecked development in low-lying areas can be detrimental. Few regulations are currently in place to reduce continued impacts on the quality and quantity of coastal habitat. One of critical stressors is freshwater inflows to the central Texas coastal estuaries, which have continued to decrease for decades. Water diversions reduce the amount of freshwater available to mix with saline waters entering from the Gulf, altering characteristic salinity gradients in the bays. The resulting salinity extremes negatively affect both plant and animal productivity and relative abundance of food resources for cranes. High salinities reduce wolfberry productivity, and salinities >35 ppt in the salt marsh can stress even those species adapted to higher salinity regimes. By altering salinity regimes, long-term reductions in freshwater inflows have changed community structure and vegetation patterns in low-latitude salt marsh communities. Even freshwater inflow pulses at certain times can affect the plant phenology and shift community dominance. Therefore, patterning freshwater inflows under managed conditions with natural cycles is as important as overall amounts of water released.

The optimum salinity range for blue crabs at various life stages has been identified as 10–25 psu, and this species can occur from fresh to ocean, salinities (35 psu). However, few studies have addressed the impact that hypersaline conditions have on blue crab abundance and distribution. Natural drought periods that are prolonged from reduced freshwater inflows can set up conditions where blue crabs are absent in high-salinity marsh ponds and sloughs and other near-shore environments. Whooping Cranes have been documented using upland habitats at a higher frequency during drought periods, although the direct link between food availability and increase in upland use has not been quantified. Studies of the bay's ecological function and responses under varying inflow conditions will be essential in identifying and recommending environmental freshwater inflow regimes that will maintain a sound ecological environment for the Guadalupe Estuary and other systems to support the expanding crane distribution.

Development along the Texas coast is currently centered northeast of the AWBP winter range in Houston and surrounding areas and south in the Lower Rio Grande area. Within the river basins that provide freshwater inflows to the Guadalupe Estuary, the area between two large municipalities of San Antonio and Austin is experiencing some of the fastest growth in the United States. At the basin scale, the impacts that water availability has on the wintering range are discussed earlier. However, another connection is becoming more important as the coast experiences unprecedented growth in second homes for people living in the upper basins and from Houston. Large tracts of coastal property have been purchased for development that converts low-lying prairie and estuarine habitats into canal developments. As a result, the limited amount of native habitat currently available for the AWBP wintering range continues to decrease. Legislative efforts to allow counties to implement zoning ordinances should be supported, and conservation design alternatives should be provided to prospective developers to minimize habitat loss and degradation.

Recreational activities by residents and visitors continue to increase as a result of development, and both direct and indirect effects are affecting crane behavior and use of remaining coastal habitat. Local efforts to increase public awareness are essential to limit airboat traffic in salt marshes as well as to keep them a respectful distance from Whooping Cranes. Waterfowl hunting is allowed within salt marsh habitat in the Matagorda Island portion of ANWRC and is legal in other salt marsh habitat in the wintering range. Increased educational efforts are needed to reduce accidental shooting of Whooping Cranes. An area within the current wintering range has been designated as a "No Sandhill Crane Hunting Zone" to prevent the potential of mistaking Whooping Cranes for Sandhill Cranes (*Grus canadensis*). As the AWBP expands along the Texas coast, the expansion of that area may need to be evaluated.

Climate changes occur at a long time scale; however, the effects have been evaluated and observed even within the last 60 years. Relative SLR along the central Texas coast has resulted in the increase of estuarine marsh and seagrass habitat and concomitant decrease in tidal flats. Within the last 30 years, black mangrove has become established as a result of fewer multiple-day freezing events and has replaced estuarine marsh and tidal flat habitats. Initial SLR models indicated a substantial decrease in estuarine marsh habitat under 1-m SLR. Preliminary modeling results from two ongoing studies indicate that sufficient habitat is currently available when additional area is included from the Nueces Estuary to Galveston Estuary to support downlisting goals. These general estimates do not take into account habitat quality and effects of SLR, two factors that are very important. Results of these studies are forthcoming, and should be considered in long-term conservation planning for AWBP recovery efforts. Managing impacts of these natural phenomena is not practical; however, the impact of current and future loss of coastal habitat from unplanned development and habitat degradation from decreases in freshwater inflow can be mitigated. A conservation design approach that plans how economic development for municipalities and industry is implemented across the coastal landscape can ensure that ecological diversity is also maintained in the current and future AWBP winter range. Balancing short-term and long-term protection of the habitat with ensuring healthy estuaries through management of freshwater inflows is essential to contribute to the complete recovery of the AWBP Whooping Crane.

References

Abell, R., Thieme, M.L., Revenga, C., Bryer, M., Kottelat, M., Bogutskaya, N., Coad, B., Mandrak, N., Contreras Baleras, S., Bussing, W., Stiassny, M.L.J., Skelton, P., Allen, G.R., Unmack, P., Naseka, A., Ng, R., Sindorf, N., Robertson, J., Armijo, E., Higgins, J.V., Heibel, T.J., Wikramanayake, E., Olson, D., López, H.L., Reis, R.E., Lundberg, R.J.,

Sabaj Pérez, M.H., Petry, P., 2008. Freshwater ecoregions of the world: a new map of biogeographic units for freshwater biodiversity conservation. BioScience 58 (5), 403–414.

Allen, R.P., 1952. The Whooping Crane. Research Report 3. National Audubon Society, New York, p. 246.

Allen, R.P., 1954. Additional data on the food of the whooping crane. Auk 71 (2), 198.

Anderson, J.B., 2007. Formation and Future of the Upper Texas Coast. Texas A&M Press, College Station, p. 163.

Anderson, J., Milliken, K., Wallace, D., Rodriguez, A., Simms, A., 2010. Coastal impact underestimated from rapid sea level rise. Eos 91(23), 205–206.

Aransas National Wildlife Refuge (ANWR), 2012. Aransas National Wildlife Refuge White-Tailed Deer, Feral Hog, and Waterfowl Hunt Plan. Aransas National Wildlife Refuge Complex, Austwell, TX, Draft Report, p. 26.

Aransas National Wildlife Refuge Complex (ANWRC), 2010. Comprehensive Conservation Plan and Environmental Assessment. U.S. Fish and Wildlife Service, Albuquerque, NM, Chapters + Appendices.

Armitage, A.R., Highfield, W.E., Brody, S.D., Louchouarn, P., 2015. The contribution of mangrove expansion to salt marsh loss on the Texas Gulf Coast. PLoS ONE 10, e01254104.

Armstrong, N.E., 1987. The Ecology of Open-Bay Bottoms of Texas: A Community Profile. U.S. Fish Wildl. Serv. Biol. Rep. 85 (7.12). U.S. Fish and Wildlife Service, Washington, DC, p. 104.

Behrens, E.W., Watson, R.L., 1973. Corpus Christi Water Exchange Pass: A Case History of Sedimentation and Hydraulics during Its First Year. U.S. Army Corps of Engineers, Coastal Research Center, Vicksburg, MS, DACW 72-72-C-0026, p. 135.

Bender, S., Shelton, S., Bender, K.C., Kalmbach, A. (Eds.), 2005. Texas Comprehensive Wildlife Conservation Strategy, 2005–2010. Texas Parks and Wildlife Department, Austin, p. 1131.

Bishop, M.A., 1984. The Dynamics of Subadult Flocks of Whooping Cranes Wintering in Texas 1978–1979 through 1982–1983. MS thesis, Texas A&M University, College Station, p. 127.

Bishop, M.A., Blankinship, D.R., 1982. Dynamics of subadult flocks of whooping cranes at Aransas National Wildlife Refuge, Texas, 1978–1981. In: Lewis, J.C. (Ed.), Proceedings of the 1981 Crane Workshop. National Audubon Society, Tavernier, FL, pp. 180–189.

Blankenship, L.H., Reeves, H.M., 1970. Mourning Dove Recoveries from Mexico. U.S. Fish Wildl. Serv. Spec. Sci. Rep. – Wildl. 135. U.S. Fish and Wildlife Service, Washington, DC.

Blankinship, D.R., 1976. Studies of whooping cranes on the wintering grounds. International Crane Workshop Proceedings, 197–206.

Block, W.M., Brennan, L.A., 1993. The habitat concept in ornithology: theory and applications. Curr. Ornithol. 11, 35–91.

Britton, J.C., Morton, B., 1989. Shore Ecology of the Gulf of Mexico. University of Texas Press, Austin, p. 87.

Brown, L.F., Brewton, J.H., McGowen, J.H., Evans, T.J., Fisher, W.L., Groat, C.G., 1976. Environmental Geologic Atlas of the Texas Coastal Zone: Corpus Christi Area. Bureau of Economic Geology, University of Texas at Austin, p. 107.

Butler, M.J., Harris, G., Strobel, B.N., 2013. Influence of whooping crane population dynamics on its recovery and management. Biol. Conserv. 162, 89–99.

Butler, M.J., Metzger, K.L., Harris, G., 2014. Whooping crane demographic responses to winter drought focus conservation strategies. Biol. Conserv. 179, 72–85.

Butzler, R.E., Davis, III, S.E., 2006. Growth patterns of Carolina wolfberry (*Lycium carolinianum* L) in the salt marshes of Aransas National Wildlife Refuge, Texas, USA. Wetlands 26 (3), 845–853.

Calabrese, A., 1969. Individual and combined effects of salinity and temperature on embryos and larvae of the coot clam *Mulinia lateralis* (Say). Biol. Bull. 137 (3), 417–428.

Callaway, J.C., DeLaune, R.D., Patrick, W.H., 1997. Sediment accretion rates from four coastal wetlands along the Gulf of Mexico. J. Coast. Res. 13 (1), 181–191.

Canadian Wildlife Service (CWS) and U.S. Fish and Wildlife Service (FWS), 2007. International Recovery Plan for the Whooping Crane, third ed. Recovery of Nationally Endangered Wildlife (RENEW). Ottawa, Ontario, and U.S. Fish and Wildlife Service, Albuquerque, NM, p. 162.

Chapman, V.J., 1975. Mangrove biogeography. In: Proceedings of the International Symposium on Biology and Management of Mangroves. University of Florida Gainesville 1, pp. 3–22.

Chavez-Ramirez, F., 1996. Food Availability, Foraging Ecology and Energetics of Whooping Cranes Wintering in Texas. PhD dissertation, Texas A&M University, College Station, p. 104.

Chavez-Ramirez, F., Hunt, H.E., Slack, R.D., Stehn, T.V., 1996. Ecological correlates of whooping crane use of fire-treated upland habitats. Conserv. Biol. 10 (1), 217–223.

Chavez-Ramirez, F., Wehtje, W., 2012. Potential impact of climate change scenarios on whooping crane life history. Wetlands 32 (1), 11–20.

Clough, J.S., Larson, E.C., 2010. Application of the Sea-Level Affecting Marshes Model (SLAMM 6) to Aransas NWR. U.S. Fish and Wildlife Service, National Wildlife Refuge System, Arlington, VA, p. 80.

Condon, E., Brooks, W.B., Langenberg, J., Lopez, D., 2018. Whooping crane shootings since 1967 (Chapter 23). In: French, Jr., J.B., Converse, S.J., Austin, J.E. (Eds.), Whooping Cranes: Biology and Conservation. Biodiversity of the World: Conservation from Genes to Landscapes. Academic Press, San Diego, CA.

Conner, W.H., Day, J.W., Baumann, R.H., Randall, J.M., 1989. Influence of hurricanes on coastal ecosystems along the northern Gulf of Mexico. Wetl. Ecol. Manage. 1 (1), 45–56.

Dahl, T.E., 2000. Status and Trends of Wetlands in the Coterminous United States 1986 to 1997. U.S. Department of the Interior, Fish and Wildlife Service, Washington, DC., p. 82.

Dahl, T.E., 2011. Status and Trends of Wetlands in the Conterminous United States 2004 to 2009. U.S. Department of the Interior, Fish and Wildlife Service, Washington, DC, p. 108.

Day, J.W., Christian, R.R., Boesch, D.M., Yáez-Arancibia, A., Morris, J., Twilley, R.R., Naylor, L., Schaffner, L., Stevenson, C., 2008. Consequences of climate change on the ecogeomorphology of coastal wetlands. Estuar. Coasts 31 (3), 477–491.

Dungan, C.F., Hamilton, R.M., Hudson, K.L., McCollough, C.B., Reece, K.S., 2002. Two epizootic diseases in Chesapeake Bay commercial clams *Mya arenaria* and *Tagellus plebeius*. Dis. Aquat. Organ. 50 (1), 67–78.

Elliott, L.F., Diamond, D.D., True, C.D., Blodgett, C.F., Pursell, D., German, D., Treuer-Kuehn, A., 2014. Ecological Mapping Systems of Texas: Summary Report. Texas Parks & Wildlife Department, Austin, p. 42.

Everitt, J.H., Yang, C., Judd, F.W., Summy, K.R., 2010. Use of archive aerial photography for monitoring black mangrove populations. J. Coast. Res. 26 (4), 649–653.

Federal Geographic Data Committee, 2013. Classification of Wetlands and Deepwater Habitats of the United States, FGDC-STD-004-2013, second ed. Wetlands Subcommittee, Federal Geographic Data Committee, and U.S. Fish and Wildlife Service, Washington, DC, p. 86.

Finkbeiner, M.J., Simons, J.D., Robinson, C., Wood, J., Summers, A., Lopez, C., 2009. Atlas of Shallow-Water Benthic Habitats of Coastal Texas: Espiritu Santo Bay to Lower Laguna Madre 2004 and 2007. NOAA Coastal Services Center, Charleston, SC, p. 60.

Fulbright, T.E., Diamond, D.D., Rappole, J., Norwine, J., 1990. The coastal sand plain of southern Texas. Rangelands 12, 337–340.

Gibeaut, J.C., White, W.A., Smyth, R.C., Andrews, J.R., Tremblay, T.R., Gutierrez, R., Hepner, T.L., Neuenschwander, A., 2003. Topographic variation of barrier island subenvironments and associated habitats. In: Davis, R.A., Sallenger, A., Howd, P. (Eds.), Coastal Sediments '03: Crossing Disciplinary Boundaries: Proceedings of the 5th International Symposium on Coastal Engineering and Science of Coastal Sediment Processes, Clearwater Beach, FL.

Gill, J.A., Norris, K., Sutherland, W.J., 2001. Why behavioural responses may not reflect the population consequences of human disturbance. Biol. Conserv. 97 (2), 265–268.

Greer, D.M., 2010. Blue Crab Population Ecology and Use by Foraging Whooping Cranes on the Texas Gulf Coast. PhD dissertation, Texas A&M University, College Station, p. 293.

Gunter, G., 1950. Distribution and abundance of fishes on the Aransas National Wildlife Refuge, with life history notes. Publ. Inst. Mar. Sci. 1 (2), 89–101.

Harris, J., Mirande, C., 2013. A global overview of cranes: status, threats and conservation priorities. Chin. Birds 4 (3), 189–209.

Hedgpeth, J.W., 1950. Notes on the marine invertebrate fauna of salt flat areas in Aransas National Wildlife Refuge. Publ. Inst. Mar. Sci. 1 (2), 103–119.

Hitchcock, D., 2011. Urban areas. In: Schmandt, J., North, G.R., Clarkson, J. (Eds.), The Impact of Global Warming on Texas, second ed. University of Texas Press, Austin, pp. 172–195.

Holland, A.F., Dean, J.M., 1977. The biology of the stout razor clam *Tagelus plebeius*: II. Some aspects of the population dynamics. Chesapeake Sci. 18 (2), 188–196.

Hunt, H.E., Slack, R.D., 1989. Winter diets of whooping and sandhill cranes in south Texas. J. Wildl. Manage. 53 (4), 1150–1154.

Hutto, R.L., 1985. Habitat selection by nonbreeding, migratory land birds. In: Cody, M.L. (Ed.), Habitat Selection in Birds. Academic Press, Orlando, FL, pp. 455–476.

International Panel on Climate Change (IPCC), 2014. Climate Change 2014: Synthesis Report. Contribution of Working Groups I, II, and III to the Fifth Assessment Report of the Intergovernmental Panel on Climate Change. Core Writing Team, R.K. Pachauri, L.A. Meyers (Eds.). IPCC, Geneva, Switzerland, p. 151.

Johnson, D.H., 1980. The comparison of usage and availability measurements for evaluating resource preference. Ecology 6 (1), 65–71.

Labuda, S.E., Butts, K.O., 1979. Habitat use by wintering whooping cranes on the Aransas National Wildlife Refuge. In: Lewis, J.C. (Ed.), 1978 Crane Workshop. Printing Service, Colorado State University, Fort Collins, pp. 152–157.

Lafever, K.E., 2006. Spatial and Temporal Winter Territory Use and Behavioral Responses of Whooping Cranes to Human Activities. MS thesis, Texas A&M University, College Station, p. 100.

Lewis, T.E., Slack, R.D., 1992. Whooping crane response to disturbances at the Aransas National Wildlife Refuge. Proceedings of the North American Crane Workshop 6, 176.

Lichvar, R.W., Butterick, M., Melvin, N.C., Kirchner, W.N., 2014. The National Wetland Plant List: 2014 update of wetland ratings. Phytoneuron, 2014-41, 1–42.

Lomovasky, B.J., Gutiérrez, J.L., Iribarne, O.O., 2005. Identifying repaired shell damage and abnormal calcification in the stout razor clam *Tagelus plebeius* as a tool to investigate its ecological interactions. J. Sea Res. 54 (2), 163–175.

Lonard, R.I., Judd, F.W., 1985. Effects of a severe freeze on native woody plants in the Lower Rio Grande Valley, Texas. Southwest. Nat. 30 (3), 397–403.

Lonard, R.I., Judd, F.W., 1991. Comparison of the effects of the severe freezes of 1983 and 1989 on native woody plants in the Lower Rio Grande Valley, Texas. Southwest. Nat. 36 (2), 213–217.

Longley, W.L., Solis, R.S., Brock, D.A., Malstaff, G., 1994. Coastal hydrology and the relationships among inflow, salinity, nutrients, and sediments. In: Longley, W.L. (Ed.), Freshwater Inflows to Texas Bays and Estuaries: Ecological Relationships and Methods for Determination of Needs. Texas Water Development Board and Texas Parks and Wildlife Department, Austin, pp. 23–72.

Mabie, D.W., Johnson, L.A., Thompson, B.C., Barron, J.C., Taylor, R.B., 1989. Responses of wintering whooping cranes to airboats and hunting activities on the Texas coast. Wildl. Soc. Bull. 17 (3), 249–257.

McKee, K.L., Mendelssohn, I.A., Materne, M.D., 2004. Acute salt marsh dieback in the Mississippi River deltaic plain: a drought-induced phenomenon? Glob. Ecol. Biogeogr. 13 (1), 65–73.

Michener, W.K., Baerwald, T.J., Firth, P., Palmer, M.A., Rosenberger, J.L., Sandlin, E.A., Zimmerman, H., 2001. Defining and unraveling biocomplexity. Bioscience 51 (12), 1018–1023.

Montagna, P.A., 2011a. Addendum to Expert Report, Opinion 4. The Aransas Project vs Shaw, unpublished report, Blackburn and Carter, Houston, TX, p. 9.

Montagna, P.A., 2011b. Coastal impacts. In: Schandt, J., Clarkson, J., North, G.R. (Eds.), The Impact of Global Warming on Texas, second ed. University of Texas Press, Austin, pp. 96–123.

Montagna, P.A., Brenner, J., Gibeaut, J., Morehead, S., 2011. Coastal impacts. In: Schmandt, J., North, G.R., Clarkson, J. (Eds.), The Impact of Global Warming on Texas, second ed. University of Texas Press, Austin, pp. 96–123.

Montagna, P.A., Li, J., 1996. Modeling and monitoring long-term change in macrobenthos in Texas estuaries. In: Final Report to the Texas Water Development Board. Technical Report TR/96-001. University of Texas at Austin, Marine Science Institute, Port Aransas, TX, p. 149.

Montagna, P.A., Kalke, R.D., 1992. The effects of freshwater inflow on meiofaunal and macrofaunal populations in the Guadalupe and Nueces estuaries, Texas. Estuaries 15 (3), 307–326.

Montagna, P.A., Gibeaut, J.C., Tunnell, Jr., J.W., 2007. South Texas climate 2100: coastal impacts. In: Norwine, J., John, K. (Eds.), The Changing Climate of South Texas 1900–2100. CREST-RESSACA, Texas A&M University-Kingsville, pp. 57–77.

Montagna, P.A., Palmer, T., 2012. Impacts of droughts and low flows on estuarine health and productivity. In: Final Report to the Texas Water Development Board,

Project for Interagency Agreement 1100011150. Harte Research Institute, Texas A&M University-Corpus Christi, p. 142.

Moulton, D.W., Dahl, T.E., Dall, D.M., 1997. Texas coastal wetlands – status and trends mid-1950s to early 1990s. U.S. Department of Interior, U.S. Fish and Wildlife Service, Albuquerque, NM, p. 32.

Moya, M., Mahoney, M., Dixon, T., 2012. Calhoun County Texas shoreline access plan. Atkins Coastal Planning and Restoration Group Report 100023672. Austin, TX, p. 62.

NatureServe, 2009. International Ecological Classification Standard: Terrestrial Ecological Classifications: Ecological Systems of Texas' Gulf Prairies and Marshes. NatureServe Central Database, Arlington, VA, p. 18.

Nelson, J.T., Slack, R.D., Gee, G.F., 1996. Nutritional value of winter foods for whooping cranes. Wilson Bull. 108 (4), 728–739.

Ning, Z.H., Turner, R.E., Doyle, T., Abdollahi, K., 2003. Preparing for a Changing Climate: The Potential Consequences of Climate Variability and Change – Gulf Coast Region. U.S. Environmental Protection Agency Gulf Coast Climate Change Assessment Council and Louisiana State University, Baton Rouge, p. 80.

Osland, M.J., Enwright, N., Day, R.H., Doyle, T.W., 2013. Winter climate change and coastal wetland foundation species: salt marshes vs. mangrove forests in the southeastern United States. Glob. Change Biol. 19 (5), 1492–1494.

Parker, R.H., 1959. Macro-invertebrate assemblages of central Texas coastal bays and Laguna Madre. Am. Assoc. Pet. Geol. Bull. 43 (9), 2100–2166.

Pfeffer, W.T., Harper, J.T., O'Neel, S., 2008. Kinematic constraints on glacier contributions to 21st century sea-level rise. Science 321 (5894), 1340–1343.

Price, W.A., 1933. Role of diastrophism in topography of Corpus Christi area Texas. Am. Assoc. Pet. Geol. Bull. 17 (8), 907–962.

Pugesek, B.H., Baldwin, M.J., Stehn, T.V., 2008. A low intensity sampling method for assessing blue crab abundance at Aransas National Wildlife Refuge and preliminary results on the relationship of blue crab abundance to whooping crane winter mortality. Proceedings of the North American Crane Workshop 10, 13–24.

Pugesek, B.H., Baldwin, M.J., Stehn, T., 2013. The relationship of blue crab abundance to winter mortality of whooping cranes. Wilson J. Ornithol. 125 (3), 658–661.

Pulich, Jr., W., 2007. Texas coastal bend. In: Handley, L., Altsman, D., DeMay, R. (Eds.), Seagrass Status and Trends in the Northern Gulf of Mexico: 1940–2002. U.S. Geological Survey Scientific Investigations Report 2006-5287 and U.S. Environmental Protection Agency 855-R-04-003, Reston, VA, pp. 41–59.

Rasser, M.K., Fowler, N.L., Dunton, K.H., 2013. Elevation and plant community distribution in a microtidal salt marsh of the Western Gulf of Mexico. Wetlands 33 (4), 575–583.

Saintilan, N., Wilson, N.C., Rogers, K., Rajkaran, A., Krauss, K.W., 2014. Mangrove expansion and salt marsh decline at mangrove poleward limits. Glob. Change Biol. 20 (1), 147–157.

Seitz, R.D., Lipcius, R.N., Seebo, M.S., 2005. Food availability and growth of the blue crab in seagrass and unvegetated nurseries of Chesapeake Bay. J. Exp. Mar. Biol. Ecol. 319 (1), 57–68.

Sherrod, C.L., Hockaday, D.L., McMillan, C., 1986. Survival of red mangrove, *Rhizophora mangle*, on the Gulf of Mexico coast of Texas. Contrib. Mar. Sci. 29, 27–36.

Sherrod, C.L., McMillan, C., 1981. Black mangrove *Avicennia germinans*, in Texas, past and present distribution. Contrib. Mar. Sci. 24, 115–131.

Sherrod, C.L., McMillan, C., 1985. The distributional history and ecology of mangrove vegetation along the northern Gulf of Mexico coastal region. Contrib. Mar. Sci. 28, 129–140.

Smith, E.H., Chavez-Ramirez, F.C., Lumb, L., Gibeaut, J., 2014. Employing the Conservation Design Approach on Sea-Level Rise Impacts on Coastal Avian Habitats along the Central Texas Coast. Gulf Coast Prairies Landscape Cooperative Program, 131 pp.

Snay, R., Cline, M., Dillinger, W., Foote, R., Hilla, S., Kass, W., Ray, J., Rohde, J., Sella, G., Soler, T., 2007. Using global positioning system-derived crustal velocities to estimate rates of absolute sea level change from North American tide gauge records. J. Geophys. Res. 112, B04409.

Stehn, T.V., 1995. Whooping Cranes during 1993–1994 Winter. Administrative Report, Aransas National Wildlife Refuge, Austwell, TX.

Stehn, T.V., Johnson, E.F., 1987. Distribution of winter territories of whooping cranes on the Texas coast. In: Lewis, J.C. (Ed.), Proceedings of the 1985 Crane Workshop, Platte River Whooping Crane Maintenance Trust, Grand Island, NE, pp. 180–195.

Stehn, T.V., Prieto, F., 2010. Changes in winter whooping crane territories and range 1950–2006. Proceedings of the North American Crane Workshop 11, 40–56.

Stevenson, J.O., Griffith, R.E., 1946. Winter life of the whooping crane. Condor 48 (4), 160–178.

Stutzenbaker, C.D., 1999. Aquatic and Wetland Plants of the Western Gulf Coast. Texas Park and Wildlife Press, Austin, p. 465.

Taylor, L.N., Ketzler, L.P., Rousseau, D., Strobel, B.N., Metzger, K.L., Butler, M.J., 2015. Observations of Whooping Cranes during Winter Aerial Surveys: 1950–2011. Aransas National Wildlife Refuge, U.S. Fish and Wildlife Service, Austwell, TX.

Thompson, B.C., George, R.R., 1987. Minimizing conflicts between migratory game bird hunters and whooping cranes in Texas. In: Lewis, J.C. (Ed.), Proceedings of the

1985 Crane Workshop, Platte River Whooping Crane Habitat Maintenance Trust, Grand Island, NE, pp. 58–68.

Tolan, J.M., 2007. El Niño-Southern Oscillation impacts translated to the watershed scale: estuarine salinity patterns along the Texas Gulf coast, 1982–2004. Estuar. Coast. Shelf Sci. 72 (1–2), 247–260.

Tremblay, T.A., Calnan, T.R., 2010. Status and Trends of Inland Wetland and Aquatic Habitats, Matagorda Bay Area. Bureau of Economic Geology, University of Texas at Austin, p. 71.

Tremblay, T.A., Calnan, T.R., 2011. Status and Trends of Inland Wetland and Aquatic Habitats, Freeport and San Antonio Bay Area. Bureau of Economic Geology, University of Texas at Austin, p. 81.

Tremblay, T.A., Vincent, J.S., Calnan, T.R., 2008. Status and Trends of Inland Wetland and Aquatic Habitats in the Corpus Christi area. Bureau of Economic Geology, University of Texas at Austin, p. 89.

Twilley, R.R., Barron, E.J., Gholz, H.L., Harwell, M.A., Miller, R.L., Reed, D.J., Rose, J.B., Siemann, E.H., Wetzel, R.G., Zimmerman, R.J., 2001. Confronting Climate Change in the Gulf Coast Region. Union of Concerned Scientists, Cambridge, MA, and Ecological Society of America, Washington, DC, p. 82.

Tunnell, Jr., J.W., 2002. The environment. In: Tunnell, Jr., J.W., Judd, F.W. (Eds.), The Laguna Madre of Texas and Tamaulipas. Texas A&M University Press, College Station, pp. 73–84.

U.S. Environmental Protection Agency (EPA), 2013. Level III ecoregions of the continental United States. U.S. EPA National Health and Environmental Effects Research Laboratory, map scale 1:7,500,000, https://www.epa.gov/eco-research/level-iii-and-iv-ecoregions-continental-united-states [3 February 2017].

Vermeer, M., Rahmstorf, S., 2009. Global sea level linked to global temperature. Proc. Natl. Acad. Sci. 106 (51), 21527–21532.

Westwood, C.M., Chavez-Ramirez, F., 2005. Patterns of food use of wintering whooping cranes on the Texas coast. Proceedings of the North American Crane Workshop 9, 133–140.

White, W.A., Tremblay, T.A., Waldinger, R.L., Calnan, T.R., 2002. Status and Trends of Wetland and Aquatic Habitats on Texas Barrier Islands: Matagorda Bay to San Antonio Bay. Bureau of Economic Geology, University of Texas at Austin, p. 66.

White, W.A., Tremblay, T.A., Waldinger, R.L., Calnan, T.R., 2006. Status and Trends of Wetland and Aquatic Habitats on Texas Barrier Islands: Coastal Bend. Bureau of Economic Geology, University of Texas at Austin, p. 64.

Wiens, J.A., 1989. Spatial scaling in ecology. Funct. Ecol. 3 (4), 385–397.

Wozniak, J.R., Swannack, T.M., Butzler, R., Llewellyn, C., Davis, III, S.E., 2012. River inflow, estuarine salinity, and Carolina wolfberry fruit abundance: linking abiotic drivers to whooping crane food. J. Coast. Conserv. 16 (3), 345–354.

Zervas, C., 2001. Sea Level Variations of the United States 1985–1999. NOAA Technical Report NOW CO-OPS 36. National Oceanic and Atmospheric Administration, Silver Spring, MD, p. 66.

14

Habitat Use by the Reintroduced Eastern Migratory Population of Whooping Cranes

Jeb A. Barzen*,**, *Anne E. Lacy**,
Hillary L. Thompson*,†, *Andrew P. Gossens**

*International Crane Foundation, Baraboo, WI, United States
**Private Lands Conservation LLC, Spring Green, WI, United States
†Clemson University, Clemson, SC, United States

INTRODUCTION

Often the habitat an organism chooses is critical to survival or reproduction (Barzen, Chapter 15, this volume; Fuller, 2012). Without habitat of adequate quality and quantity, a reintroduction effort has a low chance of success regardless of how many well-trained or healthy propagules are released (Griffith et al., 1989). Armstrong and Seddon (1987) proposed 10 key questions that often determined the success of reintroduction programs. Of these 10 questions, we focus on one for the Eastern Migratory Population (EMP) of Whooping Cranes (French et al., Chapter 1, this volume): "What habitat conditions are needed for persistence of the reintroduced population?"

A useful step in the process of reestablishing the EMP considers what habitat conditions Whooping Cranes experienced prior to their extirpation. The original breeding range of Whooping Cranes was widespread and comprised the divergent biomes of Upper Tallgrass Prairie/ Aspen Parkland, Taiga, and Gulf Coastal Plain (Allen, 1952; Austin et al., Chapter 3, this volume). Today, most of the wetlands where Whooping Cranes once nested in the Upper Tallgrass Prairie have been degraded or lost (Dahl, 1990). Wisconsin, however, has retained approximately 50% of its historic wetlands (Dahl, 1990), and these wetlands are found in a diverse array of ecoregions (Johnston, 1982; Omernik et al., 2000). So, even though less than 15% of historical Whooping Crane records occurred east of the Mississippi River, reintroduced cranes might have a reasonable chance of finding appropriate wetland habitats in Wisconsin.

Establishment of the EMP, undertaken by the Whooping Crane Eastern Partnership (WCEP) in 1999 (see French et al., Chapter 1, this volume), began with an identification of sites that might be suitable for a breeding population of Whooping Cranes in Wisconsin. Cannon (1999) summarized habitat data from the literature and collected expert judgments to establish selection criteria for potential release sites. Three sites, Crex Meadows State Wildlife Area (SWA), Necedah National Wildlife Refuge (NWR), and Horicon Marsh, as well as their surrounding areas, were considered in the final stage of the process, with Necedah NWR being chosen as the preferred release site (Appendix 1 of Cannon, 1999).

Approximately 49,800 acres (20,150 ha) of Necedah NWR and surrounding areas contain abundant, shallow emergent wetland habitat, and were considered appropriate for Whooping Cranes (Cannon, 1999). Established in 1939, Necedah NWR is a mosaic of sedge (*Carex* spp.) meadows, restored emergent marsh, open prairie, oak (*Quercus* spp.) savannah, and oak-pine (*Pinus* spp.) forest.

Release techniques used in the EMP are described elsewhere (see Hartup, Chapter 20, this volume) but the manner of release might relate to habitat selection in two ways. Philopatry, the tendency of young to return to natal areas in subsequent years (release areas in our case), is known for several crane species (Meine and Archibald, 1996). With costume-reared cranes less information is known, but Sandhill Cranes (*Grus canadensis*) released by what is now called the direct autumn release (DAR) method and by the ultralight-led (UL) method showed propensities to return to their release areas and to subsequent winter areas (Urbanek et al., 2005). Whooping Cranes had similar rates of philopatry (Maguire, 2008). A third release method, parent rearing (PR), is similar to DAR in the manner in which birds were released but PR birds were raised by captive Whooping Crane parents instead of costumed humans. Differences in

how birds learn about appropriate habitat use could, presumably, be influenced by different rearing techniques.

Second, natal dispersal distances, the distance between each bird's first established nest and its natal (or release) area (Greenwood, 1980; Johns et al. 2005), could influence how much habitat is available to returning birds. Natal dispersal distances were, however, unknown for costume-reared Whooping Cranes until recently. Of the 330 birds released into the EMP 2001–14, 70% (232 birds) have been costume-reared and led by ultralight aircraft to Florida wintering areas.

The first nest in the EMP was established in 2005 at Necedah NWR and the number of nests has increased since then, both in the refuge and outside of it (Urbanek et al., 2014b). Survival of adults has been comparable to other growing populations (Servanty et al., 2014). Reproductive rates, however, have been hampered by poor hatching success (Converse et al., 2013; King et al., 2013; Urbanek et al., 2010b), and poor chick survival (Converse et al., Chapter 8, this volume). Poor habitat quality on winter areas has been related to subsequent declines in reproductive success in other crane populations or species (Burnham et al., 2017; Gil de Weir, 2006). In addition, wetland productivity in the Glacial Lake Wisconsin Sand Plain ecoregion (Omernik et al., 2000) has been associated with lower waterfowl productivity (Baldassarre, 1978). Thus the question arises: have Necedah NWR and surrounding wetlands provided appropriate habitat for reintroduced Whooping Cranes in the EMP? If not, do *any* nesting habitats in the Upper Midwest, or do any winter areas, provide enough appropriate habitats for a self-sustaining population?

Our objectives in this chapter were to (1) present new habitat use and composition information for Whooping Cranes in the EMP during summer, considering periods when cranes were nonterritorial, territorial, and molting; (2) combine new results with previously published data to describe habitat use in the EMP during the entire annual cycle; and (3) discuss the importance of broader approaches that expose Whooping Cranes to more diverse wetland types and landscapes. Such efforts might help us to better understand the habitat needs of Whooping Cranes throughout North America.

METHODS

New habitat use data were collected for cranes studied April–September, 2011–2014, in Necedah NWR and surrounding areas (Fig. 14.1). This study area is similar to that used by Van Schmidt et al. (2014). Cranes were grouped as territorial or nonterritorial birds, and within each group cranes were further categorized according to the

presence or absence of a synchronous remigial molt occurring anytime in a summer. All flight feathers (remiges) are lost during remigial molt and replaced simultaneously. Hence, molting birds become flightless for 6–7 weeks while molted feathers regrow (Folk et al., 2008). The remigial molt does not occur every year and does not include the annual molt of contour feathers (Folk et al., 2008).

For Whooping Cranes, if an individual became territorial at any time during the summer, it was considered as being territorial for the whole summer. Most cranes became territorial early in the season (Urbanek et al., Chapter 10, this volume) so determination of territoriality was relatively simple in most years. Territorial cranes

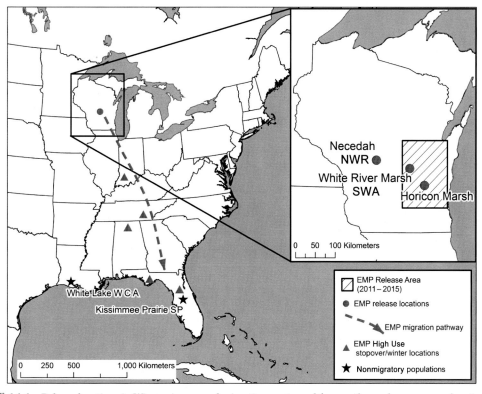

FIGURE 14.1 Release locations in Wisconsin, general migration route, and frequently used nonsummer locations for the Eastern Migratory Population (EMP) of Whooping Cranes 2001–2015. EMP release area is defined in Van Schmidt et al. (2014). Two nonmigratory populations (black stars in Louisiana and Florida) are included for reference. NWR = National Wildlife Refuge, SWA = State Wildlife Area, WCA = Wildlife Conservation Area, and SP = State Park.

associated with the same mates and defended well-defined, nonoverlapping areas from conspecifics all summer. Territory locations of individual pairs were also similar between years and usually contained nest locations. Nonterritorial birds did not always associate with the same mate, often associated with groups of three or more birds during summer, did not nest, and had home ranges that overlapped with conspecifics. Cranes undergoing remigial molt were identified by secretive behavior, presence of large numbers of remiges on the ground at night or day roosts, or observation of blood quills (new feather growth) during wing-flapping or stretching behaviors (Folk et al., 2008).

Data Collection

Most cranes introduced into the EMP (99%) and most wild-fledged chicks were outfitted with a very high frequency (VHF) transmitter mounted on the color bands of one leg. Color bands alone, or bands with a satellite-monitored transmitter, were mounted on the opposite leg (Urbanek et al., 2014a). Each year, 4–8 individuals were released and approximately half of the released birds were fitted with satellite transmitters of some kind. Nonterritorial birds were tracked with satellite transmitters (see Urbanek et al., 2014a for details). In summer, satellites typically recorded four usable locations per bird per day, once every 5th day. With nonterritorial cranes, we evaluated all available data 2011–14, starting with cranes in their second year (beginning January 1 in the year following hatch). All nonterritorial cranes were tracked until they died, their transmitters died and could not be replaced, or they became territorial. While some cranes were tracked only in their second year, some cranes were tracked in their third or fourth year as well. Since cranes undergo a remigial molt every 2–3 years (Folk et al., 2008), we expected that at least some nonterritorial cranes would become flightless during our observations.

We could not conduct roost-to-roost tracking on all territorial cranes in our study so we chose territorial cranes that represented a range of nesting habitats in the EMP 2011–12 (Fig. 14.2) and we attempted to use birds that would likely undergo the remigial molt sometime during the period that we tracked them. With VHF transmitters we tracked six territorial pairs throughout each year, one day per week.

Within each week, we tracked a territorial pair for one full day (from predawn roost to postsunset roost). Order for the study subjects was chosen randomly and, unless otherwise specified, refer to diurnal locations only. Within a day, bird location, habitat, behavior, and identity of associating birds were recorded every 1–1.5 h. If birds were hidden, bird location was determined through triangulation using VHF transmitters (Miller and Barzen, 2016; Mech, 1983) and habitat was determined later using remotely sensed data. Behavior data were not recorded for unseen birds. Habitat data were derived from the National Land Cover Database 2011 (NLCD), a land cover data set derived from remotely sensed data with a spatial resolution of 30 m for the coterminous United States (Homer et al., 2015).

Habitat Composition and Use

Home range estimates of territorial Whooping Crane pairs were calculated using Geospatial Modeling Environment (GME; Beyer, 2015), which applied the ks package in R Software programming (R Core Team, 2015). Calculations of 95% kernel density estimates (KDEs) were generated using the KDE and isopleth commands in GME with likelihood cross-validation (CVh) used to calculate smoothing parameters (Horne and Garton, 2006). Movements of mates were not independent of each other, so we displayed 95% KDE estimates using ArcGIS 10.3 for each male in a pair. To assess the composition of habitats located within each home range, we overlaid the 2011 NLCD (Homer et al., 2015) onto each home

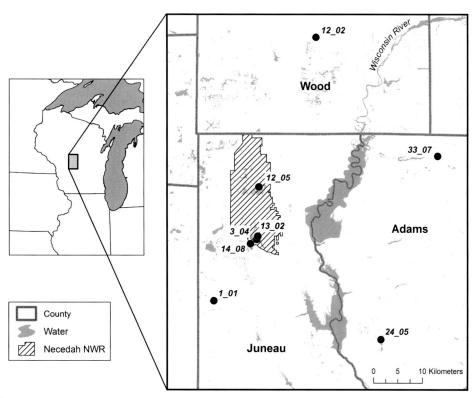

FIGURE 14.2 Location of eight nesting Whooping Crane territories in central Wisconsin that were radio-tracked 2011–12. Pairs were identified by the males of each pair, and bird ID is the bird number before the underscore and year of hatch following the underscore. Necedah National Wildlife Refuge (NWR) is cross-hatched, and water bodies are gray.

range polygon and extracted habitat types using the Extract by Mask tool in the Arc Toolbox (ESRI, 2011). Percent habitat composition was then calculated for each home range. We used nine NLCD categories: open water, developed, barren, nonforested wetland, shrub or scrub, grassland or pasture, cultivated crops, forested wetland, and herbaceous wetland. Cranberry production fields and cranberry reservoirs were classified as "cultivated crop" by NLCD, which we recategorized, after ground-truthing, to herbaceous wetland.

Home range estimates for nonterritorial Whooping Cranes were calculated using R Software programming (R Core Team, 2015). Calculation of 95% minimum convex polygon (MCP) estimates followed Calenge (2006) and

used the adehabitat HR package. We used MCP estimates for nonterritorial birds because MCP estimates are more accurate with small sample sizes and in situations where study animals range widely (Boyle et al., 2009). Location data for all individual Whooping Cranes and seasons were displayed using ArcGIS 10.3.

Habitat use reflects the actual habitats in which cranes were located as opposed to habitat composition that reflects the proportion of habitats that are located within the derived territory or home range of the bird (i.e., a measure of habitat availability). Generally habitats that are used more than expected, based on their availability, are interpreted as "preferred" (Fuller, 2012; Garshelis, 2000). Estimates of habitat use on summer areas for both territorial and nonterritorial birds were calculated by creating a 50 m radius around each crane location and determining the land cover within that area from the NLCD data layer.

Data on Whooping Cranes undergoing the remigial molt were gathered in 2011. Habitat use of 6 Whooping Cranes was a subset of the broader habitat use analysis described earlier (Lacy and McElwee, 2014). We considered our sample to consist of four independent measures of habitat use (two territorial pairs and two nonterritorial birds) because both members of each pair remained with their mates during the remigial molt. Cranes that molted were followed using the same methods as for territorial birds except that home ranges were determined before (April–May), during (June–July) and after (August–September) the remigial molt for each bird. Due to the small number of birds studied, data for both territorial and nonterritorial cranes were analyzed together using pair-wise t-tests (Sokal and Rohlf, 1981) to compare relative changes in territory or home range size before, during, and after the remigial molt within each bird.

TABLE 14.1   Home Range (km²) for Individual, Nonterritorial Cranes, by Age, Tracked April–September, 2011–14. Territory Size Estimated by 95% Minimum Convex Polygon Procedure. Individuals Were Identified by Bird Number (Preceding the Underscore) and the Hatch Year (Following the Underscore).

Bird ID	Age[a]		
	2	3	4
15_11	6,368.3	6,408.8	728.3
20_11	2,921.3	13.6	1.8
4_11	7,531.2	1,001.4	10.0
7_11	3,627.0	598.3	918.4
1_10	16,340.3	1,201.9	
14_11	9,750.5	3,063.0	
6_10	5,482.2	15,633.0	
9_10	17,549.8		
9_11	176.8		
Average	7,749.7	3,988.6	414.6
Median	6,368.3	1,201.9	369.2
SE	5,901.3	5,572.1	478.3

[a]Age: A bird began its second year on January 1 following the hatch year.

RESULTS

Home Range and Habitat Data during Summer

Though highly variable, home ranges for eight of nine nonterritorial cranes in their second year exceeded 2,900 km² and decreased substantially for most individuals as nonterritorial cranes aged (Table 14.1). The average home range of four nonterritorial cranes in their 4th year was 414.6 km². Interestingly, though large at first, not only did home ranges shrink in size as nonterritorial cranes aged, but they remained in or near a small portion their former home ranges (Fig. 14.3) and those smaller home ranges tended to be near release areas.

In contrast, even though individual territorial crane pairs used landscapes that varied from cranberry farm to the mosaic of wetlands

and restored prairies within Necedah NWR (Table 14.2), home ranges of eight territorial birds during the same time period averaged 4.58 ± 1.68 km². Within Necedah NWR boundaries, home range sizes varied greatly between pairs (range = 1.18–14.64 km²) and between years (Table 14.2). For example, male 14_08's territory was more than 3× larger in 2011 than in 2012, perhaps because it raised a chick in 2012 but not in 2011. In contrast, male 12_05 showed little change in home range size between years and did not raise a chick in either year.

Within home ranges, just under half of the locations for nonterritorial cranes were located in wetland habitats (Table 14.3) while 75% of territorial crane locations were located in wetland habitats (Table 14.4A). While we do not have home range habitat composition estimates for nonterritorial cranes (because they covered such extensive areas), home range composition

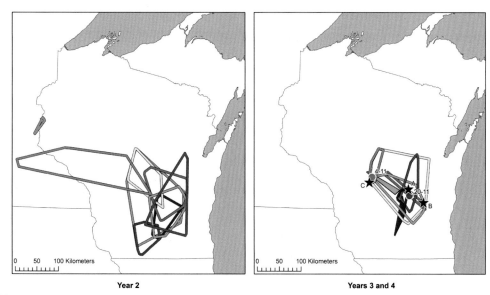

Year 2 **Years 3 and 4**

FIGURE 14.3 Home range polygons for nine nonterritorial Whooping Cranes as 2-year-olds, seven nonterritorial Whooping Cranes as 3-year-olds, and four nonterritorial Whooping Cranes as 4-year-olds (depicted as filled circles because their home range was too small to specifically describe with filled polygons). Color for each bird remained the same for all years where data existed. Bird ID is the bird number before the underscore and year of hatch follows. Birds 4_11 (blue), 7_11 (red), and 9_11 (pink) were released at White River Marsh (A); birds 14_11 (brown), 15_11 (green), and 20_11 (purple) were released at Horicon Marsh (B); and birds 1_10 (orange), 6_10 (yellow), and 9_10 (gray) were released at Necedah NWR (C).

TABLE 14.2 Summer Home Range of Individual, Territorial Whooping Cranes Tracked with Roost-to-Roost Surveillance 1 Day per Week, 2011–12, Using 95% Kernel Density Estimates. NWR, National Wildlife Refuge and SWA, State Wildlife Area. Individuals Were Identified by Bird Number (Preceding the Underscore) and the Hatch Year (Following the Underscore).

Year	Male ID of pair	Total days	Total # of locations	Locations per day	Home range area (km²)	# Polygons depicted	Landscape description
2011	12_02	22	210	9.6	2.55	8	Cranberry Farm
2012	12_02	20	242	12.1	0.78[a]	6	Cranberry Farm
2011	12_05	25	211	8.0	8.60[a]	2	Necedah NWR
2012	12_05	24	334	12.9	9.50[a]	2	Necedah NWR
2011	14_08	18	150	7.9	14.64[a]	9	Necedah NWR
2012	14_08	24	326	13.6	4.60	9	Necedah NWR
2011	33_07	23	250	10.9	0.86	2	Cranberry Farm
2012	33_07	25	349	14.0	1.16	9	Cranberry Farm
2011	24_05[b]	10	67	6.7	2.48	1	Quincy Bluff SWA
2011	3_04	24	214	8.9	5.53	1	Necedah NWR
2012	1_01	24	336	14.0	3.07	7	Private Wetland
2012	13_02	26	365	14.0	1.18	1	Necedah NWR
Ave. ± SE		22 ± 4.4	254 ± 91	11.0 ± 2.7	4.58 ± 1.68		

[a]Birds with part of their home range located more than 2 km from portion of their home range containing the nest.
[b]Bird found dead in July 2011.

TABLE 14.3   Proportion of Locations Found in Each Habitat Type (% Habitat Use) for Nonterritorial Whooping Cranes in Wisconsin, April–September, 2011–2014. Data Were Acquired with GPS Satellite-Monitored Transmitters. Sample Size Refers to the Number of Individual Birds Represented in Each Age Group but Average Number of Locations per Age Group Was 3,072. % Wetland = Herbaceous Wetland, Wooded Wetland, and Open Water Combined, While % Upland Combines the Remaining Habitats.

| | % Habitat use for nonterritorial Whooping Cranes | | | | | | | | Summary of % use | |
Age[a]	Open water	Devel-oped	Forested	Shrub or scrub	Grassland or pasture	Cultivated crop	Wooded wetland	Herbaceous wetland	% Upland	% Wetland
2 (n = 9)	24.56	0.75	2.63	1.26	5.11	44.86	2.96	18.01	55.61	45.53
3 (n = 7)	11.59	0.25	2.58	1.26	3.98	40.29	2.30	37.76	48.36	51.64
4 (n = 4)	9.51	1.90	2.25	1.01	1.79	42.00	8.23	33.31	48.95	51.05
Ave.	15.22	0.97	2.49	1.18	3.63	42.38	4.49	29.69	50.97	49.40

[a]Age follows calendar year such that the end of the first year occurred December 31 in the year the bird hatched. A bird entered its second year on January 1 of the year following its hatch.

of territorial cranes was 57.25% wetland (Table 14.4B). The majority of the locations (87%) used by nonterritorial cranes consisted of cultivated cropland (42.38%), herbaceous wetland (29.69%), and open water (15.22%), and the proportion of crane locations found in wetlands tended to increase after the second year (Table 14.3). The same three habitat types comprised the majority of locations for territorial cranes (82%) but with modest differences among the habitats used: herbaceous wetland (43.33%), open water (27.14%), and cultivated crops (11.38%; Table 14.4A). With the exception of male 1_01, the proportion of wetlands used by territorial cranes were 56.30–94.52% even though composition of home ranges for these birds was 23.74–81.18%, indicating that wetlands were used at a higher rate than expected based on their proportion found within the home range (Table 14.4). Male 1_01 and his mate never nested, even though they were clearly territorial, and 1_01 spent much of his time using grasslands associated with a nearby airport.

Home Range and Habitat Data during Remigial Molt

Habitat use was analyzed separately for cranes during the years when they underwent a remigial molt. Cranes are presumably more vulnerable to predation while molting, so they would be expected to choose habitats that would reduce this risk while flightless. All six molting birds observed began their remigial molt in the second week of June and the flightless period lasted approximately 6 weeks. Members of both pairs were flightless together but the phenology of molt was staggered by about 1 week. Flightless birds had a markedly reduced home range of 0.45 km^2 (SE = 0.12) compared to larger home range sizes before or after the remigial molt (paired t-value = 3.96, d = 5, p < 0.02; Table 14.5). Home range sizes for cranes during remigial molt were 10% of the 4.58 km^2 average territory size and less than 1% of any home range for a nonterritorial Whooping Crane (Table 14.6).

Use of wetland habitats (open water and herbaceous emergent wetland) during the remigial molt accounted for 92% of diurnal locations among all six birds on average (Table 14.5). Wetland habitat used by flightless birds also varied little among individuals (Table 14.5).

DISCUSSION

Our goal was to describe habitat use and habitat composition for reintroduced cranes in the EMP during their annual cycle. We provided

TABLE 14.4 (A) Habitat Use and (B) Habitat Composition Data from Whooping Crane Territories during Summer in Wisconsin, 2011–12, Calculated Using Roost-to-Roost Tracking. Habitat Composition Is the Proportion of Habitat Types Located in Each Home Range. Percent Habitat Use Is Frequency That Birds Were Located in Each Habitat Type within the Home Range. Summary % Wetland Is the Summation of Herbaceous Wetland, Wooded Wetland, and Open Water, While % Upland Sums the Other Habitat Types. Individuals Were Identified by Bird Number (Preceding the Underscore) and the Hatch Year (Following the Underscore).

A.

Male ID of pair	% Habitat use									Summary of use (%)	
	Open water	Developed	Barren	Forested	Shrub and scrub	Grassland and pasture	Cultivated crops	Woody wetland	Herbaceous wetland	% Upland	% Wetland
12_02	2.68	5.23	0.00	16.75	0.00	3.53	18.19	0.05	53.57	43.70	56.30
12_02	11.27	0.97	0.00	6.52	0.00	0.23	1.24	0.48	79.30	8.95	91.05
12_05	65.10	1.29	0.00	1.91	0.00	1.82	3.81	3.94	22.13	8.83	91.17
12_05	36.87	0.32	0.00	1.41	0.00	0.07	6.59	12.50	42.24	8.38	91.62
14_08	36.47	2.48	0.00	3.85	8.36	3.64	6.59	8.08	30.54	24.90	75.10
14_08	45.03	0.72	0.00	2.73	1.64	2.68	1.38	4.88	40.93	9.16	90.84
33_07	24.10	0.61	1.31	2.75	0.00	12.23	6.82	1.59	50.59	23.72	76.28
33_07	24.25	1.25	0.80	4.00	0.00	19.18	17.46	1.48	31.59	42.68	57.32
24_05[a]	49.88	0.83	0.00	15.61	0.00	3.35	7.54	7.24	15.56	27.33	72.67
3_04	19.32	1.00	0.00	0.99	0.16	0.19	3.14	6.45	68.75	5.48	94.52
1_01	0.65	4.11	0.00	6.29	0.31	17.83	57.35	2.78	10.67	85.90	14.10
13_02	10.06	2.20	0.00	1.18	0.12	1.02	6.43	4.87	74.12	10.94	89.06
Ave.	27.14	1.75	0.18	5.33	0.88	5.48	11.38	4.53	43.33	25.00	75.00

(Continued)

TABLE 14.4   (A) Habitat Use and (B) Habitat Composition Data from Whooping Crane Territories during Summer in Wisconsin, 2011–12, Calculated Using Roost-to-Roost Tracking. Habitat Composition Is the Proportion of Habitat Types Located in Each Home Range. Percent Habitat Use Is Frequency That Birds Were Located in Each Habitat Type within the Home Range. Summary % Wetland Is the Summation of Herbaceous Wetland, Wooded Wetland, and Open Water, While % Upland Sums the Other Habitat Types. Individuals Were Identified by Bird Number (Preceding the Underscore) and the Hatch Year (Following the Underscore). (*cont.*)

B.

Male ID of pair	Open water	Developed	Barren	Forested	Shrub or scrub	Grassland or pasture	Cultivated crops	Wooded wetland	Herbaceous wetland	% Upland	% Wetland
				Habitat composition (%)						Summary of composition (%)	
12_02	1.27	5.08	0.00	38.55	0.28	1.94	30.41	0.00	22.47	76.26	23.74
12_02	17.57	1.38	0.00	15.04	0.00	0.57	2.30	2.53	60.62	19.29	80.71
12_05	33.17	4.12	0.00	21.12	0.35	3.98	9.76	6.54	20.96	39.33	60.67
12_05	28.73	2.75	0.00	4.82	0.14	1.51	11.56	17.05	33.45	20.77	79.23
14_08	9.68	4.31	0.00	16.88	10.81	9.03	13.62	11.21	24.46	54.65	45.35
14_08	13.93	1.98	0.00	11.88	11.06	5.38	6.73	12.86	36.18	37.03	62.97
33_07	33.30	2.63	0.95	1.16	0.00	9.17	18.23	1.05	33.51	32.14	67.86
33_07	17.15	4.15	0.69	14.23	0.00	8.69	31.46	5.15	18.46	59.23	40.77
24_05[a]	31.46	2.56	0.00	23.41	0.00	5.61	5.00	9.92	22.03	36.59	63.41
3_04	26.06	4.55	0.00	7.85	3.89	5.13	10.05	8.57	33.89	31.48	68.52
1_01	0.26	3.55	0.26	6.45	1.03	2.58	73.51	3.08	9.29	87.37	12.63
13_02	20.73	2.22	0.00	3.90	0.46	4.67	7.57	4.28	56.16	18.82	81.18
Ave.	19.44	3.27	0.16	13.77	2.33	4.86	18.35	6.85	30.96	42.75	57.25

[a]*Bird found dead in July 2011.*

TABLE 14.5   Home Range and Habitat Use of Whooping Cranes in the EMP during 2011 for 2 Months during Their Remigial Molt (June–July), 2 Months prior to the Molt (April–May), and 2 Months after the Molt (August–September)

Bird ID[a]	Status[b]	Home range (km²)			% Wetland used during molt
		Premolt	Molt	Postmolt	
19_04	T w/12_02	3.39	0.47	24.15	92
12_02	T w/19_04	4.11	0.58	24.69	93
18_02	T w/13_02	14.98	0.30	17.58	93
13_02	T w/18_02	15.64	0.53	17.63	92
29_09	NT	70.56	0.52	159.60	81
4_08	NT	201.31	0.29	253.75	100
Ave.		51.67	0.45	82.90	92
SE		77.44	0.12	100.44	

[a]Individuals were identified by bird number (preceding the underscore) and the hatch year (following the underscore). 18_02 and 19_04 were females while 12_02, 13_02, 4_08, and 29_09 were males.
[b]T, Territorial; NT, nonterritorial. T w/12_02, paired with bird ID 12_02.

TABLE 14.6   Average Home Ranges of Nonterritorial and Territorial Whooping Cranes in the EMP throughout the Annual Cycle

# of individuals or pairs	Age[a]	Status[b]	Stage of annual cycle	Home range (km²)	Standard error	Tracking metrics[c]	Home range analysis	Source
9	2	NT	Summer	7,750	5,901	PTT, W	95% MCP	This study
7	3	NT	Summer	3,989	5,572	PTT, W	95% MCP	This study
4	4	NT	Summer	414.6	478	PTT, W	95% MCP	This study
16	>2	T	Nesting	3.68	0.95	VHS, O	95% KDE	Van Schmidt et al., 2014
8	>2	T	Nesting	4.58	1.68	VHS, RTR	95% KDE	This study
30	>1	NT	Migration	7.16	2.25	PTT, D	95% MCP	H. Thompson, unpublished data[d]
30	>1	NT	Winter	32.41	6.61	PTT, W	95% MCP	H. Thompson, unpublished data[d]

[a]Age: A bird began its second year on January 1 following the hatch year.
[b]Status catagories: nonterritorial (NT) or territorial (T).
[c]Source and frequency of tracking data, where PTT, Platform Terminal Transmitter, VHS, very high frequency; RTR, roost-to-roost tracking once a week; W, data collected every 5 days; O, data collected opportunistically (ca. once every 3 days); D, data collected every day.
[d]International Crane Foundation, WI.

new data for habitat use of Whooping Cranes in the EMP during summer and referred to other published and unpublished work to augment our understanding of habitat use in winter and migration.

Habitat Use and Habitat Composition during Summer

Nonterritorial Cranes

During their second year, 87% of male and female Whooping Cranes in the EMP returned to within 7.2 km of their release site 2002–06 (n = 56; Maguire, 2008). Following their return, however, these young birds spread out with exceptionally large home ranges that varied dramatically in size (Table 14.1) and were larger than at any other time in the annual cycle of Whooping Cranes (Table 14.6). Individuals or small groups often dispersed several hundred kilometers from their release areas after their initial return. Some dispersal movements may have been associated with unusual weather patterns during the first spring migration (Urbanek et al., 2010a). The wide-ranging and highly variable movements by most young Whooping Cranes, however, were likely representative of the population because similar large movements also occurred in nonmigratory populations where weather-related issues were absent (see Barzen, Chapter 15, this volume). Also, some young cranes in the wild, migratory AWBP have also shown extensive movements before establishing territories (Johns et al., 2005; Kuyt, 1979). This behavior has been termed "spring wandering" (Allen, 1952; Urbanek et al., 2014b), and may allow young Whooping Cranes to select habitats from a broad array of wetland types. Similar movements also occur in nonterritorial Sandhill Cranes (Hayes and Barzen, 2016).

Mean natal dispersal distance (the distance moved from location of hatch to location of first nest; Greenwood, 1980) was 19.4 km overall (17.8 km for males and 21.1 km for females; H. Thompson, unpublished data, International

Crane Foundation, WI) so, even though nonterritorial Whooping Cranes wandered widely and were likely exposed to a wide range of habitats, they still established breeding territories located relatively close to their natal areas (Fig. 14.3). Natal dispersal distances in the AWBP were similar to those in the EMP (Johns et al., 2005). The function of spring wandering, therefore, is unclear but may allow birds to (1) identify other habitats that could be used in times of severe climate alterations (such as drought), (2) identify alternative breeding habitats (Hayes, 2015), or (3) facilitate locating mates. The high degree of natal philopatry, however, appears to constrain the habitat types that are used to establish a breeding territory. The probability is low, for example, that Whooping Cranes reintroduced to the Necedah NWR will quickly (i.e., within 5–10 years) establish nest sites outside the 4,600 km^2 Glacial Lake Wisconsin Sand Plain (U.S. EPA, 2013), the ecological zone within which Necedah NWR is located. Selection of wetlands by Whooping Cranes in the near term, therefore, will mostly include ecosystems with one soil type (i.e., sand) whether or not this type of wetland is ecologically appropriate for nesting birds.

Beyond natal dispersal distance behavior, did costume rearing methods influence habitat use in the EMP? The hypothesis that introduced immature cranes had imprinted on the "viewscape" (the habitats that young, flightless Whooping Cranes could see from their rearing pens) of the habitat in which they were raised was examined by comparing the habitat composition within the home range of second-year, nonterritorial cranes with habitats of Necedah NWR (the release area) for the first six cohorts (n = 56, 2001–06; Maguire, 2008). No significant difference was detected between the habitat compositions of either area (Maguire, 2008). Nonterritorial cranes of the EMP ranged widely and yet chose habitats that provided high annual survivorship (0.974), survivorship that was comparable to similar cohorts in the AWBP (Servanty et al., 2014). It is notable that nonterritorial EMP Whooping Cranes during summer used wetlands less than did any other group of cranes studied during the same period (Tables 14.3 and 14.4A; Barzen, Chapter 15, this volume).

Territorial Cranes

Home range size of eight pairs in the EMP, calculated from weekly roost-to-roost tracking, was 4.58 km², larger than the 3.68 km² reported by Van Schmidt et al. (2014) for the same nesting population (Table 14.6). Van Schmidt et al. (2014) used an opportunistic sampling protocol, as opposed to roost-to-roost tracking, and sampled crane habitat use from April to July instead of during the April–September period used in this study. These differences in sampling period may have led to the difference in home range estimates. Home range sizes, for example, tend to increase after nest efforts have been completed (WCEP, unpublished data, International Crane Foundation, WI), a trend that also occurs in sympatric nesting Sandhill Cranes (Miller and Barzen, 2016). Our estimates of home range size of territorial cranes in the EMP were similar to the Composite Nesting Areas (CNAs) for the AWBP (3.8–4.2 km² in the Sass and Klewi River nesting areas; Kuyt, 1981b, 1993). To derive CNAs Kuyt used crane locations from several years, and did not use marked cranes. Our analyses were with marked birds within one or two years (Table 14.2).

Habitat composition of the home ranges for 16 nesting pairs of Whooping Cranes over numerous years (Van Schmidt et al., 2014) and for 8 nesting pairs over 2 years (Table 14.4B) indicated large proportions of home range territories were composed of wetlands (45.32% vs. 57.25%, respectively). Habitat use estimates in our study indicated that wetland use (75%; Table 14.4A) occurred more frequently than expected based on wetland availability within home ranges. Similarly, Van Schmidt et al. (2014) found a significant positive selection index of 0.37 within home ranges. Both studies demonstrate important selection of wetlands by Whooping Cranes, something that has been speculated on, but not quantified, for other Whooping Crane populations during summer [Kuyt, 1993; Novakowski, 1966 for the AWBP; T. Dellinger (unpublished data, Florida Fish and Wildlife Conservation Commission) for the Florida Nonmigratory Population; and S. King (unpublished data, USGS Louisiana Fish and Wildlife Cooperative Research Unit) for the Louisiana Nonmigratory Population]. Overall, wetland use by territorial Whooping Cranes was also higher than wetland use of sympatric Sandhill Cranes (46.77%; Miller and Barzen, 2016), suggesting that Whooping Cranes are more wetland dependent than Sandhill Cranes. Compared to nonterritorial cranes, the greater use of wetlands by territorial cranes (Table 14.6) is likely related to nesting and chick-rearing needs, as suggested by research at WBNP (Kuyt, 1981a; Kuyt et al., 1992). However, only one EMP breeding pair tracked in 2011–12 fledged a chick, so the use of wetlands by territorial birds in the EMP may be due to maintaining a territory with extensive wetlands, a behavior that confers the *potential* to successfully breed in any 1 year (Hayes, 2015).

Both Van Schmidt et al. (2014) and this study measured discontinuous habitat use in some but not all Whooping Crane territories. Of a composite measure of 3.68 km² for home range size of territorial cranes, 2.69 km² were for areas of the home range that were contiguous with the nest and 1.27 km² were for foraging areas (cornfields) located over 15 km from the nesting territory (Van Schmidt et al., 2014). In our study, three pairs (males 12_02, 12_05, and 14_08; Table 14.2) used separate, discontinuous foraging territories as described in Van Schmidt et al. (2014).

Molting Cranes

There is a dramatic increase in the time spent using wetland habitats and a large decrease in home range size by both territorial and nonterritorial cranes during their remigial molt (Table 14.5). These behavioral changes suggest that the brief molt (6 weeks) requires specific wetland habitats to be located within territories for territorial birds or within home ranges for nonterritorial birds.

As periodicity of the remigial molt varies for individuals (once every 2–3 years; Folk et al., 2008), the mechanism to predict when the molt might occur is less clear than in other species, such as waterfowl, that have an annual remigial molt (Baldassarre and Bolen, 2006). Habitats within territories or home ranges, as a result, must contain wetland habitat adequate to provide nutrition, chick-rearing habitat, and escape cover while flightless regardless of whether the remigial molt will occur that year.

No difference in the rate of mortality was detectable during the flightless period for cranes versus any other portion of the annual cycle for the EMP (Converse et al., Chapter 7, this volume). Of 65 confirmed mortalities that occurred April–October, 2002–15 (9.3 mortalities/month), only 6 occurred during the 6-week remigial molt (4 mortalities/month). This rough assessment suggests that habitat available for Whooping Cranes in the EMP during the remigial molt is sufficient to mitigate a vulnerable period for Whooping Cranes.

Habitat Use during Winter and Migration

Winter

High natal philopatry, well-developed territoriality, and extensive wetland use by summer birds contrast with the behavior of wintering cranes in the EMP. During the first two winters of reintroduction, all Whooping Cranes resided in Florida (Fondow, 2013; Urbanek et al., 2014a). In subsequent years, the winter distribution of this population expanded northward, and by the winter 2014/2015, only

8% of known Whooping Crane winter locations were in Florida (Teitelbaum et al., 2016; Urbanek et al., 2014a). Both studies attributed the change in winter distribution to the implementation of newer release methods, changes in winter climate, drought in winter areas, as well as associations with other Whooping Cranes. Teitelbaum et al. (2016), however, demonstrated the importance of learning that occurred among cranes to quickly establish new winter distributions in response to climatic or habitat variation. Sandhill Cranes in this flyway have shown a similar pattern of changing winter distributions over time, moving northward because of changing climate, annual variation in weather, population growth, and large-scale land use change (Lacy et al., 2015).

In Florida, pastures were the primary habitats used by EMP Whooping Cranes in winter (Fondow, 2013), whereas outside of Florida the proportion of available grain fields and winter temperature were associated with EMP Whooping Crane winter distribution (Teitelbaum et al., 2016). Further, the amount of wetland used by wintering cranes in Florida has decreased from 52.4% (2004–06; Fondow, 2013) to 32.0% in winter throughout the southeastern United States, 2014–15 (Thompson, 2018). In addition to wetland use, Thompson (2018) established that, throughout their winter distribution (Indiana, Kentucky, Tennessee, Alabama, Georgia, and Florida), Whooping Cranes used row crop agriculture fields 38.7% of the time. In comparison, diurnal winter home ranges were composed mostly of row crop agriculture fields (52.30%) and wetlands (22.60%). The use of wetlands in winter was higher than expected based on availability even though wetland use in winter by EMP Whooping Cranes was lower than wetland use found in summer areas (Table 14.4A). The lower use of row crop agriculture fields than expected from availability could be attributed to avoidance or, more likely, the high abundance of grain fields such that there were not enough cranes in the population to use all of the available habitat and thus preference was not detectable (Garshelis, 2000).

In contrast to strong territorial behavior described for wintering cranes from the AWBP (Stehn and Prieto, 2010), no evidence of territorial behavior has been seen in the EMP during winter (H. Thompson, unpublished data, International Crane Foundation, WI). Wetland use by Whooping Cranes in the AWBP also appears to be substantially higher than in the EMP (Chavez-Ramirez, 1996; Smith et al., Chapter 13, this volume). Yet even with a large difference in habitat use behavior between the two populations during winter, the overall survival rates for Whooping Cranes in the two populations were comparable (Servanty et al., 2014). No evidence exists, therefore, that heavy use of agricultural habitats confers additional threats to survival for EMP birds in winter. Wetland conditions during drought years in winter at Aransas NWR for the AWBP, however, were thought to decrease survivorship of Whooping Cranes in winter (Stehn and Haralson-Stobel, 2014) or depress reproductive rates on breeding grounds during the subsequent spring (Gil de Weir, 2006).

Migration

During migration, 14.8% of diurnal Whooping Crane locations in the EMP occurred at wetlands (H. Thompson, unpublished data, International Crane Foundation, WI), the lowest amount of wetland use in the annual cycle (Barzen, Chapter 15, this volume). Habitat use by migrants was measured from 140 stops in both spring and fall that were made by 60 birds with five locations or more at each stopover site, 2002–15. In contrast, home ranges of cranes during migration were smaller than were winter home ranges (H. Thompson, unpublished data, International Crane Foundation, WI) but larger than summer home ranges of territorial cranes (Table 14.6).

The diurnal use of wetlands by EMP Whooping Cranes during migration (14.8%) was lower than wetland use by migrating nonfamily groups of Whooping Cranes in the AWBP (30%; Howe, 1989). Johns et al. (1997), however, reported that diurnal Whooping Crane sightings in traditional stopover sites of Saskatchewan were mostly in agricultural fields during fall (85% of sightings for foraging birds) and during spring (100% of sightings for foraging birds) though birds still roosted in wetlands at night as occurred in the EMP.

Landscape-Scale Habitat Experiment on Breeding Grounds in Wisconsin

In this study, wetland habitats during summer appeared important to Whooping Cranes, especially for nesting and molting. The importance of wetlands was also illustrated for the historical population that bred in the upper tallgrass prairie/aspen parkland biome as well (Allen, 1952, Austin et al., Chapter 3, this volume; Bent, 1926, pp. 221–224). Yet, though Necedah NWR contains abundant wetland habitat, reproductive rates in the EMP remain unsustainably low and the constraints on reproduction are linked to high densities of an ornithophilic black fly species (*Simulium annulus*) that harasses nesting Whooping Cranes (Barzen et al., 2018; Converse et al., 2013; King et al., 2013; Urbanek et al., 2010b); low fledging success is due to unknown reasons (Barzen et al., 2018; Converse et al., Chapter 8, this volume). Collectively, these studies suggest that wetland quality is at least as important as wetland quantity. With low natal dispersal distances, however, it is a slow process for Whooping Cranes to move outside of the large Glacial Lake Wisconsin Sand Plain where a limitation on reproduction exists. Importantly, if Whooping Cranes succeed in moving outside the Glacial Lake Wisconsin Sand Plain, where might better quality habitat be located?

In 2010, most releases of Whooping Cranes in Necedah NWR ceased because WCEP sought to determine if appropriate wetlands were located elsewhere in Wisconsin where *S. annulus* was absent and where Whooping Cranes could be released. Van Schmidt et al. (2014) first described wetland habitat use by nesting Whooping Cranes at Necedah NWR and then identified similar patterns of habitat found in other areas of Wisconsin. WCEP then chose two release sites (White River Marsh SWA and Horicon NWR) located within a broad region of appropriate habitat in eastern Wisconsin that did not have large populations of *S. annulus* (Fig. 14.1). Wetlands in eastern Wisconsin also have more productive soils and perhaps different predator populations (Van Schmidt et al., 2014). If habitat-related issues are contributing to reproductive failure, then releasing EMP cranes into wetlands of eastern Wisconsin could more quickly expose birds to a broader array of wetlands. Thus, by comparing habitat use in these two areas we could assess habitat needs for Whooping Cranes in the temperate grassland biome more broadly (Austin et al., Chapter 3, this volume; Converse et al., Chapter 8, this volume). Conversely, if nonhabitat issues (i.e., learning or captive selection) were actually the problem, then establishing nests in eastern Wisconsin would not change reproductive success because fledging rates would remain low. The release of Whooping Cranes in eastern Wisconsin thus represents an important opportunity to discern habitat and nonhabitat hypotheses (Converse et al., Chapter 8, this volume). Between 2011 and 2015, 70 cranes were introduced into eastern Wisconsin but only two pairs had established a territory as of 2017. As birds introduced into eastern Wisconsin establish more territories, the two regions can be compared.

SUMMARY AND OUTLOOK

Whooping Cranes have been reintroduced to alternative habitats during summer in Wisconsin to identify the precise type of wetlands necessary to produce sustainable reproduction. Here, we found that wetland dependence by EMP Whooping Cranes in summer is high, especially during the remigial molt. Despite observed spring wandering of young, nonterritorial Whooping Cranes, new pairs eventually established territories close to their natal area. Low natal dispersal distances for both sexes suggested that habitat availability near release or hatching locations is most important. The importance of wetlands to Whooping Cranes has often been cited but these are the first data to quantify their relevance. In contrast, habitat use in winter has not been constrained geographically: EMP Whooping Cranes spend the winter in various habitats throughout the southeastern part of the United States Wintering cranes use wetlands much less during the daytime but wetlands are still important for roosting at night. Survival in EMP birds using these new winter habitats so far appears to be comparable to the AWBP.

The focus of future research, therefore, should discern between how habitat or nonhabitat factors can affect reproductive rates. If habitats provided to reintroduced populations are adequate, then future research should focus on nonhabitat genetic or learning hypotheses. Although it is premature to conclude that habitats are adequate in the EMP for all portions of the annual cycle, outside of ornithophilic black fly abundance, nothing yet suggests that habitats pose major barriers to reproduction or survival. Since no Whooping Crane reintroduction has yet produced a self-sustaining population (Urbanek and Lewis, 2015), understanding the role that habitat might play in the overall success of any crane reintroduction project is critical. The releases into both Necedah NWR and eastern Wisconsin provide an opportunity to examine habitat selection in a diverse array of wetlands as well as to illustrate potential nonhabitat hypotheses such as those involved with learning or genetics.

Acknowledgments

The work presented in this chapter is possible because of the efforts of many biologists and volunteers who have tracked Whooping Cranes in most places east of the Mississippi River. We thank International Crane Foundation staff, interns, and volunteers, Wisconsin Department of Natural Resources, U.S. Fish and Wildlife Service, and U.S. Geological Service staff, as well as countless volunteers and citizen scientists who have reported Whooping Crane sightings to WCEP. Lynda Garrett, Julie Langenberg, Nathan Van Schmidt, and one anonymous reviewer provided valuable comments and editing for this manuscript.

References

Allen, R.P., 1952. The Whooping Crane. Research Report 3. National Audubon Society, New York, p. 246.

Armstrong, D.P., Seddon, J.P., 1987. Directions in reintroduction biology. Trends Ecol. Evol. 23 (1), 20–25.

Austin, J.E., Hayes, M.E., Barzen, J.A., 2018. Revisiting the historic distribution and habitats of the whooping crane (Chapter 3). In: French, Jr., J.B., Converse, S.J., Austin, J.E. (Eds.), Whooping Cranes: Biology and Conservation. Biodiversity of the World: Conservation from Genes to Landscapes. Academic Press, San Diego, CA.

Baldassarre, G.A., 1978. Ecological Factors Affecting Waterfowl Production on Three Man-Made Flowages in Central Wisconsin. University of Wisconsin, Stevens Point, p. 124.

Baldassarre, G.A., Bolen, E.G., 2006. Ducks, Geese and Swans of North America. Krieger Publishing, Malabar, FL, p. 567.

Barzen, J.A., 2018. Ecological implications of habitat use by reintroduced and remnant whooping crane populations (Chapter 15). In: French, Jr., J.B., Converse, S.J., Austin, J.E. (Eds.), Whooping Cranes: Biology and Conservation. Biodiversity of the World: Conservation from Genes to Landscapes. Academic Press, San Diego, CA.

Barzen, J.A., Converse, S.J., Adler, P.H., Lacy, A.E., Gray, E., Gossens, A., 2018. Examination of multiple working hypotheses to address reproductive failure in reintroduced whooping cranes. Condor 120, 632–649.

Bent, A.C., 1926. Life histories of North American marsh birds. Smithson. Inst. U.S. Nat. Mus. Bull. 135, 490.

Beyer, H.L., 2015. Geospatial Modelling Environment (Version 0.7.4.0). Available from: http://www.spatialecology.com/gme [2 August 2018].

Boyle, S.A., Lourenco, W.C., da Silva, L.R., Smith, A.T., 2009. Home range estimates vary with sample size and methods. Folia Primatol. 80 (1), 33–42.

Burnham, J., Barzen, J., Pidgeon, A.M., Sun, B., Wu, J., Liu, G., Jiang, H., 2017. Novel foraging by wintering Siberian cranes (Leucogeranus leucogeranus) at China's Poyang Lake indicates broader changes in the ecosystem, raises new challenges for a critically endangered species. Bird Conserv. Int. 27, 204–223.

Calenge, C., 2006. The package adehabitat for the R software: a tool for the analysis of space and habitat use by animals. Ecol. Model. 197 (3–4), 516–519.

Cannon, J.R., 1999. Wisconsin whooping crane breeding site assessment. Unpublished Final Report to the Canadian-United States Whooping Crane Recovery Team, p. 112.

Chavez-Ramirez, F., 1996. Food Availability, Foraging Ecology and Energetics of Whooping Cranes Wintering in Texas. PhD dissertation, Texas A&M University, College Station, p. 104.

Converse, S.J., Royle, J.A., Adler, P.H., Urbanek, R.P., Barzen, J.A., 2013. A hierarchical nest survival model integrating incomplete temporally varying covariates. Ecol. Evol. 3 (13), 4439–4447.

Converse, S.J., Servanty, S., Moore, C.T., Runge, M.C., 2018. Population dynamics of reintroduced whooping cranes (Chapter 7). In: French, Jr., J.B., Converse, S.J., Austin, J.E. (Eds.), Whooping Cranes: Biology and Conservation. Biodiversity of the World: Conservation from Genes to Landscapes. Academic Press, San Diego, CA.

Converse, S.J., Strobel, B.N., Barzen, J.A., 2018. Reproductive failure in the Eastern Migratory Population: interaction of research and management (Chapter 8). In: French, Jr., J.B., Converse, S.J., Austin, J.E. (Eds.), Whooping Cranes: Biology and Conservation. Biodiversity of the World: Conservation from Genes to Landscapes. Academic Press, San Diego, CA.

Dahl, T.E., 1990. Wetlands Losses in the United States. 1780's to 1980's. Report to the Congress, National Wetlands Inventory, St. Petersburg, FL.

ESRI, 2011. ArcGIS Desktop: Release 10. Environmental Systems Research Institute, Redlands, CA.

Folk, M.J., Nesbitt, S.A., Parker, J.M., Spalding, M.G., Baynes, S.B., Candelora., K.L., 2008. Feather molt of nonmigratory whooping cranes in Florida. Proceedings of the North American Crane Workshop 10, 128–132.

Fondow, L.E.A., 2013. Habitat Selection of Reintroduced Migratory Whooping Cranes (Grus americana) on Their Wintering Range. Master's thesis, University of Wisconsin, Madison, p. 122.

French, Jr., J.B., Converse, S.J., Austin, J.E., 2018. Whooping Cranes past and present (Chapter 1). In: French, Jr., J.B., Converse, S.J., Austin, J.E. (Eds.), Whooping Cranes: Biology and Conservation. Biodiversity of the World: Conservation from Genes to Landscapes. Academic Press, San Diego, CA.

Fuller, R.J., 2012. Birds and Habitat: Relationships in Changing Landscapes. Cambridge University Press, Cambridge, UK, p. 542.

Garshelis, D.L., 2000. Decisions in habitat evaluation: measuring use, selection and importance. In: Boitani, L., Fuller, T. (Eds.), Research Techniques in Animal Ecology: Controversies and Consequences. Columbia University Press, New York, pp. 111–164.

Gil de Weir, K., 2006. Whooping Crane (Grus americana) Demography and Environmental Factors in a Population Growth Simulation Model. PhD dissertation, Texas A&M University, College Station, p. 159.

Greenwood, P.J., 1980. Mating systems, philopatry and dispersal in birds and mammals. Anim. Behav. 28 (4), 1140–1162.

Griffith, B., Scott, J.M., Carpenter, J.W., Reed, C., 1989. Translocation as a species conservation tool: status and strategy. Science 245 (4917), 477–480.

Hartup, B.K., 2018. Rearing and release methods for reintroduction of captive-reared whooping cranes (Chapter 20). In: French, Jr., J.B., Converse, S.J., Austin, J.E. (Eds.), Whooping Cranes: Biology and Conservation. Biodiversity of the World: Conservation from Genes to Landscapes. Academic Press, San Diego, CA.

Hayes, M.A., 2015. Dispersal and Population Genetic Structure in Two Flyways of Sandhill Cranes (*Grus canadensis*). PhD dissertation, University of Wisconsin, Madison, p. 287.

Hayes, M.A., Barzen, J.A., 2016. Dispersal patterns and pairing behaviors of non-territorial sandhill cranes. Passenger Pigeon 78 (4), 411–425.

Homer, C.G., Dewitz, J.A., Yang, L., Jin, S., Danielson, P., Xian, G., Coulston, J., Herold, N.D., Wickham, J.D., Megown, K., 2015. Completion of the 2011 National Land Cover Database for the conterminous United States: representing a decade of land cover change information. Photogr. Eng. Remote Sensing. 81 (5), 345–354.

Horne, J.S., Garton, E.O., 2006. Likelihood cross-validation versus least squares cross-validation for choosing the smoothing parameter in kernel home-range analysis. J. Wildl. Manage. 70 (3), 641–648.

Howe, M.A., 1989. Migration of radio-marked whooping cranes from the Aransas-Wood Buffalo Population: patterns of habitat use, behavior, and survival. Fish and Wildlife Technical Report 21. U.S. Department of the Interior, Fish and Wildlife Service, p. 39.

Johns, B.W., Goossen, P., Kuyt, E., Craig-Moore, L., 2005. Philopatry and dispersal in whooping cranes. Proceedings of the North American Crane Workshop 9, 117–125.

Johns, B.W., Woodsworth, E.J., Driver, E.A., 1997. Habitat use by migrant whooping cranes in Saskatchewan. Proceedings of the North American Crane Workshop 7, 123–131.

Johnston, C.A., 1982. Wetlands in the Wisconsin landscape. Wis. Acad. Rev. 29 (1), 8–11.

King, R.S., Trutwin, J.J., Hunter, T.S., Varner, D.S., 2013. Effects of environmental stressors on nest success of introduced birds. J. Wildl. Manage. 77 (4), 842–854.

Kuyt, E., 1979. Banding of juvenile whooping cranes and discovery of the summer habitat used by non-breeders. In: Lewis, J.C. (Ed.), Proceedings of the 1978 North American Crane Workshop. Colorado State University Printing Service, Fort Collins, p. 259, pp. 109–111.

Kuyt, E., 1981a. Clutch size, hatching success and survival of whooping crane chicks, Wood Buffalo National Park, Canada. In: Lewis, J.C., Masatomi, H. (Eds.), Crane Research around the World. International Crane Foundation, Baraboo, WI, pp. 126–129.

Kuyt, E., 1981b. Population status, nest site fidelity, and breeding habitat of whooping cranes. In: Lewis, J.C., Masatomi, H. (Eds.), Crane Research around the World. International Crane Foundation, Baraboo, WI, pp. 119–125.

Kuyt, E., 1993. Whooping crane, *Grus americana*, home range and breeding range expansion in Wood Buffalo National Park, 1970–1991. Can. Field Nat. 107 (1), 1–12.

Kuyt, E., Johns, B.W., Barry, J., 1992. Below average whooping crane production in Wood Buffalo National Park during drought years 1990 and 1991. Blue Jay 50 (4), 225–229.

Lacy, A.E., Barzen, J.A., Moore, D.M., Norris, K.E., 2015. Changes in the number and distribution of greater sandhill cranes in the eastern population. J. Field Ornithol. 86 (4), 317–325.

Lacy, A, McElwee, D., 2014. Observations of molt in reintroduced whooping cranes. Proceedings of the North American Crane Workshop 12, 75.

Maguire, K.J., 2008. Habitat Selection of Reintroduced Whooping Cranes, *Grus americana*, on Their Breeding Range. MS thesis, University of Wisconsin, Madison, p. 54.

Mech, L.D., 1983. Handbook of Animal Radio-Tracking. University of Minnesota Press, St. Paul.

Meine, C.D., Archibald, G.A., 1996. The Cranes: Status Survey and Conservation Action Plan. International Union for the Conservation of Nature, Gland, Switzerland, and Cambridge, UK, p. 294.

Miller, T.P., Barzen, J.A., 2016. Habitat selection by breeding sandhill cranes in central Wisconsin. In: Proceedings of the North American Crane Workshop 13, 1–12.

Novakowski, N.S., 1966. Whooping crane population dynamics on the nesting grounds, Wood Buffalo National Park, Northwest Territories, Canada. Canadian Wildlife Service Report Series 1, Ottawa, Canada. 20 pp.

Omernik, J.M., Chapman, S.S., Lillie, R.A., Dumke, R.T., 2000. Ecoregions of Wisconsin. Trans. Wis. Acad. Sci. Arts Lett. 88, 77–103.

R Core Team, 2015. R: A Language and Environment for Statistical Computing. R Foundation for Statistical Computing, Vienna, Austria. http://www.R-project.org/ [2 August 2018].

Servanty, S., Converse, S.J., Bailey, L.L., 2014. Demography of a reintroduced population: moving toward management models for an endangered species, the whooping crane. Ecol. Appl. 24 (5), 927–937.

Smith, E.H., Chavez-Ramirez, F., Lumb, L., 2018. Winter habitat ecology, use, and availability for the Aransas-Wood Buffalo Population of whooping cranes (Chapter 13). In: French, Jr., J.B., Converse, S.J., Austin, J.E. (Eds.), Whooping Cranes: Biology and Conservation. Biodiversity of the World: Conservation from Genes to Landscapes. Academic Press, San Diego, CA.

Sokal, R.R., Rohlf, F.J., 1981. Biometry. W.H. Freeman New York, p. 859.

Stehn, T.V., Haralson-Strobel, C.L., 2014. An update on mortality of fledged whooping cranes in the Aransas/Wood Buffalo Population. Proceedings of the North American Crane Workshop 12, 14–50.

Stehn, T.V., Prieto, F., 2010. Changes in winter whooping crane territories and range 1950–2006. Proceedings of the North American Crane Workshop 11, 40–56.

Teitelbaum, C.S., Converse, S.J., Fagan, W.F., Bohning-Gaese, K., O'Hara, R.B., Lacy, A.E., Mueller, T., 2016. Experience drives innovation of new migration patterns of whooping cranes in response to global change. Nat. Commun. 7 (12793). doi:10.1038/ncomms12793.

Thompson, H.L., 2018. Characteristics of whooping crane home ranges during the nonbreeding season in the Eastern Migratory Population. MS thesis, Clemson University, Clemson, SC.

Urbanek, R.P., Duff, J.W., Swengel, S.R., Fondow, L.E.A., 2005. Reintroduction techniques: post-release performance of sandhill cranes (1) released into wild flocks and (2) led on migration by ultralight aircraft. Proceedings of the North American Crane Workshop 9, 203–211.

Urbanek, R.P., Fondow, L.E.A., Zimorski, S.E., 2010a. Survival, reproduction, and movements of migratory whooping cranes during the first seven years of reintroduction. Proceedings of the North American Crane Workshop 11, 124–132.

Urbanek, R.P., Lewis, J.C., 2015. Whooping crane (*Grus americana*). In: Rodewald, P.G. (Ed.), The Birds of North America. Cornell Lab of Ornithology, Ithaca, NY. Retrieved from the Birds of North America Online: https://birdsna.org/Species-Account/bna/species/whocra/introduction. [2 August 2018] doi:10. 2173/bna. 153.

Urbanek, R.P., Szyszkoski, E.K., Zimorski, S.E., 2014a. Winter distribution dynamics and implications to a reintroduced population of migratory whooping cranes. J. Fish Wildl. Manage. 5 (2), 340–362.

Urbanek, R.P., Szyszkoski, E.K., Zimorski, S.E., Fondow, L.E.A., 2018. Pairing dynamics of reintroduced migratory whooping cranes (Chapter 10). In: French, Jr., J.B., Converse, S.J., Austin, J.E. (Eds.), Whooping Cranes: Biology and Conservation. Biodiversity of the World: Conservation from Genes to Landscapes. Academic Press, San Diego, CA.

Urbanek, R.P., Zimorski, S.E., Fasoli, A.M., Szyszkoski, E.K., 2010b. Nest desertion in a reintroduced population of migratory whooping cranes. Proceedings of the North American Workshop 11, 133–144.

Urbanek, R.P., Zimorski, S.E., Szyszkoski, E.K., Wellington, M.M., 2014b. Ten-year status of the eastern migratory whooping crane reintroduction. Proceedings of the North American Crane Workshop 12, 33–42.

U.S. Environmental Protection Agency (EPA), 2013. Level III and IV ecoregions of the continental United States. EPA, National Health and Environmental Effects Research Laboratory, Corvallis, OR, map scale 1:3,000,000. http://www.epa.gov/wed/pages/ecoregions/level_iii_iv.htm [2 August 2018].

Van Schmidt, N.D., Barzen, J.A., Engels, M.J., Lacy, A.E., 2014. Refining reintroduction of whooping cranes with habitat use and suitability analysis. J. Wildl. Manage. 78 (8), 1404–1414.

Ecological Implications of Habitat Use by Reintroduced and Remnant Whooping Crane Populations

Jeb A. Barzen

International Crane Foundation, Baraboo, WI, United States;
Private Lands Conservation LLC, Spring Green, WI, United States

INTRODUCTION

Allen (1952) described Whooping Crane populations as historically occupying divergent landscapes that are now called biomes. Yet, within each biome, the habitats used occurred within narrow ecological boundaries. Close adherence to a particular area within a biome suggests habitat specialization, while habitat use between broadly different biomes suggests plastic habitat preferences. It appears that characteristics highly important to survival or reproduction may not be related to the biomes in which these habitats can be found. Based on historic observations, characterization of habitats used suggests four elements as being important to Whooping Cranes (Austin et al., Chapter 3, this volume): (1) open habitat structure and moderate topographic variability; (2)

high densities of wetland complexes; (3) hydrological regimes providing reliable conditions for nesting, brood rearing, and flightless adults; and (4) high productivity due to fertile soils, hydrological pulsing, periodic inflow of nutrients, or other periodic perturbations. Importantly, providing all habitat needs simultaneously may require large territories that encompass diverse habitats (criterion 2).

Required habitat diversity can be measured in many ways. The reintroduction of any migratory Whooping Crane population, for example, will cross temperature and moisture gradients that differ from the Great Plains distribution of the Aransas-Wood Buffalo Population (AWBP; Carpenter, 1940). The historic distribution of breeding Whooping Cranes in the Midwest (Allen, 1952) suggests that Whooping Cranes can generally adapt to those environmental conditions, but it is less clear if released cranes of the Eastern Migratory Population (EMP), derived from the AWBP, can adapt to environmental variation inherent in these new, albeit historical, geographical regions. Reintroduction efforts thus become broad experiments on habitat requirements that relate to an entire species, to each individual population, and to individuals as they respond to their environment.

In addition to spatial variation, habitat conditions can vary temporally as well. Since the late 1800s, historical wetland conditions in the Midwest have changed extensively (Tiner, 1984), making it difficult to understand how habitats today may affect cranes. Iowa, for example, held the highest density of historical Whooping Cranes nesting records (Allen, 1952) but has since lost >98% of its wetlands to drainage (Tiner, 1984) and lost associated grasslands to row crop agriculture. Given such extensive temporal change, reintroduction efforts should test, rather than assume, whether hypothesized environmental conditions truly suffice as appropriate Whooping Crane habitat. If habitats are found lacking, adjustments to habitat conditions should be made through releases in new areas (Van Schmidt et al., 2014) or through behavioral modification.

Examining the ecological ramifications of habitat use by any species, especially one that has been reduced to such low numbers as the Whooping Crane and has been eliminated from a large component of its historical range, is also challenging. For example, absence of use for a particular habitat when population numbers are low may mean the habitat is undesirable or it may mean that there are not enough animals available to use the otherwise good habitat. Though the ultimate factors that influence resource selection must invariably affect either survival or natality (Cody, 1985; Fuller, 2012b), there are many proximate environmental, demographic, and behavioral mechanisms through which resource selection is modified. A brief summary of these mechanisms is useful here.

Environmentally, traditional habitat use can occur at multiple geographical scales: (1) among food items within a specific habitat, (2) among habitats within a home range, (3) among home ranges within a landscape, and (4) among landscapes within a region (Johnson, 1980). Though habitat use estimates have been made at all geographic levels for Whooping Cranes, studies have focused mostly on estimating habitat use within home ranges and regions. By summarizing habitat use measured within home ranges and regions for multiple populations, I hope to provide insights that can improve management action and ecological understanding.

When comparing habitat use in different Whooping Crane populations, demographic characteristics, mostly related to crane density, are relevant as they influence use of available habitats (Garshelis, 2000). Though demographic modifiers of habitat selection may proximately determine how an individual increases or maintains ultimate fitness through habitat use, they can also alter our interpretation of the habitat use that we measure (Fuller, 2012a). Habitat selection of the AWBP at its nadir of 15 individuals (Stevenson and Griffith, 1946), for example, will differ greatly from habitat use of the current population containing well over 300 cranes, which will differ from habitat use of a population of 1,100 cranes at the estimated carrying capacity for the winter area (Stehn and Prieto, 2010).

Behavioral responses by individuals can also modify habitat selection in Whooping Cranes. Breeding Whooping Cranes are territorial, long-lived individuals that tend to pair for multiple years (Urbanek et al., 2014b; Urbanek and Lewis, 2015). Whooping Cranes also exhibit an unusual resource defense mating system (Greenwood, 1980) where natal dispersal distances between males and females are similar and generally short (Johns et al., 2005; H. Thompson, unpublished data, International Crane Foundation, WI). Short natal dispersal distances, in turn, tend to limit the overall area for which habitats can be considered available and thus influence habitat selection (Barzen et al., Chapter 14, this volume; Garshelis, 2000).

Behavior of conspecifics can also influence habitat selection. With Whooping Cranes, the formation of long-term pair bonds means that one member of the pair will necessarily follow the other pair member, influencing habitat selection. On winter areas (Stehn and Prieto, 2010) and on summer areas (Urbanek et al., 2010a) the gender choosing territories appears to be the male, suggesting that females should be the greater dispersers (Greenwood, 1980), but little evidence of female-biased natal dispersal behavior exists in Whooping Cranes. Members of a group can also influence habitat selection in the wild (Mueller et al., 2013) or in captivity (Maguire, 2008).

The goal of this chapter, therefore, is to examine the collective characterizations of remnant and reintroduced populations of Whooping Cranes across divergent landscapes to provide a more complete picture of habitat associations over the full range of environmental gradients that Whooping Cranes currently occupy (Fuller, 2012a). To accomplish this, habitat use data for four of five populations were examined. Whooping Cranes reintroduced to Grays Lake, Idaho (Grays Lake Population, GLP), had few habitat use data to examine (Ellis et al., 1992). I present habitat use for each stage of the annual cycle (summer, winter, and migration) and compare populations within each stage as well as comparing Whooping Crane data with sympatric populations of Sandhill Cranes (*Grus canadensis*) where relevant. Ecological implications of habitat use are discussed for each population and for each annual cycle stage.

NOTES ON METHODS

Primary data reviewed in this chapter are presented as home range size (km²), natal dispersal distance (km), and habitat use (%) for the AWBP, EMP, the Louisiana Nonmigratory Population (LNMP), and the Florida Nonmigratory Population (FNMP), as well as two populations of Sandhill Cranes (Eastern Population of Greater Sandhill Cranes, *G. c. tabida*, EP; and the nonmigratory Florida Sandhill Crane, *G. c. pratensis*) obtained from published and unpublished research projects. For Sandhill Cranes, most data that were compared to Whooping Cranes came from the EP. Within a year, the annual cycle was subdivided into arrival on breeding grounds, nesting, flightless molt, chick rearing (these first four categories were collectively called summer), fall migration, winter, and spring migration.

Where home range estimates were cited, the methods used to calculate home range size in various studies differed. For Whooping Cranes, both 95% Kernel Density Estimates (KDEs) and Minimum Convex Polygon (MCP) estimates have been used frequently (Boyle et al., 2009). Analysis determining home range size in the LNMP used Brownian bridge movement models (Horne et al., 2007) and 50% utilization distribution contours to represent the seasonal core–use areas (Pickens et al., 2017). The 50% Brownian model estimates calculate smaller home range sizes when compared to 95% KDE or MCP estimates. Variation in interpretations related to each technique used to estimate home range was beyond the scope of this summary.

Time periods used for each home range estimate varied among studies and were identified in the tables. For both home range and habitat use studies, sample size is problematic because not all individuals represent independent samples. Where cranes commingled, the number of groups or pairs was noted instead of the number of individuals. Because estimates from studies among populations and decades were so disparate, only qualitative comparisons of home range or habitat use were made between various portions of the annual cycle. Home range size was interesting to examine because it forms a behavioral boundary that defines habitat availability at one spatial scale and because it often reflects resource abundance or distribution within groups of similar individuals (Hutto, 1985).

Natal dispersal distance, defined by Greenwood (1980) and applied to Whooping Cranes (Johns et al., 2005), is the linear distance moved by an individual between the area where it was hatched and the area where it established its first nest as an adult. As with home range size, linking nesting territory and natal locations for the same crane helps define habitat availability on a landscape or regional scale. In effect, a crane natal dispersal distance represents a minimum lifetime assessment of habitat availability for an individual (Hayes, 2015; Hayes and Barzen, 2016).

Habitat use data represent the percentage of crane locations found in various habitat types. A similar measure, habitat composition, is the proportion of habitat that occurred within the boundaries of a home range. Both measures are presented here. Analysis to see if habitats were actually used more or less frequently than they were available, based on the overall distribution of habitat types located within a specified spatial scale, is often used to infer selection for or against specific habitats (Garshelis, 2000). Most Whooping Crane studies investigating habitat use did not compare use and availability, so few conclusions regarding habitat selection were possible. Finally, because habitat use was compared among crane populations, all habitat categories that cranes used were consolidated into wetland (i.e., emergent marsh, shallow open water, shallow riverine areas, inundated agricultural areas) versus all other habitat groups. Only wetland use is reported here. In the LNMP, wetland designation of agricultural habitats was dynamic and difficult to determine (Foley, 2015). A rice (*Oryza sativa*) field, for example, is a wetland-like habitat (periodically flooded) but could be dry or wet, depending upon the stage of crop development. For the purposes of this chapter we assumed rice and crawfish fields were static wetland habitat types and fallow fields were static upland habitats. Greater differentiation than wetland versus upland would make habitat comparisons among populations too difficult to use here. Finally, non-wetland

use of nighttime roosts is rare (Urbanek and Lewis, 2015) so comparisons made here reflect only daytime habitat use.

Territories – home ranges that are constrained by intraspecific interactions among pairs or family groups and create mostly nonoverlapping use areas – are a focus of this chapter because they determine the spacing between crane groups more precisely than do home ranges. Territories were demarcated areas that a pair or family group (parents with associated chicks) repeatedly defended from conspecifics for a significant portion of the annual cycle, such as winter or summer periods. Defense of territories is achieved through threat displays and active or passive displacement of conspecifics. With long-lived cranes (Urbanek and Lewis, 2015) the same basic territory is defended across multiple years (Bonds, 2000; Hayes, 2015; Stehn, 1992; Stehn and Johnson, 1987). Winter or summer territories provide resources that are clumped and economically defendable (Brown, 1964, 1969; Brown and Orians, 1970) and in cranes likely include food and/or habitat structure that provides safety from predators. In summer territories, additional resources defended within a territory include the nest and chick-rearing habitats (Miller and Barzen, 2016; Van Schmidt et al., 2014) and parallel Type 2 territories described in waterfowl (Anderson and Titman, 1992, p. 259).

RESULTS AND DISCUSSION

Summer Period

In summer I first looked at spacing mechanisms through examination of differences in home range size for territorial and nonterritorial cranes among populations, and then examined natal dispersal distance. The flightless molt in Whooping Cranes presents a unique challenge and is treated separately at the end of this section. Home range and natal dispersal distance define the amount of area over which cranes select habitats at different spatial scales. Within these behavioral boundaries of habitat availability, extrinsic factors like drought and flood were considered in both home range and habitat use sections as they modify habitat use, which follows the home range section.

Spacing Mechanisms

TERRITORIAL CRANES

The home range of territorial individuals formed the boundaries of habitats a pair or family group of cranes used and included summer resources that were important for survival and reproduction. As such, territorial behavior in cranes appeared to greatly influence habitat selection, but this aspect of resource use has never been tested directly even though it appeared likely that sexually mature Whooping Cranes did not breed until they obtained a territory (Kuyt, 1979, 1993). The importance of a territory to breeding cranes was described in detail for Sandhill Cranes (Hayes, 2015).

The composite nesting areas (CNAs) of Kuyt (1981b, 1993) for the AWBP contained clumped nest locations and reflected the total use of habitats by pairs or family groups over multiple years (usually a minimum of 3 years). CNAs also varied in their boundaries, presumably based on environmental conditions, such as drought, which varied between years (Kuyt, 1995). The size of CNAs therefore approximated territory area. At high population densities, however, variation in territory boundaries can be limited by interactions among territorial pairs as occurred with breeding Sandhill Cranes (Hayes, 2015). Within Wood Buffalo National Park (WBNP), CNAs averaged 4.1 km^2 and varied from 1.3 to 18.9 km^2 (Kuyt, 1993; Table 15.1). Environmental conditions (but not territory size) were qualitatively related to productivity in the AWBP: Novakowski (1966) suggested climate influenced productivity, while Kuyt et al. (1992) suggested that climate-related hydrological conditions such as pond depth were the drivers for productivity because low

water levels would lead to higher predation of chicks. In the WBNP, some CNAs might be large simply because no other Whooping Cranes are nesting nearby and presumably there is no cost for defense (Kuyt, 1993), so care must be taken in overinterpreting home range size as an indicator of habitat quality (Garshelis, 2000).

In two studies for the EMP, territories averaged 3.68 km^2 (Van Schmidt et al., 2014) and 4.58 km^2 (Barzen et al., Chapter 14, this volume), similar in size to the AWBP's CNAs, larger than territories of sympatric Sandhill Cranes in similar habitat conditions, and larger than territories of nonmigratory Whooping Cranes in Florida (Folk et al., 2005). No estimates of territory size were available for the LNMP. The similar size of home ranges for reintroduced and remnant populations suggests that spacing behavior in reintroduced populations is likely normal and that habitat requirements for individual Whooping Crane territories occur over a large expanse. It is possible that the four habitat characteristics that are thought important to Whooping Cranes (Austin et al., Chapter 3, this volume) must occur within each Whooping Crane home range, requiring nesting territories to be larger than nesting territories of conspecific Sandhill Cranes that do not rely on wetlands to the same extent (Miller and Barzen, 2016; Table 15.1). Minimum territory size reported by any nesting study for Whooping Cranes was 0.78 km^2 (Barzen et al., Chapter 14, this volume).

Natal dispersal distance was relatively short in all Whooping Crane populations so far studied (Table 15.2). Dispersion from the historic breeding range at WBNP has been limited (Kuyt, 1993) and has not expanded beyond the primary nesting area of the park (Johns et al., 2005), so little can be learned about how Whooping Cranes respond to novel breeding habitats in the AWBP. Interestingly, though average natal dispersal distances in all populations of Whooping Cranes are relatively short, they are over 4 times longer than sympatric migratory or nonmigratory male Sandhill Cranes but not very different for female

TABLE 15.1   Estimates of Average Home Range Size (km^2) for Two Reintroduced Populations of Whooping Cranes (Eastern Migratory Population, EMP; Louisiana Nonmigratory Population, LNMP), the Remnant Migratory Population (Aransas-Wood Buffalo Population, AWBP) of Whooping Cranes, and the Eastern Population of Sandhill Cranes (EP). NA = Data Not Applicable, n = Sample Size of Nesting Territories (nt), Pairs (p), Groups (g), or Individuals (i). Age Is Defined by Calendar Year Where a Bird Becomes 2 Years Old on January 1 Following the Hatch Year and Ages 1 Year Each Subsequent January.

| | Reintroduced populations | | Remnant population | Sandhill Crane |
| | EMP | LNMP | AWBP | EP |
Life cycle stage	Average ± SE (*n*)	Average ± SE (*n*)	Average ± SE (*n*)	Average ± SE (*n*)
Summer – Territorial Birds	3.68 ± 0.95 (16 nt)[a]		4.1 ± 1.4 (13 nt)[b]	1.275 ± 0.20 (12 nt)[c]
Summer – Territorial Birds	4.58 ± 1.68 (8 nt)[d]			2.847 ± 0.60 (12 nt)[c]
Summer – Nonterritorial Birds		55 ± 7 (53 i)[e]		
2nd Year	7,750 ± 5,901 (10 i)[d]		70 (7 i)[f]	284.43 ± 86.62 (30 i)[g]
3rd Year	3,989 ± 5,572 (7 i)[d]			86.34 ± 20.46 (15 i)[g]
4th Year	414.6 ± 478 (4 i)[d]			31.34 ± 10.32 (8 i)[g]
Summer – Molting Birds	0.45 ± 0.12 (2 p & 2 i)[d]			NA
Staging Birds	7.16 ± 2.25 (30 i)[h]			35.4 ± 39.25 (6 i)[i]
Winter – Territorial Birds			2.10 (66 p)[j]	
Winter – Territorial Birds			1.63 ± 1.68 (142 p)[k]	
Winter – Nonterritorial Birds	0.29 ± 0.12 (14 g)[l]	54 ± 5 (53 i)[m]		48.7 ± 64.29 (6 i)[n]
Winter – Nonterritorial Birds	2.04 ± 1.23 (18 g)[l]			
Winter – Nonterritorial Birds	32.41 ± 6.61 (30 i)[o]			

[a]*Van Schmidt et al., 2014. Data from April to July includes off-territory habitat use (1.27 km^2) where no territory was defended.*

[b]*Kuyt, 1993. Mean home range from non-isolated composite nesting territories containing at least one marked bird measured for more than 3 years. Data collected mostly June–August.*

[c]*Miller and Barzen, 2016. Smaller estimate of March–July while larger estimate of March–October.*

[d]*Barzen et al., Chapter 14, this volume, April–September.*

[e]*Pickens et al., 2017. Summer months, May–July, included five territorial pairs.*

[f]*Kuyt, 1979. This is an approximation based on resightings of color-banded birds, all of which used the same area of 14 × 5 km.*

[g]*Hayes and Barzen, 2016. Data from April to September.*

[h]*H. Thompson, unpublished data, International Crane Foundation, WI. Data from 46 stopovers made by 30 birds consisting of at least five locations in eight states of Iowa, Indiana, Illinois, Kentucky, Tennessee, Arkansas, Alabama, and Georgia during migration.*

[i]*Thompson and Lacy, 2016. Includes 30 bird-stops on 19 migrations through Wisconsin, Indiana, Kentucky, Tennessee, and Georgia made by six birds in spring and fall that lasted more than 3 days.*

[j]*Stehn and Prieto, 2010. Estimate is for all banded and unbanded territorial pairs or family groups in 1 year (2006). This estimate is the total area occupied by territorial birds divided by the number of territories, so there is no error estimate.*

[k]*Bonds, 2000. Estimated as a 5-year running average for all territories using only banded birds (66% of total population).*

[l]*Fondow, 2013. Data from Table 15.2, winter 2004/2005, and Table 15.3, winter 2005/2006.*

[m]*Pickens et al., 2017. Winter months, November–January.*

[n]*Thompson and Lacy, 2016. Includes nine winter records for six independent birds within Tennessee, Indiana, Georgia, and Florida. Winter records are based on a minimum of 30 locations per winter record.*

[o]*Thompson (2018), data from 37 winter records made by 30 independent individuals having at least 30 locations each among six states of Indiana, Kentucky, Tennessee, Alabama, Georgia, and Florida during winter.*

TABLE 15.2   Natal Dispersal Distance in Whooping Crane and Comparable Sandhill Crane Populations (Whooping Cranes: FNMP, Florida Nonmigratory Population; EMP, Eastern Migratory Population; AWBP, Aransas-Wood Buffalo Population; Sandhill Cranes: EP, Eastern Population of Greater Sandhill Cranes, *Grus canadensis tabida*; FL, Florida Nonmigratory Subspecies, *G. c. pratensis*)

Metric	Whooping Cranes		Sandhill Cranes		
	FNMP	EMP	AWBP	EP	FL
Male, Ave. ± SE (km)	28.0 ± 34.3	24.55 ± 49.72	16.8 ± 16.6	2.3 ± 0.4	3.9 ± 2.8
Female, Ave. ± SE (km)	40.0 ± 41.2	27.52 ± 53.56	16.2 ± 10.5	10.7 ± 4.0	11.6 ± 11.2
No. males, no. females	90, 91	57, 55	31, 30	21, 14[a]	12, 16
Natal dispersal range (km)[b]	0.1–135.5	0.6–357.4	0.3–54.8	0.1–57.4	0.4–48.3
Percent >2 × SE[c]	18.2	2.67	18.0[d]	5.7	No data
Source	T. Dellinger, unpublished data[e]	H. Thompson, unpublished data[f]	Johns et al., 2005	Hayes and Barzen, 2016	Nesbitt et al., 2002

[a] Hayes, 2015.
[b] Range is combined for males and females.
[c] Percentage of individuals with natal dispersion over 2 times the standard error.
[d] B. Johns, unpublished data, Saskatoon, SK.
[e] Florida Fish and Wildlife Conservation Commission.
[f] International Crane Foundation, WI.

Sandhill Cranes (Table 15.2). Natal dispersal distances of either sex in the FNMP appeared larger than any other Whooping Crane population, perhaps because wetland complexes were widely scattered and exposed to drought conditions during the reintroduction, creating both spatial and temporal variation in wetland abundance. Though most cranes did not have large natal dispersal distances, a few cranes in each population (<20%) dispersed more than twice the standard error of their average natal dispersal distance, and some individuals moved as much as 357 km (Table 15.2).

Short natal dispersal behavior concentrated cranes in habitat complexes that were similar to their natal or reintroduction sites and may have been adaptive where survival or reproduction could be improved by local cranes adjusting to specific environmental conditions (e.g., food availability for chicks or predator risk), called Optimal Discrepancy by Anderson

et al. (1992). Allen (1952) suggested that different breeding populations of Whooping Cranes remained fairly isolated. Limited gene flow between local populations, and thus the differentiation of local genotypes, has been demonstrated within the EP Sandhill Cranes by Hayes (2015). For reintroduced populations, however, short natal dispersal distances will tend to maintain any potential errors in site selection because cranes are less likely to disperse from the reintroduction area (Barzen et al., Chapter 14, this volume). Rare but long natal dispersal distances can affect gene flow to the extent that metapopulations that are separated by as little as 100 km (2 × SE, Table 15.2) may be genetically distinct (Hayes, 2015). In addition, even though EP Sandhill Cranes breed in areas of high crane densities (Barzen et al., 2016), natal dispersal distances are still short (Hayes and Barzen, 2016). This is a situation that remains true even though nonterritorial cranes cannot breed unless they find a territory (Hayes, 2015), suggesting that the mechanisms that determine natal dispersal distance may well be genetically fixed.

NONTERRITORIAL CRANES

Within the EMP, LNMP, and perhaps within the AWBP, home ranges of nonterritorial Whooping Cranes in summer were 10–1,000 times larger than those of territorial Whooping Cranes, meaning that the amount of habitat available to cranes in these two social groups differed markedly (Table 15.1). From limited data, home ranges of FNMP nonterritorial cranes were also large (T. Dellinger, unpublished data, Florida Fish and Wildlife Conservation Commission). Home range size estimates from nonterritorial cranes in all populations except the EMP and LNMP, however, were sparse. The only estimate of home range size from WBNP comes from seven marked (but not telemetered), nonterritorial cranes that occupied an area of approximately 70 km² (Kuyt, 1979), 20 times the size of the average CNA. Though this was adjacent to the breeding areas, Kuyt felt that the nonterritorial cranes of various ages intentionally occupied habitat that was unused by territorial cranes.

Nonterritorial cranes of the EMP were tracked using VHF radio and satellite telemetry, providing more accurate estimates of home range sizes. Average territory size for nonterritorial cranes in the EMP was substantially larger than those of territorial cranes and decreased with age (Table 15.1). Large home ranges of nonterritorial EMP Whooping Cranes in summer seemed to reflect extensive sampling of available habitats located well away (>100 km) from natal or release areas. As 2-year-olds, for example, a small proportion of cranes in the EMP traveled to adjoining states of Minnesota, Iowa, Illinois, and Michigan (>300 km from release/natal areas; Barzen et al., Chapter 14, this volume). Though some of these large movements may have been caused by idiosyncrasies of migration by costume-reared cranes (Urbanek et al. 2010a),

most spring wanderings appeared comparable in size to the AWBP.

In the LNMP, 50% core-use areas ranged from 4.7 to 438 km² with an overall mean of 53.2 ± 2.9 km² and seasonal means of 65 ± 7 km² (spring, February–April), 55 ± 7 km² (summer, May–July), 39 ± 2 km² (fall, August–October), and 54 ± 5 km² (winter, November–January; Pickens et al., 2017). In addition to differences in home range metrics used, Pickens et al. (2017) combined age groups and breeding status of cranes to estimate home range size, so they were difficult to compare with other populations (Table 15.1). As with the EMP, some nonterritorial cranes in the LNMP temporarily dispersed from Louisiana and used wetlands approximately 500 km west of their White Lake release site (Pickens et al., 2017).

Termed "spring wandering" (Allen, 1952; Urbanek et al., 2014b), extensive home ranges of subadults decreased most between a crane's third and fourth year (Table 15.1). Though Kuyt (1979) found a large proportion of marked, nonterritorial cranes utilizing WBNP during summer, numerous summer records of nonterritorial Whooping Cranes occurred well outside (>500 km) of WBNP as well (Johns et al., 2005). Yet, even with extensive wandering, cranes from any population studied had remarkably short natal dispersal distances for either sex (Table 15.2). Except for the EMP, maximum natal dispersal distances did not exceed 136 km (Table 15.2). Extensive natal dispersal distance (>2 × SE) in the EMP, though rare (<3%), is possible. The relationship between broadly wandering nonterritorial and highly philopatric territorial Whooping Cranes in summer is poorly understood. Perhaps survival of these long-lived individuals (Urbanek and Lewis, 2015) improves when cranes have knowledge of a landscape that offers safe habitat alternatives when environmental extremes occur. Though sympatric EP Sandhill Cranes have smaller home ranges than do EMP Whooping Cranes, the pattern found in home range size for different age groups was similar (Table 15.1). This pattern, therefore, is not likely an artifact of reintroduction.

Habitat Use

Determination of home range makes habitat use easier to evaluate because it defines a spatial scale over which habitat use is measured (Johnson, 1980). For territorial Whooping Cranes, nesting cranes used wetlands a reported 100% of the time at WBNP, but this estimate was not supported with data (Allen, 1956; Novakowski, 1966). The high density of wetlands, associated with small river systems, is unique to the crane nesting area of WBNP and suggests that a preference for this dynamic landscape may occur on a regional basis (Kuyt, 1981b; Timoney et al., 1997). Since little upland habitat exists in the nesting area at WBNP (Olson + Olson Planning Design Consultants, 2003), however, it is difficult to evaluate wetland preference by territorial cranes there, especially within home ranges.

In the EMP, Whooping Cranes from eight nesting territories were located in wetlands 75% of the time (Barzen et al., Chapter 14, this volume) and wetlands were used by EMP Whooping Cranes more than expected based on their availability (Van Schmidt et al., 2014). Without understanding wetland preference at WBNP by territorial cranes, however, it is difficult to evaluate any biological significance between 100% wetland use hypothesized at WBNP and 75% wetland use measured in the EMP. The frequency that wetlands were used by territorial Whooping Cranes in the EMP was 45% higher than wetland use by territorial EP Sandhill Cranes (Table 15.3), a species that appears less wetland dependent than Whooping Cranes (Armbruster, 1987; Miller and Barzen, 2016).

Fifty percent of locations for nonterritorial Whooping Cranes in the LNMP were found in a variety of wetlands during summer (Pickens et al., 2017) but they used only crawfish fields (agricultural wetlands) more than expected while other wetland types (rice fields and natural marsh) were used to the same degree as their availability. Nonterritorial cranes in the EMP used wetlands about as much as did cranes in the LNMP (Table 15.3) during summer. A small sample of nonterritorial Whooping Cranes in the FNMP, however, used wetlands about 70% of the time (T. Dellinger, unpublished data, Florida Fish and Wildlife Conservation Commission).

Currently, more wetlands occur in WBNP than there are cranes available to use them. As of 2003, only 14% of the area of appropriate wetlands appeared to be used by nesting cranes (Olson + Olson Planning Design Consultants, 2003; Timoney, 1999). It is not clear, however, if rivers were incorporated into the habitat models used to measure availability of nesting habitat. Having many wetland basins that vary in area and depth, and are located within each home range of a territorial family group, is thought important to Whooping Cranes (Austin et al., Chapter 3, this volume; Kuyt, 1995; Kuyt et al., 1992). River systems, snow melt, groundwater flow (Timoney et al., 1997), and particularly spring floods (Kuyt, 1981b), create hydrological conditions (i.e., both pulsing and predictable occurrence of water) for cranes to successfully breed each year, albeit in different parts of the CNA as habitat conditions vary annually. In droughts, for example, Whooping Cranes shifted their CNA toward river systems (Kuyt, 1981a) and annual production of young was correlated with yearly mean water depth at nest sites (Kuyt, 1981a; Kuyt et al., 1992).

Is there a problem with the lower proportion of wetland use by territorial Whooping Cranes in the EMP? Adult survival at all stages of the life cycle in the EMP appear as high as or higher than survival found in the AWBP (Servanty et al., 2014) but EMP natality rates are still low (Urbanek et al., 2014b) with poor hatching success (Barzen et al., 2018a; Converse et al., 2013; King et al., 2013; Urbanek et al., 2010b) and low chick fledging rates (Barzen et al., 2018a; Converse et al., Chapter 8, this volume), largely being responsible for the current

TABLE 15.3 Proportion of Wetland Habitats Used by Two Reintroduced Populations of Whooping Crane (Eastern Migratory Population, EMP, and Louisiana Nonmigratory Population, LNMP), the Remnant Migratory Population of Whooping Cranes (Aransas-Wood Buffalo Population, AWBP), and the Eastern Population of Sandhill Cranes (EP) during the Day. NA = Data Not Applicable; n = Sample Size of Nesting Territories (nt), Pairs (p), Groups (g), or Individuals (i).

| | Reintroduced populations | | Remnant population | Sandhill Crane |
| | EMP | LNMP | AWBP | EP |
Life cycle stage	% Wetland (n)	% Wetland (n)	% Wetland (n)	% Wetlands (n)
Summer – Territorial Birds	75.0 (8 nt)[a]		Mostly[b]	30.6 (12 nt)[c]
Summer – Nonterritorial Birds	49.4 (9 i)[a]	50.0 (53 i)[d]		25.7 (36 i)[e]
Summer – Molt	92.0 (2 pr and 2 i)[f]			NA
Staging Birds	14.8 (60 i)[g]		30, 67 (7) i[h]	35.7 (6 i)[i]
Winter – all birds combined	52.4 (32 g)[j]	61.0 (53 i)[k]	89.0 (128 i)[l]	29.7 (17 i)[m]
	32 (30 i)[n]		91.2 (122 i)[l]	59.1 (6 i)[o]

[a]*Barzen et al., Chapter 14, this volume, April–September.*
[b]*Novakowski, 1966. No supporting data are provided. This is the only reference that makes any attempt to measure habitat use by nesting cranes in the AWBP.*
[c]*Miller and Barzen, 2016. March–October.*
[d]*Pickens et al., 2017. Summer (May–July), includes five nesting pairs.*
[e]*International Crane Foundation, WI, unpublished data. From Briggsville, Wisconsin. Nonterritorial cranes that were in their second, third, and fourth years. A bird entered its second year on January 1 of the year following its hatch. Data collected April–September.*
[f]*Barzen et al., Chapter 14, this volume, June–July.*
[g]*H. Thompson, unpublished data, International Crane Foundation, WI. During migration 60 cranes made 140 stops with five or more locations per stop, in eight states of Iowa, Indiana, Illinois, Kentucky, Tennessee, Arkansas, Alabama, and Georgia.*
[h]*Howe, 1989. Seven birds were marked with radios but 27 cranes were associated with marked birds. Howe split estimates of habitat use by family groups and non-family groups where foraging, non-family groups were found in wetlands 30% of the time and foraging family groups were found in wetlands 67% of the time.*
[i]*Thompson and Lacy, 2016. Includes 30 bird-stops on 19 migrations through Wisconsin, Indiana, Kentucky, Tennessee, and Georgia made by six birds in spring and fall that lasted more than 3 days.*
[j]*Fondow, 2013. Data from two winters combined in Florida.*
[k]*Pickens et al., 2017. Winter months (November–January).*
[l]*Chavez-Ramirez, 1996. Results not from telemetry but from aerial survey for most of the winter population. Estimates of habitat use provided separately for two winters. Number of individuals estimated by multiplying total population size for that year by proportion of population detected.*
[m]*D. Aborn, unpublished data. Hiwassee State Wildlife Area, Tennessee.*
[n]*Thompson, 2018. These data include 37 winter records made by 30 independent individuals having at least 30 locations each among six states of Indiana, Kentucky, Tennessee, Alabama, Georgia, and Florida.*
[o]*Thompson and Lacy, 2016. Includes nine winter records for six independent birds within Tennessee, Indiana, Georgia, and Florida. Winter records are based on a minimum of 30 locations per winter record.*

lack of population growth in the EMP (Converse et al., Chapter 7, this volume). Upland habitat use could be higher in the EMP because black fly disturbance in wetlands harassed cranes enough for them to seek upland areas, away from black flies, but within their territory (Urbanek et al., 2010b). Upland habitat use, however, did not diminish when black flies were experimentally reduced and nest success improved (J. Barzen, unpublished data, Private Lands Conservation LLC, WI; Barzen et al., Chapter 14, this volume). Foraging in upland agricultural fields, mostly corn (*Zea mays*) fields and cranberry (*Vaccinium macrocarpon*) levies, during spring (Van Schmidt et al., 2014) may meet energetic needs during a period of net negative energy consumption (Fitzpatrick, 2016) when wetland food availability is low. Agricultural areas were unavailable to

AWBP cranes at Wood Buffalo National Park. The influence of habitat use and condition on chick survival has not yet been examined.

Though 75% of all locations for EMP cranes were found in wetlands, it is possible that wetland use by territorial pairs might still be too low for Whooping Cranes to nest successfully. In Iowa, where the highest density of historical breeding records occurred (Allen, 1956), J.W. Preston described a Whooping Crane nest in 1893 (Bent, 1926, p. 221). Paraphrasing, the nest was located in the middle of a shallow wetland dominated by emergent vegetation and located approximately 1.6 km from any shore, making the wetland approximately 8 km², large enough to contain a Whooping Crane home range in its entirety, potentially allowing for 100% wetland use. Wetlands used by Whooping Cranes in Wisconsin were large enough to contain more than one Whooping Crane territory, so use of upland areas by cranes did not appear to be constrained by wetland size. It is possible, however, that habitat imprinting formed behavioral patterns of habitat selection that caused costume-reared cranes to use both wetlands and uplands as adults (Maguire, 2008). If the habitat use that costume-reared cranes learned was inappropriate, then the higher proportion of upland use in Wisconsin could be problematic. Upland use by Whooping Cranes in the GLP appeared related to high chick mortality, but this situation was complicated by Whooping Crane chicks being raised by Sandhill Cranes (Ellis et al., 1992), a situation not found in Wisconsin.

Though all four habitat characteristics identified by Austin et al. (Austin et al., Chapter 3, this volume), open topography, high density of wetlands, reliable water conditions, and highly productive wetlands, are thought to exist at a regional scale in summer areas of WBNP, EMP, FNMP, and LNMP (Barzen et al., Chapter 14, this volume; Dellinger, Chapter 9, this volume; King et al., Chapter 22, this volume; Kuyt, 1993), it is not clear if these four characteristics occur within each home range of individual pairs

for each study area. Of the 37 CNAs at WBNP, home ranges were thought to contain all the key characteristics (Kuyt, 1993, 1995). Each EMP territory met most of the four criteria for nesting Whooping Cranes except that wetland productivity may be insufficient within the Central Sand Plains (Baldassarre, 1978; Kadlec, 1962; Keys et al., 1995). Lower productivity of wetlands in territories could depress chick provisioning rates of easily digestible protein that is necessary to maintain rapid growth (Wellington et al., 1996). Dragonfly naiads were the primary foods fed to Whooping Crane chicks for the first few days after hatch at WBNP (Bergeson et al., 2001). Low wetland productivity in the Central Sands Plains can be mitigated through conducting drawdowns (Mitsch and Gosselink, 2000; Murkin et al., 2000) but, if poorly timed, drawdowns might eliminate reliable water levels for molting cranes (see "Molt" section). Alternatively, releases in east-central Wisconsin might allow cranes to select wetlands that have similar vegetation and open water characteristics to those found at Necedah National Wildlife Refuge (NWR) but are located on soils that have higher productivity, have no large populations of ornithophilic black flies, and may have different types and numbers of predators (Van Schmidt et al., 2014). Since few nest territories have yet been established in east-central Wisconsin, results of releases there are inconclusive (Barzen et al., Chapter 14, this volume).

The value of maintaining reliable water conditions for Whooping Cranes is perhaps demonstrated by nesting records of Whooping Cranes in Florida. There, drought conditions (i.e., low winter water elevation in ponds) were linked to production of smaller eggs, lower hatch success, and lower fledgling success in the following spring (Spalding et al., 2009). In Florida, only 4 out of 10 years produced water levels in the winter that were high enough to allow good subsequent reproduction (Spalding et al., 2009). In Louisiana, only a few territories have yet occurred but habitat composition found within nesting territories appeared to be mostly wetlands (S. King, unpublished data, USGS Louisiana Cooperative Fish and Wildlife Research Unit).

Of course, the paucity of habitat use data from historical studies means that other important habitat characteristics may remain unknown. The ability of black flies to harass Whooping Cranes and limit hatching appears important but was unreported until recently (Urbanek et al., 2010b). Predators or disease may also be important habitat components since, once black flies were experimentally reduced, chick survival was still unsustainable even though hatch success improved (Barzen et al., 2018a; Converse et al., Chapter 8, this volume). Aspects of habitat quality at this level of detail have not been studied at WBNP, but Kuyt et al. (1992) believed that drought conditions rendered Whooping Crane chicks more susceptible to predation. Interactions among predators, habitat, and chick survival have only begun to be studied in the EMP (Converse et al., Chapter 8, this volume) and may be of particular importance for the overall success of the species.

Molt

Whooping Cranes have a poorly understood, synchronous remigial molt (Allen, 1952; Barzen et al., Chapter 14, this volume; Folk et al., 2008; Gee and Russman, 1996; Urbanek and Lewis, 2015), where adults become flightless for 6–7 weeks while new remiges grow (Folk et al., 2008). Further, unlike the annual molt of contour feathers, the synchronous remigial molt occurs once every 2–3 years (Folk et al., 2008). Habitat use appears to change dramatically when flightless individuals, presumably more vulnerable to predation while incapable of flight, restrict their home range and remain mostly in wetlands (Barzen et al., Chapter 14, this volume). Folk et al. (2008) described the eight individuals studied in the FNMP as becoming more secretive during the flightless molt and concentrating habitat use in dense emergent vegetation and large marshes. Wetland use comprised 92% of Whooping Crane locations during the flightless

molt in the EMP (Barzen et al., Chapter 14, this volume) as compared to 75% wetland use by territorial cranes that were not replacing remiges. Further, home range for two molting pairs in the EMP was reduced from 3.75 km² and 15.31 km², respectively, in the two months prior to molting to 0.52 km² and 0.42 km² during the molt (Barzen et al., Chapter 14, this volume).

High wetland productivity (criterion 4) and reliable wetland conditions (criterion 3; Austin et al., Chapter 3, this volume), when located within summer territories, can often depend upon opposing environmental conditions, which might necessitate high wetland diversity (criterion 2). Wetland productivity depends upon nutrients that become available either through water level drawdowns that allow aerobic respiration or through flooding from rivers that provide nutrient input (Mitsch and Gosselink, 2000). Reliable areas of inundated marsh that provide water depth amenable for wading, however, can be reduced by either drawdown or flood. The importance of providing both opposing habitat characteristics simultaneously through having a high wetland diversity within territories becomes more understandable when the flightless molt is considered. Though molting cranes have not been studied at WBNP, summer habitat use during mid-summer, when molt would likely occur, suggests that the habitats necessary to maintain high chick and adult survival include diverse wetlands that contain appropriate water depths throughout a variety of climatic conditions. It is not clear if this diversity of habitat characteristics is sufficiently available to reintroduced cranes. Still, mortality during the remigial molt does not appear disproportionate to the time span over which molt occurs (Barzen et al., Chapter 14, this volume). Importantly, the need for flooded areas of emergent vegetation during molt may impose restrictions for long-lived, highly territorial individuals that prevent easy adaptation to human-altered habitats such as occurs for Whooping Cranes during migration and winter, at least in the EMP (see "Migration" section).

Winter Period

For migratory Whooping Cranes, winter habitats are located thousands (AWBP) or hundreds (EMP) of kilometers distant from summer areas. Conversely, winter habitats are often located in or near summer territories for nonmigratory populations even though nonmigratory cranes were not strongly territorial in winter, if at all (S. King, unpublished data, USGS Louisiana Cooperative Fish and Wildlife Research Unit; T. Dellinger, unpublished data, Florida Fish and Wildlife Conservation Commission). Even between the two migratory populations, distinct differences in winter spacing and habitat use behavior occur. Adult cranes that defend territories in summer at WBNP also defend territories in winter at Aransas NWR (Allen, 1952; Bonds, 2000; Stehn and Johnson, 1987; Stehn and Prieto, 2010; Stevenson and Griffith, 1946) whereas no territorial behavior has been documented in the EMP during winter (Fondow, 2013; H. Thompson, unpublished data, International Crane Foundation, WI; Urbanek et al., 2014a). Here I describe winter spacing and habitat use behavior for each population and then discuss differences in how cranes use winter habitats to inform future needs in a changing world.

Winter Spacing and Habitat Use Behavior

Most of our knowledge regarding Whooping Crane use of winter areas arises from the study of migratory populations. The remnant AWBP has been studied on winter grounds for over 70 years (Smith et al., Chapter 13, this volume), while wintering cranes in the EMP have been studied since reintroduction efforts began in 2001 (Barzen et al., Chapter 14, this volume).

REMNANT POPULATION

For Whooping Cranes in winter, intraspecific behaviors, such as spacing between territorial and nonterritorial cranes, as well as habitat use, illustrate what cranes need from their environment and how habitat management of winter habitats might directly improve survival or, indirectly, reproduction (Allen, 1952; Chavez-Ramirez, 1996; Gil de Weir, 2006; Stevenson and Griffith, 1946). Habitats and locations where Whooping Cranes forage have been the focus of most research in Aransas NWR (Smith et al., Chapter 13, this volume). Few studies, however, followed marked individuals, used telemetry to gain unbiased habitat use samples, or measured habitat availability. Still, data related to use of winter habitat by Whooping Cranes of the AWBP are rich, even if incomplete.

Along the Texas coast, territorial cranes tended to be older individuals that had established breeding territories in summer areas (Stehn and Prieto, 2010), while nonterritorial cranes tended to be young individuals that had not yet paired or attempted breeding (Bishop, 1984; Bishop and Blankinship, 1982; Stehn, 1997). Nonterritorial cranes of the AWBP appeared to utilize much of the winter habitats that were not occupied by territorial adults (Bishop, 1984; Bishop and Blankinship, 1982).

Estimates of winter home range size for territorial cranes in the AWBP were half that of summer territories at WBNP or in Wisconsin (Table 15.1). Qualitative estimates suggested that home ranges of nonterritorial cranes were considerably larger than conspecific territorial cranes in winter (Bishop and Blankinship, 1982). Territories of family groups or pairs ranged from an average of 3.88 km² on San Jose Island to an average of 1.09 km² on Aransas NWR during winter 2006/2007. Estimates of territory size using only marked individuals (Bonds, 2000) were 53% smaller on average than territory sizes reported by Stehn and Prieto (2010) for the same winter. Though differences in the two territory estimates were thought due to different methodologies used (T. Stehn, unpublished data, Aransas Pass, TX), this assertion has not been tested.

Many factors likely influence territory size. For example, territories located on the end of a linear array of territories tended to be larger than territories located in the middle of that string

(Stehn and Prieto, 2010). From 1950 to 2006, territory size in wintering Whooping Cranes of the AWBP also declined significantly as the population grew (Stehn and Prieto, 2010). The inverse relationship between population density and territory size fit predictions of territorial behavior proposed by Brown (1969). Territory size will tend to be larger where populations have not yet filled all high-quality habitat because territory size is not constrained by conspecifics (i.e., Level 1). As population densities grow, some individuals will occupy suboptimal habitats as primary habitats become saturated (Level 2). Stehn and Prieto (2010) described winter crane densities that likely occurred between Levels 1 and 2. Except in years where food abundance was low, there was little evidence that significant numbers of floating pairs or family groups were unable to find territories because both optimal and suboptimal territories were occupied (Level 3; Brown, 1969; Stehn and Prieto, 2010). Though not cited, both Stehn and Johnson (1987), as well as Stehn and Prieto (2010) used Brown's reasoning to estimate winter carrying capacity along the Texas coast. As yet, population density of the AWBP in winter does not appear to be resource limited (Butler et al., 2013).

Habitat quality is another factor influencing territorial behavior. Home ranges of color-banded Whooping Cranes utilizing winter territories consisted primarily of open water and emergent vegetation in salt marsh habitats (50–90%; Bonds, 2000). Similar conclusions, but without the use of marked individuals, were reached by other researchers (Allen, 1952; Chavez-Ramirez, 1996; Greer, 2010; Labuda and Butts, 1979; Stevenson and Griffith, 1946), with >89% use of wetlands in winter reported by Chavez-Ramirez (1996; Table 15.3). None of these research efforts, however, used telemetry to sample habitat use in an unbiased manner. Lack of telemetry data might result in undersampling habitat use outside of territories when unmarked territorial and nonterritorial cranes, each group having different habitat needs, mix.

In addition, wetland use varied directly among years with variation in food abundance (Chavez-Ramirez, 1996; Greer, 2010). In years where the abundance of blue crabs, a primary food, for example, was extremely low, Whooping Crane survival was lower (Pugesek et al., 2013). Butler et al., (2014) also found lower winter crane survival as drought severity increased, and Gil de Weir (2006) calculated a positive relationship between freshwater inflow and Whooping Crane survival, presumably because freshwater inflow is positively related to blue crab abundance in salt marsh habitats (Longley, 1994).

Even without precise telemetry data, habitat use by cranes outside of territories was still linked with Whooping Crane territory locations. Winter territories that contained primary food sources like blue crabs, and were occupied by cranes for many decades, were located near freshwater sources for drinking or alternate food sources (e.g., acorns) even though these additional resources were not part of any territory (Smith et al., Chapter 13, this volume). Freshwater habitats were not economically defensible (Brown, 1964) and cranes did not defend them from other cranes though a hierarchy among individuals within crane groups mutually using freshwater habitats was observed (Stehn and Prieto, 2010). Discontinuous habitat use, where cranes flew daily from territories to upland areas to obtain fresh water, was likely required of most territorial cranes (T. Stehn, unpublished data, Aransas Pass, TX). Discontinuous habitat use by some territorial, breeding Whooping Cranes in the EMP has also been described, so this behavior in winter was not an anomaly (Barzen et al., Chapter 14, this volume; Van Schmidt et al., 2014).

Use of recently burned areas created other foraging opportunities that some (though not all) territorial cranes utilized on a discontinuous basis (Bishop, 1984; Chavez-Ramirez, 1996; Chavez-Ramirez et al., 1996; Westwood and Chavez-Ramirez, 2005). Whooping Cranes were found more frequently in upland areas when drought conditions were severe (Butler et al., 2014). The purpose for these movements is unknown but was likely important for augmenting food acquisition when food availability within a territory was limited. Discontinuous movements could also have been for social or other non-food-related reasons. Though the degree to which social behavior may influence habitat use might vary with crane density, social behavior should not oscillate dramatically between years as might food availability. Habitat use outside of territories, therefore, suggests that the abundance and distribution of foods may sometimes change crane foraging patterns enough to alter the degree to which territoriality is expressed (Allen, 1952).

Winter food abundance was linked to mortality and winter territories were organized around available foods that varied annually in abundance. In contrast, other resources, such as open habitat structure that may provide safety from predation, did not vary annually and were therefore of secondary importance for influencing territory size (Chavez-Ramirez, 1996). As compared to summer territories, smaller territories in winter suggested that there is greater behavioral flexibility in meeting crane needs given that nest and chick-rearing habitats are not required. Both Chavez-Ramirez (1996) and Greer (2010), however, argued that cranes were opportunistic foragers but they did not specify at what spatial scale. Opportunistic habitat use behavior between home ranges by nonterritorial Whooping Cranes in winter likely did occur. Food abundance and distribution in home ranges of territorial birds must be clumped and economically defendable, both of which mean they are predictable to the individual (Brown, 1969), so cranes would not likely be opportunistically foraging between territories except to the extent that discontinuous habitat use occurred. Further, the reuse of winter territory locations by individual pairs from one year to the next (Stehn, 1997) suggests that food abundance is largely predictable and adequate even if substantial variation in food occurs among years (Allen, 1952; Chavez-Ramirez, 1996;

Pugesek et al., 2013). Given repeated, annual winter territory use by long-lived cranes, it is possible that long-term productivity of a territory is as important as habitat quality within any single year. A similar situation is known from Sandhill Cranes where maintaining a breeding territory over the long term is the best measure of lifetime reproductive success (Hayes, 2015). A countervailing condition is that the cost of attempting to maintain a territory in winter, even when food conditions are poor, can be lower survival (Pugesek et al., 2013).

REINTRODUCED POPULATIONS

With the importance of winter territorial behavior in the AWBP, why does similar behavior not occur in reintroduced populations even when summer territorial behavior occurs in both populations? First, winter wetland use in the AWBP is high compared to the EMP and the LNMP (Table 15.3). In contrast, daily use of wetlands by EMP cranes was similar to use of wetlands by sympatric EP Sandhill Cranes in winter (Table 15.3). Territorial behavior in EP Sandhill Cranes has not been observed in winter (D. Aborn, unpublished data, University of Tennessee, Chattanooga; S. Nesbitt, unpublished data, Gainesville, FL).

Compared to the distinct, small size of winter territories in the AWBP, winter home ranges of Whooping Cranes from reintroduced populations were large (Table 15.1) and overlapped (Thompson, 2018). Primary winter habitats in the EMP used by foraging cranes included upland agricultural fields (Barzen et al., Chapter 14, this volume), while rice and crawfish fields (both wetland-like habitats usually) were used more than expected based on availability in the LNMP (Pickens et al., 2017). In addition, overall winter use of wetland habitats occurred approximately 61% of the time in the LNMP (Pickens et al., 2017). Food in agricultural fields is typically distributed uniformly over a large area, and food availability can be altered throughout the winter by foraging flocks, reducing the likelihood of territorial behavior forming because these foods are not economically defendable in time or space (Brown, 1964).

Regionally, winter habitat use changed dramatically in the EMP over 13 winters. During the first few winters home ranges of cranes were small and restricted to Florida (Table 15.1). Cranes used wetlands 52% of the time (Table 15.3). Subsequently, the winter distribution of EMP Whooping Cranes expanded north (Teitelbaum et al., 2016; Urbanek et al., 2014a) and changed rapidly because cranes learned from conspecifics in the population (Mueller et al., 2013; Teitelbaum et al., 2016). Concomitant with northward expansion of winter areas, wetland use declined to 32% (Table 15.3). Interestingly, the shift in winter distribution over a decade is similar to a shift northward over several decades made by the sympatric EP Sandhill Cranes that winter in the same region (Lacy et al., 2015). Over similar time periods, however, no shifts have occurred with the winter distribution of the AWBP or of the LNMP. Novel agricultural habitats used in winter by the EMP and LNMP still approximated key habitat characteristics identified by Austin et al. (Austin et al., Chapter 3, this volume), though modified anthropogenically; agricultural fields provided open habitat, predictability, and high energy foods while reservoirs, man-made ponds, remnant wetlands, or agricultural wetlands provided shallow, open waters for night roosting, as well as high-energy foods, at least in some cases (Foley, 2015).

Ecological implications of the novel winter distribution in the EMP, or novel habitat use by the EMP and LNMP, are still unclear. Whooping Cranes in the EMP return to breeding areas in a pattern, adjusted for different nesting latitudes, that is similar to cranes in the AWBP (Fitzpatrick et al., Chapter 12, this volume). Whooping Cranes in both populations arrive on breeding grounds when it is still cold and habitats are thawing. Initial nest attempts begin before food becomes readily available, but hatching coincides with broad food abundance as spring advances. The role of energetics in poor hatch success, however, has not been ruled out (Converse et al., Chapter 8, this volume). Compared to Whooping Cranes, sympatric Sandhill Cranes utilize a capital breeding system (a reliance on stored body reserves for breeding; Krapu et al., 1985) and have similar winter distributions (Lacy et al., 2015) but show no sign of being unable to acquire energy reserves in winter habitats even though their population has increased dramatically over the last 60 years (Lacy et al., 2015). Survival of EMP adults during winter is also similar to AWBP cranes (Servanty et al., 2014). Though no direct evidence suggests that the extensive winter distribution and associated habitat use in the EMP is problematic, the EMP does not yet have sustainable reproduction, so problems with winter use of novel habitats cannot yet be ruled out. Further exploration of novel winter habitat use is warranted. Reintroduction efforts in the LNMP are too early yet to judge if use of novel habitats will result in sustainable populations.

Changes in Winter Crane Behavior and Future Winter Habitat Management

Spacing behavior of cranes can be used to predict future challenges with winter distribution for the AWBP. In winter, nonterritorial subadult or adult cranes consisted of 29–55% of the overall winter population in the nine winters studied during 1950–2006 (Stehn and Prieto, 2010), similar to estimates for the size of the nonbreeding flock at WBNP during 1968–79 (37.2–60%; Kuyt, 1981b), so no demographic evidence suggests that AWBP cranes are having difficulty establishing winter territories, the condition that defines Level 3 of territoriality. In addition, behavioral evidence suggests that all known floating pairs eventually established territories (Stehn and Johnson, 1987). Stehn and Prieto (2010) estimated that 160 potential territories, of any quality, remain unfilled at Aransas NWR while the Texas Coast might possibly provide habitat for an additional 320 territories.

Their hypothesis fits the theoretical framework that, until Level 3 is reached for the coastal population of wintering cranes, expansion of the winter range will predominantly occur within the Texas coast rather than outside of it.

The management challenge related to Stehn and Prieto's (2010) carrying capacity estimate is that, when major changes in habitat conditions happen rapidly (e.g., hurricane, drought, or chemical spill), exploratory behavior by cranes may be incapable of changing rapidly enough to avoid increased mortality rates associated with reduced habitat quality (even if temporarily). This perhaps happened in winter 2008 when a projected 23 Whooping Cranes died (a 60-year high) during a winter with abnormally low blue crab populations (Stehn and Haralson-Strobel, 2014). Alternatively, Butler et al. (2014) argued that cranes did not die but were undetected because they utilized new, unmonitored areas. With either case, mortality or movement, cranes responded to a change in habitat quality and this change may have been directly (i.e., starvation) or indirectly (i.e., higher predation rates when using new winter habitats) related to survival.

The degree of crane exploration away from territories in relation to low food abundance has been well documented (Bishop, 1984; Butler et al., 2014; Chavez-Ramirez, 1996; Greer, 2010; Harrell and Bidwell, 2013) and additional exploration did not always lead to increased mortality. Exploratory flights represented shifts between salt marsh habitats and upland burned areas (Chavez-Ramirez et al., 1996) or shifts between currently used salt marsh habitats and novel salt marsh habitats (Harrell and Bidwell, 2014). Some exploration outside the Texas coast occurred in the dry winter of 2012/2013 when six adults and two accompanying juveniles utilized Granger Lake, a reservoir located 280 km northwest of Aransas NWR (Harrell and Bidwell, 2013). Habitat use at Granger Lake was, however, temporary, lasting only one winter (Harrell and Bidwell, 2014; W. Harrell, unpublished data, Aransas National Wildlife Refuge, TX).

In contrast to extensive winter territorial behavior by cranes of the AWBP, territorial behavior in the EMP may never have begun because original winter distributions did not occur in regions where food distribution (Fondow, 2013) allowed territoriality to develop. In addition, winter distributions in the EMP changed rapidly as individual cranes learned from each other how to exploit broadly distributed foods (predominantly waste corn) on extensive agricultural fields (Teitelbaum et al., 2016). To the degree that the winter distribution in the EMP is beneficial (this is currently unclear), it is important to determine if lessons from crane distribution during winter for the EMP can be applied to the AWBP. Once suitable, alternate winter habitats are identified by cranes, dramatic shifts in population abundance can occur within a few years (Urbanek et al., 2014a) through social transferal of regional habitat use behavior (Mueller et al., 2013).

Response by the AWBP winter flock to longer-term change in habitat condition, such as mangrove expansion (Chavez-Ramirez and Wehtje, 2012; Smith et al., Chapter 13, this volume), is more difficult to predict. Climate models predict increasing drought and rising sea levels for the Texas coast, and these changes, if climate change predictions are accurate, might lead to a decline in blue crab production and salt marsh area, lowering the current carrying capacity of the Texas coast for Whooping Cranes (Chavez-Ramirez and Wehtje, 2012; Smith et al., 2014). Determining to what degree Whooping Cranes in the AWBP will utilize new winter habitats that are located away from the Texas coast therefore becomes critical. Before conclusions can be made, however, current unknowns must still be resolved: Is winter habitat use of the EMP truly beneficial? Do appropriate alternate winter habitats exist outside of the Texas coast? Can artificial management of winter foods replace or augment natural salt marsh habitats in the AWBP (Shields and Benham, 1969)?

In the LNMP, long-term environmental changes may be less of a concern. Though coastal deltaic marshes in Louisiana (i.e., those formed directly by the Mississippi River) are eroding rapidly due to subsidence and sea-level rise (Barras et al., 2003), subsidence rates and wetland loss are much lower in the coastal marshes of the Chenier Plain where Whooping Cranes are being reintroduced (King et al., Chapter 22, this volume). The effects of sea-level rise on coastal marshes of the Chenier Plain are unknown and will depend upon the rate of sea-level rise relative to mineral and organic accretion (S. King, unpublished data, USGS Louisiana Cooperative Fish and Wildlife Research Unit; Kirwan et al. 2016).

Migration Period

Fall and spring migrations occur during 17% of the annual cycle of Whooping Cranes but safe passage between disjunct summer and winter regions is critical to population viability. Melvin and Temple (1982) described dynamic migration behavior for Sandhill Cranes as encompassing staging areas, traditional stopover areas, and nontraditional stopover areas. Staging areas occurred within the first 20% of the migration path (within one day's flight) where cranes remained for several weeks. Traditional stopover areas occurred in mid-migration where individuals stopped for days to weeks, and nontraditional stopovers tended to be opportunistic pauses in migration for cranes to rest or wait until adverse weather conditions changed. What cranes do when they pause during migration determines habitat use behavior, such that staging and traditional stopover habitats must provide abundant, high-energy foods for fat acquisition, along with safety and fresh water, while habitats used by migrants during nontraditional stopovers may only provide habitats safe from predators and contain fresh water. With minor modification, this system of describing migration dynamics is used here.

AWBP

Allen (1952) reported extensive use of the Big Bend stretch of the Platte River (i.e., the central Platte) as a traditional stopover area in both spring and fall. He also identified the northern portion of the Sand Hills of Nebraska as being used by fall migrants while the nearby Niobrara River was used by spring migrants. Rainwater Basin wetlands south of the central Platte were also used, mostly by spring migrants. Current use of the central Platte has been evaluated by many and the region remains classified as critical habitat for Whooping Cranes (National Research Council, 2005). Austin and Richert (2005) also documented the continued use of the Rainwater Basins and Sandhills of Nebraska by migrating Whooping Cranes.

Johnson and Temple (1980) compiled the first contemporary summary of Whooping Crane habitats used during migration since Allen (1952). They described migration habitats as being 100% wetlands (palustrine wetlands, inland salt marsh, reservoir, stock pond, and shallow river) used for night roosting and 70% of foraging sites being used as agricultural, similar to what was found through following telemetered non-family groups of Whooping Cranes (Howe, 1989; Table 15.3). Foraging areas were typically within 1–2 km of night roosts but could occur within 22–27 km (Johnson and Temple, 1980). Austin and Richert (2005) summarized information from >1,000 site-evaluation records collected 1977–99 to describe habitat use patterns by Whooping Cranes in the United States. Similar to earlier descriptions, most night roost sites occurred in palustrine or riverine wetlands and daytime feeding sites were predominantly in upland areas (78%), mostly in agricultural crops. Use of riverine roosts occurred mainly in Nebraska (Platte, Loup, and other rivers). Two studies found regional differences in habitat use among social groups, where families tended to use palustrine wetlands in the Sand Hills of Nebraska while non-family groups tended to use riverine wetlands in the Rainwater

Basin or along the central Platte (Austin and Richert, 2005; Howe, 1989).

Fall migration occurred in three phases: (1) a 2–3-day flight from WBNP to southern Saskatchewan, (2) staging in Saskatchewan for 1–5 weeks, and (3) rapid flight to the Texas Gulf coast in 7–10 days using nontraditional stopping areas (Kuyt, 1992). The pause in Saskatchewan was likely needed to acquire fat reserves for migration or for a hedge against sudden inclement weather. Family groups or breeding pairs made the return spring migration in 10–11 days without any significant stops (Kuyt, 1992). Subsequent data suggested that some Whooping Cranes used the Platte River as traditional stopovers in spring (National Research Council, 2005). Besides the wetlands used in southern Saskatchewan staging areas (Johns et al., 1997), Whooping Cranes chose nontraditional stopover wetlands randomly along their fall or spring migration route (Howe, 1989), selecting among wetlands located toward the end of daily migratory flights (Moore and Aborn, 2000). All cranes roosted at night in wetlands, but during the day 70% of areas used by non-family groups were upland agriculture, while family groups used agricultural habitats only 33% of the time (Howe, 1989). Using Whooping Crane sightings rather than telemetry, Johns et al. (1997) found that wetlands in Saskatchewan were used exclusively for night roosts while upland agricultural fields were predominantly used for diurnal foraging in spring (100%) and fall (85%). Regionally, Allen's description of migration habitats focused mainly on areas in Nebraska, while current data suggests that Saskatchewan appears to contain the most frequent stopping locations, with 68% of the fall and 44% of the spring migration records being located there (Johns, 1992; Johns et al., 1997). Shifts from staging in Nebraska to staging in Saskatchewan, to the extent that this shift is biologically meaningful and not due to sampling differences between the two studies, suggests that cranes have some flexibility regarding where they establish staging areas.

EMP

Migration behavior by Whooping Cranes in the EMP resulted in completing the main fall or spring passage in 3–5 days (H. Thompson, unpublished data, International Crane Foundation, WI), approximately half the time spent in migration by the AWBP. Staging areas for the EMP, if they occurred at all, were adjacent to summer areas (Urbanek et al., 2014a). Migration behavior of the EMP was similar to sympatric EP Sandhill Cranes (Thompson and Lacy, 2016). The distance between summer and winter areas in the EMP was as little as 600 km (H. Thompson, unpublished data, International Crane Foundation, WI) as opposed to 3,800 km between WBNP and Aransas NWR, so extensive use of traditional stopovers to acquire fat for long, extended migrations may not be as important to cranes in the EMP.

Habitat use in the EMP was similar to that of the AWBP in that agricultural fields were used extensively during the day (H. Thompson, unpublished data, International Crane Foundation, WI). Diurnal wetland use for Whooping Crane migrants occurred less frequently than at any other time in the annual cycle and was similar to use by EP Sandhill Crane migrants. Nocturnal wetland use, however, was as high in the EMP (H. Thompson, unpublished data, International Crane Foundation, WI) as it was in the AWBP (Austin and Richert, 2005; Howe, 1989; Johns et al., 1997) and in EP Sandhill Cranes (Thompson and Lacy, 2016).

Emerging Threats and Opportunities for Future Habitat Use

A large portion of Whooping Crane summer habitat remains currently unused at WBNP (Olson + Olson Planning Design Consultants, 2003) and impacts from climate change on summer habitat conditions may not be significant for Whooping Cranes (Chavez-Ramirez and Wehtje, 2012). The future of important crane habitats in winter or staging areas for the AWBP, however, is less certain. Long-term tension between agricultural and conservation uses of wetlands (Dahl, 1990) is expected to continue and will be exasperated by human population growth and climate change. New pressures from energy development (primarily construction of wind turbines) are arising on migration areas as well but may be avoidable through proper planning and ecological input (Belaire et al., 2013). Development pressures on winter areas are extensive (Smith et al., 2014) and future impacts due to climate change may be substantial (Chavez-Ramirez and Wehtje, 2012; Smith et al., Chapter 13, this volume). What might happen to Whooping Cranes if conflicts in land use in migration or winter habitats are unresolvable and the quality of habitats declines significantly?

Chavez-Ramirez (1996) estimated that net energy balance in Whooping Cranes varied greatly between winters. If cranes leave winter habitats with lower energy reserves (such as in winter 1993/1994) than obtained in normal winters, they might stop longer during spring migration to accumulate more fat. Alternative responses would be to over-summer on the Texas coast or delay winter departure until adequate fat reserves have been acquired (Chavez-Ramirez, 1996). Over-summering cranes at Aransas NWR are rare (Stehn and Prieto, 2010) and would prevent reproduction, leaving only extension of time spent using winter or migration habitats as options for cranes attempting to nest during the subsequent summer.

In fall, Whooping Cranes in the AWBP already have a tradition of leaving their breeding grounds and then staging in Saskatchewan where high-energy foods (wheat, *Triticum aestivum*) are acquired. Howe (1989) quotes Kuyt as saying that, in wet years at WBNP, individual Whooping Cranes that stay longer in the park shorten their staging period in Saskatchewan. Kuyt's observation suggests that endogenous reserves that are acquired for fall migration can be acquired on either summer or fall habitats. In the variable environments where Whooping Cranes live, such plasticity seems warranted, especially

where large, unpredictable expanses separate breeding and winter grounds as occurs in the AWBP. It might be possible, then, that similar behavioral flexibility can occur during the transition from winter to spring as well. Acquiring fat on spring staging or traditional stopover areas already occurs with Sandhill Cranes in the Central Flyway along the central Platte (Krapu et al., 1984, 2011). Subadult Whooping Cranes also stop in Saskatchewan for longer periods during spring than do breeding cranes (Johns et al., 1997; Kuyt, 1992), so this behavioral flexibility is possible, but no tradition, as yet, exists for breeding Whooping Crane adults.

Linking resources acquired on winter areas with subsequent reproduction on summer areas has been examined in Whooping Cranes through energetics studies (Chavez-Ramirez, 1996) and correlative studies that compare winter habitat conditions and subsequent summer reproductive rates (Gil de Weir, 2006). Butler et al. (2014), however, found no relationship between recruitment (proportion of hatch-year cranes in the entire population) and a drought severity index of the preceding winter. Gil de Weir's (2006) model, however, included a more complex array of variables for winter grounds (freshwater inflow, temperature, and net evaporation) than found in the Palmer Hydrological Drought Severity Index (used by Butler et al., 2014) alone. The capital breeding model has been described for Sandhill Cranes (Krapu et al., 1985). This model follows the basic model of capital breeding waterbirds where energy is acquired on winter or staging areas and is then used to fuel migration, egg formation, and perhaps incubation, as well as survival (Alisauskas and Ankney, 1992; Stephens et al., 2009). In years of poor winter habitat condition Siberian Cranes (*Leucogeranus leucogeranus*), for example, had lower reproduction in the subsequent breeding season (Burnham et al., 2017).

The primary cost of extending spring migration after cranes encounter poor winter habitat conditions is a shortening of time available to fledge young on breeding areas in the taiga's short summer. Compared to the EMP (Urbanek et al., 2014b), cranes in the AWBP have little time available for renesting in most years (Kuyt, 1981a), so the cost of delaying nest initiation could be substantial for the AWBP unless the length of the nesting season expanded with warming temperatures. An important data gap regarding this hypothesis is to describe what individuals do during spring migration with known habitat quality conditions during winter in either the EMP or the AWBP. Experimentation with artificial feeding could test if cranes might become territorial (in the EMP), improve body condition (in both the EMP and AWBP), or alter departure dates on winter areas. Some flexibility in where reproductive energy reserves are acquired has been described in other capital breeding species, such as Canvasback (*Aythya valisineria*; Barzen, 1989; Barzen and Serie, 1990).

Habitat use in Whooping Cranes contrasts a high degree of flexibility seen among individual Whooping Cranes with substantial habitat or behavioral specificity seen between populations. Understanding this contrast allows a clearer interpretation of the interactions between the crane (through behavior, genetics, and population density) and habitat (through intrinsic and extrinsic mechanisms). Specifically, this means that, where populations occur at carrying capacity, if winter habitats decline in quality or quantity, an increase in use of migration habitats is to be expected or demographic rates will change. Linkage of winter and migration habitats is, therefore, critical to informing future management actions necessary to safeguard Whooping Cranes in the future.

SUMMARY AND OUTLOOK

Though Whooping Cranes historically nested in ecosystems as varied as taiga, upper tallgrass prairie, and coastal marsh, they also appeared to need specific habitat components located within each of these biomes. In summer, territorial cranes had small territories and used wetlands extensively during both day and night. In contrast, nonterritorial cranes had overlapping, large home ranges, and utilized wetlands to a lesser extent. Compared to all populations of Whooping Cranes, quantified wetland use was most extensive during the flightless molt and home ranges were the smallest size measured anywhere. What is less clear is how productive wetlands used by cranes need to be and to what extent wetlands need to retain water within relatively small Whooping Crane territories, especially in years when the flightless molt occurs. In the LNMP the role of wetlands appears to be important to nesting cranes, but, for nests that occur in agricultural wetlands, water permanence is not solely a function of crane management. Research to address these knowledge gaps is critical. Further, wetlands or wetland complexes that exceed 4 km² (the average territory size across populations) and have the high productivity required by Whooping Cranes are uncommon. For reintroductions to succeed, wetland restoration and management actions in proposed nesting areas will likely be needed.

In winter Whooping Crane territories are smaller than in summer and appear based primarily on the abundance of food. In the high-quality habitats of Aransas NWR, territories are stable in all but years of lowest food abundance. Extensive winter territoriality appears to have led AWBP cranes to develop a strong tradition that is absent in other Whooping Crane populations. In contrast, EMP cranes feed extensively in agricultural areas during winter and were not territorial. No population appears to have saturated available winter habitat. Research needs for cranes in winter should examine if use of agricultural fields in the EMP allows for sustainable survival and reproduction, as well as whether similar habitats could fulfill energy and nutritional needs of cranes in the AWBP. For the AWBP, it is also important to determine if appropriate winter habitats exist outside of the Texas coast. Conversely, it would also be important to better understand the role of territorial quality in shaping winter crane distributions and to determine how quickly such behavior can be altered if food abundance is low in any one winter or if habitat quality declines on a longer-term basis.

Even though migration occurs during a brief portion of the annual cycle, habitats used by migrating cranes provide the critical bridge between summer and winter habitats and are therefore equally important to overall population viability. Habitat use for daytime foraging appears to be the most flexible of any period of the annual cycle for both migratory populations. Dependence on agricultural areas for food, though viable at present, assumes that agricultural practices will not change in a manner that would be detrimental to cranes. This is a difficult assumption, at least in some cases (e.g., Krapu et al., 2004), and should be examined more closely. As important, habitats used by staging cranes may provide some flexibility for cranes to adjust to problems encountered in summer or winter areas, but this hypothesis needs to be tested.

Acknowledgments

This chapter has benefited from the exchange of data and ideas by many individuals: Mark Bidwell and Brian Johns (retired), Canadian Wildlife Service; Hillary Thompson, Clemson University; Felipe Chavez-Ramirez, Gulf Coast Bird Observatory; Tim Dellinger, Ryan Druyor, Andrew Hayslip, and Steve Nesbitt (retired), Florida Fish and Wildlife Conservation Commission; Betsy Didrickson, Andrew Gossens, Anne Lacy, Dorn Moore, Karis Ritenour, and Elizabeth Smith, International Crane Foundation; Phillip Vasseur, Louisiana Department of Wildlife and Fisheries; Bret Collier, Louisiana State University School of Renewable Natural Resources; David Aborn, University of Tennessee-Chattanooga; Matt Hayes, University of Wisconsin-Madison; Sammy King and Aaron Pearse, U.S. Geological Survey; and Wade Harrell, Kris Metzger, and Tom Stehn (retired), U.S. Fish and Wildlife Service. Editorial assistance and comments on the manuscript were provided by Lynda Garrett (U.S. Geological Survey), Pat Gonzalez (Elsevier), Sammy King, Julie Langenberg (International Crane Foundation), Tom Stehn, and an anonymous reviewer.

References

Alisauskas, R.T., Ankney, C.D., 1992. The cost of egg laying and its relationship to nutrient reserves in waterfowl. In: Batt, B.D.J., Afton, A.D., Anderson, M.G., Ankney, C.D., Johnson, D.H., Kadlec, J.A., Krapu, G.L. (Eds.), Ecology and Management of Breeding Waterfowl. University of Minnesota Press, Minneapolis, pp. 30–61.

Allen, R.P., 1952. The Whooping Crane. Research Report 3. National Audubon Society, New York, 246 pp.

Allen, R.P., 1956. A Report on the Whooping Crane's Northern Breeding Grounds: A Supplement to Research Report Number 3. National Audubon Society, New York, 31 pp.

Anderson, M.G., Rhymer, J.M., Rohwer, F.C., 1992. Philopatry, dispersal, and the genetic structure of waterfowl populations. In: Batt, B.D.J., Afton, A.D., Anderson, M.G., Ankney, C.D., Johnson, D.H., Kadlec, J.A., Krapu, G.L. (Eds.), Ecology and Management of Breeding Waterfowl. University of Minnesota Press, Minneapolis, pp. 365–395.

Anderson, M.G., Titman, R.D., 1992. Spacing patterns. In: Batt, B.D.J., Afton, A.D., Anderson, M.G., Ankney, C.D., Johnson, D.H., Kadlec, J.A., Krapu, G.L. (Eds.), Ecology and Management of Breeding Waterfowl. University of Minnesota Press, Minneapolis, pp. 251–289.

Armbruster, M.J. 1987. Habitat Suitability Index Models: Greater Sandhill Crane. U.S. Fish and Wildlife Service. Biological Report 82(10.140), 26 pp.

Austin, J.E., Hayes, M.A., Barzen, J.A., 2018. Revisiting the historic distribution and habitats of the whooping crane (Chapter 3). In: French, Jr., J.B., Converse, S.J., Austin, J.E. (Eds.), Whooping Cranes: Biology and Conservation. Biodiversity of the World: Conservation from Genes to Landscapes. Academic Press, San Diego, CA.

Austin, J.E., Richert, A.L., 2005. Patterns of habitat use by whooping cranes during migration: summary from 1977–1999 site evaluation data. Proceedings of the North American Crane Workshop 9, 79–104.

Baldassarre, G.A., 1978. Ecological Factors Affecting Waterfowl Production on Three Man-Made Flowages in Central Wisconsin. MS thesis, University of Wisconsin, Stevens Point, 124 pp.

Barras, J., Beville, S., Britsch, D., Hartley, S., Hawes, S., Johnston, J., Kemp, P., Kinler, Q., Martucci, A., Porthouse, J., Reed, D., Roy, K., Sapkota, S., Suhayda, J., 2003. Historical and projected coastal Louisiana land changes: 1978–2050. United States Geological Survey Open File Report 03-334, 39 pp.

Barzen, J.A., 1989. Patterns of nutrient acquisition in canvasbacks during spring migration. MS thesis, University of Grand Forks, ND, 74 pp.

Barzen, J.A., Serie, J., 1990. Nutrient reserve dynamics of breeding canvasbacks. Auk 107 (1), 75–85.

Barzen, J.A., Converse, S.J., Adler, P.H., Lacy, A., Gracy, E., Gossens, A., 2018a. Examination of multiple working hypotheses to address reproductive failure in reintroduced whopping cranes. Condor 120, 632–649.

Barzen, J.A., Lacy, A., Thompson, H.L., Gossens, A.P., 2018b. Habitat use by the reintroduced Eastern Migratory Population of whooping cranes (Chapter 14). In: French, Jr., J.B., Converse, S.J., Austin, J.E. (Eds.), Whooping Cranes: Biology and Conservation. Biodiversity of the World: Conservation from Genes to Landscapes. Academic Press, San Diego, CA.

Barzen, J.A., Su, L., Lacy, A.E., Gossens, A.P., Moore, D.M., 2016. High nest density for sandhill cranes of central Wisconsin. Proceedings of the North American Crane Workshop 13, 13–24.

Belaire, J.A., Kreakie, B.J., Keitt, T., Minor, E., 2013. Predicting and mapping potential whooping crane stopover habitat to guide site selection for wind energy projects. Conserv. Biol. 28 (2), 541–550.

Bent, A.C., 1926. Life Histories of North American Marsh Birds. United States National Museum Bulletin, Government Printing Office, Washington, DC, p. 135.

Bergeson, D.G., Bradley, M., Holroyd, G.L., 2001. Food items and feeding rates for wild whooping crane colts in Wood Buffalo National Park. Proceedings of the North American Crane Workshop 8, 36–39.

Bishop, M.A., 1984. The dynamics of sub-adult flocks of whooping cranes wintering in Texas, 1978–1979 through 1982–1983. MS thesis, Texas A&M University, College Station, 127 pp.

Bishop, M.A., Blankinship, D.R., 1982. Dynamics of sub-adult flocks of whooping cranes at Aransas National Wildlife Refuge, Texas, 1978–1981. In: Lewis, J.C. (Ed.), Proceedings of the 1981 Crane Workshop. National Audubon Society, Tavernier, FL, pp. 180–189.

Bonds, C.J., 2000. Characterization of banded whooping crane winter territories from 1992–93 to 1996–97 using GIS and remote sensing. PhD thesis, Texas A&M University, College Station, 106 pp.

Boyle, S.A., Lourenco, W.C., da Silva, L.R., Smith, A.T., 2009. Home range estimates vary with sample size and methods. Folia Primatol. 80 (1), 33–42.

Brown, J.L., 1964. The evolution of diversity in avian territorial systems. Wilson Bull. 76 (2), 160–169.

Brown, J.L., 1969. Territorial behavior and population regulation in birds: a review and re-evaluation. Wilson Bull. 81 (3), 293–329.

Brown, J.L., Orians, G.H., 1970. Spacing patterns in mobile animals. Annu. Rev. Ecol. Syst. 1, 239–262.

Burnham, J., Barzen, J., Pidgeon, A.M., Sun, B., Wu, J., Liu, G., Jiang, H., 2017. Novel foraging by wintering Siberian cranes (*Leucogeranus leucogeranus*) at China's Poyang Lake indicates broader changes in the ecosystem and raises new challenges for a critically endangered species. Bird Conserv. Int. 27, 204–223.

Butler, M.J., Harris, G., Strobel, B.N., 2013. Influence of whooping crane population dynamics on its recovery and management. Biol. Conserv. 162, 89–99.

Butler, M.J., Metzger, K.L., Harris, G., 2014. Whooping crane demographic responses to winter drought focus conservation strategies. Biol. Conserv. 179, 72–85.

Carpenter, J.R., 1940. The grassland biome. Ecol. Monogr. 10 (4), 617–684.

Chavez-Ramirez, F., 1996. Food Availability, Foraging Ecology and Energetics of Whooping Cranes Wintering in Texas. PhD dissertation, Texas A&M University, College Station, 104 pp.

Chavez-Ramirez, F., Hunt, H.E., Slack, R.D., Stehn, T.V., 1996. Ecological correlates of whooping crane use of fire-treated upland habitats. Conserv. Biol. 10 (1), 217–223.

Chavez-Ramirez, F., Wehtje, W., 2012. Potential impact of climate change scenarios on whooping crane life history. Wetlands 32 (1), 11–20.

Cody, M.L., 1985. Habitat Selection in Birds. Academic Press, New York, 558 pp.

Converse, S.J., Moore, C.T., Folk, M.J., Runge, M.C., 2013. A matter of tradeoffs: reintroduction as a multiple objective decision. J. Wildl. Manage. 77 (6), 1145–1156.

Converse, S.J., Servanty, S., Moore, C.T., Runge, M.C., 2018. Population dynamics of reintroduced whooping cranes (Chapter 7). In: French, Jr., J.B., Converse, S.J., Austin, J.E. (Eds.), Whooping Cranes: Biology and Conservation. Biodiversity of the World: Conservation from Genes to Landscapes. Academic Press, San Diego, CA.

Converse, S.J., Strobel, B.N., Barzen, J.A., 2018. Reproductive failure in the Eastern Migratory Population: interaction of research and management (Chapter 8). In: French, Jr., J.B., Converse, S.J., Austin, J.E. (Eds.), Whooping Cranes: Biology and Conservation. Biodiversity of the World: Conservation from Genes to Landscapes. Academic Press, San Diego, CA.

Dahl, T.E., 1990. Wetlands losses in the United States, 1780's to 1980's. National Wetlands Inventory, St. Petersburg, FL, Report to the Congress PB-91-169284/XAB.

Dellinger, T.A., 2018. Florida's nonmigratory whooping cranes (Chapter 9). In: French, Jr., J.B., Converse, S.J., Austin, J.E. (Eds.), Whooping Cranes: Biology and Conservation. Biodiversity of the World: Conservation from Genes to Landscapes. Academic Press, San Diego, CA.

Ellis, D.H., Lewis, J.C., Gee, G.F., Smith, P.G., 1992. Population recovery of the whooping crane with emphasis on reintroduction efforts: past and future. Proceedings North American Crane Workshop 6, 142–150.

Fitzpatrick, M.J., 2016. Mechanistic models and tests of whooping crane energetics and behavior locally and at landscape scales: implications for food requirements, migration, conservation strategies and other bird species. PhD dissertation, University of Wisconsin-Madison, 255 pp.

Fitzpatrick, M.J., Mathewson, P.D., Porter, W.P., 2018. Ecological energetics in whooping cranes in the Eastern Migratory Population (Chapter 12). In: French, Jr., J.B., Converse, S.J., Austin, J.E. (Eds.), Whooping Cranes: Biology and Conservation. Biodiversity of the World: Conservation from Genes to Landscapes. Academic Press, San Diego, CA.

Foley, C.C., 2015. Wading bird food availability in rice fields and crawfish ponds of the Chenier Plain of southwest

Louisiana and southeast Texas. MS thesis, Louisiana State University, Baton Rouge, 76 pp.

Folk, M.J., Nesbitt, S.A., Parker, J.M., Spalding, M.G., Baynes, S.B., Candelora, K.L., 2008. Feather molt of nonmigratory whooping cranes in Florida. Proceedings of the North American Crane Workshop 10, 128–132.

Folk, M.J., Nesbitt, S.A., Schwikert, S.T., Schmidt, J.A., Sullivan, K.A., Miller, T.J., Baynes, S.B., Parker, J.M., 2005. Breeding biology of re-introduced non-migratory whooping cranes in Florida. Proceedings of the North American Crane Workshop 9, 105–109.

Fondow, L.E.A., 2013. Habitat selection of reintroduced migratory whooping cranes (Grus americana) on their wintering range. MS thesis, University of Wisconsin, Madison, 122 pp.

Fuller, R.J., 2012a. The bird and its habitat: an overview of concepts. In: Fuller, R.J. (Ed.), Birds and Habitat: Relationships in Changing Landscapes. Cambridge University Press, Cambridge, UK, pp. 3–36.

Fuller, R.J., 2012b. Habitat quality and habitat occupancy by birds in variable environments. In: Fuller, R.J. (Ed.), Birds and Habitat: Relationships in Changing Landscapes. Cambridge University Press, Cambridge, UK, pp. 37–62.

Garshelis, D.L., 2000. Decisions in habitat evaluation: measuring use, selection and importance. In: Boitani, L., Fuller, T.K. (Eds.), Research Techniques in Animal Ecology: Controversies and Consequences. Columbia University Press, New York, pp. 111–164.

Gee, G.F., Russman, S.E., 1996. Reproductive physiology. In: Ellis, D.H., Gee, G.F., Mirande, C.M. (Eds.), Cranes: Their Biology, Husbandry, and Conservation. National Biological Service, Washington, DC, and International Crane Foundation, Baraboo, WI, pp. 123–136.

Gil de Weir, K., 2006. Whooping crane (Grus americana) demography and environmental factors in a population growth simulation model. PhD dissertation, Texas A&M University, College Station, 159 pp.

Greenwood, P.J., 1980. Mating systems, philopatry and dispersal in birds and mammals. Anim. Behav. 28 (4), 1140–1162.

Greer, D.M., 2010. Blue crab population ecology and use by foraging whooping cranes on the Texas Gulf Coast. Texas, PhD dissertation, A&M University, College Station, 288 pp.

Harrell, W., Bidwell, M., 2013. Report on Whooping Crane Recovery Activities (2012 Breeding Season–2013 Spring Migration). Unpublished Report to Whooping Crane Recovery Team, 90 pp.

Harrell, W., Bidwell, M., 2014. Report on Whooping Crane Recovery Activities (2013 Breeding Season–2014 Spring Migration). Unpublished Report to Whooping Crane Recovery Team, 78 pp.

Hayes, M.A., 2015. Dispersal and population genetic structure in two flyways of sandhill cranes (Grus canadensis). PhD dissertation, University of Wisconsin, Madison, 287 pp.

Hayes, M.A., Barzen, J.A., 2016. Dispersal patterns and pairing behaviors of non-territorial sandhill crane. Passenger Pigeon 78 (4), 411–425.

Horne, J.S., Garton, E.O., Krone, S.M., Lewis, J.S., 2007. Analyzing animal movements using Brownian bridges. Ecology 88 (9), 2354–2363.

Howe, M.A., 1989. Migration of radio-marked whooping cranes from the Aransas-Wood Buffalo Population: patterns of habitat use, behavior, and survival. Fish and Wildlife Technical Report 21, U.S. Fish and Wildlife Service, Department of the Interior, 39 pp.

Hutto, R.L., 1985. Habitat selection by non-breeding migratory land birds. In: Cody, M.L. (Ed.), Habitat Selection in Birds. Academic Press, New York, pp. 455–476.

Johns, B.W., 1992. Preliminary identification of whooping crane staging areas in prairie Canada. In: Wood, D.A. (Ed.), Proceedings of the 1988 North American Crane Workshop, State of Florida Game and Freshwater Fish Commission Nongame Wildlife Program Technical Report 12, 61–66.

Johns, B.W., Goossen P., Kuyt E., and Craig-Moore, L., 2005. Philopatry and dispersal in whooping cranes (Grus americana). Proceedings of the North American Crane Workshop 9, 117–125

Johns, B.W., Woodsworth, E.J., and Driver, E.A. 1997. Habitat use by migrant whooping cranes in Saskatchewan. Proceedings of the North American Crane Workshop 7, 123–131.

Johnson, D.H., 1980. The comparison of usage and availability measurements for evaluating resource preference. Ecology 61 (1), 65–71.

Johnson, K.A., Temple, S.A., 1980. The Migratory Ecology of the Whooping Crane (Grus americana). Unpublished Report, Prepared for U.S. Fish and Wildlife Service by University of Wisconsin, Madison, 89 pp.

Kadlec, J.A., 1962. Effects of a drawdown on a waterfowl impoundment. Ecology 43 (2), 267–281.

Keys, Jr., J., Carpenter, C., Hooks, S., Koenig, F., McNab, W.H., Russell, W., Smith, M.L., 1995. Ecological units of the eastern United States – first approximation (CD-ROM). U.S. Department of Agriculture, Forest Service, Atlanta, GA, GIS coverage in ARCINFO format, selected imagery, and map unit tables.

King, R.S., Trutwin, J.J., Hunter, T.S., Varner, D.M., 2013. Effects of environmental stressors on nest success of introduced birds. J. Wildl. Manage. 77 (4), 842–854.

King, S.L, Selman, W., Vasseur, P., Zimorski, S.E., 2018. Louisiana nonmigratory whooping crane reintroduction (Chapter 22). In: French, Jr., J.B., Converse, S.J., Austin, J.E (Eds.), Whooping Cranes: Biology and Conservation. Biodiversity of the World: Conservation from Genes to Landscapes. Academic Press, San Diego, CA.

Kirwan, M.L., Temmerman, S., Skeehan, E.E., Guntenspergen, G.R., Fagherazzi, S., 2016. Overestimation of marsh vulnerability to sea level rise. Nat. Clim. Change 6, 253–260.

Krapu, G.L., Facey, D.E., Fritzell, E.K., Johnson, D.H., 1984. Habitat use by migrant sandhill cranes in Nebraska. J. Wildl. Manage. 48 (2), 407–417.

Krapu, G., Iverson, G., Reinecke, K., Boise, C., 1985. Fat deposition and usage by arctic-nesting sandhill cranes during spring. Auk 102 (2), 362–368.

Krapu, G.L., Brandt, D.A., Cox, Jr., R.R., 2004. Less waste corn, more land in soybeans, and the switch to genetically modified crops: trends with important implications for wildlife management. Wildl. Soc. Bull. 32 (1), 127–136.

Krapu, G.L., Brandt, D.A., Jones, K.L., Johnson, D.H., 2011. Geographic distribution of the mid-continent population of sandhill cranes and related management applications. Wildl. Monogr. 175, 1–38.

Kuyt, E., 1979. Banding of juvenile whooping cranes and discovery of the summer habitat used by nonbreeders. In: Lewis, J.C. (Ed.), Proceedings of the 1978 North American Crane Workshop. Colorado State University Printing Service, Fort Collins, pp. 109–111.

Kuyt, E., 1981a. Clutch size, hatching success and survival of whooping crane chicks, Wood Buffalo National Park, Canada. In: Lewis, J.C., Masatomi, H. (Eds.), Crane Research around the World. International Crane Foundation, Baraboo, WI, pp. 126–129.

Kuyt, E., 1981b. Population status, nest site fidelity, and breeding habitat of whooping cranes. In: Lewis, J.C., Masatomi, H. (Eds.), Crane Research around the World. International Crane Foundation, Baraboo, WI, pp. 119–125.

Kuyt, E. 1992. Aerial tracking of whooping cranes migrating between Wood Buffalo National Park and Aransas National Wildlife Refuge, 1981–1984. Canadian Wildlife Service, Occasional Paper 74, 53 pp.

Kuyt, E., 1993. Whooping crane, *Grus americana*, home range and breeding range expansion in Wood Buffalo National Park, 1970–1991. Can. Field Nat. 107 (1), 1–12.

Kuyt, E., 1995. The nest and eggs of the whooping crane, *Grus americana*. Can. Field Nat. 109 (1), 1–5.

Kuyt, E., Johns, B.W., Barry, S.J., 1992. Below average whooping crane production in Wood Buffalo National Park during drought years 1990 and 1991. Blue Jay 50 (4), 225–229.

Labuda, S.E., Butts, K.O., 1979. Habitat use by wintering whooping cranes on the Aransas National Wildlife Refuge. In: Lewis, J.C. (Ed.), Proceedings of the 1978 North American Crane Workshop. Colorado State University Printing Service, Fort Collins, pp. 151–157.

Lacy, A.E., Barzen, J.A., Moore, D.M., Norris, K.E., 2015. Changes in the number and distribution of greater sandhill cranes in the eastern population. J. Field Ornithol. 86 (4), 317–325.

Longley, W.L. (Ed.), 1994. Freshwater Inflows to Texas Bays and Estuaries: Ecological Relationships and Methods for Determination of Needs. Texas Water Development Board and Texas Parks and Wildlife Department, Austin.

Maguire, K.J., 2008. Habitat selection of reintroduced whooping cranes, *Grus americana*, on their breeding range. MS thesis, University of Wisconsin, Madison, 54 pp.

Melvin, S.M., Temple, S.A., 1982. Migration ecology of sandhill cranes: a review. In: Lewis, J.C. (Ed.), Proceedings of the 1981 North American Crane Workshop. National Audubon Society, Tavernier, FL, pp. 73–87.

Miller, T.P., Barzen, J.A., 2016. Habitat selection by breeding sandhill cranes in central Wisconsin. Proceedings of the North American Crane Workshop 13, 1–12.

Mitsch, W.J., Gosselink, J.G., 2000. Wetlands, third ed. John Wiley & Sons, New York.

Moore, F.R., Aborn, D.A., 2000. Mechanisms of *en route* habitat selection: how do migrants make habitat decisions during stopover? Stud. Avian Biol. 20, 34–42.

Mueller, T., O'Hara, R.B., Converse, S.J., Urbanek, R.P., Fagan, W.F., 2013. Social learning of migratory performance. Science 341 (6149), 999–1002.

Murkin, H.R., van der Valk, A.G., Clark, W.R., 2000. Prairie Wetland Ecology: The Contribution of the Marsh Ecology Research Program. Iowa State University Press, Ames, 413 pp.

National Research Council, 2005. Endangered and Threatened Species of the Platte River. National Research Council of the National Academy of Sciences, Washington, DC, 336 pp.

Nesbitt, S.A., Schwikert, S., Folk, M.J., 2002. Natal dispersal in Florida sandhill cranes. J. Wildl. Manage. 66 (2), 349–352.

Novakowski, N.S., 1966. Whooping Crane Population Dynamics on the Nesting Grounds, Wood Buffalo National Park, Northwest Territories, Canada. Canadian Wildlife Service Report Series 1, Ottawa, Canada, 20 pp.

Olson + Olson Planning Design Consultants, 2003. Final Report: Whooping Crane Potential Habitat Mapping Project. Environment Canada and Canadian Wildlife Service, Ottawa.

Pickens, B.A., King, S.L., Vasseur, P., Zimorski, S.E., Selman, W., 2017. Seasonal movements and multiscale habitat selection of whooping crane (*Grus americana*) in natural and agricultural wetlands. Waterbirds 40 (4), 322–333.

Pugesek, B.H., Baldwin, M.J., Stehn, T., 2013. The relationship of blue crab abundance to winter mortality of whooping cranes. Wilson J. Ornithol. 125 (3), 658–661.

Servanty, S., Converse, S.J., Bailey, L.L., 2014. Demography of a reintroduced population: moving toward management models for an endangered species, the whooping crane. Ecol. Appl. 24 (5), 927–937.

Shields, R.H., Benham, E.L., 1969. Farm crops as food supplements for whooping cranes. J. Wildl. Manage. 33 (4), 811–817.

Smith, E.H., Chavez-Ramirez, F., Lumb, L., Gibeaut, J., 2014. Employing the Conservation Design Approach on Sea-Level Rise Impacts on Coastal Avian Habitats along the Central Texas Coast. Final Report Submitted to: Gulf Coast Prairies Landscape Conservation Cooperative, 131 pp.

Smith, E.H., Chavez-Ramirez, F., Lumb, L., 2018. Wintering habitat ecology, use, and availability for the Aransas-Wood Buffalo Population of whooping cranes (Chapter 13). In: French, Jr., J.B., Converse, S.J., Austin, J.E. (Eds.), Whooping Cranes: Biology and Conservation. Biodiversity of the World: Conservation from Genes to Landscapes. Academic Press, San Diego, CA.

Spalding, M.G., Folk, M.J., Nesbitt, S.A., Folk, M.L., Kiltie, R., 2009. Environmental correlates of reproductive success for introduced resident whooping cranes in Florida. Waterbirds 32 (4), 538–547.

Stehn, T.V., 1992. Re-pairing of whooping cranes at Aransas National Wildlife Refuge. In: Wood, J.D. (Ed.), Proceedings of the 1988 Crane Workshop. Florida Game and Freshwater Commission, Nongame Wildlife Program Technical Report 12, 185–188.

Stehn, T.V., 1997. Pair formation by color-marked whooping cranes on the wintering grounds. Proceedings of the North American Crane Workshop 7, 24–28.

Stehn, T.V., Haralson-Strobel, C.L., 2014. An update on mortality of fledged whooping cranes in the Aransas/Wood Buffalo Population. Proceedings of the North American Crane Workshop 12, 43–50.

Stehn, T., Johnson, E.F., 1987. Distribution of winter territories of whooping cranes on the Texas coast. In: Lewis, J.C. (Ed.), Proceedings of the 1985 North American Crane Workshop. Platte River Whooping Crane Maintenance Trust, Grand Island, NE, pp. 180–195.

Stehn, T.V., Prieto, F., 2010. Changes in winter whooping crane territories and range 1950–2006. Proceedings of the North American Crane Workshop 11, 40–56.

Stephens, P.A., Boyd, I.L., McNamara, J.M., Houston, A.I., 2009. Capital breeding and income breeding: their meaning, measurement, and worth. Ecology 90 (8), 2057–2067.

Stevenson, J.O., Griffith, R.E., 1946. Winter life of the whooping crane. Condor 48, 160–178.

Teitelbaum, C.S., Converse, S.J., Fagan, W.F., Bohning-Gaese, K., O'Hara, R.B., Lacy, A.E., Mueller, T., 2016. Experience drives innovation of new migration patterns of whooping cranes in response to global change. Nat. Commun. 7. doi:10.1038/ncomms12793, Article 12793.

Thompson, H., 2018. Characteristics of whooping crane home ranges during the non-breeding season in the Eastern Migratory Population. MS thesis, Clemson University, Clemson, SC.

Thompson, H., Lacy, A., 2016. Winter and migratory habitat use of six eastern greater sandhill cranes. Proceedings of the North American Crane Workshop 13, 47–53 pp.

Timoney, K., 1999. The habitat of nesting whooping cranes. Biol. Conserv. 89 (2), 189–197.

Timoney, K., Zoltai, S.C., Goldsborough, L.G., 1997. Boreal diatom ponds: a rare wetland associated with nesting whooping cranes. Wetlands 17 (4), 539–551.

Tiner, Jr., R.W., 1984. Wetlands of the United States: Current Status and Recent Trends. National Wetlands Inventory, U.S. Fish and Wildlife Service, Department of the Interior, Washington, DC, 76 pp.

Urbanek, R.P., Fondow, L.E.A., Zimorski, S.E., 2010a. Survival, reproduction, and movements of migratory whooping cranes during the first seven years of reintroduction. Proceedings of the North American Crane Workshop 11, 124–132.

Urbanek, R.P., Lewis, J.C., 2015. Whooping crane (Grus americana). In: Rodewald, P.G. (Ed.), The Birds of North America Online. Cornell Lab of Ornithology, Ithaca, NY. Retrieved from the Birds of North America. Available from: http://bna.birds.cornell.edu/bna/species/153. [2 August 2018] doi:10.2173/bna.153.

Urbanek, R.P., Szyszkoski, E.K., Zimorski, S.E., 2014a. Winter distribution dynamics and implications to a reintroduced population of migratory whooping cranes. J. Fish Wildl. Manage. 5 (2), 340–362.

Urbanek, R.P., Zimorski, S.E., Fasoli, A.M., Szyszkoski, E.K., 2010b. Nest desertion in a reintroduced population of migratory whooping cranes. Proceedings of the North American Workshop 11, 133–141.

Urbanek, R.P., Zimorski, S.E., Szyszkoski, E.K., Wellington, M.M., 2014b. Ten-year status of the eastern migratory whooping crane reintroduction. Proceedings of the North American Crane Workshop 12, 33–42.

Van Schmidt, N.D., Barzen, J.A., Engels, M.J., Lacy, A.E., 2014. Refining reintroduction of whooping cranes with habitat use and suitability analysis. J. Wildl. Manage. 78 (8), 1404–1414.

Wellington, M., Burke, A., Nicolich, J.M., O'Malley, K., 1996. Chick rearing. In: Ellis, D.H., Gee, G.F., Mirande, C.M. (Eds.), Cranes: Their Biology, Husbandry, and Conservation. Department of the Interior, National Biological Service, Washington, DC, and International Crane Foundation, Baraboo, WI, pp. 77–95.

Westwood, C.M., Chavez-Ramirez, F., 2005. Patterns of food use of wintering whooping cranes on the Texas coast. Proceedings of the North American Crane Workshop 9, 133–140.

CAPTIVE BREEDING AND WHOOPING CRANE HEALTH

Whooping Crane chick with puppet

Advances in Conservation Breeding and Management of Whooping Cranes (*Grus americana*)

Sandra R. Black*, Kelly D. Swan**

*Calgary Zoological Society, Calgary, AB, Canada
**Centre for Conservation Research, Calgary Zoological Society, Calgary, AB, Canada

INTRODUCTION

From the first Whooping Crane egg collections in Wood Buffalo National Park in 1967 to the release of the species to a fourth reintroduction site (Louisiana) in 2011 (French et al., Chapter 1, this volume; King et al., Chapter 22, this volume), the captive population of Whooping Cranes has served as an insurance population, a source for reintroduction efforts, and fertile ground for research. Captive Whooping Cranes and the reintroduction of their offspring to the wild have also captured the interest and imagination of the general public (Mooallem, 2009; Roberge, 2014). However, 2017 marked 50 years of investment in captive breeding and 41 years of investment in reintroduction, yet the Whooping Crane remained Endangered on the International Union for Conservation of Nature (IUCN) Red List (IUCN, 2015) and in the United States and

Canada (Canadian Wildlife Service and U.S. Fish and Wildlife Service, 2005), and no selfsustaining reintroduced populations had been established (Converse et al., Chapter 8, this volume). Given that conservation breeding and translocation programs are on the rise (Brichieri-Colombi and Moehrenschlager, 2016; Collar and Butchart, 2014; Swan et al., 2016), and effective management of these – frequently expensive – programs will be increasingly necessary to meet conservation goals, it is crucial that the lessons learned from long-term breeding programs are communicated to program managers. Here, we present a synthesis of the major techniques utilized and advancements made over the course of the Whooping Crane captive breeding program. We identify ongoing challenges and areas for future research, with a focus on genetic management, optimizing the production of birds for release, and maintaining a controlled yet naturalistic captive environment.

Whooping Cranes: Biology and Conservation
http://dx.doi.org/10.1016/B978-0-12-803555-9.00016-5

Overview of the Captive Population

After the near extinction of the Whooping Crane in the early 1940s, there was a push to develop husbandry and propagation techniques in the interest of creating an insurance population of Whooping Cranes in captivity (Canadian Wildlife Service and U.S. Fish and Wildlife Service, 2005). Beginning in 1947, the last known bird from the nonmigratory population in Louisiana was paired sequentially with several injured birds originating from the Aransas-Wood Buffalo Population (AWBP). The intent was to form a breeding pair and produce offspring, thereby preserving some genetic representation of the Louisiana population (Barrett and Stehn, 2010). Although Whooping Cranes had been held in captivity previously in both North America and Europe (Allen, 1952), this was the first focused effort to breed the species and protect its genetic diversity for the future. Although these earliest pairs did successfully lay eggs and produce the first captive-bred chick in 1950, none of the offspring survived to reproduce and the genetic material of the Louisiana lineage was lost (Barrett and Stehn, 2010). The subsequent development of the captive flock – including early advocates, egg collections from Wood Buffalo National Park, establishment of the initial program at Patuxent Wildlife Research Center (PWRC) in Laurel, Maryland, in the mid-1960s, early successes and challenges, and the division of the flock among the five extant breeding centers – has been well documented elsewhere (e.g., Barrett and Stehn, 2010; Doughty, 1989; Kuyt, 1996).

The International Recovery Plan for Whooping Cranes (Canadian Wildlife Service and U.S. Fish and Wildlife Service, 2005) specifies that a genetically stable captive population of Whooping Cranes (at least 153 individuals, including 50 breeding pairs by 2010) should be maintained to ensure against the extinction of the species and to provide individuals for release into reintroduced populations. As of December 2016, the captive population included 153 individual birds and 56 breeding pairs housed at five breeding institutions (PWRC; International Crane Foundation, Baraboo, WI; Calgary Zoo, Calgary AB, Canada; Audubon Nature Institute, New Orleans, LA; San Antonio Zoo, San Antonio, TX), with an additional 10 nonbreeding birds held at other zoological institutions for public education and display (Lowry Park Zoo, Tampa, FL; White Oak Conservation, Yulee, FL; Homasassa Springs Wildlife State Park, Homasassa, FL; Stone Zoo, Stoneham, MA; Smithsonian National Zoo, Washington, DC; Jacksonville Zoo, Jacksonville, FL; Milwaukee County Zoo, Milwaukee, WI; Sylvan Heights Bird Park, Scotland Neck, NC). All Whooping Cranes are under the joint stewardship of Canada and the United States through a Memorandum of Understanding between the Canadian Wildlife Service and the U.S. Fish and Wildlife Service as the lead agencies for each nation (Barrett and Stehn, 2010).

GENETIC MANAGEMENT

Maintaining Genetic Diversity

The global population of Whooping Cranes dropped to just 22 individuals in 1942 (Canadian Wildlife Service and U.S. Fish and Wildlife Service, 2005), and the severity of this genetic bottleneck is evident in the species' low allelic diversity and low heterozygosity in the major histocompatibility complex (relative to the closely related and vastly more numerous Sandhill Crane, *Grus canadensis*; Jarvi et al., 2001). Haplotype diversity of mitochondrial DNA was also reduced by two-thirds following the bottleneck (Glenn et al., 1999). Careful management of the captive population of Whooping Cranes is essential to retain the remaining genetic diversity, prevent genetic drift, and maximize production of birds for release to the wild.

To date, the genetic management of captive Whooping Cranes has largely depended on known and assumed founder relationships, based on knowledge of the nests and wild pairs from which the flock derived. This information, kept in a historic studbook, has served as the primary source for ranking and pairing birds according to their relative genetic diversity or value. However, over the past two decades, research to create and refine microsatellite DNA profiles for individual cranes, in order to better understand founder relationships, has provided new information for genetic management and pairing decisions (Jones et al., 2002, 2010). These decisions are ultimately the responsibility of the International Whooping Crane Recovery Team, which incorporates the latest research and also enlists the expertise of a captive management team consisting of a research geneticist and experienced staff from each captive center, including veterinarians, flock managers, and curators.

Management of captive populations for the purpose of reintroduction will often involve trade-offs between what is good for the captive population and what is best for the wild, reintroduced population. For instance, Mirande et al. (1996a) advocate for rapid expansion of the captive Whooping Crane population while also maintaining an equal sex ratio and equal contribution of each breeding pair to the next generation, in order to maximize genetic diversity. They further advise that the age structure of the population should be maintained for stability and inbreeding should be minimized. While these actions are important to maintain genetic diversity in captivity, the avoidance of inbreeding may have negative trade-offs for the reintroduction program overall (Converse et al., 2012). The numbers of chicks produced for reintroduction may be reduced when successful (and therefore overrepresented) pairs are held back to decrease inbreeding, despite only weak evidence that individual genotype or inbreeding has a negative effect on post-release survival of reintroduced Whooping Cranes (Converse et al., 2012) and no evidence that it influences reproductive performance in captivity (Brown et al., 2017; Smith et al., 2011).

The need for a formal program to optimize genetic management has grown with the number of Whooping Cranes in captivity. Recognizing this need, the International Whooping Crane Recovery Team reached out to the Association of Zoos and Aquariums (AZA) in early 2015, in the interest of developing a Species Survival Plan (SSP) for Whooping Cranes. SSPs were first developed by the AZA in the early 1980s, with the aim of coordinating the breeding and management of select species held by member institutions and ensuring sustainable demographics over the long term (Association of Zoos and Aquariums, 2014; Conway, 1982; Meritt, 1980). The establishment of an SSP for Whooping Cranes required the appointment of a Regional Studbook Keeper to maintain histories of all captive Whooping Cranes and an SSP Coordinator to lead and support the program. Following a master planning meeting in August 2016, a Population Analysis & Breeding and Transfer Plan for Whooping Cranes was completed (Association of Zoos and Aquariums, 2017); this plan outlines goals for the captive population, describes the demographic and genetic status of the population, and makes recommendations for suitable breeding pairs and transfers between captive facilities. Future advancement of the SSP may include refinement of the genetic information analyzed to better describe relatedness of birds in the population (K. Boardman, International Crane Foundation, personal communication).

Pairing of Birds

Behavioral compatibility of Whooping Crane pairs is important for reproductive success (Derrickson and Carpenter, 1987). Nevertheless, pairing and artificial insemination (AI) decisions have primarily been driven by the objectives of maximizing allele retention and minimizing inbreeding. All birds in the population are assigned a genetic value, and breeding efforts are focused on those individuals with the highest genetic value, that is, the lowest mean kinship (MK) values (Ballou and Lacy, 1995; Montgomery et al., 1997). The most robust method for selecting breeding pairs in long-lived animals is to use a dynamic MK selection process, where once a pair is formed, their potential offspring are added to the MK list, and the MK rankings are recalculated before new pairs are selected (Ivy and Lacy, 2012). Such an MK ranking approach was found to minimize inbreeding and slow the loss of genetic diversity relative to random mating or selection for a particular trait (Ivy and Lacy, 2012; Willoughby et al., 2015). These practices have been challenging to apply within the captive Whooping Crane population, as the MK values for individual birds vary, sometimes substantially, when measured using the historic studbook information only versus a studbook that integrates genetic information to determine relatedness of founders (Jones et al., 2002). Microsatellite information has identified some individual Whooping Cranes with a lower overall genetic value, but who carry rare alleles that may be lost if not captured in the subsequent generations; the relative MK ranking of these birds within the breeding population increased when this genetic information was incorporated.

As the complexity of managing this population of endangered Whooping Cranes increases, and given new resources available through the SSP program, other (concurrent) strategies to maintain the greatest gene diversity in the population could include mating birds of similar MK (based on either the historic studbook or a studbook integrating genetic value) to avoid mixing rare and common alleles, and producing offspring with the lowest MK values possible. When a valuable pair of birds (defined by low and well-matched MKs; Association of Zoos and Aquariums, 2017) fails to reproduce, less valuable pairings may still have merit. The potential adverse linking of rare and common alleles may be minimized when parental output

is continually managed to minimize MK in the population (Ivy and Lacy, 2012).

One possible result of focusing on genetic management rather than mate compatibility in Whooping Cranes, however, is that many pairs remain nonreproductive or produce only infertile eggs. In addition, pairs that successfully reproduce in captivity do so at a later age than their wild counterparts (Mirande et al., 1996b). Unfortunately, successfully pairing birds is not a simple process. While pairing can begin as early as the second or third year, new pairings are often ephemeral (Swengel et al., 1996). Furthermore, some cranes that have been reared together or paired at a young age may appear behaviorally compatible, but nonetheless fail to lay (Swengel et al., 1996). The reproductive success of pairs can be maximized by pairing individuals of similar age, behavior, and condition (Mirande et al., 1996a). Older birds may intimidate younger birds (when managers attempt to pair birds of dissimilar age), and stress and disturbance may interfere with pair bond formation (S. Black, personal communication). Overly aggressive birds, usually males, can pose a threat to their mates. Nelson et al. (1995) published preliminary information on early pair behaviors, such as unison calling, dancing, and strutting and other synchronous activities that may be predictive of later reproductive success. Brown et al. (2016) found that the frequency of specific behaviors (e.g., unison calling, marching, and copulation or copulation attempts) and the overall time spent performing reproductive behaviors were significantly higher in successful than unsuccessful breeding pairs.

Allowing greater choice of mates among captive Whooping Cranes may result in pairs that are more compatible and therefore more productive. At the present time, the majority of Whooping Crane pairs are matched genetically before any testing for compatibility begins. Historically, many of these genetically based pairings have laid infrequently, produced infertile eggs, or remained nonreproductive (upublished data, 1995–2016, Calgary

Zoological Society; Canadian Wildlife Service and U.S. Fish and Wildlife Service, 2005). More choice in mate selection may offer advantages to the Whooping Crane program, although there would be a cost in space and resources. One potential method, considered by the International Recovery Team, would create groups of 6–10 currently unpaired or unsuccessfully paired cranes in large enclosures, allowing new pairs to form. These potential pairs would be identified by behavior and removed before they became aggressive or otherwise adversely affected other birds through territoriality (Kepler, 1976 in White, 2000). Another proposed method involves allowing a single female to encounter two or more males, each in his own adjacent pen, through fencing. Over time, the female might display preference for one male by spending time in proximity to his pen. Further behavioral pairing techniques would be used for this self-selected pair. This method was successfully implemented with several species of cranes at both PWRC and facilities managed by the Smithsonian National Zoo (S. Derrickson, Smithsonian Conservation Biology Institute, personal communication). While the results vary greatly, research on wild avian species indicates that mate choice can increase the success of reproduction via improved quality of territory, reduced risk of disease, and increased parental effort (Moller and Jennions, 2001). There may also be effects on the genetic fitness of the offspring as measured by heterozygosity of MHc alleles and the attractiveness of male offspring (Chargé et al., 2014; Jones and Ratterman, 2009). While not all benefits accruing to free-ranging birds will be available in a captive environment, there is promise in using mate choice as a tool to improve fitness in captive populations. The goals of doing so should be clear, and outcomes should be carefully evaluated. Increasing the ability of female Whooping Cranes to select sexual partners within the constraints of responsible genetic management must be a priority research focus for the captive breeding centers in the next few years.

Artificial Insemination

Among the earliest innovations in Whooping Crane captive breeding was the use of AI to stimulate fertile egg production and, ultimately, to achieve desired genetic pairings. First employed in 1969, AI was responsible for all captive chick production until 1991, when natural breeding was first confirmed in captivity (Canadian Wildlife Service and U.S. Fish and Wildlife Service, 2005). Although semen collection and AI had long been used in the breeding of domestic birds, their application to endangered species recovery was novel. Gee and Mirande (1996) give an excellent overview of the practice of AI for crane species. Using an adapted massage technique, most males will adjust to handling for semen collection; in practice, calmer males are more effective donors, producing larger and more consistent samples. During insemination of females, deposition of semen samples directly into the female's oviduct, rather than cloaca, increases the production of fertile eggs. As with males, calmer trained females are more likely to evert their cloaca, thus facilitating this placement without potentially invasive cloacal manipulation (Gee and Mirande, 1996).

Swengel and Tuite (1997) reviewed practical techniques for the use of AI in 10 crane species, including Whooping Cranes, and determined that fertility rates increased with samples of higher sperm density and motility, and with multiple insemination events during the 4–7 days preoviposition. Jones and Nicolich (2001) found through paternity determinations that fertilizations occurred as late as 6 days post insemination, lending credence to the possibility of sperm storage in this species. Evaluation of seminal quality by Brown et al. (2017) indicated that volume and sperm concentration peaked in the midseason; seminal quality was similar across age groups. Additionally, there was no discernible effect of male inbreeding coefficient on sperm quality. This information can be used to maximize the effectiveness of AI procedures.

AI has allowed flock managers to genetically pair birds that are not behaviorally paired, and to increase fertility and fecundity in pairs that are not naturally fertile. If greater mate choice were allowed, AI might also be increased to produce desirable genetic pairings where behavioral pairs do not constitute such pairings. Cryopreservation of samples could facilitate the use of AI for genetic management, but cryopreservation technologies are still in development (see Songsasen et al., Chapter 17, this volume). Some captive pairs of Whooping Cranes do successfully breed without intervention. Nicholich et al. (2001) reviewed the breeding success of eight pairs of captive Whooping Cranes that were not manipulated for AI, but allowed to breed naturally. While fertility rates were slightly lower than those for pairs where AI was used, natural breeding for pairs with good flight capability is now standard at several of the captive centers. Furthermore, it bears consideration whether Whooping Cranes that are the most productive AI donors or recipients may have heritable behavioral profiles that are less adaptive for release to the wild, thereby producing offspring with lower chance of survival or breeding success. However, no empirical analyses of this question have been conducted.

PRODUCTION OF BIRDS FOR RELEASE

Maximizing Egg Production, Fertility, and Hatch Rates

Whooping Cranes are long-lived, are slow to reach sexual maturity, lay few eggs (usually two in a single annual clutch), and have relatively long periods of parental dependence and high adult survival (Canadian Wildlife Service and U.S. Fish and Wildlife Service, 2005). These characteristics have implications for population growth both in the wild and in captivity. In the wild, chicks hatch asynchronously, because incubation begins with the first egg (Kuyt, 1995, 1996). In many instances, the older chick kills the younger chick posthatch (siblicide); this behavior may be associated with resource competition (Bergeson et al., 2001; Boyce et al., 2005; Lewis, 2001). Additionally, one or both chicks may be lost to predation or other causes of mortality prior to fledging (Bergeson et al., 2001). Low fledge rates have been implicated in the slow intrinsic growth rates of both the AWBP and reintroduced populations (Converse et al., Chapter 8, this volume; Wilson et al., 2016). By comparison, in captivity, managers can increase egg and chick production by removing eggs from nests, thus stimulating the female to lay additional clutches (Mirande et al., 1996b). Further, protection from predators and ample food resources ensure that captive-bred chicks fledge at higher rates than those in the wild (Ellis and Gee, 2001). However, the captive population is subject to other limitations, including reduced fertility and hatch rates relative to the AWBP, the need for surrogate and artificial incubators, and behavioral challenges.

Although the survival of hatched offspring is higher in captivity, captive Whooping Cranes tend to breed later in life than their wild counterparts (Canadian Wildlife Service and U.S. Fish and Wildlife Service, 2005). Additionally, disturbances occurring in captive environments (e.g., human presence or construction activity) have been associated with reduced egg production. Further, some captive Whooping Crane pairs never produce fertile eggs and are obligate AI donors or recipients (Gee and Mirande, 1996). Brown et al. (2016) examined fecal hormone samples of captive Whooping Cranes and found that females that laid eggs have higher levels of estrogen and lower levels of progestogens relative to females that do not. Laying females also exhibited higher levels of reproductive behaviors, such as unison calling and copulation attempts. Glucocorticoid hormones are often used as a measure of behavioral stress, and did not differ significantly between the two groups. While it does not appear that stress responses differ between laying and nonlaying pairs, females that produce eggs have higher levels of gonadotropic hormones, which affect follicular development and ovulation (Brown et al., 2016). This work will allow future research to evaluate impacts of management and husbandry practices on the hypothalamic–pituitary–gonadal axis (Songsasen et al., Chapter 17, this volume).

Even when a given Whooping Crane pair successfully mates and lays eggs in captivity, those eggs may not be fertile. The fertility of captive-produced Whooping Crane eggs can vary widely between years, pairs, and breeding facilities. For example, from 1990 to 1999, mean fertility rates for naturally fertile Whooping Crane pairs at PWRC ranged from 40% to 94% (Nicholich et al., 2001). The average fertility rate across all pairs was 65%. Estimates of egg fertility in the AWBP range from 73% to 95% (Ellis and Gee, 2001; Kuyt, 1995 in Brown et al., 2016); thus there is room for improvement in captive fertility rates. AI is widely used to increase fertility of Whooping Cranes in captivity, but it can be difficult to obtain high-quality samples from all males, and fertility rates nonetheless remain below those seen in the wild.

Egg fertility is presently the greatest limiting factor in the captive production of Whooping Cranes, but the hatch rates of fertile eggs are also lower than desired. Smith et al. (2011) report that the percentages of fertile eggs hatching at PWRC, the International Crane Foundation (ICF), and Calgary Zoo (CZ) between 1990 and 2005 were 89%, 73.8%, and 57%, respectively. By comparison, the observed hatching success of eggs in the AWBP (between 1967 and 1974) was 95% (Erickson and Derrickson, 1981). Smith et al. (2011) used 15 years of captive production data to assess the impacts of several factors on hatching success of fertile eggs. These factors included facility, age of the dam, inbreeding

measures of both parents, year laid, sequence order of the egg (first laid, second laid, etc.), and the amount of time an egg was naturally incubated. Their results strongly confirmed subjective experience that the longer an egg is incubated by the crane parents or surrogates (as opposed to artificial incubators), the greater the hatching success. Given that artificial incubators can incubate far more eggs than crane pairs (and are more economical than caring for surrogate incubating birds), the question remains as to how they can be improved to achieve hatch rates comparable to that of natural incubation. Smith et al. (2012) used specially designed data logging eggs to collect information on incubation conditions (temperature, humidity, rotation, and light levels) in crane nests and in artificial incubators. Their devices showed that the temperature and humidity settings traditionally used in artificial incubators are quite different from those recorded in crane nests. This underscored the need to maximize the duration of natural incubation and to thoroughly review incubator settings and practices. Research is currently underway to identify the specific incubation conditions that predict hatch success, such that the operation (and potentially design) of artificial incubators can be improved (unpublished data, S. Converse et al., 2016, U.S. Geological Survey).

Embryonic death during incubation is a common finding in captive-laid Whooping Crane eggs and is the primary cause of decreased hatchability. A retrospective study of 44 Whooping Crane embryonic deaths identified malposition, hemorrhage, and incubation failure as the most common conditions associated with embryonic death in captivity, and determined that none of the sources of mortality appeared to be preventable (Letoutchaia et al., 2010). At times, embryonic death during incubation is the result of adverse parental behaviors, such as egg breaking. Of 15 species of captive cranes held at ICF, Puchta et al. (2008) found that Whooping Cranes had the highest percentage of broken eggs at 13%. Behavioral observations at CZ suggest that most egg destruction is deliberate and egg break-ing behavior correlates with decreased pair bond behaviors, pairs that have been produc-tive but have not raised a chick for several years, and disturbance by staff checking nests (V. Edwards, Calgary Zoological Society, per-sonal communication). Efforts to counteract this behavior include the provision of larger and more secluded pen space, decreased visits by staff where possible, and increased distance between breeding pairs. Additionally, some full-term crane chicks will simply fail to hatch for unknown reasons (e.g., Converse et al., 2010), and others may require human assistance during the hatching process.

Preparing Chicks for Reintroduction

Egg production, fertility, and hatching are just the first of several critical hurdles in the Whooping Crane conservation breeding pro-gram. Chick rearing and preparation for release comprise a substantial portion of the resources expended by ICF and PWRC, and care dur-ing these stages may have an impact on the post-release performance of Whooping Cranes. (Converse et al., Chapter 8, this volume; Hartup, Chapter 20, this volume). For instance, one con-sequence of maximizing captive egg production is that there are not enough Whooping Crane pairs available to rear all the chicks that hatch. This need has primarily been met by human caretakers in crane costumes (the costumes are intended to prevent imprinting). However, Olsen et al. (2014) found that costume-raised Whooping Crane chicks spent less time being vigilant and foraging – behaviors that are known to be beneficial for post-release survival (Kreger et al., 2004) – than did parent-raised chicks. Similarly, initial techniques developed for hand-raising Sandhill Crane chicks for ultralight aircraft-led migrations resulted in problematic tameness in released birds (Duff et al., 2001).

The relationship between captive rearing conditions and subsequent performance of Whooping Cranes in the wild is not fully under-stood, however (Converse et al., Chapter 8, this volume). Dinets (2015) suggests that there may be an advantage to releasing relatively naïve, costume-reared birds to the wild because not all behaviors learned from captive parents will be suitable for the novel habitats they face upon release. Indeed, research on Mississippi Sandhill Cranes (G. c. pulla) determined that postrelease survival rates of hand-raised chicks (at ages 6 months and 1–5 years) were higher than those for parent-raised chicks (Ellis et al., 2001). In the same study, the post-release reproductive suc-cess (measured by age at first egg or first hatch) did not differ between hand-raised and parent-raised birds. Importantly, however, recruitment rates for wild-hatched chicks were twice as high when at least one parent was wild-hatched, compared to when both parents were captive-hatched (Ellis et al., 2001; Hereford et al., 2014). These findings underscore the challenges of pro-ducing captive-reared birds that can successfully fledge their own chicks in the wild, and although several hundred captive-raised chicks have been released to the wild in four separate reintro-duction programs (French et al., Chapter 1, this volume), many questions about the long-term influence of the captive environment remain.

HUSBANDRY

Captive Environments

Whooping Cranes are highly territorial, and while habitat has not been a limiting factor on the breeding grounds for wild cranes, aggressive encounters have been observed between pairs occupying a territory and subadults that wander into proximity (Johns et al., 2005). With this ter-ritoriality in mind, pens for captive Whooping Cranes should be designed to minimize audi-tory and visual interactions among pairs. Visual barriers and distance from other breeding pairs may assist in reducing aggression and stress in breeding birds (Mirande et al., 1997). Breeding pens at most of the captive facilities are arranged in linear rows, and breeding pairs are separated from each other by vacant pens and visual barriers. While the logistics of caring for large numbers of birds are simplified using this design, the effects of stress related to social dominance may negatively impact breeding outcomes. Early researchers noted increased adrenal size and decreased gonad size in subordinate individuals in domestic fowl (Flickinger, 1961). In other non-cooperatively breeding avian species, similar to Whooping Cranes, subordinate animals show higher plasma levels of stress-related hormones than their dominant counterparts (Rohwer and Wingfield, 1981; Schwabl et al., 1988). In many species, breeding rates are controlled at moderate population densities, most likely via social inhibition of less dominant breeders (Eccard et al., 2011). While the relationship between stress, dominance, and reproductive suppression is complex and not fully understood, reproductive rates are generally lower for subordinates (Creel, 2001). The proximity of most captive Whooping Crane pairs to other breeding pairs and their unison calling may decrease reproductive activity, particularly in younger or less confident pairs. At the same time, the more distant presence of other pairs may serve as a positive stimulus for competition and breeding behavior (Swengel et al., 1996).

In addition to providing a secure territory, increasing the naturalistic components of the captive environment may also increase reproductive success. Natural ponds, or some form of running water (Mirande et al., 1997; Swengel and Carpenter, 1996), and high grasses for cover can provide security and foraging opportunities for natural and provisioned food items. Occupational enrichment in the form of varied food items (e.g., aquatic and terrestrial invertebrates, baked potatoes, and melons; V. Edwards, Calgary Zoological Society, personal communication) may also stimulate natural behaviors and thereby support successful reproduction (Carlstead and Shepherdson, 1994).

Research is underway at PWRC to determine how naturalistic environments with wetland components affect Whooping Crane breeding behaviors, reproductive hormone production, and reproductive output (Songsasen et al., Chapter 17, this volume).

It is important also to consider environmental conditions that may trigger reproduction. For species that breed at middle to high latitudes, annual photoperiodic cycle is often the most important factor in synchronizing reproductive activities (Farner, 1986 in Mirande et al., 1996b; Saint-Jalme, 2002; Songsasen et al., Chapter 17, this volume). Given that all living Whooping Cranes descend from a population that breeds at a high latitude (~59 degrees north) with wide seasonal variation in daylight hours, captive managers at lower latitudes (where changes in day length are not as extreme; PWRC, ICF, San Antonio Zoo, and Audubon Center for Research of Endangered Species) will often use photoperiod lights to mimic the daylight hours of WBNP (Mirande et al., 1996b). For example, when artificial lighting up to 170 lux per pen was used to lengthen the daylight hours at PWRC in the middle of February, egg laying occurred from 8 to 16 days earlier than when no lights were used (Olsen, 2014). This allowed eggs to be laid at lower temperature and higher humidity, which help to mitigate the decrease in reproduction that has been associated with higher environmental temperatures (Mirande et al., 1996b). Additional research is required to investigate this and other potential triggers for reproduction in captivity (e.g., rainfall and ambient temperature), but some observations and hypotheses are outlined by Mirande et al. (1996b).

Equally important to re-creating natural environmental conditions is minimizing the impact of "unnatural" disturbances often associated with the operation of breeding facilities. Disturbance by human and vehicular traffic, construction activities, or movement of cranes to new pens or between facilities can have negative effects on breeding success (Mirande et al., 1997). As such, necessary disruptions should be timed to begin after breeding and chick rearing and to end well in advance of the onset of hormonal changes in January. Ultimately, maintaining a captive environment that closely approximates that of a free-ranging bird should increase the likelihood of successful reproduction and may also reduce opportunities for incidental captive selection (i.e., genetic adaptation to captivity over successive generations; Frankham, 2008; Frankham et al., 1986; Williams and Hoffman, 2009; see later).

Nutrition and Diet

The functions of diet for animals in human care go beyond providing optimum nutrition for support of growth and physiologic functions to providing enrichment and promotion of natural behaviors. While it can be difficult to replicate wild diets for captive animals, a sound knowledge of food choices made by animals in the wild can contribute positively to the development of complete diets for their captive counterparts.

Several researchers have contributed to the information available on the wild diet of Whooping Cranes. Duxbury and Holroyd (1996) examined the proportions of stable isotopes in the feathers of AWBP cranes and demonstrated that these birds were feeding high in the trophic food web, including a large proportion of fish and marine invertebrates. These findings are compatible with those of Hunt and Slack (1989), who examined feces of wintering Whooping Cranes in Texas and determined that blue crab and clams are the most important winter diet items. Other foods of importance include wolfberry, acorns, snails, fish, and insects (Hunt and Slack, 1989).

In other work, Stocking et al. (2008) noted that captive Whooping Cranes had daily energy intakes that were highest in March, during the preoviposition cycle, and in October and November, during the season when their wild counterparts are migrating. At CZ, the highest body weights for both male and female adult cranes are recorded in October and November

(S. Black, personal communication). Energy intake also increased in captive cranes with increasing and decreasing temperatures, indicating that, as in other homeothermic species, energy expenditure is required for temperatures above and below a thermal neutral zone (Fitzpatrick et al., Chapter 12, this volume; Stocking et al., 2008). This information allows Whooping Crane managers to more closely approximate annual feeding cycles, as well as provide appropriate enrichment food items.

Despite the omnivorous nature of wild Whooping Cranes, the largest component of captive diets is a pelleted feed derived from plants and developed from formulations initially used for game birds (Swengel and Carpenter, 1996). This allows a cost-effective, palatable, and easily accessible food to be used year-round. Managers may consider adding supplemental invertebrate and fish protein sources to more closely replicate wild diets. Higher protein formulations of the pelleted diet are used in winter and spring to support reproduction (Swengel and Carpenter, 1996).

In an effort to improve the pelleted formulations, Nelson et al. (1997) milled 30% of a natural food item (clam, crab, acorn, or wolfberry) into the standard pelleted feed for Whooping Cranes. The resultant feeds were not as palatable and resulted in a decrease in assimilation of energy. Rather than milling supplemental food items into the pelleted feed, managers provide whole food items before and during the breeding season to assist in providing additional protein and energy for the physiological expenditures of breeding. Smelt, aquatic invertebrates, insects, and commercial poultry flushing formulations (supplemental pellets with high protein and calcium levels) have all been used by the various captive breeding facilities (J. San Miguel, San Antonio Zoo, personal communication; S. Black, personal communication). Feeding of wild insects and other plant and animal material by the parents when chicks are parent-raised may be one factor that protects hatchlings from developmental limb abnormalities (Kelley and Hartup, 2008). Finally, inclusion of more natural diet items, particularly for birds destined for release, is likely to support normal foraging and food choices in the wild (Moehrenschlager and Lloyd, 2016). In a reintroduction program for the Capercaillie (*Tetrao urogallus*), initially high mortality levels of released birds were thought to be due to a combination of predation and unfamiliarity with wild food sources (Siano et al., 2006).

Certain food additives may offer other health benefits for breeding birds. Musculoskeletal problems, including osteoarthritis, are the second most common source of morbidity in captive Whooping Cranes (Fitzpatrick et al., Chapter 12, this volume; Hartup et al., 2010). Poor joint health may compromise reproductive success by affecting breeding and incubating postures, and thereby decreasing fertility and hatchability of naturally incubated eggs (Gabel and Mahan, 1996). The addition of a nutraceutical containing minerals, vitamins, collagen, chondroitin sulfate, and other nutrients known to support joint and connective tissue health was trialed in cranes (Bauer et al., 2014). Early results showed promise in increasing mobility in affected joints, but further research is required.

THE SPECTER OF CAPTIVE SELECTION

The practice of breeding endangered species in captivity is fraught with challenges, including high financial costs, risks of disease transmission (in both captive and reintroduced populations), and the difficulties of managing the demography and genetics of extremely small populations (Bowkett, 2009; Snyder et al., 1996). Even when conservation breeding programs are successfully established, some of the greatest challenges will remain, including the potential for captive selection. In captivity, the selective pressure of limited food resources is eliminated, and rates of illness and injury may be much lower due to

husbandry controls and treatments (Frankham et al., 1986). Further, some heritable personality traits, such as tameness or boldness, may confer a reproductive advantage in captivity but have negative fitness consequences in the wild (Smith and Blumstein, 2008). For example, the fitness of released steelhead trout (measured by reproductive output) decreased by 37.5 percent for each prior generation in captivity (Araki et al., 2007).

To date, reintroduced populations of Whooping Cranes have demonstrated poor reproductive success, and there are concerns that captive selection may be a contributing factor (Converse et al., Chapter 8, this volume). Pressure to maximize reproductive output from the captive breeding flock may inadvertently result in selection for traits such as docility (e.g., if males that readily provide semen samples become overrepresented in the population) or low vigilance (e.g., if pairs that have high tolerance for human disturbance produce more offspring than those that are more easily stressed and/or are more territorial). One strategy to minimize adaptation to captivity is to limit the number of generations spent in human care (Williams and Hoffman, 2009); however, Whooping Cranes have already progressed through several captive generations. Thus, resumption of egg collection from the AWBP, to introduce new, wild-adapted genetics to the captive population, may be warranted. Collection and cryopreservation of semen from these wild-sourced birds (for future AI) may also assist in mitigating captive selection, as has been accomplished for the endangered black-footed ferret (*Mustela nigripes*; Howard et al., 2016). Captive selection may also be minimized by concerted efforts of managers to apply no breeding selection other than that needed to maximize genetic diversity (Frankham et al., 1986). Fortunately, the new SSP program for Whooping Cranes should make more tools available to identify and address potential selection issues.

Much as the identification of microsatellite DNA improved the accuracy of MK ranking for the Whooping Crane over the past 2 decades, the accessibility of new genomic tools will allow us to collect data across the entire genome in years to come. The role of captive selection will be elaborated at the genome level, and genome-wide association studies and quantitative trait loci analyses can help to identify the genes responsible for individual fitness (Steiner et al., 2013). The use of these next-generation sequencing techniques is allowing the discovery of genes responsible for adaptation in an increasing number of species (Stapley et al., 2010). The loci responsible for behavioral adaptation to captivity in Whooping Cranes, as well as the type of genetic variation responsible for the changes, may well be discernible in the near future using these techniques. By identifying such loci and the individuals carrying alleles that are of optimum value to a wild genotype, more informed breeding decisions can be made with reference to the management of adaptation to captivity.

CONSERVATION CONTRIBUTIONS OF CAPTIVE BREEDING

The challenges discussed in this chapter – maintaining genetic diversity; pairing birds; maximizing egg production, fertility, and hatch rates; preparing chicks for reintroduction; and providing proper nutrition and captive habitat – are not unique to Whooping Cranes. All avian conservation breeding programs must address these challenges, and effective captive breeding requires a careful balance between the strategic use of management interventions (e.g., AI or reclutching) and maintenance of conditions that promote survival and breeding in both captivity and the wild. While the success of reintroductions and other conservation translocations can be difficult to evaluate (Fischer and Lindenmayer, 2000; Moehrenschlager et al., 2013; Robert et al., 2015), it has been estimated that only 11% of avian reintroduction programs result in self-sustaining wild populations (or

unsupported populations with greater than 500 individuals; Beck et al., 1994). Additionally, animal translocations that depend on captive source populations tend to fare worse than those that use wild source populations (Fischer and Lindenmayer, 2000; Griffith et al., 1989; Rummel et al., 2016; Wolf et al., 1996).

Fundamentally, captive breeding and reintroduction efforts have yet to improve the conservation status of Whooping Cranes. The species remains Endangered under the U.S. Endangered Species Act, Canadian Species at Risk Act, and IUCN Red List of Endangered Species. However, the establishment and longevity of the Whooping Crane captive breeding program has been a success in its own right – from 178 eggs collected in the wild (Barrett and Stehn, 2010), there are now approximately 161 Whooping Cranes in captive breeding facilities and 148 in reintroduced populations in North America (unpublished data, W. Harrell, U.S. Fish and Wildlife Service, and M. Bidwell, Environment and Climate Change Canada). Objective 2 of the International Whooping Crane Recovery Plan ("Maintain a genetically stable captive population to ensure against the extinction of the species"; Canadian Wildlife Service and U.S. Fish and Wildlife Service, 2005) has generally been met, and the reintroduced populations, although not yet self-sustaining (Converse et al., Chapter 7, this volume), do provide additional insurance in the event of catastrophic losses in the AWBP. Furthermore, innovative breeding and rearing methods developed for Whooping Cranes have informed the field of reproductive biology in general (Wildt et al., 2010) and have led the way for many other avian conservation programs (D'Elia, 2010). For example, the successful use of AI to breed captive Whooping Cranes is considered to have "drastically changed the playing field for captive breeding of endangered avifauna" (D'Elia, 2010). The steady progress and collective determination of captive breeders and reintroduction managers are reasons for optimism.

SUMMARY AND OUTLOOK

The establishment of the captive breeding population of Whooping Cranes involved many advancements in crane husbandry and propagation, as exceptionally low numbers of Whooping Cranes in the wild compelled captive breeders to maximize egg and chick production however possible. A combination of methods, including AI, reclutching, surrogate incubation, and costume rearing, ultimately facilitated the growth of the captive population to just over 160 individuals distributed among five captive breeding facilities in Canada and the United States.

While many long-used techniques are still employed in the captive breeding program, increasing attention is being paid to more natural methods of breeding and rearing Whooping Cranes, in response to the poor reproductive success of captive-bred cranes in the wild. A study comparing the performance of costume-reared and parent-reared Whooping Cranes (in the wild) is currently underway (Converse et al., Chapter 8, this volume), and the possibilities of introducing mate choice and providing or simulating larger "territories" in captivity are under discussion. These and any other management alternatives should be implemented as part of controlled experiments.

Questions about the potential role of captive selection need to be addressed also; if the present captive population is not capable of producing birds that will reproduce successfully in the wild, the implications for the reintroduction program are substantial. Ideally, captive selection pressures (should they exist) would be identified and mitigated, but if this were not possible, reintroduced individuals could be sourced from the wild (i.e., the AWBP), or resources could be redirected toward other actions, such as increased habitat protection.

Ultimately, great strides have been made in captive management of Whooping Cranes over the 5 decades they have been bred in captivity, but several obstacles remain. Low fertility, lack of mate choice and mate compatibility,

aging flocks, and operational challenges limit the number of chicks produced for release into the wild (Songsasen et al., Chapter 17, this volume). However, the collaborative nature of the many government agencies, nonprofit organizations, and academic institutions working to recover the Whooping Crane is perhaps the species' greatest asset. With continued research and careful management, the captive breeding population of Whooping Cranes will continue to protect the species from extinction.

References

Allen, R.P., 1952. The Whooping Crane. Research Report 3. National Audubon Society, New York.

Araki, H., Cooper, B., Blouin, M.S., 2007. Genetic effects of captive breeding cause a rapid, cumulative fitness decline in the wild. Science 318 (5847), 100–103.

Association of Zoos and Aquariums, 2014. Species Survival Plan® (SSP) Program Handbook. Association of Zoos and Aquariums, Silver Spring, MD, 85 pp.

Association of Zoos and Aquariums, 2017. Population Analysis and Transfer Plan-Whooping Crane (*Grus americana*) AZA Species Survival Program® Green Program. Population management Center, Association of Zoos and Aquariums, Silver Spring, MD, 36 pp.

Ballou, J.D., Lacy, R.C., 1995. Identifying genetically important individuals for management of genetic diversity in pedigreed populations. In: Ballou, J.D., Fooze, T.J., Gilpin, M. (Eds.), Population Management for Survival and Recovery. Columbia University Press, New York, pp. 76–111.

Barrett, C., Stehn, T.V., 2010. A retrospective of whooping cranes in captivity. Proceedings of the North American Crane Workshop 11, 166–179.

Bauer, K.L., Dierenfeld, E.S., Hartup, B.K., 2014. Evaluation of a nutraceutical joint supplement in cranes. Proceedings of the North American Crane Workshop 12, 27–32.

Beck, B., Rapaport, L., Price, M.S., Wilson, A.C., 1994. Reintroduction of captive-born animals. In: Olney, P.J., Mace, G., Feistner, A. (Eds.), Creative Conservation: Interactive Management of Wild and Captive Animals. Springer, New York, pp. 265–286.

Bergeson, D.G., Johns, B.W., Holroyd, G.L., 2001. Mortality of whooping crane colts in Wood Buffalo National Park, Canada, 1997–99. Proceedings of the North American Crane Workshop 8, 6–10.

Bowkett, A.E., 2009. Recent captive-breeding proposals and the return of the Ark concept to global species conservation. Conserv. Biol. 23 (3), 773–776.

Boyce, M.S., Lele, S.R., Johns, B.W., 2005. Whooping crane recruitment enhanced by egg removal. Biol. Conserv. 126 (3), 395–401.

Brichieri-Colombi, T.A., Moehrenschlager, A., 2016. Alignment of threat, effort, and perceived success in North American conservation translocations. Conserv. Biol. 30 (6), 1159–1172.

Brown, M.E., Converse, S.J., Chandler, J.N., Crosier, A.L., Lynch, W., Wildt, D.E., Keefer, C.L., Songsasen, N., 2017. Time within reproductive season, but not age or inbreeding coefficient, affects seminal and sperm quality in the whooping crane (*Grus americana*). Reprod. Fertil. Dev. 29 (2), 294–306.

Brown, M.E., Converse, S.J., Chandler, J.N., Shafer, C., Brown, J.L., Keefer, C.L., Songsasen, N., 2016. Female gonadal hormones and reproductive behaviors as key determinants of successful reproductive output of breeding whooping cranes (*Grus americana*). Gen. Comp. Endocrinol. 230, 158–165.

Canadian Wildlife Service and U.S. Fish and Wildlife Service, 2005. International Recovery Plan for the Whooping Crane. Recovery of Nationally Endangered Wildlife (RENEW), Ottawa, and U.S. Fish and Wildlife Service, Albuquerque, NM, 162 pp.

Carlstead, K., Shepherdson, D., 1994. Effects of environmental enrichment on reproduction. Zoo Biol. 13 (5), 447–458.

Chargé, R., Teplitsky, C., Sorci, G., Low, M., 2014. Can sexual selection theory inform genetic management of captive populations? a review. Evol. Appl. 7 (9), 1120–1133.

Collar, N.J., Butchart, S.H.M., 2014. Conservation breeding and avian diversity: chances and challenges. Int. Zoo Yearb. 48 (1), 7–28.

Converse, S.J., Chandler, J.N., Olsen, G.H., Shafer, C.C., 2010. Evaluating propagation method performance over time with Bayesian updating: an application to incubator testing. Proceedings of the North American Crane Workshop 11, 110–117.

Converse, S.J., Royle, J.A., Urbanek, R.P., 2012. Bayesian analysis of multi-state data with individual covariates for estimating genetic effects on demography. J. Ornithol. 152 (2), 561–572.

Converse, S.J., Servanty, S., Moore, C.T., Runge, M.C., 2018. Population dynamics of reintroduced whooping cranes (Chapter 7). In: French, Jr., J.B., Converse, S.J., Austin, J.E. (Eds.), Whooping Cranes: Biology and Conservation. Biodiversity of the World: Conservation from Genes to Landscapes. Academic Press, San Diego, CA.

Converse, S.J., Strobel, Bradley N., Barzen, Jeb A., 2018. Reproductive failure in the Eastern Migratory Population: interaction of research and management (Chapter 8). In: French, Jr., J.B., Converse, S.J., Austin, J.E. (Eds.), Whooping Cranes: Biology and Conservation. Biodiversity of the World: Conservation from Genes to Landscapes. Academic Press, San Diego, CA.

Conway, W.G., 1982. The Species Survival Plan: tailoring long-term propagation species by species. Proceedings of the 1981 AAZPA Annual Conference, Wheeling, WV, 1982, pp. 6–11.

Creel, S., 2001. Social dominance and stress hormones. Trends Ecol. Evol. 16 (9), 491–497.

D'Elia, J., 2010. Evolution of avian conservation breeding with insights for addressing the current extinction crisis. J. Fish Wildl. Manage. 1 (2), 189–210.

Derrickson, S.R., Carpenter, J.W., 1987. Behavioral management of captive cranes – factors influencing propagation and reintroduction. In: Archibald, G.W., Pasquier, R.F. (Eds.), Proceedings of the 1983 International Crane Workshop. International Crane Foundation, Baraboo, WI, pp. 493–511.

Dinets, V., 2015. Can interrupting parent-offspring cultural transmission be beneficial? the case of whooping crane reintroduction. Condor 117 (4), 624–628.

Doughty, R.W., 1989. Return of the Whooping Crane. University of Texas Press, Austin.

Duff, J.W., Lishman, W.A., Clark, D.A., Gee, G.F., Sprague, D.T., Ellis, D.H., 2001. Promoting wildness in sandhill cranes conditioned to follow an ultralight aircraft. Proceedings of the North American Crane Workshop 8, 115–121.

Duxbury, J.M., Holroyd, G.L., 1996. The Determination of the Diet of Whooping Cranes on the Breeding Ground: A Stable Isotope Approach. Interim Report of the Canadian Wildlife Service, Ottawa, Ontario, Canada, 20 pp.

Eccard, J.A., Jokinen, I., Ylönen, H., 2011. Loss of density-dependence and incomplete control by dominant breeders in a territorial species with density outbreaks. BMC Ecol. 11 (1), 16.

Ellis, D.H., Gee, G.F., 2001. Whooping crane egg management: options and consequences. Proceedings of the North American Crane Workshop 8, 17–23.

Ellis, D.G., Gee, G.F., Olsen, G.H., Hereford, S.G., Nicolich, J.M., Thomas, N.J., Nagendran, M., 2001. Minimum survival rates for Mississippi sandhill cranes: a comparison of hand rearing and parent rearing. Proceedings of the North American Crane Workshop 8, 80–84.

Erickson, R.C., Derrickson, S., 1981. The whooping crane. In: Lewis, J.C. (Ed.), Crane Research around the World. International Crane Foundation, Baraboo, WI, pp. 103–134.

Farner, D.S., 1986. Generation and regulation of annual cycle in migratory passerine birds. Am. Zool. 26 (3), 493–501.

Fischer, J., Lindenmayer, D.B., 2000. An assessment of the published results of animal relocations. Biol. Conserv. 96 (1), 1–11.

Fitzpatrick, M.J., Mathewson, P.D., Porter, W.P., 2018. Ecological. Energetics of whooping cranes in the Eastern Migratory Population (Chapter 12). In: French, Jr., J.B., Converse, S.J., Austin, J.E. (Eds.), Whooping Cranes: Biology and Conservation. Biodiversity of the World: Conservation from Genes to Landscapes. Academic Press, San Diego, CA.

Flickinger, G.L., 1961. Effect of grouping on adrenals and gonads of chickens. Gen. Comp. Endocrinol. 1 (4), 332–340.

Frankham, R., 2008. Genetic adaptation to captivity in species conservation programs. Mol. Ecol. 17 (1), 325–333.

Frankham, R., Hemmer, H., Ryder, O.A., Cothran, E.G., Soulé, M.E., Murray, N.D., Snyder, M., 1986. Selection in captive populations. Zoo Biol. 5 (2), 127–138.

French, Jr., J.B., Converse, S.J., Austin, J.E., 2018. Whooping cranes past and present (Chapter 1). In: French, Jr., J.B., Converse, S.J., Austin, J.E. (Eds.), Whooping Cranes: Biology and Conservation. Biodiversity of the World: Conservation from Genes to Landscapes. Academic Press, San Diego, CA.

Gabel, R.R., Mahan, T.A., 1996. Incubation and hatching. In: Ellis, D.H., Gee, G.F., Mirande, C.M. (Eds.), Cranes: Their Biology, Husbandry and Conservation. Department of the Interior, National Biological Service, Washington, DC, and International Crane Foundation, Baraboo, WI, pp. 59–76.

Gee, G.F., Mirande, C.M., 1996. Special techniques, part A: crane artificial insemination. In: Ellis, D.H., Gee, G.F., Mirande, C.M. (Eds.), Cranes: Their Biology, Husbandry and Conservation. Department of the Interior, National Biological Service, Washington, DC, and International Crane Foundation, Baraboo, WI, pp. 205–217.

Glenn, T.C., Stephan, W., Braun, M.J., 1999. Effects of a population bottleneck on whooping crane mitochondrial DNA variation. Conserv. Biol. 13 (5), 1097–1107.

Griffith, B., Scott, J.M., Carpenter, J.W., Reed, C., 1989. Translocation as a species conservation tool: status and strategy. Science 245 (4917), 477–480.

Hartup, B.K., 2018. Rearing and release methods for reintroduction of captive-reared whooping cranes (Chapter 20). In: French, Jr., J.B., Converse, S.J., Austin, J.E. (Eds.), Whooping Cranes: Biology and Conservation. Biodiversity of the World: Conservation from Genes to Landscapes. Academic Press, San Diego, CA.

Hartup, B.K., Niemuth, J.N., Fitzpatrick, B., Fox, M., Kelley, C., 2010. Morbidity and mortality of captive whooping cranes at the International Crane Foundation: 1976–2008. Proceedings of the North American Crane Workshop 11, 183–185.

Hereford, S., Grazia, T., Billodeaux, L., 2014. Effect of rearing technique on age at first reproduction of released Mississippi sandhill cranes. Proceedings of the North American Crane Workshop 12, 92.

Howard, J.G., Lynch, C., Santymire, R.M., Marinari, P.E., Wildt, D.E., 2016. Recovery of gene diversity using long-term cryopreserved spermatozoa and artificial insemination in the endangered black-footed ferret. Anim. Conserv. 19 (2), 102–111.

Hunt, H.E., Slack, R.D., 1989. Winter diets of whooping and sandhill cranes in south Texas. J. Wildl. Manage. 53 (4), 1150–1154.

IUCN, 2015. The IUCN Red List of Threatened Species. Version 2015.3. Available from: http://www.iucnredlist.org.

Ivy, J.A., Lacy, R.C., 2012. A comparison of strategies for selecting breeding pairs to maximize genetic diversity retention in managed populations. J. Hered. 103 (2), 186–196.

Jarvi, S.I., Miller, M.M., Gotol, R.M., Gee, G.F., Briles, W.E., 2001. Evaluation of the major histocompatibility complex (MHc) in cranes: applications to conservation efforts. Proceedings of the North American Crane Workshop 8, 223.

Johns, B.W., Goossen, J.P., Kuyt, E., Craig-Moore, L., 2005. Philopatry and dispersal in Whooping Cranes. Proceedings of the North American Crane Workshop 9, 117–125.

Jones, A.G., Ratterman, N.L., 2009. Mate choice and sexual selection: what have we learned since Darwin? Proc. Natl. Acad. Sci. 106 (Suppl. 1), S10001–S10008.

Jones, K.L., Glenn, T.C., Lacy, R.C., Pierce, J.R., Unruh, N., Mirande, C.M., Chavez-Ramirez, F., 2002. Refining the whooping crane studbook by incorporating microsatellite DNA and leg-banding analyses. Conserv. Biol. 16 (3), 789–799.

Jones, K.L., Henkel, J.R., Howard, J.J., Lance, S.L., Hagen, C., Glenn, T.C., 2010. Isolation and characterization of 14 polymorphic microsatellite DNA loci for the endangered whooping crane (Grus americana) and their applicability to other crane species. Conserv. Genet. Resour. 2 (1), 251–254.

Jones, K.L., Nicolich, J.M., 2001. Artificial insemination in captive whooping cranes: results from genetic analyses. Zoo Biol. 20 (4), 331–342.

Kelley, C., Hartup, B.K., 2008. Risk factors associated with developmental limb abnormalities in captive whooping cranes. Proceedings of the North American Crane Workshop 10, 119–124.

Kepler, C.B., 1976. Dominance and dominance-related behavior in the whooping crane. In: Lewis, J.C. (Ed.), Proceedings of the 1975 International Crane Workshop. Oklahoma State University Publishing and Printing, Stillwater, pp. 177–196.

King, S.L., Selman, W., Vasseur, P.L., Zimorski, S.E., 2018. Louisiana nonmigratory whooping crane reintroduction (Chapter 22). In: French, Jr., J.B., Converse, S.J., Austin, J.E. (Eds.), Whooping Cranes: Biology and Conservation. Biodiversity of the World: Conservation from Genes to Landscapes. Academic Press, San Diego, CA.

Kreger, M.D., Estevez, I., Hatfield, J.S., Gee, G.F., 2004. Effects of rearing treatment on the behavior of captive whooping cranes (Grus americana). Appl. Anim. Behav. Sci. 89 (3–4), 243–261.

Kuyt, E., 1995. The nests and eggs of the whooping crane, *Grus americana*. Can. Field Nat. 109 (1), 1–5.

Kuyt, E., 1996. Reproductive manipulation in the whooping crane, *Grus americana*. Bird Conserv. Int. 6 (1), 3–10.

Letoutchaia, J.N., Maguire, K., Hartup, B.K., 2010. Causes of embryonic death in captive whooping cranes. Proceedings of the North American Crane Workshop 11, 192.

Lewis, J.C., 2001. Increased egg conservation – is it essential for recovery of whooping cranes in the Aransas Wood Buffalo Population? Proceedings of the North American Crane Workshop 8, 1–5.

Meritt, Jr., D.A., 1980. A Species Survival Plan for AAZPA. AAZPA Annual Conference Proceedings 1980, 69–75.

Mirande, C.M., Carpenter, J.W., Burke, A.M., 1997. The effect of disturbance on the reproduction and management of captive cranes. Proceedings of the North American Crane Workshop 7, 56–61.

Mirande, C.M., Gee, G.F., Burke, A., Sheppard, C., 1996a. Genetic management. In: Ellis, D.H., Gee, G.F., Mirande, C.M. (Eds.), Cranes: Their Biology, Husbandry and Conservation. Department of the Interior, National Biological Service, Washington, DC, and International Crane Foundation, Baraboo, WI, pp. 45–57.

Mirande, C.M., Gee, G.F., Swengel, S.R., Whitlock, P., 1996b. Egg and semen production. In: Ellis, D.H., Gee, G.F., Mirande, C.M. (Eds.), Cranes: Their Biology, Husbandry and Conservation. Department of the Interior, National Biological Service, Washington, DC, and International Crane Foundation, Baraboo, WI, pp. 175–183.

Moehrenschlager, A., Lloyd, N.A., 2016. Release considerations and techniques to improve conservation translocation success. In: Jachowski, D.S., Millspaugh, J.J., Angermeier, P.L., Slotow, R. (Eds.), Reintroduction of Fish and Wildlife Populations. University of California Press, Oakland, pp. 245–280.

Moehrenschlager, A., Shier, D.M., Moorhouse, T.P., Stanley Price, M.R., 2013. Righting past wrongs and ensuring the future. In: Macdonald, D.W., Willis, K.J. (Eds.), Key Topics in Conservation Biology 2. Wiley-Blackwell, Oxford, pp. 405–429.

Moller, A.P., Jennions, M.D., 2001. How important are direct fitness benefits of sexual selection? Naturwissenschaften 88 (10), 401–405.

Montgomery, M.E., Ballou, J.D., Nurthen, R.K., England, P.R., Briscoe, D.A., Frankham, R., 1997. Minimizing kinship in captive breeding programs. Zoo Biol. 16 (5), 377–389.

Mooallem, J., 2009. Rescue flight. *New York Times*. Available from http://nyti.ms/1N7CHzL.

Nelson, J.T., Gee, G.F., Slack, R.D., 1997. Food consumption and retention time in captive whooping cranes (*Grus americana*). Zoo Biol. 16 (6), 519–531.

Nelson, J.T., Small, C.R., Ellis, D.H., 1995. Quantitative assessment of pair formation behavior in captive whooping cranes (*Grus americana*). Zoo Biol. 14 (2), 107–114.

Nicholich, J.M., Gee, G.F., Ellis, D.H., Hereford, S.G., 2001. Natural fertility in whooping cranes and Mississippi sandhill cranes at Patuxent Wildlife Research Center. Proceedings of the North American Crane Workshop 8, 170–177.

Olsen, G., 2014. Photoperiod and nesting phenology of whooping cranes at two captive facilities. Proceedings of the North American Crane Workshop 12, 100.

Olsen, G.H., Matthews, L., Converse, S., 2014. Comparison of behaviors of crane chicks that were parent reared and reared by costumed humans. Proceedings of the North American Crane Workshop 12, 101.

Puchta, S., Putnam, M.S., Maguire, K., 2008. Egg breakage by captive cranes at the International Crane Foundation. Proceedings of the North American Crane Workshop 10, 170.

Roberge, J.M., 2014. Using data from online social networks in conservation science: which species engage people the most on Twitter? Biodiversity Conserv. 23 (3), 715–726.

Robert, A., Colas, B., Guigon, I., Kerbiriou, C., Mihoub, J.B., Saint-Jalme, M., Sarrazin, F., 2015. Defining reintroduction success using IUCN criteria for threatened species: a demographic assessment. Anim. Conserv. 18 (5), 397–406.

Rohwer, S., Wingfield, J.C., 1981. A field study of dominance, plasma levels of luteinizing hormone and steroid hormones in wintering Harris' sparrows. Zeits. Tierpsychol. 57 (2), 173–183.

Rummel, L., Martínez–Abraín, A., Mayol, J., Ruiz-Olmo, J., Mañas, F., Jiménez, J., Gomez, J.A., Oro, D., 2016. Use of wild–caught individuals as a key factor for success in vertebrate translocations. Anim. Biodiv. Conserv. 39 (2), 207–219.

Saint-Jalme, M., 2002. Endangered avian species captive propagation: an overview of functions and techniques. Avian Poult. Biol. Rev. 13 (3), 187–202.

Schwabl, H., Ramenofsky, M., Scwabl-Benzinger, I., Farner, D.S., Wingfield, J.C., 1988. Social status, circulating levels of hormones, and competition for food in winter flocks of the white-throated sparrow. Behaviour 107 (1), 107–121.

Siano, R., Barlein, F., Exo, K.M., Herzog, S.A., 2006. Überlebensdauer, Todesursachen und Raumnutzung gezüchteter Auerhuhner (*Tetrao urogallus* L.), ausgewildert im Nationalpark Harz. Survival, causes of death and spacing of captive Capercaillies (*Tetrao urogallus* L.) released in the Harz Mountains National Park. Vogelwarte 44, 145–158.

Smith, B., Blumstein, D., 2008. Fitness consequences of personality: a meta analysis. Behav. Ecol. 19 (2), 448–455.

Smith, D.H., Converse, S.J., Gibson, K.W., Moehrenschlager, A., Link, W.A., Olsen, G.H., Maguire, K., 2011. Decision analysis for conservation breeding: maximizing production for reintroduction of whooping cranes. J. Wildl. Manage. 75 (3), 501–508.

Smith, D.H., Moehrenschlager, A., Christensen, N., Knapik, D., Gibson, K., Converse, S.J., 2012. Archive eggs: a research and management tool for avian conservation breeding. Wildl. Soc. Bull. 36 (2), 342–349.

Snyder, N.F., Derrickson, S.R., Beissinger, S.R., Wiley, J.W., Smith, T.B., Toone, W.D., Miller, B., 1996. Limitations of captive breeding in endangered species recovery. Conserv. Biol. 10 (2), 338–348.

Songsasen, N., Converse, S.J., 2018. Reproduction and reproductive strategies relevant to management of whooping cranes ex situ (Chapter 17). In: French, Jr., J.B., Converse, S.J., Austin, J.E. (Eds.), Whooping Cranes: Biology and Conservation. Biodiversity of the World: Conservation from Genes to Landscapes. Academic Press, San Diego, CA.

Stapley, J., Reger, J., Feulner, P.G.D., Smadja, C., Galindo, J., Ekblom, R., Bennison, C., Ball, A.D., Beckerman, A.P., Slate, J., 2010. Adaptation genomics: the next generation. Trends Ecol. Evol. 25 (12), 705–712.

Steiner, C.C., Putnam, A.S., Hoeck, E.A., Ryder, O.A., 2013. Conservation genomics for threatened animal species. Annu. Rev. Anim. Biosci. 1, 261–281.

Stocking, J.J., Putnam, M.S., Warning, N.B., 2008. A year-long study of food consumption by captive whooping cranes at the International Crane Foundation. Proceedings of the North American Crane Workshop 10, 178.

Swan, K.D., McPherson, J.M., Seddon, P.J., Moehrenschlager, A., 2016. Managing marine biodiversity: the rising diversity and prevalence of marine conservation translocations. Conserv. Lett. 9 (4), 239–251.

Swengel, S.R., Archibald, G.W., Ellis, D.H., Smith, D.G., 1996. Behavior management. In: Ellis, D.H., Gee, G.F., Mirande, C.M. (Eds.), Cranes: Their Biology, Husbandry and Conservation. Department of the Interior, National Biological Service, Washington, DC, and International Crane Foundation, Baraboo, WI, pp. 105–122.

Swengel, S.R., Carpenter, J.W., 1996. General husbandry. In: Ellis, D.H., Gee, G.F., Mirande, C.M. (Eds.), Cranes: Their Biology, Husbandry and Conservation. Department of the Interior, National Biological Service, Washington, DC, and International Crane Foundation, Baraboo, WI, pp. 31–43.

Swengel, S.R., Tuite, M.L., 1997. Recent advances in scheduling strategies and practical techniques in crane artificial insemination. Proceedings of the North American Crane Workshop 7, 46–55.

White, J.L., 2000. Management of captive whooping cranes (*Grus americana*) to improve breeding behaviour and success. M.Sc. thesis, University of Calgary.

Wildt, D.E., Comizzoli, P., Pukazhenthi, B., Songsasen, N., 2010. Lessons from biodiversity—the value of nontraditional species to advance reproductive science, conservation, and human health. Mol. Reprod. Dev. 77 (5), 397–409.

Williams, S.A., Hoffman, E.A., 2009. Minimizing genetic adaptation in captive breeding programs: a review. Biol. Conserv. 142 (11), 2388–2400.

Willoughby, J.R., Fernandez, N.B., Lamb, M.C., Ivy, J.A., Lacy, R.C., Dewoody, J.A., 2015. The impacts of inbreeding, drift and selection on genetic diversity in captive breeding populations. Mol. Ecol. 24 (1), 98–110.

Wilson, S., Gil-Weir, K.C., Clark, R.G., Robertson, G.J., Bidwell, M.T., 2016. Integrated population modeling to assess demographic variation and contributions to population growth for endangered whooping cranes. Biol. Conserv. 197, 1–7.

Wolf, C.M., Griffith, B., Reed, C., Temple, S.A., 1996. Avian and mammalian translocations: update and reanalysis of 1987 survey data. Conserv. Biol. 10 (4), 1142–1154.

Reproduction and Reproductive Strategies Relevant to Management of Whooping Cranes Ex Situ

Nucharin Songsasen*, Sarah J. Converse**,†, Megan Brown*

*Smithsonian Conservation Biology Institute, National Zoological Park, Front Royal, VA, United States
**U.S. Geological Survey, Patuxent Wildlife Research Center, Laurel, MD, United States
†U.S. Geological Survey, Washington Cooperative Fish and Wildlife Research Unit, School of Environmental and Forest Sciences (SEFS) & School of Aquatic and Fishery Sciences (SAFS), University of Washington, Seattle, WA, United States

INTRODUCTION

The captive (ex situ) Whooping Crane (*Grus americana*) population was established in the late 1960s at Patuxent Wildlife Research Center (PWRC), Laurel, Maryland (Black and Swan, Chapter 16, this volume; French et al., Chapter 1, this volume; Kuyt, 1996). Founders were added through the mid-1990s via collection of one of two fertile eggs from nests at Wood Buffalo National Park, Canada. Currently, there are approximately 160 birds maintained ex situ. Until 2018, most of the birds were housed in five major breeding centers (PWRC; the International Crane Foundation, ICF; Calgary Zoo; Audubon Species Survival Center; and San Antonio Zoo), with a small number displayed at zoos without breeding programs.

The captive breeding program has a dual purpose: insurance against extinction of the species (Canadian Wildlife Service and U.S. Fish and Wildlife Service, 2005) and a source of offspring for reintroduction efforts (French et al., Chapter 1, this volume). However, reproduction of captive birds housed in breeding centers is poor (Black and Swan, Chapter 16, this volume). For example, in 2013, 10 out of 26 behavioral pairs housed at PWRC laid 22 eggs, only 5 of which were fertile (Harrell, 2014). During the same year, 14 of 19 pairs housed at ICF produced 42 eggs, of which 23 were fertile. In addition to a low fertility rate, captive birds also exhibit delayed onset of egg laying (7 years vs. 5 years in situ; Ellis et al., 1996). Poor reproductive success presents as an impediment to the sustainability of the ex situ population that, in turn, limits its contribution to the recovery of the species. To improve the contribution of the captive population to Whooping Crane recovery, in terms of both the quantity and the quality of birds produced, there is a need to improve understanding of the basic reproductive biology of the species, to identify the underlying cause(s) of poor reproduction, and to develop management actions to improve reproduction. In addition, reproductive technologies could contribute to improved genetic management of the population, which is especially critical given the genetic bottleneck through which Whooping Cranes passed in the early 20th century (Austin et al., Chapter 3, this volume).

In this chapter, we begin by reviewing Whooping Crane reproductive biology and current knowledge of factors affecting reproductive success. We also discuss captive management strategies and reproductive technologies with the potential to enhance reproductive capacity in Whooping Cranes. We close with a discussion of future research priorities.

WHOOPING CRANE REPRODUCTIVE BIOLOGY

To understand, and to improve management of, reproduction in captive Whooping Cranes, it is beneficial to understand the reproductive biology of the species in the wild. Whooping Cranes are socially monogamous and generally mate for life, although pairs have been known to separate if they are not reproductively successful (Swengel et al., 1996). Also, extra-pair copulations have been reported in the wild (Dellinger et al., 2013), but whether or how often

extra-pair fertilizations occur is not known. Based on data from birds in the Aransas-Wood Buffalo Population (AWBP), Whooping Cranes form pair bonds at ~4–5 years of age (Ellis et al., 1996), though earlier pairing and breeding are well documented in reintroduced populations where better information on bird age is available (Dellinger, Chapter 9, this volume; Urbanek et al., Chapter 10, this volume).

In captivity, by contrast, while pairing is attempted when birds are as young as 2 years old, first successful egg production rarely occurs when the birds are younger than 5 years old (Ellis et al., 1992). Of females monitored at PWRC between 1975 and 1993, only about half (53%) of 7-year-olds had begun breeding, and it was not until 10 years of age that all females had done so (Canadian Wildlife Service and U.S. Fish and Wildlife Service, 2005).

As in other seasonally breeding birds, reproduction in Whooping Cranes is photoperiod dependent. Whooping Cranes in the AWBP construct nests after arrival at their northern breeding ground and usually lay two eggs (48–60 h apart) in late April through mid-May (Canadian Wildlife Service and U.S. Fish and Wildlife Service, 2005; Johnsgard, 1983; Urbanek and Lewis, 2015). Birds in the EMP at Necedah National Wildlife Refuge generally produce nests beginning in early April and failed pairs may renest and incubate these second nests through mid-June (Converse et al., 2013), while the nesting season of the Florida Nonmigratory Population (FNMP), when nesting occurred in that population, began in January and extended to May (Folk et al., 2005). The earlier onset of nesting results in a longer breeding season. Because of the longer breeding season, birds breeding at lower latitudes may produce additional clutches if the first clutch fails (Converse et al., Chapter 8, this volume; Dellinger, Chapter 9, this volume). However, in the AWBP, Whooping Cranes typically lay only one clutch, even if the first clutch is lost (Mirande et al., 1996). Eggs are incubated by both members of the pair for 28–34 days; both egg laying and hatching are asynchronous. With suitable habitat conditions (e.g., high food availability and low predator density), crane pairs may hatch and raise two chicks (Archibald and Lewis, 1996; Urbanek and Lewis, 2015) though only approximately 10% of family groups observed in winter at Aransas National Wildlife Refuge include two chicks (Erickson, 1975; Kuyt, 1987). In the EMP, no pair has successfully raised two chicks to fledging. If an egg does not hatch because it is infertile or damaged, the pair may continue to incubate the egg for up to 50 additional days (Archibald and Lewis, 1996).

Cranes in captivity also respond to photoperiod. Cranes at ICF housed in pens with no artificial lighting start nesting 2 weeks later than birds housed at PWRC. The presence and number of artificial lights also influence nesting period in the Whooping Crane (Olsen, 2011). At PWRC, birds housed in a pen equipped with two lights laid eggs 10 days earlier than those housed in a pen with one light and 16 days earlier than those with no light. However, at ICF, the difference between lights on a pen and no lights was only 8 days (Olsen, 2011).

Whooping Cranes housed at PWRC begin laying eggs in April (Gee, 1983). Copulation generally occurs 2–5 weeks before the onset of egg production (Gee and Russman, 1996). By contrast, female birds housed at the Audubon Species Survival Center in Louisiana can lay first eggs as early as February and egg production can last until early June (Mirande et al., 1996). Captive birds lay clutches of sometimes one and more frequently two eggs. After the first egg is laid, the second egg can be expected 2–3 days later and clutches are usually collected soon after completion. Whooping Cranes may produce two to three additional clutches if they are not allowed to incubate their clutches (Brown et al., 2015).

Reproductive Endocrinology

There is limited information on reproductive endocrinology specific to Whooping Cranes. As in other vertebrates, generally, reproduction in crane is controlled by the hypothalamo-pituitary gonadal (HPG) axis. The primary responsibility of the hypothalamus is to coordinate the activation and inhibition of the axis by releasing neuropeptides into the pituitary (Bedecarrats, 2015). These neuropeptides bind to specific targeted receptors in the anterior pituitary, which results in the release of follicle-stimulating hormone (FSH) and luteinizing hormone (LH), which, in turn, stimulates gametogenesis and the release of gonadal hormones, including estrogen and progesterone in females and testosterone in males (Bedecarrats, 2015).

The impact of photoperiod on breeding birds is mediated through the HPG axis. Whooping Cranes are long-day breeders, and reproductive organs recrudesce in response to increasing day length (Archibald and Lewis, 1996). As in other birds, photo-stimulation of cranes promotes the production of thyrotropin-stimulating hormone (TSH), which converts thyroxin (T4) into triiodothyronine (T3). T3 stimulates the secretion of gonadotropin-releasing hormone 1 (GnRH 1) from the hypothalamus that, in turn, up regulates the production of gonadotropins from the anterior pituitary and gonadal hormones from the gonad (Sharp et al., 1998; Tsutsui et al., 2012). Continued increases in day length past a critical point (normally >12 h) decrease GnRH1 production, resulting in down regulation of gonadotropins and gonadal hormones. Thus, birds enter a refractory period when the gonad decreases in size and becomes quiescent (Sharp et al., 1998; Tsutsui et al., 2012). This refractory period lasts until the next annual photo-stimulation cycle begins (Sharp et al., 1998). Within the same bird species, the onset of reproductive season varies depending on the latitude (Baker, 1939). Birds living at a low latitude begin their breeding season sooner than those at high latitudes (Baker, 1939). Baker estimated that the breeding season will start 3 days later with each additional degree of latitude (Baker, 1939). However, the onset of gonadal regression (or photorefraction) occurs at the same photoperiod, regardless of latitude. Therefore, the reproductive season of birds living at a lower latitude is longer than individuals living at a higher latitude (Baker, 1939), which accounts for the previously discussed differences in breeding season length across the range of Whooping Cranes. As noted, Whooping Cranes in captivity also respond to photoperiod changes (Mirande et al., 1996). There may be sex differences in the response to photoperiod. In other avian species, reproductive recrudescence in the male is directly linked to photoperiod, whereas female birds require additional stimuli, such as temperature, food supply, a nesting site, and behavioral interaction with a prospective mate to trigger the final stage of egg production (El Halawani et al., 1984; Pollock and Orosz, 2002).

Semen Characteristics and Factors Influencing Sperm Production

The morphology of bird spermatozoa varies greatly among species, but can be classified into two basic types: a simple morphology typical of nonpasserine birds (e.g., cranes) and a more complex morphology found in passerines. Within the same species, there are also variations in sperm size, and this has been suggested to be driven by sperm competition processes (Bennison et al., 2015). For the Sandhill Crane (*Grus canadensis*), sperm head length varied among subspecies, with the Greater (*G. canadensis tabida*) and Florida Sandhill Cranes (*G. canadensis pratensis*) having longer (13.8 ± 0.3 and 13.0 ± 0.3 μm, respectively) head length than Lesser Sandhill Cranes (*G. canadensis canadensis*; 9.8 ± 1.3 μm). Greater Sandhill Cranes have longer head length than the Mississippi subspecies (*G. canadensis pulla*; 9.0 ± 1.1 μm) (Sharlin et al., 1979). Furthermore, sperm head length is correlated with fertility in that the Greater and Florida subspecies have higher fertility rates (46–60%) than Mississippi (23%) and Lesser

Sandhill Cranes (7%) (Sharlin et al., 1979). There is no information about the impact of sperm head length on male fertility in the Whooping Crane. Given that AI in Whooping Cranes often mixes semen from more than one male, future studies aiming at investigating the impact of sperm head length on male fertility are warranted.

Semen volume varies greatly between individuals and can differ among species (Table 17.1), individuals of the same species, and ejaculates from the same male (Brown et al., 2017; Chen et al., 2001). For captive Whooping Cranes, previous work suggests that peak semen production occurs between March 30th and April 26th (Mirande et al., 1996).

Brown et al. (2017) examined a variety of factors with potential to influence semen production in 29 Whooping Cranes housed at PWRC. The study documented substantial variation in semen characteristics among individual males and among ejaculates within the same individual (Table 17.1). On average, Whooping Crane males ejaculate about 60 μl, containing nearly 15 million total spermatozoa, of which 45% are motile. Crane ejaculates are slightly alkaline with an osmolality of ~300 mOsm. Approximately 35% of spermatozoa within a given ejaculate are structurally abnormal; the abnormalities primarily are related to head deformity. This proportion in sperm deformities is similar to that reported in the Sandhill Crane, a species that has not experienced a population bottleneck (Chen et al., 2001). Furthermore, there is no direct relationship between inbreeding coefficient for individual Whooping Crane males and semen characteristics. This is rather surprising as loss of heterozygosity has been associated with low numbers of sperm with structurally normal morphology in many mammalian species (Fitzpatrick and Evans, 2009; Ryan et al., 2003).

Recently, comparison of semen characteristics among samples obtained during early (March 30 and April 1), mid (April 21 and 28), and late (May 10 and 11) breeding season has

TABLE 17.1 Semen Characteristics of Crane Species (Family Gruidae Reported in the Literature) (NA = Data Not Available)

Species

Common name	Scientific name	Volume (µl)	Density[a]	Concentration[b]	% Motility	% Normal morphology
Black-necked	Grus nigricollis	74.0[c]	9.9[c]	NA	70.7[c]	NA
Blue	Anthropoides paradisea	20.0[c]	7.5[c]	NA	35.0[c]	NA
Brogla	Grus rubicunda	47.1[c]	7.1[c]	NA	44.9[c]	NA
Common	Grus grus	62.1[c]	7.2[c]	NA	52.6[c]	NA
Demoiselle	Anthropoides virgo	100.0[c]	5.0[c]	NA	60.0[c]	NA
Hooded	Grus monacha	62.9[c]	9.0[c]	NA	63.8[c]	NA
Red-crowned	Grus japonensis	84.4[c]	5.8[c]	NA	56.0[c]	NA
Sandhill	Grus canadensis	52.9[c]		1.9[e]	62.6[c]	59.1[e]
Sarus	Grus antigone	90.2[c]		350.0[f]	60.0[f]	54.7[f,i]
Siberian	Grus leucogeranus	43.7[c]	8.6[c]	NA	91.3[e]	NA
Wattled	Bugeranus carunculatus	40.8[c]	5.5[c]	NA	43.6[c]	NA
White-naped	Grus vipio	61.3[c]		621.2[g]	66.8[g]	
Whooping	Grus americana	61.1[h]		180.4[h]	44.6[h]	65.4[h]

[a]"Density" is assessed on a 12-point scale based on an established procedure used at the International Crane Foundation. A higher number indicates a category with more sperm in the sample.

[b]Concentration reported as 10^6 sperm/mL.

[c]International Crane Foundation artificial insemination records from 1992 to 2015.

[d]Chen et al., 2001.

[e]Maksudov, G., Panchenko, V.G., 2002. Production of interspecies hybrid of cranes by artificial insemination with frozen semen. Izvestiia Akademii Nauk SSSR Seriya Biologicheskaya 2, 243–247.

[f]Kongbuntad, W., Kaewmad, P., Tanomtong, A., 2013. Semen quality and artificial insemination of Eastern Sarus crane Grus antigone shappii Linn. In captive condition in the Nakhon Ratchasima Zoo, Thailand. World Appl. Sci. J. 28 (1), 145–152.

[g]Panyaboriban et al., 2016.

[h]Brown et al., 2017.

shown that sperm output (i.e., total sperm per ejaculate) increases as breeding season progresses (Brown et al., 2017). However, the same study demonstrates that the percentage of structurally normal sperm is highest in samples collected early in the breeding season (Brown et al., 2017). Interestingly, time within breeding season does not affect sperm motility (Brown et al., 2017). The increase in structurally abnormal spermatozoa toward the end of the breeding season does not appear to compromise fertility, perhaps because of the offset of accelerated spermatogenesis that continues to increase numbers of morphologically normal spermatozoa per ejaculate through the season's end. The influence of time within breeding season on total sperm per ejaculate has also been observed in the Sandhill Crane where sperm output increases as breeding season progresses (Chen et al., 2001). However, unlike the Whooping Crane, in the Sandhill Crane, total number of sperm per ejaculate increases from the seasonal onset, plateaus midseason, and then is sustained for the remainder of the reproductive interval (Chen et al., 2001). As most captive and wild Whooping Crane eggs are laid during mid and late breeding season, increased semen production as the season progresses ensures that there are sufficient sperm for successful fertilization.

Egg Production

In female birds, FSH stimulates follicle growth and estrogen production by theca interstitial cells of small follicles (Ottinger and Baskt, 1995; Paster, 1991; Pollock and Orosz, 2002). Estrogen stimulates secondary sex characteristics and reproductive organ development, as well as vitellogenesis by signaling the liver to produce calcium-binding vitellogenin, which is critical for eggshell production (Norris, 2006; Pollock and Orosz, 2002). Progesterone is produced by granulosa cells of large growing follicles within the ovary (Pollock and Orosz, 2002). As the ovum grows, progesterone levels continue to increase and signal LH secretion (Norris, 2006), which, in turn, upregulates the conversion of androgen produced by the theca interna layer into estrogen, resulting in ovulation. LH secretion begins to elevate when females nest, and the concentration reaches peak levels at the onset of egg production and immediately declines after eggs are laid (Joyner, 1990; Ottinger and Baskt, 1995). After ovulation, steroid hormone production decreases until the growth of a new follicle cohort (Ottinger and Baskt, 1995).

A recent study compared gonadal and adrenal hormone levels between successful and unsuccessful Whooping Crane pairs (Brown et al., 2016). Successful females excreted higher mean fecal estrogen metabolite levels, but lower progesterone metabolites than did unsuccessful females. However, there were no differences detected in gonadal hormone production between successful and nonsuccessful males, as measured using fecal hormone metabolites. Differences in gonadal hormone production in females do not appear to be associated with stress, as mean fecal glucocorticoid metabolites do not differ among individual cranes.

FACTORS INFLUENCING REPRODUCTIVE PERFORMANCE

Genetic Diversity

For many wildlife species, subspecies, and populations, there is a relationship between loss of genetic variation and reproductive fitness (Fitzpatrick and Evans, 2009; Ryan et al., 2003). In mammals, loss of heterozygosity has been associated with a high incidence of pleomorphic sperm (Asa et al., 2007; Pukazhenthi et al., 2001; Roldan et al., 1998; Wildt et al., 1982), reduced litter size (Rabon and Waddell, 2010), and high neonatal mortality (Facemire et al., 1995; Roelke et al., 1993). For birds, low gene diversity has also been associated with poor reproductive performance in both wild and captive individuals (Swinnerton et al., 2004). Specifically, low heterozygosity decreases fertility and survival of captive and wild endangered Pink Pigeons (*Columba mayeri*) (Swinnerton et al., 2004), although free-living birds are more susceptible to inbreeding depression than ex situ counterparts (Swinnerton et al., 2004). In particular, poor reproduction and survival are observed only in highly inbred (inbreeding coefficient [F] $\geq$ 0.25) birds maintained ex situ, while moderate inbreeding levels ($0 < F < 0.25$) can negatively influence wild birds (Swinnerton et al., 2004). Inbreeding has also been shown to negatively affect hatchability in the Great Reed Warbler (*Acrocephalus arundinaceus*; Bensch et al., 1994) and Blue Tit (*Parus caeruleus*; Kempenaers et al., 1996) living in the wild. Genetic similarity between parents also has been shown to affect clutch size and fledging success in these species (Bensch et al., 1994; Kempenaers et al., 1996). In Lesser Kestrels (*Falco naumanni*), gene diversity has been shown to impact egg production; females with high heterozygosity lay more eggs than homozygous individuals (Ortego et al., 2007). A study in European Shags (*Phalacrocorax aristotelis*) has demonstrated a strong effect of genetic diversity on reproductive success of female, but not male, birds (Velando et al., 2015). Specifically, levels of homozygosity are negatively correlated with the number of chicks produced.

A recent study in Whooping Cranes housed at PWRC did not demonstrate a direct relationship between inbreeding coefficient and semen characteristics during the breeding season (Brown et al., 2017). Smith et al. (2011) also reported no association of inbreeding on hatching success in the ex situ Whooping Crane population. Furthermore, Converse et al. (2012) reported only weak evidence of an inbreeding effect on post-release survival of Whooping Cranes produced and reared in captivity. Because the Whooping Crane experienced a severe population bottleneck in the early 1940s and the captive population is descended from 16 or fewer founders, there is very little variation in inbreeding level among pairs (Ellis et al., 1996). Low overall gene diversity within the population may explain the lack of an apparent impact of inbreeding on reproductive success in this species.

Extrinsic Factors

The reproductive cycle of Whooping Cranes is highly dependent on several extrinsic factors (Mirande et al., 1996). As described previously, increasing day length stimulates the activation of the neuroendocrine cascade in preparation for reproduction (Ball and Ketterson, 2008; Farner and Wingfield, 1980). In addition to photoperiod, other factors, including light intensity, temperature, rainfall, open water, nest size, and food availability have been shown to influence crane reproduction (Fig. 17.1; Mirande et al., 1996; Olsen, 2011).

Changes in weather patterns can affect reproductive success of Whooping Cranes. In the FNMP, an extended period of drought was identified as a probable cause of low hatching success, possibility due to limited food sources (Folk et al., 2005). Similarly, reduced chick survival during dry years also has been observed in the WBNP population (Kuyt, 1992).

In the FNMP, several reproductive parameters, including fertility and hatching success, have been shown to be positively correlated with the amount of rainfall and water levels during winter months (i.e., prebreeding season; Spalding et al., 2009). Delayed nesting was observed in a dry year, and the resulting hatching success was poor (Spalding et al., 2009). However, other environmental parameters, including relative humidity, precipitation, soil temperature, and solar radiation during the incubation period have not been documented to affect hatching success.

Mate Selection and Breeding Behavior

Because a main goal of ex situ management is to maintain genetic diversity, crane pairs are selected to minimize a population's average kinship (Ivy and Lacy, 2012). Although this breeding strategy has been widely used in ex situ management of various wildlife species, there is evidence that pairing in this way may have negative impacts on reproductive outcomes, especially in socially monogamous species (Black and Swan, Chapter 16, this volume; Dixon et al., 2003; Griffith et al., 2011; Ihle et al., 2015; Schaedelin and van Dongen, 2015).

Studies in other avian species, such as Zebra Finches (*Taeniopygia guttata*), have shown that breeding pairs resulting from free mate choice achieve higher reproductive success (fertility and chick survival) than pairs that did not have a choice of mates (Ihle et al., 2015). In the Pine Siskin (*Spinus pinus*), the presence in the spring of a potential mate stimulates LH production, facilitated yolk deposition, and appearance of a brood patch in the paired female compared to an unpaired control (Ball and Ketterson, 2008). These responses were more pronounced when the pair had high affiliation, as determined by increased performance of courtship behavior. In the Canvasback Duck (*Aythya valisineria*), females show preference to a particular male and lay fertile egg only if they have formed

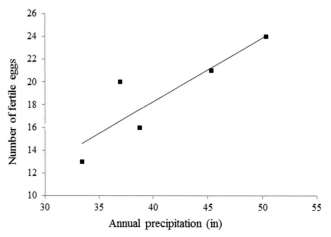

FIGURE 17.1   Relationship between annual precipitation (2007–11) and the number of fertile eggs produced in the following year, from breeding pairs housed at the Patuxent Wildlife Research Center.

a strong pair bond with that individual (Bluhm et al., 1984). There is little information about the influence of mate choice on reproductive capacity of cranes in breeding centers. However, a recent study that compared reproductive behavior between laying and nonlaying Whooping Crane pairs demonstrated that laying pairs exhibited a greater rate of reproductive behavior prior to laying, including marching, unison calling, and copulation attempts, than did their nonlaying counterparts. Members of laying pairs spent more time in physical proximity as well (Brown et al., 2016). This finding suggests that behavioral compatibility of breeding pairs may play an important role in reproductive performance in Whooping Cranes.

MANAGEMENT STRATEGIES AND TECHNOLOGIES FOR ENHANCING REPRODUCTIVE PERFORMANCE

Environmental Management

Like many avian species, Whooping Crane reproduction depends on environmental cues. Hence, providing an ex situ environment that mimics the natural habitat may improve reproductive performance in individuals maintained at breeding centers. Wild Whooping Cranes depend on wetland ecosystems for nesting, overwintering, and migratory stopovers (Ellis et al., 1996; Spalding et al., 2009; Urbanek and Lewis, 2015). Although wild cranes rely on wetlands, captive cranes are mostly housed in dry outdoor pens to reduce potential health risks associated with pathogens present in standing water (Ellis et al., 1996). However, there is evidence suggesting that reproduction of captive cranes may depend on water availability. First, captive-bred Whooping Cranes in a reintroduced population exhibit low reproductive success when water levels are low (Spalding et al., 2009). Second, retrospective analysis of egg fertility in breeding pairs housed at PWRC during 2007–11 indicates that the total number of fertile eggs produced is related to annual precipitation in the previous year (Fig. 17.1). Third, it has been shown that crane pairs in a pen containing a pool with flowing water show increased egg production (Mirande et al., 1996). Seasonal pen flooding also has been shown to increase pair interaction and foraging behavior (Mirande et al., 1996). Ongoing research at PWRC is exploring the influence of enclosure environment on egg production in Whooping

Cranes. Preliminary findings suggest that moving a nonlaying crane pair into an enclosure with a large pond may increase egg production, and the concentration of fecal estrogen metabolites appears to be higher when the pair is housed in a pen with a pond compared to a dry pen.

Artificial Insemination

Sandhill Cranes were the first crane species for which artificial insemination (AI) was implemented, in 1969. The practice was later adapted for other crane species, including the Whooping Crane (Gee, 1983). AI has been used extensively in Whooping Crane reproductive management in all breeding centers. An excellent review on techniques involved in crane AI, including semen collection and insemination, was published by Gee and Mirande (1996). The main objectives of using AI in captive Whooping Cranes are to (1) achieve genetically desirable pairing, (2) augment reproductive performance of crane pairs that fail to lay or do not reliably lay fertile eggs, and (3) enable reproduction in cranes that are physically or behaviorally impaired (e.g., improperly imprinted birds) (Black and Swan, Chapter 16, this volume; Canadian Wildlife Service and U.S. Fish and Wildlife Service, 2005). AI in cranes is an involved process: birds that participate should be properly trained to reduce stress associated with handling, and AI should be performed by well-trained personnel with whom cranes are familiar (Gee and Mirande, 1996). When fresh semen is used, the percentage of fertile eggs obtained is often higher than via natural breeding (Gee and Mirande, 1996). Some breeding centers, such as the Audubon Species Survival Center, rely heavily on this technology to produce fertile eggs (Harrell and Bidwell, 2016).

The fertility rate achieved through AI using fresh semen depends on several factors, including sperm quality (concentration and motility), the location at which sperm are deposited (vagina vs. cloaca), the time between insemination and oviposition, and the sample volume (Black and Swan, Chapter 16, this volume; Gee and Mirande, 1996). When sperm are deposited in the cloaca, AI should be performed more frequently and timed to closely match oviposition (Gee and Mirande, 1996). The optimal volume of sample to use for each insemination depends on sperm concentration and the capacity of the female reproductive tract to retain the semen. Large-volume insemination may result in a portion of sperm being expelled from the vagina, which in turn reduces fertility. More frequent insemination may be necessary when using samples with low sperm concentration or of poor quality. AI with fresh semen has been routinely used in Whooping Crane management during the past 2 decades, but frozen–thawed semen has not been utilized in this species due to the poor quality of frozen–thawed samples (see later).

Sperm Cryopreservation

The value of sperm cryopreservation in ex situ management of wildlife species has been extensively discussed (Pukazhenthi et al., 2011; Pukazhenthi and Wildt, 2004; Wildt, 2003). For birds, semen cryopreservation has been studied mostly in domestic poultry, including chickens, turkeys, ducks, and guinea fowl (Blesbois et al., 2008). Protocols used in domestic poultry have been adapted to many nondomestic birds with varying results (Table 17.2). Dimethylacetamide and dimethylsulfoxide are the two main cryoprotectants used for cryopreserving sperm of wild birds (Table 17.2). Percent motility of frozen–thawed sperm ranges from 5% to 40%. For crane species, sperm cryopreservation has primarily been studied in the Sandhill Crane (Blanco et al., 2011, 2012; Gee et al., 1985), and the protocol developed for this species has been adapted for the endangered Mississippi Sandhill Crane, resulting in the production of live chicks. A recent study on the White-naped Crane (*Grus vipio*) demonstrated that two-step cooling was superior to

TABLE 17.2   Summary of Freezing and Thawing Protocols Used for Semen in Wild Avian Species and Survival of Thawed Sperm. Every Study Here That Attempted AI with Thawed Samples Resulted in the Production of a Chick. DMA, Dimethylacetamide; DMSO, Dimethylsulfoxide; LN2, Liquid Nitrogen; AI, Artificial Insemination.

| Species | | | | | Thawing survival | | AI |
Common name	Order	Cryoprotectant	Method and rate	Thawing temperature	Motility (%)	Viability (%)	performed
Imperial Eagle	Accipitriformes[a]	DMA	50°C/min	4°C		70	No
Bonelli's Eagle	Accipitriformes[a]	DMA	1°C/min (4° to −20°), 2°C/min (−20° to −70°), then plunged in LN2	4°C		65	No
Golden Eagle	Accipitriformes[a]	DMA	1°C/min (4° to −20°), 2°C/min (−20° to −70°), then plunged in LN2	4°C		60	No
White Backed Vulture	Accipitriformes[b]	DMSO	1°C/min (24° to 4°), 6°C/min (4° to −85°), then plunged into LN2	37°C	33.5		No
Emu	Casuariiformes[c]	DMA	11°C/min (5° to −140°), then plunged into LN2	5°C	29.8	42.8	No
Blue Rock Pigeon	Columbiformes[d]	DMSO	1°C/min (4° to −20°), 8°C/min (−20° to −80°), then plunged in LN2	37°C	39.2		Yes
Peregrine Falcon	Falconiformes[a]	DMA	1°C/min (4° to −20°), 2°C/min (−20° to −70°), then plunged in LN2	4°C		50	No
Pheasant spp.	Galliformes[e]	DMA	Vitrification	60°C	5-20		Yes

Sandhill Crane	*Gruiformes*[f]	DMSO	1°C/min (4° to –20°), 50°C/min (–20° to –80°), then plunged into LN2	0.5°C		50	Yes
Houbara Bustard	*Gruiformes*[g]	DMA	Vitrification	60°C	24.9	36.4	Yes
King Penguin	*Sphenisciformes*[h]	DMSO	Placed on dry ice for 5 min then plunged into LN2	35°C	36.8	37.6	No
Megellanic Penguin	*Sphenisciformes*[i]	DMSO + Ethylene Glycol	6°C/min (21° to 4°), 51°C/min (4° to –80°), then plunged into LN2	50°C		77.8	No

[a]Blanco, J.M., Gee, G., Wildt, D.E., Donoghue, A.M., 2000. Species variation in osmotic, cryoprotectant, and cooling rate tolerance in poultry, eagle, and peregrine falcon spermatozoa. Biology of Reproduction. 63 (4). 1164–1171.

[b]Umapathy, G., Sontakke, S., Reddy, A., Ahmed, S., Shivaji, S., 2005. Semen characteristics of the captive Indian white-backed culture (Gyps bengalensis). Biology of Reproduction 73 (5): 1039–1045.

[c]Sood, S., Malecki, I.A., Tawang, A., Martin, G.B., 2012. Survival of emu (Dromaius novaehollandiae) sperm preserved at subzero temperatures and different cryoprotectant concentrations. Theriogenology 78 (7): 1557–1569.

[d]Sontakke, S.D., Umapathy, G., Sivaram, V., Kholkute, S.D., Shivaji, S., 2004. Semen characteristics, cryopreservation, and successful artificial insemination in the blue rock pigeon (Columba livia). Theriogenology 62 (1–2): 139–153.

[e]Saint Jalme, M., Lecoq, R., Seigneurin, F., Blesbois, E., Plouzeau, E., 2003. Cryopreservation of semen from endangered pheasants: the first step towards a cryobank for endangered avian species. Theriogenology 59 (3–4): 875–888.

[f]Gee et al., 1985.

[g]Hartley, P.S., Dawson, B., Lindsay, C., McCormick, P., Wishart, G., 1999. Cryopreservation of houbara semen: a pilot study. Zoo Biology 18 (2): 147–152.

[h]O'Brien, J.K., Robeck, T.R., 2014. Semen characterization, seasonality of production, and in vitro sperm quality after chilled storage and cryopreservation in the king penguin (Aptenodytes patagonicus). Zoo Biology 33 (2): 99–109.

[i]O'Brien, J.K., Steinman, K.J., Montano, G.A., Dubach, J.M., Robeck, T.R., 2016. Chicks produced in the Magellanic penguin (Spheniscus magellanicus) after cloacal insemination of frozen-thawed semen. Zoo Biology 35 (4): 326–338.

the one-step method in maintaining postthaw sperm motility (Panyaboriban et al., 2016). The same study also showed that thawing condition impacted sperm cryosurvival: warming samples in a 37°C water bath for 30 s resulted in a higher percentage of motile sperm than warming in a 4°C water bath for 1 min. Interestingly, although these authors reported significant reduction in sperm motility and viability postthawing (in comparison to freshly collected samples), cryopreservation procedures had no effect on the integrity of acrosomal membranes, which are essential for sperm–egg interaction.

Cryopreservation has not been used routinely in reproductive and genetic management for any crane species. Currently, the quality of cryopreserved sperm is low, but the ability to successfully cryopreserve Whooping Crane spermatozoa and establish germplasm banks for this species will certainly enhance genetic management of the ex situ population, as it allows production of offspring from behaviorally incompatible or geographically separated, but genetically desirable, pairs. Furthermore, the capacity to cryopreserve semen samples from the current population has the potential for reintroducing valuable genetics into the future population, as has been demonstrated in the black-footed ferret (*Mustela nigripes*; Santymire et al., 2014). More research is needed to improve cryopreservation practices if the method is to be used more widely.

SUMMARY AND OUTLOOK

To improve reproductive success of ex situ Whooping Cranes, a better understanding of the basic reproductive biology of the species is needed. While some data have been generated, especially at the gonadal level (Brown et al., 2016), a priority now is to understand (1) the mechanisms regulating ovarian function, yolk deposition, and follicle hierarchy development and (2) sensitivity of endocrine pathways

and reproductive behavior to male presence. Having this knowledge would improve understanding of cause(s) of poor reproduction in female cranes that will lead to the development of mitigation strategies, such as hormone therapy to enhance fertility in the most genetically valuable individuals. Better understanding the influence of males (their presence and behavior) on endocrine and reproductive behavior of females will help determine (1) whether males are needed for exogenous hormone treatment to stimulate HPG axis and egg production in a nonlaying female, (2) how pair bonding influences endocrine patterns or egg production in cranes, and (3) whether delayed reproduction in females is linked to the lack of opportunity to choose a mate. The latter two questions are extremely important, as mate choice is very restrictive in captive Whooping Cranes; the effects of a more naturalistic formation of pairs should be explored.

Because reproductive capacity of Whooping Cranes is highly dependent on environmental cues, future research examining the impact of the captive environment (e.g., water availability, the presence or density of conspecifics) on reproductive outcomes in birds at breeding centers would be beneficial. The findings from such studies will aid managers in developing improved husbandry for this species.

There are no agreed-upon sperm cryopreservation protocols for Whooping Cranes. There is a need for research to investigate the influence of various procedural factors (e.g., cryoprotectant type and concentration, freezing and warming rates) on the survival of cryopreserved Whooping Crane sperm. Future research also should explore the usefulness of cold storage for short-term preservation of crane spermatozoa, as such technology would aid genetic exchange among breeding centers.

Since the 1960s when the first Whooping Crane eggs were brought to PWRC, several hundred chicks have been produced and released into the wild. Despite this tremendous success, poor reproductive performance of ex situ birds has been an ongoing challenge that limits the contribution of this population to species recovery. Recent studies have shed some light on the potential causes of poor reproduction in captive cranes. However, more studies are needed for crane managers to effectively establish strategies to overcome this challenge and to ensure long-term sustainability of ex situ populations, as well as recovery of the wild Whooping Crane.

Acknowledgments

The authors thank Bryant Tarr of the International Crane Foundation for providing artificial insemination record data to be incorporated into this chapter. The authors also acknowledge Dr. Glen Olsen for his continuing support.

References

Archibald, G.W., Lewis, J.C., 1996. Crane biology. In: Ellis, D.H., Gee, G.F., Mirande, C.M. (Eds.), Cranes: Their Biology, Husbandry, and Conservation. National Biological Service, Washington, DC, and International Crane Foundation, Baraboo, WI, pp. 1–29.

Asa, C., Miller, P., Agnew, M., Rebolledo, J.A.R., Lindsey, S.L., Callahan, M., Bauman, K., 2007. Relationship of inbreeding with sperm quality and reproductive success in Mexican gray wolves. Anim. Conserv. 10 (3), 326–331.

Austin, J.E., Hayes, M.A., Barzen, J.A., 2018. Revisiting the historic distribution and habitats of the whooping crane (Chapter 3). In: French, Jr., J.B., Converse, S.J., Austin, J.E. (Eds.), Whooping Cranes: Biology and Conservation. Biodiversity of the World: Conservation from Genes to Landscapes. Academic Press, San Diego, CA.

Baker, J.R., 1939. The relation between latitude and breeding seasons in birds. Proc. Zool. Soc. Lond. A 108 (4), 557–582.

Ball, G.F., Ketterson, E.D., 2008. Sex differences in the response to environmental cues regulating seasonal reproduction in birds. Philos. Trans. R. Soc. Lond. B 363 (1490), 231–246.

Bedecarrats, G.Y., 2015. Control of the reproductive axis: balancing act between stimulatory and inhibitory inputs. Poult. Sci. 94 (4), 810–815.

Bennison, C., Hemmings, N., Slate, J.M., Birkhead, T., 2015. Long sperm fertilize more eggs in a bird. Proc. R. Soc. B 282 (1799), 20141897.

Bensch, S., Hasselquist, D., von Schantz, T., 1994. Genetic similarity between parents predicts hatching failure: nonincestuous inbreeding in the great reed warbler? Evolution 48 (2), 317–326.

Black, S.R., Swan, K.D., 2018. Advances in conservation breeding and management of whooping cranes (Chapter 16). In: French, Jr., J.B., Converse, S.J., Austin, J.E. (Eds.), Whooping Cranes: Biology and Conservation. Biodiversity of the World: Conservation from Genes to Landscapes. Academic Press, San Diego, CA.

Blanco, J.M., Long, J.A., Gee, G., Wildt, D.E., Donoghue, A.M., 2011. Comparative cryopreservation of avian spermatozoa: benefits of non-permeating osmoprotectants and ATP on turkey and crane sperm cryosurvival. Anim. Reprod. Sci. 123 (3–4), 242–248.

Blanco, J.M., Long, J.A., Gee, G., Wildt, D.E., Donoghue, A.M., 2012. Comparative cryopreservation of avian spermatozoa: effects of freezing and thawing rates on turkey and sandhill crane sperm cryosurvival. Anim. Reprod. Sci. 131 (1–2), 1–8.

Blesbois, E., Grasseau, I., Seigneurin, F., Mignon-Grasteau, S., Saint Jalme, M., Mialon-Richard, M.M., 2008. Predictors of success of semen cryopreservation in chickens. Theriogenology 69 (2), 252–261.

Bluhm, C.K., Phillips, R.E., Burke, W.H., Gupta, G.N., 1984. Effects of male courtship and gonadal steroids on pair formation, egg-laying, and serum LH in canvasback ducks (Aythya valisineria). J. Zool. 204 (2), 185–200.

Brown, M.E., Converse, S.J., Chandler, J.N., Crosier, A.C., Lynch, W., Wildt, D.E., Keefer, C.L., Songsasen, N., 2017. Time within reproductive season, but not age or inbreeding coefficient, affects seminal and sperm quality in the whooping crane (Grus americana). Reprod. Fertil. Dev. 29, 294–306.

Brown, M.E., Converse, S.J., Chandler, J.N., Shafer, C., Brown, J.L., Keefer, C.L., Songsasen, N., 2016. Female gonadal hormones and reproductive behaviors as key determinants of successful reproductive output of breeding whooping cranes (Grus americana). Gen. Comp. Endocrinol., 230–231.

Brown, M.E., Converse, S.J., Keefer, C.L., Songsasen, N. 2015. What makes my nest best? The effect of captive environment on whooping crane stress and reproduction. In: 5th Conference of the International Society of Wildlife Endocrinology, October 12–14, Berlin, Germany, p. 24, Abstract #94.

Canadian Wildlife Service and U.S. Fish and Wildlife Service, 2005. International Recovery Plan for the Whooping Crane. Recovery of Nationally Endangered Wildlife (RENEW), Ottawa, and U.S. Fish and Wildlife Service, Arbuquerque, NM, p. 162.

Chen, G., Gee, G.F., Nicolich, J.M., Taylor, J.A., 2001. The effects of semen collection on fertility in captive, naturally fertile, sandhill cranes. Proceedings of the North American Crane Workshop 8, 185–194.

Converse, S.J., Royle, J.A., Adler, P.H., Urbanek, R.P., Barzen, J.A., 2013. A hierarchical nest survival model integrating incomplete temporally-varying covariates. Ecol. Evol. 3 (13), 4439–4447.

Converse, S.J., Royle, J.A., Urbanek, R.P., 2012. Bayesian analysis of multiple state data with individual covariates for estimating genetic affects on demography. J. Ornithol. 152 (Suppl. 2), S561–S572.

Converse, S.J., Strobel, B.N., Barzen, J.A., 2018. Reproductive failure in the Eastern Migratory Population: the interaction of research and management (Chapter 8). In: French, Jr., J.B., Converse, S.J., Austin, J.E. (Eds.), Whooping Cranes: Biology and Conservation. Biodiversity of the World: Conservation from Genes to Landscapes. Academic Press, San Diego, CA.

Dellinger, T.A., 2018. Florida's nonmigratory whooping cranes (Chapter 9) In: French, Jr., J.B., Converse, S.J., Austin, J.E. (Eds.), Whooping Cranes: Biology and Conservation. Biodiversity of the World: Conservation from Genes to Landscapes. Academic Press, San Diego, CA.

Dellinger, T., Folk, M.J., Spalding, M.G., 2013. Copulatory behavior of non-migratory whooping cranes in Florida. Wilson J. Ornithol. 125 (1), 128–133.

Dixon, A., Harvey, N., Patton, M., Setchell, J., 2003. Behavior and reproduction. In: Holt, W.V., Pickard, A.R., Rodger, J.C., Wildt, D.E. (Eds.), Reproductive Science and Integrated Conservation. Cambridge University Press, Cambridge, pp. 24–41.

El Halawani, M.E., Burke, W.H., Millam, J.R., Fehrer, S.C., Hargis, B.M., 1984. Regulation of prolactin and its role in gallinaceous bird reproduction. J. Exp. Zool. 232 (3), 521–529.

Ellis, D.H., Gee, G.F., Mirande, C.M., 1996. Cranes: Their Biology, Husbandry and Conservation. National Biological Service, Washington DC, and International Crane Foundation, Baraboo, WI.

Ellis, D.H., Lewis, J.C., Gee, G.F., Smith, D.G., 1992. Population recovery of the whooping crane with emphasis on reintroduction efforts: past and future. In: Stahlecker, D.W. (Ed.), Proceedings of the Sixth North American Crane Workshop, Oct. 3–5, 1991, Regina, Sask. North American Crane Working Group, Grand Island, NE, pp. 142–150.

Erickson, R.C., 1975. Captive breeding of whooping cranes at the Patuxent Wildlife Research Center. In: Martin, R.D. (Ed.), Breeding Endangered Peciesin Captivity. Academic Press, New York, pp. 99–114.

Facemire, C.F., Gross, T.S., Guillette, L.J., 1995. Reproductive impairment in the Florida panther – nature or nurture. Environ. Health Perspect. 103, 79–86.

Farner, D.S., Wingfield, J.C., 1980. Reproductive endocrinology of birds. Annu. Rev. Physiol. 42, 457–472.

Fitzpatrick, J.L., Evans, J.P., 2009. Reduced heterozygosity impairs sperm quality in endangered mammals. Biol. Lett. 5 (3), 320–323.

Folk, M.J., Nesbitt, S.A., Schwikert, S.T., Schmidt, J.A., Sullivan, K.A., 2005. Breeding biology of re-introduced non-migratory whooping cranes in Florida. Proceedings of the North American Crane Workshop 9, 105–109.

French, Jr., J.B., Converse, S.J., Austin, J.E., 2018. Whooping cranes past and present (Chapter 1). In: French, Jr., J.B., Converse, S.J., Austin, J.E. (Eds.), Whooping Cranes: Biology and Conservation. Biodiversity of the World: Conservation from Genes to Landscapes. Academic Press, San Diego, CA.

Gee, G.F., 1983. Crane reproductive physiology and conservation. Zoo Biol. 2 (3), 199–213.

Gee, G.F., Baskt, M.R., Sexton, S.F., 1985. Cryogenic preservation of semen from the greater sandhill crane. J. Wildl. Manage. 49 (2), 480–484.

Gee, G.F., Mirande, C.M., 1996. Special technique, part A: crane artificial insemination. In: Ellis, D.H., Gee, G.F., Mirande, C.M. (Eds.), Cranes: Their Biology, Husbandry, and Conservation. National Biological Service, Washington, DC, and International Crane Foundation, Baraboo, WI, pp. 205–217.

Gee, G.F., Russman, S.E., 1996. Reproductive physiology. In: Ellis, D.H., Gee, G.F., Mirande, C.M. (Eds.), Cranes: Their Biology, Husbandry, and Conservation. National Biological Service, Washington, DC, and International Crane Foundation, Baraboo, WI, pp. 123–136.

Griffith, S.C., Pryke, S.R., Buttemer, W.A., 2011. Constrained mate choice in social monogamy and the stress of having an unattractive partner. Proc. R. Soc. Lond. B 278 (1719), 2798–2805.

Harrell, W., 2014. Report on Whooping Crane Recovery Activities (2013 Breeding Season – 2014 Spring Migration). U.S. Fish and Wildlife Service, Aransas, TX.

Harrell, W., Bidwell, M., 2016. Report on Whooping Crane Recovery Activities (2015 Breeding Season – 2016 Spring Migration). U.S. Fish and Wildlife Service, Aransas, TX, 22 pp.

Ihle, M., Kempenaers, B., Forstmeier, W., 2015. Fitness benefits of mate choice for compatibility in a socially monogamous species. PLoS Biol. 13 (9), e1002248.

Ivy, J.A., Lacy, R.C., 2012. A comparison of strategies for selecting breeding pairs to maximize genetic diversity retention in managed populations. J. Hered. 103 (2), 186–196.

Johnsgard, P.A., 1983. Cranes of the World: 5. Comparative Reproductive Biology. Indiana University Press, Bloomington.

Joyner, K.L., 1990. Avian theriogenology. In: Ritchie, B.W., Harrison, G.J. (Eds.), Avian Medicine: Principles and Application. H. B. D. International, Brentwood, pp. 749–804.

Kempenaers, B., Adriaensen, F., Noordwijk, A.J.V., Dhondt, A.A., 1996. Genetic similarity, inbreeding and hatching failure in blue tits: are unhatched eggs infertile? Proc. R. Soc. Lond. B 263 (1367), 179–185.

Kuyt, E., 1987. Whooping crane migration studies, 1981–82. In: Archibald, G., Pasquier, R. (Eds.), Proceedings of the 1983 International Crane Workshop. International Crane Foundation, Baraboo, WI, pp. 371–379.

Kuyt, E. 1992. Aerial radio-tracking of whooping cranes migrating between Wood Buffalo National Park and Aransas National Wildlife Refuge, 1981–1984. Canadian Wildlife Service, Occasional Paper 74, 53.

Kuyt, E., 1996. Reproductive manipulation in the whooping crane Grus americana. Bird Conserv. Int. 6 (1), 3–10.

Mirande, C.M., Gee, G.F., Burke, A., Whitlock, P., 1996. Egg and semen production. In: Ellis, D.H., Gee, G.F., Mirande, C.M. (Eds.), Cranes: Their Biology, Husbandry, and Conservation. National Biological Service, Washington, DC, and International Crane Foundation, Baraboo, WI, pp. 45–57.

Norris, D.O., 2006. Vertebrate Endocrinology. Elsevier Science, Oxford, UK.

Olsen, G. 2011. Photoperiod and nesting phenology of whooping cranes at two captive facilities. Proceedings of the North American Crane Workshop 13–16 March, Grand Island, NE, p. 100.

Ortego, J., Calabuig, G., Cordero, P.J., Aparicio, J.M., 2007. Egg production and individual genetic diversity in lesser kestrels. Mol. Ecol. 16 (11), 2383–2392.

Ottinger, M.A., Baskt, M.R., 1995. Endocrinology of the avian reproductive system. J. Avian Med. Surg. 9 (4), 242–250.

Panyaboriban, S., Pukazhenthi, B., Brown, M.E., Crowe, C., Lynch, W., Singh, R.P., Songsasen, N., 2016. Influence of cooling and thawing conditions and cryoprotectant concentration on frozen-thawed survival of white-naped crane (*Antigone vipio*) spermatozoa. Cryobiology 73 (2), 209–215.

Paster, M.B., 1991. Avian reproductive endocrinology. Vet. Clin. North Am. 21 (6), 1343–1359.

Pollock, C.G., Orosz, S.E., 2002. Avian reproductive anatomy, physiology and endocrinology. Vet. Clin. North Am. 5 (3), 441–474.

Pukazhenthi, B.S., Hagedorn, M., Comizzoli, P., Songsasen, N., Wildt, D.E., 2011. Cryopreserving endangered species gametes, embryos and gonadal tissue: challenges, successes and future directions. Cryobiology 63 (3), 308–309.

Pukazhenthi, B.S., Wildt, D.E., 2004. Which reproductive technologies are most relevant to studying, managing and conserving wildlife? Reprod. Fertil. Dev. 16 (1–2), 33–46.

Pukazhenthi, B.S., Wildt, D.E., Howard, J., 2001. The phenomenon and significance of teratospermia in felids. J. Reprod. Fertil. 57, 423–433.

Rabon, Jr., D.R., Waddell, W., 2010. Effects of inbreeding on reproductive success, performance, litter size, and survival in captive red wolves (*Canis rufus*). Zoo Biol. 29 (1), 36–49.

Roelke, M.E., Martenson, J.S., O'Brien, S.J., 1993. The consequence of demographic reduction and genetic depletion in the endangered Florida panther. Curr. Biol. 3 (6), 340–350.

Roldan, E.R., Cassinello, J., Abaigar, T., Gomendio, M., 1998. Inbreeding, fluctuating asymmetry, and ejaculate quality in an endangered ungulate. Proc. R. Soc. Lond. B 265 (1392), 243–248.

Ryan, K.K., Lacy, R.C., Margulis, S.W., 2003. Impacts of inbreeding on components of reproductive success. In: Holt, W.V., Pickard, A.R., Rodger, J.C., Wildt, D.E. (Eds.), Reproductive Science and Integrated Conservation. Cambridge University Press, Cambridge, pp. 82–96.

Santymire, R.M., Livieri, T.M., Branvold-Faber, H., Marinari, P.E., 2014. The black-footed ferret: on the brink of recovery? Adv. Exp. Med. Biol. 753, 119–134.

Schaedelin, F.C., van Dongen, W.F.D., Wagner, R.H., 2015. Mate choice and genetic monogamy in a biparental, colonial fish. Behav. Ecol. 26 (3), 782–788.

Sharlin, J.S., Shaffner, C.S., Gee, G.F., 1979. Sperm head length as a predictor of fecudity in the sandhill crane, *Grus canadensis*. J. Reprod. Fertil. 55 (2), 411–413.

Sharp, P.J., Dawson, A., Lea, R.W., 1998. Control of luteinizing hormone and prolactin secretion in birds. Comp. Biochem. Physiol. C 119 (3), 275–282.

Smith, D.H.V., Converse, S.J., Gibson, K.W., Moehrenschlager, A., Link, W.A., Olsen, G.H., Maguire, K., 2011. Decision analysis for conservation breeding: maximizing production for reintroduction of whooping cranes. J. Wildl. Manage. 75 (3), 501–508.

Spalding, M.G., Folk, M.J., Nesbitt, S.A., Folk, M.L., Kiltie, R., 2009. Environmental correlates of reproductive success for introduced resident whooping cranes in Florida. Waterbirds 32 (4), 538–547.

Swengel, S.R., Archibald, G.W., Ellis, D.H., Smith, D.G., 1996. Behavior management. In: Ellis, D.H., Gee, G.F., Mirande, C.M. (Eds.), Cranes: Their Biology, Husbandry, and Conservation. National Biological Service, Washington, DC, and International Crane Foundation, Baraboo, WI, pp. 105–117.

Swinnerton, K.J., Groombridge, J.J., Jones, C.G., Burn, R.W., Mungroo, Y., 2004. Inbreeding depression and founder diversity among captive and free-living populations of the endangered pink pigeon *Columba mayeri*. Anim. Conserv. 7 (4), 353–364.

Tsutsui, K., Ubuka, T., Bentley, G.E., Kriegsfeld, L., 2012. Gonadotropin-inhibitory hormone (GnIH): discovery, progress and prospect. Gen. Comp. Endocrinol. 177 (3), 305–314.

Urbanek, R.P., Lewis, J.C., 2015. Whooping crane (*Grus americana*). In: Rodewald, P.G. (Ed.), The Birds of North America Online. Cornell Lab of Ornithology, Ithaca, NY. Retrieved from the Birds of North America Online: https://birdsna.org/Species-Account/bna/species/whocra/introduction. doi:10.2173/bna.153.

Urbanek, R.P., Szyszkoski, E.K., Zimorski, S.E., Fondow, L.E.A., 2018. Pairing dynamics of reintroduced migratory whooping cranes (Chapter 10). In: French, Jr., J.B., Converse, S.J., Austin, J.E. (Eds.), Whooping Cranes: Biology and Conservation. Biodiversity of the World: Conservation from Genes to Landscapes. Academic Press, San Diego, CA.

Velando, A., Barros, A., Moran, P., 2015. Heterozygosity-fitness correlations in a declining seabird population. Mol. Ecol. 24 (5), 1007–1018.

Wildt, D.E., 2003. The role of reproductive technologies in zoos: past, present and future. Int. Zoo Yearb. 38 (1), 111–118.

Wildt, D.E., Bass, E.J., Chakraborty, P.K., Wolfle, T.L., Stewart, A.P., 1982. Influence of inbreeding on reproductive performance, ejaculate quality and testicular volume in the dog. Theriogenology 17 (4), 445–452.

Health of Whooping Cranes in the Central Flyway

Barry K. Hartup

International Crane Foundation, Baraboo, WI, United States

INTRODUCTION

The North American Whooping Crane (*Grus americana*) has successfully returned from the brink of extinction, but remains endangered despite over 40 years of concerted conservation action (Canadian Wildlife Service and U.S. Fish and Wildlife Service, 2005). Managing threats to Whooping Crane health will be an important component of ongoing efforts to recover the species. Whooping Cranes frequent agricultural lands, interact with domestic animals, and are exposed to a variety of physical hazards (e.g., power lines) throughout the Central Flyway migratory corridor. Oil and natural gas extraction and industrial shipping in the midst of winter habitat present potential risks to individual and population health and welfare that may negatively affect species recovery. In addition, the Whooping Crane Health Advisory Team (which advises the International Whooping Crane Recovery Team) has long sought information on the health status of the Aransas-Wood Buffalo Population (AWBP) to discern disease risks associated with reintroduction

of captive-bred Whooping Cranes into nearby Louisiana. There is concern that reintroduced cranes from Louisiana could mix with the AWBP at migratory stopover or wintering sites and transmit parasites or disease agents with negative outcomes for the AWBP. Knowledge of the health status of source populations is also a necessary component in conducting proper risk assessments for locating reintroductions or conservation translocations (International Union for Conservation of Nature/Species Survival Commission, 2013), but has never been obtained for the Whooping Crane. Indeed, "little is known about the importance of diseases or parasites as mortality factors for wild Whooping Cranes" (Canadian Wildlife Service and U.S. Fish and Wildlife Service, 2005, p. 23).

Virtually all published veterinary information on Whooping Cranes has come from captive, *ex situ* populations or free-ranging, captive-bred reintroduced cranes in the eastern United States, which are outside of the core range of the species (Cole et al., 2009; Hartup et al., 2010a; Hoar et al., 2007; Olsen et al., 1996, 1997; Olsen et al., Chapter 19, this volume; Spalding et al., 2008).

Wild Whooping Cranes are presumably susceptible to a variety of infectious and toxicological diseases, but evidence of disease-related mortality is infrequently documented. Avian mycobacteriosis is known to affect the species, and infectious bursal disease virus is presumed to have resulted in significant morbidity within two reintroduction cohorts in Florida (Canadian Wildlife Service and U.S. Fish and Wildlife Service, 2005; Candelora et al., 2010; Hartup et al., 2010b). Human impacts on the environment and increased movement of biological materials around the globe are resulting in more frequent exposure to pathogens of possible significance to Whooping Cranes. For example, West Nile virus first emerged in North America in 1999 and there is no information available on the exposure of AWBP Whooping Cranes to this pathogen, despite its study in captive Whooping Cranes (Hartup, 2008).

Studies of various threatened and endangered birds have focused on population health status and have addressed similar issues of concern (Deem et al., 2005; Ortiz-Catedral et al., 2011). Uhart et al. (2006) assessed the Greater Rhea (*Rhea americana*) to establish baseline health information useful in translocation and reintroduction schemes involved in *ex situ* production in Argentina. The concerns for this population centered on potential negative interaction (i.e., transmission of infectious agents) between intensively farmed and wild rheas. The authors determined reference information for hematology and plasma biochemistry, and evaluated the exposure of wild rheas to several infectious disease agents through the use of serological testing. Smith et al. (2008) conducted similar work with Humboldt Penguins (*Sphenicus humboldti*) in Peru, capitalizing on ongoing field studies as part of long-term population monitoring. Padilla et al. (2006) conducted a health assessment in a diverse Galapagos island avian community believed to be at risk due to human population expansion and potential introduction of diseases from domestic and peridomestic avian sources.

In each of these studies, veterinary evaluations of individual birds were aggregated to provide baseline population data to enhance conservation efforts in various ways (Karesh and Cook, 1995).

This chapter presents new data gleaned from a multiyear field study, and provides an initial assessment of the health of the migratory population of Whooping Cranes in the Central Flyway. These data illustrate several differences between wild Whooping Cranes and *ex situ* or reintroduced populations that were previously unknown. The potential implications from key disease risks to future crane health and conservation are highlighted.

FIELD METHODS

Whooping Cranes were captured for satellite telemetry tagging, color banding, and health assessment under U.S. Fish and Wildlife Service and Texas Parks and Wildlife Department permits. Thirty-eight adult cranes were captured at Aransas National Wildlife Refuge (NWR) during winters between December 2009 and February 2014 with a self-activated or remotely triggered leg snare followed by manual restraint within 30 s (Pearse et al., Chapter 6, this volume). A cloth hood was placed on each bird to reduce visual stimuli and stress, and facilitate the tagging and examination procedures.

The physical examination consisted of rapid visual inspection, palpation, auscultation, and photography of abnormalities with emphasis on evaluation of the head and oral cavity, feather condition and presence of ectoparasites, extremities, and the respiratory tract. Pectoral muscle scores were determined on a 1–5 scale to provide a measure of body condition (Olsen et al., 1996). Up to 12 mL of blood was collected from the right jugular vein and placed into sterile tubes for analysis of individual clinical pathology measures and antibodies to diseases of concern, among others. Thin blood smears were made to provide hematological measures of inflammation, anemia, or blood parasitism. Cloacal swabs for virus isolation were immersed in viral transport media per USGS National Wildlife Health Center (NWHC, Madison, WI) protocol, and a second swab was collected and placed into bacterial transport media for subsequent inoculation of selective media. Sampling protocols were approved by a University of Wisconsin Institutional Animal Care and Use Committee.

All samples were stored on wet ice in the field until return to controlled conditions. Following centrifugation, aliquots of whole blood, serum, and plasma were placed into cryovials. The vials and the cloacal swabs were stored at −20°C for 2–14 days, and then held at −80°C until analysis. Thin blood films were fixed in methanol and later stained with Wright's stain at the University of Wisconsin Veterinary Care Clinical Pathology laboratory (Madison, WI). Slides exhibiting significant artifact (e.g., large numbers of unidentifiable cells, thrombocyte clumping, or poor staining) were not included in the analyses. Red blood cell morphology, polychromasia (1–4 scale; a measure of active red blood cell regeneration), and the presence of hemoparasites were noted during the 100 cell white blood cell (WBC) differential counts and total WBC estimations. Last, smaller sample sizes in some test results were due to insufficient volumes of blood, unpredictable quality control concerns not mentioned earlier, or specific commercial laboratory constraints beyond the author's control.

Samples from Captive Cranes

Hematology and serum biochemistry tabulations have been available for captive Whooping Cranes since the 1970s and are a well-integrated diagnostic tool in the health management of this population (Olsen et al., 1996, 2001). No trace element references from blood have been determined for Whooping Cranes and only limited information from tissues is available (Lewis et al., 1992). I conducted a comparison of wild Whooping Crane results to reference intervals determined from a contemporary representative sample of captive adult cranes at the International Crane Foundation (ICF, Baraboo, WI). This analysis accounts for differences in diet, underlying disease, and age structure between the populations. Thin blood smears and serum from 30 adult, clinically healthy Whooping Cranes (15M:15F) were collected during the October 2012 flock health check at ICF. The mean age of this population was 20 years (median = 22), with a range of 3–45 years. All samples were processed and analyzed similarly to those from wild Whooping Cranes in terms of content, technique, materials, and laboratory methods. Data were used for reference interval determination using Excel freeware Reference Value Advisor v2.1, National Veterinary School, Toulouse, France, available at http://www.biostat.envt.fr/spip/ in accordance with the American Society of Veterinary Clinical Pathology guidelines (Friedrichs et al., 2012; Geffré et al., 2011). Reference intervals represent a range of values for a physiological measurement in healthy animals from a reference population used to determine a possible pathological state in an individual. The captive Whooping Crane population is appropriate for this purpose because repeated, long-term assessment of health status among known individuals is possible, and lends validity to the measures of interest, albeit under ideal conditions (e.g., ad libitum food, water, shelter, etc.).

RESULTS AND DISCUSSION

A total of 38 subadult or adult Whooping Cranes were sampled at Aransas NWR and surrounding private lands in Texas. One crane was sampled in each of the first two winters of the capture effort, and 36 in the subsequent three winters. The sex ratio of the sampled cranes was one (19 M:19 F).

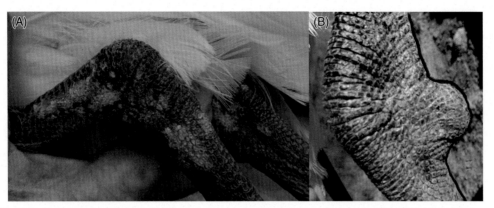

FIGURE 18.1　Integumentary lesions observed in adult Whooping Cranes at Aransas NWR. (A) Multiple depigmented epidermal scars in an adult whooping crane. (B) A 1.4 × 1.7 × 1.1 cm (w × l × d) raised cutaneous mass at the cranial aspect of the right tarsus. *Photo credit: Barry Hartup.*

All cranes were visibly normal and in generally good condition and no injuries were identified during any examination. Respiratory and cardiac auscultations were within normal limits, and response to stimuli and release was unremarkable. The mean weight ± SE of males was 6755 ± 135g (range 5563–7603g). The mean weight of females was 6158g ± 112g (range 5040–6906g). Pectoral muscle scores for males and females were similar (each median = 3, range 2–4).

Notable physical findings included depigmented epidermal scars in five cranes, found on the unfeathered portion of the legs at the tarsal joint (Fig. 18.1A). Raised soft tissue masses of two general types were observed: three cranes had cutaneous lesions on the crown or proximate to the commissure of the mouth, each 3 mm in diameter or smaller, and two cranes had masses at the tarsus 4 mm in diameter or greater in size (Fig. 18.1B). These lesions were consistent with cutaneous chondromas (benign cartilaginous tumors) diagnosed from the necropsy of another Aransas NWR Whooping Crane in January 2014 and in Sandhill Cranes (*Grus canadensis*) from Florida (unpublished data, J. Lankton, USGS NWHC; Forrester and Spalding, 2003; Nesbitt et al., 2005). A cutaneous 1.8 × 2.2 cm swelling

consistent with pododermatitis (e.g., "bumblefoot," a chronic inflammatory foot condition) of the third toe was observed in one crane. Poor feather condition was rarely observed, but was noted in some second-year cranes and in one recaptured individual in its third year with poor flight feather condition. Poor feather condition is often a sign of malnourishment, systemic illness, or contact with an external substance that is damaging to the feather structure.

Hematology

Assessment of blood smears from AWBP cranes provided the following core complete blood count measurements: estimated WBC (Fudge, 2000), relative (%) heterophils, lymphocytes, eosinophils, monocytes and basophils, ratio of heterophils to lymphocytes (H:L; a stress-related measure), and estimated thrombocytes (the clotting cells; Samour, 2008) (Table 18.1). Wild AWBP cranes had greater WBC and lower H:L ratios (due to lower % heterophils and greater % eosinophils) compared to captive adults. These modest differences likely resulted from exposure of the wild cranes to a wider diversity of potential pathogens, particularly parasites, which typically induce an increase in eosinophils. The captive

TABLE 18.1   Comparison of Hematologic Values between Aransas-Wood Buffalo Population (AWBP; Samples Collected 2009–14) and Captive (Samples Collected 2012) Whooping Cranes. Reference Intervals Are the Expected Range of 95% of Observations of Each Analyte from a Healthy Captive Population of Whooping Cranes (See Text for Methodology).

Variables	AWBP adults ($n = 28$)		Captive adults ($n = 27$)	
	Mean ± SE (median)	Range	Mean ± SE (median)	Reference interval
White blood cells (#/µL)[a]	18,932 ± 1,810 (18,375)	2,600–45,250	12,995 ± 805 (12,750)	6,526–23,800
Heterophil/lymphocyte ratio[a]	0.39 ± 0.03 (0.41)	0.10–0.81	0.53 ± 0.04 (0.56)	0.12–1.01
Heterophils (%)[a]	22.2 ± 1.5 (23.5)	6–34	28.8 ± 1.5 (29)	11–43
Lymphocytes (%)	59.8 ± 1.7 (59.5)	39–76	57.9 ± 1.9 (55)	41–85
Eosinophils (%)[a]	13.5 ± 1.6 (11)	4–37	9.1 ± 0.8 (9)	0–18
Monocytes (%)	4.0 ± 0.5 (4)	1–10	4.0 ± 0.6 (4)	0–10
Basophils (%)	0.5 ± 0.2 (0)	0–3	0.1 ± 0.1 (0)	0–1
Thrombocytes (no./µL)	17,960 ± 2,368 (15,840)	99–41,148	13,192 ± 1,208 (12,420)	1,417–27,622

[a] Statistical difference between AWBP and captive Whooping Cranes (Student's t test, P < 0.05).

flock at ICF is closely monitored for endoparasites and managed to reduce worms, while pen occupancy is managed to reduce the soil burden of pathogens. AWBP cranes are exposed to a wider array of parasites that are only occasionally observed in captivity (unpublished data, M. Bertram, Texas A&M University). The lower H:L ratio may also suggest wild cranes had lower baseline stress levels associated with younger age. In Song Sparrows (*Melospiza melodia*), higher H:L ratios were found in older individuals (Losdat et al., 2016). In this study, the birds selected from captivity (median age of 22 years), were likely older, on average, than the sample of wild cranes of unknown age (based on Gil-Weir et al., 2012).

The relative WBC estimates are higher than reference norms published by Olsen et al. (1996, 2001). Those studies were based on differential WBC counts from blood stored in the anticoagulant tri-potassium ethylenediaminetetraacetic acid (K_3EDTA) for extended periods of time before smears were made. Mauer et al. (2010) showed that these techniques resulted in lower estimates of lymphocytes and may bias subsequent determinations, such as WBC (downward) and H:L ratios (upward). The data presented here are based on blood samples with no exposure to anticoagulant, subjected to constant handling and preparation techniques, and thus represent an improved set of reference material for the species under both captive and free-ranging conditions.

Serum Biochemistry

All available serum samples were thawed and submitted to Marshfield Labs (Marshfield, WI) for analysis of standard avian biochemistry profiles (Table 18.2). Seven values in wild cranes were significantly different than in captive cranes. The AWBP cranes had lower alkaline phosphatase levels than captive adults. Though low alkaline phosphatase has been linked to dietary zinc deficiency (Hochleithner, 1994), blood zinc levels of all of the crane samples were considered adequate (see later). Greater creatinine kinase levels in captive cranes were attributed to housing management in the 24 h prior to annual health check sampling (kept indoors, which may result in pacing); the increased

TABLE 18.2   Comparison of Serum Biochemistry Values between AWBP (Samples Collected 2009–14) and Captive (Samples Collected 2012) Whooping Cranes. Reference Intervals Were Calculated as Described in Text.

Analytes	AWBP adults (*n* = 38)		Captive adults (*n* = 25)	
	Mean ± SE (median)	Range	Mean ± SE (median)	Reference interval
Glucose (mg/dL)	241 ± 8 (232)	187–411	257 ± 7 (250)	199–343
Aspartate aminotransferase (U/L)	299 ± 17 (276)	158–692	318 ± 18 (297)	199–621
Alanine aminotransferase (U/L)	48 ± 2 (48)	24–71	44 ± 4 (40)	25–62
Alkaline phosphatase (U/L)[a]	105 ± 12 (74)	27–368	231 ± 17 (202)	116–417
Creatine kinase (U/L)[a]	113 ± 6 (109)	59–225	215 ± 49 (133)	38–688
Lactate dehydrogenase (U/L)	256 ± 23 (221)	98–742	200 ± 16 (181)	90–424
Cholesterol (mg/dL)	174 ± 6 (168)	116–317	166 ± 7 (162)	103–256
Calcium (mg/dL)	9.8 ± 0.1 (9.8)	8.6–11.3	9.8 ± 0.1 (9.8)	9.1–10.6
Phosphorus (mg/dL)[a]	3.5 ± 0.2 (3.1)	2.1–7.2	4.3 ± 0.2 (4.2)	2.4–6.2
Sodium (mmol/L)	147 ± 1 (147)	140–156	149 ± 1 (149)	142–155
Potassium (mmol/L)[a]	3.2 ± 0.1 (3.2)	2.1–4.4	4.3 ± 0.1 (4.4)	2.8–5.9
Chloride (mmol/L)	106 ± 1 (106)	101–116	107 ± 0.4 (107)	103–112
Bicarbonate (mmol/L)[a]	22 ± 1 (19)	11–42	16 ± 1 (17)	12–21
Anion gap (mmol/L)[a]	23 ± 1 (25)	4–39	30 ± 1 (29)	21–38
Uric acid (mg/dL)[a]	9.4 ± 1.0 (8.8)	3.0–42.8	5.8 ± 0.4 (6.0)	2.2–9.8

[a] *Statistical difference between AWBP and captive Whooping Cranes (Student's t test, $P < 0.05$).*

physical activity is thought to be responsible for this slight variation, and increases are not indicative of myopathy or other pathologic process.

The differences in inorganic phosphorus, potassium, and uric acid are considered to be generally reflective of dietary differences between the AWBP (varied wild food items including high-protein invertebrates) and the captive cranes fed an ad libitum pelleted diet (Zeigler Bros. Inc., Gardners, PA). Sample handling was similar and quality was considered excellent with minimal lipemia or hemolysis that may have affected phosphorus or potassium levels. Significant hyperuricemia in one crane (42.8 mg/dL, a significant outlier) may have indicated renal disease, dehydration, an undefined toxicity, or bacterial infection.

The higher levels of bicarbonate in AWBP compared to captive cranes are consistent with several individuals affected with mild metabolic alkalosis, or elevated tissue pH, at the time of capture (however, this was not confirmed by blood gas analysis). Alkalosis is also reflected by many low anion gap measures compared to captive adults, and the lower potassium levels. Bicarbonate is an important component of the pH buffering system and maintenance of acid–base homeostasis. Though hypoalbuminemia (low levels of blood protein) may be associated with increased bicarbonate levels, AWBP cranes exhibited similar albumin levels to captive cranes (see later). Disease seems unlikely to be causally linked to these observations due to normal clinical presentation and hematological results, but mild dehydration due to limited access to fresh water during the prevailing drought conditions during the study may be. The increased levels of uric acid in AWBP cranes are also supportive of mild dehydration in several individuals at the time of capture.

Blood Protein Analysis

Protein electrophoresis was performed on serum samples according to current standards for birds by the Division of Comparative Pathology, University of Miami Miller School of Medicine (Miami, FL; Hausmann et al., 2015). Blood proteins, including albumin and globulins, are important measures of metabolism, inflammation, infection, clotting, and immune response, and changes may be reflective of disease processes of concern for Whooping Cranes, such as mycobacteriosis and aspergillosis (Hartup and Schroeder, 2006; Melillo, 2013). The wild adult Whooping Cranes ($n = 22$) had greater concentrations of gamma globulins than the captive adult cranes held at ICF, but most values were within calculated reference intervals from the captive cranes (Table 18.3). The higher gamma globulin level observed in the AWBP cranes was likely due to greater antigenic exposure to potential infectious agents in the wild, but resulted in only a modest immunologic response. On average, AWBP cranes exhibited a 20% increase in gamma globulins compared to captive cranes.

Trace Element Analysis

Lithium heparinized whole blood from AWBP and captive Whooping Cranes was submitted to the Wisconsin Veterinary Diagnostic Laboratory (WVDL, Madison, WI) for trace element analysis by inductively coupled plasma mass spectrometry direct dilution for fluids (Table 18.4). Adult AWBP cranes were observed with greater arsenic, boron, lead, and selenium levels, and lower iron, molybdenum, nickel, phosphorus, potassium, and zinc levels than captive Whooping Cranes. These numerical differences were minimal and most values bounded the reference intervals determined from the captive cranes and hence were unlikely to be of clinical consequence. The medians observed were similar to available reference information on bird and mammalian species for which normal blood trace element analyses are available, and no trend for either deficiency or potentially toxic elevation of a trace element at the population level was identified (Jansen and Nijboer, 2003; Mertz, 1986, 1987;

TABLE 18.3   Comparison of Protein Electrophoresis Values from Serum Samples of AWBP (Samples Collected 2009–14) and Captive Whooping Cranes (Information Provided in Hausmann et al., 2015). Reference Intervals Were Calculated as Described in Text.

Analytes	AWBP adults ($n = 22$)		Captive adults ($n = 30$)	
	Mean ± SE (median)	Range	Mean ± SE (median)	Reference interval
Total protein (g/dL)	3.6 ± 0.1 (3.6)	2.8–4.6	3.7 ± 0.1 (3.8)	2.3–5.1
Albumin/globulin ratio	1.3 ± 0.04 (1.2)	1.0–1.7	1.3 ± 0.04 (1.3)	0.9–1.7
Prealbumin (g/dL)	0.19 ± 0.01 (0.17)	0.12–0.26	0.17 ± 0.01 (0.18)	0.06–0.29
Albumin (g/dL)	1.85 ± 0.08 (1.71)	1.38–2.66	1.93 ± 0.08 (1.89)	1.02–2.85
Alpha 1 (g/dL)	0.17 ± 0.01 (0.17)	0.11–0.26	0.18 ± 0.03 (0.17)	0.08–0.29
Alpha 2 (g/dL)	0.62 ± 0.02 (0.60)	0.42–0.80	0.59 ± 0.02 (0.60)	0.38–0.80
Beta (g/dL)	0.46 ± 0.2 (0.48)	0.27–0.62	0.54 ± 0.03 (0.52)	0.21–0.86
Gamma (g/dL)[a]	0.36 ± 0.01[a](0.37)	0.24–0.47	0.30 ± 0.01[a](0.31)	0.17–0.44

[a] Statistical difference between AWBP and captive Whooping Cranes (Student's t test, P < 0.05).

TABLE 18.4   Comparison of Trace Element Values from Whole Blood Samples of AWBP (Samples Collected 2009–14) and Captive (Samples Collected 2012) Whooping Cranes. Reference Intervals Were Calculated as Previously Described.

Analytes	AWBP adults (n = 38)		Captive adults (n = 27)	
	Mean ± SE (median)	Range	Mean ± SE (median)	Reference interval
Arsenic (μg/mL)[a]	0.18 ± 0.02 (0.12)	0.021–0.54	0.04 ± 0.002 (0.04)	0.03–0.06
Boron (μg/mL)[a]	0.11 ± 0.01 (0.11)	0.02–0.24	0.03 ± 0.002 (0.02)	<0.02–0.05
Cadmium (μg/mL)	na	<0.02	na	<0.01
Calcium (μg/mL)	57 ± 1 (57)	49–68	58 ± 1 (59)	49–64
Chromium (μg/mL)	na	<0.20	na	<0.20
Cobalt (μg/mL)	na	<0.01	na	<0.01
Copper (μg/mL)	0.21 ± 0.01 (0.22)	0.093–0.28	0.23 ± 0.01 (0.23)	0.19–0.27
Iron (μg/mL)[a]	319 ± 8 (330)	240–430	365 ± 7 (360)	307–460
Lead (μg/mL)	0.018 ± 0.004 (0.01)	0.003–0.15	na	<0.01
Magnesium (μg/mL)	73 ± 2 (67)	58–99	73 ± 1 (74)	66–83
Manganese (μg/mL)	0.015 ± 0.001 (0.015)	0.01–0.025	0.014 ± 0.001 (0.014)	0.001–0.022
Molybdenum (μg/mL)[a]	0.05 ± 0.01 (0.04)	0.004–0.17	0.09 ± 0.01 (0.09)	0.05–0.16
Nickel (μg/mL)	na	<0.20	na	<0.20
Phosphorus (μg/mL)[a]	640 ± 10 (640)	530–790	717 ± 11 (710)	594–840
Potassium (μg/mL)[a]	1541 ± 28 (1514)	1300–2000	1674 ± 25 (1700)	1404–1944
Selenium (μg/mL)[a]	0.38 ± 0.01 (0.36)	0.27–0.59	0.30 ± 0.01 (0.30)	0.23–0.37
Sodium (μg/mL)	2068 ± 27 (2050)	1800–2400	2041 ± 21 (2100)	1814–2268
Zinc (μg/mL)[a]	2.4 ± 0.1 (2.6)	1.6–3.3	2.7 ± 0.1 (2.7)	2.2–3.6

[a] *Statistical difference between AWBP and captive Whooping Cranes (Student's t test, $P < 0.05$).*

Puls, 1994). Zinc levels of AWBP Whooping Cranes were lower than those reported for reintroduced cranes from Florida (mean = 3.9, range 0.54–17 ug/mL, $n = 33$; Forrester and Spalding, 2003), but several cases of confirmed metal ingestion and presumed zinc toxicosis were included in the Florida samples.

Individual AWBP cranes, however, did show blood concentrations of arsenic, boron and selenium over captive crane maxima. The source of these elements is unknown, but may be affected by human activities that increase biological availability in food and water, such as dumping of industrial wastes and discharges of agricultural drainage (Ohlendorf, 1996). Each of these elements has negative effects on reproductive performance in poultry: high dietary exposure is generally associated with decreased egg production, poor embryonic development, or reduced hatchability (Puls, 1994). Boron and selenium have been shown to have similar effects in mallards (Heinz et al., 2012), but no interaction effects between the two elements were observed (Stanley et al., 1996). Tissue selenium levels from two AWBP cranes collected in 1989 were in the range associated with reproductive impairment in birds, but the toxic or harmful levels for these elements are not known for Whooping Cranes (Lewis et al., 1992). Diagnosis of selenium toxicosis is complicated by interactions with other elements, such as mercury (which was not quantified in this study), and relies on determination of tissue concentrations, histological changes in liver, and environmental sampling (Franson, 1999). Blood levels alone may not adequately assess the potential for selenium toxicity or toxicity of other elements of biological significance. Ongoing monitoring, through either tissue or egg sampling of freshly retrieved remains, seems warranted based on these results.

Microbiology

Cloacal swabs were collected from 35 cranes and submitted to Marshfield Laboratories for gram stain evaluation ($n = 7$) and aerobic culture with identification of dominant bacterial isolates ($n = 35$). Gram stains showed the cranes' enteric bacteria consisted of 60–90% gram positive bacteria and 10–40% gram negative bacteria, consistent with its omnivorous diet. *Escherichia coli* was isolated from 21 samples (60%). One isolate of *Salmonella litchfield* (group C2–C3) was made, but no *Campylobacter* sp. was isolated from any sample. *Salmonella* sp. and *Campylobacter* sp. are potentially zoonotic bacteria of public health significance.

A complete list of bacterial isolates was established for 16 of the 35 cloacal swab samples (Table 18.5). The results show general similarity to the enteric bacteria of captive Whooping Cranes, with high prevalences of gram positive cocci, coliforms, and gram negative bacilli (Hoar et al., 2007). The low prevalence of *Salmonella* is consistent with recent observations of captive juvenile Whooping Cranes at ICF (1%; Keller and Hartup, 2013), but annual prevalence has varied considerably in the Whooping Crane captive propagation program where several different strains have been isolated. *Salmonella* is most commonly a transient infection in Whooping Cranes, but acute infections have been associated with enteritis and lethargy in captive Whooping Crane chicks.

Most significant was the lack of *Campylobacter* sp. in cultures from wild cranes. This organism is commonly found in captive Whooping Cranes across the North American breeding centers and in wild Sandhill Cranes of the Central Flyway. Waterfowl and Sandhill

TABLE 18.5 Types and Frequencies of Gram Positive and Gram Negative Microorganisms Isolated from Cloacal Swabs of 16 Whooping Cranes in the Aransas-Wood Buffalo Population, 2009–14

Microorganism	N (%)
Gram positive	
Arthrobacter aurescens	1 (6)
Coryneform bacilli	4 (25)
Curtobacterium sp.	1 (6)
Enterococcus sp.	2 (12)
Enterococcus avium	1 (6)
Enterococcus faecalis	2 (12)
Enterococcus casselflavus	2 (12)
Enterococcus gallinarum	5 (31)
Gram + bacilli	10 (62)
Rothia sp.	2 (12)
Staphylococcus sp. coagulase negative	3 (19)
Staphylococcus aureus	2 (12)
Staphylococcus hyicus	1 (6)
Staphylococcus intermedius group	1 (6)
Staphylococcus sciuri	4 (25)
Streptococcus viridans group	6 (38)
Streptomyces sp.	1 (6)
Gram negative	
Bacillus sp. (multiple morphologic types)	6 (38)
Citrobacter sp.	1 (6)
Enterobacter aerogenes	1 (6)
Enterobacter cloacae complex	5 (31)
Enterobacter cancerogenus	1 (6)
Escherichia coli	10 (62)
Gram – bacilli	2 (12)
Klebsiella oxytoca	2 (12)
Pantoea dispersa	1 (6)
Pseudomonas sp.	1 (6)
Serratia sp.	1 (6)

Cranes are known to have asymptomatic infections of *Campylobacter* sp. and have the potential to spread the bacterium in the environment. Sandhill Cranes from Alaska in the summer of

2008 were tied to the spread of the bacteria to pea fields; ingested raw peas sickened 99 people. *Campylobacter jejuni* was recovered from 71% of Sandhill Crane fecal samples collected in the central Platte River, in areas designated as critical habitat and visited by Whooping Cranes (Vogel et al., 2013). In Texas, Whooping Cranes may only be exposed to limited numbers of *Campylobacter* bacteria at managed freshwater sites at Aransas NWR or game feeders on private land where there is frequent contact among concentrations of Sandhill Cranes and Whooping Cranes.

A limitation of the data is that traditional aerobic culture and biochemical identification of bacterial isolates only determines a microorganism's presence under specific laboratory conditions. Anaerobic and more fastidious bacterial organisms are not quantified, and detailed comparisons of abundance are not reliable. Next-generation sequencing of 16S rRNA has recently allowed the determination of relative abundances of bacterial phyla in the gut microbiota of endangered Red Crowned Cranes (*Grus japonensis*; Xie et al., 2016). Wild Red Crowned Cranes had distinct compositions of gut microbiota and lower bacterial diversity compared to captive and reintroduced adults. Many wild adult Red Crowned Cranes had greater relative abundances of Proteobacteria (gram negative organisms) compared to captive and reintroduced adults, whose microbiota showed greater relative abundances of Firmicutes (gram positive organisms) and Fusobacteria (anaerobic organisms). The authors noted that captive and reintroduced Red Crowned Cranes are limited by bacterial disease outbreaks and lower survival compared to wild cranes that are likely associated with these differences in the gut microbiota. A similar study of wild, captive, and reintroduced Whooping Cranes may also be of benefit, as reintroduction of Whooping Cranes also faces similar infectious disease challenges (Keller and Hartup, 2013).

Virology

Cranes were tested for antibodies to one or more viruses of concern: inclusion body disease of cranes (herpesvirus), infectious bursal disease virus type 2 (birnavirus), West Nile virus (flavivirus), avian influenza (orthomyxovirus), Newcastle disease virus (paramyxovirus), and a novel herpesvirus that was isolated from a dead Whooping Crane at the Aransas NWR in winter 2007–08 and has been maintained at the NWHC.

None of the AWBP adults were seropositive for inclusion body disease of cranes ($n = 38$; by serum neutralization, NWHC). No Whooping Crane (wild or captive) has tested positive for exposure to this virus since an outbreak at ICF in 1978 (Docherty and Henning, 1980). No AWBP cranes were seropositive for the novel herpesvirus ($n = 27$; by serum neutralization, NWHC).

Twenty-six of 37 cranes (70%) were seropositive for infectious bursal disease virus type 2 (by serum neutralization, University of Georgia Poultry Diagnostic and Research Center, Athens, Georgia). The geometric mean titer of the sample population was 35.8 (range 1–256; greater titer values equate with stronger immune responses and higher concentration of antibodies in serum, and cutoff values are used by test laboratories to define a "positive" versus "negative" sample). These data seem most similar to those described by Spalding et al. (2008) for the Florida Nonmigratory Population (FNMP), where 75% of cranes exposed to the Florida landscape for several years were seropositive, and prevalence of positive titers increased with age of the cranes. In contrast, 15% of adults at ICF in 2002 were found to be seropositive (Hartup et al., 2004). Despite the high prevalence of positive titers, no virus has been isolated in the laboratory from a crane, nor has an AWBP crane been diagnosed with the wasting syndrome described in the FNMP and attributed to infectious bursal disease virus (Spalding et al., 2008). The disease, however, typically manifests among juvenile cranes, and would be hard to detect without constant monitoring of the Wood Buffalo nesting grounds. The source of exposure in the AWBP may be from within the Whooping Crane flock or exposure to another avian source. For example, a serosurvey in Florida and Georgia showed that seropositive Sandhill Cranes and Wild Turkeys (*Meleagris gallopavo*) shared habitat with FNMP Whooping Cranes and may have been a source of virus exposure (Candelora et al., 2010).

Two of 28 cranes (7%) were seropositive for exposure to West Nile virus, with antibody titers of 160 and 1280 respectively (by serum neutralization, NWHC). During West Nile virus surveillance at ICF in 2000–04, 6 of 52 (11.5%) Whooping Cranes tested from the ICF captive flock were seropositive, but no clinical cases of West Nile encephalitis were observed (Hartup, 2008). No cases of West Nile encephalitis have been diagnosed in Whooping Cranes, despite the documentation of several cases in captive Sandhill Cranes and seroprevalence >20% in free-ranging Sandhill Cranes (Hansen et al., 2008; Hartup, 2008).

No cranes were seropositive for avian influenza virus (*n* = 9; agar gel immunodiffusion, WVDL) or Newcastle disease virus (*n* = 9; by hemagglutination inhibition, WVDL). More extensive testing for exposure to these diseases in Whooping Cranes from the reintroduced Eastern Migratory Population also has shown no evidence of exposure to birds nearing release (unpublished data, B. Hartup, ICF). In addition, all cloacal swabs tested for the presence of avian influenza DNA by polymerase chain reaction (PCR) technique were negative (*n* = 37; NWHC).

Parasitology

Collection of fecal samples from captured cranes was too infrequent to lead to meaningful analysis. Instead, fecal samples from the AWBP were obtained during the winters of 2012–13 and 2013–14 through monthly, noninvasive collection at excavated upland freshwater sites at the Aransas NWR. This sampling strategy was designed to better estimate the parasite diversity and seasonal aspects of endoparasitism in the AWBP.

The protozoan parasites *Eimeria gruis* and *E. reichenowi* were observed from 26% of fecal samples (*n* = 328) evaluated by microscopy and confirmed by PCR (Fig. 18.2; Bertram et al., 2015a). Though the proportion of positive fecal samples varied with date of collection (range 8–59%), the proportion of positive samples did not vary significantly between sampling sites, between the two years, or suggest a distinct seasonal peak of shedding. The results were similar to a limited sampling of Whooping Cranes from the 1970s which showed 32% of fecal samples contained *Eimeria* (Forrester et al., 1978). *Eimeria* are of considerable concern due to the frequent occurrence of an extraintestinal form of parasitism known as disseminated

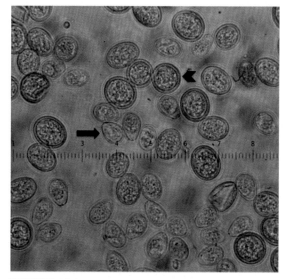

FIGURE 18.2 Coccidian parasites observed during fecal flotation, 500X, from adult Whooping Crane samples collected at Aransas NWR. The smaller, pear-shaped oocysts are consistent with *Eimeria gruis* (arrow) and the larger, round to oval oocysts are consistent with *E. reichenowi* (arrowhead). *Source: From Bertram, M.R., Hamer, G.L., Snowden, K.F., Hartup, B.K., Hamer, S.A., 2015a. Coccidian parasites and conservation implications for the endangered whooping crane (Grus americana). PLoS One, 10(6), p. e0127679.*

visceral coccidiosis (Olsen et al., 1996). This form results in widespread inflammation and the development of granulomatous lesions in affected organs and tissues. *Eimeria* infections are potentially fatal, and have impacted fledging rates and overall productivity of captive Whooping Cranes that did not receive antiparasitic drugs. It is not known if *Eimeria* is a limiting factor to the AWBP.

Trematode eggs were detected in 11% ($n = 63$) of fecal sedimentation samples, and trematode-specific PCR and DNA sequences commonly confirmed urinary tract flukes of the genus *Tanasia* (Bertram et al., 2015b). Unfortunately, potential contamination of samples by soil dwelling species of nematodes prevented estimation of the prevalence of parasitic infections from this phylum in Whooping Cranes.

Microscopic evaluation showed 20% ($n = 35$) of AWBP blood smears contained a parasite morphologically consistent with *Haemoproteus* sp. Parasitemias were quite mild, with 1–6 infected cells/10,000 observed (all captive cranes were negative for hemoparasites, $n = 29$). Molecular detection of parasitic DNA by PCR provided a much greater estimate of the prevalence of infection in the AWBP: 68% ($n = 37$) tested positive for Haemosporidia DNA (personal communication, M. Bertram, Texas A&M University). Additional analysis suggests that these parasites belong to a novel clade within the apicomplexan subclass Haemosporidia. In addition, 3% of the AWBP blood samples were positive for *Leucocytozoon* sp. DNA. No AWBP blood samples were positive for microfilarial nematodes by microscopy ($n = 35$).

Many instances of blood parasite infection have no significant detrimental effect on the health of avian hosts, but there are several notable exceptions (Clark et al., 2009). *Haemoproteus* and *Leucocytozoon* are generally considered to be of little pathogenic threat to Whooping Cranes, but individual morbidity in birds with high parasitemias have been diagnosed in other crane species (especially *Leucocytozoon* in Red Crowned Cranes). Whether either parasite influences the fitness or reproductive success of the AWBP remains unknown.

SUMMARY AND OUTLOOK

The AWBP Whooping Cranes sampled during the winters of 2009–10 through 2013–14, as described here, exhibited generally normal physical condition for free-ranging birds. The complete evaluation of clinical pathology, however, elucidated important differences from captive Whooping Cranes: eosinophilias and hyperglobulinemia consistent with exposure to a greater array and density of parasites or other antigens, serum biochemistry values reflective of a varied natural diet and stressful environmental conditions (possible fluid stress due to limited freshwater availability), and previously poorly described prevalence of endo- and hemoparasites. These baseline data are important for the interpretation of health and productivity assessments in future. Several potential disease risks were also identified that may not result in severe morbidity or mortality by themselves, but many are recognized as important cofactors for other pathogens or may become pathogenic under favorable environmental conditions or with compromise of the host. For example, recent findings on the differences of gut microbiota of Red Crowned Cranes from different populations (Xie et al., 2016) raise questions for Whooping Cranes: is the gut microbiota of wild Whooping Cranes similarly distinct compared to captive or reintroduced cranes, and is morbidity associated with any of the bacterial communities over another?

The potential impact of viruses and suspected viral-associated conditions on the AWBP is difficult to discern. First, cutaneous tumors with ulcerated epithelium and underlying viral particles were described from the FNMP (Forrester and Spalding, 2003). The virus was never identified, nor its role in the pathogenesis of the tumors discerned. But the tumors are common in the AWBP (as high as 5% of the cranes); the prevalence of similar lesions in Florida Sandhill Cranes was estimated at 0.1% (Nesbitt et al., 2005). Perhaps not all of the cases observed in the AWBP sample would be expected to become complicated or life-threatening. Currently, there is a gap in knowledge of the underlying etiology and morbidity associated with this condition. Second, over 10% of sampled adults from the AWBP tested positive for exposure to West Nile virus, but no clinical case of West Nile-associated encephalitis has been described in Whooping Cranes in Canada or the United States. Glial cell aggregates (inflammatory brain lesions), however, were described in Sandhill Cranes exposed to one mosquito dose of West Nile virus under laboratory conditions, but that did not exhibit clinical illness (Olsen et al., 2009). The authors suggested that "stressed, compromised, or unhealthy Sandhill Cranes could develop clinical illness or die when exposed to a complicating factor such as West Nile virus." Third, the majority of the sampled adults from the AWBP had antibody titers consistent with exposure to infectious bursal disease virus type 2. Testing of prefledging juveniles from Wood Buffalo National Park revealed a similar pattern to that observed in the FNMP: antibody titers of adults are significantly greater than those of younger cranes (data not shown). There appears to be repeated exposure to the virus after hatching and in subsequent seasons, but the infectious bursal disease virus strain in the AWBP is generally nonpathogenic except perhaps under unusual circumstances (e.g., as observed in captivity; Hartup and Sellers, 2008). Yet, the virus has not been isolated or characterized in the laboratory, and the diagnosis remains uncertain. These three examples show additional work is needed to diagnose and understand the pathophysiology of virus-associated threats to wild Whooping Cranes.

Coccidiosis is a recognized threat to captive Whooping Cranes. The species' small brood size and the defense of large territories in the Wood Buffalo National Park natal area likely limit exposure of juveniles to quantities of coccidian oocysts sufficient to cause severe disease. In comparison, the loss of wetlands along the migration corridor has concentrated birds at stopover sites, thereby increasing the risk of disease exposure and pathogen transmission. For example, avian cholera epizootics occur regularly among waterfowl in the Central Flyway that share sites with Whooping Cranes (Friend, 1999). As the wintering Whooping Crane population expands, the provision of additional suitable habitat and managed water resources at Aransas NWR will be required to help limit the negative host effects and enhanced pathogen transmission among overconcentrated cranes, especially in times of extreme environmental fluctuation.

Recovery actions continue to promote study and recognition of existing threats, such as habitat loss and degradation, disease, mortality from power lines, and loss of genetic diversity (Canadian Wildlife Service and U.S. Fish and Wildlife Service, 2005). The ability to discern the effects of disease on AWBP health and productivity have thus far been extremely limited; nearly all health knowledge on the species has come from the study of captive cranes and reintroduced populations at the margin of the species' range, likely with far different host-disease agent-environment dynamics. Core protected areas exist for both the breeding and wintering grounds of Whooping Cranes, but are subject to outside pressures, including habitat transformation, resource extraction, and pollution. Efforts to define the magnitude and potentially limit the threat of disease in these transforming landscapes cannot be accomplished without periodic monitoring and assessment of the health status and exposure risks of the AWBP.

Acknowledgments

The author extends gratitude to F. Chavez-Ramirez, D. Brandt, A. Pearse, D. Baasch, G. Wright, J. Rempel,

W. Wehtje, L. Craig-Moore, W. Harrell, B. Strobel, and FWS interns for field assistance and logistical support during work in Texas. Special thanks also go to the Aransas NWR project leaders, staff and volunteers, and T. Stehn for original project approval. The following institutions provided significant project support: USGS Northern Prairie Research Center, The Crane Trust, Platte River Recovery Implementation Program, and the University of Wisconsin Companion Animal Fund. Thanks also to H. Ip and USGS National Wildlife Health Center staff for ongoing diagnostic service support. ICF veterinary staff and interns played critical roles in data analysis, including A. Aeschbach and K. Shultz.

References

Bertram, M.R., Hamer, G.L., Snowden, K.F., Hartup, B.K., Hamer, S.A., 2015a. Coccidian parasites and conservation implications for the endangered whooping crane (*Grus americana*). PLoS ONE 10 (6), e0127679.

Bertram, M.R., Hamer, G.L., Snowden, K.F., Hartup, B.K., Rech, R., Hensel, M., Hamer, S.A., 2015b. Worms of the wild whoopers: characterization of helminths in endangered whooping cranes (*Grus americana*) and sympatric sandhill cranes (*Grus canadensis*). American Association of Veterinary Pathologists, 60th annual meeting, July 11–14, 2015, Boston, p. 93.

Canadian Wildlife Service and U.S. Fish and Wildlife Service, 2005. International Recovery Plan for the Whooping Crane. Recovery of Nationally Endangered Wildlife (RENEW), Ottawa, and U.S. Fish and Wildlife Service, Albuquerque, NM, p. 162.

Candelora, K.L., Spalding, M.G., Sellers, H.S., 2010. Survey for antibodies to infectious bursal disease virus serotype 2 in wild turkeys and sandhill cranes of Florida, USA. J. Wildl. Dis. 46 (3), 742–752.

Clark, P., Boardman, W.S.J., Raidal, S.R., 2009. Atlas of Clinical Avian Hematology. Wiley-Blackwell, West Sussex, UK.

Cole, G.A., Thomas, N.J., Spalding, M., Stroud, R., Urbanek, R., Hartup, B.K., 2009. Postmortem evaluation of reintroduced migratory whooping cranes in eastern North America. J. Wildl. Dis. 45 (1), 29–40.

Deem, S.L., Noss, A.J., Cuéllar, R.L., Karesh, W.B., 2005. Health evaluation of free-ranging and captive blue-fronted Amazon parrots (*Amazona aestiva*) in the Gran Chaco, Bolivia. J. Zoo Wildl. Med. 36 (4), 598–605.

Docherty, D.E., Henning, D.J., 1980. The isolation of a herpesvirus from captive cranes with an inclusion body disease. Avian Dis. 24 (1), 278–283.

Forrester, D.J., Carpenter, J.W., Blankinship, D.R., 1978. Coccidia of whooping cranes. J. Wildl. Dis. 14 (1), 24–27.

Forrester, D.J., Spalding, M.G., 2003. Parasites and diseases of wild birds in Florida. University Press of Florida, Gainseville.

Franson, J.C., 1999. Selenium. In: Friend, M., Franson, J.C. (Eds.), Field Manual of Wildlife Diseases: General Field Procedures and Diseases of Birds. U.S. Geological Survey, Biological Resources Division, Madison, WI, pp. 335–336.

Friedrichs, K.R., Harr, K.E., Freeman, K.P., Szladovits, B., Walton, R.M., Barnhart, K.F., Blanco-Chavez, J., 2012. ASVCP reference interval guidelines: determination of de novo reference intervals in veterinary species and other related topics. Vet. Clin. Pathol. 41 (4), 441–453.

Friend, M., 1999. Avian cholera. In: Friend, M., Franson, J.C. (Eds.), Field Manual of Wildlife Diseases: General Field Procedures and Diseases of Birds. U.S. Geological Survey, Biological Resources Division, Madison, WI, pp. 75–92.

Fudge, A.M., 2000. Laboratory Medicine: Avian and Exotic Pets. W.B. Saunders, Philadelphia.

Geffré, A., Concordet, D., Braun, J.P., Trumel, C., 2011. Reference Value Advisor: a new freeware set of macroinstructions to calculate reference intervals with Microsoft Excel. Vet. Clin. Pathol. 40 (1), 107–112.

Gil-Weir, K.C., Grant, W.E., Slack, R.D., Wang, H.-H., Fujiwara, M., 2012. Demography and population trends of whooping cranes. J. Field Ornithol. 83 (1), 1–10.

Hansen, C.H., Hartup, B.K., Gonzalez, O.D., Lyman, D.E., Steinberg, H., 2008. West Nile encephalitis in a captive Florida sandhill crane. Proceedings of the North American Crane Workshop 10, 115–118.

Hartup, B.K., 2008. Surveillance for West Nile virus at the International Crane Foundation 2000–2004. Proceedings of the North American Crane Workshop 10, 111–114.

Hartup, B.K., Niemuth, J.N., Fitzpatrick, B., Fox, M., Kelley, C., 2010a. Morbidity and mortality of captive whooping cranes at the International Crane Foundation: 1976–2008. Proceedings of the North American Crane Workshop 11, 183–185

Hartup, B.K., Olsen, G.H., Sellers, H.S., Smith, B., Spalding, M., 2004. Serologic evidence of infectious bursal disease virus exposure in captive whooping cranes. In: Proceedings of the American Association of Zoo Veterinarians. American Association of Wildlife Veterinarians, Wildlife Disease Association Joint Conference, San Diego, CA, p. 71.

Hartup, B.K., Schroeder, C.A., 2006. Protein electrophoresis in cranes with presumed insect bite hypersensitivity. Vet. Clin. Pathol. 35 (2), 226–230.

Hartup, B.K., Sellers, H.S., 2008. Serological survey for infectious bursal disease virus exposure in captive cranes. Proceedings of the North American Crane Workshop 10, 173–174

Hartup, B.K., Spalding, M.G., Thomas, N.J., Cole, G.A., Kim, Y.J., 2010b. Thirty years of mortality assessment in whooping crane reintroductions: patterns and implications. Proceedings of the North American Crane Workshop 11, 204.

Hausmann, J.C., Cray, C., Hartup, B.K., 2015. Comparison of serum protein electrophoresis values in wild and captive whooping cranes (*Grus americana*). J. Avian Med. Surg. 29 (3), 192–199.

Heinz, G.H., Hoffman, D.J., Klimstra, J.D., Stebbins, K.R., 2012. A comparison of the teratogenicity of methylmercury and selenomethionine injected into bird eggs. Arch. Environ. Contam. Toxicol. 62 (3), 519–528.

Hoar, B.M., Whiteside, D.P., Ward, L., Inglis, G.D., Morck, D.W., 2007. Evaluation of the enteric microflora of captive whooping cranes (*Grus americana*) and sandhill cranes (*Grus canadensis*). Zoo Biol. 26 (2), 141–153.

Hochleithner, M., 1994. Biochemistries. In: Ritchie, B.W., Harrison, G.J., Harrison, L.R., (Eds.), Avian Medicine: Principles and Application. Wingers Publishing, Lake Worth, FL, pp. 223–245.

International Union for Conservation of Nature/Species Survival Commission, 2013. Guidelines for Reintroductions and Other Conservation Translocations, Version 1.0. IUCN Species Survival Commission, Gland, Switzerland.

Jansen, W.L., Nijboer, J., 2003. Zoo Animal Nutrition: Tables and Guidelines. European Zoo Nutrition Center, Amsterdam, Netherlands.

Karesh, W.B., Cook, R.A., 1995. Applications of veterinary medicine to *in situ* conservation efforts. Oryx 29 (4), 244–252.

Keller, D., Hartup, B.K., 2013. Reintroduction medicine: whooping cranes in Wisconsin. Zoo Biol. 32 (6), 600–607.

Lewis, J.C., Drewien, R.C., Kuyt, E., Sanchez, Jr., C., 1992. Contaminants in habitat, tissues, and eggs of whooping cranes. Proceedings of the North American Crane Workshop 6, 159–165

Losdat, S., Arcese, P., Sampson, L., Villar, N., Reid, J.M., 2016. Additive genetic variance and effects of inbreeding, sex and age on heterophil to lymphocyte ratio in song sparrows. Funct. Ecol. 30 (7), 1185–1195.

Mauer, J., Reichenberg, B., Kelley, C., Hartup, B.K., 2010. The effects of anticoagulant choice and sample processing time on hematologic values of juvenile whooping cranes. Proceedings of the North American Crane Workshop 11, 105–109

Melillo, A., 2013. Applications of serum protein electrophoresis in exotic pet medicine. Vet. Clin. North Am. Exot. Anim. Pract. 16 (1), 211–225.

Mertz, W., 1986. Trace Elements in Human and Animal Nutrition, fifth ed. Academic Press, Orlando, Florida.

Mertz, W., 1987. Trace elements in human and animal nutrition, fifth ed. Academic Press, San Diego, CA.

Nesbitt, S.A., Spalding, M.G., Schwikert, S.T., 2005. Injuries and abnormalities of sandhill cranes captured in Florida. Proceedings of the North American Crane Workshop 9, 15–20.

Ohlendorf, H.M., 1996. Selenium. In: Fairbrother, A., Locke, L., Hoff, G.L. (Eds.), Noninfectious Diseases of Wildlife. second ed. Iowa State University Press, Ames, pp. 128–140.

Olsen, G.H., Carpenter, J.W., Langenberg, J.A., 1996. Medicine and surgery. In: Ellis, D.H., Gee, G.F., Mirande, C.M. (Eds.), Cranes: Their Biology, Husbandry, and Conservation. Hancock House, Blaine, WA, pp. 137–174.

Olsen, G.H., Hartup, B.K., Black, S., 2018. Health and disease treatment in captive and reintroduced whooping cranes (Chapter 19). In: French, Jr., J.B., Converse, S.J., Austin, J.E. (Eds.), Whooping Cranes: Biology and Conservation. Biodiversity of the World: Conservation from Genes to Landcapes. Academic Press, San Diego, CA.

Olsen, G.H., Hendricks, M.M., Dressler, L.E., 2001. Hematological and serum chemistry norms for sandhill and whooping cranes. Proceedings of the North American Crane Workshop 8, 178–184

Olsen, G.H., Miller, K., Docherty, D.E., Bochsler, V.S., Sileo, L., 2009. Pathogenicity of West Nile virus and response to vaccination in sandhill cranes (*Grus canadensis*) using a killed vaccine. J. Zoo Wildl. Med. 40 (2), 263–271.

Olsen, G.H., Taylor, J.A., Gee, G.F., 1997. Whooping crane mortality at Patuxent Wildlife Research Center, 1982–95. Proceedings of the North American Crane Workshop 7, 243–248

Ortiz-Catedral, L., Prada, D., Gleeson, D., Brunton, D.H., 2011. Avian malaria in a remnant population of red-fronted parakeets on Little Barrier Island, New Zealand. N.Z. J. Zool. 38 (3), 261–268.

Padilla, L.R., Whiteman, N.K., Merkel, J., Huyvaert, K.P., Parker, P.G., 2006. Health assessment of seabirds on Isla Genovesa, Galápagos Islands. Ornithol. Monogr. 60, 86–97.

Pearse, A.T., Brandt, D.A., Hartup, B.K., Bidwell, M., 2018. Mortality in Aransas-Wood Buffalo Whooping Cranes: Timing, Location, and Causes (Chapter 6). In: French, Jr., J.B., Converse, S.J., Austin, J.E. (Eds.), Whooping Cranes: Biology and Conservation. Biodiversity of the World: Conservation from Genes to Landcapes. Academic Press, San Diego, CA.

Puls, R., 1994. Mineral Levels in Animal Health, Diagnostic Data, second ed. Sherpa International, Clearbrook, British Columbia, Canada.

Samour, J., 2008. Avian Medicine, second ed. Mosby, St. Louis, MO.

Smith, K.M., Karesh, W.B., Majluf, P., Paredes, R., Zavalaga, C., Hoogesteijn Reul, A., Stetter, M., Braselton, W.E., Puche, H., Cook, R.A., 2008. Health evaluation of free-ranging Humboldt penguins (*Spheniscus humboldti*) in Peru. Avian Dis. 52 (1), 130–135.

Spalding, M., Sellers, H.S., Hartup, B.K., Olsen, G.H., 2008. A wasting syndrome in released whooping cranes in Florida associated with infectious bursal disease titers. Proceedings of the North American Crane Workshop 10, 176 pp.

Stanley, T.R., Smith, G.J., Hoffman, D.J., Heinz, G.H., Rosscoe, R., 1996. Effects of boron and selenium on mallard reproduction and duckling growth and survival. Environ. Toxicol. Chem. 15 (7), 1124–1132.

Uhart, M., Aprile, G., Beldomenico, P., Solís, G., Marull, C., Beade, M., Carminati, A., Moreno, D., 2006. Evaluation of the health of free-ranging greater rheas (*Rhea americana*) in Argentina. Vet. Rec. 158 (9), 297–303.

Vogel, J.R., Griffin, D.W., Ip, H.S., Ashbolt, N.J., Moser, M.T., Lu, J., Beitz, M.K., Ryu, H., Santo Domingo, J.W., 2013. Impacts of migratory sandhill cranes (*Grus canadensis*) on microbial water quality in the central Platte River, Nebraska, USA. Water Air Soil Pollut. 224, 1576.

Xie, Y., Xia, P., Wang, H., Yu, H., Giesy, J.P., Zhang, Y., Mora, M.A., Zhang, X., 2016. Effects of captivity and artificial breeding on microbiota in feces of the red crowned crane (*Grus japonensis*). Sci. Rep. 6, 33350.

Health and Disease Treatment in Captive and Reintroduced Whooping Cranes

Glenn H. Olsen, Barry K. Hartup**,
Sandra R. Black†*

*U.S. Geological Survey, Patuxent Wildlife Research Center, Laurel, MD, United States
**International Crane Foundation, Baraboo, WI, United States
†Calgary Zoological Society, Calgary, AB, Canada

INTRODUCTION

The International Union for Conservation of Nature (IUCN) Red List categorizes 11 of the 15 species of cranes (Gruidae) as vulnerable, endangered, or critically endangered (www.iucnredlist.org; accessed 7 December 2016). There are captive breeding programs at zoos and other institutions for all of these crane species. In North America there are captive breeding and release programs for endangered Whooping Cranes (*Grus americana*) to support reintroduction programs.

Information relevant to Whooping Crane health has been gleaned through studies of health risks and mortality factors in reintroduced populations. However, the majority of the available information on Whooping Crane medicine comes from the experience of the

veterinarians and biologists working in captive breeding programs. Other information comes from the many zoological institutions that display but do not breed Whooping Cranes. There are several important disease entities, as well as common injuries, which can affect Whooping Cranes in reintroduced and captive settings. Since 1992, the Whooping Crane Health Advisory Team (WCHAT) has advised the International Whooping Crane Recovery Team on disease risks and health protocols applicable to reintroduced and captive settings.

In the past 20 years, multidisciplinary collaborations led by the U.S. Fish and Wildlife Service (USFWS) and the Canadian Wildlife Service have undertaken three large Whooping Crane reintroduction programs, including the efforts to establish the Florida Nonmigratory Population (FNMP), the Eastern Migratory Population (EMP), and the Louisiana Nonmigratory Population (LNMP). Two of the programs, FNMP and EMP, have existed long enough that studies of mortality causes on a reasonably large sample size of birds have been possible. Because all birds released in each population were given leg bands and most had a working radio transmitter, carcasses were readily located after mortality events, and necropsies were performed or some final diagnosis as to the cause of mortality was elucidated. Necropsies were performed at the Department of Infectious Diseases and Pathology, College of Veterinary Medicine, University of Florida, the U.S. Geological Survey (USGS) National Wildlife Health Center in Wisconsin, and the USFWS National Wildlife Forensic Laboratory in Oregon.

In the FNMP, 58% of 186 mortalities were caused by predation, primarily by bobcats (*Lynx rufus*) and American alligators (*Alligator mississippiensis*) (Spalding et al., 2004). Other documented causes of mortality included trauma (7.5%), and infectious diseases (7.5%), while nearly 27% of mortalities were undetermined (Spalding et al., 2004). There were several infectious diseases documented in the FNMP, including aspergillosis, eastern equine encephalitis, disseminated visceral coccidiosis (DVC), and a wasting syndrome associated with an infectious bursal disease-like virus (Spalding et al., 2004).

In the EMP, causes of all mortalities documented at necropsy were similar, with predation (47%), trauma (12%), and degenerative disease (6%) documented at necropsy in 17 cases (Cole et al., 2009). Thirty-five percent of necropsies did not determine cause of death in this initial survey. The single case of degenerative disease was associated with exertional myopathy after the crane was captured and translocated to another area. Five nonfatal traumatic injuries were documented in the EMP, including three utility line collisions, gunshots, and impact traumas (Cole et al., 2009).

In addition to the studies of mortality causes in these two reintroduced Whooping Crane populations, there has been an extensive history of developing captive Whooping Crane populations to support release programs and a resulting wealth of information on crane health and medicine to support these captive populations. Early efforts at captive propagation began when the last remnant nonmigratory Louisiana Whooping Crane was captured in 1950 and paired with birds from the remnant Aransas-Wood Buffalo Population (AWBP) at the Audubon Zoo in New Orleans, Louisiana (Black and Swan, Chapter 16, this volume). Unfortunately, there was not enough known about nutrition, health, or diseases at that time to maintain a captive population of cranes, let alone produce birds for reintroduction programs. No descendants survive from the remnant nonmigratory Louisiana Whooping Cranes (Barrett and Stehn, 2010).

The next major effort at starting a captive population resulted in the establishment of the program at Patuxent Wildlife Research Center, Laurel, Maryland (Patuxent). In late 1964, a lone male young-of-the-year Whooping Crane severely fractured its wing and was taken to the College of Veterinary Medicine, Colorado State University, Fort Collins. Attempts were made to repair the fracture but the wing was later amputated (Dr. Richard Slemons, personal communications; 19 November 2015) (Fig. 19.1). The Whooping Crane, which was named Canus, was later moved to Patuxent, and went on to produce 186 offspring (http://whoopingcrane.com/cranus-186-whooping-crane-descendents/ posted 25 February 2013, 12:06 am, accessed 17 January 2017). The bulk of the breeding population at Patuxent was composed of individuals collected as eggs from the AWBP (Olsen, unpublished data).

Patuxent had a history of rearing birds in captivity for research dating back to the establishment of the research facility in 1936 (Olsen, 2016). In addition to personnel skilled in captive animal care, Patuxent employed several veterinarians, including the first director, Dr. Richard Morley. By 1964, Patuxent had a record of research and publication on wildlife disease

FIGURE 19.1   Whooping Crane, photographed at Colorado State University, where attempts were made to repair the fracture but the wing was later amputated (Dr. Richard Slemons, personal communications, 19 November 2015). This Whooping Crane, named Canus, was later moved to Patuxent, and went on to produce 186 offspring (http://whoopingcrane.com/cranus-186-whooping-crane-descendents/ posted 25 February 2013, 12:06 am, accessed 17 January 2017).

issues (Olsen, 2016). Patuxent went on to establish the largest captive crane population in the world, numbering over 300 birds in the 1990s, and breeding endangered Whooping Cranes and Mississippi Sandhill Cranes (*Antigone canadensis pulla*) for reintroduction programs.

In 1989, a portion of the Patuxent flock was moved to the International Crane Foundation (ICF) in Baraboo, Wisconsin, to establish a second breeding population. The Devonian Wildlife Conservation Centre of the Calgary Zoo, Calgary, Alberta, received Whooping Cranes in 1992. The establishment of these two new breeding centers was the direct result of a mycotoxin epizootic that occurred in 1989 at Patuxent (Olsen et al., 1995). Establishing additional captive flocks that were physically separate was undertaken to ensure that a catastrophic disease would not destroy the captive Whooping Crane program.

Coupled with the effort to study crane health in the context of reintroduction programs was a well-documented health program at the captive breeding centers, including Patuxent, ICF, and Calgary Zoo, and the smaller breeding populations at the Audubon Zoo and San Antonio Zoo, San Antonio, Texas. Drawing on multiple journal papers on single disease issues and empirical research, a summary chapter on health management and diseases in cranes was first published in 1986 (Carpenter, 1986), followed by a book devoted to crane husbandry and reintroduction including two chapters on crane medicine (Olsen et al., 1996; Olsen and Langenberg, 1996). Other general avian medical textbooks have included chapters summarizing crane medicine (Olsen, 2000, 2009; Olsen and Carpenter, 1997; MacLean and Beaufrere, 2015) and all arose as a direct result of the increased emphasis on reintroduction of cranes, especially Whooping Cranes, in North America. Whooping Cranes are also one of four species featured in a chapter on conservation medicine in a recent textbook (Olsen et al., 2016).

Thus, a wealth of information has been developed over the years to support the management of health in captive and reintroduced Whooping Cranes. In this chapter, we review health care and treatment for Whooping Cranes in these populations. We concentrate on health conditions that can affect the suitability of birds for reintroduction.

CAPTURE, SEDATION, AND ANESTHESIA

Capture

Capturing wild cranes is often important in reintroduction programs to allow for treatment, relocation, or other management of birds. Alpha-chloralose ($C_6H_{11}CL_3O_6$) (Sigma Chemical Company, St. Louis, MO) is an oral sedative that has been useful for capturing wild cranes. Alpha-chloralose acts to depress the brain's control centers (Balis and Munroe, 1964). One field technique is to bait the cranes into a site using whole corn. Once the target cranes are feeding regularly on the corn, alpha-chloralose is added at 0.39–0.48 g/280 mL whole corn (1 cup) (Hayes et al., 2003). To be adequately sedated, a crane needs to eat 140–280 mL of the corn/alpha-chloralose mixture. Ataxia may be seen in 20–30 min: the crane may have difficulty walking and flying, or not attempt to fly at all. Often the sedated crane will lie down in sternal recumbency or be poorly responsive to stimuli. Some type of temporary holding facility is needed if using alpha-chloralose for field captures of wild cranes as recovery can take 12–24 h. Fluid therapy, such as lactated Ringer's solution (Lactated Ringers Injection, Hospira, Inc., Lake Forest, IL) administered subcutaneously, is helpful in significantly shortening recovery times (Hartup et al., 2014). With alpha-chloralose, there is the potential for a crane to consume some medication and become mildly sedated, but not enough to be captured. There is the potential for the crane to hurt itself or be killed while in such a state, and careful monitoring of all treated cranes is advisable.

Cranes are susceptible to exertional myopathy when handled improperly and allowed to struggle during handling. In one study (Hayes et al., 2003) of Sandhill Cranes there was a 3.7% morbidity and 1.6% mortality rate caused by exertional myopathy. Another study (Businga et al., 2007) describes treatment for exertional myopathy in Sandhill Cranes with dexamethasone (1–2 mg/kg given by subcutaneous injection every 12 h for 1–2 days), selenium/vitamin E (0.06 mg/kg given as an intramuscular injection initially and repeated on day 7 of treatment), lactated Ringer's solution (60–180 mL per crane given subcutaneously every 12 h for 2–5 days), and nutritional support using gavage feeding of 30–120 mL every 12 h. The first step to reduce struggling is to place a hood over an individual's head, leaving the bill and nares unobstructed so that the crane can breathe easily, or pant if it becomes overheated.

Sedation

Several sedatives can aid in calming agitated cranes but require a few minutes to take effect after being administered. Diazepam sodium (Diazepam Injectable, Hospira, Inc., Lake Forest, IL; any use of trade, firm, or product names is for descriptive purposes only and does not imply endorsement by the U.S. government) at 0.5–1.0 mg/kg is effective; midazolam (Akoron, Inc., Lake Forest, IL) also works well at 0.5–1.0 mg/kg, but can cause a crane to be unsteady on its feet, especially at the highest dose (Olsen, unpublished data). Diazepam effects may last 4–6 h and midazolam up to 24 h; length of effect must be considered prior to use. If the crane is suffering from injury or exertional myopathy, longer sedation may be useful. If cranes are to be released immediately after handling, however, sedative effects that last for hours are inadvisable. Midazolam can be reversed with flumazenil (0.02 mg/kg, Hikma Pharmacutica, Portugal) given by intramuscular injection (Abou-Madi, 2001; Machin and Caulkett, 1998).

An ultrashort-acting benzodiazepine, triazolam, was administered experimentally in Sandhill Cranes (0.1–0.15 mg/kg) and was effective at decreasing both heart rates and the level of struggling in handled birds (Black and Whiteside, 2003). Effects were less marked in birds with more apprehensive personalities, while mild ataxia was seen in the calmest birds. Triazolam effects were most noticeable for 1–2 h.

Anesthesia

General anesthesia is useful when complete immobilization is desired, such as for surgery or radiographs. General anesthesia can be administered in either an inhalant or injectable format, or a combination of the two. The most common inhalant anesthetics used in avian species are isoflurane (Isoflurane, Piramal Critical Care, Inc., Bethlehem, PA) and sevoflurane (Sevoflurane, Piramal Critical Care, Inc., Bethlehem, PA) (Jones, 2009). These are given first by face mask for induction (vaporizer setting of 4–5%) (Olsen, 2009). For maintenance (1–3%) with oxygen flow rate of 1–2 L/min. the crane may be intubated with a noncuffed or cuffed soft silicon endotracheal tube to prevent tracheal damage. Because birds have complete tracheal rings, unlike the partial rings in mammals, the trachea in birds cannot expand when the cuff on the endotracheal tube is inflated. At least one Whooping Crane died due to a granulomatous lesion that developed in the trachea in the location where an endotracheal tube cuff was inflated (Olsen, unpublished data). Inflating endotracheal tube cuffs should be avoided to prevent resultant necrosis of the mucosa and subsequent development of granuloma in the area.

Injectable anesthetic agents may offer some advantage over inhalant anesthetics, especially for field situations, as no bulky vaporizer, oxygen regulator, or oxygen tank is required. Propofol (PropoFlo Injectable, Abbott Laboratories, North Chicago, IL) is given intravenously at 1.0–1.5 mg/kg to effect (Rupiper et al., 2000) and can be maintained with subsequent small doses or a constant infusion also given intravenously. As with inhalant anesthetics, an open or patent airway should be maintained by intubation. Respiratory depression or complete apnea can occur at any point during general anesthesia. Supplemental oxygen or even ambient air delivered with an Ambu bag should be readily available for artificial ventilation. Another useful intravenous anesthetic agent, alphaxolone (Alfaxan, Jurox, Inc., Kansas City, MO), has been used in avian species for over a decade and has recently become available in the United States. The recommended dose is 6.5–7.0 mg/kg given by intravenous injection, which provides a smooth and rapid induction (13–26 s). The maximum anesthetic effect lasts only 5–6 min without additional doses given periodically (Bailey et al., 1999; Olsen, 2009). Cranes are generally standing with no visible signs of ataxia in as little as 20–33 min following the last dose given (Bailey et al., 1999).

SURGERY

Techniques used for rehabilitating injured cranes are many and varied (Olsen, 2000). Recovery from surgery can be protracted. Therefore, not many surgical interventions are used on pre-release or wild cranes. Nevertheless, many surgical procedures have been used on captive cranes, especially those Whooping Cranes considered genetically valuable to the breeding and reintroduction programs. Here we focus on surgery used to implant VHF radio transmitters in Sandhill Crane and Whooping Crane chicks and management of orthopedic problems that have the potential to impact releasability of captive raised Whooping Cranes.

Radio Implant Surgery

One surgical procedure used specifically for field studies of wild crane chicks is the subcutaneous implantation of small (<3 g) VHF radio transmitters (Olsen, 2004; Spalding et al., 2001). This technique has been used in field studies of both Mississippi Sandhill Cranes and Whooping Cranes. Crane chicks are captured on or near the nests, anywhere between 1 and 12 days of age (older chicks have been implanted but the technique is complicated in a more mature chick due to increased difficulty of capturing and safely restraining the chick). The chick is given a subcutaneous injection of bupivacaine (2 mg/kg, Bupivacaine Injection, Hospira, Inc., Lake Forest, IL) in the caudal dorsal cervical skin. A 1:10 or 1:20 dilution of bupivacaine and sterile water may be used, to increase injection volume and allow perfusion of an area along the dorsal neck of about 2.5 cm with the local anesthetic (Olsen, unpublished data). After allowing 5–10 min for the local anesthetic to take effect, the surgery site is plucked and disinfected and an incision 1.0–1.5 cm long is made. The subcutaneous tissue caudal to the incision and between the scapulae is undermined using blunt dissection to create space to place the small transmitter. The skin is perforated from the outside with an 18 or 20 gauge 1½ in. needle so that the tip of the needle is visible in the cranial portion of the incision. The tip of the antenna is then passed up and out through the needle. The needle is removed and, pulling on the antenna end, the body of the transmitter is manipulated into the subcutaneous space. The transmitter should be totally under the skin and not visible at the surgery site. The skin is sutured in a simple interrupted pattern using a small diameter absorbable monofilament suture material, such as 4-0 polydiaxaone (PDS II, Ethicon, Inc., Johnson & Johnson, Somerville, NJ) or 4-0 poliglecaprone (Monocryl, Ethicon, Inc., Johnson & Johnson, Somerville, NJ). A small amount of subcutaneous fluid (lactated Ringer's solution, 3–5 mL/100 g body weight) is given, injected subcutaneously away from the surgery site. The crane chick can be immediately returned to the nest or area where it was found.

In one study in Mississippi, no observed morbidity or mortality events were found to be associated with radio implant surgery and all chicks were successfully reunited with their parents (Olsen, 2004). The transmitter should be removed when the chick is captured for banding at or around fledging. This can be done by making a small incision next to the antenna exit site and pulling the transmitter out by the antenna. The incision can be sealed with several drops of tissue glue and the incision edges pinched shut until the glue sets. If the transmitter cannot be retrieved, experience with captive Sandhill Cranes at Patuxent suggests the transmitter will be rejected or fall out within the first year (Olsen, unpublished data).

ORTHOPEDIC TREATMENTS

Orthopedic problems that develop in captivity can result in a Whooping Crane chick becoming unfit for release. Whooping Crane chicks grow from 140 g at hatch to 5–6 kg at fledging in 90 days, and rapid growth of the long bones is associated with a significant occurrence of bone abnormalities. Often considerable amounts of time and other resources are expended to treat orthopedic problems in young Whooping Cranes that eventually cannot be released or may have to be euthanized. Specific orthopedic conditions were identified by Olsen and Langenberg (1996) in a publication focused on crane chick medicine, and include lateral rotation of the distal end of the carpometacarpus, rotations and linear toe abnormalities, and rotational/angular pelvic limb deformities. In one retrospective study, 18% (31/179) of captive Whooping Crane chicks had carpal deformities, and 29% (51/179) had pelvic limb deformities (Kelley and Hartup, 2008). Kelley and Hartup (2008) showed that (1) increased risk of toe and leg deformities was associated with hand rearing, (2) lower relative weight change in week 1 increased risk of toe deformities, (3) higher relative weight change in week 2 increased risk of leg deformities, (4) female chicks and chicks from third clutch eggs were at increased risk of carpal deformities, and (5) an increased risk of carpal deformities was observed in chicks from eggs collected at Wood Buffalo National Park and in chicks with preexisting or concurrent deformities (Kelley and Hartup, 2008). Hand rearing was highly correlated with increased incidence of toe and leg deformities compared to parent-reared chicks. One reason for this may be that chicks with developmental toe deformities exercise less, thus contributing to development of one or both of the other problems. Identifying, treating, and resolving toe problems quickly may help keep the chick in good condition for release. Twisted or bent toes can be returned to a more normal conformation using several corrective procedures. Splints made from wooden dowels and low-tack adhesive strapping or packing tape may be applied to individual toes for 48 h (Fig. 19.2) (Olsen and Langenberg, 1996). Another technique is to create a snowshoe-like structure from light cardboard and to tape this to the bottom of the foot using the low-tack adhesive tape (Fig. 19.3). The snowshoe is also removed after 48 h (Olsen and Langenberg, 1996).

In one study, Olsen (1994) found that 32% (7/22) of orthopedic injuries occurred in crane chicks between 9 and 12 weeks of age. A range from 5 to 16 weeks of age accounted for 64% (14/22) of all orthopedic injuries in cranes. There was a high incidence of long bone fractures found in crane chicks between 9 and 12 weeks of age, which corresponds to the time of epiphyseal closure of the long bones. For reintroduction programs, this has led to the recommendation not to ship crane chicks between 8 and 12 weeks of age. In the EMP, to avoid leg injuries, chicks for ultralight-led reintroductions were normally shipped to Wisconsin before 8 weeks of age, while chicks for direct autumn release and parent-reared chicks were shipped after 12 weeks of age. Chicks for the LNMP are shipped when older than 16 weeks of age.

FIGURE 19.2 A Whooping Crane chick with splints on its toes to correct a curving or deviation of the toes from the normal straight position. The splint is made of a wooden dowel and low-tack adhesive packing tape. The splint is left in place for 2 days and then removed because of the rapid growth seen in young chicks. If needed, another splint can be placed on the problem toes, but generally one splint in place for 2 days is sufficient to see correction of the problem.

FIGURE 19.3   A Whooping Crane chick with both splints on the toes of one foot and a snowshoe on the other foot. The snowshoe is made from note cardboard or similar weight material and is attached using low-tack adhesive packing tape, being careful to keep all toes straight. The snowshoe is used when the entire foot is curled and the chick is not placing the foot properly. Snowshoes remain in place for 2 days and then are removed. If need be, and the foot is still curled, a second snowshoe can be placed on the foot after a few hours' rest for the chick.

For reintroduction programs, fractures of wing limb bones, as compared to leg bones, more likely heal sufficiently for release, depending on timing with respect to migration or release, with none of the 19 cranes in one retrospective study dying or being euthanized due to wing bone fractures (Olsen, 1994). Fractures of pelvic bones did result in euthanasia or death (n = 15 of 19). The implications for reintroduction programs are twofold. Crane chicks that have wing injuries may be treated successfully and may continue in the reintroduction program, but crane chicks with pelvic limb injuries most likely will not be suitable for release. For Whooping Cranes already released in a reintroduction program, any major limb injury may render the crane unsuitable for rerelease after capture and treatment. Even treating wing injuries, in which a successful outcome is more likely, can require extended hospitalization of 29–93 days (Olsen, 1994).

INFECTIOUS DISEASES

The following material on infectious diseases is not intended to cover all the infectious diseases reported in cranes, most of which are reported from captive crane populations (Hartup et al., 2010a, 2010b). Instead, we review a few diseases in each category (bacterial, fungal, parasitic, and viral) that have broader implications for reintroduction programs. Some diseases may result in a captive crane becoming unfit for release. Diseases may occur after release and cause morbidity or mortality in the reintroduced population of Whooping Cranes. In addition, infectious diseases from captive crane facilities have the potential to spread to wild crane populations. One study (Hartup, Chapter 18, this volume) reports on the incidence of diseases found in the remnant AWBP.

Bacterial Diseases

A number of bacterial diseases can pose a threat to crane health, particularly in reintroduced populations where there are limited treatment options. Management of the habitat used by reintroduced populations can be critical for avoiding epizootics. For captive and reintroduced populations, bacterial diseases of primary concern include diseases caused by *Mycobacteria* and *Salmonella*. Both bacteria cause diseases of zoonotic importance, and testing has been required prior to shipping Whooping Cranes for release or between captive breeding facilities.

Mycobacteriosis

Avian tuberculosis (*Mycobacterium avium*) has been found in wild cranes in North America but is not reported in captive Sandhill Cranes or Whooping Cranes. Snyder et al. (1997) reported on a 5-year-old Whooping Crane from the AWBP that was found debilitated, with a large, palpable midcoelomic splenic mass and a 2 cm diameter mass in a chronically prolapsed cloaca. *M. avium* was isolated from the cloacal mass after surgical removal and the diagnosis was confirmed with a DNA-specific probe. A 12-month treatment program with rifampin (45 mg/kg once daily, Rifampin Injection, Pfizer, Inc., New York, NY) and ethambutol (30 mg/kg once daily, Ethambutol, Lupin Pharmaceuticals, Baltimore, MD) was ineffective at controlling clinical signs. After 16 weeks of azithromycin treatment (40 mg/kg once daily, Azithromycin, Greenstone, LLC, Peapack, NJ), the infection appeared to successfully resolve. Ten months later, radiographic findings, weight loss, and elevated white blood cell counts suggested avian tuberculosis was again active. The crane was again treated with azithromycin (20 mg/kg once daily to start, then increased to 40 mg/kg once daily in food). During the 16-week course of treatment the crane improved (Snyder et al. 1997). However, when ethambutol was added to the treatment, the crane had a fatal adverse reaction. *M. avium* could not be isolated from any tissues at necropsy, suggesting treatment may have ultimately been effective. The required long course of treatment makes treatment of wild cranes challenging at best. This case illustrates the potential risks associated with Whooping Cranes acquiring this disease in the wild. The ease of contagion among birds reinforces the need for quarantine coupled with adequate disease screening when bringing wild birds into captivity. *M. avium* can be zoonotic and especially poses a risk for immunocompromised humans (Friend and Franson, 1999).

Salmonellosis

In Japan, Maeda et al. (2001) isolated 29 strains of *Salmonella* spp. from 420 crane fecal samples from mixed flocks of Hooded Cranes (*Grus monacha*), White-naped Cranes (*G. vipio*), Common Cranes (*G. grus*), Siberian Cranes (*G. leucogeranus*), Demoiselle Cranes (*Anthropoides virgo*), and Sandhill Cranes. In captive reintroduction programs, *Salmonella typhimurium* was isolated from one Whooping Crane chick showing symptoms of diarrhea, lethargy, and weight loss, and from several healthy older colts at Patuxent (Olsen, unpublished data). Numerous other strains of *Salmonella* have been isolated from the feces of healthy juvenile cranes being tested as part of a pre-reintroduction health screening; however, shedding is often sporadic. In 2002, a nonpathogenic strain of *S. typhimurium* was isolated from a Whooping Crane chick at Patuxent (Olsen, unpublished data). Other nonpathogenic strains isolated from Whooping Cranes include *S. montevido, S. panama, S.infanis, S. muenster,* and *S. lexington* (Olsen, unpublished data). These findings are not unusual given the modes of transmission of *Salmonella*, which include direct contact with feces, contaminated feed, insects, rodents, or feral birds; treatment is not indicated for these cases. Some forms of *Salmonella* can be zoonotic, and precautions to prevent human exposure should be taken.

Campylobacteriosis

Several species of *Campylobacter* have been isolated from cranes. *Campylobacter jejuni* has been reported from both captive and free-living cranes (Hoar et al., 2007; Langenberg et al., 1996; Lu et al., 2013). Langenberg et al. (1996) isolated *C. jejuni* in 19 of 21 crane chicks of three species and 8 of 55 adult cranes of 10 species. In addition, *C. jejuni* was isolated from several Florida Sandhill Crane (*A. c. pratensis*) chicks. *C. enteritis* has not been isolated from cranes. However, a novel *Campylobacter* species, *C. canadensis*, was isolated from captive Whooping Cranes (Inglis et al., 2007). *Campylobacter* is probably part of the normal flora of cranes, especially Sandhill Cranes, and does not require treatment in the absence of clinical signs (Lu et al., 2013). Campylobacter infections are of zoonotic concern, and personnel handling cranes should take precautions to avoid being infected (Lu et al., 2013).

Other Bacteria

To assist in investigating diseases or testing and certifying cranes for release in reintroduction programs, it is useful to know what normal bacterial flora to expect. Hoar et al. (2007) collected aerobic and anaerobic cloacal swabs from captive, clinically normal Whooping Cranes and Sandhill Cranes. Gram-positive cocci, coliforms, and gram-negative bacilli were the most common isolates. *Escherichia coli*, *Enterococcus faecalis*, and *Streptococcus* Group D were the most frequent individual species isolated. *Campylobacter* spp. was isolated from five Whooping Cranes. Significant differences exist in the normal flora of these two closely related crane species, which has implications for the management of these birds, often held in proximity in captivity (Hoar et al., 2007). Knowing pathogenic versus normal bacterial flora can assist the veterinary clinician and biologist in determining the suitability of cranes for reintroduction programs and the type of prerelease testing that should be performed.

Bacterial eye infections are reported in captive crane chicks and can cause withdrawal of some chicks from release programs due to permanent eye damage and loss of vision in the affected eye. Miller et al. (1995) described bacterial isolates from the inferior conjunctival cul-de-sac of 48 clinically normal captive cranes of three species: Whooping Cranes, Florida Sandhill Cranes, and Siberian Cranes. *Corynebacterium* spp., *Enterobacter* spp., and coagulase negative *Staphylococcus* spp. were among the most frequent isolates. *Pseudomonas* spp. was isolated but *P. aeruginosa* was not found. *P. aeruginosa* has been associated with pathogenic corneal ulcerations resulting in permanent corneal damage in young Whooping Cranes at ICF (Hartup, unpublished data) and Patuxent (Olsen, unpublished data). Other species of bacteria isolated from normal crane eyes included *E. coli*, *Klebsiella* spp., *Pasteurella* spp., *Aeromonas* spp., *Acinetobacter baumanni*, *Serratia fonticola*, *Citrobacter amalonaticus*, and *Alcaligenes faecalis* in the gram-negative category. Among gram-positive organisms isolated, in addition to the aforementioned *Staphylococcus* and *Corynebacterium*, were *Streptococcus* spp., *Enterococcus* spp., and *Bacillus* spp. (Miller et al., 1995).

Fungal Diseases

Fungal diseases cause significant issues for crane captive breeding and reintroduction programs. Some fungal organisms, such as *Candida* and *Aspergillus* sp., can be directly pathogenic to cranes, while others may produce toxins in grains used in feed and cause death and debilitation through ingestion of the toxin.

Candidiasis

Candida spp. is occasionally isolated from lesions on the bill or within the oral cavity (Olsen, 2009). These lesions can be treated successfully with topical nystatin cream (Nystatic, TARO Pharmaceuticals, Ltd., Hawthorn, NY) or clotrimazole cream (Clotrimazole Cream, Major Pharmaceuticals, Livonia, MI). There are no known cases of *Candida* spp. causing lesions that led to a crane being eliminated from a reintroduction program.

Aspergillosis

Aspergillosis is a common fungal disease of the respiratory tract of many avian species, including cranes. The most common isolate is *Aspergillus fumigatus*, though other species of *Aspergillus* have been infrequently isolated from cranes. Aspergillosis can cause acute death or chronic infections, and is most common in young cranes (Keller and Hartup, 2013). The organism is often associated with granuloma formation in the airway passages, most notably in the trachea and at the tracheal bifurcation, both of which can lead to airway obstruction. Traditional plain radiographs have not been found to be of diagnostic value. However, computed tomography scans have proven useful in distinguishing these masses (Schwarz et al., 2016). While there are no fully reliable premortem tests for aspergillosis, weight loss, lethargy, and vocal changes are clinical signs. Several plasma-based tests can be used in combination, including antibody and antigen titers, *Aspergillus* sp. PCR, plasma galactomann levels, and plasma protein electrophoresis, with increased suspicion related to peaks in beta globulins. Exposure to *Aspergillus* organisms can start in the egg, when spores from contaminated nesting material or bedding, or a contaminated incubator can penetrate the eggshell and result in embryo mortality or decreased hatchability. Posthatch infected hatchlings may be seen; these chicks are weak and may die as young as 9 days old from aspergillosis (Olsen and Clubb, 1997). *Aspergillus* spp. organisms are ubiquitous in the environment and are strongly associated with accumulation of moist organic material.

Effective preventive therapy for aspergillosis can be accomplished with itraconazole (10 mg/kg twice daily, reduced to once daily after several weeks; Itraconazole, Patriot Pharmaceuticals, LLC, Horsham, PA; Hartup, unpublished data). For treatment, itraconazole with terbinafine (10–20 mg/kg once daily in the evening to avoid potential appetite suppression; Terbinafine, Harris Pharmaceutical, Inc., Fort Myers, FL) may be used for several weeks. A bolus of Amphotericin B (Amphotericin B for Injection USP, Xgen Pharmaceuticals, Inc., Big Flats, NY), given intravenously at 1.5 mg/kg once and repeated in 4 weeks, may also be added to the treatment (Hartup, unpublished data). Refractory or severe cases can be treated with voriconazole (5 mg/kg SID, Voriconazole, Greenstone, LLC, Peapack, NJ), and the bird monitored for clinical signs of hepatotoxicity. Clotrimazole nebulization (made by various compounding pharmacies) has also been used for treatment (Olsen, 2009). Oral medication can often be given in a smelt or grape to an individual chick. Treatment can be effective, but respiratory sequelae are common, even after treatment. Associated limited exercise tolerance may render birds unsuitable for reintroduction.

Mycotoxicosis

Mycotoxins are produced by a group of fungal organisms, primarily *Fusarium graminearum*, that grow on corn, peanuts, and other crops. Multiple large-scale mortality events in wild Sandhill Cranes have been caused by mycotoxins (Roffe et al., 1989; Windingstad et al., 1989). Predominant clinical signs include inability to hold the neck erect, submandibular edema, and inability to fly. At necropsy, multiple muscle hemorrhages may be seen. These mycotoxin epizootics in Sandhill Cranes were attributed to moldy peanuts, associated with cold, wet weather.

A mycotoxin epizootic among captive cranes at Patuxent sickened 240 of 300 cranes and caused 15 deaths, including two Mississippi Sandhill Cranes that were part of a reintroduction effort (Olsen et al., 1995). Clinical signs were nonspecific, and included weakness, necrosis of the oral mucous membranes, depression, dehydration, ataxia, and recumbency, followed by death. Serum uric acid levels were elevated on clinical pathology testing. Gross pathology findings included dehydration, atrophy of fat, renal insufficiency, and splenic atrophy. *Fusarium* spp. was isolated from the pelleted feed, and two mycotoxins, T2 and deoxynivalenol (vomitoxin), were identified (Olsen et al., 1995). Testing of each batch of grain-based feed is highly recommended for crane breeding and reintroduction programs. Because of the danger of mycotoxins, the public should be discouraged from feeding cranes with grain tossed on the ground.

Parasitic Diseases

Control of parasitic diseases is important for Whooping Crane programs for two reasons. Some parasitic diseases, especially when amplified in the confines of captive pens, can lead to lethal infections (Olsen et al., 1996; Olsen and Langenberg, 1996). Enteric coccidial infections, which develop into DVC, were found to limit the survival of Whooping Crane chicks until the cause of the disease was discovered and control measures instituted (Olsen and Langenberg, 1996).

Other parasitic diseases are found in the captive setting but may not exist in the wild, making efforts to eliminate parasites prior to release a critical component of prerelease health care.

Protozoa

DVC is undoubtedly the most important parasitic disease in cranes. DVC is caused by two protozoan parasites, *Eimeria gruis* and *Eimeria riechenowi*, and has been recorded in free-ranging Sandhill and Whooping Cranes (Bertram et al., 2015; Hartman et al., 2010; Parker and Duszynski, 1986). *E. gruis* and *E. riechenowi* have also been identified in four species of cranes in captivity in North America, including Sandhill and Whooping Cranes. DVC has been found in a wild White-naped Crane (*Grus vipio*) in Korea (Kwon et al., 2006). This White-naped Crane died of phosphamidon poisoning with the additional gross findings of disseminated white nodules on the serosa of the proventriculus, gizzard, and intestines, and the parenchyma of the liver, spleen, and heart. Asexual coccidial stages were observed on histopathological examination but the species of *Eimeria* was not identified because no oocysts were found in the feces or intestinal contents.

Monoclonal antibodies have been used to study *Eimeria* in crane chicks (Augustine et al., 1998, 2001). Monoclonal antibodies elicited against the *Eimeria* spp. of chickens and turkeys were found to cross-react with sporozoites and various stages in the development of *Eimeria gruis* in Sandhill Crane chicks (Augustine et al., 1998, 2001). Results indicated that the typical 7-day life cycle of coccidia parasites does not appear to apply to the two *Eimeria* of cranes. Fourteen days is the minimum length of time before various stages are found in the jejunum, liver, and lungs, and longer periods are required before DVC lesions occur on other internal organs. In one study at Patuxent, peak shedding in the feces occurred 17 days post inoculation for *E. gruis* and 21–22 days for *E. riechenowi* (Olsen, unpublished data). Based on

these findings, morbidity and mortality from DVC may not occur until the chick is a minimum of 3–4 weeks old for individuals exposed soon after hatching. In the wild, chick disappearances are often attributed to predation, when the underlying cause of the morbidity leading to predation or the mortality itself may be DVC. Mortality studies of crane chicks are needed to assess the effects of diseases such as DVC on chick survival.

In captive crane colonies, amprolium (Corid, Merial, Ltd., Duluth, GA) added to drinking water was initially used to control coccidia infections. However, some resistance to amprolium in the form of increasing fecal parasite counts was noted at Patuxent during one year (Olsen, unpublished data). Two other coccidiastatic medications, monensin (Elanco Animal Health, Indianapolis, IN) and clazuril (Biotain Pharma Company, Ltd, Yiamen City, Fujian Province, China) were tested and found efficacious (Carpenter et al., 1992). Commercially available monensin mixed with feed at 90 g/ton was chosen as the method for controlling coccidia in the captive reintroduction programs, where DVC can pose a problem. This treatment continues to be efficacious as monitored by fecal parasite counts and the absence of DVC-associated morbidity and mortality (Olsen, unpublished data). However, obtaining monensin-impregnated feed has become more challenging because monensin is highly toxic to horses and many feed mills are reluctant to work with the medication. Toltrazuril (Baycox, Bayer, Inc.) has been used to eliminate severe coccidial infections in Whooping Crane chicks (Olsen, unpublished data), but may be limited in application due to limited availability. In the future, a new coccidiastat will need to be tested and used if DVC is to be prevented in the captive-rearing facilities that support reintroduction efforts.

Avian malarial blood parasites, including *Haemoproteus antigonis*, *H. balearicae*, and *Leucocytozoon grusi*, have been diagnosed in cranes with associated severe anemia (hematocrit as low as 13% in one case) and death of wild Florida Sandhill Crane chicks (Dusek et al., 2004). In the FNMP, no avian malarial parasites were found to be associated with chick mortality. *Haemaproteus* infections have caused illness with severe anemia in captive Mississippi Sandhill Crane chicks. The chicks were treated successfully with a simple blood transfusion and atovaquone proquantil (20 mg/kg PO SID for 3 days, Prasco Laboratories, Cincinnati, OH) (MacLean and Beaufrere, 2015). Both *Heamoproteus* spp. and *Leucocytozoon* spp. have been documented in raptors in Louisiana (Olsen and Gaunt, 1985) and were associated with raptor mortality. While there has been no associated mortality during the first 4 years of the program, malarial blood parasites may be of concern in the LNMP and should be monitored.

In one study (Hartup, Chapter 18, this volume) 20% ($n = 35$) of blood smears from Whooping Cranes in the AWBP were positive for hemoparasites consistent with *Haemoproteus* sp. morphology. The parasitemia found ranged from 1 to 6 cells per 10,000 cells observed. Using molecular detection methods, 68% ($n = 37$) of the AWBP Whooping Cranes were positive for *Haemosporidia* DNA and 3% of samples were positive for *Leucocytozoan* spp. DNA. Hemoparasites have not been identified in captive Whooping Cranes at ICF (Hartup, unpublished data), Patuxent (Olsen, unpublished data), nor the Calgary Zoo (Black, unpublished data).

Hexamita spp. has been reported to cause catarrhal necrotizing enteritis and death in Demoiselle Cranes (Ippen et al., 1980; Zwart et al., 1985) and in two captive-reared Florida Sandhill Cranes that were to be released in Florida (Spalding et al., 1994). The Florida Sandhill Cranes were raised at Patuxent and shipped to Florida in February 1991, where they were released into a large pen. One crane was found dead in March 1991 and the other in April 1991. No unusual behavior or clinical signs were noted prior to death. *Hexamita* spp. had not been diagnosed in the crane flock at Patuxent prior to

when these mortalities occurred, nor has it since. Spalding et al. (1994) suggest that *Hexamita* sp. is a potential pathogen of cranes under stress (e.g., from shipping). Testing for the presence of *Hexamita* in fecal samples and/or prophylactic therapy is recommended prior to shipping cranes.

Helminths

Whooping Cranes and Sandhill Cranes in captivity have been diagnosed with spiny-headed worms (Acanthocephalans) (Olsen et al., 1996), gapeworm (*Cyanthostoma coscorobae*, *Syngamus* spp.) (Carpenter, 1993; Olsen et al., 1996), *Capillaria* spp., and *Ascaridia* spp. (Olsen et al., 1996). Acanthocephalans have caused severe morbidity and mortality in captive-reared Whooping Crane chicks, including perforation of the intestines with resulting severe peritonitis (Olsen, 2009; Olsen et al., 1996).

In the most complete study of helminth parasites in reintroduced Whooping Cranes, 27 birds found dead up to 28 months post-release in the FNMP were necropsied (Spalding et al., 1996). The nine species of nematodes identified were: *Ascaridia pterophora*, *Strongyloides* sp., *Eucoleus obtusiusscula*, *Dispharynx nasuta*, *Hystrichis tricolor*, *Capillaria* sp., *Contracaecum multipapillatum*, *Cyanthostoma variegatum*, and *Physaloptera* sp. In addition, three trematode species (*Brachylaema* sp., *Lyperorchis lyperochis*, and *Philopthalmus gralli*) and two species of acanthocephalans (*Centrorhynchus kuntzi* and *Southwellina* sp.) were found (Spalding et al., 1996). All Whooping Cranes released in the FNMP underwent a 60-day quarantine in a clean facility (not previously used by captive cranes) and received anthelminthic treatment and testing of feces for parasites during quarantine. Despite these precautions, a relatively large population of parasites was found in these birds at necropsy. As the authors point out, three of the parasites (*A. pterophora*, *E. obtusiuscula*, and *C. variegatum*) had never been reported for wild Sandhill Cranes in Florida (Forrester et al., 1974, 1975) and may have been introduced along with the Whooping

Cranes. One of the parasites, *A. pterophora*, had previously only been reported from gruiformes in Brazil (Cristofaro and Feijo, 1976) and from a Sandhill Crane reintroduced in Florida in 1990 (Spalding et al., 1996). In most of these cases, parasitism was an incidental finding (Spalding et al., 1996), with most mortalities due to bobcat predation and 24 of 27 birds in very good body condition. The three birds with low body fat levels had greater numbers of parasitic species than the other Whooping Cranes (Spalding et al., 1996).

Arthropods

In the Spalding et al. (1996) study, one necropsied Whooping Crane was found to have a very low number of biting lice (*Heleonomus assimilis*). No pathological problems associated with the lice were identified, and only two lice were found. The bird died 10 months after release from unrelated processes. Given the amount of time post-release, the Whooping Crane may have acquired the lice post-release rather than in the captive facility.

Captive cranes may be exposed to environmental conditions significantly different from those of wild cranes. One issue that has plagued individual cranes at some of the breeding centers is an apparent hypersensitivity to fly (Diptera) bites (Hartup and Schroeder, 2006). While many treatments have been used to treat and mitigate this condition, including nonsteroidal anti-inflammatory drugs, topical steroids and antibiotics, and parenteral antibiotics, only Tylosin tartrate administered in drinking water has been found to correlate with significant clinical benefit (Kelman and Hartup, 2014).

Viral Diseases

Inclusion Body Disease of Cranes

In 1978, there was an epizootic of a herpes virus, causing morbidity and mortality, in captive cranes at ICF. Once identified and based on histologic findings, the virus was named inclusion body disease of cranes virus (IBDC; Docherty and Henning, 1980). IBDC was believed to have been introduced to the collection by cranes imported from Europe, as the disease had not been previously reported in captive or wild cranes in North America (Docherty, 1987, 1999). IBDC was controlled and eventually eliminated at ICF through a combination of strict quarantine of infected animals and repeated testing (Docherty, 1999). Concern exists that IBDC may be reintroduced to North America in the future by migratory Lesser Sandhill Cranes (*A. c. canadensis*). As the breeding range of this subspecies continues to spread west in Siberia, individuals may come into contact with Eurasian crane species (Bysykatova et al., 2010). This could result in crossover infection with IBDC, and the potential for this virus to be carried to North American by Lesser Sandhill Cranes migrating across the Bering Strait.

IBDC causes lethargy, loss of appetite, diarrhea, and death. During the epizootic in a mixed flock of 51 cranes, 18 mortalities occurred, including Sandhill Cranes, Red-crowned Cranes (*Grus japonensis*), Blue Cranes (*Anthropoides paradise*), and Hooded Cranes (*Grus monacha*) (Docherty and Henning, 1980). Pathology was characterized by liver and spleen enlargement with small (1 mm diameter) yellow-white lesions present in both, and hemorrhage in the thymus and intestines. Characteristic intranuclear inclusion bodies were described in the liver and spleen (Docherty, 1987, 1999). Whooping Cranes were not present at the institution when this epizootic occurred, but, given the broad crane species range affected, it is assumed Whooping Cranes would also be susceptible to IBDC, though crane species are known to vary in their response to IBDC (Docherty and Romaine, 1983).

To date, this has been the only outbreak in North America, and the risk of this disease spreading from a captive facility to wild cranes in North America is carefully managed. All Whooping Cranes designated to be reintroduced are tested for IBDC (antibody serology) two weeks before shipment during a period of quarantine from other cranes. In addition, all captive Whooping Cranes in flocks producing offspring for reintroduction are tested semi-annually for IBDC.

Eastern Equine Encephalitis

Eastern equine encephalitis virus (EEE) is an arbovirus spread by mosquitoes, primarily *Culiseta melanura* (Morris, 1989; Pagac et al., 1992). EEE occurs in eastern North American, sections of Central and South America, and the Caribbean, where the enzootic cycle is well established between native bird species and mosquitoes (Ritchie, 1995). The virus is found in a large number of bird species, with mortality usually limited to introduced species, such as Ring-necked Pheasants (*Phasianus colchicus*), House Sparrows (*Passer domesticus*), and Chukar Partridges (*Alectoris chukar*) (Ritchie, 1995). Once a bird recovers from infection, immunity is believed to be lifelong with antibody titers rising when challenged again with the virus (Ritchie, 1995). There are vaccines available to provide protective immunity in horses and humans, and these are also used to protect Whooping Cranes (Olsen, 2010; Olsen et al., 1997, 2005).

In 1984, a deadly epizootic of EEE occurred in Whooping Cranes at Patuxent, killing 7 of 35 birds (Dein et al., 1986). Another deadly epizootic occurred among captive Mississippi Sandhill Cranes (Young et al., 1996) in Louisiana. In the course of the clinical disease in cranes, death can be acute with a lack of clinical signs or following brief neurological signs such as ataxia. Because the disease is endemic to native species in North America, research to test various vaccination programs was initiated on Sandhill Cranes (Clark et al., 1987). In the original vaccination trials, a human EEE vaccine (Salk Institute, Government Service Division, Swiftwater, PA) was compared with an equine vaccine; the human vaccine produced higher antibody titers post vaccination (Clark et al., 1987). Olsen (2010) and Olsen et al. (1997, 2005) have tested various vaccines using antibody titers to measure potential effectiveness. Both Sandhill and Whooping

Cranes in captive breeding centers, as well as all birds slated for release, have been safely vaccinated. Currently, adult Whooping Cranes held in facilities where EEE is endemic receive an annual EEE vaccination with a 1 mL intramuscular injection of products, such as Vetera EWT (Boehringer Ingelheim Vetmedica, Inc., St. Joseph, MI) or Equiloid Innovator (Zoetis, Inc., Kalamazoo, MI).

There have been no vaccination-challenge studies in either Sandhill Cranes or Whooping Cranes to test the efficacy of vaccination as a protective agent for EEE. However, one retrospective study (Olsen et al., 1997) looked at known exposure cycles documented in other birds, including northern bobwhite quail (*Colinus virginianus*), and in mosquitoes to determine years when captive Whooping Cranes would have been exposed. No Whooping Cranes that were previously vaccinated for EEE died from the disease in those peak years, even though an approximate 20% mortality rate could have been expected based on the 1984 epizootic of EEE. The authors concluded that EEE vaccination of Whooping Cranes was efficacious (Olsen et al., 1997).

EEE does not occur in most of the areas used by the AWBP. However, EEE does occur in Wisconsin and Louisiana where Whooping Crane reintroductions are occurring and in other southeastern states used by wintering Whooping Cranes in reintroduced populations. All Whooping Crane chicks in reintroduction programs receive two doses of EEE vaccination prior to release, at 35 and 56 days of age. The potential impact of EEE on reintroduced populations cannot be known until unvaccinated wild-hatched chicks are found in higher numbers in the populations.

West Nile Virus

Since its first documented occurrence in North America in 1999, West Nile virus (WNV) has been of great concern to wildlife rehabilitation and reintroduction programs (Steele et al., 2000). The disease was first identified in the 1930s in Africa, southern Europe, and the Middle East (McLean and Ubico, 2007). In the intervening years, associated avian mortality has been commonly reported, along with occasional infections and mortalities in many mammalian species, including horses and humans. However, no crane mortalities were reported before 1999.

During the 1999 outbreak of WNV in New York City, three species of cranes at the Bronx Zoo were exposed, and five of the six individual cranes developed antibody titers indicating exposure, but no clinical disease (Calle et al., 2000). However, as WNV spread across North America, occasional crane mortalities have been documented. The Audubon Center for Research on Endangered Species in New Orleans, Louisiana, lost seven chicks in a reintroduction program and had some morbidity among adult Mississippi Sandhill Cranes. One adult Mississippi Sandhill Crane died at White Oak Plantation, Florida; a Demoiselle Crane died at the Milwaukee County Zoo, Wisconsin; three Sandhill Cranes died at Patuxent (one adult, two young-of-the–year); and one Sandhill Crane was lost at ICF (Hansen et al., 2008).

Two challenge studies were conducted using Sandhill Cranes (Olsen et al., 2008, 2009) to study the effects of WNV in cranes and evaluate possible prevention of the disease with vaccines. In the first study, adult Sandhill Cranes were vaccinated with a killed-virus WNV vaccine manufactured for horses, then challenged with 5,000 plaque-forming units of a WNV isolate from crows in New York State. The most significant findings were marked weight loss in challenged birds as compared to controls and shedding of the virus by unvaccinated challenged birds. In addition, among challenged Sandhill Cranes, the vaccinated birds produced significantly higher antibody titers than did unvaccinated Sandhill Cranes. Sandhill Cranes challenged with the virus, whether vaccinated or not, all exhibited lower hematocrits (30%) as compared to unchallenged control cranes (44%) at day 10 post challenge

(Olsen, 2011). Mean white blood cell counts in challenged birds, vaccinated and unvaccinated, rose to 20,000–25,000 between day 7 and day 14, while unchallenged control cranes remained at 7,000–10,000 (Olsen, 2011; Table 19.1).

A second experiment used juvenile Sandhill Cranes (3–4 months old) as subjects, and administered either a killed virus WNV vaccine ($n = 6$), a recombinant DNA vaccine ($n = 6$), or no vaccine ($n = 6$), The results were similar as with adult cranes: elevated antibody titers by day 14 and elevated white blood cell counts on days 7–14 (Olsen et al., 2008). However, in the juvenile unvaccinated cranes there was a 33% mortality rate ($n = 6$) from WNV infection versus 0% mortality in the two vaccinated groups (Olsen et al., 2008; Table 19.2). After initial testing of safety and efficacy of the vaccines in Sandhill Cranes, vaccine testing in Whooping Cranes ($n = 9$), including young Whooping Cranes, was undertaken, but no challenge studies were conducted on Whooping Cranes (Olsen et al., 2003).

Currently, for all released Whooping Cranes, young birds are given two doses of a combination EEE–WNV vaccine (such as West Nile-Innovator + EWT, Zoetis, Inc., Kalamazoo, MI, USA) 3–4 weeks apart, starting at approximately 35 days of age. No deaths of Whooping Crane chicks have been attributed to WNV in the reintroduction program before or after this vaccination program was established, but no challenge studies have been done in Whooping Crane chicks. At ICF and Patuxent, captive adult Whooping Cranes develop antibody titers to natural infections with no morbidity or mortality from WNV, and thus are not vaccinated for WNV (Hartup, 2008; Olsen et al., 2009).

If there is 33% mortality in Sandhill Crane chicks, as demonstrated by Olsen et al. (2008), there could be undetected mortality among wild-hatched crane chicks. Any long-term effect of WNV on Sandhill Crane or Whooping Crane populations has not been studied. However, there has been no WNV-associated morbidity or mortality in captive Whooping Cranes

TABLE 19.1 Titers, White Blood Cell (WBC) Counts, and Survival of Sandhill Crane Chicks Challenged with 1 Mosquito Dose (5,000 Plaque-Forming Units) of West Nile Virus (WNV), USGS National Wildlife Health Center, Madison, WI, February 2002. For Vaccinated Challenged Sandhill Cranes, Titers Were Significantly Higher ($P = 0.048$, $F = 1.59$) Than the Titers of the Unvaccinated Challenged Sandhill Cranes.

Crane number	Vaccinated	Challenged	Titer day 0	Titer day 14	Titer day 42	Survived to day 42	WBC count maximum	Day of maximum WBC count
003	Yes	No	<1:5	<1:5	<1:160	Yes	10,119	3
017	Yes	No	<1:5	<1:5	<1:10	Yes	8,573	7
001	Yes	Yes	<1:5	1:10,240	1:10,249	Yes	26,527	7
028	Yes	Yes	<1:5	1:640	1:5,120	Yes	38,177	21
060	Yes	Yes	<1:5	1:2,560	1:2,560	Yes	21,153	7
053	Yes	Yes	<1:5	1:10,240	1:5,120	Yes	30,147	8
061	Yes	Yes	<1:5	1:2,480	1:10,240	Yes	17,247	10
113	No	Yes	<1:5	1:320	1:2,560	Yes	33,257	7
004	No	Yes	<1:5	1:320	1:1,280	Yes	20,923	21
041	No	Yes	<1:5	1:1,280	1:10,240	Yes	27,264	8
055	No	Yes	<1:5	1:640	1:2,560	Yes	22,999	10
065	No	Yes	<1:5	1:640	>1:2,560	Yes	29,265	10

Modified from: Olsen, G.H., Miller, K.J., Docherty, D.E., Bochsler, V.S., Sileo, L., 2009. Pathogenicity of West Nile virus and response to vaccination in Sandhill Cranes (Grus canadensis) using a killed vaccine. J. Zoo Wildl. Med., 40 (2), 263–271.

422

19. HEALTH AND DISEASE TREATMENT IN CAPTIVE AND REINTRODUCED WHOOPING CRANES

TABLE 19.2 Results from Vaccinating Sandhill Crane Chicks with Either FD Equine West Nile Vaccine or CDC Recombinant DNA Vaccine, PVAX-WN-1 Starting on Day 7 of Age. Chicks Were Then Challenged at 4 Months of Age with 5,000 Plaque-Forming Units of Live WNV (Day 0 of Challenge). Five Thousand Plaque-Forming Units of WNV Represents Approximately One Mosquito Bite Dose of the Virus. Sandhill Cranes Vaccinated with FD Vaccine Showed Similar Protection, Responses to Challenge, and Results as Did Adult Sandhill Cranes (See Table 19.1) When Administered the Same FD Vaccine. Sandhill Crane Chicks Administered the CDC Vaccine Showed Higher Titers from Day 0 of Challenge to Day 41 and Had Much Lower WBC Counts Post Challenge Than Did the FD Vaccinated and Nonvaccinated Sandhill Crane Chicks.

Sandhill Crane number	Vaccinated	Challenged	Titer day 0 post challenge	Titer day 14 post challenge	Titer day 41 post challenge	WBC day 0 post challenge	WBC day 10 post challenge	WBC maximum	Day of maximum WBC
45	No	No	<1:40	<1:20	<1:20	9,348	11,236	11,236	10
49	No	Yes	<1:20	>1:1,280	>1,280	11,494	27,301	36,457	24
52	No	Yes	<1:20	1:640	>1:640	13,511	21,079	29,932	1
46	FD	No	<1:20	<1:20	<1:20	11,323	20,674	20,674	10
51	FD	Yes	<1:20	>1:1,280	>1:2,560	5,141	20,963	26,386	7
53	FD	Yes	<1:20	<1:640	<1:80	6,594	7,213	11,927	3
48	CDC	Yes	1:2,560	1:1,280	>1:640	18,526	16,271	18,717	3
50	CDC	Yes	>1:2,560	>1:640	1:1,280	6,154	7,379	14,764	5
54	CDC	Yes	1:1,280	1:2,560	>1:2,560	13,924	13,735	14,269	7
55	CDC	Yes	1:2,560	1:2,560	>1:1,280	21,455	32,585	33,135	3

CDC, Centers for Disease Control recombinant DNA vaccine; FD, Fort Dodge equine WNV vaccine; WBC, White blood cell count.

before or after vaccination programs were initiated (Olsen, unpublished data). This may indicate that Sandhill Cranes are more susceptible to illness caused by WNV than are Whooping Cranes.

Infectious Bursal Disease

Captive-reared Whooping Cranes reintroduced into the FNMP in 1997–98 and again in 2001–02 had higher than expected morbidity and mortality. Seroconversion to infectious bursal disease virus (IBD) serotype 2 was considered a probable cause of the mortality (Candelora et al., 2010). The disease was described as a wasting syndrome characterized by diarrhea, lethargy, subcutaneous hemorrhage, emaciation, dehydration, immunosuppression associated with small bursas, and death (Candelora et al., 2010). The disease affected primarily younger Whooping Cranes, with 8 of 10 released cranes having titers >1:128 also having clinical signs. Older Whooping Cranes in the same habitat were unaffected.

In a retrospective study of Whooping Cranes at ICF and Patuxent, a small number of samples collected between 1995 and 2003 were positive for IBD (4 of 32 adults, 4 of 54 juveniles; Hartup et al., 2004b, 2005; Hartup and Sellers, 2008). None of the captive cranes with titers had clinical signs seen in reintroduced Whooping Cranes. In another study of wild Sandhill Cranes and Turkeys (*Meleagris gallopavo*) in central Florida and southern Georgia, 46% (50/108) of the Sandhill Cranes and 6% (33/596) of the Turkeys were serologically positive (Candelora et al., 2010). No incidents of similar disease have been detected in any reintroduced Whooping Crane populations aside from the FNMP, or in the AWBP. The source of the infection in FNMP Whooping Cranes, whether from captivity, or from Turkeys or Sandhill Cranes, remains unknown (Spalding et al., 2008). Reintroduced Whooping Cranes were serologically tested by virus neutralization assay (Candelora et al., 2010) for IBD approximately 2 weeks prior to shipping to a release location (see Fig. 19.4; Keller and Hartup, 2013).

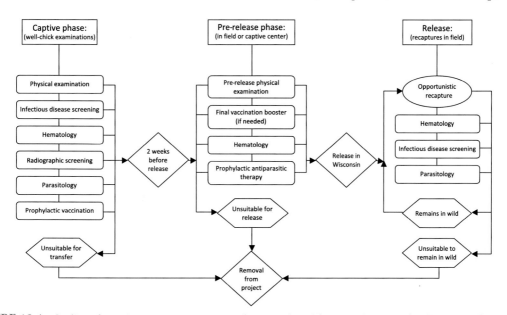

FIGURE 19.4 Outline of veterinary management procedures used on Whooping Cranes in the Wisconsin release project (Eastern Migratory Population, EMP), 2001–17. *Source: Chart modified from Keller, D.L., Hartup, B.K., 2013. Reintroduction medicine: Whooping Cranes in Wisconsin. Zoo Biol., 32 (6), 600–607.*

CHALLENGES PRESENTED BY CRANE ANATOMY AND PHYSIOLOGY

Crane anatomy is uniquely characterized by the convoluted trachea that is enclosed in the bones of the sternum in the *Grus* genus. The long-looping trachea of cranes can result in several medical issues that affect the suitability of young captive cranes for release in reintroduction programs (Schwarz et al., 2016). At times mucus or fluids build up in the trachea leading to severe dyspnea (Schwarz et al., 2016). Cranes, especially young captive cranes, can develop *Aspergillus* granulomas in the trachea, often in the loop or at the tracheal bifurcation. Either of these conditions can render a captive crane unfit for reintroduction programs.

Cranes also are characterized by long legs and long wings coupled with a relatively heavy body (4–7 kg depending on species), which makes them slow fliers, slow at maneuvering, and subject to collisions with man-made objects such as power lines. Cranes rely on thermals for lift and gliding flight as much as possible, especially during migration, and migratory behavior is closely tied to wind direction and strength. Even when reintroducing Whooping Cranes that flap-fly behind ultralight aircraft (as opposed to the natural soaring and gliding migration flight), wind direction and speed become important factors in determining which days are suitable for leading the cranes on migratory flights.

Cranes undergo complete primary and secondary feather molts in the summer every 2–4 years. During this period of time, which can last up to 6 weeks, they are flightless and at much greater risk of predation than when capable of flight.

Stress levels and associated fitness may vary throughout the year. In a study of wild nonbreeding Sandhill Cranes near Briggsville, Wisconsin, mean fecal corticosterone levels were found to be highest after spring migration compared to levels found during summer and early fall (Hartup and Gutwillig, 2014). In other studies examining reintroduction programs with captive Sandhill Cranes and Whooping Cranes following ultralight aircraft, researchers found the highest fecal corticosterone levels when cranes were shipped from rearing facilities to training areas, and after handling for banding and health examinations, all considered acute stress responses (Hartup et al., 2004a, 2005). Hartup et al. (2004a) found slight increases in fecal corticosterone during the first few days of migration due to increased physical demands from longer flights and the natural annual rhythm of migration. In Sandhill Cranes trained to follow ultralights, mean fecal corticosterone levels during training were 109.5 ± 7.5 ng/g, while during the first 3 weeks of ultralight-led migration, mean fecal corticosterone levels increased to a high of 210.5 ± 21.6 ng/g. By the fourth week of migration the fecal corticosterone levels had fallen to 118.0 ± 14.5 ng/g, similar to levels found during training (Hartup et al., 2004a). For Whooping Cranes, average fecal corticosterone levels during training were 85.7 ± 3.6 ng/g. This did not change significantly during ultralight-led migration (88.0 ± 4.1), except for an increase following a severe storm and escape of the Whooping Cranes following the storm (251.9 ± 21.9 ng/g; Hartup et al., 2005).

Cranes are long-lived species, occasionally living over 40 years in captivity. Older female cranes may still defend territory and have a mate, but fail to produce fertile eggs or to produce any eggs at all. This may result in unproductive pairs retaining prime breeding habitat. For captive Whooping Cranes, this can result in scarce pen space being occupied by nonreproductive Whooping Cranes. This poses a management issue to be addressed in the future, possibly placing older, nonreproductive Whooping Cranes with cooperating zoos for display.

SUMMARY AND OUTLOOK

There are several topics of veterinary medical concern for the rearing and release of Whooping Cranes. What is known has been summarized in this chapter, but some areas have not been fully researched, including IBDC, EEE, WNV, infectious bursal disease, and avian influenza.

Though not covered in this chapter, among noninfectious disease issues, nutrition is often discussed. There were two reports in the literature on captive crane diets published 35 years ago (Serafin, 1980, 1982) but little information since on Whooping Crane diets in captivity. When Whooping Cranes began abandoning nests at Necedah National Wildlife Refuge, one hypothesis was that the returning adults did not have enough energy to sustain them through egg laying and incubation (Runge et al., 2011). Nutrition of released Whooping Cranes has not been adequately explored. There is one study in progress examining the upper gastrointestinal tract contents of Whooping Cranes from the AWBP and the EMP. Preliminary results indicate that Whooping Cranes in the two populations are consuming similar foodstuffs, though not the same exact species (Chavez-Ramirez, 1996; Neri, MS thesis in preparation). Further work in the area of nutrition of both captive and released Whooping Cranes should be undertaken.

Captive centers continue to vaccinate captive Whooping Cranes in eastern North America on an annual basis for EEE, with Whooping Crane chicks treated twice prior to release. After release, they are unlikely to be vaccinated again, although we do not know how long immunity from vaccination for this disease lasts. Revaccinating released Whooping Cranes when they are captured for radio transmitter changes could be considered. A serosurvey of captured Whooping Cranes could help to determine the extent of postrelease exposure and immunity, if any. There has been no evidence of reintroduced Whooping Cranes dying of EEE in the necropsy reports available on EMP birds (Hartup and Yaw, unpublished data).

The effects of WNV on Whooping Crane populations are not well understood. We know that WNV can kill up to a third of unvaccinated Sandhill Crane chicks between the ages of 3 and 4 months (Olsen et al., 2008). In Whooping Cranes, this may be an unsustainable loss.

In reintroduced populations, notably the EMP, most chicks do not survive the first few months (Converse et al., Chapter 8, this volume) but the causes of mortality in chicks are generally not known. Chicks predominantly die early in the summer while WNV tends to peak later in summer in Wisconsin. Similar to EEE, a serosurvey would be informative in determining the degree of exposure of various populations. WNV-related morbidity may play a role in increasing mortality or predation of older chicks, with loss of the carcass through scavenging. Further study of the potential effects of WNV on Whooping Crane reintroductions is needed.

Infectious bursal disease was implicated in the deaths of Whooping Cranes in the FNMP, with considerable research completed on potential sources of infection and routes of transmission (Candelora et al., 2010; Spalding et al., 2004, 2008). To date, however, the causative agent has not been confirmed. Preventive management will be incomplete until the virus is fully characterized. In addition, the specificity of the serological test is unknown. Further research into this disease and the value of continued testing in the captive and reintroduced populations should be a high priority.

References

Abou-Madi, N., 2001. Avian anesthesia. Vet. Clin. North Am. Exot. Anim. Pract. 4 (1), 147–167.

Augustine, P.C., Klein, P.N., Danforth, H.D., 1998. Use of monoclonal antibodies against chicken coccidia to study invasion and early development of *Eimeria gruis* in Florida sandhill crane (*Grus canadensis*). J. Zoo Wildl. Med. 29 (1), 21–24.

Augustine, P., Olsen, G., Danforth, H., Gee, G., Novilla, M., 2001. Use of monoclonal antibodies development against chicken coccidia (*Eimeria*) to study invasion and development of *Eimeria riechenowi* in Florida sandhill cranes (*Grus canadensis*). J. Zoo Wildl. Med. 32 (1), 65–70.

Bailey, T.A., Toosi, A., Samour, J.H., 1999. Anesthesia of cranes with alphaxolone-alphadolone. Vet. Rec. 145 (3), 84–85.

Balis, G., Munroe, R., 1964. The pharmacology of chloralose: a review. Psychopharmacologia 6 (1), 1–30.

Barrett, C., Stehn, T.V., 2010. A retrospective of whooping cranes in captivity. Proceedings of the North American Crane Workshop 11, 166–179.

Bertram, M.R., Hamer, G.L., Snowden, K.F., Hartup, B.K., Hamer, S.A., 2015. Coccidian parasites and conservation implications for the endangered whooping cranes (*Grus americana*). PLoS ONE 10 (6), e0127679.

Black, S.R., Swan, K.D., 2018. Advances in conservation breeding and management of whooping cranes (*Grus americana*) (Chapter 16). In: French, Jr., J.B., Converse, S.J., Austin, J.E. (Eds.), Whooping Cranes: Biology and Conservation. Biodiversity of the World: Conservation from Genes to Landcapes. Academic Press, San Diego, CA.

Black, S.R., Whiteside, D.W., 2003. Short term tranquilization of sandhill cranes (*Grus canadensis*) using triazolam. 52nd Annual Wildlife Disease Association Conference, 82–83.

Businga, N.K., Langenberg, J., Calson, L., 2007. Successful treatment of capture myopathy in three wild greater sandhill cranes (*Grus canadensis tabida*). J. Avian Med. Surg. 21 (4), 294–298.

Bysykatova, I., Sleptsov, S., Vasiliev, N., 2010. Current status of lesser sandhill cranes in Yakutia. Proceedings of the North American Crane Workshop 11, 198.

Calle, P.P., Ludwig, G.V., Smith, J.F., Raphael, B.L., Clippinger, T.L., Rush, E.M., McNamara, T., Manduca, R., Linn, M., Turrell, M.J., Schoepp, R.J., Larsen, T., Mangiafico, J., Steele, K.E., Cook, R.A., 2000. Clinical aspects of West Nile virus infection in a zoological collection. Proceedings of the AAZV and IAAAM Joint Conference, 92–96.

Candelora, K.L., Spalding, M.G., Sellers, H.S., 2010. Survey for antibodies to infectious bursal disease virus serotype 2 in wild turkeys and Sandhill cranes of Florida, USA. J. Wildl. Dis. 46 (3), 742–752.

Carpenter, J.W., 1986. Cranes. In: Fowler, M.E. (Ed.), Zoo, Wild Animal Medicine, second ed. W.B. Saunders, Philadelphia, PA, pp. 315–326.

Carpenter, J.W., 1993. Infectious and parasitic diseases of cranes. In: Fowler, M.E. (Ed.), Zoo, Wild Animal Medicine: Current Therapy 3. W.B. Saunders, Philadelphia, PA, pp. 229–237.

Carpenter, J.W., Novilla, M.N., Hatfield, J.S., 1992. The safety and physiological effects of the anticoccidial drugs Monensin and Clazuril in Sandhill cranes (*Grus canadensis*). J. Zoo Wildl. Med. 23 (2), 214–221.

Chavez-Ramirez, F., 1996. Food Availability, Foraging Ecology, and Energetics of Whooping Cranes Wintering in Texas. PhD dissertation, Texas A&M University, College Station, p. 104.

Clark, G.C., Dein, F.J., Crabbs, C.L., Carpenter, J.W., Watts, D.M., 1987. Antibody response of sandhill and whooping cranes to an eastern equine encephalitis virus vaccine. J. Wildl. Dis. 23, 539–544.

Cole, G.A., Thomas, N.J., Spalding, M., Stroud, R., Urbanek, R.P., Hartup, B.K., 2009. Postmortem evaluation of reintroduced migratory whooping cranes in eastern North America. J. Wildl. Dis. 45 (1), 29–40.

Converse, S.J., Strobel, B.N., Barzen, J.A., 2018. Reproductive failure in the Eastern Migratory Population: interaction of research and management (Chapter 8). In: French, Jr., J.B., Converse, S.J., Austin, J.E. (Eds.), Whooping Cranes: Biology and Conservation. Biodiversity of the World: Conservation from Genes to Landcapes. Academic Press, San Diego, CA.

Cristofaro, R., Feijo, L.M.F., 1976. Contribucao ao estudo da fauna helmintologica do Estado Mato Grosso. Atas Soc. Biol. Rio de Janeiro 18, 53–57.

Dein, F.J., Carpenter, J.W., Clark, G.C., Montali, R.J., Crabbs, C.L., Tsai, T.F., Docherty, D.E., 1986. Mortality of captive whooping cranes caused by eastern equine encephalitis virus. J. Am. Vet. Med. Assoc. 189 (9), 1006–1010.

Docherty, D.E., 1987. Inclusion body disease of cranes. In: Friend, M. (Ed.), Field Guide to Wildlife Diseases, vol. 1, General Field Procedures and Diseases of Migratory Birds. Resource Publication 167, U.S. Department of the Interior, Fish and Wildlife Service, Washington, DC, pp. 128–134.

Docherty, D.E., 1999. Inclusion body disease of cranes. In: Friend, M., Franson, J.C. (Eds.), Field Manual of Wildlife Diseases: General Field Procedures and Diseases of Birds. Information and Technology Report 1999-001, U.S. Department of the Interior, U.S. Geological Survey, Biological Resources Division, Washington, DC, pp. 153–156.

Docherty, D.E., Henning, D.J., 1980. The isolation of a herpes virus from captive cranes with an inclusion body disease. Avian Dis. 24 (1), 278–283.

Docherty, D.E., Romaine, R.I., 1983. Inclusion body disease of cranes: a serological follow-up to the 1978 die-off. Avian Dis. 27 (3), 830–835.

Dusek, R.J., Spalding, M.G., Forrester, D.J., Greiner, E.C., 2004. *Haemoproteus balearicae* and other blood parasites of free-ranging Florida sandhill crane chicks. J. Wildl. Dis. 40 (4), 682–687.

Forrester, D.J., Bush, A.O., Williams, Jr., L.E., 1975. Parasites of Florida Sandhill Cranes, *Grus canadensis pratensis*. J. Parasitol. 61 (3), 547–548.

Forrester, D.J., Bush, A.O., Williams, Jr., L.E., Weiner, D.J., 1974. Parasites of greater sandhill cranes (*Grus canadensis tabida*) on their wintering grounds in Florida. Proc. Helminthol. Soc. Washington 41 (1), 55–59.

Friend, M., Franson, J.C., 1999. Aspergillosis. Field Manual of Wildlife Diseases: Birds (Chapter 13). U.S. Geological Survey, Biological Resources Division, II Series, Washington DC, pp. 129–133.

Hansen, C.H., Hartup, B.K., Gonzalea, O.D., Lyman, D.E., Steinberg, H., 2008. West Nile encephalitis in a captive Florida sandhill crane. Proceedings of the North American Crane Workshop 10, 115–118.

Hartman, S., Reichenberg, B., Fanke, J., Lacy, A.E., Hartup, B.K., 2010. Endoparasites of greater sandhill cranes in south-central Wisconsin. Proceedings of the North American Crane Workshop 11, 186–188.

Hartup, B.K., 2008. Surveillance for West Nile virus at the International Crane Foundation 2000–2004. Proceedings of the North American Crane Workshop 10, 111–114.

Hartup, B.K., 2018. Health of whooping cranes in the Central Flyway (Chapter 18). In: French, Jr., J.B.,

Converse, S.J., Austin, J.E. (Eds.), Whooping Cranes: Biology and Conservation. Biodiversity of the World: Conservation from Genes to Landcapes. Academic Press, San Diego, CA.

Hartup, B.K., Gutwillig, A., 2014. Seasonal fecal corticosterone measurements in Wisconsin sandhill cranes. Proceedings of the North American Crane Workshop 12, 90.

Hartup, B.K., Niemuth, J.N., Fitzpatrick, B., Fox, M., Kelley, C., 2010a. Morbidity and mortality of captive whooping cranes at the International Crane Foundation 1976–2008. Proceedings of the North American Crane Workshop 11, 183–185.

Hartup, B.K., Olsen, G.H., Czekala, N.M., 2005. Fecal corticoid monitoring in whooping cranes (*Grus americana*) undergoing reintroduction. Zoo Biol. 24 (1), 15–28.

Hartup, B.K., Olsen, G.H., Czekala, N.M., Paul-Murphy, J., Langenberg, J.A., 2004a. Levels of fecal corticosterone in sandhill cranes during a human-led migration. J. Wildl. Dis. 40 (2), 267–272.

Hartup, B.K., Olsen, G.H., Sellers, H.S., Smith, B., Spalding, M. 2004b. Serological evidence of infectious bursal disease virus exposure in captive whooping cranes. Proceedings of the Joint Conference of the American Association of Zoo Veterinarians, Wildlife Disease Association, and American Association of Wildlife Veterinarians, p. 71.

Hartup, B.K., Schneider, L., Engels, J.M., Hayes, M.A., Brazen, J.A., 2014. Capture of sandhill cranes using alpha-chloralose: a 10-year follow-up. J. Wildl. Dis. 50 (1), 143–145.

Hartup, B.K., Schroeder, C.A., 2006. Protein electrophoresis in cranes with presumed insect bite. Vet. Clin. Pathol. 35 (2), 226–230.

Hartup, B.K., Sellers, H.S., 2008. Serological survey for infectious bursal disease virus exposure in captive cranes. Proceedings of the North American Crane Workshop 10, 173–174.

Hartup, B.K., Spalding, M.G., Thomas, N.J., Cole, G.A., Kim, Y.J., 2010b. Thirty years of mortality assessment in whooping crane reintroductions: patterns and implications. Proceedings of the North American Crane Workshop 11, 204.

Hayes, M.A., Hartup, B.K., Pittman, J.M., Brazen, J.A., 2003. Capture of sandhill cranes using alpha-chloralose. J. Wildl. Dis. 39 (4), 859–868.

Hoar, B.M., Whitesie, D.P., Ward, L., Inglis, G.D., Morck, D.W., 2007. Evaluation of the enteric microflora of captive whooping cranes (*Grus americana*) and sandhill cranes (*Grus canadensis*). Zoo Biol. 26 (2), 141–153.

Inglis, G.D., Hoar, B.M., Whiteside, D.P., Morck, D.W., 2007. *Campylobacter canadensis* sp. nov., from captive whooping cranes in Canada. Int. J. Syst. Evol. Microbiol. 57 (11), 2636–2644.

Ippen, V.R., Wisser, J., Zwart, P., 1980. Ein weiterer Beitrag zur Hexamitiasis der Jungfernkraniche (*Anthropides virgo*). Milu 5 (1/2), 293–297.

Jones, A.K., 2009. The physical examination. In: Tully, Jr., T.N., Dorrestein, G.M., Jones, A.K. (Eds.), Handbook of Avian Medicine, second ed. Saunders Elsevier, Edinburgh, UK, pp. 56–76.

Keller, D.L., Hartup, B.K., 2013. Reintroduction medicine: whooping cranes in Wisconsin. Zoo Biol. 32 (6), 600–607.

Kelley, C., Hartup, B.K., 2008. Risk factors associated with developmental limb abnormalities in captive whooping cranes. Proceedings of the North American Crane Workshop 10, 119–124.

Kelman, A., Hartup, B.K., 2014. Tylosin tartrate promotes resolution of insect bite hypersensitivity reactions in captive cranes. Proceedings of the North American Crane Workshop 12, 73–74.

Kwon, Y.K., Jeon, W.J., Kang, M.I., Kim, J.H., Olsen, G.H., 2006. Disseminated visceral coccidiosis in a wild white-naped crane (*Grus vipio*). J. Wildl. Dis. 42 (3), 712–714.

Langenberg, J., Spalding, M., Ramer, J. 1996. *Campylobacter jejuni* in captive and free-living cranes. Proceedings of the American Association of Zoo Veterinarians Annual Conference, pp. 487-489.

Lu, J., Ryu, H., Vogel, J., Domingo, J.S., Ashbolt, N.J., 2013. Molecular detection of *Campylobacter* spp. and fecal indicator bacteria during the northern migration of sandhill cranes (*Grus canadensis*) at the central Platte River. Appl. Environ. Microbiol. 79 (12), 3762–3769.

Machin, K.L., Caulkett, N.A., 1998. Investigation of injectable anesthetic agents in mallard ducks (*Anas platyrhynchos*): a descriptive study. J. Avian Med. Surg. 12 (4), 255–262.

MacLean, R.A., Beaufrere, H., 2015. Gruiformes (cranes, limpkins, rails, gallinules, coots, bustards). In: Miller, R.E., Fowler, M.E. (Eds.), Fowler's Zoo and Wild Animal Medicine, vol. 8, Elsevier, St. Louis, MO, pp. 155–164.

Maeda, Y., Tohya, Y., Nakagami, Y., Yamashita, M., Sugimura, T., 2001. An occurrence of *Salmonella* infection in cranes at the Izumi Plains, Japan. J. Vet. Med. Sci. 63 (8), 943–944.

McLean, R.G., Ubico, S.R., 2007. Arboviruses in birds. In: Thomas, N.J., Hunter, D.B., Atkinson, C.T. (Eds.), Infectious Diseases of Wild Birds. Blackwell Publishing, Oxford, UK, pp. 17–62.

Miller, P.E., Langenberg, J.A., Hartmann, F.A., 1995. The normal conjunctival aerobic bacterial flora of three species of captive cranes. J. Zoo Wildl. Med. 26 (4), 545–549.

Morris, C.D., 1989. Eastern equine encephalomyelitis. In: Monath, T.P. (Ed.), The Arboviruses: Epidemiology, Ecology, 3, CRC, Boca Raton, FL, pp. 1–20.

Neri, H., in preparation. Characterization of the diets of wild whooping cranes (*Grus americana*) and reintroduced whooping cranes, MS thesis, Hood College, Frederick, MD.

Olsen, G.H., 1994. Orthopedics in cranes: pediatrics and adults. Semin. Avian Exot. Pet Med. 3 (2), 73–80.

Olsen, G.H., 2000. Cranes (Chapter 9). In: Tully, T.N., Lawton, M.P.C., Dorrestein, G.M. (Eds.), Avian Medicine. Butterworth Heinemann, Oxford, UK, pp. 215–227.

Olsen, G.H., 2004. Mortality of Mississippi sandhill crane chicks. J. Avian Med. Surg. 18 (4), 269–272.

Olsen, G.H., 2009. Cranes. In: Tully, Jr., T.N., Dorrestein, G.M., Jones, A.K. (Eds.), Handbook of Avian Medicine, second ed. Saunders Elsevier, Edinburgh, UK, pp. 243–257.

Olsen, G.H., 2010. Whooping crane titers in response to eastern equine encephalitis immunization. Proceedings of the North American Crane Workshop 11, 180–183.

Olsen, G.H., 2011. Clinical pathology results from cranes with experimental West Nile virus infection. Proc. Assoc. Avian Vet. 32, 321–324.

Olsen, G.H., 2016. Wildlife disease studies at Patuxent. In: Perry, M.C. (Ed.), The History of Patuxent – America's Wildlife Research Story. U.S. Geological Survey Circular 1422, pp. 213–219.

Olsen, G.H., Carpenter, J.W., 1997. Cranes. In: Altman, R.B., Clubb, S.L., Dorrestein, G.M., Quesenberry, K.E. (Eds.), Avian Medicine and Surgery. W.B. Saunders, Philadelphia, PA, pp. 973–991.

Olsen, G.H., Carpenter, J.W., Gee, G.F., Thomas, N.J., Dein, F.J., 1995. Mycotoxin-induced disease in captive whooping cranes (Grus americana) and sandhill cranes (Grus canadensis). J. Zoo Wildl. Med. 26 (4), 569–576.

Olsen, G.H., Carpenter, J.W., Langenberg, J.A., 1996. Medicine and surgery. In: Ellis, D.H., Gee, G.F., Mirande, C.M. (Eds.), Cranes, Their Biology, Husbandry and Conservation. U.S. Department of the Interior, National Biological Service, Washington, DC, and International Crane Foundation, Baraboo, WI, pp. 137–174.

Olsen, G.H., Clubb, S.L., 1997. Embryology, incubation, and hatching. In: Altman, R.B., Clubb, S.L., Dorrestein, G.M., Quesenberry, K.E. (Eds.), Avian Medicine, Surgery. W.B. Saunders, Philadelphia, PA, pp. 54–71.

Olsen, G.H., Crosta, L., Gartrell, B.D., Marsh, P.M., Stringfield, C.E., 2016. Conservation of avian species (Chapter 23). In: Speer, B.L. (Ed.), Current Therapy in Avian Medicine and Surgery. Elsevier, St. Louis, MO, pp. 719–748.

Olsen, G.H., Gaunt, S.D., 1985. Effect of hemoprotozoal infections on rehabilitation of wild raptors. J. Am. Vet. Med. Assoc. 187 (11), 1204–1205.

Olsen, G.H., Kolski, E., Hatfield, J.S., Docherty, D.E., 2005. Whooping crane titers to eastern equine encephalitis vaccinations. Proceedings of the North American Crane Workshop 9, 21–23.

Olsen, G.H., Langenberg, J.A., 1996. Veterinary techniques for rearing crane chicks. In: Ellis, D.H., Gee, G.F., Mirande, C.M. (Eds.), Cranes: Their Biology, Husbandry, Conservation. U.S. Department of the Interior, National Biological Service, Washington, DC, and International Crane Foundation, Baraboo, WI, pp. 95–104.

Olsen, G.H., Miller, K.J., Docherty, D.E., Bochsler, V., 2008. Safety of West Nile virus vaccines in crane chicks. Proceedings of the North American Crane Workshop 10, 175.

Olsen, G.H., Miller, K.J., Docherty, D.E., Bochsler, V.S., Sileo, L., 2009. Pathogenicity of West Nile virus and response to vaccination in sandhill cranes (Grus canadensis) using a killed vaccine. J. Zoo Wildl. Med. 40 (2), 263–271.

Olsen, G.H., Miller, K.J., Docherty, D., Sileo, L., 2003. West Nile virus vaccination and challenge in sandhill cranes (Grus canadensis). Proceedings Association of Avian Veterinarians, 123–124.

Olsen, G.H., Turell, M.J., Pagac, B.B., 1997. Efficacy of eastern equine encephalitis immunization in whooping cranes. J. Wildl. Dis. 33, 312–315.

Pagac, B.B., Turell, M.J., Olsen, G.H., 1992. Eastern equine encephalomyelitis virus and Culiseta melanura activity at the Patuxent Wildlife Research Center, 1985–1990. J. Am. Mosq. Control Assoc. 8 (3), 328–330.

Parker, B.B., Duszynski, D.D., 1986. Coccidiosis of sandhill cranes (Grus canadensis) wintering in Mexico. J. Wildl. Dis. 22 (1), 25–35.

Ritchie, B.W., 1995. Avian Viruses: Function and Control. Wingers Publishing, Lake Worth, FL, p. 525.

Roffe, T.J., Stroud, R.K., Windingstad, R.M., 1989. Suspected fusariomycotoxicosis in sandhill cranes (Grus canadensis): clinical and pathological findings. Avian Dis. 33 (3), 451–457.

Runge, M.C., Converse, S.J., Lyons, J.E., 2011. Which uncertainty? Using expert elicitation and expected value of information to design an adaptive program. Biol. Conserv. 144 (4), 1214–1223.

Rupiper, R.J., Carpenter, J.W., Mahima, T.Y., 2000. Formulary. In: Olsen, G.H., Orosz, S.E. (Eds.), Manual of Avian Medicine. Mosby, St. Louis, MO, pp. 553–589.

Schwarz, T., Kelley, C., Pinkerton, M.E., Hartup, B.K., 2016. Computed tomographic anatomy and characteristics of respiratory aspergillosis in juvenile whooping cranes. Vet. Radiol. Ultrasound 57 (1), 16–23.

Serafin, J.A. 1980. Influence of dietary energy and sulfur amino acid levels upon growth and development of young sandhill cranes. Proceedings of the American Association of Zoo Veterinarians Annual Conference, p. 30.

Serafin, J.A., 1982. The influence of diet composition upon growth and development of sandhill cranes. Condor 84 (4), 427–434.

Snyder, S.B., Richard, M.J., Meteyer, C.U., 1997. Avian tuberculosis in a whooping crane: treatment and outcome. Proceedings of the North American Crane Workshop 7, 253–255.

Spalding, M.G., Erlandsen, S.L., Nesbit, S.A., 1994. Hexamita-like sp. associated with enteritis and death in captive Florida sandhill cranes (Grus canadensis pratensis). J. Zoo Wildl. Med. 25 (2), 281–285.

Spalding, M.G., Kinsella, J.M., Nesbitt, S.A., Folk, M.J., Foster, G.W., 1996. Helminth and arthropod parasites of experimentally introduced whooping cranes in Florida. J. Wildl. Dis. 32 (1), 44–50.

Spalding, M.G., Nesbitt, S.A., Schwikert, S.T., Dusek, R.J., 2001. The use of radio transmitters to monitor survival of sandhill crane chicks. Proceedings of the North American Crane Workshop 8, 213–215.

Spalding, M.G., Sellers, H.S., Hartup, B.K., Olsen, G.H., 2004. Infectious bursal disease virus associated with a wasting syndrome in released whooping cranes in Florida. In: Baer, C.K. (Ed.), Proceedings of the American Association of Zoo Veterinarians. American Association of Wildlife Veterinarians, Wildlife Disease Association Joint Conference, San Diego, CA, pp. 74–75.

Spalding, M.G., Sellers, H.S., Hartup, B.K., Olsen, G.H., 2008. A wasting syndrome in released whooping cranes in Florida associated with infectious bursal disease titers. Proceedings of the North American Crane Workshop 10, 176.

Steele, K.E., Linn, M.J., Schoepp, R.J., Komar, N., Geisbert, T.W., Manduca, R.M., Calle, P.P., Rapheal, B.L., Clippinger, T.L., Larsen, T., Smith, J., Lanciotti, R.S., Panella, N.A., McNamara, T.S., 2000. Pathology of fatal West Nile virus infections in native and exotic birds during the 1999 outbreak in New York City, New York. Vet. Pathol. 37 (3), 208–224.

Windingstad, R.M., Cole, R.J., Nelson, P.E., Roffe, T.J., George, R.R., Dorner, J.W., 1989. *Fusarium* mycotoxins from peanuts suspected as a cause of sandhill crane mortality. J. Wildl. Dis. 25 (1), 38–46.

Young, L.A., Citino, S.B., Seccareccia, V., Munson, L., Nichols, D. 1996. Eastern equine encephalomyelitis in an exotic avian collection. Proceedings of the Annual Conference of the Association of Avian Veterinarians, pp. 163–165.

Zwart, P., Vroege, C., Borst, G.H.A., Poelma, F.G., Truijens, E.H.A., 1985. Hexamitiasis bei frisch importierten Jungfernkranichen (*Anthropoides virgo*). Erkrankungen der Zootiere: Verhandlungsbericht des 27, Internationalen Symposiums uber die Erkrankungen der Zootiere 27, 241–244.

REINTRODUCTION AND CONSERVATION

Migrating Whooping Cranes following ultralight aircraft

Rearing and Release Methods for Reintroduction of Captive-Reared Whooping Cranes

Barry K. Hartup

International Crane Foundation, Baraboo, WI, United States

INTRODUCTION

Recovery planning for the Whooping Crane (*Grus americana*) has emphasized three primary strategies: (1) habitat protection to foster growth and long-term viability of the Aransas-Wood Buffalo Population (AWBP), (2) maintenance of a demographically and genetically viable ex situ population as a hedge against extinction of the species in the wild, and (3) reintroduction of geographically distinct, self-sustaining wild populations to reduce the probability of catastrophic loss of the species and facilitate down-listing from endangered to threatened (Canadian Wildlife Service and U.S. Fish and Wildlife Service, 2005).

The translocation of wild breeding adults or juveniles from the AWBP has not been used as a reintroduction strategy with Whooping Cranes, despite the applicability of this approach for reintroduction of other taxa (Jones and Merton, 2012). Capture and transfer of live wild cranes is extremely costly and challenging, and

the physical and psychological stressors of this operation would likely be detrimental to translocated individuals (Parker et al., 2012). Instead, the translocation of 216 AWBP eggs, along with 73 eggs from the Patuxent Wildlife Research Center (PWRC) captive population, was used to create a population in the Rocky Mountain flyway between 1975 and 1989 (Grays Lake Population, GLP). The attempt, which included cross-fostering by Sandhill Cranes, resulted in few breeding attempts and high mortality; the GLP peaked at only 31 individuals, and went extinct after 27 years (Canadian Wildlife Service and U.S. Fish and Wildlife Service, 2005; Ellis et al., 1992a). The current recovery plan assumes the use of specially managed offspring from the captive breeding population for reintroduction efforts.

This chapter reviews the development of techniques for crane rearing and reintroduction in North America, and how they have been adapted for use in three attempts to create populations of Whooping Cranes distinct from the

Whooping Cranes: Biology and Conservation
http://dx.doi.org/10.1016/B978-0-12-803555-9.00020-7

AWBP. Initial rates of assimilation of the juvenile cranes to the release area – the probability of survival 1 year following release or the start of fall migration (i.e., for the period approximately 6–18 months of age) – are the most commonly available measures in the relevant literature, and are used here to provide a metric for comparison among reintroduction methods. Long-term success of reintroduction is determined by population viability (Converse et al., Chapter 8, this volume), which is a function of first-year survival as well as longer-term survival and reproduction, for example, as measured by lifetime reproductive contribution (Canessa et al., 2016). Combining predicted viability and economic assessment will allow for thorough evaluation of reintroduction methods. Initial survival estimates, however, are the most thoroughly studied and finest-scale information available to reflect how different release methods may function and impact prospects for reintroduction success, and thus help reflect the potential efficacy of

reintroduction from ex situ sources as a conservation tool for Whooping Crane recovery.

PARENT REARING AND RELEASES

Both parent-reared (by true or conspecific foster parents) and hand-reared Whooping Cranes have been produced for use in reintroduction projects based on work with Sandhill Cranes and other crane species (Nagendran et al., 1996). The perceived need to release large numbers of Whooping Cranes while growing the captive breeding population, however, has often limited the use of parent-reared chicks in reintroduction projects. Relatively few pairs of captive Whooping Cranes with chick rearing experience are available in any given year (a trait thought necessary for high chick survival) and they are generally managed to rear one chick (Fig. 20.1), though two are possible (Wellington et al., 1996). Dedicating valuable breeding adults to chick

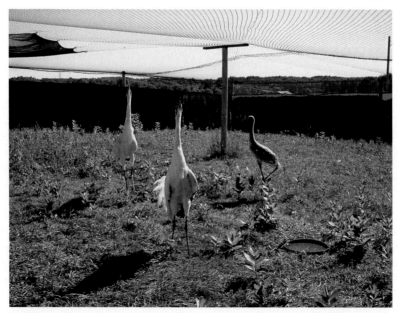

FIGURE 20.1   Typical captive environment of parent-reared Whooping Cranes at the International Crane Foundation. *Source: Photo credit: Barry Hartup.*

rearing might also preclude their production of additional clutches of eggs, thus lowering annual production.

Cross fostering with adults of common crane species (e.g., Sandhill Cranes) results in inappropriate imprinting and socialization, most notably confirmed by GLP, and as observed in other species in captivity (Mahan and Simmers, 1992). In addition, concerns have been expressed regarding the overall large monetary investment of maintaining a breeding population for parent rearing (Urbanek and Bookhout, 1992). Captive propagation costs for Whooping Cranes were estimated at >$1.7 million in 2006 (Canadian Wildlife Service and U.S. Fish and Wildlife Service, 2005). The commitment in the International Recovery Plan and by various institutions to grow the captive population for other purposes, however, has ultimately provided perhaps 10–15 available pairs for parent rearing, well below target allocations typically put forth by reintroduction project planners in the past decade. Tension remains whether to maximize egg production for releases using hand-rearing and make more rapid progress toward numerical population targets, or commit to a longer process using fewer parent-reared chicks, especially if they are potentially more adaptable and may exhibit better reproductive success.

Extensive work refined the basic rearing and release techniques with nonmigratory, parent-reared Mississippi Sandhill Crane chicks in the 1980s (Ellis et al., 1992b; Wellington et al., 1996). Briefly, fertile or hatching eggs were provided to captive, foster Sandhill Crane pairs, or rarely single cranes, for incubation, hatching, and rearing. Adoption methods using live chicks were used occasionally to compensate for illness or deaths of chicks, or new chick availability. The parent-reared chicks were separated from the adults in mid-October and housed in a net-covered community pen until transfer to Mississippi. So-called soft releases were conducted following a 30-day acclimatization period in a large, open pen at the release site, where birds were restrained from flight through the use of brails (Ellis and Dein, 1996) and provided pelleted food. Once debrailed, chicks could come and go at will. Assimilation to the release area was achieved by the majority of cranes: 1-year survival was 0.65 (SE = 0.04; $n = 142$) among parent-reared cranes released between 1980 and 1992 (Ellis et al., 1992b, 2001c).

Few studies of captive-bred, parent-reared cranes reintroduced in a migratory context have been conducted. An initial trial was conducted with 11 after-hatch-year (AHY, >January 1 of year following hatch; in this study birds were 1–3 years old) parent-reared Greater Sandhill Cranes released at Grays Lake NWR (Drewien et al., 1982). Birds were released directly following transfer from PWRC. Despite socializing with wild cranes, the captive cranes spent less time in vigilant behavior than wild cranes and only one formed a discernible association with a wild counterpart. One of the 11 cranes was resighted following intensive searches on wintering grounds. The authors hypothesized that "younger" cranes would be more suitable for release programs, and "that preconditioning them at the release site might increase survival" (Drewien et al., 1982). A follow-up study with 21 AHY cranes used a 4–6-day acclimation period before birds left a protective pen (Bizeau et al., 1987). Upon release, most cranes associated with their penmates and not wild Sandhill Cranes, and only four reintroduced cranes were believed to survive their first migration. Bizeau et al. (1987) recommended releasing birds singly to enhance socialization and integration with wild cranes.

HAND-REARING AND RELEASES

Hand-rearing cranes has been considered more efficient for raising large numbers of captive cranes, despite increased housing densities, additional labor costs, and close human contact that might impair appropriate imprinting or initial survival in a wild environment (Wellington et al., 1996). In 2011, hand rearing, releasing, and monitoring reintroduced Whooping Cranes was estimated to cost $113,886/crane for 32 cranes released that year (W. Brooks, U.S. Fish and Wildlife Service, personal communication). Despite these apparent high costs, the supply and use of Whooping Crane chicks of captive origin has been sustained since the early 1990s, and followed decades of technique development with Sandhill Cranes.

Hatch-year (HY; up to December 31 of a bird's HY) Sandhill Cranes hand-reared by conventional means were originally believed to be unsuitable for release (Derrickson and Carpenter, 1987; Nesbitt, 1979). Studies in the 1970s and 1980s removed or disguised human contact with chicks (through the use of puppets, taxidermy brood models, and costumed caretakers), and applied these techniques to both nonmigratory and migratory reintroductions (Archibald and Archibald, 1992; Ellis et al., 1992b; Horwich, 1989; Horwich et al., 1992; Urbanek and Bookhout, 1992). The typical results of these efforts were groups of young cranes socialized with each other, with behavioral attachment to costumes and captive conspecifics. Cranes were exposed to captive or controlled field environments, depending on release application, ranging from 2 to 6 months. The release strategies for hand-reared Sandhill Cranes centered on soft releases conducted over the course of weeks to assist with acclimation to the release area. The release pens allowed exposure to common food items and free-ranging cranes, while imposing limitations on flight. Some cranes were translocated to help facilitate migration or manage migration problems (Urbanek and Bookhout, 1992).

The 1-year post-release survival rate of costume-reared, nonmigratory Mississippi Sandhill Cranes was greater (0.77, SE = 0.06; n = 56) than that of parent-reared cranes (0.68, SE = 0.05; n = 76; a subset of the n = 142 birds reported on in "Parent Rearing and Releases") released in the same area

and over the same time period (Ellis et al., 2001c). Urbanek and Bookhout (1992) reported 1-year post-release survival of 0.84 (SE = 0.06; n = 38) from costume-reared Sandhill Cranes released into a wild migratory population in Upper Michigan. Each study, however, conducted releases of naïve HY cranes into areas with free-ranging adult conspecifics, whether derived from previous reintroductions or a wild, predator-wary native population. Additional work with other groups of hand-reared Sandhill Cranes indicated that initial survival was improved if cranes were released singly or in pairs into free-ranging populations (Ellis et al., 2001d). Urbanek et al. (2005a, 2005b) described the successful individual releases of eight HY cranes originally reared as a group onto Sandhill Crane staging areas in Central Wisconsin.

SPECIAL APPLICATION: GUIDED MIGRATION

The reintroduction of Whooping Cranes to establish a migratory population presents special challenges: the lack of an existing migratory population of conspecifics with an established migratory route, management goals to overwinter at a predetermined destination, and technical challenges to maintain appropriate imprinting and behavioral development. A partial solution to many of the outstanding concerns facing reintroduction of migratory cranes came through the application of guided migration. These techniques were developed using slow-flying light aircraft or ground vehicles with Canada Geese (*Branta canadensis*), Trumpeter Swans (*Cygnus buccinators*), and Sandhill Cranes (Duff, Chapter 21, this volume; Clegg and Lewis, 2001; Clegg et al., 1997; Duff et al., 2001; Ellis et al., 1997, 2001a; Lishman et al., 1997). Ellis et al. (2001b) gleaned several important lessons from these studies: (1) juvenile cranes can be led hundreds of kilometers south; (2) juvenile cranes will travel 75 km or more in a single flight behind vehicles; (3) most juveniles will return north the following spring; (4) cranes will return to the general area where they first flew and began their migration (akin to natal site fidelity); (5) cranes will return to the same or nearby wintering area the following autumn without assistance, and (6) the cranes' subsequent migration routes are more direct than their training route. A general model had emerged: cranes could be safely manipulated in their HY to follow a motorized vehicle to a specific southern destination and return on their own.

Horwich (2001) recommended several general procedures be followed in the reintroduction of migratory Whooping Cranes: (1) use as soft a release as possible, (2) use fledged HY cranes, (3) use costumes to control chick behavior through conditioning, (4) promote conspecific breeding, (5) capitalize on developmental periods (such as behavioral reattachment to parent models near fledging) to enhance training and releases, (6) consider habitat site imprinting (train and rear at the intended breeding site, promoting natal site fidelity), (7) avoid human contact and habituation, (8) use a parent model (costume) for migration, and (9) minimize costs. The majority of these recommendations stemmed from the challenge to balance behavioral conditioning in the migration trials with aspects that promoted wildness, to minimize the chances of mistaken geographic orientation, to maximize sexual imprinting on conspecifics, and to promote high survival post-release.

An experimental trial with Sandhill Cranes, utilizing many of these procedures, consisted of a guided migration behind ultralight aircraft between Wisconsin and Florida in 2000 (Urbanek et al., 2005a, 2005b). Eleven of 13 cranes completed the route and were managed in a soft release fashion at an upland site to help promote survival over the winter. Nine of the 11 that departed on northward migration the following spring were observed in Wisconsin the following summer, and 1-year survival of the guided migration group from the start of guided migration was 0.75 (SE = 0.13; n = 12;

1 bird that escaped early in the migration is not included). This rate is comparable to those observed in previous Sandhill Crane release projects. All birds exhibited normal foraging, roosting, social association, and human avoidance behaviors, according to the authors. The risks inherent in guided migration included: collision with aircraft; maladaptive behavior with regard to aircraft training (i.e., flies away); traumatic accidents common to confinement; and disease among concentrated, immunologically naïve juveniles held in pens. Refinements (i.e., protocols for aircraft training, husbandry, and preventive veterinary care, among others) were expected to improve the number of birds reaching the intended terminus of the guided migration, survive their first winter, and disperse to breeding grounds the following summer.

CASE STUDY – FLORIDA NONMIGRATORY RELEASES

Beginning with birds hatched in 1993, the Florida Fish and Wildlife Conservation Commission reintroduced Whooping Cranes into central Florida to establish a nonmigratory population (Florida Nonmigratory Population, FNMP) (Nesbitt et al., 1997). Cranes were reared in captivity under established methods that included a mixture of parent-reared and costume-reared cranes (15% and 85%, respectively, 1993–1995; Nesbitt et al., 1997) socialized into groups during a pretransfer quarantine period. Most of the cranes were less than 1 year old, but many were of the AHY age class when transferred for release. Cranes were transferred to Florida in groups of 2–14 and brought to acclimation pens after dark to reduce stress. The cranes were immediately given a physical examination, banded, fitted with radio transmitters, and brailed to prevent flight out of the 18 m × 18 m pen. A 30-day soft release, similar to that used for Mississippi Sandhill Cranes, was favored to acclimatize the cranes to the site and to allow adequate quarantine of the birds to prevent introduction of disease agents to the landscape.

Survival of cranes released from 1993 to 1995 was 0.42 (SE = 0.07; n = 52) for 1 year post-release (Nesbitt et al., 1997). Fourteen of 30 (47%) deaths during this period occurred within 30 days of release. However, the second-year annual survival rate (0.81, SE = 0.09; n = 22) was similar to that of wild Florida Sandhill Cranes (0.87; Nesbitt et al., 1997). The point estimate for 1-year post-release survival of parent-reared cranes (0.13, SE = 0.12, 95% CI = 0.00, 0.53; n = 8) was lower than survival of costume-reared cranes (0.48, SE = 0.08; 95% CI = 0.32, 0.63; n = 44) though they are not statistically different. Groups with mixed rearing histories were not as socially cohesive after release as those with similar rearing history (all costume-reared) or of smaller size (6–8 individuals) (Nesbitt et al., 1997).

Cranes hatched in 1993 and 1994 tended to roost on dry ground, use heavily vegetated areas, and be frequently killed by bobcats following release (Nesbitt et al., 1997). Changes in captive rearing protocols and facilities prior to transfer provided water for the birds to roost in, improved unobstructed views of surrounding landscapes, and stressed management that included active learning with food items similar to those in Florida. Stricter limits on pretransfer flight experience were used in an attempt to discourage dispersal at the release site. Release sites were also modified to reduce cover for bobcats, and a transition to use of private lands was made to expand management flexibility in times of water level and local habitat change.

By the spring of 2000, 208 cranes had been released and the effects of the management changes instituted in 1995 could be evaluated (Nesbitt et al., 2001). For the period 1996–2000, the average 1-year survival rate had increased to 0.50 (SE = 0.03; n = 208). Prolonged drought, predation by alligators, ingestion of metal objects, and increased prevalence of coccidiosis together were believed to keep survival low. Drought influenced mortalities during flightless molting

periods and increased dispersal, including the long-distance migration of a pair of cranes (Nesbitt et al., 2001).

An additional 81 cranes were released from 2001 to 2005; a total of 289 cranes were released for the FNMP during 1993–2005. The total population began to decline from the peak of 85 individuals at the end of 2003 (Folk et al., 2008). The cranes showed poor long-term survival and productivity (Converse et al., Chapter 8, this volume; Moore et al., 2012). By late 2008, the population numbered 30 individuals, and no males had survived past 10 years of age due to high rates of predation and power line strikes (Folk et al., 2010). Nine chicks had fledged from 31 documented hatchings in 68 nest attempts. Cyclical drought was believed to impact nesting success; rainfall and water levels in shallow marshes did not meet thresholds for productivity in 6 of 10 years between 1999 and 2008 (Spalding et al., 2009). In addition, incubation lapses by the reintroduced cranes – leading to embryonic death and incubation failure – were recorded by researchers in 2010, and were hypothesized to be a by-product of the cranes' (artificial) captive upbringing (Folk et al., 2014). Plans for future releases were withdrawn by recovery planners in September 2008.

CASE STUDY – EASTERN MIGRATORY RELEASES

The failure of the FNMP reintroduction, a desire to establish another geographically distinct population from the AWBP, and the refinement of guided migration motivated interest in reintroduction of a migratory population of Whooping Cranes into the eastern United States (Eastern Migratory Population, EMP) by the International Whooping Crane Recovery Team. Ultralight aircraft were chosen to help establish a migratory route between Wisconsin and the Gulf coast of Florida using costume-reared Whooping Cranes (Duff, Chapter 21, this volume). The cranes were habituated to aircraft sounds and sight, and operant conditioning with food rewards was used to induce following of aircraft. Though the FNMP reintroduction involved maintaining the release candidates at captive breeding centers for 6–9 months prior to transfer to the release site, the ultralight training program accelerated the transfer of chicks to the field in less than 2 months. Urbanek et al. (2005a, 2005b) provide an excellent overview of the techniques used in 2001, the project's first year.

For the period 2001–10, training and departure for migration occurred from the Necedah NWR in central Wisconsin. Ninety-six percent (150 of 156) of cranes completed guided migration to one of two Florida winter destinations. Survival to 1 year from the start date of the guided migration was 0.76 (SE = 0.03; n = 156). Mortalities were caused by traumatic injury (e.g., aircraft strike, automobile collision, conspecific aggression) and predation, and occurred throughout the migratory range and annual cycle. A severe storm at the Florida release pen was associated with the deaths of 17 birds. A few cranes did not migrate northwest along the route they were trained on and ended well east of the migration origin; these birds were captured and translocated to Wisconsin (Zimorski and Urbanek, 2010). Nevertheless, despite the inherent risks of artificial migration, maladaptation of Whooping Cranes to conditions of captivity, and environmental uncertainty, the husbandry and soft release methodology was associated with high first-year survival and appropriately oriented migratory behavior compared to any previous reintroduction effort (Mueller et al., 2013; Servanty et al., 2014; Urbanek et al., 2014).

Beginning in 2005, the EMP releases using guided migration were supplemented with the release of juvenile cranes in the autumn of their first year (termed direct autumn release, or DAR). Protocols very similar to those of Horwich (1989) and Urbanek and Bookhout (1992) were followed to hand-rear cranes in captivity using costumed caretakers (Wellington and Urbanek, 2010). The cranes were transferred at <1–2 months of age from the International Crane Foundation (ICF) to a chick rearing facility in the field, and costumed caretakers led chicks to wetlands to forage, exercise, and loaf throughout most days (Fig. 20.2). A large covered pen

FIGURE 20.2 Active exposure of juvenile Whooping Crane chicks to native habitats is a core pre-release activity in the direct autumn release method used in the Eastern Migratory Population reintroduction. *Source: Photo credit: Tom Lynn.*

was used to acclimate the cranes to unsheltered conditions and provide nighttime protection, but birds were allowed to take flights during daylight following fledging and would return to the holding pen at night. In late October, the chicks were released singly, in pairs, or in small groups near adult Whooping Cranes, or if not available, near staging Sandhill Cranes. DAR Whooping Cranes frequently reunited shortly after release, but ultimately accompanied free-ranging cranes on migration or departed south by themselves. With the hard release, the cranes were no longer provisioned with pelleted food or managed for protection from predators or other influences.

From 2005 to 2010, 44 Whooping Cranes were released by the DAR method in addition to those released with guided migration from the same refuge. Similar to other release projects, behavioral and environmental conditions produced numerous challenges. For example, the hard release technique resulted in unexpected local movements and selection of habitats or locations that resulted in interventions, which increases the risk of traumatic injury from handling and transport. Later challenges included incorrect orientation of southward migration (due south instead of to the southeastern United States) or return migration north; exclusive association with Sandhill Cranes, especially by lone birds; and losses from predation, power line collision, gunshot, and traumatic injury (Urbanek et al., 2014). Translocation was used as a management tool to correct some of the migration orientation challenges (Zimorski and Urbanek, 2010). The 1-year survival rate from release was 0.68 (SE = 0.07; n = 44) during this period. Though initial survival was lower than for costume-reared Sandhill Cranes released in autumn (0.84, SE = 0.06; above), the Whooping Crane releases were conducted in areas with far fewer free-ranging adults or certainty of staging behavior, perhaps crucial elements for predator recognition and avoidance, as well as efficient migratory orientation and safe roosting site selection.

Since 2013, a small number of parent-reared Whooping Crane chicks (i.e., reared by captive foster parents) have also been released at Necedah NWR (G.H. Olsen, U.S. Geological Survey, unpublished data; Wellington et al., 1996). Parent-reared HY juveniles were separated from the captive adults at 3–4 months of age, socialized briefly and housed in pens with water features for roosting experience, and then transferred to the release site and penned singly within the existing territories of adult Whooping Crane pairs. This method has taken advantage of the modest number of reintroduced territorial pairs without young of the year. Chicks were provided with food and water in the temporary pens; no introduction to local habitat was conducted prior to hard release that occurred after about 4 days. The goal was to provide a protected "introduction" of the chick to the intended allo-parents and to create opportunity for social bond formation that would carry through to migration and possibly subsequent wintering and spring migration. Four chicks were released in both 2013 and 2014, and three were released in 2015. Four of the 11 cranes died within 1 month after release, never forming substantial associations with adult Whooping Cranes despite numerous encounters. Two of these birds were hit by cars; two were depredated. The other seven birds successfully migrated to various southern destinations with Whooping Cranes, five of them with allo-parents. The 1-year survival post-release of this small sample was 0.64 (95% CI = 0.35, 0.88; see also Converse et al., Chapter 8, this volume).

Managers of the EMP shifted the site of costume-reared crane releases >80 km SE in 2011 in order to enhance range expansion and promote site fidelity in areas with fewer of the parasitic black flies believed to be impeding reproductive success at Necedah NWR (Converse et al., Chapter 8, this volume). The costume-rearing methods for both release projects remained unchanged, but the shift resulted in the separation of the aircraft-guided migration (White River Marsh State Wildlife Area) and DAR projects (Horicon NWR) from occurring at the same refuge. The two locations are separated by

48 km, and Whooping Cranes originally from the Necedah NWR releases infrequently used the region.

Initially, no adult Whooping Cranes were present in eastern Wisconsin to release DAR birds with; the chicks dispersed independently or affiliated with Sandhill Cranes. Later, interaction with older aircraft-guided and DAR cranes (representing a mix of geographic orientations, either Necedah or eastern Wisconsin) occurred at wintering sites or along migration. Site-specific challenges also necessitated modification to the prerelease phase of the DAR project at Horicon NWR. The chick facilities at Necedah NWR were used for field rearing from approximately 1.5 to 2 months of age until fledging before transfer to Horicon NWR in 2011 and 2012. In 2013, chicks were managed exclusively at the ICF prior to transfer to Horicon NWR at fledging age. Extended field exposure was provided for the birds at Horicon NWR in all years, as well as overnight protection in a covered pen, until the hard release of small groups similar to earlier releases.

From 2011 to 2014, 52 cranes were released in eastern Wisconsin to become a part of an expanded EMP (29 via guided migration, 23 via DAR; there was no DAR group in 2014). The 1-year survival rates from the start of guided migration and direct release were 0.90 (SE = 0.06) and 0.52 (SE = 0.10), respectively. Despite encouraging results in 2011 and 2012 among DAR cranes, all but one of the 2013 cranes showed delayed migration, which increased risk of predation, traumatic injury, and disease-related mortality.

CASE STUDY – LOUISIANA NONMIGRATORY RELEASES

Southwestern Louisiana was the last remaining site of wild resident Whooping Cranes in the United States and was under consideration as a candidate site for reintroduction for many years (Gomez, 1992; Gomez et al., 2005). Whooping Cranes have been released since 2011 at the White Lake Wetlands Conservation Area in southwest Louisiana with the goal of creating a self-sustaining population (King et al., Chapter 22, this volume). A short-term objective was to exceed the 0.50 average 1-year survival rate of the FNMP following 1995 (Nesbitt et al., 2001).

The project implemented a soft release strategy similar to that used with the FNMP with modifications for local conditions. The cranes were costume-reared by established methods at PWRC. No parent-reared birds were included in the release groups. After shipping to the release site, the birds were initially held for 3–4 weeks in a 70-foot-diameter net-covered pen in order to acclimate to the environment (Fig. 20.3). Use of netting to restrain flight meant brails were unnecessary during the acclimation. The pen

FIGURE 20.3   A fully enclosed pen used to acclimatize juvenile Whooping Cranes for a soft release in Louisiana. *Source: Photo credit: Louisiana Department of Wildlife and Fisheries.*

FIGURE 20.4   The large 1.5 acre release pen used in soft release of nonmigratory Whooping Cranes in Louisiana. *Source: Photo credit: Louisiana Department of Wildlife and Fisheries.*

was larger (3848 ft.^{2}) than the portable pen used for FNMP acclimation periods and release (~3100 ft.^{2}) (Nesbitt et al., 2001), and it was expanded to a 100-foot-diameter structure in 2011 to accommodate larger cohorts of cranes. The cranes were banded and given physical examinations within a few days of arrival, rather than immediately as in Florida. The covered pen was built within an open, rectangular 1.5 acre release pen containing multiple feeding platforms, similar to that described by Urbanek and Bookhout (1992) for use with migratory Sandhill Crane reintroduction (Fig. 20.4). The Whooping Cranes were provided supplemental food until they began to forage regularly outside the enclosure. Mammalian predator control at the release site was accomplished by live trap and removal, and by manipulating water levels to deter terrestrial approach; alligators were uncommon in the immediate vicinity of the pen.

The first release group consisted of 10 cranes hatched in 2010 and transferred to Louisiana in February 2011 (similar to delayed AHY release in Florida). Following less than 4 weeks of acclimation, the cranes were allowed into the large release pen. The cranes began roosting at night in areas outside the pen and using the interior and feed dispensers during daylight hours, opposite to the intended management objective. Food was removed from the pen after 8 weeks. Regional drought at this time was then associated with a dispersal of the cranes to east Texas, but most birds later settled in rice and crawfish agricultural areas north of the release site in Louisiana (Louisiana Department of Wildlife and Fisheries, unpublished data).

The second release group consisted of 16 cranes that were transferred in December 2011. The cranes were held for nearly 4 weeks in the acclimation pen prior to moving to the release pen. Less effort was made to encourage birds to roost at night in the release pen, and they did not do so. Supplemental food was provided until March 2012 to facilitate the convalescence of a bird following surgery to repair a fracture sustained during handling. Release groups of 14 and 10 cranes were transferred to Louisiana in December 2012 and 2013. Both groups were allowed 3 weeks of acclimation and several months' access to supplemental food from the large release pen. Behavior of the cranes upon release was similar to behavior of previous groups through the end of the year. The fifth release group consisted of 14 cranes transferred in early December 2014 to the acclimation pen and given access to the open release pen after 3.5 weeks. Supplementary feeding was managed creatively to minimize interaction and aggression from older cranes visiting the release pen. Food was withdrawn entirely after 11 weeks.

Overall 1-year survival post-release of the 40 cranes released from 2010 to 2013 was 0.64 (SE = 0.06; n = 64; Louisiana Department of Wildlife and Fisheries, unpublished data). First-year survival of Whooping Cranes released in Louisiana has been greater than that observed in the FNMP releases. The FNMP 1-year survival rate for 100 cranes released in the first 4 years was 0.50 (SE = 0.05; Nesbitt et al., 2001). Mortality factors of LNMP cranes are similar to other reintroduced Whooping Cranes (predation, disease, power line strike), but the prevalence of gunshot-related deaths (24%, n = 6) is greater than in any other project. Shootings were associated with dispersal to agricultural prairie north of the reintroduction site, and possibly complicated by human habituation or high visibility (and hence easy targets for vandals; see Condon et al., Chapter 23, this volume).

Another challenge in Louisiana includes wandering of yearling cranes into east and north Texas in proximity to the migratory corridor of the AWBP. The site of the Louisiana reintroduction is well within the subadult wandering distance from the natal or release site recently documented in Whooping Cranes. WBNP subadults have been observed ranging 200 km or more northwest of Wood Buffalo National Park, and EMP subadults have ranged between 780 (west) and 1370 km (east) from the Wisconsin release site (Urbanek et al., 2010). Seasonal differences in range use and natal site fidelity will likely keep the LNMP and AWBP separate for many years, until the populations grow and the likelihood of significant behavioral interaction increases. Overlap with southward migrating birds from Wood Buffalo seems the likeliest location of contact.

SUMMARY AND OUTLOOK

Methods to reintroduce captive-reared Whooping Cranes to sites in the United States have resulted in assimilation of most cranes to the landscape for the year following release (Table 20.1). The 1-year survival rates are similar to, but generally below, those of experimental trials with Sandhill Cranes. Both parent- and hand-reared Whooping Cranes have been released, though vastly more of the latter due to efficiencies of scale achievable with hand-rearing, and techniques have resulted in a high probability of appropriate sexual imprinting. Many of the releases have equaled or exceeded the 1-year survival rates of previous attempts, but all have been limited by the lack of an extant, high-density, free-ranging adult population of Whooping Cranes within which to release birds, unlike many of the Sandhill Crane projects.

The greatest initial rates in survival with reintroduction of young cranes to a site without Whooping Cranes has been achieved through the application of guided migration combined with soft release at a southern terminus. This method reduces predation risk through the use of enclosures, and uncontrolled migratory movements are limited through behavioral conditioning. Foraging skill development is likely prolonged because exposure to summer habitat and food resources is limited to small pen sites and provisioning with an artificial diet. Though most yearling cranes the following spring are successful in their return, several did exhibit naiveté, especially in small groups or alone, because an experienced guide was often missing (Mueller et al., 2013). This same process occurs with DAR cranes, but active pre- and post-fledging learning is used to prepare them for release. The direct release method relies on experienced birds to provide guidance and

TABLE 20.1   First-Year Survival Rates of Whooping Cranes Released by Costume Rearing and Parent Rearing and Release Methods, across Three Different Reintroduction Efforts (EMP, Eastern Migratory Population; FNMP, Florida Nonmigratory Population; LNMP, Louisiana Nonmigratory Population)

Rearing			Release				Survival	References
Method	Location	No.	Location	Year	Migration[a]	Method[b]		
Mixed	Multiple	52	FNMP	1993–95	n/a	Soft	0.40	Nesbitt et al., 1997
Mixed	Multiple	208	FNMP	1996–2000	n/a	Soft	0.50	Nesbitt et al., 2001
Costume	PWRC	156	EMP	2001–10	Guided	Soft	0.76	WCEP[c]
Costume	PWRC	29	EMP	2011–14	Guided	Soft	0.90	WCEP
Costume	ICF	44	EMP	2005–10	Self	Hard	0.68	WCEP
Costume	ICF	23	EMP	2011–14	Self	Hard	0.52	WCEP
Parent	PWRC	8	EMP	2013–14	Self	Hard	0.62	WCEP
Costume	PWRC	40	LNMP	2010–13	n/a	Soft	0.64	LDWF[d]

[a] First fall migration method; Guided = juvenile cranes are managed to fly in stages behind ultralight aircraft, Self = juvenile cranes migrate on own or in presence of experienced Whooping Cranes or Sandhill Cranes.

[b] Release method; Soft = technique involving a period of protection and provisioning at a specific release site to encourage acclimation and gradual dispersal, Hard = sudden release following minimal acclimation at the specific release site (may or may not have been preceded by in situ acclimation at other site at the same refuge prior to release).

[c] Whooping Crane Eastern Partnership annual reports, available at: http://www.bringbackthecranes.org/.

[d] Louisiana Department of Wildlife and Fisheries, unpublished data.

some degree of protection on migration, and perhaps more successfully if they are Whooping Cranes versus Sandhill Cranes due to the potential risk of hybrid pairings between lone birds. As originally designed, DAR provides a lower-cost, logistically feasible method to add cranes to an experienced migratory population; recent results in eastern Wisconsin suggest lower survival when DAR is used to colonize a new site. The active in situ learning components of DAR (frequent exposure to diverse wetland habitat beginning at an early age) could certainly also be adapted to use with a nonmigratory reintroduction project, such as in Louisiana, which now has a core population of free-ranging cranes to add to, but does not transfer cranes to the release area until 6 months of age.

The release methodology used with captive parent-reared Whooping Cranes also shows promise, but the hard release techniques used with single juveniles and potential allo-parents may benefit from refinement. Extending the acclimation time of the juveniles and food baiting the

adults to the release site might be one option to induce greater interaction; and beginning earlier and capitalizing on the developmental reattachment period near fledging described by Horwich (1989) may result in a higher rate of association by migration. Unfortunately, producing effective numbers of releasable parent-reared cranes will require reorienting management and reducing production from captive breeding centers. The potential greater adaptability of these birds is hypothesized to provide an increase in breeding success in the future (Converse et al., Chapter 8, this volume), which may be limited among hand-reared Whooping Cranes. Unfortunately, no example exists to support the contention that parent-reared cranes outperform genetically equivalent hand-reared counterparts in habitats currently used. The Mississippi Sandhill Crane program continues to release birds after 35 years partly because neither releases of parent-reared nor costume-reared cranes have resulted in a self-sustaining population despite intensive habitat management.

Though releases have resulted in the addition of many Whooping Cranes to eastern North America over the past 15 years, additional analysis is urgently needed to elucidate the critical factors associated with unsatisfactory long-term natural recruitment and population persistence (Armstrong and Seddon, 2008). These areas represent important aspects for improvement and validation of reintroduction as an effective conservation measure for the species.

References

Archibald, K., Archibald, G., 1992. Releasing puppet-reared sandhill cranes into the wild: a progress report. In: Wood, D.A., (Ed.), Proceedings of the 1988 North American Crane Workshop. Nongame Wildlife Program Technical Report 12, Florida Game Fresh Water Fish Commission Tallahassee, pp. 251–254.

Armstrong, D.P., Seddon, P.J., 2008. Directions in reintroduction biology. Trends Ecol. Evol. 23 (1), 20–25.

Bizeau, E.G, Schumacher, T.V., Drewien, R.C., Brown, W.M., 1987. An experimental release of captive–reared greater sandhill cranes. In: Lewis, J.C. (Ed.), Proceedings of the 1985 Crane Workshop. Platte River Whooping Crane Maintenance Trust, Grand Island, NE, 78–88.

Canadian Wildlife Service and U.S. Fish and Wildlife Service, 2005. International recovery plan for the whooping crane. Recovery of Nationally Endangered Wildlife (RENEW), Ottawa, and U.S. Fish and Wildlife Service, Albuquerque, NM, p. 162.

Canessa, S., Guillera-Arroita, G., Lahoz-Monfort, J.J., Southwell, D.M., Armstrong, D.P., Chadès, I., Lacy, R.C., Converse, S.J., 2016. Adaptive management for improving species conservation across the captive-wild spectrum. Biol. Conserv. 199, 123–131.

Clegg, K.R., Lewis, J.C., 2001. Continuing studies of ultralight aircraft applications for introducing migratory populations of endangered cranes. Proceedings of the North American Crane Workshop 8, 96–108.

Clegg, K.R., Lewis, J.C., Ellis, D.H., 1997. Use of ultralight aircraft for introducing migratory crane populations. Proceedings of the North American Crane Workshop 7, 105–113.

Condon, E., Brooks, W.B., Langenberg, J., Lopez, D., 2018. Whooping crane shootings since 1967 (Chapter 23). In: French, Jr., J.B., Converse, S.J., Austin, J.E. (Eds.), Whooping Cranes: Biology and Conservation. Biodiversity of the World: Conservation from Genes to Landscapes. Academic Press, San Diego, CA.

Converse, S.J., Strobel, B.N., Barzen, J.A. Reproductive failure in the eastern migratory population: the interaction of research and management (Chapter 8). In: French, Jr., J.B., Converse, S.J., Austin, J.E. (Eds.), Whooping Cranes: Biology and Conservation. Biodiversity of the World: Conservation from Genes to Landscapes. Academic Press, San Diego, CA.

Derrickson, S.R., Carpenter, J.W., 1987. Behavioral management of captive cranes – factors influencing propagation and reintroduction. In: Archibald, G.W., Pasquier, R.F. (Eds.), Proceedings of the 1983 International Crane Workshop. International Crane Foundation, Baraboo, WI, 493–511.

Drewien, R.C., Derrickson, S.R., Bizeau, E.G., 1982. Experimental release of captive parent-reared greater sandhill cranes at Grays Lake Refuge, Idaho. In: Lewis, J.C. (Ed.), Proceedings of the 1981 Crane 580 Workshop. National Audubon Society, Tavernier, FL, pp. 99–111.

Duff, J.W., 2018. The operation of an aircraft-led migration: goals, successes, challenges 2001 to 2015 (Chapter 21). In: French, Jr., J.B., Converse, S.J., Austin, J.E. (Eds.), Whooping Cranes: Biology and Conservation. Biodiversity of the World: Conservation from Genes to Landscapes. Academic Press, San Diego, CA.

Duff, J.W., Lishman, W.A., Clark, D.A., Gee, G.F., Ellis, D.H., 2001. Results of the first ultralight-led sandhill crane migration in eastern North America. Proceedings of the North American Crane Workshop 8, 109–114.

Ellis, D.H., Clauss, B., Watanabe, T., Curt, R.M., Kinloch, M., Ellis, C.H., 1997. Results of an experiment to lead cranes on migration behind motorized ground vehicles. Proceedings of the North American Crane Workshop 7, 114–122.

Ellis, D.H., Clauss, B., Watanabe, T., Curt, R.M., Shawkey, M., Mummert, D.P., Sprague, D.T., Ellis, C.H., Trahan, F.B., 2001a. Results of the second (1996) experiment to lead cranes on migration behind a motorized ground vehicle. Proceedings of the North American Crane Workshop 8, 122–126.

Ellis, D.H., Dein, F.J., 1996. Special techniques part E: flight restraint. In: David, H.E., George, F.G., Claire, M.M. (Eds.), Cranes: Their 596 Biology, Husbandry, and Conservation. Department of the Interior, National Biological Service, Washington, DC, and International Crane Foundation, Baraboo, WI, pp. 241–244.

Ellis, D.H., Gee, G.F., Clegg, K.R., Duff, J.W., Lishman, W.A., Sladen, W.J.L., 2001b. Lessons from the motorized migrations. Proceedings of the North American Crane Workshop 8, 139–144.

Ellis, D.H., Gee, G.F., Olsen, G.H., Hereford, S.G., Nicolich, J.M., Thomas, N.J., Nagendran, M., 2001c. Minimum survival rates for Mississippi sandhill cranes: a comparison of hand-rearing and parent-rearing. Proceedings of the North American Crane Workshop 8, 80–84.

Ellis, D.H., Lewis, J.C., Gee, G.F., Smith, D.G., 1992a. Population recovery of the Whooping Crane with emphasis on reintroduction efforts: past and future. Proceedings of the North American Crane Workshop 6, 142–150.

Ellis, D.H., Mummert, D.P., Urbanek, R.P., Kinloch, M., Mellon, C., Dolbeare, T., Ossi, D., 2001d. The one-by-one method for releasing cranes. Proceedings of the North American Crane Workshop 8, 225.

Ellis, D.H., Olsen, G.H., Gee, G.F., Nicolich, J.M., O'Malley, K.E., Nagendran, M., Hereford, S.G., Range, P., Harper, W.T., Ingram, R.P., Smith, D.G., 1992b. Techniques for rearing and releasing nonmigratory cranes: lessons from the Mississippi sandhill crane program. Proceedings of the North American Crane Workshop 6, 135–141.

Folk, M.J., Dellinger, T., Baynes, S., Chappell, K., Spalding, M., 2014. Status of the Florida resident flock of whooping cranes. Proceedings of the North American Crane Workshop 12, 86.

Folk, M.J., Nesbitt, S.A., Parker, J.M., Spalding, M.G., Baynes, S.B., Candelora, K.L., 2008. Current status of nonmigratory whooping cranes in Florida. Proceedings of the North American Crane Workshop 10, 7–12.

Folk, M.J., Rodgers, J.A., Dellinger, T.A., Nesbitt, S.A., Parker, J.M., Spalding, M.G., Baynes, S.B., Chappell, M.K., Schwikert, S.T., 2010. Status of nonmigratory whooping cranes in Florida. Proceedings of the North American Crane Workshop 11, 118–123.

Gomez, G.M., 1992. Whooping cranes in southwest Louisiana: history and human attitudes. Proceedings of the North American Crane Workshop 6, 19–23.

Gomez, G.M., Drewien, R.C., Courville, M.L., 2005. Historical notes on whooping cranes at White Lake, Louisiana: the John J. Lynch interviews, 1947–1948. Proceedings of the North American Crane Workshop 9, 111–116.

Horwich, R.H., 1989. Use of surrogate parental models and age periods in a successful release of hand-reared sandhill cranes. Zoo Biol. 8 (4), 379–390.

Horwich, R.H., 2001. Developing a migratory whooping crane flock. Proceedings of the North American Crane Workshop 8, 85–95.

Horwich, R.H., Wood, J., Anderson, R., 1992. Release of sandhill crane chicks hand-reared with artificial stimuli. In: Wood, D.A. (Ed.), Proceedings of the 1988 North American Crane Workshop. Nongame Wildlife Program Technical Report 12, Florida Game Fresh Water Fish Commission, Tallahassee, pp. 255–262.

Jones, C.G, Merton, D.V., 2012. A tale of two islands: the rescue and recovery of endemic birds in New Zealand and Mauritius. In: Ewen, J.G., Armstrong, D.P., Parker, K.A., Seddon, P.J., (Eds.), Reintroduction Biology: Integrating Science and Management. Wiley-Blackwell, Oxford, UK, pp. 33–72.

King, S.L, Selman, W., Vasseur, P., Zimorski, S.E., 2018. Louisiana nonmigratory whooping crane reintroduction (Chapter 22). In: French, Jr., J.B., Converse, S.J., Austin, J.E (Eds.), Whooping Cranes: Biology and Conservation. Biodiversity of the World: Conservation from Genes to Landscapes. Academic Press, San Diego, CA.

Lishman, W.A., Teets, T.L., Duff, J.W., Sladen, W.J.L., Shire, G.G., Goolsby, K.M., Bezner, W.A.K., Urbanek, R.P., 1997. A reintroduction technique for migratory birds: leading Canada geese and isolation-reared sandhill cranes with ultralight aircraft. Proceedings of the North American Crane Workshop 7, 96–104.

Mahan, T.A., Simmers, B.S., 1992. Social preference of four cross-foster reared sandhill cranes. Proceedings of the North American Crane Workshop 6, 114–119.

Moore, C.T., Converse, S.J., Folk, M.J., Runge, M.C., Nesbitt, S.A., 2012. Evaluating release alternatives for a long-lived bird species under uncertainty about long-term demographic rates. J. Ornithol. 152 (Suppl. 2), S339–S353.

Mueller, T., O'Hara, R.B., Converse, S.J., Urbanek, R.P., Fagan, W.F., 2013. Social learning of migratory performance. Science 341 (6149), 999–1002.

Nagendran, M., Urbanek, R.P., Ellis, D.H., 1996. Special techniques, part D: reintroduction techniques. In: David, H.E., George, F.G., Claire, M.M. (Eds.), Cranes: Their Biology, Husbandry, and Conservation. Department of the Interior, National Biological Service, Washington, DC, and International Crane Foundation, Baraboo, WI, pp. 231–240.

Nesbitt, S.A., 1979. Notes on the suitability of captive-reared sandhill cranes for release into the wild. In: Lewis, J.C. (Ed.), Proceedings of the 1978 Crane Workshop. Colorado State University, pp. 85–88.

Nesbitt, S.A., Faulk, M.J., Sullivan, K.A., Schwikert, S.T., Spalding, M.G., 2001. An update of the Florida whooping crane release project through June 2000. Proceedings of the North American Crane Workshop 8, 62–73.

Nesbitt, S.A., Folk, M.J., Spalding, M.G., Schmidt, J.A., Schwikert, S.T., Nicolich, J.M., Wellington, M., Lewis, J.C., Logan, T.H., 1997. An experimental release of whooping cranes in Florida – the first three years. Proceedings of the North American Crane Workshop 7, 79–85.

Parker, K.A., Dickens, M.J., Clarke, R.H., Lovegrove, T.G., 2012. The theory and practice of catching, holding, moving and releasing animals. In: Ewen, J.G., Armstrong, D.P., Parker, K.A., Seddon, P.J. (Eds.), Reintroduction Biology: Integrating Science and Management. Wiley-Blackwell, Oxford, UK, pp. 105–137.

Servanty, S., Converse, S.J., Bailey, L.L., 2014. Demography of a reintroduced population: moving toward management models for an endangered species, the whooping crane. Ecol. Appl. 24 (5), 927–937.

Spalding, M.G., Folk, M.J., Nesbitt, S.A., Folk, M.L., Kiltie, R., 2009. Environmental correlates of reproductive success for introduced resident whooping cranes in Florida. Waterbirds 32 (4), 538–547.

Urbanek, R.P., Bookhout, T.A., 1992. Development of an isolation-rearing/gentle release procedure for reintroducing migratory cranes. Proceedings of the North American Crane Workshop 6, 120–130.

Urbanek, R.P., Duff, J.W., Swengel, S.R., Fondow, L.E.A., 2005a. Reintroduction techniques: post-release performance of sandhill cranes (1) released into wild flocks and (2) led on migration by ultralight aircraft. Proceedings of the North American Crane Workshop 9, 203–211.

Urbanek, R.P., Fondow, L.E.A., Satyshur, C.D., Lacy, A.E., Zimorski, S.E., Wellington, M., 2005b. First cohort of migratory whooping cranes reintroduced to eastern North America: the first year after release. Proceedings of the North American Crane Workshop 9, 213–223.

Urbanek, R.P., Fondow, L.E.A., Zimorski, S.E., 2010. Survival, reproduction, and movements of migratory whooping cranes during the first seven years of reintroduction. Proceedings of the North American Crane Workshop 11, 124–132.

Urbanek, R.P., Zimorski, S.E., Szyszkoski, E.K., Wellington, M.M., 2014. Ten-year status of the eastern migratory whooping crane reintroduction. Proceedings of the North American Crane Workshop 12, 33–42.

Wellington, M.M., Burke, A., Nicolich, J.M., O'Malley, K., 1996. Chick rearing. In: David, H.E., George, F.G., Claire, M.M. (Eds.), Cranes: Their Biology, Husbandry, and Conservation. Department of the Interior, National Biological Service, Washington, DC, and the International Crane Foundation, Baraboo, WI, pp. 95–104.

Wellington, M.M., Urbanek, R.P., 2010. The direct autumn release of whooping cranes into the Eastern Migratory Population: a summary of the first three years. Proceedings of the North American Crane Workshop 11, 215.

Zimorski, S.E., Urbanek, R.P., 2010. The role of retrieval and translocation in a reintroduced population of migratory whooping cranes. Proceedings of the North American Crane Workshop 11, 216.

21

The Operation of an Aircraft-Led Migration: Goals, Successes, Challenges 2001 to 2015

Joseph W. Duff

Operation Migration Inc., Port Perry, ON, Canada;
Operation Migration USA, Niagara Falls, NY, United States

INTRODUCTION

All of the Whooping Cranes (*Grus americana*) that survive today are the descendants of only 16 individuals that wintered in the southern United States in the early 1940s and nested in or near the Wood Buffalo National Park in northern Canada. Thanks to extensive conservation efforts, including habitat protection and controls on hunting, the population recovered from that bottleneck (Glenn et al., 1999) and has grown at a rate of approximately 4% per year (Converse et al., Chapter 7, this volume), reaching a record estimated 308 birds in 2014 (Harrell, 2014). Despite encouraging growth, this small population faces many pressures, primarily at the southern terminus at the Aransas National Wildlife Refuge (NWR) where reduced freshwater inflow, human encroachment, invasive species, rising sea levels, land subsidence, chemical spills, avian disease, and/or catastrophic weather events could threaten their

habitat (Canadian Wildlife Service and U.S. Fish and Wildlife Service, 2005). The International Whooping Crane Recovery Team (IWCRT), composed of Canadian and U.S. members, proposed establishing additional, discrete, populations to safeguard the species from extinction.

In the first attempt to reintroduce Whooping Cranes in a migratory situation, eggs were placed in the nests of Sandhill Cranes (*Grus canadensis*) in Grays Lake NWR, Idaho, beginning in 1975 (Ellis et al., 1992). The project ended in 1985 because of low recruitment, due in part to cross imprinting of Whooping Cranes on Sandhill Cranes.

From 1993 to 1995, a Canadian nongovernmental organization (NGO) Operation Migration (OM), conducted preliminary studies using Canada, Geese (*Branta canadensis*) and Trumpeter Swans (*Cygnus buccinator*) as research surrogates. OM is based in Canada, where it is a registered charity. It is also a 501(c)(3) nonprofit in the

United States. The two corporations are operated by one management team and governed by one board of directors. Over the 15-year duration of the project, OM contributed over US$10 million in privately sourced funds to help fulfill the mandate of the IWCRT and the USFWS.

The purpose of those early experiments was to develop a method of instilling migration behavior in nidifugous birds that could eventually be applied to endangered species (Lishman et al., 1997). The birds were imprinted on the pilots and conditioned to follow modified ultralight aircraft (the aircraft-guided or UL method). From 1995 to 1998, four UL migration experiments were conducted with Sandhill Cranes to determine if the technique could be applied to a crane species. The results of each study were presented at the annual meeting of the IWCRT, and by 1998 OM was able to demonstrate that Sandhill Cranes could be isolation-reared while being conditioned to follow the ultralight aircraft, and taught to migrate along a safe route to preselected wintering grounds. The following spring, the Sandhill Cranes initiated the northern migration without the aid of the aircraft guide and returned to the core reintroduction area while avoiding humans and selecting what appeared to be appropriate habitat (Duff et al., 2000).

The success of these studies led, in 1999, to the formation of a consortium of nine federal, state, and private agencies collectively known as the Whooping Crane Eastern Partnership (WCEP). Most of the year 2000 was required for the partnership to obtain multijurisdictional, regulatory approval for the Eastern Migration Population (EMP). This included the issuance of a Nonessential Experimental Population (NEP) designation under the U.S. Endangered Species Act (Federal Register 66 (123):33903–33917) for Whooping Cranes that would be released into the eastern flyway. The NEP designation reduced the status of the EMP cranes from endangered to threatened, allowing for greater flexibility in the management of the reintroduced Whooping Cranes. The legal range of the NEP covers 20 American states. Consensus was also obtained from the Canadian Wildlife Service and the Provinces of Ontario and Manitoba. While this lengthy approval process was underway, a test of the complete protocol was undertaken using nonendangered Sandhill Cranes. The 1900 km migration route was finalized along with prearranged stopover locations at 80–100 km intervals.

In 2001, WCEP began the second attempt to reintroduce migratory Whooping Cranes. After more than a month of imprinting and conditioning the Whooping Crane chicks, the NEP Final Rule was published in the Federal Register just 2 days before the birds were scheduled for transport to the reintroduction site in Wisconsin.

TRAINING METHODS AND HOUSING FACILITIES

Early Imprinting and Conditioning

Captive-produced Whooping Crane eggs, as well as eggs salvaged from abandoned nests at Necedah NWR after 2007, were reared for UL release. Eggs were hatched in incubators at the U.S. Geological Survey (USGS) Patuxent Wildlife Research Center (Patuxent). Digital recordings of a Whooping Crane brood call were played to pipping eggs to simulate parent/chick communications. A recording of the aircraft engine was also played to the chicks, which was effective in reducing the chick's fear of the aircraft. Recorded natural marsh sounds were used to mask the noise of human activity. In addition to the recorded brood call, hatchlings were provided with taxidermied brood models to help ensure they identified with conspecifics. The chicks were penned next to adult Whooping Cranes intended to act as sexual imprint models (Horwich, 1989).

Isolation rearing methods were used from the time the eggs hatched in April and May until the young birds left the winter pen site the following spring (Horwich, 2001). Three OM pilots were in charge of caring for the birds and ensuring observance of the protocol. They managed the early imprinting, the training in Wisconsin, and the migration. Only a limited number of people were allowed near the birds at any time. All refrained from talking, and each wore a costume designed to disguise the human form. It consisted of a loose-fitting, white smock that covered the handlers from shoulders to midcalf. OM handlers wore white plastic construction helmets that were covered with a white fabric veil, which included a reflective, plastic visor so the birds could not see the handler's eyes. Long sleeves covered their hands and they wore dark rubber boots. Handlers also carried a digital recorder that could play the brood call and a hand puppet that looked like the head of an adult crane. Efforts were made to disguise all equipment that was needed to care for the birds, such as the pens, tools, feed buckets, cameras, and radios. No vehicles, lawn mowers, tractors, power tools, or similar equipment was allowed near the birds. The pens and equipment that the birds were allowed to see were painted in natural earth tones and/or draped with camouflage netting to obscure their shape. When heavy maintenance was needed on the pens or the runway such as mowing, the birds were led away from the area to minimize their exposure to human activity (Operation Migration Protocol, 2001). Attempts were made to remove all opportunities for the birds to become familiar with people and their equipment, reasoning that once released into the wild, the birds would be naturally wary of those unfamiliar anthropogenic environments (Duff et al., 2000). During the 15 years of the project, WCEP removed only three UL birds from the project due to habituation to humans or proximity to human-built structures (see "Outcomes and Discussion" for details).

When the chicks were old enough to leave their pens, they were exposed to the aircraft for the first time. Soon thereafter, a circular pen was used to begin conditioning the chicks to follow the aircraft (Fig. 21.1). A wire fabric enclosure, 10 m in diameter and 0.5 m tall, was built in an isolated area at Patuxent. It was designed to protect the chicks from the wheels and propeller of the aircraft as it taxied around the outside perimeter. Chicks were placed inside this circular pen and encouraged to follow the aircraft by means of food (mealworms) dispensed from a crane head puppet, which had an extension long enough to reach from the aircraft to the ground inside the enclosure. By that stage, the chicks were familiar with the sight and sound of the stationary aircraft: however, they often appeared fearful when it first began to move. With patience and practice, the chicks began to follow the aircraft around the circle pen. The chicks were conditioned in the circle pen for the first time at a mean age of 8.41 days (SE m 0.346; range, 7–10.9 days; WCEP Annual Reports 2002–13). As the chicks became stronger and larger, the circular pen was replaced with a long, low fence. The chicks ran on one side, while the aircraft taxied on the other. While at

FIGURE 21.1 An Operation Migration (OM) pilot trains two Whooping Crane chicks, about 2 weeks old, to follow the ultralight aircraft trike in the circle pen at Patuxent Wildlife Research Center, Laurel, MD. *Source: Photo credit: Operation Migration.*

Patuxent, the prefledged chicks were conditioned with the aircraft an average of 7.77 h (SE m 0.776, range 3.55–11.56 h) (Operation Migration, unpublished data). The birds often appeared fearful of the large aircraft wing over their heads. It was cumbersome to operate in high winds and limited the number of training opportunities. Therefore, it was replaced with a small, imitation wing during the early training. That substitute helped prepare the birds for the transition to the real wing when they arrived in Wisconsin.

Whooping Cranes in the wild typically lay two eggs (Allen, 1952). Chicks are agonistic with siblings, requiring that hand-reared birds be trained individually in the early stages of development to avoid aggression injuries. Depending on the temperament of the individual, aggressive behavior ends early or can continue for months. Typically, after a week or two the chicks could be socialized into small groups based on age and compatibility (Operation Migration, unpublished reports). Establishing groups of chicks allowed the handlers to train more than one bird at a time, greatly reducing the workload. With annual cohorts of chicks as large as 23 individuals and training sessions lasting approximately 20 min, conditioning the birds individually took many hours every day, especially when it could not be conducted during the heat of midafternoon. Prior to translocation to Wisconsin, all of the birds were socialized into as many as three groups. This was the first step in creating a single cohesive flock that would eventually follow the aircraft.

Summer Training, Fledging, and Acclimation to the Reintroduction Site

The goals of the summer training were to maintain the birds' affinity to the aircraft throughout the fledging process so they would follow it in flight, as well as on the ground. Also, during that extended time spent at the training areas, the birds became familiar with the introduction site from the air.

Prior to fledging, and when the birds were old enough to tolerate shipping, they were loaded into custom-fabricated containers and airlifted by private aircraft to Wisconsin. Relocation by private aircraft provided direct flights and shorter travel times and allowed us to use smaller airports closer to the reintroduction sites. The average age at shipping was 48.17 days (SE m, 0.88; range 36.7–56.7 days). As many as three groups were shipped per year, for a total of 30 groups, over 15 years (mean shipping date 27 June, range 12 June–29 July). An average of 13.9 birds was shipped per season (SE m 1.47; range 6–23 birds). The most severe injury resulting from shipping was a broken blood feather that caused only minor blood loss.

Summer training sessions with the aircraft generally lasted an hour and involved two handlers and one pilot. On nonflying days, one person typically took 15–30 min to check the birds in the morning and the evening. The birds spent an average of 100.2 days at the reintroduction site (SE m 2.08 days; range 77–121 days). The birds were trained every morning at sunrise when the weather allowed. The mean number of training days was 53.6 (SE m 8.87 days; range 40–69 days). Pilots began by simply taxiing the aircraft up and down the runway while the birds followed. Eventually the birds began to make elongated steps, then short hops until they were able to fly at a low height, in ground effect. When flying close to the ground, induced drag is dramatically reduced, making flight easier (Cui and Zhang, 2010). When birds are raised in the group, it is difficult to determine exactly when individual birds are able to fly. OM staff considered all birds to have fledged when the entire group was able to fly beyond ground effect for a distance of ~50 m at an altitude of ~10 m. The average fledging date was 12 August at 94 days of age. The longest premigration flights averaged 31.2 min (SE m 2.27; range 18–47 min).

During years when the cohort consisted of more than one group of chicks, the birds had to be socialized into one flock prior to the migration. Each group had its own dominance structure that had to be combined with the dominance structure of the other group(s). To minimize aggression during this transition, OM staff found it was best to wait until all the birds had fledged before creating a single social group. The pilots would then lead the middle age group to where the youngest birds were housed. The pen was temporarily divided to keep the two groups separate, yet allow them to see each other and interact through a chain link fence. Initially the two groups were trained separately even though they were at the same site. After training, all the birds were allowed to forage on the runway. Eventually they would begin to compete for leadership. Handlers allowed these important confrontations to take place but did intervene if aggressive interactions made injury a possibility. After a week or so, the birds would be trained together, and eventually the pen divider separating them would be removed. Later, the oldest group would be moved to join the other two. The older birds were generally larger, with a well-established dominance structure; however, they were disadvantaged by being outnumbered and in an unfamiliar pen. This appeared to help balance their size advantage and eased the transition to one large cohort.

From 2001 to 2009, the cranes underwent a premigration health examination and were fitted with permanent marking bands and leg-mounted tracking devices. All of the birds carried VHF units (short-range RF transmitters). A few birds were fitted with Platform Terminal Transmitters (PTT) to communicate with satellites. Others wore Global System for Mobile (GSM) units, which operate using the Cell Communication Network. Those remotely trackable, more advanced units were not used on all of the birds because of cost. Capturing and holding the birds for examination and band fitting was invasive and sometimes resulted in muscle strain and lameness. The birds often appeared distrustful of the handlers afterward and were reluctant to follow the aircraft for periods of up to 2 weeks In 2010, the WCEP Vet Team agreed to end the premigration medical examinations. Temporary tracking devices that could be quickly clipped

onto the legs of the birds were thereafter used during the migration phase of the project. That procedure eliminated the need to hold the birds and shortened their acclimation to the leg bands and tracking devices. Permanent units were fitted to the birds once they arrived at the wintering grounds when their affinity to the handlers and the aircraft was no longer critical.

Beginning in 2009, the birds were monitored continuously using a remote camera that transmitted live coverage of the pen and surrounding area. A network of volunteers, working on a schedule, logged into the program remotely to operate the camera. They could pan, tilt, and zoom the camera to follow the training or watch the interaction of the birds in the pen. This feed was broadcast live at www.ustream.tv. Camera operators could contact the team members, who could access the live stream on smart phones if a concern was raised. The camera operated constantly, switching to infrared at night. A version of the camera also covered the migration portion of the reintroduction project, including inflight, streaming video when the signal allowed.

Pens and Facilities at Necedah National Wildlife Refuge, 2001–10

The IWCRT and the USFWS selected the Necedah NWR in central Wisconsin as the initial reintroduction site for the EMP. Necedah NWR was used for ultralight-led releases from 2001 to 2010. Over that period, four pens and training facilities were constructed in closed areas of the refuge (Fig. 21.2).

1. Site 1/East Site: On the east side of Rynearson Pool One; Lat 44.0534713, Long −90.1496800
2. Site 2/West Site: On the west side of Rynearson Pool One; Lat 44.0673562, Long −90.1658463.
3. Site 4/North Site: South of Coaver Rd; Lat 44.0783262, Long −90.1670986.
4. Site 5/Canfield Site: South of Speedway Rd; Lat 44.0952229, Long −90.1910115.

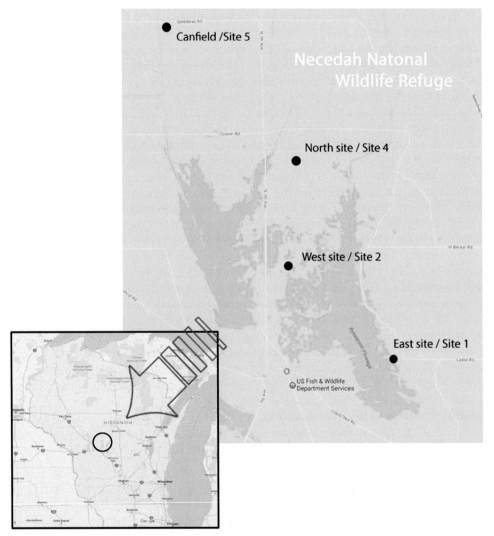

FIGURE 21.2   Location of training pen sites at Necedah National Wildlife Refuge, Necedah, WI. *(Note: Site 3 not used for UL training.)*

Each site consisted of a pen built on dry land using post and stringer construction and clad with sheet steel or vertical boards so the birds could not see out from three sides. This provided a visual barrier if it became necessary to shield the birds from outside activity.

The dry pens were approximately 25 m in diameter and were rounded so birds could not be cornered during aggressive encounters within the flock. In the center, a 3 m by 3 m covered feeding station was divided by a chain link fence to prevent one dominant bird from monopolizing the two feeders that were hung in the shelter. The birds were provided with food ad libitum. Their food consisted of a custom blended, pelletized crane chow provided by Patuxent. In addition, treats such as mealworms or grapes were used as rewards to assist in the training.

Cornstalks with cobs attached, pumpkins, squash, and other natural foods were provided periodically as enrichment items in the pens (Newberry, 1995). Adjacent to the dry pen was another enclosure built of chain link fence covering approximately 1,000 m² and providing a view of the surrounding marsh. That pen was kept flooded with freshwater pumped in daily to provide a clean roosting site. Chain link fencing was dug 0.5 m into the ground around both pens to discourage burrowing predators. Both enclosures were top-netted and protected by three strands of electric fencer wire. Over the course of 15 years, the pens were never breached by a predator resulting in injury or loss of any birds. Next to each pen enclosure, was a turf runway approximately 30 m wide and 250 m long from which to operate the aircraft.

Each group was housed at one or more of these Necedah NWR facilities. The average group size was 6.33 birds (SE m 1.55; range 3–10 birds). Birds were initially grouped together by age. Similar developmental stages allowed them to participate in the training at the same rate. Birds that were too young to follow the aircraft in concert with the others would otherwise turn back to the pen site due to fatigue. Once established, that behavior was often difficult to correct. The pen sites were approximately 2 km apart, which allowed up to three pilots to train the three groups at the same time, thus taking advantage of the relatively short training window associated with calm, early-morning winds. Birds were later socialized to form one cohesive group as previously described. Over 10 years, 209 Whooping Crane chicks were reintroduced at Necedah NWR using the UL method.

Pens and Facilities at White River Marsh State Wildlife Area, 2011–15

In 2010, the IWCRT recommended that no more Whooping Cranes should be reintroduced at the Necedah NWR due to low reproductive success. Poor reproduction was attributable, in part, to harassment of incubating birds by blood-feeding black flies (Simuliidae), which appear to cause nest abandonment (Converse et al., 2013; Urbanek et al., 2010a). The reintroduction site was moved to the White River Marsh State Wildlife Area (WRMSWA) in 2011. One pen site was used and was located at Lat 43.9063555, Long −89.1102375. It was constructed using similar methods to the pens at Necedah NWR and included a dry pen (150 m²), a wet pen (400 m²), and a grass runway 250 m long. Over 5 years, 37 Whooping Crane chicks were reintroduced at WRMSWA.

MIGRATION

How well the birds were socialized, their flight endurance, and their willingness to follow the aircraft were all factors that determined when the migration could begin. All of the necessary equipment was prepared and the migration team assembled prior to a target departure date. The actual day the migration began was dictated by the weather (mean departure date 10 October, range 28 September–17 October). During the trips south, the pilots were able to fly with the birds on an average of 20.87 days; (SE m 0.97; range 19–26 days). However, due to poor flying conditions and other variables, it took an average of 84.3 days (SE m 6.85; range 48–130 days) to complete the migration.

Original Route, 2001–07

The IWCRT and the USFWS selected Chassahowitzka NWR on the mid Gulf coast of Florida as the initial wintering site for the UL birds. OM developed the original route to follow the basic flight path of the Sandhill Cranes that migrate between Wisconsin and Florida. The route was modified to avoid overflying built-up areas such as Chicago and through controlled airspace around busy airports. From Necedah NWR, the birds were led due south, passing to the west of the Chicago metropolitan area.

The intent was to stop at an important Sandhill Crane staging area at Jasper-Pulaski Fish and Wildlife Area (JP), in northwestern Indiana. However, that was reconsidered because it required a drastic course change south of Chicago. Because of the JP location, OM staff was concerned that the birds might fail to make that course correction on their return migration and end up on the east side of Lake Michigan. It was believed that the route should be simple, making it easier for the birds to return on their own. Additionally, our arrival at JP with cranes following our low-flying aircraft could have disturbed the many Sandhill Cranes that use the area for feeding and resting.

From Indiana, the route led south and over the Appalachian Mountains in Kentucky and Tennessee, then through Georgia and into Florida. In total, it covered seven states and almost 2000 km. The route included 28 potential stopovers at 80–100 km intervals. With the exception of Muscatatuck NWR in Indiana and the Hiwassee State Wildlife Refuge in Tennessee, the stopovers were on private land.

The Appalachian Mountains presented a substantial obstacle during the migration. Cranes that normally soar in thermals to climb had to use powered flight (flap flying) behind the aircraft in order to clear the high ridges. Often they turned back due to exhaustion. Limited access roads and mountainous terrain made it difficult to track birds that dropped out. The forests and rocks left few landing options if the birds or pilots encountered a problem, and mountain flying conditions can be challenging. Even gentle winds can become problematic when they tumble over mountain peaks.

Modified Route, 2008–15

In 2008, OM acquired funding to explore a new route and look for options that allowed us to lead the birds around the Appalachian Mountains instead of over them and thereby increase the safety margin for the pilots and the birds (Fig. 21.3). To develop the route, the path was first flown in a conventional aircraft to lay out the course over flat, open land avoiding built-up areas and controlled airspace. It was flown again to sight potential stopover locations along that path. Then it was driven twice to assess the stopover locations, track down the landowners, and obtain their cooperation. That process took more than a month. The new route led south from Wisconsin, through Illinois, Kentucky, and Tennessee, then around the southern end of the Appalachian Mountains in Alabama. From there, the route led over the western region of Georgia and into Florida. The migration along the old route was often delayed while waiting for the lower winds and higher ceiling needed to clear mountain ridges. Avoiding those obstacles had the potential to shorten the duration of the migration; however, year-to-year weather variations made that determination impossible. Improvements in safety for the birds and the pilots made the change worthwhile.

Stopovers

The new route included up to 26 possible stopovers at 80–100 km intervals, all on private property. Not all of the stopovers were used each year based on whether headwinds slowed their progress or tailwinds pushed birds and aircraft along, allowing them to skip sites. The pen locations at each stopover were selected to provide isolation from buildings, roads, and the sounds of human activity. Pens were positioned next to a flat, open area where the aircraft could land and take off while remaining clear of power lines and utility poles. Open areas were selected to encourage the birds to make similar habitat choices once released. It was not possible to assemble the pens in wetland habitat and still be able to operate the aircraft. That meant that the birds were not able to water-roost for the duration of the migration; however, they were able to water-roost in the release pen over the winter.

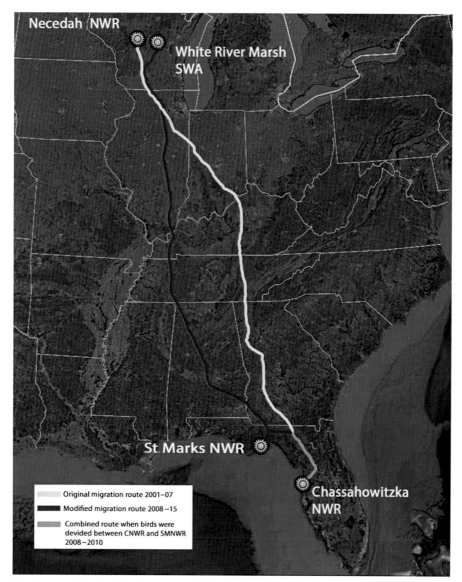

FIGURE 21.3   Routes of the ultralight-led migration from Wisconsin to Florida.

Portable Pens

OM fabricated two identical, portable pens built on custom trailers that could be set up by two people in ~30 min. Each carried 18 interlocking steel-frame panels measuring 1.8 m high and 3 m long. The panels were covered with nonabrasive wire mesh. When assembled, they connected to the back of the trailer, thereby creating a circular enclosure ~15 m in diameter. The first panels on either side of the back of the trailer were solid and, together with the trailer itself, created a windbreak, providing shelter in high winds. The enclosed trailer, painted in camouflage colors, also carried 120 L of fresh water and a supply of crane chow. The pen was top-netted and protected by three strands of electric fencer wire. Two strands were around the base of the pen and the third was set up 20 m out from the enclosure. Braided cable was used that was nearly invisible, making it more effective against predators. Most wild predators are wary of strange structures that show up in their territory. Feral dogs were a greater concern because they are familiar with human environments. In 15 years, these mobile pens were never breached by a predator.

The two travel pen trailers were identical and were used to leapfrog, with one pen occupied by the birds while the other waited at the next site. Occasionally, good weather allowed the pilots to skip one or more stopover sites. The ground crew would then have to clean and disassemble the pen just vacated by the birds, and relocate it to the new site. On those occasions, after the pilots had landed at the new site, they would keep the birds isolated in a preselected hiding place such as over a hill or behind a tree line. The pen was set up while the pilots held the birds out of sight so they did not witness this human activity. If the pilots were able to skip stopovers and cover 200 km or so, that meant the ground crew would have to drive well over that distance to reach the new site by road. Often the birds were held for several hours at a time.

Aircraft

Depending on the number of birds that were led south in a year, OM used up to four aircraft (minimum two) during the annual migrations. The aircraft used are commonly referred to as "trikes" because of the three-wheeled fuselage. They use a wing similar in design to a hang glider. The pilot moves the entire wing to maneuver the aircraft in a control system known as weight shift.

The primary challenge in flying with birds is finding an aircraft able to fly slowly enough to

match the bird's speed, yet remain stable and safe. Whooping Cranes fly at a cruise speed of 60 km/h (personal observations, confirmed using GPS), but the aircraft must be able to fly more slowly (50 km/h), allowing the birds to catch up if they fall behind.

In simplified terms, an airfoil must maintain a minimum forward speed in order to generate enough lift to keep the aircraft flying. If the airfoil is slowed below that critical speed, it will begin to fall in what is known as a wing stall [Federal Aviation Administration (FAA) Aviation Manuals]. That transition from flying to falling can occur with little warning and is dangerous, particularly at low altitude. As an example, the birds needed only a few steps to take off; however, the aircraft must reach its minimum flight speed to get airborne. That generally placed the aircraft 50 m ahead of the birds. If they could not close that distance, the birds would often turn back to the pen. The pilots regularly flew close to the stall speed while turning the aircraft to allow the birds to cut the corner and catch up. Although the aircraft performance matched the flight abilities of Whooping Cranes, a wider flight performance envelope would have increased the effectiveness of the training and the safety margin. However, no such aircraft was available. Other considerations making weight-shift Light Sport Aircraft (LSA) appropriate for leading Whooping Cranes included:

- Open configuration allows the birds to see the costumed pilot.
- The "pusher" configuration (engine and propeller at the back) made it possible to protect the birds from the propeller using a custom-fabricated prop guard.
- Unlike conventional aircraft control systems with many components, the wing of a trike pivots on one moving part, decreasing the threat of mechanical failure.
- They are intuitive and simple to fly, with forgiving stall characteristics.
- They can be quickly disassembled for easy transport.

Initially, OM used Cosmos, Phase II aircraft equipped with 19 m^2 wings and powered by 50 hp engines. In compliance with an FAA-issued exemption, OM purchased North Wing, Apache trikes equipped with 20 m^2 wings and powered by the same engines. Both of these aircraft types had custom-built prop guards, GPS navigation systems, aviation communication radios, and amplifiers to broadcast the brood call loud enough to be heard over the sound of the engine. All the aircraft were registered to OM, in the FAA's LSA category.

Although the aircraft used by OM are commonly referred to as ultralights, they are categorized by the FAA as LSA. They are limited to recreational activities only and cannot be used for any commercial purpose except to provide flight instruction. They are operated by pilots holding a Light Sport Pilot Certificate or higher. In 2010, OM was questioned by the FAA regarding the commercial use of our aircraft. The FAA initially agreed with OM's explanation that the staff was paid for their other bird-related duties and that the limited flying time was volunteered. However, the FAA later determined that a portion of the job responsibilities of OM staff pilots was to fly with the birds. In the FAA's judgment, that constituted the commercial use of an LSA. The FAA continued to work with OM and provided long-term exemptions with the provision that the pilots upgrade their licenses from Light Sport Pilot Certificates to Private Pilot Licences and that OM purchase Special LSAs. These machines are similar in design but, rather than being owner-maintained, all work on these new aircraft must be performed and endorsed by an FAA-certified mechanic.

In 15 years of flying with Whooping Cranes, OM led birds over 27,000 km, often with as many as four aircraft. In addition, the aircraft were used to condition and exercise the birds on 1,287 days. In all of that time airborne, no accidents or incidents occurred that legally required reporting to the FAA or the National Transportation Safety Board (NTSB). The only accident occurred when the volunteer top-cover pilots exhausted the fuel supply in their Cessna 182 and performed an emergency landing in a plowed field. Fortunately, no injuries resulted although the aircraft was severely damaged.

Migration Team

Eight to 12 staff and volunteers made up the migration team. Up to four pilots and aircraft (minimum two) led the birds. The lead pilot was selected on a rotating basis and the others flew in the chase positions. The job of the chase pilots was to assist if the birds were reluctant to follow, broke up into smaller groups, or landed away from the pens. The lead pilot for the day would land next to the isolated pen and signal for the release of the birds. Two experienced handlers released the birds from the pen at the beginning of each migration leg. During the flight, two experienced trackers attempted to stay within visual or radio range while following the aircraft and the birds on the roads below. Their job was to track and collect dropout birds or to assist if weather or mechanical issues arose. An outreach representative was positioned at prearranged flyover sites, which gave the public opportunities to watch the birds and aircraft fly overhead. Often, 50 or more people gathered in the early morning cold to watch. Other team members assisted with cleaning and assembling the travel pens and moving the camp to the next stopover location. In addition to their primary assignments, team members drove one of the OM accommodation motor homes or one of the trucks pulling the equipment trailers.

From 2001 to 2008, OM benefited from the services of volunteer top-cover pilots who donated their time and the use of their Cessna 182 aircraft for the entire migration period. With one volunteer acting as pilot and the other as observer, they circled 150 m above the trikes and kept track of any wayward birds. When the OM pilots were flying with large groups of birds, having an additional airborne observer was helpful, particularly if the flock broke up and was following multiple trikes. The top-cover pilots communicated with Air Traffic Control using an FAA service known as Flight Following, to help keep the OM aircraft clear of controlled airspace. The top-cover pilots could also fly ahead to check weather conditions or to provide the ground crew with directions to the locations where dropout birds had landed.

Leading Whooping Cranes in Flight

Conditioning the birds to follow the aircraft required many months of early imprinting, ground taxi exercise, and flying to extend their endurance. Successfully leading them on their first migration depended on the birds' inclination to follow the pilots and aircraft. The birds appeared more willing to follow the aircraft when all of the handlers worked to build trust with them. The birds became wary and reluctant to follow when they were captured and held for health examinations or banding, herded aggressively into or out of the pens, or made to be fearful of the handlers for any reason. Within each group there appeared to be a dominance structure that evolved as the chicks matured (Piper, 1997). Aggressive individuals would periodically challenge the costumed handlers, who had to carefully regulate their responses. A submissive reaction could lead to increased aggression from the challenger with various consequences, including injury to the handler, the bird, or both. Birds that displayed dominance over the handlers appeared more independent and were less likely to follow. Too aggressive a response from the costumed handler, and the attacking bird could become fearful and less likely to follow (personal observation). A variation of the dominance structure was also noticed when flying. Aggressive birds would occasionally attempt to force their way ahead of the aircraft and assume the leadership of the flock, often taking them off the intended course. The pilot had to speed ahead to retake the lead

and reassert dominance while being mindful of the slow birds in the back of the formation that could be left behind. Despite the efforts of the handlers, the birds' willingness to follow the aircraft was tenuous and could be disrupted by many factors, including temperature, visibility, humidity, turbulence, density altitude, wind direction, and the duration of the birds' stay at one location. There were occasions when the pilots had to spend more than an hour attempting to lead the birds away from the pen before getting them on course for the next stopover.

Whooping Cranes have high aspect ratio wings (ratio of span to mean chord) that are effective for soaring (Alvarez et al., 2001). During long-distance flights, Whooping Cranes often gain altitude by soaring in thermals or rising warm air caused by differential solar heating of the earth's surface (Oke, 1987). Between those thermal flights, they will often glide or fly in lines or chevrons to take advantage of wingtip vortices produced by the birds in the lead (Hedenstrom, 1993; Kerlinger and Moore, 1989). In this manner, they can cover long distances without expending much energy, as long as the sun keeps producing thermals and the wind blows in the right direction.

The aircraft used by OM did not have the fuel endurance or the agility to soar as effectively as cranes, so the pilots avoided turbulent, soaring conditions during midday and instead flew with the birds in the calm, colder air of early morning (Fig. 21.4). In these calm conditions, the birds learned to fly at the wingtips of the aircraft in order to benefit from the vortices generated by induced drag (Green, 1995). In smooth air, when the wing was stable, the lead bird would glide within one meter of the wingtip, flapping occasionally to keep up or to reposition. The first bird in line immediately behind the wingtip did the least work while successive birds received less assistance the farther back they were in the order. As long as those smooth conditions prevailed, the pilots could lead the birds for up to the 3.5 h fuel endurance of the aircraft. If the

FIGURE 21.4   Flock of birds flying behind the UL in early morning. *Source: Photo credit: Operation Migration.*

aircraft wing was unstable in turbulent conditions, the birds expended more energy adjusting to that movement than they saved by surfing on it. After ~45 min in rough air that forced them to engage in powered flight for much of the time, they began to show signs of fatigue (personal observation). They would pant with beaks open and splay their toes for increased cooling. They would sometimes move from wingtip to wingtip appearing to look for an easier spot to fly. Eventually, they would fall behind or begin to descend. Once they lost their position on the wingtip vortices, they had to work much harder to keep up, adding to their fatigue. The lead pilot could descend to collect a bird in these situations, but that often meant dropping back into the rough air closer to the surface and increasing the risk of losing all of the birds. Instead, the chase pilot would move in. Once close enough, the tired bird would generally move into position directly behind the chase aircraft's wingtip. From there it could garner the benefit from the vortices and rest. Occasionally, birds were so exhausted that they would ignore the chase plane and glide down to land. The chase pilot would then guide the ground crew to the bird's landing

site. If the conditions allowed, the chase pilot might land with the bird and take off again after it recovered. On flight days when the weather deteriorated and several birds dropped out, all the ground crew members participated in collecting the birds and relocating them to the next stopover site. Only a few birds per year flew the entire migration without dropping out and having to be transported in a crate.

The aircraft used by OM were legal to fly only at or after sunrise (FAA – Visual Flight Rules). The calm conditions of early morning last only until the sun begins to heat the earth's surface, causing thermal activity. That meant the flight window was short, never lasting more than 3 or 4 hours. Ideal flying conditions for leading birds included overcast ceilings to inhibit the generation of thermal activity, low humidity to minimize fog, and winds that were calm on the surface with a slight (~16 km/h) tailwind aloft.

Cold air is denser and improves the performance of the wings of both the aircraft and the birds. Because it is compressed, it contains less moisture and more oxygen for breathing (FAA Handbook of Aeronautical Knowledge). Additionally, the birds did not overheat on cold flights. Because many factors affected the flight performance of the birds, it was difficult to quantify the benefits of cold air. However, in temperatures above 15°C, their flight endurance was reduced by as much as 50%, especially in humid conditions (personal observation). OM pilots flew in temperatures as low as −10°C.

Over the years, the pilots developed a number of methods designed to encourage the birds to follow the aircraft. As an example, when the birds turned back to the pen after takeoff, the pilots could retake the lead by powering ahead of the dominant bird and redirecting the flock on course. Alternatively, the aircraft could be maneuvered in such a way as to break up the flight order of the birds to remove the most dominant bird from the lead position. Often, with a different bird in the front of the formation, directly behind the aircraft wingtip, the other birds would follow more efficiently. The pilots could also break up the flock into smaller groups, with each aircraft leading a few birds. All flight maneuvers were coordinated over the aviation radios, which were monitored by the ground and tracking crews. Once the birds took off with the aircraft, the ground crew that released them hid inside the pen trailer to avoid causing a distraction. When all attempts to prevent the birds from returning to the pen failed, the pilots had the option of directing the ground crew to exit the trailer, cover themselves with large plastic tarpaulins, and walk around the pen area. The birds appeared fearful of those unfamiliar, moving shapes and they would not land (personal observation). That gave the pilots another opportunity to collect the birds and attempt to lead them back on course. Occasionally, the pilots would intentionally lead a group of reluctant birds back over the pen, allowing them to see the tarp-covered handlers from the air. That appeared to encourage the birds to follow the aircraft. Once on course, the pilots would begin to climb, typically at no more than 15 m per min, leveling off as needed to allow the birds to rest. As the morning became warmer, the associated

thermal activity also began to rise. A slow continuous climb up to 1200–1500 m often ensured smoother conditions and better opportunity for the birds to benefit from the wingtip vortices.

OVERWINTERING

Chassahowitzka NWR

The IWCRT and the USFWS initially selected Chassahowitzka NWR on the Gulf coast of Florida as the wintering site for the EMP. That site was used exclusively until 2007. During those years, a number of adult birds that had already been taught to migrate often stopped at the release pen at Chassahowitzka NWR during their fall migration. Because they were older, larger, and sometimes aggressive, the WCEP winter monitoring team thought they posed a risk to the chicks. Because there were adult birds at the Chassahowitzka pen site in 2006, OM pilots were asked to short stop the migration 40 km to the northeast at a temporary site called Halpata Tastanaki Preserve near Dunnellon, Florida. The birds were held at that site from 19 December 2006 to 12 January 2007 until the older birds cleared the pen site at Chassahowitzka.

Later that winter, some of the older birds returned to the Chassahowitzka NWR site. To keep the chicks separated from the older birds, the WCEP winter monitoring team held the chicks in a top-netted pen. On 2 February 2007, a winter storm of unpredicted severity struck central Florida. Storm-driven tides raised the water level in the pen and 17 of the 18 birds drowned. The one surviving bird managed to escape the pen but was later killed by a predator at the Halpata site.

St. Marks NWR

The loss of 17 birds in one event prompted the WCEP to consider other options. After extensive research, St. Marks NWR (SMNWR) was selected, in part because of its open marsh, low vegetation, limited human use, and lack of storm-driven tides. St. Marks NWR is on the Gulf coast of the Florida panhandle, south of Tallahassee.

From 2008 until 2010, the flocks were divided between Chassahowitzka NWR and St. Marks NWR. From 2012 to 2015, all the birds wintered at St. Marks NWR.

Overwintering Facilities and Management

Large release pens were constructed in areas closed to the public at both Chassahowitzka (Lat 28.7294195, Long −82.6464937) and St. Marks (Lat 30.1069221, Long −84.2835622). Three-meter-tall plastic fence enclosed approximately 1.43 ha of open marsh. The pens included a covered feeding station, fresh water supply, and areas for the birds to water-roost. The pen were protected by three strands of electric fencer wire but were not top netted.

On average, the migration terminated at the wintering site on 24 December (range 23 November–6 February). When the birds arrived, they were held in a top-netted portion of the larger pen for an acclimation period, and while tracking devices and permanent leg bands were attached. Thereafter they were released to the larger open pen. They were able to fly out into the marsh during the day to forage. In the evening, the birds were called back into the pen using an amplified brood call. If the birds failed to return to the safety of the pen in the evening, the OM winter monitoring team would locate them and attempt to lead or to flush them back into the pen. If they would not follow and were not in good roosting habitat, team members would often spend the night with them to ensure their safety.

In this manner, the birds underwent a soft release. They were checked at least twice daily, once in the morning and once in the evening prior to roost. Although the birds were provided with crane chow ad libitum, efforts were made to minimize human contact. During this time,

they also learned to use thermals when flying on their own. In the spring, the birds made an unassisted northern migration, leaving in March or April, and thereafter were considered fully released.

OUTCOMES AND DISCUSSION

Effectiveness of the UL Release Method

From 2001 to 2015, 167 young-of-year Whooping Cranes were released into the EMP using the UL method. Three birds (1.8%) were removed from the project due to their encroachment into anthropogenic environments:

- In 2009, after frequenting an ethanol plant in Wisconsin to feed on spilled corn, male 10_07 (designating the 10th bird hatched in year 2007) was removed from the population because his behavior appeared to be encouraging other Whooping Cranes to do the same.
- Male 5_01 repeatedly flew into the Homosassa Springs Wildlife State Park in Florida to associate with a captive female Whooping Crane. After several attempts to discourage this behavior, the bird was brought into captivity in 2011 and eventually paired with that female.
- Male 1_01, the oldest bird in the EMP, was captured in 2014 after he could not be discouraged from using areas close to the active runways at Volk Field Air National Guard Base in Wisconsin.

Only one bird was not retrieved after dropping out of the migration. In 2011, the birds turned back or dropped out during a failed attempt to leave the first stopover location in Wisconsin. They landed in fields when the wind picked up and the air became rough. The radio tracking device on number 2_11 (hatch year 2011) failed and she could not be located despite extensive ground and aerial searches. Later that day, she was observed with a large flock of Sandhill Cranes. Typically, the birds could easily be approached by costumed handlers; in that case, however, she would flush with the Sandhill Cranes when the handlers approached. That female completed the migration to Florida with Sandhill Cranes and returned to central Wisconsin in the spring of 2012. She was never recaptured for rebanding and was last reported in Florida in April 2013. Her fate was not determined with certainty, but she is presumed dead.

Weather conditions affected the duration of the migration, and over 15 years migration times increased from 48 to 115 days (Fig. 21.5). On average, it took 20.8 flying days (averaging 32.55 accumulated flight hours) to complete. However, it took an average of 75 days to get those good flying mornings. The decisions to

fly or not were made by the pilots. It is possible that they gradually became more discriminating about the weather and opted for more optimal conditions. However, based on comparisons of journal entries, weather history, and the observations of long-term crewmembers, that did not appear to be the case. Poor flying conditions, particularly winds from the south, increased in frequency over the period of 2001 and 2015. That could be attributable to cyclical weather patterns or climate change.

In 2014, the early onset of winter weather caused delays in the migration. By mid-November, the team had covered only 83 km in 36 days due to consistent winds from the south. That prompted the decision by WCEP to approve the relocation of the birds south of that weather system to the halfway point of the migration route in

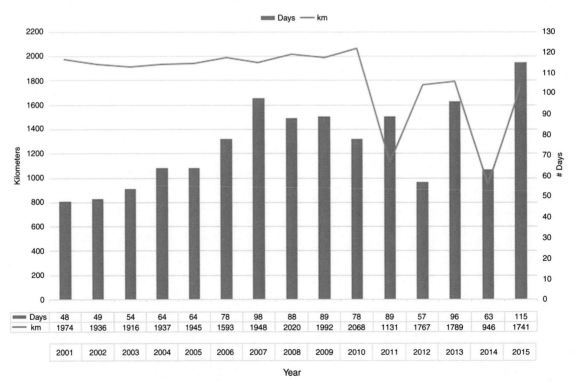

	2001	2002	2003	2004	2005	2006	2007	2008	2009	2010	2011	2012	2013	2014	2015
Days	48	49	54	64	64	78	98	88	89	78	89	57	96	63	115
km	1974	1936	1916	1937	1945	1593	1948	2020	1992	2068	1131	1767	1789	946	1741

FIGURE 21.5 Length and duration of ultralight-led migration, 2001–15. In 2011, the migration was terminated at Wheeler NWR in northern Alabama; in 2014, migration was delayed in the northern portions of the route so the birds were transported to Tennessee, from where the migration flights continued to Florida.

Tennessee. From there, rapid progress was made and the remaining 910 km to St. Marks NWR were covered in 16 days. This unprecedented move caused a break in the birds' knowledge of the route south and inhibited their ability to return to the reintroduction site in Wisconsin. WCEP developed a monitoring and intervention protocol, and OM established a team to track the birds north. One bird successfully followed an experienced adult back to the Wisconsin reintroduction site. The remaining five were monitored for several weeks. In southern Illinois, they appeared to become sedentary and ignored favorable flying conditions as if their northern migration was complete. In May 2015, they were captured and relocated to the WRMSWA reintroduction site. The following fall they successfully migrated to Florida.

Demography and Challenges with Reproductive Success

For birds reintroduced at Necedah NWR via the UL method, annual survivorship from hatching at captive centers through the transport to Necedah (mean arrival date 27 June), UL-led migration south (mean migration arrival date 24 December), and subsequent springtime soft-release averaged 85% (95% CI = 74–94%; Servanty et al., 2014). One hundred percent of released individuals were then confirmed moving north on their first spring migration. Thereafter, annual survival of UL birds exceeded 92%, regardless of age or state (i.e., unpaired, paired, nonbreeding, and breeding; Servanty et al., 2014). Moreover, birds involved in the UL program that later paired, established territories, and attempted to reproduce did so unilaterally with conspecifics – heterospecific interactions with Sandhill Cranes remained unremarkable and required no intervention.

The reintroduction effort generated promising results for establishing a viable population. Demographic parameters, including survival, habitat selection, philopatry, pair bonding, territoriality, and preincubation reproductive behavior, broadly met or exceeded our objectives and reported vital rates in growing populations of cranes (Converse et al., Chapter 7, this volume; Servanty et al., 2014; Urbanek et al., 2011). However, recruitment fell surprisingly short of compensating for apparent mortality. Such low recruitment appeared to be caused, in part, by the presence of blood-feeding black flies (Simuliidae) that caused nest abandonment during incubation (Converse et al., Chapter 8, this volume; Urbanek et al., 2010a).

In addition to nest abandonment, prefledging chick mortality was also high at Necedah NWR (Converse et al., Chapter 8, this volume). Between 2001 and 2010, 132 Whooping Cranes were reintroduced at Necedah NWR using the UL method. As of December 2015 there were 27 reproductive pairs using that habitat. There have been 161 nesting pair-years producing 64 chicks; however, only nine of those chicks survived to December 1 of the year of hatching. Beginning in 2005, WCEP began the direct autumn release (DAR) reintroduction method (Hartup, Chapter 20, this volume). Between 2005 and 2010, 43 juveniles were raised in captivity and released in the fall in proximity to experienced Whooping Cranes and/or Sandhill Cranes. These DAR birds account for a portion of the reproductive pairs and nesting numbers listed earlier. The nine chicks that survived to fledge, however, were the product of breeding pairs in which both birds were introduced using the UL method.

The low fecundity in the EMP prompted the IWCRT to end releases at Necedah NWR until the cause of the problem was identified and mitigated. In response, the WCEP identified a wetland complex that had low densities of the black fly species thought to be associated with nest abandonment. The WRMSWA is part of that wetland complex, located 86 km east of Necedah NWR. At this writing, the 37 birds that were reintroduced at WRMSWA using the UL method were still too young to evaluate whether birds reintroduced at WRMSWA will demonstrate better breeding success than those reintroduced at Necedah NWR.

In the first 5 years of the reintroduction program at Necedah NWR, 70 birds were released compared to only 37 that were released during the first 5 years at WRMSWA. The smaller group sizes were due in part to lower reproduction rates at some of the captive breeding centers and the IWCRT's decision to approve the reintroduction of a nonmigratory population in Louisiana (LNMP) (King et al., Chapter 22, this volume). Beginning in 2011, all of the chicks produced in captivity were divided between the LNMP and the EMP.

Education and Outreach

The image of modern aircraft leading endangered birds along a safe migration route generated worldwide interest. It was covered by national media outlets like CBS's *60 Minutes*, ABC's *20/20*, the Discovery Channel, and *National Geographic*. It has been the subject of several documentaries and covered by hundreds of media outlets such as the *Washington Post*, the *Chicago Tribune*, and the *New York Times*. An OM aircraft is on permanent display at the Smithsonian National Air and Space Museum and another at Disney's Animal Kingdom in Florida. It has been featured in at least 11 books and referenced in several education textbooks. In terms of education and public interest, live cameras, blogs updated daily, an extensive social media network, and an online education program that reached over a million students a year made the EMP one of the most recognized wildlife reintroductions in U.S. history.

SUMMARY AND OUTLOOK

At the end of the 2015 field season, Region 3 of the USFWS ended the use of the UL method for reintroducing birds into the EMP. They hypothesized that birds raised by people do not learn nurturing skills that might normally be taught by their parents and that lack of attentiveness was the cause of nest abandonment and high prefledge mortality (Converse et al., Chapter 8, this volume; Fasbender et al., 2018). The UL method was deemed too artificial. In recent years, many of the Whooping Cranes in the EMP wintered as far north as Goose Pond in Indiana (39°N latitude; see also Mueller et al., Chapter 11, this volume). Had the use of the aircraft-led release method been continued, the migration route could easily have been shortened to use wintering locations farther north.

To date, causal links between low recruitment at Necedah NWR and the artificiality of the UL method or hand or costume rearing are unsubstantiated by data. This paucity of data obfuscates effective management and policy formation. For example, black flies have been linked to poor reproduction (Converse et al., Chapter 8, this volume). However, apart from a 3-year experimental application of the natural bacterium *Bacillus thuringiensis israelensis* (*Bti*), there have been relatively few resources dedicated to verifying the apparent correlation and/or simply reducing Necedah NWR's ubiquitous black flies as a precautionary measure to facilitate breeding productivity in the Whooping Cranes on the refuge. Additionally, parental inattentiveness may lead to high prefledge chick mortality; however, this mortality could also be the result of unusually high predation pressure(s) or other environmental factors at Necedah NWR. Considering the investment made by WCEP to reintroduce birds at Necedah NWR and the value of that resource, it is my opinion that efforts should be made to understand the problem and mitigate it with habitat management and/or black fly population suppression. The relocation of the reintroduction site to WRMSWA will ultimately allow for testing of the hypothesis that reproductive success is related to habitat (Converse et al., Chapter 8, this volume). Future releases into the EMP using parent-reared release techniques (Hartup, Chapter 20, this volume) may assist in testing the costume-rearing hypothesis.

Many wildlife reintroduction projects have noted significant differences in the viability of wild-sourced animals over those produced by captive propagation programs (Fischer and Lindenmayer 2000; Roche et al., 2008). Whooping Crane reintroduction projects beginning with the Florida Nonmigratory Population (FNMP) (Folk et al., 2008) and including the EMP and the LNMP have used birds from captive sources. Although it is still too early to evaluate recruitment in the LNMP, both the FNMP and the EMP have suffered from low fecundity. It has been hypothesized that the genetic composition of the reintroduced populations has been influenced by captive selection, leading to poor reproductive success post-release (Converse et al., Chapter 8, this volume). Considering the cost of operating the five Whooping Crane captive breeding programs in North America and the investments made by the many agencies that have been reintroducing these birds, further study of the effects of captive selection is warranted.

Acknowledgments

This chapter could not have been written without the important work of the Whooping Crane Eastern Partnership, which was established in 1999 to reintroduce a migratory population of Whooping Cranes to eastern North America. The nine founding members are the Canada-U.S. Whooping Crane Recovery Team, the U.S. Fish and Wildlife Service, USGS Patuxent Wildlife Research Center, USGS National Wildlife Health Center, the Wisconsin Department of Natural Resources, Operation Migration Inc., the International Crane Foundation, the National Fish and Wildlife Foundation, and the Natural Resources Foundation of Wisconsin.

OM is grateful for the generosity of so many landowners who hosted the birds and our migration team for so many years. That was no simple task. The area where the birds were secluded on their property was closed to visitors, including the owners themselves. We often parked our aircraft, motor homes, and vehicles on their property, and used electricity, water, and internet connections. Our arrival was dependent on the weather and was as unpredictable as the length of our stay. We are forever grateful to the late Terry Kohler for his generous financial support and the 32 round-trip flights his aircraft and crew made between Maryland and Wisconsin to safely deliver 209 Whooping Crane colts. We owe a great debt to our supporters who not only provided financial sustenance but also moral support when it was really needed. We are grateful to corporate sponsors like the National Fish and Wildlife Foundation/Southern Company partnership and Disney's Worldwide Conservation Fund. We are also thankful to the staff at Disney's Animal Kingdom who provided veterinary care for the birds and winter monitoring assistance. In addition, there are too many volunteers to mention individually who generously assisted with training and migrations, dedicating weeks or months of their lives away from home.

Leading the birds south with an aircraft is not a simple undertaking. It required specialized equipment and piloting skills honed over many years of flying with birds. The people flying the aircraft were not only pilots but also experienced bird handlers, responsible for their own safety and the well-being of the birds, both in the air and on the ground. Together with the OM ground crew, they had to understand and manipulate the dominance structure of the birds and develop methods to mitigate bad habits. They had to learn how to encourage the birds to follow the aircraft and influence their behavior without causing stress, while keeping them isolated from human encounters. The OM crew had the stubborn dedication that kept them flying in adverse conditions or standing knee deep in a cold marsh for hours at a time. OM is extremely proud of our team of people who worked closely with the birds and made this project a success, despite the self-sacrifice that was required.

References

Allen, R.P., 1952. The Whooping Crane. Research Report 3. National Audubon Society, New York, p. 246.

Alvarez, J.C., Meseguer, J., Meseguer, E., Perez, A., 2001. On the Role of the Alula in the Steady Flight of Birds. Ardeola 48 (2), 161–173.

Canadian Wildlife Service and U.S. Fish and Wildlife Service, 2005. International Recovery Plan for the Whooping Crane. Recovery of Nationally Endangered Wildlife (RENEW), Ottawa, and U.S. Fish and Wildlife Service, Albuquerque, NM.

Converse, S.J., Royle, J.A., Adler, P.H., Urbanek, R.P., Barzen, J.A, 2013. A hierarchical nest survival model integrating incomplete temporally varying covariates. Ecol. Evol. 3 (13), 4439–4447.

Converse, S.J., Servanty, S., Moore, C.T., Runge, M.C., 2018. Population dynamics of reintroduced whooping cranes (Chapter 7). In: French, Jr., J.B., Converse, S.J., Austin, J.E. (Eds.), Whooping Cranes: Biology and Conservation. Biodiversity of the World: Conservation from Genes to Landscapes. Academic Press, San Diego, CA.

Converse, S.J., Strobel, B.N., Barzen, J.A., 2018. Reproductive failure in the Eastern Migratory Population: the interaction of research and management. In: French, Jr., J.B., Converse, S.J., Austin, J.E. (Eds.), Whooping Cranes: Biology and Conservation. Biodiversity of the World: Conservation from Genes to Landscapes. Academic Press, San Diego, CA.

Cui, E., Zhang, X., 2010. Ground effect aerodynamics. In: Richard, B., Wei, S. (Eds.), Encyclopedia of Aerospace Engineering. John Wiley & Sons, Chichester, West Sussex, UK, pp. 245–256.

Duff, J.W., Lishman, W.A., Clark, D.A., Gee, G.F., Sprague, D.T., Ellis, D.H., 2000. Promoting wildness in sandhill cranes conditioned to follow an ultralight aircraft. Proceedings of the North American Crane Workshop 8, 115–121.

Ellis, D.H., Lewis, J.C., Gee, G.F., Smith, D.G., 1992. Population recovery of the whooping crane with emphasis on reintroduction efforts: past and future. Proceedings of the North American Crane Workshop 6, 142–150.

Fasbender, P., Staller, D., Harrell, W., Brooks, B., Strobel, B., Warner, S., 2018. The Eastern Migratory Population of Whooping Cranes: FWS Vision for the Next 5-year Strategic Plan. Available from: https://www.fws.gov/midwest/whoopingcrane/pdf/FWS5YrVisionDoc09222015.pdf; accessed 15 January 2017.

Fischer, J., Lindenmayer, D.P., 2000. An assessment of published results of animal relocation. Biol. Conserv. 96 (1), 1–11.

Folk, M.J., Nesbitt, S.A., Parker, J.M., Spalding, M.G., Baynes, S.B., Candelora, K.L., 2008. Current status of nonmigratory whooping cranes in Florida. Proceedings of the North American Crane Workshop 10, 7–12.

Glenn, T.C., Stephan, W., Braun, M.J., 1999. Effects of a population bottleneck on whooping crane mitochondrial DNA variation. Conserv. Biol. 13 (5), 1097–1107.

Green, S.I., 1995. Wing Tip Vortices. In: Green, S.I. (Ed.), Fluid Vortices. Fluid Mechanics and Its Applications, vol. 30. Kluwer Academic Publishers, Dordrecht, Netherlands, pp. 427–469.

Harrell, W., 2014. Report on Whooping Crane Recovery Activities (2013 Breeding Season – 2014 Spring Migration). U.S. Fish and Wildlife Service. Available at: https://www.google.ca/search?q=Harrell%2C+Wade.+Report+on+Whooping+crane+Recovery+Activities+2013+Breeding+season&oq=Harrell%2C+Wade.+Report+on+Whooping+crane+Recovery+Activities+2013+Breeding+season+&aqs=chrome..69i57.41630j0j7&sourceid=chrome&ie=UTF-8, accessed 22 March 2018.

Hartup, B.K., 2018. Rearing and release methods for reintroduction of captive-reared whooping cranes (Chapter 20). In: French, Jr., J.B., Converse, S.J., Austin, J.E. (Eds.), Whooping Cranes: Biology and Conservation. Biodiversity of the World: Conservation from Genes to Landscapes. Academic Press, San Diego, CA.

Hedenstrom, A., 1993. Migration by soaring or flapping flight in birds: the relative importance of energy cost and speed. Philos. Trans. R. Soc. B 342 (1302), 353–361.

Horwich, R.H., 1989. Use of surrogate parental models and age periods in a successful release of hand-reared sandhill cranes. Zoo Biol. 8 (4), 379–390.

Horwich, R.H., 2001. Developing a migratory whooping crane flock. Proceedings of the North American Crane Workshop 8, 85–95.

Kerlinger, P., Moore, F.R., 1989. Atmospheric structure and avian migration. In: Current Ornithology, vol. 6. Plenum Press, New York, pp. 109–142.

King, S.L., Selman, W., Vasseur, P.L., Zimorski, S.E., 2018. Louisiana nonmigratory whooping crane reintroduction (Chapter 22). In: French, Jr., J.B., Converse, S.J., Austin, J.E. (Eds.), Whooping Cranes: Biology and Conservation. Biodiversity of the World: Conservation from Genes to Landscapes. Academic Press, San Diego, CA.

Lishman, W.A., Teets, T.I., Duff, J.W., Sladen, W.J.L., Shire, G.G., Goolsby, K.M., Bezner Kerr, W.A., Urbanek, R.P., 1997. A reintroduction technique for migratory birds: leading Canada geese and isolation-reared sandhill cranes with ultralight aircraft. Proceedings of the North American Crane Workshop 7, 96–104.

Mueller, T., Teitelbaum, C.S., Fagan, W.F., Converse, S.J., 2018. Movement ecology of reintroduced migratory whooping cranes (Chapter 11). In: French, Jr., J.B., Converse, S.J., Austin, J.E. (Eds.), Whooping Cranes: Biology and Conservation. Biodiversity of the World: Conservation from Genes to Landscapes. Academic Press, San Diego, CA.

Newberry, R.C., 1995. Environmental enrichment increasing the biological relevance of captive environments. Appl. Anim. Behav. Sci. 44 (2–4), 229–243.

Oke, T.R., 1987. Boundary Layer Climates, second ed. Methuen, London.

Operation Migration Protocol 2001, Available at: http://operationmigration.org/protocol.asp; accessed 3 March 2018.

Piper, W.H., 1997. Social dominance in birds, early findings and new horizons. In: Nolan, Jr., V., Ketterson, E.D., Thompson, C.F. (Eds.), Current Ornithology, vol. 14, Plenum Press, New York, pp. 125–187.

Roche, E.A., Cuthbert, F.J., Arnold, T.W., 2008. Relative fitness of wild and captive-reared piping plovers: does egg salvage contribute to recovery of the endangered Great Lakes population? Biol. Conserv. 141 (12), 3079–3088.

Servanty, S., Converse, S.J., Bailey, L.L., 2014. Demography of a reintroduced population: moving toward management models for an endangered species, the whooping crane. Ecol. Appl. 24 (5), 927–937.

Urbanek, R.P., Zimorski, S.E., Fasoli, A.M., Szyszkoski, E.K., 2010a. Nest desertion in a reintroduced population of migratory whooping cranes. Proceedings of the North American Crane Workshop 11, 133–141.

Urbanek, R.P., Zimorski, S.E., Szyszkoski, E.K., Wellington, M.M., 2011. Ten-year status of the eastern migratory whooping crane reintroduction. Proceedings of the North American Crane Workshop 12, 33–42.

Louisiana Nonmigratory Whooping Crane Reintroduction

Sammy L. King*, Will Selman**, Phillip L. Vasseur[†],
Sara E. Zimorski[‡]

*U.S. Geological Survey, Louisiana Cooperative Fish and Wildlife Research Unit,
Baton Rouge, LA, United States
**Biology Department, Millsaps College, Jackson, MS, United States
[†]Louisiana Department of Wildlife and Fisheries, Rockefeller Wildlife Refuge,
Grand Chenier, LA, United States
[‡]International Crane Foundation, Baraboo, WI, United States;
Louisiana Department of Wildlife and Fisheries, Gueydan, LA, United States

INTRODUCTION

Historically, it is speculated that Whooping Cranes (*Grus americana*) were more abundant during the winter in Louisiana than anywhere else in North America (Allen, 1952). Louisiana supported both resident and migratory Whooping Cranes. As with the trends range-wide, the numbers of resident and migratory Whooping Cranes began to decline in the late 1800s; the last Whooping Crane was removed from Louisiana in 1950.

In February 2011, the first of several releases of Whooping Cranes, in their former habitat, was carried out in Louisiana as part of a reintroduction program. The objectives of this chapter are to (1) discuss the history of Whooping Cranes in Louisiana, (2) describe the historic and current conditions of the ecological systems that they historically used, and (3) identify potential factors that may limit the success of the Louisiana reintroduction program.

HISTORY OF WHOOPING CRANES IN LOUISIANA

The most important historical Whooping Crane winter habitats in Louisiana were the Great Southwest Prairies (hereafter, coastal prairies) and coastal marshes of the Chenier Plain of southwestern Louisiana (Allen, 1952) (Fig. 22.1). Most of what is known about the historical presence of Whooping Cranes in southwest Louisiana comes from the detailed interviews conducted in 1947 by Mr. John Lynch, a U.S. Fish and Wildlife Service biologist. The interviews were of local people

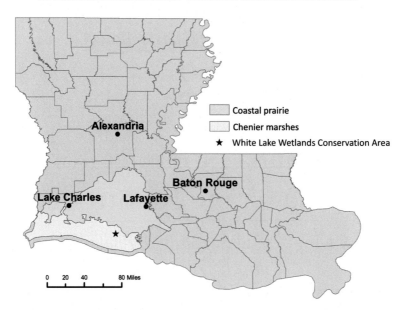

FIGURE 22.1   The coastal prairie and coastal marshes of Louisiana. Data are from U.S. Geological Survey and the Louisiana Department of Transportation and Development. *Source: Figure by Whitney Kroschel.*

(particularly trappers and individuals who lived near marsh habitats) in southwestern Louisiana, who shared local ecological knowledge with Lynch, specifically the habitats Whooping Cranes used and when they were observed through-out the annual cycle (summary by Gomez et al., 2005). Based on the information Lynch gathered, which was thereafter communicated to Mr. R.P. Allen, Allen (1952) noted important habitat types in Louisiana as coastal prairie, panicum (*Panicum hemitomon*) or paille-fine marsh, prairie swale and prairie marsh, sea rim and brackish marsh, and sawgrass (*Cladium jamaicense*) and deep marsh. Sawgrass and deep marsh habitats were identified as the least important of these habitats, and sea rim marshes supported only migratory Whooping Cranes. Furthermore, it appears from Lynch's interviews that nests of Whooping Cranes were found only in freshwater marshes and coastal prairie wetlands (Gomez et al., 2005).

Although Allen (1952) stated that Whooping Cranes were not reported in the region until 1880, journals from the Freeman and Custis

expedition of 1806 (April–July) reported that Whooping Cranes were "very abundant" throughout the Red River Valley (Flores, 1984). Similarly, it was reported that Whooping Cranes were abundant in the decades after the Civil War (Gomez et al., 2005). McIlhenny (1943) noted that prior to extensive human settlement of the coastal prairie, "Quite large flocks (of Whooping Cranes) were in evidence all winter to anyone riding over the open range." According to Lowery (1974), Mr. Vernon Bailey, a respected naturalist from southwestern Louisiana, reported that Whooping Cranes remained abundant near Iowa Station (i.e., coastal prairie habitats) in Calcasieu Parish as late as 1899. In contrast to Vernon Bailey's observations, Mr. O'Neil Nunez (as recorded by Mr. J. Lynch) reported that he saw his last Whooping Crane nest around 1897, but as a boy had seen 10–12 nests per year in platins (i.e., nearly circular ponds often associated with meander scars in the coastal prairie; Landry, 1973). Gomez (1992) provides a more complete summary of the history of Whooping Cranes in Louisiana.

The demise of the Whooping Crane in Louisiana, and throughout its range, was especially rapid in the late 1800s to early 1900s (Allen, 1952). Drainage of the Midwest prairie breeding grounds for migratory Whooping Cranes, shooting for food and museum collection, egg collection, and the rapid transformation of habitats on the coastal prairie (see later) all impacted migratory and resident Whooping Cranes in Louisiana (Gomez, 1992; McNulty, 1966). The last observation of migratory Whooping Cranes in Louisiana was in 1918 when a rice farmer shot 12 Whooping Cranes feeding on rice near his thresher (Allen, 1952). The last documented successful breeding of Whooping Cranes in the resident population occurred in 1939 when Mr. John Lynch observed two prefledged young during an aerial survey near White Lake (Lynch, 1984). In August 1940, a hurricane deluged the area, and prolonged flooding for several months dispersed the remaining individuals; only six birds returned at a later date. One of the birds, an injured female, was captured by a farmer about 80 km north of White Lake and was given to the Audubon Zoo in New Orleans. She lived at the zoo until 1948 when she was transferred to Aransas National Wildlife Refuge (ANWR), situated along the Texas Gulf coast, where she was kept in a pen. In 1950, there was only one Whooping Crane left in Louisiana's marshes; a decision was made to capture the bird and bring it to ANWR to potentially breed with other wild Whooping Cranes. On 11 March 1950, Mr. Lynch and Mr. Nick Schexnayder, superintendent of Audubon's nearby Paul J. Rainey Wildlife Sanctuary, used a helicopter piloted by Mr. L.L. "Mac" McCombie to capture the last Whooping Crane from the marshes of the current White Lake Wetlands Conservation Area (WLWCA; Gomez, 1992). The bird was transported to ANWR and placed in a pen. She responded poorly to captivity, was attacked by Whooping Cranes (from the

Aransas-Wood Buffalo Population, AWBP), and was released to the wild, where she died a few months later (McNulty, 1966). The female Louisiana bird from the Audubon Zoo was able to mate in captivity with other Whooping Cranes from the AWBP before dying in 1965. She produced four offspring that lived 12 or more years. With their deaths, the last known direct link to the Louisiana resident population was lost.

Louisiana Reintroduction

Only 27 years after the capture of the last bird at White Lake, efforts were initiated to reestablish a resident population of Whooping Cranes in the same area (Allender and Archibald, 1977; Gomez, 1992). In 1977, Dr. George Archibald, cofounder of the International Crane Foundation, and Dr. John Allender, the former curator of birds at the Audubon Zoo in New Orleans, wrote a proposal to the U.S. Fish and Wildlife Service to reestablish a resident population of Whooping Cranes in Louisiana. The Grays Lake Population experimental reintroduction (French et al., Chapter 1, this volume) was ongoing at the time and neither the U.S. Fish and Wildlife Service nor the Louisiana Department of Wildlife and Fisheries (LDWF) showed interest in the proposal (Dr. George Archibald, International Crane Foundation, personal communication). During several visits to Louisiana in the 1980s, Dr. Archibald advocated for the proposal to no avail. In 1990, Dr. Archibald joined the Whooping Crane Recovery Team (WCRT) and continued efforts to convince the U.S. Fish and Wildlife Service and LDWF to reintroduce Whooping Cranes in Louisiana. However, the Florida Nonmigratory Population reintroduction effort that began in 1993 (Dellinger, Chapter 9, this volume) was of primary interest to the WCRT. In regard to Louisiana, there were concerns over the proximity of Louisiana to ANWR, the potential intermixing of AWBP and reintroduced birds, and thus the possibility of introducing diseases or undesirable behaviors into the remnant population. As the Florida reintroduction struggled during an intense and prolonged drought (Spalding et al., 2009), the vast wetlands of Louisiana received more serious thought from the WCRT and the potential value of Louisiana was reinforced by site visits (Dr. George Archibald, International Crane Foundation, personal communication). LDWF administrators warmed to the idea, and at the behest of the WCRT funded a project to equip migratory Sandhill Cranes (*Grus canadensis*) with satellite transmitters to determine if they migrated through the Platte River. The WCRT was concerned that reintroduced Whooping Cranes would overlap with migratory Sandhill Cranes and introduce infectious bursal disease to the population, which could then be transmitted to the AWBP during the Sandhills' stopover at the Platte (Spalding et al., 2008, 2010). Some of the marked Sandhill Cranes did in fact migrate through the Platte River area (an area used also by migrating AWBP birds: King et al., 2011), but risks over disease dissipated when it was determined that the AWBP had already been exposed to infectious bursal disease. In 2007, the WCRT requested a study on food availability in the coastal marshes in and around the White Lake Wetlands Conservation Area (Fig. 22.2). The WCRT decided to move forward with the reintroduction, and on 16 February 2011 the first cohort of Whooping Cranes was transported to Louisiana from Patuxent Wildlife Research Center. In March, the cohort was released from a top-netted pen at WLWCA, where Lynch and colleagues had captured the last Louisiana Whooping Crane in 1950.

HISTORIC AND CURRENT HABITAT CONDITIONS

Habitat suitability and quality impact the success of any wildlife reintroduction (Seddon, 2013). In general, the wetland habitats in

Louisiana may be more productive and expansive than those found in any previous Whooping Crane release sites (Foley, 2015; Kang and King, 2013a, 2013b, 2014; Visser et al., 2000). However, the landscape has changed since Whooping Cranes were last present, and many factors could potentially hinder a successful reintroduction. In this section, we describe the historic and current habitat conditions in the region. In a subsequent section, we will discuss the potential implications of habitat characteristics and historic habitat changes on the success of the Louisiana reintroduction.

Climate

The climate of southwestern Louisiana is somewhat typical of low-latitude coastal habitats in North America. Precipitation reflects an east/west gradient, with Lafayette and Alexandria receiving about 154 cm/year and Lake Charles, the westernmost large city in the coastal prairie region, 146.1 cm/year (all weather data accessed 10 September 2015 at http://www.usclimate-data.com for respective cities). Daily average temperatures are approximately 15–25°C. The average date of the first freeze is highly variable spatially but ranges from about 6 November to 16 December. The average date of the last freeze is 30 January in the extreme southern Chenier Plain and early to mid-March in the northern Chenier Plain. Hurricanes and tropical storms are common in the region (Wallace and Anderson, 2010), with the hurricane season running from 1 June to 30 November.

Coastal Prairie

The Chenier Plain of southwestern Louisiana is within the Western Gulf Coastal Plain (Fig. 22.1) (Fenneman, 1928), and can be further subdivided into the coastal marshes (Fig. 22.2) and the coastal prairies (Autin et al., 1991). These two regions support vastly different habitats and have been subjected to different

(A)

FIGURE 22.2 (A) The paille-fine (*Panicum hemitomon*) marshes of White Lake historically supported a resident population of Whooping Cranes; the last bird was captured on 11 March 1950. (B) Interspersed in the paille-fine are ponds of various sizes and depths. The White Lake Wetlands Conservation Area, owned by the Louisiana Department of Wildlife and Fisheries, encompasses the historic capture site and supports a release pen for the Louisiana reintroduction.

human development pressures and habitat changes. The coastal prairie was formed from marine and fluvial deposits during a series, or sequence, of rising and falling sea levels during the Pleistocene (Autin et al., 1991). The coastal prairie contains many relict natural levees and channel scars reflecting the past occupation by the Red River.

The coastal prairie ranges from about 1.5 m in elevation near the Gulf coast to about 16 m in the north (Allen et al., 2001). Although generally flat but with gradually sloping hills and valleys, the landscape was historically filled with *marais* (shallow, water-filled depressions), platins, and mima or pimple mounds. The pimple mounds were low, circular mounds from 2 to 20 m in diameter and from a few cm to over 1.5 m in height (Allen et al., 2001). The origin of these features is still debated but may be related to differential erosion and/or fluvial processes (Smeins et al., 1992; Wiseman et al., 1979).

Soils in the coastal prairie region of Louisiana are primarily marine-derived with a loess cap over much of the eastern quarter of the coastal prairie (Heinrich, 2008; Vidrine et al., 2001). Coastal prairie soils have dense clay subsoils and generally are of low relief; thus they have poor surface and internal drainage; the clay-pan is typically >20–40 cm deep. Winter and spring often have ponded or flooded conditions, whereas during summer the soils can become droughty (Vidrine, 2010).

The coastal prairie region historically was composed of many small prairies or coves that were separated by floodplain forests (Post, 1940). Vidrine (2010) noted that the region was a mixture of dry and wet prairies. This tallgrass prairie system extended across about 1,000,000 ha in parts of present-day Acadia, Allen, Calcasieu, Cameron, Evangeline, Iberia, Jefferson Davis, Lafayette, St. Landry, St. Martin, and Vermilion parishes (Allen et al., 2001; Post, 1940). Allain et al. (2006) found 594 plant species, of which 254 species were considered prairie-specific species; this number of species exceeded that of other tallgrass prairies (Smeins et al., 1992). Little bluestem (*Schizachyrium scoparium*) and Indiangrass (*Sorghastrum nutans*) were the dominant species overall, but switchgrass (*Panicum virgatum*) and eastern gamagrass (*Tripsacum dactyloides*) were dominant in low areas.

In the coastal prairie, sheetflow and ponding of water in marais and platins created a wet prairie for much of the year (Vidrine, 2010). Vidrine (2010) noted that springs, emanating from shallow groundwater, were also common. Fires, both natural and anthropogenically induced, were also common on the native coastal prairie (Allen and Vidrine, 1989; Allen et al., 2001; Vidrine, 2010). Fire and hydrology were the dominant forces shaping the highly diverse grasslands. The combination of wet, flooded conditions and

droughty, often cracking soils created a harsh environment for woody plants and limited their distribution to the floodplains of numerous bayous that ran through the region.

Coastal Prairie Conversion

The coastal prairie was sparsely populated and relatively unaltered until the mid-1800s (Meacham, 1986; Vidrine, 2010). Although rice was grown in the platins, rice agriculture was not a commercial enterprise. Rapid transformation of the region occurred with Louisiana leading the nation in rice production by 1918 (Meacham, 1986). Less than 1% of native coastal prairie remains (Allain et al., 2006), as it was rapidly replaced by rice fields, and more recently by crawfish aquaculture (Fig. 22.3). Roadways, ditches, agricultural levees, and fence lines have disrupted sheetflow, and leveling of farm fields has removed marais, platins, and mima mounds.

In 2015, about 182,186 ha of rice were grown in Louisiana (U.S. Department of Agriculture, 2015), with the greatest concentration in the coastal prairie region (Steve Linscombe, LSU AgCenter, Rice Experiment Station, personal communication). Crawfish farming began in Louisiana in 1949 (LaCaze, 1968) and about 4,049 ha of crawfish farms existed by the mid-1960s (https://www.lsuagcenter.com/en/our_offices/research_stations/Aquaculture/Features/extension/Classroom_Resources/History+of+Crawfish+Aquaculture+in+Louisiana.htm; accessed 5 November 2015). As of 2014, there were about 48,583 ha of crawfish farms, and crawfish revenues are considered essential by many rice farmers to sustain economic viability of their farms.

Rice and crawfish are often grown in rotation with each other and with other crops. Rice is grown in shallower water (5–10 cm) during the warmer months (May–September), whereas crawfish are grown in deeper water (20–50 cm) and generally cooler weather (October–June) (Huner et al., 2002). Crawfish and rice production creates a diversity of wetland habitats

FIGURE 22.3 The coastal prairie of Louisiana has been converted from tallgrass prairie to an agriculture-based ecosystem dominated by rice agriculture and crawfish aquaculture east of Lake Charles; cattle grazing (not shown) is more common in the coastal prairie south and west of Lake Charles. (A) Rice fields in various stages of production. (B) Crawfish aquaculture pond following a drawdown at the end of the crawfish season (July 2015). Note the number and diversity of wading bird species using this habitat. *Source: Photos by George Archibald and Sammy L. King.*

that have ample food resources throughout the annual cycle and support over 260 species of birds (Foley, 2015; Huner et al., 2002). Huner et al. (2002) provide a detailed explanation of the various management schemes of rice and crawfish.

Coastal Marshes

The coastal marshes of the Chenier Plain were formed in the late Holocene, and the region is classified as a microtidal, storm-dominated coast. It is located west and downdrift of the Mississippi River deltaic plain (McBride et al., 2007). McBride et al. (2007) noted that the coastal marshes are composed primarily of mud deposits supporting marsh and interspersed with thin sand and shell beach ridges known as cheniers.

The Chenier Plain region is composed of two major hydrologic basins: the Mermentau Basin and the Calcasieu-Sabine (Louisiana Coastal Wetlands Conservation and Restoration Task Force and the Wetlands Conservation and Restoration Authority, 1998). Ponds and lakes are exceptionally common throughout the Louisiana coastal zone (Chabreck, 1971), with four major lakes in the Chenier Plain region: Calcasieu, Sabine, Grand, and White lakes. The lakes are about 1–2.5 m in depth (Louisiana Coastal Wetlands Conservation and Restoration Task Force, 1998).

There are ~3,085 km² of coastal marshes in the Chenier Plain region (Field et al., 1991) and the marsh types have been mapped periodically since 1949 (O'Neil, 1949). Visser et al. (2000) found that marshes dominated by paille-fine (17%) and freshwater bulltongue (*Sagittaria lancifolia*; 11%) accounted for ~28% of the marsh area. These freshwater marshes have average annual salinities <0.5 ppt, and the large area in these marshes may be particularly important for Whooping Cranes that are known to successfully rear chicks in only freshwater habitats (see Gomez et al., 2005). Another 51% of the area is oligohaline marsh types (average annual salinity between 0.5 and 5 ppt), with 88% of oligohaline marshes dominated by wiregrass (*Spartina patens*).

Although there are vast quantities of marsh in the region, plant species composition does fluctuate with salinity, and is particularly pronounced after hurricanes and drought (Glasgow and Ensminger, 1957; Valentine, 1976). Increased salinities following hurricanes can kill freshwater and oligohaline vegetation, thus resetting succession by opening up dense, rank vegetation and improving habitat for waterbirds. Hurricanes Audrey (1957), Rita (2005), and Ike (2008) are some of the more recent

hurricanes that have had a substantial impact on coastal marshes of the Chenier Plain.

Fire is also an important natural process, and management tool, in coastal marshes. Lynch (1941) described three types of burns in coastal marshes: (1) cover burns, (2) root burns, and (3) peat burns. Cover burns are of lower intensity and can be used for a variety of purposes, including to open the marsh up for waterbirds and to reduce the probability of root and peat burns. Root burns often occur during droughts in peaty marsh areas such as paille-fine, wiregrass, and sawgrass marshes. In the absence of fire, these vegetation types build up tremendous litter, which can be ignited by lightning, creating an intense fire that kills the plant's rhizomes. Fires that ignite the peat layer can burn for months or years. Cover burns are not harmful to wildlife but their effect on openings is short lived. In contrast, root and peat burns are initially detrimental to wildlife but the openings in the marsh can last several years.

Changes in Coastal Marshes

Human population density in the Chenier Plain was historically low and remains so today: changes in the habitat are primarily caused by hydrologic changes, as well as oil and gas development. Oil and gas development began in the 1920s in the coastal region (Lindstedt et al., 1991) and is still a regional economic driver, although most oil development is now offshore. The development of oil canals for boat access to oil development infrastructure was believed to have contributed to marsh die-offs (Glasgow and Ensminger, 1957) because they rapidly drained rainwater from the marsh and allowed saline water to move deeper into the marsh than would have occurred prior to canal construction. The construction of the Old Intracoastal Waterway from 1912 to 1924 was the first of numerous hydrologic changes (Louisiana Department of Wildlife and Fisheries, 2011). The Old Intracoastal connected White and Grand lakes and disrupted southerly sheetflow of water across the marsh. The Gulf Intracoastal Waterway (GIWW) was constructed from 1925 to 1944, replaced the Old Intracoastal Waterway, and lies near the intersection of the coastal prairie and coastal marshes. This location coincides with the last nesting location of the resident population of Whooping Cranes described by John Lynch (Allen, 1952). The GIWW also facilitated exploration of Chenier Plain marshes, including those of the White Lake area that previously had been largely protected because of their remoteness.

Hydrologic changes in the basin are not limited to canals. Three major water control structures were created in the 1950s to complete the Mermentau Basin Project. The objectives of the project were to provide a freshwater resource for farming activities on the coastal prairie and to maintain water levels high enough to support navigation. The structures, and associated hydrologic modifications, converted the White Lake area from an estuarine system characterized by variable salinities and fluctuating water levels (Gunter and Shell, 1958) to a predominantly freshwater system with little salinity and water level variation. The construction of Highway 82 from Pecan Island to Grand Chenier in 1958 further disrupted sheetflow in the marsh, thereby reducing freshwater inputs to marshes south of Highway 82 and minimizing salinity pulses to marshes north of the highway. Although some effort has been made to improve flow under Highway 82, it is still far less than historical conditions.

Although hydrologic modifications began in the early 1900s, so did broad-scale conservation efforts. E.A. McIlhenny was instrumental in acquiring several large properties – State, Rockefeller, and Marsh islands – in the region to serve as wildlife refuges (Louisiana Department of Conservation, 1929). By 1912, the state owned nearly 69,000 ha of coastal marshes for the preservation of wildlife resources (Louisiana Department of Conservation, 1929). Since that

time, several other large national wildlife refuges have been established, including Lacassine (13,779 ha), Cameron Prairie (3,788 ha), and Sabine National Wildlife Refuges (49,020 ha). Furthermore, the Audubon Paul J. Rainey Wildlife Sanctuary (10,522 ha) is also a large tract of coastal marsh in the region, and the largest land holding for this nonprofit organization. Coastal marsh tract sizes for private landowners are often greater than several hundred hectares in size.

Waterbird Populations

Prior to habitat conversion, the wildlife populations of the coastal marshes and coastal prairie were diverse and abundant. Mottled Ducks (*Anas fulvigula*), Sandhill Cranes, and Whooping Cranes were some of the many common avian species that nested in the coastal prairie (Gomez et al., 2005; McIlhenny, 1943). Today, over 260 species of birds use the rice/crawfish complex of the coastal prairie (Huner et al., 2002). Fleury and Sherry (1995) found that colonial wading bird populations in Louisiana increased dramatically following the expansion of crawfish aquaculture in the state, particularly for those species that are crawfish specialists (eight species).

We are unaware of complete avian surveys of coastal marshes in the Chenier Plain, and any species other than Whooping Cranes and Mississippi Sandhill Cranes (*G. c. pulla*) that have been extirpated. Figgins (1923) noted that Sandhill Cranes nested in the marshes of Louisiana's Chenier Plain. Currently, the marshes support a diverse assemblage and high abundance of waterbirds (Pickens and King, 2014). Sherry and Chavez-Ramirez (2005) suggested that wading bird habitat use is a good surrogate for measuring quality of Whooping Crane winter habitat. Other studies indicate that the number and size of wading bird rookeries are correlated with the abundance and quality of habitat within a certain distance (varies by species) from the rookery (Hafner, 2000). Rookery data from the Louisiana Natural Heritage Program indicate that there were 24 active rookeries surveyed in 2008 of an unknown total in Cameron and Vermilion parishes. These rookeries supported an estimated 50,701 wading birds of 14 species. Thus, in a general context, the region appears exceptionally productive for waterbirds and creates a sense of optimism regarding habitat quality for Whooping Cranes (Kang and King, 2014).

WHOOPING CRANE REINTRODUCTION STATUS AND CHALLENGES

As of June 2016, a total of 75 Whooping Cranes (43 females; 32 males) have been released in Louisiana. Annual cohorts have ranged in size from 10 to 16 birds. There were also a maximum of 38 adult/subadult Whooping Cranes in the Louisiana population, plus one wild-hatched chick. The entire 2010 cohort is deceased. Annual survival estimates are about 83.4 ± 0.03% across all birds and cohorts. Gunshot ($n = 10$) is the leading known cause of death (see later), and predation ($n = 6$) is the second largest known mortality agent. However, the remains of an additional seven birds have been recovered, of which six are suspected to be a result of predation (although disease cannot be ruled out).

A total of 16 nests have been attempted, including nests from at least five pairs in 2016. The first successful nest was hatched in 2016. Twins were hatched from a nest in a crawfish pond; both chicks survived for 1 month, but only one chick remains alive at this time. Whooping Cranes continue to use a wide range of habitats, with most locations occurring in crawfish, rice, and coastal marshes.

The proximate factors leading to the demise of the resident population of Whooping Cranes in Louisiana were likely shooting for food and incidental take from fur trapping, but ultimately their demise was a result of increased human activity in the region (Allen, 1952). The development of the Intracoastal Waterway facilitated human activity in the coastal marshes during a period when rates of human subsistence hunting in the region were high. Subsistence hunting is no longer an issue, but the marsh does receive much greater human disturbance overall than would have occurred during the 1800s when Whooping Cranes were abundant in the region.

Whether the conversion of the coastal prairie grasses to rice per se was the cause of the decline of birds on the coastal prairie is difficult to determine. That is a possibility but along with this conversion came much higher human population densities and human activity, including hunting and other human-caused mortality (Allen, 1952; Vidrine, 2010). Furthermore, even if rice conversion had been primarily responsible for the decline, rice agriculture and the former coastal prairie landscape have continued to change.

Another characteristic of breeding habitats within the core historic range of the Whooping Crane is high densities of food in breeding wetlands (Austin et al., Chapter 3, this volume; Bataille and Baldassarre, 1993). The long growing season, abundant rainfall, and fertile soils found within southwest Louisiana support high levels of wetland productivity. Crawfish aquaculture ponds (Foley, 2015) and coastal marsh ponds (Kang and King, 2013a, 2013b) have abundant Whooping Crane food resources during the breeding season. Several food resource studies were conducted to evaluate the efficacy of WLWCA as a potential reintroduction site (Kang and King, 2013a, 2013b, 2014). These studies indicated that the freshwater marshes at WLWCA and the intermediate brackish and saline marshes at Rockefeller Wildlife Refuge support abundant invertebrate and fish prey. The types and density of food resources vary seasonally and spatially across marsh types; however, food availability does not appear to be a limiting factor.

Food availability, however, is not the only factor that defines successful breeding habitat. In the coastal prairie region, Whooping Crane nesting must be compatible with crop activities. Rice will likely not be a major nesting habitat simply because rice is not even planted until late April–May and is not mature until late in the breeding season (e.g., June–July); therefore, a nesting substrate is absent. Crawfish ponds, which may have perennial wetland plants and rice from the previous growing season as a forage crop for crawfish, are shallowly flooded (about 0.5 m) throughout the winter and spring and provide suitable nesting substrates and abundant food resources (Foley, 2015). Whooping Cranes nesting in crawfish ponds, however, must adapt to the near-daily presence of crawfishermen and other farm operations. Based on nesting attempts to date in crawfish ponds (11 of 16 nests), this does not appear to be an issue, as birds appear habituated to normal crawfish farming operations; they continue normal incubation behaviors even while crawfishermen pass closely by the nest in boats (all authors, personal observation). However, the timing of drawdown of the crawfish pond relative to egg hatch and chick development is potentially significant. Farmers typically draw down their ponds in May and June to facilitate crawfish reproduction and to negate water quality issues that can develop in shallow water during summer. It typically takes about 30 days for a Whooping Crane egg to hatch and another 90 days for the chick to fly (Lewis, 1995). Thus, having a reliable water source within walking distance and through habitat that minimizes predation is necessary for fledging to occur. Luckily, on many crawfish farms, flooded rice will be present during the drawdown period for crawfish ponds. However, lack of reliable water in areas managed for crawfish remains a potential concern for Louisiana reintroduction efforts.

In coastal marshes, both the spatial extent and overall productivity suggest that Whooping Cranes will do quite well in Louisiana's marsh habitats. However, similar to the coastal prairie, the marsh has changed since the 1800s when Whooping Cranes were more abundant. Hydrologic alterations within the Chenier Plain region have made the marshes a more stable freshwater environment (Gunter and Shell, 1958). Prior to the hydrologic alterations, salinity levels would have fluctuated more frequently. Hydrologic variability is critical to supporting long-term productivity in wetlands (Euliss et al., 2004), and the stabilization of salinity levels and water levels (within their ability to do so) and other hydrologic modifications described previously have facilitated the development of dense, rank vegetation in many areas that is unsuitable for Whooping Cranes (S.L. King, personal observation). This is particularly restrictive during flightless periods, such as molt and chick rearing, when birds can move about only on the ground. The marshes are still highly productive, but they are different from historical conditions. Further, the temporal and spatial distribution of resources during critical life history events may prove to be important. For example, at White Lake, ponds in the freshwater marsh are highly productive, but during low water conditions the ponds are isolated from each other by tall (>1 m), dense stands of paille-fine. For a flightless chick or adult, moving from pond to pond during this period would be a high predation risk.

As in all coastal systems, sea-level rise will be an issue (Church et al., 2013; Smith et al., Chapter 13, this volume). However, predictions based solely on estimates of sea-level rise and landscape elevation are poor predictors of actual coastal impacts (Anderson et al., 2014). These simplistic models do not account for changes in sediment accommodation, accretion (including organic), and erosion. Furthermore, predicting coastal impacts is complicated due to spatial variability in these processes. The ability of coastal systems to accrue and store sediment, however, has mostly been negatively impacted due to dams, reduced river discharges to estuaries, interruption of long-shore and cross-shore sand transport, and subsidence due to fluid extraction (Anderson et al., 2014).

While the wet climate in Louisiana is conducive to high ecosystem productivity, a potential negative effect of the climate is that torrential rainfalls can cause flash flooding within virtually any month. This could be particularly problematic during nesting season if rapidly rising floodwaters overtop nests or if cold spring rains kill newly hatched chicks. Flooding of nests should be less of an issue in crawfish ponds where pond design and water control structures allow water depth to be maintained at a desired level. The marshes, however, are potentially subject to rapid rises in water levels. In fact, one nest in 2015 was submerged at WLWCA following ~25.4 cm of rainfall within a 4-day period.

Reproduction and survival rates sufficient to sustain a self-reproducing population will be necessary for the reintroduction effort to succeed. While the habitat appears to be highly productive and in adequate quantities, nonhabitat factors, including captive rearing techniques and founder effects, could also limit reintroduction success (Converse et al., Chapter 8, this volume). Inadequate incubation behavior and poor chick rearing success have been observed in the reintroduced Eastern Migratory Population (EMP; Converse et al., Chapter 7, this volume; Converse et al., Chapter 8, this volume). These behaviors may be related to captive rearing techniques and, if so, could affect the outcome of the Louisiana reintroduction, as Whooping Cranes released in Louisiana to date have been reared similarly. Although captive rearing techniques are a plausible reason for poor reproductive performance in reintroduced birds, habitat quality may also be totally or partially responsible. Poor wetland conditions and insufficient stimulation of the neuroendocrine system of Whooping Cranes due to intense drought affected reproductive activity in the Florida reintroduced population (Spalding et al., 2009). High black fly abundance and low wetland productivity characterize the breeding habitat of the EMP (Baldassarre, 1978; King et al., 2015), and black fly abundance is a strong predictor of nest abandonment in the EMP (Converse et al., 2013; Converse et al., Chapter 8, this volume). Moore et al. (2012) through population modeling of the Florida resident population found that reintroduction success was highly sensitive to reproductive vital rates (see also Converse et al., Chapter 7, this volume). When vital rates from the AWBP were used in the model, the reintroduction was successful regardless of the number of birds released. In contrast, when observed rates from the Florida population were used in the model, the reintroduction failed regardless of the number of birds released. The vital rates necessary to allow a sustainable population in Louisiana are unknown, but if they are sensitive to habitat quality, then Louisiana may fare better than previous reintroductions. The negative effects of any non-habitat issues, however, could exceed the benefits of improved habitat conditions. Time is necessary to distinguish these effects.

Fischer and Lindenmayer (2000) reviewed the literature and found that most avian reintroductions are not successful. Of the projects that were successful, the three most common traits were a wild source, a large number of released animals ($n > 100$), and the cause of the original decline being removed. Whooping Crane reintroductions have always used eggs from captive birds. It may be possible in the future that eggs from AWBP birds could be brought into captivity and reared either by parent cranes at the captive facilities or using the costume/puppet rearing technique (Converse et al., Chapter 8, this volume). However, releasing birds captured from the AWBP is not currently a strong possibility (Dr. George Archibald, WCRT, personal communication). Getting to 100 individuals as rapidly as possible is a goal of the reintroduction based on Fischer and Lindenmayer (2000), but it is complicated by limitations in the captive rearing process. A finite number of eggs can be produced in a given year and the labor and facility requirements to rear and train chicks are substantial. Furthermore, simultaneous reintroduction efforts reduce the number of chicks available in any given year. However, it is hoped that in the next 5–7 years or earlier the Louisiana population will reach 100 birds.

Finally, shooting of Whooping Cranes in Louisiana and elsewhere is affecting all populations of Whooping Cranes (Condon et al., Chapter 23, this volume). A total of six shooting incidents that eliminated 10 birds (eight directly killed; two euthanized due to injuries sustained) are known to have occurred; shootings of Louisiana birds were not limited to Louisiana, as two individuals were shot in Texas. These shootings included the first pair of Whooping Cranes to exhibit signs of nesting (i.e., constructed nest platform the previous spring) in the reintroduced population in Louisiana. Shootings have been common throughout the history of Whooping Cranes (Allen, 1952; McNulty, 1966) and at least 25 Whooping Cranes have been shot and killed in at least eight states since 2009 (Condon et al., Chapter 23, this volume). In Louisiana, the first shooting deaths of reintroduced Whooping Cranes prompted a massive and multipronged public education campaign that continues to target multiple audiences, including middle and high school teachers, radio and television audiences, and travelers, through the use of billboards. Reward signs, in both English and Spanish, have been placed throughout southwest Louisiana warning people to not shoot Whooping Cranes and encouraging individuals to contact law enforcement with information about shootings. LDWF received a multiyear grant from Chevron Corporation to support this program, and millions of people have been reached. It is estimated that 850,000 travelers were exposed to the billboards in 2014–15 alone. A survey of 2,165 licensed hunters in 2015 found that 56% had heard about the Whooping Crane program with television advertisements

being the most commonly cited source of information (Louisiana Department of Wildlife and Fisheries, 2014). In spite of these efforts, a pair of birds were shot and killed in east Texas in January 2016 and another pair met the same fate in Louisiana in May 2016. These birds were not killed accidentally in the sport of hunting, but were the deliberate target of criminal activity. The ultimate success of the Louisiana reintroduction may depend on solving this all too common problem.

SUMMARY AND OUTLOOK

Louisiana historically supported both resident and migratory Whooping Cranes. Although Louisiana continues to support vast amounts of productive natural and agricultural wetlands, the structure and processes of these wetlands have changed since Whooping Cranes last occupied the region. The reintroduction is only in its sixth year, but Whooping Cranes are using both the agricultural wetlands and marshes of Louisiana. The combined annual survival is $83.4 \pm 0.03\%$, and nesting has been initiated with the first successful nest in 2016. The success of this reintroduction will likely hinge on the answers to three questions: (1) Can captive-reared Whooping Cranes reproduce at a great enough rate to sustain a viable population? (2) Will crawfish farming operations be compatible with Whooping Crane reproduction in crawfish ponds? (3) Can shootings of Whooping Cranes be reduced or eliminated? In regard to reproduction rates, to date no reintroduced populations of captive-reared Whooping Cranes have demonstrated reproductive rates capable of sustaining a viable population (Converse et al., Chapter 7, this volume). If this is partially or wholly a habitat issue, then the quality and amount of Louisiana's wetlands give reason for optimism. However, if this is a behavioral issue related to the captive-rearing process (Converse et al., Chapter 8, this volume), then the Louisiana reintroduction may follow the same path as its predecessors.

Whooping Cranes are attracted to crawfish aquaculture ponds, and most of the nests have been located in crawfish ponds. Crawfish farmers typically draw down ponds in June or July, and, depending on nesting phenology, this could strand flightless chicks in areas without adequate wetland habitat. However, there are large amounts of alternative wetland habitats in the region (Foley, 2015), and, while worth mentioning, it may not be as significant an issue as the other two.

Finally, the level of gunshot mortality in Louisiana and throughout all populations of Whooping Cranes is discouraging. Public education efforts in Louisiana have been extensive, but shootings of Whooping Cranes have been deliberate and continue even following the massive education and outreach campaign. Thus, the challenge on how to effectively address deviant behavior that leads to shooting of Whooping Cranes is perplexing. Yet, our efforts to address the problem with whatever means possible cannot be deterred, as the success of this project may well be dependent on solving this issue.

References

Allain, L., Smith, L., Allen, C., Vidrine, M.F., Grace, J.B., 2006. A floristic quality assessment system for the coastal prairie of Louisiana. Proceedings of the 19th North American Prairie Conference 19, 118.

Allen, C.M., Vidrine, M.F., 1989. Wildflowers of the Cajun Prairie. Louisiana Conservation. 41, 20–25.

Allen, C.M., Vidrine, M.F., Borsari, B., Allain, L., 2001. Vascular flora of the Cajun Prairie of southwestern Louisiana. Proceedings of the North American Prairie Conference 17, 35–41.

Allen, R.P., 1952. The Whooping Crane. Research Report 3. National Audubon Society, New York.

Allender, J., Archibald, G., 1977. The Preliminary Proposal Regarding the Reestablishing of Resident Whooping Cranes in Louisiana. Audubon Park Zoological Garden, New Orleans, LA, p. 25; unpublished report.

Anderson, J.B., Wallace, D.J., Simms, A.R., Rodriguez, A.B., Milliken, K.T., 2014. Variable response of coastal environments of the northwestern Gulf of Mexico to sea-level rise and climate change: implications for future change. Mar. Geol. 352, 348–366.

Austin, J.E., Hayes, M.A., Barzen, J.A., 2018. Revisiting the historic distribution and habitats of the whooping crane (Chapter 3). In: French, Jr., J.B., Converse, S.J., Austin, J.E. (Eds.), Whooping Cranes: Biology and Conservation. Biodiversity of the World: Conservation from Genes to Landscapes. Academic Press, San Diego, CA.

Autin, W.J., Burns, S.F., Miller, B.J., Saucier, R.T., Snead, J.I., 1991. Quaternary geology of the Lower Mississippi Valley. In: Morrison, R.B. (Ed.), Quaternary Nonglacial Geology: Conterminous U.S. Geological Society of America, vol. K-2. Geology of North America, Boulder, CO, pp. 547–582.

Baldassarre, G.A., 1978. Ecological factors affecting waterfowl production on three man-made flowages in central Wisconsin. MS thesis, University of Wisconsin, Stevens-Point, p. 124.

Bataille, K.J., Baldassarre, G.A., 1993. Distribution and abundance of aquatic macroinvertebrates following drought in three prairie pothole wetlands. Wetlands 13 (4), 260–269.

Chabreck, R.H., 1971. Ponds and lakes of the Louisiana coastal marshes and their value to fish and wildlife. Proceedings of the Annual Conference of the Southeastern Association of Fish and Wildlife Agencies 25, 206–215.

Church, J.A., Clark, P.U., Cazenave, A., Gregory, J.M., Jevrejeva, S., Levermann, A., Merrifield, M.A., Milne, G.A., Nerem, R.S., Nunn, P.D., Payne, A.J., Pfeffer, W.T., Stammer, D., Unnikrishnan, A.S., 2013. Sea level change. In: Stocker, T.F., Qin, D., Plattner, G.K., Tignor, M., Allen, S.K., Boschung, J., Nauels, A., Xia, Y., Bex, V., Midgley, P.M. (Eds.), Climate Change 2013: The Physical Science Basis Contribution of Working Group I to the Fifth Assessment Report of the Intergovernmental Panel on Climate Change. Cambridge University Press, Cambridge, UK, and New York, pp. 1137–1216.

Condon, E., Brooks, W.B., Langenberg, J., Lopez, D., 2018. Whooping crane shootings since 1967 (Chapter 23). In: French, Jr., J.B., Converse, S.J., Austin, J.E. (Eds.), Whooping Cranes: Biology and Conservation. Biodiversity of the World: Conservation from Genes to Landscapes. Academic Press, San Diego, CA.

Converse, S.J., Royle, J.A., Adler, P.H., Urbanek, R.P., Barzen, J.A., 2013. A hierarchical nest survival model integrating incomplete temporally varying covariates. Ecol. Evol. 3 (13), 4439–4447.

Converse, S.J., Servanty, S., Moore, C.T., Runge, M.C., 2018. Population dynamics of reintroduced Whooping Cranes (Chapter 7). In: French, Jr., J.B. Converse, S.J., Austin, J.E. (Eds.), Whooping Cranes: Biology and Conservation. Biodiversity of the World: Conservation from Genes to Landscapes. Academic Press, San Diego, CA.

Converse, S.J., Strobel, B.N., Barzen, J.A. Reproductive failure in the eastern migratory population: the interaction of research and management (Chapter 8). In: French, Jr., J.B., Converse, S.J., Austin, J.E. (Eds.), Whooping Cranes:

Biology and Conservation. Biodiversity of the World: Conservation from Genes to Landscapes. Academic Press, San Diego, CA.

Dellinger, T.A., 2018. Florida's nonmigratory whooping cranes (Chapter 9). In: French, Jr., J.B., Converse, S.J., Austin, J.E. (Eds.), Whooping Cranes: Biology and Conservation. Biodiversity of the World: Conservation from Genes to Landscapes. Academic Press, San Diego, CA.

Euliss, N.H., LaBaugh, J.W., Fredrickson, L.H., Mushet, D.M., Laubhan, M.K., Swanson, G.A., Winter, T.C., Rosenberry, D.O., Nelson, R.D., 2004. The wetland continuum: a conceptual framework for interpreting biological studies. Wetlands 24 (2), 448–458.

Fenneman, N.M., 1928. Physiographic divisions of the United States. Ann. Assoc. Am. Geogr. 18 (4), 261–353.

Field, D.W., Reyer, A.J., Genovese, P.A., Shearer, B.D., 1991. Coastal wetlands of the United States: an accounting of a valuable national resource. Special National Oceanic and Atmospheric Administration 20th Anniversary Report. National Oceanic and Atmospheric Administration and U.S. Fish and Wildlife Service, Washington, DC.

Figgins, J.D., 1923. The breeding birds of the vicinity of Black Bayou and Bird Island, Cameron Parish, Louisiana. Auk 40 (4), 666–677.

Fischer, J., Lindenmayer, D.B., 2000. An assessment of published results of animal relocations. Biol. Conserv. 96 (1), 1–11.

Fleury, B.E., Sherry, T.W., 1995. Long-term population trends of colonial wading birds in the southern United States: the impact of crayfish aquaculture on Louisiana populations. Auk 112 (3), 613–632.

Flores, D.L., 1984. Southern Counterpart to the Lewis and Clark: The Freeman and Custis Expedition of 1806. University of Oklahoma Press, Norman.

Foley, C.C., 2015. Wading bird food availability in rice fields and crawfish ponds of the Chenier Plain of southwest Louisiana and southeast Texas. MS thesis, Louisiana State University, Baton Rouge, p. 76.

French, Jr., J.B., Converse, S.J., Austin, J.E., 2018. Whooping cranes past and present (Chapter 1). In: French, Jr., J.B., Converse, S.J., Austin, J.E. (Eds.), Whooping Cranes: Biology and Conservation. Biodiversity of the World: Conservation from Genes to Landscapes. Academic Press, San Diego, CA.

Glasgow, L.L., Ensminger, A., 1957. A marsh deer "die-off" in Louisiana. J. Wildl. Manage. 21 (2), 245–247.

Gomez, G.M., 1992. Whooping cranes in southwest Louisiana: history and human attitudes. Proceedings of the North American Crane Workshop 6, 19–23.

Gomez, G.M., Drewien, R.C., Courville, M.L., 2005. Historical notes on whooping cranes at White Lake, Louisiana: the John J. Lynch interviews, 1947–1948. Proceedings of the North American Crane Workshop 9, 111–116.

Gunter, G., Shell, Jr., W.E., 1958. A study of the estuarine area with water-level control in the Louisiana marsh. Proc. Louisiana Acad. Sci. 21, 5–34.

Hafner, H., 2000. Heron nest site conservation. In: Kushlan, J.A., Hafner, H. (Eds.), Heron Conservation. Academic Press, San Diego, CA, pp. 201–217.

Heinrich, P.V., 2008. Loess map of Louisiana. Public Information Series 12. Louisiana Geological Survey, Baton Rouge.

Huner, J.V., Jeske, C.W., Norling, W., 2002. Managing agricultural wetlands for waterbirds in the coastal regions of Louisiana, USA. Waterbirds 25, 66–78, Special Publication 2.

Kang, S.R., King, S.L., 2013a. Effects of hydrologic connectivity on aquatic macroinvertebrate assemblages in different marsh types. Aquat. Biol. 18 (2), 149–160.

Kang, S.R., King, S.L., 2013b. Effects of hydrologic connectivity and environmental variables on nekton assemblage in a coastal marsh system. Wetlands 33 (2), 321–334.

Kang, S.R., King, S.L., 2014. Suitability of coastal marshes as whooping crane foraging habitat in southwest Louisiana, USA. Waterbirds 37 (3), 254–263.

King, R.S., McKann, P.C., Gray, B.R., Putnam, M.S., 2015. Host-parasite behavioral interactions in a recently introduced whooping crane population. J. Fish Wildl. Manage. 6 (1), 220–226.

King, S.L., Pierce, A.R., Hersey, K., Winstead, N., 2011. Migration patterns and movements of sandhill cranes wintering in central and southwestern Louisiana, USA. Proceedings of the 11th North American Crane Workshop 11, 57–61.

LaCaze, C.G., 1968. Louisiana Wildlife and Fisheries Commission: Biennial Report 1966–1967. http://archive.org/stream/louisianawildlif12depa/louisianawildlif12depa_djvu.txt [8 August 2018].

Landry, Jr., W.E., 1973. The distribution of platins: circular surface depressions on the prairie of southwest Louisiana. MA thesis, University of Southwestern Louisiana, Lafayette.

Lewis, J.C., 1995. Whooping crane (Grus americana). In: Poole, A., Gill, F. (Eds.), The Birds of North America 153. Academy of Natural Sciences, Philadelphia, and American Ornithologists' Union, Washington, DC.

Lindstedt, D.M., Nunn, L.L., Holmes, Jr., J.C., Willis, E.E., 1991. History of oil and gas development in coastal Louisiana. Resource Information Series 7. Louisiana Geological Survey, Baton Rouge.

Louisiana Coastal Wetlands Conservation and Restoration Task Force and the Wetlands Conservation and Restoration Authority, 1998. Coast 2050 toward a Sustainable Coastal Louisiana. Louisiana Department of Natural Resources, Baton Rouge, p. 161. Available from: https://www.doi.gov/sites/doi.gov/files/migrated/deepwaterhorizon/adminrecord/upload/Louisiana-Coastal-Wetlands-Conservation-and-Restoration-Task-Force-and-the-Wetlands-Coast-2050-Toward-a-Sustainable-Coastal-Louisiana-1998.pdf [8 August 2018].

Louisiana Department of Conservation, 1929. The creating of the wild life refuges in Louisiana. Ninth Biennial Report of the Department of Conservation of the State of Louisiana, pp. 132–139.

Louisiana Department of Wildlife and Fisheries, 2011. Rockefeller Wildlife Refuge Management Plan. Rockefeller Wildlife Refuge, Grand Chenier, LA.

Louisiana Department of Wildlife and Fisheries, 2014. 2014 Louisiana Whooping Crane Report: 1 June 2014 to 30 June 2015. Louisiana Department of Wildlife and Fisheries, Coastal and Non-game Resources, Baton Rouge.

Lowery, Jr., G.H., 1974. Louisiana Birds. Louisiana State University Press, Baton Rouge, p. 651.

Lynch, J.J., 1941. The place of burning in management of the Gulf coast wildlife refuges. J. Wildl. Manage. 5 (4), 454–457.

Lynch, J.J., 1984. A field biologist. In: Hawkins, A.S., Hanson, R.C., Nelson, H.K., Reeves, H.M. (Eds.), Flyways: Pioneering Waterfowl Management in North America. U.S. Fish and Wildlife Service, Washington, DC, pp. 35–40.

McBride, R.A., Taylor, M.J., Byrnes, M.R., 2007. Coastal morphodynamics and Chenier-Plain evolution in southwestern Louisiana USA: a geomorphic model. Geomorphology 88 (3–4), 367–422.

McIlhenny, E.A., 1943. Major changes in the bird life of southern Louisiana during sixty years. Auk 60 (4), 541–549.

McNulty, F., 1966. The Whooping Crane: The Bird That Defies Extinction. E.P. Dutton, New York.

Meacham, Jr., S.W., 1986. The origin and evolution of the southwestern Louisiana rice region, 1880–1920. PhD dissertation, University of Tennessee, Knoxville, pp. 255.

Moore, C.T., Converse, S.J., Folk, M.J., Runge, M.C., Nesbitt, S.A., 2012. Evaluating release alternatives for a long-lived bird species under uncertainty about long-term demographic rates. J. Ornithol. 152 (Suppl. 2), S339–S353.

O'Neil, T., 1949. The muskrat in the Louisiana coastal marshes. Louisiana Wildlife and Fisheries Commission, New Orleans.

Pickens, B.A., King, S.L., 2014. Multiscale habitat selection of wetland birds in the northern Gulf coast. Estuar. Coast. 37 (5), 1301–1311.

Post, L.C., 1940. The rice country of southwestern Louisiana. Geogr. Rev. 30 (4), 574–590.

Seddon, P.J., 2013. The new IUCN guidelines highlight the importance of habitat quality to reintroduction success – reply to White et al. Biol. Conserv. 164, 177.

Sherry, D.A., Chavez-Ramirez, F., 2005. Use of wading birds as indicators of potential whooping crane wintering habitat. Proceedings of the North American Crane Workshop 9, 127–132.

Smeins, F.E., Diamond, D.D., Hanselka, C.W., 1992. Coastal prairie. In: Coupland, R.T. (Ed.), Ecosystems of the World 8A: Natural Grasslands. Elsevier, New York, pp. 269–290.

Smith, E.H., Chavez-Ramirez, F., Lumb, L., 2018. Wintering habitat ecology, use, and availability for the Aransas-Wood Buffalo Population of whooping cranes (Chapter 13). In: French, Jr., J.B., Converse, S.J., Austin, J.E. (Eds.), Whooping Cranes: Biology and Conservation. Biodiversity of the World: Conservation from Genes to Landscapes. Academic Press, San Diego, CA.

Spalding, M.G., Folk, M.J., Nesbitt, S.A., Folk, M.L., Kiltie, R., 2009. Environmental correlates of reproductive success for introduced resident whooping cranes in Florida. Waterbirds 32 (4), 538–547.

Spalding, M.G., Folk, M.J., Nesbitt, S.A., Kiltie, R., 2010. Reproductive health and performance of the Florida flock of introduced whooping cranes. Proceedings of the North American Crane Workshop 11, 142–155.

Spalding, M.G., Sellers, H.S., Hartup, B.K., Olsen, G.H., 2008. A wasting syndrome in released whooping cranes in Florida associated with infectious bursal disease. Proceedings of the North American Crane Workshop 10, 176.

U.S. Department of Agriculture, 2015. Louisiana Acreage Report. Available at: http://www.nass.usda.gov/Statistics_by_State/Louisiana/Publications/Crop_Releases/Acreage/2015/laacreage15.pdf; accessed 8 September 2015.

Valentine, J.M., 1976. Plant succession after saw-grass mortality in southwestern Louisiana. Proceedings of the Southeastern Association of Fish and Wildlife Agencies 30, 630–640.

Vidrine, M.F., 2010. The Cajun Prairie: A Natural History. Malcolm Francis Vidrine, Eunice, LA.

Vidrine, M.F., Fontenot, W.R., Allen, C.M., Borsari, B., Allain, L., 2001. Prairie Cajuns and the Cajun Prairie: a history. Proceedings of the 17th North American Prairie Conference 220–224.

Visser, J.M., Sasser, C.E., Chabreck, R.H., Linscombe, R.G., 2000. Marsh vegetation types of the Chenier Plain, Louisiana, USA. Estuaries 23 (3), 318–327.

Wallace, D.J., Anderson, J.B., 2010. Evidence of similar probability of intense hurricane strikes for the Gulf of Mexico over the late Holocene. Geology 38 (6), 511–514.

Wiseman, D.E., Weinstein, R.A., Glander, W.P., Landry, L.A., 1979. Environment and Settlement on the Southwestern Louisiana Prairies: A Cultural Resources Survey in the Bell City Watershed. Coastal Environments, Baton Rouge, LA.

Whooping Crane Shootings since 1967

Elisabeth Condon*, William B. Brooks**,
Julie Langenberg*, Davin Lopez[†]

*International Crane Foundation, Baraboo, WI, United States
**U.S. Fish and Wildlife Service, Jacksonville, FL, United States
†Wisconsin Department of Natural Resources, Madison, WI, United States

INTRODUCTION

Shootings are having an impact in all remnant and reintroduced Whooping Crane populations. Shootings contribute to 19% and 24% of known mortality the Eastern Migratory Population (EMP) and the Louisiana Nonmigratory Population (LNMP), respectively (Harrell, 2014), two populations that are struggling to become self-sustaining. In the Aransas-Wood Buffalo Population (AWBP), shootings contribute to 20% of known mortality (Stehn and Haralson-Strobel, 2014). Whooping Cranes are a long-lived species with low adult mortality; this significant contributor to Whooping Crane mortality is likely having an impact at the population level, especially in the reintroduced populations.

In 1918, the Migratory Bird Treaty Act (MBTA) made the shooting of a migratory bird, including a Whooping Crane, illegal. In 1967, the Whooping Crane was listed as a Federally Endangered Species under the Endangered Species Preservation Act of 1966, the precursor to the Endangered Species Act (ESA). The Whooping Crane was then "grandfathered in" to the Endangered Species Act of 1973. Whooping Cranes are listed as a "Game Bird" under the bird conventions between the U.S., Mexico, and Canada, although there has never been a Whooping Crane hunting season under modern hunting regulations. In this chapter, we examine the record since the 1967 Endangered Species listing of the Whooping Crane – describing numbers of shootings over time and among populations, examining seasonal and geographic patterns in losses, and comparing situations in relation to hunting and nonhunting shooting cases. We discuss patterns of shooting incidents in the AWBP, the Florida Nonmigratory Populations (FNMP), the LNMP, and the EMP.

Muth and Bowe (1998) define poaching as "any act that intentionally contravenes the laws and regulations established to protect wild, renewable resources, such as plants, mammals, birds, insects, reptiles, amphibians, fish, and shellfish. It is important to note that this definition purposely excludes law violators who unintentionally or accidentally contravene acts and regulations." They note that hunters are excluded from this definition of poaching, unless they are violating hunting regulations on purpose. In this chapter, we make a distinction between hunting-related Whooping Crane shooting incidents and those that are unrelated to hunting. This distinction is useful in that efforts to reduce shootings may require different strategies for hunters who accidentally shoot Whooping Cranes while hunting a different species and people who shoot Whooping Cranes unrelated to hunting activities.

Hunters who accidentally shoot Whooping Cranes may not be poaching by Muth and Bowe's definition, but people who shoot them outside of a hunting season are considered poachers. The commercial market for the millinery trade that provided financial motivation for shooting Whooping Cranes no longer exists, but there may be other motivations for shooting a Whooping Crane. Muth and Bowe categorize many different motivations for poaching, including several that might apply to those who shoot Whooping Cranes: recreational poaching, trophy poaching, poaching as thrill killing, poaching to protect self and property, poaching as rebellion, poaching as traditional right of use, disagreement with specific regulations, and gamesmanship (Muth and Bowe, 1998). We discuss how these motivations may be applied to modern cases of Whooping Crane shootings later in this chapter.

WHOOPING CRANES KILLED BEFORE THE 1966 ENDANGERED SPECIES PRESERVATION ACT

When Robert Porter Allen wrote *The Whooping Crane* in 1952, he identified shootings as a major cause of decline of the Whooping Crane population. Allen scoured records to find all known locations of Whooping Crane killings. He did not specify if the killings resulted from shootings or from a different cause, but we can assume that many of these birds died from gunshot. Allen found 389 killings recorded between 1722 and 1952, the year of the publication, but speculated that the loss of Whooping Cranes was vastly underreported. These occurred in 16 states, in addition to Canada and Mexico. Date of the incident was available for only 190 of the 389 killings. A large number of museum specimens lack records (Hahn, 1963), suggesting that they probably were not collected by museum employees and may have been donated to the museums by hunters or other nonprofessional collectors. Location of many of the killing incidents indicates that many of the shootings took place at key staging areas during migration. A large driver for these shootings was the millinery trade; Whooping Crane feathers, like those of many white birds, were popular on the feather market and helped drive commercial collectors (Hornaday, 1913; Hughes, 2008).

Unregulated hunting was one of the primary causes for decline of the Whooping Crane (Canadian Wildlife Service and U.S. Fish and Wildlife Service, 2005). When Allen wrote *The Whooping Crane*, he did not make a distinction between legal hunting of Whooping Cranes that would have occurred before their protection under the Migratory Bird Treaty Act in 1918, and purposeful or accidental shooting of Whooping Cranes after 1918, which would have all been illegal. Although he compiled records of 389 Whooping Crane killings from gunshot from Colonial times to 1948, he did not provide information about the circumstances surrounding each case. It is not clear what motivated the majority of Whooping Crane shootings. Regardless of the reason for shooting Whooping Cranes, the large number of shootings has been identified as one of the primary causes for decline of the Whooping Crane.

WHOOPING CRANE SHOOTINGS SINCE 1967

Methods

We started with dates and locations of known shooting cases that had been previously compiled by the U.S. Fish and Wildlife Service (USFWS), and expanded upon that record. Public information available for each case is highly variable. For this chapter the authors submitted Freedom of Information Act (FOIA) requests to the USFWS and the U.S. Department of Justice (USDOJ) and were able to obtain information on nine cases. Other information was gleaned from news reports and personal communications with people involved in investigations. In some cases very little information exists in the public record. One confirmed case took place in Saskatchewan, Canada; we did not make a request for information to the Canadian Wildlife Service but compiled information from news outlets for that case. We combined all information into a database (Appendix).

Summary of Shootings across Populations

Although the Endangered Species Preservation Act of 1966, the precursor to the Endangered Species Act of 1973, conferred an additional level of protection to the Whooping Crane, there have been 27 recorded instances of shootings since the 1967 listing of the Whooping Crane as a federally endangered species, resulting in 37 total Whooping Crane fatalities. Here we summarize available information on shooting cases to date, and discuss the possible causes and apparent differences in cases between the four current Whooping Crane populations.

Whooping Crane shootings have been documented across all four existing free-ranging populations (Fig. 23.1). The Grays Lake Population (GLP) in the intermountain West no longer has Whooping Cranes and did not have a recorded shooting incident during its existence (1975–2002). In the EMP, shootings account for 19% of all mortality where the cause of death could be determined (Harrell, 2014). For the AWBP, the percentage of mortality caused by

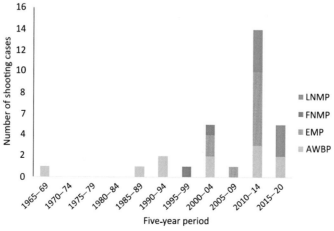

FIGURE 23.1 Number of confirmed shooting cases between 1967 and 2016 in the four wild populations. AWBP, Aransas-Wood Buffalo Population; FNMP, Florida Nonmigratory Population; LNMP, Louisiana Nonmigratory Population; EMP, Eastern Migratory Population.

shootings is about 20% of known causes of mortality (Stehn and Haralson-Strobel, 2014). The LNMP shooting incidents account for 24% of known mortality (Harrell, 2014).

The first documented shooting of a Whooping Crane after the 1967 listing as an endangered species occurred near Aransas Bay, Texas, in January 1968. The next documented shooting mortality occurred in 21 years later in 1989. The estimated shooting rate across all populations for 1989–2016 is 1.33 Whooping Cranes per year. In recent years, we document an accelerating trend in the number of Whooping Crane shootings (Fig. 23.1). Between 2011 and 2016 there have been 15 shooting incidents causing the loss of 21 Whooping Cranes from all populations combined, a rate of 4.2 birds per year over a 5-year period; 86% of Whooping Crane deaths from gunshot in that period occurred in reintroduced populations (EMP and LNMP) (Table 23.1 and Appendix).

In the 15 years since the initiation of the reintroduction of the EMP in 2001, there have been 10 documented shootings, compared to nine documented shooting cases in the AWBP since 1967 (Table 23.1). These result in a difference of nearly one Whooping Crane documented being shot per year in the EMP compared with one every five years in the AWBP (Table 23.1). The FNMP was established in 1993 and has had two known shooting incidents. As of the most recent winter survey of the AWBP in 2014–15, there are about 300 Whooping Cranes in that population. This estimate accounts for close to 70% of all Whooping Cranes in the wild, but only 20% of confirmed shooting incidents have occurred in that population since 2011.

Shootings in the AWBP

As previously mentioned, the first shooting incident in the AWBP after the ESA listing occurred in 1968, but no more were documented until 1989. A total of nine known Whooping Cranes have been taken in the six shooting cases in the AWBP, spread across three U.S. states and one Canadian province (Kansas, South Dakota, Texas, and Saskatchewan). Six incidents in the AWBP are related to hunting (see later section for details). The three non-hunting-related shootings were documented in two states, Texas and South Dakota, and one province, Saskatchewan.

Shootings in the EMP

The first confirmed shooting case in the EMP took place in July 2004 in Oceana County, Michigan. In this case, the necropsy of a 1-year-old Whooping Crane determined that it had been shot in the foot and later died of other causes. Since this first case, Whooping Crane shootings in the EMP have been documented in an additional five states: Alabama, Georgia, Indiana, Kentucky, and Wisconsin. Of the 10 cases in the EMP since the first birds were released in 2001, eight cases have been unrelated to a hunting season; the two cases unrelated to hunting are discussed in the hunting-related shootings section of this chapter.

TABLE 23.1   Number of Shooting Cases and Whooping Crane Deaths from Gunshot in the Four Wild Populations from 1967 to 2016: the AWBP, the FNMP, the LNMP, and the EMP

Population	Number of shooting cases	Number of Whooping Cranes shot	Number of cases/year	Number of Whooping Crane deaths from shootings/year
AWBP (since 1967)	9	10	0.18	0.20
FNMP (est. 1993)	2	3	0	0.13
EMP (est. 2001)	10	14	0.67	0.93
LNMP (est. 2011)	6	10	1.2	2

For the five cases in the EMP in which a perpetrator was identified, situational analysis of the crime does not support a conclusion that the bird was shot because it was a Whooping Crane. Instead, it appears that the Whooping Crane happened to be an easy target for someone who shoots nongame wildlife for recreation or because the shooter mistook a Whooping Crane for another species. In one EMP case in which a shooter was identified, a man shot a Whooping Crane in a wheat field in July 2013. The perpetrator claimed that he mistook the Whooping Crane for an albino Sandhill Crane, a species for which there is no hunting season in Wisconsin and for which he otherwise had no permit to shoot. In another shooting in January 2011, a man shot an adult and a juvenile Whooping Crane on a peanut (*Arachis hypogaea*) farm in Cherokee County, Alabama. The perpetrator may have been squirrel hunting on the farm when he shot two Whooping Cranes that were standing near a pond on the farm. He claimed that he did not know what the birds were at the time of the shooting. In both cases a rifle was used to conduct the shooting, as is true for the eight non-hunting-related shooting cases in which we could identify the type of weapon used (eight cases).

In many cases perpetrators claimed to not know that they had shot a Whooping Crane, mistaking it for another species. However, under the MBTA it is illegal to hunt migratory birds with a rifle rather than a shotgun, except under specific licenses; in some states, nonmigratory game birds, such as Wild Turkeys (*Meleagris gallopavo*) can be hunted with a rifle. Thus, the shooters were in violation of the MBTA law even if the claim of aiming for another bird was true.

Shootings in the FNMP

The FNMP project began in 1993, and the last release of captive birds into that population occurred in 2004. Whooping Cranes in the FNMP do not migrate. Although no longer an active reintroduction project, a small number of remnant birds (<12) remain as of 2016. Two shooting incidents have been documented in this population, in May 1999 and November 2000, resulting in the deaths of three Whooping Cranes; neither case was related to hunting. No information except cause of death is available on the May 1999 case. The other incident, which occurred in St. Augustine, St. Johns County, Florida, in November 2000, was perpetrated by an 18-year-old male who was accompanied by a 14-year-old male. It is unclear from USFWS reports whether the 14-year-old was involved in the shootings. The perpetrators shot two Whooping Cranes from their truck and continued driving. The two also were accused of vandalizing homes with a slingshot and steel shot on the same day. The 18-year-old was charged and pled guilty to both killing endangered wildlife and shooting into an occupied dwelling.

Shootings in the LNMP

The most recent Whooping Crane reintroduction project began in 2011 in the White Lake Wetlands Conservation Area, Vermilion Parish, Louisiana. In the 5 years since that time, there have been six confirmed shooting incidents in the population that have taken a total of 10 birds, resulting in a rate of 1.2 Whooping Crane deaths from shootings per year. This is the highest rate of any of the four existing Whooping Crane populations. Nearly one in four Whooping Cranes in the LNMP has died from gunshot (Harrell and Bidwell, 2013).

In October 2011, 2 months after the inception of the Louisiana reintroduction project, two Whooping Cranes were shot in a rice (*Oryza sativa*) field in Jefferson-Davis Parish. The two cranes were shot by two teenagers, ages 13 and 16 years, who turned themselves in to the authorities after seeing news reports of the shootings. They claimed that they thought the Whooping Cranes were Snow Geese (*Chen caerulescens*), although there was no concurrent Snow Goose hunting season.

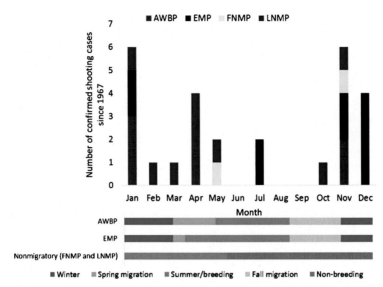

FIGURE 23.2   Seasonal occurrence of Whooping Crane shooting cases, divided into the four wild populations: the AWBP, the FNMP and LNMP, and the EMP. Timing of Whooping Crane breeding and migration status is represented by horizontal bars.

In January 2016, an 18-year-old man shot and killed two Whooping Cranes near Beaumont, Texas. Those Whooping Cranes were part of a small number of LNMP birds that spend part of the year in eastern Texas. This case is unique among modern shooting cases in that there is substantial evidence that the man preplanned the shooting and fully intended to shoot a pair of Whooping Cranes, knowing their Endangered Species status (Smith, 2016). Although the man identified strongly as a hunter, this was not a hunting accident. He pled guilty to an Endangered Species Act violation.

As of 2016, there has yet to be information made public about whether suspects have been identified for the other four cases in the LNMP (April 2013, February 2014, November 2014, and May 2016).

Seasonal and Geographic Differences between Shootings in the AWBP, EMP, and LNMP

For the two migrating Whooping Crane populations (AWBP and EMP), 33% of shooting incidents have occurred during migration, compared to 8% in the summer and 64% in the winter (Fig. 23.2). Allen (1952) found that 66% of all reported kills from 1722 to 1952 took place during migration. A similar proportion (60%) of Whooping Cranes have been lost to shooting in the AWBP since 1967: five shooting incident taking a total of six Whooping Cranes during fall and spring migration combined, four incidents taking four Whooping Cranes during the winter in the AWBP in the same period, and no recorded instances of people shooting Whooping Cranes during the summer (Fig. 23.2).

While the AWBP is experiencing a proportion of shootings during migration similar to what Allen recorded from 1722 to 1952, in the other migratory flock of Whooping Cranes (the EMP), 79% of all shooting cases have occurred during the wintering period. The wintering period is hard to define for the EMP, since EMP Whooping Cranes move fluidly in the wintering area throughout the winter period (see Mueller et al., 2018), but Allen's cutoff date of November 15th makes for a helpful comparison with the AWBP. In the EMP the wintering area is much larger than the wintering area for the AWBP. The wintering area for the AWBP is centered on Aransas National Wildlife Refuge, although recently the Whooping Crane winter use area has been expanding outward. The winter use area for the AWBP is still confined to the Gulf coast of Texas, primarily the Coastal Bend area. In contrast, birds in the EMP during the wintering period, defined by Allen (1952) as November 15 to March 14, can be found in eight different states, four of which (Alabama, Georgia, Indiana, and Kentucky) had Whooping Crane shooting cases during the winter months (Fig. 23.2).

Differences in known shooting rates could be attributed to many different factors. Monitoring levels are much higher in the reintroduced populations, which could mean that we are detecting a higher proportion of shooting events in the reintroduced populations than in the AWBP. All of the Whooping Cranes in the EMP and LNMP are banded, and a substantial number of them have radio, cellular, or satellite telemetry devices. Only 10–15% of the AWBP cranes are banded and less than 5% have satellite telemetry devices (Table 23.2).

TABLE 23.2 Number of Whooping Crane Shooting Incidents by State or Province and Population since 1967: the AWBP, the FNMP, the LNMP, and the EMP

State	AWBP	EMP	FNMP	LNMP
TX	6			1
LA				5
IN		4		
AL		2		
FL			2	
GA		1		
KS	1			
KY		1		
MI		1		
Saskatchewan	1			
SD	1			
WI		1		

Relationship of Hunting to Whooping Crane Shootings since 1967

Of the 25 known shooting incidents that have taken place since 1967 across all populations, seven have been confirmed as related to hunting activities: six in the AWBP, and one in the EMP. Two cases in the AWBP were related to a Snow Goose (*Chen caerulescens*) hunting season, three were related to Sandhill Crane (*Grus canadensis*) hunting, and in one case we were unable to determine which species the hunter was intending to shoot. In one case where a bird was shot during a waterfowl hunt on private land (AL: Dec. 2004), the perpetrator was never identified. One case in the EMP (IN: Dec. 2013) was suspected to be related to hunting activities, but authorities failed to identify a suspect and we could not confirm this information (Table 23.3).

We cannot know what the intention was for the hunters who have shot Whooping Cranes. In most cases that involved hunters, the hunters were already in violation of a separate hunting law. In the first two recorded cases of Whooping Crane shootings that were related to Snow Goose hunting seasons, there were no details available on whether the perpetrator was following all other rules and regulations and made an identification mistake.

In November 2003, a waterfowl hunter shot a Whooping Crane during fall migration in Ellis County, Texas, but the waterfowl hunter was

TABLE 23.3 Hunter-Related Shooting Cases in the Wild Populations: the AWBP, the FNMP, the LNMP, and the EMP

Population	Hunter related		
	No	Yes	Unknown
AWBP	3	6	
EMP	8	1	1
FNMP	2		
LNMP	5		
Total	19	7	1

in violation of multiple laws at the time, making it difficult to classify this case as a hunting accident and not intentional poaching (the perpetrator was hunting outside waterfowl hunting season and had no permit to hunt in the area). In November of the following year a group of seven hunters shot two Whooping Cranes in Stafford County, Kansas. The hunters shot the Whooping Cranes before legal hunting hours. It is not clear whether they thought they were shooting at Sandhill Cranes or at waterfowl, but the light was probably too low to make a correct identification.

The next case of a hunter-related Whooping Crane shooting occurred in January 2012 in Calhoun County, Texas, but there is little information available about this case other than that it was related to hunting. If the perpetrator had been hunting Sandhill Cranes, he would have been illegally hunting Sandhill Cranes in a closed zone. In January 2013 a hunter shot and killed a Whooping Crane in Aransas County, Texas. The perpetrator claimed he was intending to shoot a Sandhill Crane but instead shot a juvenile Whooping Crane. This took place in the closed zone for Sandhill Crane hunting.

In the AWBP, these cases of Whooping Crane shootings related to hunting, by hunters already in violation of other hunting laws, are pertinent due to Section 7 of the Endangered Species Act. This section states that federal agencies must consult with the U.S. Fish and Wildlife Service about activities that might endanger or threaten an endangered species. The USFWS has determined that incidental taking of Whooping Cranes due to legal hunting activities is allowed at a rate of one per decade, starting from 2011. Since then, there have been two cases related to hunting activities in the AWBP, although whether the hunters were legally hunting is questionable, since in both cases the hunters were in violation of several hunting regulations in addition to shooting a Whooping Crane.

None of the incidents in Louisiana appear to have been related to hunting. There is no Sandhill Crane hunting season in Louisiana, and none of the shootings were concurrent with a waterfowl hunting season.

In three out of seven cases that were confirmed to be related to hunting activities, the hunters intended to shoot Sandhill Cranes. As of 2016, 17 states allow Sandhill Crane hunting, and several more are discussing adding a Sandhill Crane season. Many opponents of Sandhill Crane hunting cite accidental taking of Whooping Cranes as a major reason why the state should not allow a Sandhill Crane season. We cannot determine from this limited data whether Sandhill Crane hunting is a major threat to Whooping Cranes in areas where it is allowed. The data here show that 11% of shooting cases are attributed to Sandhill Crane hunting, and in all of these cases the perpetrators were in violation of other hunting laws while attempting to take a Sandhill Crane. In four out of seven hunting-related cases, the intended target species was not determined. In two shooting cases that were related to hunting, the perpetrators turned themselves in to the authorities, which has never happened in a shooting case that was unrelated to hunting. In three hunter-related cases it is unknown whether the hunter turned himself in, in one case a hunter was reported by others, and in one case a suspect was never identified. In six out of seven Whooping Crane shooting cases related to hunting activities, perpetrators were identified, compared with non-hunter-related cases, where a perpetrator was identified for 10 out of 18 shooting cases. Non-hunting-related cases may be harder to detect; people who shoot Whooping Cranes unrelated to a hunting season may try harder to evade the authorities, as evidenced by the fact that none of the non-hunting-related shooters turned themselves in.

Another factor that could possibly contribute to lower shooting rates in the AWBP is the longer prevalence of legal Sandhill Crane hunting in the Central Flyway states. Kansas, Montana, North Dakota, South Dakota, Oklahoma, Wyoming, and Texas all have had Sandhill Crane hunting seasons since the 1960s or 1970s and have Whooping Cranes that either winter or migrate through. Tennessee and Kentucky (EMP states) also have Sandhill Crane seasons, although both of those states instituted Sandhill Crane seasons within the last 5 years. In some of the states that have a Sandhill Crane season, Sandhill Crane hunters must take an identification test to show that they know how to properly distinguish between a Sandhill Crane and both adult and juvenile Whooping Cranes. A heightened level of awareness among hunters may contribute to overall higher levels of awareness of Whooping Cranes in rural communities, but the effect of Sandhill Crane hunting on awareness and shootings of Whooping Cranes is unclear.

Relationship of Shootings to Habitat and Area Protection

For the 18 cases in which we could determine what type of habitat the crane(s) were in at the time of the shooting, 11 took place in agricultural fields, 1 took place on a riverbank, and 6 took place in wetlands. All hunter-related shootings took place either in wetlands or in unknown locations. Non-hunting-related shootings took place in agricultural fields, on a riverbank, and in wetlands.

Whooping Cranes spend large amounts of time in agricultural fields, especially during migration and the wintering period (Barzen et al., Chapter 14, this volume; Belaire et al., 2014; Howe, 1989). Agricultural fields are easily accessible to the public, especially if they are near a major road. In addition, Whooping Cranes are highly visible in agricultural fields as they tend to use them only when the fields are fallow or have low vegetation, providing no cover for the birds. These factors may make Whooping Cranes highly vulnerable to poaching while they are in agricultural fields.

In the 15 cases where the location of a Whooping Crane shooting could be defined as public or private land, 12 shooting cases took place on private land, and three took place on public land. In the AWBP one event was confirmed to take place on private land and two took place on public land; six other events could not be categorized. In the EMP six shooting cases took place on private land, one took place on public land, and three could not be categorized by land ownership. In the FNMP one case took place on public land, and the other case could not be categorized by land ownership. In the LNMP, four cases took place on private land, and two cases could not be categorized by land ownership.

The difference in availability of public and private land in each of the populations could account for some differences in shooting rates on public versus private lands. As of 2016 the EMP has about 105 birds, and during the winter they are spread throughout a large region and most diurnal habitat use is on unprotected land (Mueller et al., 2018). In contrast, the AWBP has about 300 birds that are concentrated in a smaller area along the Gulf coast of Texas. Much of the AWBP wintering area is protected, although Whooping Cranes in the AWBP continue to expand away from the core wintering area (Smith et al., Chapter 13, this volume).

Age and Sex of Whooping Cranes Shot

Of the 33 Whooping Cranes that have been shot since 1987, 25 were adults, 6 were juveniles, and 6 were of unknown age. Birds were classified as juveniles if they were under 1 year of age (and usually based on the continued presence of rusty-colored juvenile feathers), and adults if they were 1 year of age or older. It may be easier to mistake a first-year Whooping Crane for a Sandhill Crane due to some similarity in coloration, despite differences in size. Both of the juveniles shot in the AWBP were classified as unintentional shootings by a Sandhill Crane hunter, although they were both shot in the closed zone for Sandhill Crane hunting.

All Whooping Cranes in the EMP, LNMP, and FNMP are banded; therefore, sex, age, and other information are available for all individuals in those populations. No information on age was available to us for the FNMP shooting cases. Of the Whooping Cranes shot in the EMP, 29% were juveniles. The ratio of juveniles to adults in the EMP has fluctuated over the years, depending on how many juveniles have been released into the population. Typically juveniles represent no more than 20% of the population. The number of juvenile Whooping Cranes shot in the EMP is skewed by a single incident in Georgia in 2010 that took three juveniles at the same time. All of the cranes shot in the LNMP were adults. A larger proportion of the Whooping Cranes in Louisiana are juveniles compared with adults, since the population currently numbers only about 40 birds, and 11 juveniles were released there in 2016.

Of the 37 Whooping Cranes that have been shot since 1967, 12 were female, 12 were male, and 13 were of unknown sex. The sex ratio of shot birds varied among populations: the AWBP had eight birds whose sex was not reported along with one male and one female, the EMP had an even sex ratio, the FNMP had two males shot and one bird of unknown sex, and the LNMP had four birds whose sex was not reported, and five females and three males shot.

There is no sexual dimorphism between male and female Whooping Cranes, making it unlikely that one sex would be a better target than the other. Since there is no strong pattern toward male or female shootings, it is unlikely that shootings are having an impact on breeding dynamics, other than the disruption that a shooting causes when the surviving member of the pair is forced to find a new mate.

Analysis of Shooting Events

A perpetrator was identified in 17 of 27 cases (63%). Because the offer of a financial reward for information leading to a conviction may influence reporting and identification of a perpetrator, we used institutional and media records to record data on the 11 studied cases where rewards were offered. There were often many contributors to an individual case's reward funds, including local and national conservation and animal welfare nonprofits, but in some cases the names of the contributing organizations, refuge affiliates, state and federal resource agencies, and individuals were not available to us. As of 2016, 25 organizations and several individual donors have contributed reward money in various shooting cases. Rewards ranged from $2,500 to $23,250.

Rewards did help identify a suspect in some cases. For example, in the November 2009 case of a Whooping Crane shot in Cayuga County, Indiana, a citizen tip in response to a $2,500 reward led to identification of a suspect. In many cases a reward was not needed because perpetrators were identified right away or turned themselves in.

Demographics of Perpetrators

For 10 cases there is information available on the demographics of the perpetrators. All identified perpetrators were white males with an average age of 27.6 years old; the youngest perpetrator was 13 years old, and the oldest 48. Most people who commit wildlife crimes in the United States are white and male, so this was not surprising (Crow et al., 2013). Some of the perpetrators had prior convictions, such as a citation for driving under the influence of alcohol. Several of the perpetrators were illegally in possession of the gun they used to shoot a Whooping Crane. In six cases the perpetrators also conducted concurrent crimes, such as vandalism of property, or shooting of other non-game wildlife or game wildlife out of season.

Case Prosecutions

Of the 17 cases in which a perpetrator was identified, 12 cases (44% of the total number of cases) resulted in prosecutions; three of these cases were related to hunting activities. In nine cases we were able to identify which law was used to prosecute the perpetrator. The two federal laws used to prosecute people who have shot a Whooping Crane are the Endangered Species Act (ESA) and the Migratory Bird Treaty Act (MBTA), while three cases were prosecuted under state laws in Florida, Indiana, and Louisiana (Table 23.4).

The Endangered Species Act (ESA) of 1973 provides the harshest punishments for the taking of a Whooping Crane of any applicable statute, with civil penalties of up to $25,000 per violation, and criminal penalties of up to $50,000, and 1 year imprisonment per violation. According to the ESA Section 3(19), the term "take" means to harass, harm, pursue, hunt, shoot, wound, kill, trap, capture, or collect, or to attempt to engage in any such conduct. The ESA has been used for prosecution in two recent cases. A South Dakota man shot a Whooping Crane in Hand County in April 2012. He was also found to be in violation of the Migratory Bird Treaty Act and witness

TABLE 23.4 Sentencing from Nine Whooping Crane Shooting Cases That Resulted in Successful Prosecutions from 1967 to 2016 in the Wild Populations: the AWBP, the FNMP and LNMP, and the EMP. Information Was Pulled from News Reports and Freedom of Information Act Requests to the U.S. Fish and Wildlife Service (USFWS) and the U.S. Department of Justice (USDOJ). These Cases Represent Only Those for Which We Were Able to Identify the Law That Was Used for Prosecution. One Case, a Shooting That Took Place in Kansas in 2004, Resulted in a Federal Prosecution, but We Could Not Identify Which Law Was Used. In 18 Cases a Perpetrator Was Not Identified, or the Case Did Not Go to Court. The Law Column Indicates Which Law Was Used for Prosecution in Each Case: Migratory Bird Treaty Act (MBTA), Endangered Species Act (ESA), or State Law.

Law	State	Date	Hunter related	Fine	Jail time	Probation	Community service	Hunting privileges revoked
ESA	TX	Jan. 2016	No	$25,850	No	5 years	200 h	5 years
ESA and MBTA	SD	Apr. 2012	No	$85,000	30 days	2 years	Yes, hours unknown	2 years
MBTA	TX	Nov. 2003	Yes	$2,000	6 months	No	None	Indefinitely
MBTA	AL	Jan. 2011	No	$425 (never collected)	No	No	None	No
MBTA	TX	Jan. 2012	Yes	Unknown	No	Unknown	Unknown	Unknown
MBTA	TX	Jan. 2013	Yes	$5,000	No	1 year	$10,000 community service payment to Friends of Aransas and Matagorda Island National Wildlife Refuges	No
MBTA	WI	Jul. 2013	No	$2,000	No	Unknown	None	2 years
State	FL	Nov. 2000	No	No	75 days	2.5 years	200 h	Unknown
State	IN	Nov. 2009	No	$1 + $504.50 in court fees	No	Unknown	None	Unknown
State	LA	Oct. 2011	No	$0	No	None	Unknown number of hours at White Lake Wetlands Conservation Area	No

tampering. His punishment included an $85,000 fine, 30 days in jail, 2 years' probation, and hunting privileges revoked for 2 years. It is likely that the charge of witness tampering increased his punishment. The second case took place in Texas in January 2016. The sentencing administered included a $25,850 fine, 5 years' probation during which time the perpetrator cannot own a gun or hunt or fish in any state, and 200 h of community service.

Most shootings have occurred in reintroduced populations, which are considered experimental, nonessential populations under the ESA. This distinction only impacts accidental shootings. If an accidental shooting were to occur in a nonessential population, it would not be prosecuted under the ESA, whereas an intentional shooting in a nonessential population could be prosecuted under the ESA. In theory, it is easier to prosecute under the ESA in the AWBP rather than in the reintroduced populations because of the rules regarding accidental shootings in nonessential populations. However, in the AWBP three cases have gone to court and they were all prosecuted under the MBTA (Table 23.4).

One possible obstacle to ESA prosecution is the McKittrick Policy, which was set in 1998. Prior to *U.S. v. McKittrick*, it was not required for prosecutors to prove that a defendant knew that he or she was taking an endangered species and was in violation of the ESA (Newcomer et al., 2010). In the case, McKittrick shot a Gray Wolf near Red Lodge, Montana, in 1995. After McKittrick was convicted of illegally taking an animal listed as endangered pursuant to the ESA, the case went through several appeals until reaching the Supreme Court. The Supreme Court appeal had to do with the jury instruction on the definition and understanding of "knowingly." The Supreme Court declined the case, but in a government brief regarding the appeal, the government claimed that the jury instruction failed to fully explain the meaning of the term "knowingly."

The result in practice is that the burden falls on the government to prove that a defendant knew that he or she was shooting an endangered species in order to be prosecuted under the ESA. Whether this was the original intent of the ESA or not, it makes prosecution under the ESA very difficult in cases of illegal taking of an endangered species. This potentially complicates the information we have about past shooting events. Although there have likely been cases where a person has made a genuine identification mistake, there may have been instances where a person who shot a Whooping Crane merely stated that he was intending to shoot another, nonendangered target (e.g., a Snow Goose) to avoid an ESA prosecution. This could apply to both hunting- and non-hunting-related shooting cases.

The January 2016 case in Texas involved the shooting of two LNMP birds that were part of a nonessential population. The case was initially tried under the Migratory Bird Treaty Act, but was refiled under the ESA after more evidence was gathered. The prosecution may have felt that they had sufficient evidence to prove intent, but regardless, the perpetrator pled guilty to violating the ESA. Specifically, he pled guilty to possession, rather than taking, but all violations of the ESA are subject to the same punishment. The International Crane Foundation (ICF), local conservation groups, and individuals wrote letters to the judge and prosecuting attorney, encouraging them to consider a harsh sentence in order to serve as an effective deterrent. As described previously, the sentencing included a $25,850 fine, 200 h of community service, and 5 years' probation during which the man cannot own firearms or hunt or fish in all states.

If another shooting case were to occur in eastern Texas with an LNMP bird, it is important to note that the Endangered Species Act requires that the populations be given a geographic boundary. Technically, when an LNMP bird flies outside of its designated boundary and past the AWBP boundary, its population status changes from a nonessential population to an essential population. This could potentially affect whether or not the ESA could be used for prosecution of a shooting case of an LNMP bird in Texas. Since the LNMP continues to spend part of the year in Texas, we may yet see another shooting case similar to the January 2016 shooting of LNMP Whooping Cranes in Texas.

The law most commonly invoked for Whooping Crane shooting cases is the MBTA, which has been used in the prosecution of six cases. Whooping Cranes have been protected under the MBTA since its inception in 1918. "The Migratory Bird Treaty Act makes it illegal for anyone to take, possess, import, export, transport, sell, purchase, barter, or offer for sale, purchase, or barter, any migratory bird, or the parts, nests, or eggs of such a bird except under the terms of a valid permit issued pursuant to Federal regulations" (16 U.S.C. 703-712). The MBTA allows for a fine of up to $10,000 per violation or 1 year in jail, or both. Punishments under the MBTA have ranged from a $425 fine (AL: Jan. 2011) to a $5,000 fine and 1 year's probation (TX: Jan. 2013).

If a case is not federally prosecuted, some cases have been prosecuted under state law. In two cases there have been successful prosecutions under state law for a Whooping Crane shooting. In one case (IN: Nov. 2009) the prosecution resulted in a $1 fine. In another case (FL: Nov. 2000) the punishment was more severe, resulting in 75 days in jail and 2.5 years of probation.

In summary, punishments for Whooping Crane shooting cases have run from very small (a $1 fine) to very large (an $85,000 fine plus jail time, probation, and revocation of hunting privileges) (Table 23.4). Some of this variation is related to the laws used for prosecution. The most severe punishment was given out for a combined ESA and MBTA prosecution along with witness tampering, while the least severe was for a state law prosecution. The choices

and perspectives of the district attorney and the judge involved in each case play a very significant role in the variation seen in the severity of penalties in these cases. As an example, two cases that were both prosecuted under the MBTA (AL: Jan. 2011 and TX: Jan. 2013) had very different sentencing results (Table 23.4). Judges, district attorneys, and juries may not be familiar with the conservation history of Whooping Cranes prior to being assigned to a Whooping Crane shooting case. For the case in South Dakota in 2012, the presiding judge took it upon himself to contact the ICF to get an estimate for the monetary value of a Whooping Crane, which ICF estimated at $85,000. The judge decided to fine the defendant that amount, which the defendant is paying in regular small increments in support of Whooping Crane conservation activities.

The authors believe that providing prosecutors and judges a succinct, compelling summary of the conservation history and investment for Whooping Cranes could help get more severe penalties in prosecuted cases. Putting such information in the hands of local advocates, ready to oversee investigations and prosecutions of these shooting cases when they occur, could also be effective in encouraging more severe penalties upon conviction. Educating district attorneys in key Whooping Crane areas before another shooting case occurs may be worthwhile as well. This could include sending packets of information but could also include field trips to view Whooping Cranes in the wild and tours of facilities where Whooping Cranes are raised in captivity, so that people in the criminal justice system can understand the resources and labor that go into Whooping Crane conservation efforts.

Other Factors Possibly Affecting Shooting Cases

The relatively high level of Whooping Crane shootings in Louisiana may be attributed to local cultural views of Whooping Cranes and poaching. Historically, Whooping Cranes in Louisiana were viewed as an agricultural pest and a source of food (Gomez, 1991). In the 1970s, locals were interviewed concerning a possible reintroduction project, and it was met with mixed opinions. Among those who opposed reintroducing Whooping Cranes to Louisiana, some disliked the idea of an increased federal presence to protect Whooping Cranes, and thought that local people may be asked to make changes to their land to accommodate Whooping Cranes. There were also concerns about how the presence of Whooping Cranes might impact hunting and fishing, which are important to Cajun culture (Gomez, 1991).

The recent reintroduction of Whooping Cranes to Louisiana may have brought up feelings of resentment toward the government. In addition, poaching is considered a folk crime by some rural people in Louisiana. It is culturally acceptable in some places, and may even be seen as a tradition and an entertainment (Forsyth et al., 1997).

EFFORTS TO REDUCE WHOOPING CRANE SHOOTINGS

Reducing Non-Hunting-Related Shootings

Situational crime prevention (SCP) is a set of strategies used to reduce opportunities for crimes to occur. SCP has been successfully applied to reducing the occurrence of many different crimes, from home burglaries to terrorism, and has been applied to wildlife crimes such as poaching (Clarke, 1997; Pires and Clarke, 2012). One of the underlying theories of SCP is routine activity theory, which dictates that three circumstances have to align in order for a crime to occur: a desirable target or victim, a lack of guardianship of that target, and a motivated offender. Routine activity theory focuses on the situation

surrounding crimes, and also focuses on the idea that criminals commit crimes that intersect with the activities of their daily routines (Cohen and Felson, 1979). Wildlife crimes tend to occur while people are already engaged in legitimate outdoor activities and an opportunity for poaching presents itself (Crow et al., 2013). As far as we know, non-hunting-related shootings of Whooping Cranes fit into the mold of routine activity theory; these are not cases of premeditated crimes. As in the case of the two Florida teenagers who shot two Whooping Cranes in November 2000, these were crimes of opportunity; they were in the middle of a streak of vandalism, they had vandalized at least six homes with slingshots and steel shot, and when they saw the large white birds in a field, they chose to shoot them.

SCP provides a set of 25 strategies used to reduce crimes. Many of these strategies could be applied to Whooping Crane shootings, such as marking more birds with color bands, especially in the AWBP, where very few birds are banded. This shows ownership of the birds, which reduces motivation to shoot them, similar to the effect of cow branding on reducing cattle rustling. Educating the public about Whooping Cranes fits into several SCP crime reduction strategies.

A case of a non-hunting-related shooting might provide an example of how education and awareness might help reduce the rate of Whooping Crane shootings. The 25-year-old male who shot a Whooping Crane in Hand County, South Dakota, admitted in a USFWS interview to shooting hawks and waterfowl out of season on other occasions but maintained that he had never killed an eagle. This man had a reputation in his community for "riding and shooting," meaning he would drive in the countryside and shoot at whatever animals he came across. Why would a man who had no qualms about shooting hawks and other birds feel the need to point out that he had never shot an eagle? In the United States there is tremendous social pressure against shooting Bald Eagles (*Haliaeetus leucocephalus*). We presume that this came about through decades of social marketing, designed to make Americans identify with Bald Eagles as the national symbol (Wendell et al., 2002). In addition, many cases of Bald Eagle shootings have been prosecuted to the fullest extent of the law, and these cases have been highly publicized, creating an effective deterrent for potential shooters.

Allen (1952) suggested that losses of Whooping Cranes due to "other than natural causes" could be solved by a public education campaign: "It is believed that these losses can be cut by 50% and perhaps this can only be accomplished by continued education through popular articles and similar publicity calculated to reach 'the man with the gun'" (Allen, 1952, p. 86). Allen did not have social science data to substantiate his belief that people who shot Whooping Cranes simply lacked education about the species; it was just a suggestion based on his knowledge about the number of Whooping Cranes being shot each year. Increased awareness of Whooping Cranes and the potential punishments for shooting one fit into several strategies for crime reduction under SCP. For example, if more people know about Whooping Cranes and want to protect them in a community that has Whooping Cranes on the landscape, this could increase guardianship of the species, thereby increasing the risks to individuals committing the crime of shooting a Whooping Crane. Another strategy that education on Whooping Cranes can address is neutralizing peer pressure. Using slogans such as "I give a Whoop!" and "Texas cares about Whooping Cranes!" could create pressure against shooting them. (Clarke and Eck, 2005)

Several organizations are currently educating the public about Whooping Cranes in an effort to reduce shootings. As of 2016, the Louisiana Department of Wildlife and Fisheries (LDWF) has been conducting a statewide awareness campaign for several years in an effort to reduce Whooping Crane shootings. This campaign has consisted of billboards, 30-second radio and television public service announcements, programs for children, posters, and other outreach activities. Following the example of the LDWF campaign and other campaigns designed to generate pride in endangered species, the ICF is leading efforts to develop pride campaigns for Whooping Cranes. A pride campaign addresses all three parts of the crime triangle under routine activity theory. When more people are aware that Whooping Cranes are on the landscape and feel pride that they have Whooping Cranes nearby, this increases guardianship of the species. Hopefully, this will cause more people to be willing to prevent and report Whooping Crane shootings, which will deter likely offenders. In addition, creating pride in Whooping Cranes reduces motivation to shoot them and reduces their attractiveness as a target.

The ICF, with the help of many partners, has initiated a pride campaign in a pilot community in northern Alabama that will use public service announcements on television and radio, billboards, a pledge campaign, a Whooping Crane mascot, hunter education, traditional and social media outreach, and K-12 education. ICF is working with Auburn University staff, who have developed pre- and postcampaign surveys to evaluate the impact of pride campaign activities on the public. In the future the pride campaign efforts will expand to new pilot regions, including southern Indiana, central Wisconsin, and the Gulf coast of Texas. Each of these projects will incorporate survey or focus group evaluations to determine which activities and messages are the most effective at increasing pride in Whooping Cranes.

Another potential strategy for this pride campaign is engagement of citizen scientists in which the public is encouraged to participate in Whooping Crane monitoring through a website called "Whooper Watcher." Reports can be submitted fairly easily through a user-friendly web-based reporting platform. This could help citizens feel engaged in Whooping Crane conservation and increase guardianship for the birds sharing the landscape with the local community. It will also aid conservation efforts by developing more frequent and widespread information on Whooping Crane locations, which could "remove the excuses" for illegal shootings by "posting instructions" and "alerting the public conscience," according to the SCP framework (Clarke and Eck, 2005).

Reducing Hunting-Related Accidental Shootings

Although most of the past cases have been unrelated to a hunting season, efforts to educate waterfowl and Sandhill Crane hunters in areas where Whooping Cranes also occur are underway. After the shooting of two Whooping Cranes in Kansas by a group of seven hunters in 2004, the state of Kansas made changes to hunting regulations in an attempt to decrease the likelihood of a future Whooping Crane shooting event. Legal hunting hours now start later in the morning, and the start of hunting season has been moved to a later date to allow time for more Whooping Cranes to pass through the state on migration before the hunting season starts. In addition, Sandhill Crane hunters must pass an online identification test. Some other states also include Whooping Cranes in their hunter education courses. Including Whooping Cranes in hunter education courses serves to remove excuses for shooting them, another strategy for crime prevention in the SCP framework (Clarke and Eck, 2005).

As of 2016, efforts are underway to make Whooping Cranes a more prominent part of national online hunter education courses that are used across many states for hunter

safety certification, such as hunter-ed.com, beasafehunter.org, and HUNTERcourse.com. We are also working to reach out to individual state hunter education programs. Ducks Unlimited, Inc., and other hunting organizations are reaching out to their members, asking them to be alert for Whooping Cranes and to report any potential Whooping Crane shootings to the authorities (Cooper, 2015).

These efforts are designed to curb the potential for accidental shootings related to a hunting season, but they could also provide an avenue into rural communities. Hunting is popular in rural communities, and the January 2016 shooting of a Whooping Crane demonstrates that some people who identify as hunters have little regard for hunting regulations. The perpetrator's social media indicates that he identifies strongly as a hunter, and he aspires to be a champion duck caller. This is directly contradicted by his decision to poach two Whooping Cranes. Continued education of hunters through online and classroom hunter education courses, along with hunter outreach events, could help prevent both accidental and intentional shootings.

SUMMARY AND OUTLOOK

It is very likely that the number of Whooping Crane shootings exceeds the 27 confirmed cases examined in this chapter. Even without accounting for additional shootings, Whooping Crane shootings are having a significant impact on the growth of wild Whooping Crane populations, especially in the reintroduced populations. Establishing self-sustaining populations outside of the AWBP will depend partly on adult survival in those populations. Partners involved with the International Whooping Crane Recovery Program need to address the threat of shootings in all wild populations in order to ensure the recovery of the species.

More shootings appear to be occurring in the reintroduced populations than in the AWBP. Higher shooting rates in the reintroduced populations could be attributed to higher levels of monitoring, but there may be other factors involved. The vast majority of the Whooping Cranes in the reintroduced populations in 2016 were raised in captivity. Although aviculturists at Whooping Crane breeding facilities go to great lengths to isolate Whooping Crane chicks from humans, captive rearing may be producing Whooping Cranes with less fear of humans, and thus less wariness of humans in general, consequently increasing their accessibility for a lethal shooting event.

People who live in the areas occupied by Whooping Cranes in the AWBP have always shared a landscape with these birds. They may be more familiar with Whooping Cranes, may be better able to identify one, and may know more about the conservation history of this species than people who share landscapes with the reintroduced populations. This difference may explain the higher ratio of non-hunting-related shootings to hunting-related shootings in the reintroduced populations.

Without more social science and wildlife crime research on endangered species issues, we are lacking crucial information. Surveys should be designed to find out what would help people understand the conservation value of the species and take pride in the fact that Whooping Cranes are living in their communities. It is likely that different biases toward Whooping Cranes exist in different communities. For example, in Wisconsin many farmers have negative feelings toward Sandhill Cranes, which cause crop damage in certain parts of the state (Hygnstrom and Craven, 1985). Such negative bias toward cranes may have played a role in the July 2013 shooting of a Whooping Crane in Wisconsin; the perpetrator stated that he thought he had shot an albino Sandhill Crane. A study conducted in Louisiana found that some negative attitudes toward Whooping Cranes might stem from a fear of federal intervention due to the status of the species as federally endangered (Gomez, 1991). Having access to interviews with those successfully prosecuted for Whooping Crane shootings might also contribute information about motivations of the perpetrators, and inform about possible mitigation activities to stop these crimes.

From examining FOIA requested documents, it is clear that investigators take Whooping Crane shootings seriously. It appears that judges and district attorneys often do not understand the conservation value of this species. As conservationists, we need to work closely with key members of the criminal justice system, and we must communicate to them with language that they are likely to understand. This includes assigning a monetary value to Whooping Cranes, but there may be other ways to raise the level of importance of Whooping Cranes in the eyes of people in the criminal justice system. Advocacy groups, such as conservation non-profit organizations, should work with judges and attorneys to frame the Whooping Crane conservation story in language and materials that will appeal to the court. The January 2016 shooting case in Texas demonstrates that outreach to the court can result in a more severe punishment.

Conservationists need to address the threat of shootings to Whooping Cranes. While it is unlikely that we could ever completely eliminate the threat, a reduction of the shooting rate, in combination with ongoing efforts to reduce other threats and increase productivity on the breeding grounds, will aid recovery efforts for all wild Whooping Crane populations. Whooping Crane conservation efforts rely heavily on public support, and a campaign to raise awareness about Whooping Cranes will benefit many aspects of the recovery plan.

References

Allen, R.P., 1952. The Whooping Crane. Research Report 3. National Audubon Society, New York.

Barzen, J.A., Lacy, A., Thompson, H.L., Gossens, A.P., 2018. Habitat use by the reintroduced Eastern Migratory Population of whooping cranes (Chapter 14). In: French, Jr., J.B., Converse, S.J., Austin, J.E. (Eds.), Whooping Cranes: Biology and Conservation. Biodiversity of the World: Conservation from Genes to Landscapes. Academic Press, San Diego, CA.

Belaire, J.A., Kreakie, B.J., Keitt, T., Minor, E., 2014. Predicting and mapping potential whooping crane stopover habitat to guide site selection for wind energy projects. Conserv. Biol. 28 (2), 541–550.

Canadian Wildlife Service and U.S. Fish and Wildlife Service, 2005. International recovery plan for the whooping crane. Recovery of Nationally Endangered Wildlife (RENEW), Ottawa, and U.S. Fish and Wildlife Service, Albuquerque, NM.

Clarke, R.V., 1997. Situational Crime Prevention, second ed. Harrow and Heston, Guilderland, NY.

Clarke, R.V., Eck, J.E., 2005. Crime Analysis for Problem Solvers in 60 Small Steps. U.S. Department of Justice, Center for Problem-Oriented Policing, Washington, DC. Available from: http://www.cops.usdoj.gov/pdf/CrimeAnalysis60Steps.pdf.

Cohen, L.E., Felson, M., 1979. Social change and crime rate trends: a routine activity approach. Am. Sociol. Rev. 44 (4), 588–608.

Cooper, A., 2015. Hunters can help one of our rarest birds. Ducks Unlimited. Available from: www.ducks.org/conservation/national/hunters-can-help-one-of-our-rarest-birds; accessed 12 November 2016.

Crow, M.S., Shelley, T.O., Stretesky, P.B., 2013. Camouflage-collar crime: an examination of wildlife crime and characteristics of offenders in Florida. Deviant Behav. 34 (8), 635–652.

Forsyth, C., Gramling, R., Wooddell, G., 1997. The game of poaching: folk crimes in southwest Louisiana. Soc. Nat. Resour. 11 (1), 25–38.

Gomez, G.M., 1991. Whooping cranes in southwest Louisiana: history and human attitudes. Proceedings of the North American Crane Workshop 6, 19–23.

Hahn, P., 1963. Where Is That Vanished Bird? – An Index to the Known Specimens of the Extinct and Near Extinct North American Species. Royal Ontario Museum, University of Toronto, Canada.

Harrell, W., 2014. Report on Whooping Crane Recovery Activities (2013 Breeding Season – 2014 Spring Migration). U.S. Fish and Wildlife Service. Available from: https://www.fws.gov/uploadedFiles/WC%20Recovery%20Activities%20Report_SeptApril%202014_Sub4.pdf.

Harrell, W., Bidwell, M., 2013. Report on Whooping Crane Recovery Activities (2012 Breeding Season – 2013 Spring Migration). Canadian Wildlife Service and U.S. Fish and Wildlife Service. Available from: https://www.fws.gov/uploadedFiles/WCRecoveryActivitiesReport_SeptApril2013_24Sept2013_Sub_508.pdf.

Hornaday, W.T., 1913. Our Vanishing Wild Life: Its Extermination and Preservation. Charles Scribner's Sons, New York.

Howe, M.A., 1989. Migration of Radio-Marked Whooping Cranes from the Aransas-Wood Buffalo Population: Patterns of Habitat Use, Behavior, and Survival. Fish and Wildlife Technical Report 21. U.S. Fish and Wildlife Service, Washington, DC.

Hughes, J.M., 2008. Cranes: A Natural History of a Bird in Crisis. Firefly Books, Richmond Hill, ON.

Hygnstrom, S.E., Craven, S.R., 1985. State funded wildlife damage programs: the Wisconsin experience. In: Bromley, P.T. (Ed.), Proceedings of the Second Eastern Wildlife Damage Control Conference, North Carolina State University, Raleigh, NC, pp. 234–242.

Mueller, T., Teitelbaum, C.S., Fagan, W.F., Converse, S.J., 2018. Movement ecology of reintroduced migratory whooping cranes (Chapter 11). In: French, Jr., J.B., Converse, S.J., Austin, J.E. (Eds.), Whooping Cranes: Biology and Conservation. Biodiversity of the World: Conservation from Genes to Landscapes. Academic Press, San Diego, CA.

Muth, R.M., Bowe, Jr., J.F., 1998. Illegal harvest of renewable natural resources in North America: toward a typology of the motivations for poaching. Soc. Nat. Resour. 11 (1), 9–24.

Newcomer, E., Palladini, M., Jones, L., 2010. The Endangered Species Act v. The United States Department of Justice: how the Department of Justice derailed criminal prosecutions under the Endangered Species Act. Anim. Law 17 (2), 251–271.

Pires, S., Clarke, R.V., 2012. Are parrots CRAVED? An analysis of parrot poaching in Mexico. J. Res. Crime Delinq. 49 (1), 122–146.

Smith, S., 2016. Ruffled feathers. Texas Monthly, September. Available from: http://www.texasmonthly.com/articles/whooping-cranes-texas/.

Smith, E.H., Chavez-Ramirez, F., Lumb, L., 2018. Wintering habitat ecology, use, and availability for the Aransas-Wood Buffalo Population of whooping cranes (Chapter 13). In: French, Jr., J.B., Converse, S.J., Austin, J.E. (Eds.), Whooping Cranes: Biology and Conservation. Biodiversity of the World: Conservation from Genes to Landscapes. Academic Press, San Diego, CA.

Stehn, T.V., Haralson-Strobel, C.L., 2014. An update on mortality of fledged whooping cranes in the Aransas/Wood Buffalo Population. Proceedings of the North American Crane Workshop 12, 43–50.

Wendell, M.D., Sleeman, J.M., Kratz, G., 2002. Retrospective study of morbidity and mortality of raptors admitted to Colorado State University Veterinary Teaching Hospital during 1995 to 1998. J. Wildl. Dis. 38 (1), 101–106.

APPENDIX

Whooping Crane Shooting Cases between 1967 and 2016 Compiled from News Reports and Freedom of Information Act Requests to the USFWS and the USDOJ in the Four Wild Populations: the AWBP, the FNMP, the LNMP, and the EMP.

Date	State	No. cranes shot	Population	Life cycle status	Age class	Hunter related?	Perpetrator identified?	Prosecution	Sentencing
1968	TX	1	AWBP	Wintering		Yes	Yes	Unknown	
Jan. 1989	TX	1	AWBP	Wintering	Adult	Yes	Yes	Unknown	$21,000 fine
Apr. 1990	SK	1	AWBP	Spring migration	Adult	No	Yes	Unknown	
Apr. 1991	TX	1	AWBP	Spring migration		No	Yes	Unknown	$23,100 fine
May 1999	FL	1	FNMP	Nonmigratory		No	Unknown	Unknown	
Nov. 2000	FL	2	FNMP	Nonmigratory		No	Yes	State	75 days' jail time, 2.5 years' probation, 200 h community service, loss of driving privileges
Nov. 2003	TX	1	AWBP	Fall migration		Yes	Yes	MBTA	6 months' jail time, $2,000 fine, hunting privileges revoked forever, $8,200.50 in civil penalties from the state of Texas (Texas Parks and Wildlife Department determined that this is the "replacement" cost of a Whooping Crane)
July 2004	MI	1	EMP	Summer	Adult	No	No		
Nov. 2004	KS	2	AWBP	Fall migration	Adult	Yes	Yes	MBTA	$3,000 fine for each hunter, 2 years' probation, hunting privileges revoked for 2 years, 50 h of community service, required to attend a hunter education course
Dec. 2004	AL	1	EMP	Wintering	Adult	Unknown	No		
Nov. 2009	IN	1	EMP	Wintering	Adult	No	Yes	State/juvenile	$1 fine, $504.50 in court fees
Dec. 2010	GA	3	EMP	Wintering	Juvenile	No	Yes	MBTA	
Jan. 2011	AL	2	EMP	Wintering	Juvenile & adult	No	Yes	MBTA	$425 fine, which was never collected
Oct. 2011	LA	2	LNMP	Nonmigratory	Adults (from summer 2010)	No	Yes	MBTA	Probation for unknown duration, unknown number of community service hours to be served at White Lake Wetlands Conservation Area

(Continued)

Whooping Crane Shooting Cases between 1967 and 2016 Compiled from News Reports and Freedom of Information Act Requests to the USFWS and the USDOJ in the Four Wild Populations: the AWBP, the FNMP, the LNMP, and the EMP. (cont.)

Date	State	No. cranes shot	Population	Life cycle status	Age class	Hunter related?	Perpetrator identified?	Prosecution	Sentencing
Dec. 2011	IN	1	EMP	Wintering	Adult	No	No		
Jan. 2012	TX	1	AWBP	Wintering	Juvenile	Yes	Yes	MBTA	Unknown
Jan. 2012	IN	1	EMP	Wintering	Adult	No	Yes		$5,000 fine (paid to the International Crane Foundation), 3 years' probation, hunting privileges revoked for 3 years, 120 h at Goose Island Fish and Wildlife area community service, other sentencing: 3 years' probation during which period he may not hunt, nor possess or use a firearm or alcohol
Apr. 2012	SD	1	AWBP	Spring migration	Adult	No	Yes	ESA and MBTA	30 days to be served on weekends jail time, $85,000 fine, 2 years' probation, hunting privileges revoked for 2 years, 40 h community service, $25 assessment to the Victim Assistance Fund
Jan. 2013	TX	1	AWBP	Wintering	Juvenile	Yes	Yes	MBTA	$5,000 fine, 1 year's probation, $10,000 community service payment to Friends of Aransas and Matagorda Island National Wildlife Refuges
Apr. 2013	LA	1	LNMP	Nonmigratory	Adult	No	No		
July 2013	WI	1	EMP	Summer	Adult	No	Yes	MBTA	$2,000 fine, hunting privileges revoked for 2 years, $500 MBTA fine, and $1,500 to the International Crane Foundation
Nov. 2013	KY	2	EMP	Wintering	Adult	No	No		
Dec. 2013	IN	1	EMP	Wintering	Adult	Unknown	No		
Feb. 2014	LA	2	LNMP	Nonmigratory	Adult	No	No		
Nov. 2014	LA	1	LNMP	Nonmigratory	Adult	No	No		
Jan. 2016	TX	2	LNMP	Nonmigratory	Adult	No	Yes	ESA	$25,850 fine split between Texas Parks and Wildlife Department and the International Crane Foundation, hunting and fishing privileges revoked for 5 years, gun ownership revoked for 5 years, 200 hours community service
May 2016	LA	2	LNMP	Nonmigratory	Adult	No	No		

Future of Whooping Crane Conservation and Science

Sarah J. Converse*,**, *John B. French, Jr.**,
Jane E. Austin†

*U.S. Geological Survey, Patuxent Wildlife Research Center, Laurel, MD, United States
**U.S. Geological Survey, Washington Cooperative Fish and Wildlife Research Unit,
School of Environmental and Forest Sciences (SEFS) & School of Aquatic and Fishery
Sciences (SAFS), University of Washington, Seattle, WA, United States
†U.S. Geological Survey, Northern Prairie Wildlife Research Center,
Jamestown, ND, United States

INTRODUCTION

Few other birds have the grandeur and beauty of Whooping Cranes and few attract so much interest from the public (Bowker and Stoll, 1988; Duff, Chapter 21, this volume; Loomis and White, 1996; Stoll and Johnson, 1984). The species also evokes fascination for the biologists that study it, as demonstrated by the dedication of the large number of individuals who contributed to this volume, and the even larger number that contribute to research and conservation efforts for this species. Several factors make this bird unique: it is the world's rarest crane, North America's only endemic crane, and the tallest bird in North America. Ultimately, it is the beauty of this species, its reliance on habitats that are both remote and threatened, and the many unknowns of its biology and ecology that hold our attention.

The National Audubon Society's Research Report #3, *The Whooping Crane*, by Robert Porter Allen (1952), provided the last comprehensive source on Whooping Crane biology and management. In this volume, we have endeavored to summarize the current state of knowledge on Whooping Cranes to provide a comprehensive contemporary source on the biology, ecology, and management of this rare species. This volume provides an update to Allen (1952), and reviews new areas of research. We specifically concentrate on the growing body of work in the past 20 years in the areas of population biology and demography, habitat selection, behavior and social structure, disease and health, and Whooping Crane conservation.

Our purpose, in addition to summarizing current advances in understanding the biology and ecology of the species, is to support and galvanize continued action dedicated to the conservation challenges of Whooping Crane restoration, in much the same way that Robert Porter Allen's work did in his day. While the outlook for the species has improved substantially since 1952, it is still rare and continues to receive legal protection as an endangered species in both the United States and Canada.

Any pathway to recovery is likely to involve contributions from Whooping Crane populations of three distinct types, which span the captive-wild spectrum (Canessa et al., 2016). The first of these is the remnant Aransas-Wood Buffalo Population (AWBP), the only non-reintroduced wild population, which has been growing slowly but steadily since the 1950s. The second is the set of three reintroduced populations, though only two of these, the Eastern Migratory Population (EMP) and the Louisiana Nonmigratory Population (LNMP), continue to receive releases of captive-reared birds, while the Florida Nonmigratory Population (FNMP) is all but extirpated. Finally, the third of these types is the captive population, which is distributed among several captive breeding centers in the United States and Canada. The captive population has long been the source of individuals for reintroduction. The future role of the captive population is uncertain and in flux as we complete this volume.

In this chapter, we consider these three distinct types of populations. Each has an important role to play in recovery, and research efforts on all three types of populations have yielded important information about Whooping Cranes. The information needs and conservation challenges are different in each, so we provide a summary, based on the chapters in this volume, of critical information needs. Finally, we consider integration across the captive-wild spectrum (Canessa et al., 2016) and how these populations can fit together to achieve recovery.

THE REMNANT POPULATION

The AWBP is essential to species recovery. It is the largest population, the only population without captive-produced members, and the only one that is increasing due to in situ reproduction. The chapters in this volume that are focused on the AWBP (Austin et al., Chapter 3, this volume; Condon et al., Chapter 23, this volume; Hartup, Chapter 18, this volume; Pearse et al., Chapter 6, this volume; Smith et al., Chapter 13, this volume; Strobel and Butler, Chapter 5, this volume; Wilson and Bidwell, Chapter 4, this volume) identify key information needs relevant to conservation of this population, and the species overall. These fall within two general areas: (1) information needed to acquire, sustain, or restore habitat for an expanding population in the face of anthropogenic stresses, and (2) information on the factors influencing survival and reproduction. This information will be important in determining the conservation actions that can most effectively support continued population growth and viability.

Information Needed to Sustain or Restore Habitat

Sustaining or restoring habitat conditions for the AWBP is critical to supporting a growing population. Habitat has traditionally been thought to be less limiting on the remote breeding grounds than on the wintering grounds (e.g., Canadian Wildlife Service and U.S. Fish and Wildlife Service, 2005). Whooping Cranes are expanding their breeding range beyond the protection of Wood Buffalo National Park (NP), which exposes them to risks associated with oil and gas, and other development (Wilson and Bidwell, Chapter 4, this volume). The type, extent, and magnitude of risks that cranes may encounter outside of Wood Buffalo NP are unclear and deserve further investigation to inform conservation decisions. Expanding the critical habitat, as designated under Canada's

Species at Risk Act, beyond Wood Buffalo NP to include the current breeding range would ensure that protections afforded under the Act are extended to the entire breeding population.

More research is needed on wetland habitat and ecological processes on the breeding grounds (Timoney, 1999). Little is known about how the hydrological and productivity features identified by Austin et al. (Chapter 3, this volume) relate to AWBP reproductive success and survival. And related to this, information is needed to identify the areas outside Wood Buffalo NP that provide the ecological conditions needed to sustain new breeding territories. Addressing these questions will provide not only a better understanding of ecological factors influencing vital rates, but also a basis for better assessing risks from human activities and climate change.

Along the migration corridor, power lines and wind farms are expanding (Loss, 2016) while wetland losses continue (Dahl, 2011; Doherty et al., 2013; Johnson et al., 2012). Substantial uncertainty remains about the risks from power lines and wind farms, and their effect on habitat use (Loss, 2016; Pearse et al., 2016). Understanding the spatial extent of energy infrastructure relative to the areas used by Whooping Cranes during migration will be key to guiding conservation. Loss of wetlands in some areas along the migration corridor may concentrate cranes on remaining habitat with other waterbirds, leading to higher risks of disease exposure and pathogen transmission (e.g., coccidiosis, avian cholera epizootics; Hartup, Chapter 18, this volume). Recent assessments of mortality during migration are lower than earlier estimates (Pearse et al., Chapter 6, this volume), but continued monitoring and assessment are warranted, especially as land use along the migration corridor continues to change.

Landscape and habitat issues are most challenging on the wintering grounds. The Guadalupe-San Antonio River basin encompasses the main wintering range of the AWBP.

Not only is the wintering range small compared to the population's range during the rest of the year, but also the wintering range is undergoing great alteration due to changes in land use and climate. The growth of the AWBP means that more territory is needed for winter habitat. Areas currently used by the AWBP are increasingly overtaken by human development and degraded by human water demands, sea-level rise, and other climate-associated effects (e.g., invasion of salt marsh habitat by mangroves; Strobel and Butler, Chapter 5, this volume; Smith et al., Chapter 13, this volume). Freshwater inflows to the basin, a key driver of the coastal wetland ecosystem, have been declining for decades due in part to increasing diversions for human use (Longley et al., 1994). The resulting higher salinity, exacerbated during drought, affects key estuarine food resources at Aransas National Wildlife Refuge (NWR), particularly blue crab (*Callinectes sapidus*) populations. However, few studies have addressed the impact of hypersaline conditions on blue crabs. Such information will be key in recommending freshwater flow regimes that will sustain the functioning of the estuarine ecosystems and thus food availability for wintering cranes (Smith et al., Chapter 13, this volume).

Sea-level rise is one of the primary concerns for future Whooping Crane winter range along the Texas coast (Chavez-Ramirez and Wehtje, 2012; Harris and Mirande, 2013; Smith et al., Chapter 13, this volume). Sea-level rise predictions indicate that potential Whooping Crane habitat is likely to decrease along barrier island marshes (Smith et al., Chapter 13, this volume). As sea-level rise continues, updated analyses will be needed to understand what management actions are needed to conserve adequate habitat to support the population. Smith et al. (Chapter 13, this volume) discuss strategies that can be implemented now, given existing knowledge, to protect wintering habitat in the face of predicted sea-level rise.

Information on the Factors Influencing Reproduction and Survival

Historical and ongoing surveys of the AWBP, combined with advancements in population modeling, have improved our understanding of the population dynamics and ecology of Whooping Cranes. Using an integrated population model, Wilson et al. (2016) demonstrated that the variability in the growth of the AWBP was driven most by variability in fledging rate, followed by variability in breeding propensity (proportion of breeding-age females that nest in a year). These results, along with new mortality data for fledged juveniles on the breeding grounds (Pearse et al., Chapter 6, this volume), bring into focus the importance of vital rates on the breeding grounds to population viability. The analyses presented by Wilson and Bidwell (Chapter 4, this volume) lend further support to the alternative-prey hypothesis, which posits that cyclic fluctuations in boreal forest predators play an important role in the reproductive output of Whooping Cranes (Boyce et al., 2005; Butler et al., 2013). Further information is needed on how shifting boreal predator–prey communities influence productivity and survival in the AWBP, including which predators are involved and the role of environmental conditions. Such information will be valuable for assessing how landscape and climate change on the breeding grounds might inform ongoing recovery efforts (e.g., Butler et al., 2017).

Survival during migration and winter remain serious issues for Whooping Cranes, with primary concerns focused on anthropogenic mortality risks, such as power line collisions, oil spills, and shootings. Weather (e.g., hurricanes) and climate (drought) events have the potential to cause a marked increase in winter mortality, while causing long-term changes to winter habitat, although predicting effects is not straightforward. Reduced survival would have profound impacts on population growth (Wilson and Bidwell, Chapter 4, this volume). Condon et al. (Chapter 23, this volume) consider shooting mortality, and identify it as a more problematic source of mortality in the reintroduced populations than in the remnant population. Anthropogenic sources of mortality have the greatest potential to be addressed through conservation and management actions, and actions to reduce anthropogenic mortality are more feasible during winter than in the migration corridor because of the constrained geographic size of the wintering area. Furthermore, the benefits of mitigating risks on wintering areas could be greater because birds reside there longer (Pearse et al., Chapter 6, this volume).

Conservation and management activities directed toward reducing winter mortality, especially those sources of mortality related to drought conditions, will need to be identified and tested to determine efficacy and cost. As Whooping Cranes expand into unprotected areas of the Guadalupe-San Antonio River basin or beyond, they will likely be exposed to greater risks from energy development, pollution, habitat loss or degradation, human disturbance, or shooting. The type and significance of such risks to Whooping Cranes need further investigation to inform the selection of management actions over space and time.

Winter surveys have long provided critical information on the abundance of the AWBP. The winter survey protocol will be critical to population research and decision making going forward. The recently developed protocol was designed to provide a rigorous assessment of population size, including an estimate of uncertainty, and to provide for more detailed spatial analyses of distribution and habitat use (Strobel and Butler, Chapter 5, this volume). As more powerful analytical tools and environmental data become available, these data will be invaluable for guiding conservation strategies. Strobel and Butler (Chapter 5, this volume) envision that in the future the protocol will need to be revised to address evolving conservation challenges for the AWBP.

Finally, a key component to sustaining a growing AWBP is crane health throughout the annual cycle. The risk of diseases such as West Nile virus, infectious bursal disease, avian influenza, and avian cholera deserve further research to better understand implications for Whooping Crane health and productivity, and to identify management actions to mitigate risks (Hartup, Chapter 18, this volume). Changing habitat and climatic conditions will likely alter risks of disease exposure and transmission as well as the health effects on cranes; those effects may be particularly serious during periods of extreme environmental conditions such as drought on the wintering grounds. Hence, periodic monitoring and assessment of health in the AWBP will be important for developing appropriate actions to reduce risks. However, the ability to discern the health status and effects of disease on AWBP health and productivity has thus far been limited (Hartup, Chapter 18, this volume), and more attention to useful assessment methods is needed.

REINTRODUCED POPULATIONS

Reintroduction has been a major component of recovery planning for Whooping Cranes for decades. Indeed, more than 40 years have passed since the initiation of the first reintroduction effort at Grays Lake NWR in 1975. However, four reintroductions have yet to lead to successful establishment of a viable population. The two reintroduced populations currently receiving releases (EMP and LNMP) offer hope, but substantial questions remain regarding which management strategies would make successful reintroduction more likely, and even if reintroduction is a viable conservation strategy for Whooping Cranes.

Eight chapters in this volume consider reintroduced populations, including the EMP (Barzen et al., Chapter 14, this volume; Converse et al., Chapter 8, this volume; Duff, Chapter 21, this volume; Fitzpatrick et al., Chapter 12, this volume; Mueller et al., Chapter 11, this volume; Urbanek et al., Chapter 10, this volume), the FNMP (Dellinger, Chapter 9, this volume), the LNMP (King et al., Chapter 22, this volume), and several integrate information across multiple populations (Barzen, Chapter 15, this volume; Condon et al., Chapter 23, this volume; Converse et al., Chapter 7, this volume; Hartup, Chapter 20, this volume; Olsen et al., Chapter 19, this volume). Information needs identified therein can be summarized into two major areas. First, information is needed to understand and combat the major threat to reintroduction success, reproductive failure. Second, if reproductive failure can be overcome, additional information will be needed to inform decisions necessary to sustain a growing reintroduced population, including information on long-term habitat availability, health, and anthropogenic mortality.

Information on the Causes of Reproductive Failure

For the EMP, the reintroduced population about which we have the most information, it is well documented that the major challenge to reintroduction success is reproductive failure. Reproductive failure was also a serious problem in the FNMP (Dellinger, Chapter 9, this volume), and contributed substantially to failure of that reintroduction effort (Converse et al., Chapter 7, this volume). It is too early to know whether reproduction will be a major challenge in the LNMP, although early indications are mixed. Virtually all authors here agree that identifying the causes of reproductive failure must be a primary focus of research and monitoring activities concerning the EMP (Barzen et al., Chapter 14, this volume; Converse et al., Chapter 7, this volume; Converse et al., Chapter 8, this volume; Urbanek et al., Chapter 10, this volume). The likely hypotheses for reproductive failure in reintroduced populations of cranes can be categorized based on whether poor reproduction would be predicted to affect only first-generation birds (e.g., rearing and release effects; Hartup, Chapter 20, this volume) or would be predicted to affect multiple generations (e.g., habitat effects; Barzen et al., Chapter 14, this volume). A third category is the captive selection hypothesis (Converse et al., Chapter 8, this chapter), which explains poor reproduction based on selection for traits in captivity that are nonadaptive in the wild. If this is the primary mechanism affecting reproductive performance, performance may or may not improve in subsequent generations post-release, depending on population growth, the remaining genetic variability, and the strength of the selection pressure, all of which can affect the rate of population change due to natural selection. Depending on the mechanism behind reproductive failure, the implications are that, to be successful, managers must (1) improve the methods with which birds are raised and released, (2) improve the choice of habitats in which birds are released, or (3) choose birds with more favorable genetic characteristics to release.

Learning more about the factors influencing reproduction may require observational or experimental approaches. One challenge is that some of the proposed mechanisms leading to reproductive failure may interact. While evidence is strong that black flies are associated with nest failure in the EMP, this does not indicate that learned or genetic traits are not important (Converse et al., Chapter 8, this volume). Traits of the birds themselves may be the underlying cause of reproductive failure due to poor nest attendance or poor chick care, though the response manifests itself as a response to environmental factors.

In the coming years, nesting in eastern Wisconsin – or elsewhere in areas with different characteristics than Necedah NWR – will provide opportunities for learning about hypotheses explaining the habitat effects on reproduction (Barzen et al., Chapter 14, this volume). The

presence of birds in the EMP reared and released using multiple methods (parent-reared, costume-reared and released in autumn, costume-reared and trained behind ultralights) will allow evaluation of hypotheses about the effects of rearing and release strategies. Learning about captive selection effects could be attained by releasing individuals hatched from eggs produced in a population that has not experienced captive selection, that is, the AWBP. Initiating such an action will require international cooperation and must ultimately be evaluated by the International Whooping Crane Recovery Team.

Information Needs to Support Long-Term Growth

Over the long term, if challenges with reproduction are successfully managed, we may have an expanding EMP. Though currently it is too early to evaluate, adequate demographic performance in the LNMP will result in a growing population (King et al., Chapter 22, this volume). With growing populations, the availability of high-quality habitats will be increasingly important in determining population trajectory. There is still much to learn about the quality of different habitats used by reintroduced Whooping Cranes (Barzen, Chapter 15, this volume; Fitzpatrick et al., Chapter 12, this volume). Both summering and wintering habitat requirements are still relatively poorly understood, even in the EMP, which has received the most study. Furthermore, understanding the flexibility of populations in their response to changing climate and land use will be important to understand the habitats available to birds (Mueller et al., Chapter 11, this volume). It is striking that birds in the EMP have survived in a wide variety of winter habitats and notably displayed quite different behavior in winter compared to the AWBP. Understanding which habitats reintroduced Whooping Cranes need, and how a changing environment will influence the availability of those habitats, will have an important impact on determining areas to protect and how to manage those areas.

Disease is not a challenge currently in reintroduction programs, but the risk of a known or unknown disease emerging as a threat to the populations cannot be discounted. As in the AWBP, research on infectious diseases – including inclusion body disease of cranes, eastern equine encephalitis virus, West Nile virus, infectious bursal disease, and avian influenza – is needed (Olsen et al., Chapter 19, this volume) and such work may be more feasible in the reintroduced populations.

A known source of mortality in reintroduced Whooping Cranes – illegal shooting (Condon et al., Chapter 23, this volume) – can be reduced only by managing human behavior. Therefore, social science research that can help managers understand the behavior of individuals who engage in illegal shooting, and how to design education or enforcement programs to combat the behavior, is needed. A review of the prosecution of Whooping Crane shooting (Condon et al., Chapter 23, this volume) indicates that the perspectives and decisions of the district attorney or the judge involved in each case play significant roles in legal outcomes. Hence, one strategy to deter shooting may be to educate local advocates and key members of the criminal justice system about Whooping Crane conservation efforts and importance.

CAPTIVE POPULATION

Captive Whooping Cranes have been an important focus of conservation efforts for 50 years (Black and Swan, Chapter 16, this volume). Biologists from institutions housing Whooping Cranes have participated in almost all aspects of crane conservation, including development of methods for reintroduction, undertaking critical research, and providing a large proportion of the members of the International Whooping Crane Recovery Team to guide overall recovery efforts. The purpose of the captive population is to act as a hedge against extinction and to provide propagules for reintroduction efforts (Canadian Wildlife Service and U.S. Fish and Wildlife Service, 2005). After many decades of successful maintenance of a population in captivity, it is apparent that the most critical information needs involve the question of how to produce the quantity and quality of birds necessary to support reintroduction efforts.

Whooping Cranes do not breed readily in captivity and this presents challenges to captive management and reintroduction (Black and Swan, Chapter 16, this volume), though developments in the management of breeding show promise to improve productivity (Songsasen et al., Chapter 17, this volume). Most institutions make use of artificial insemination to improve low fertility, though this is costly and does not fully address the challenges of captive reproduction. Certain behaviors, such as courtship and mate choice, are difficult to facilitate in captivity for such a large bird, given space constraints. Overall, pens are quite small when compared to territories in the wild, and usually lack the vegetation and water bodies characteristic of wetland habitats. Songsasen et al. (Chapter 17, this volume) note the importance of environmental cues for reproduction, and recent research points to a positive effect of larger pens that include ponds (Brown, 2017). More experiments evaluating alternative approaches to captive management would be beneficial, with the goal of enhancing reproduction in captive Whooping Cranes and producing adequate numbers of chicks to support reintroduction.

Reintroduction programs have depended on the development and testing of rearing and release techniques often undertaken at the captive centers (Hartup, Chapter 20, this volume). As the reintroduced populations have struggled to reproduce in the wild, attention has turned toward how the chicks are reared in captivity (Converse et al., Chapter 8, this volume; Hartup, Chapter 20, this volume). It is important that managers of captive cranes work with the biologists studying reintroduced birds to identify the conditions in captivity that might affect reproduction in the wild, and work to improve captive management accordingly. From a practical standpoint, however, the cost of captive breeding for a large animal like a Whooping Crane is substantial, so considerable thought will be needed to identify the most useful modifications that can be made to the captive environment given the resources available.

Early in the establishment of captive populations, much, if not all, of the productivity was retained to build up captive populations, and Whooping Cranes were bred in a manner designed to maintain genetic diversity. Black and Swan (Chapter 16, this volume) point out that the balance between maintaining diversity and numbers in the captive population and producing chicks for release is a challenge that continues today. Recent development of a more formal Species Survival Plan (SSP) under Association of Zoos and Aquariums guidelines will add greater structure to decisions about how to best strike that balance across all captive institutions. The goals of the SSP may be periodically revisited by the Recovery Team to ensure that the captive population is used most efficiently to advance recovery goals of the species in the wild.

WHOOPING CRANES ACROSS THE WILD-CAPTIVE SPECTRUM

The existing downlisting criteria for Whooping Cranes from endangered to threatened reflect the greater safety from extinction afforded by multiple independent populations compared to just one population (Canadian Wildlife Service and U.S. Fish and Wildlife Service, 2005). Under these criteria, if there are fewer populations, the size of each must be larger. This underscores the fundamental interaction of wild, reintroduced, and captive populations in conservation of Whooping Cranes.

For the downlisting and eventual recovery of Whooping Cranes to be affected solely based on the AWBP, fundamental challenges will have to be addressed, primarily around habitat limitations on growth and changing habitat in the face of land use and climate change (Smith et al., Chapter 13, this volume). Downlisting and recovery will require relatively high confidence in the security of this population, if it is the only self-sustaining population. Alternatively, if reintroduced populations can be successfully established, the downlisting and recovery of the species are not dependent solely on the status of the AWBP. However, after four attempts at reintroduction – including two that are ongoing – success remains elusive (Converse et al., Chapter 7, this volume). Reintroduced populations have demonstrated poor survival and reproductive success; more recently, survival has improved but reproductive success remains far from adequate in the EMP. The relatively young age of the LNMP leaves room for hope that this population may demonstrate success (King et al., Chapter 22, this volume). Ongoing attempts to address reproductive failure in the EMP through management interventions also provide hope for success.

If reintroduction is to be successful, the captive population is likely to be a key contributor. The FNMP, EMP, and LNMP have been derived entirely from the captive population, and the Grays Lake Population (GLP) was derived partially from the captive population. But simply continuing to provide propagules from the captive population may not be enough. Credible hypotheses for reproductive failure in reintroduced populations explain this failure based on inadequate learning imparted by the early rearing experiences of captive-bred birds (Converse et al., Chapter 8, this volume). Continuing efforts to learn about the relationship between captive breeding methods and post-release success of birds, and to modify management accordingly, could be necessary for eventual success of reintroductions (Hartup, Chapter 20, this volume).

Much remains to be learned about the behavior and ecology of reintroduced birds, and prospects for successful establishment of reintroduced populations using captive-bred birds are uncertain. The potential for the AWBP to contribute directly to reintroduced populations is an intriguing idea that may receive increasing attention in the coming years.

An important operating principle set out by the International Whooping Crane Recovery Team was to place reintroduced populations where they would not encounter birds from the remnant AWBP (French et al., Chapter 1, this volume). Mostly, this was to prevent possible transmission of disease from captive-raised animals to the AWBP. However, birds from the EMP and LNMP have encountered, and may continue to encounter, birds from the AWBP. Transmission of seriously debilitating diseases from captive-bred to wild birds has not been observed (Hartup, Chapter 18, this volume; Olsen et al., Chapter 19, this volume). One intriguing question is whether management interventions designed to increase interaction between these populations could have conservation benefits (i.e., through the exchange of genetic information), and whether these benefits would outweigh the potential risks.

A final, and critical, way in which these populations interact is through the transfer of knowledge gained in one population to decision making in another population. Each type of population offers different opportunities for learning. The AWBP offers the best information available on how Whooping Crane populations would have functioned before numbers declined substantially with Euro-American settlement of the Great Plains, which resulted in a severe genetic bottleneck (Krajewski, Chapter 2, this volume). The AWBP can be difficult to study in its remote breeding habitat, and extreme caution is practiced in field research to avoid undue negative impacts. The reintroduced populations offer fascinating opportunities to understand how Whooping Cranes behave in novel

environments, and, because so many individuals are marked and the populations are monitored intensively, they provide large data sets. However, reintroduced populations are affected simultaneously by so many factors, including the captive environments of origin, novel habitats, and interactions between these, that it can be hard to disentangle the impacts. Captive populations offer extremely up-close observation and the best opportunities to learn about health and physiology, but there is substantial uncertainty about the relationship between bird ecology and behavior in both captive and wild settings.

Long-term guidance for the management of these populations across the captive-wild spectrum is currently being developed through an updated planning process (Miller et al., 2017), and will include both population modeling and evaluation of management strategies. Much of the information in this volume can contribute to that process.

The state of knowledge about Whooping Cranes today is impressive, especially for an endangered species. While much uncertainty remains, there is a wealth of information summarized in this volume that can support ongoing management, and a dedicated Whooping Crane conservation community exists to apply it. As conservation work proceeds into the future, and the Recovery Team implements updated management strategies, recovery and delisting may well be within our grasp. However, for Whooping Cranes, the risks from habitat limitation, anthropogenic and climatic stressors, and shootings may require long-term management. This type of management is not qualitatively different from the management needed to maintain healthy populations of many wildlife species on a human-dominated planet. We remain optimistic that, with the combined efforts and support of researchers, managers, conservationists, and the public, Whooping Cranes will continue to grace the wetlands of North America well into the future.

References

Allen, R.P., 1952. The Whooping Crane. Research Report 3. National Audubon Society, New York.

Austin, J.E., Hayes, M.A., Barzen, J.A., 2018. Revisiting the historic distribution and habitats of the whooping crane (Chapter 3). In: French, Jr., J.B., Converse, S.J., Austin, J.E. (Eds.), Whooping Cranes: Biology and Conservation. Biodiversity of the World: Conservation from Genes to Landscapes. Academic Press, San Diego, CA.

Barzen, J.A., 2018. Ecological implications of habitat use by reintroduced and remnant whooping crane populations (Chapter 15). In: French, Jr., J.B., Converse, S.J., Austin, J.E. (Eds.), Whooping Cranes: Biology and Conservation. Biodiversity of the World: Conservation from Genes to Landscapes. Academic Press, San Diego, CA.

Barzen, J.A., Lacy, A., Thompson, H.L., Gossens, A.P., 2018. Habitat use by the reintroduced Eastern Migratory Population of whooping cranes (Chapter 14). In: French, Jr., J.B., Converse, S.J., Austin, J.E. (Eds.), Whooping Cranes: Biology and Conservation. Biodiversity of the World: Conservation from Genes to Landscapes. Academic Press, San Diego, CA.

Black, S.R., Swan, K.D., 2018. Advances in conservation breeding and management of whooping cranes (Chapter 16). In: French, Jr., J.B., Converse, S.J., Austin, J.E. (Eds.), whooping cranes: Biology and Conservation. Biodiversity of the World: Conservation from Genes to Landscapes. Academic Press, San Diego, CA.

Bowker, J., Stoll, J., 1988. Use of dichotomous choice nonmarket methods to value the whooping crane resources. Am. J. Agric. Econ. 70 (2), 372–381.

Boyce, M.S., Lele, S.R., Johns, B.W., 2005. Whooping crane recruitment enhanced by egg removal. Biol. Conserv. 126 (3), 395–401.

Brown, M.E. 2017. Biology and management of reproduction in captive cranes. Dissertation, University of Maryland, College Park.

Butler, M.J., Harris, G., Strobel, B.N., 2013. Influence of whooping crane population dynamics on its recovery and management. Biol. Conserv. 162, 89–99.

Butler, M.J., Metzger, K.L., Harris, G.M., 2017. Are whooping cranes destined for extinction? Climate change imperils recruitment and population growth. Ecol. Evol. 7, 2821–2834.

Canadian Wildlife Service and U.S. Fish and Wildlife Service, 2005. International Recovery Plan for the Whooping Crane. Recovery of Nationally Endangered Wildlife (RENEW), Ottawa, and U.S. Fish and Wildlife Service, Albuquerque, NM, p. 162.

Canessa, S., Guillera-Arroita, G., Lahoz-Monfort, J.J., Southwell, D.M., Armstrong, D.P., Chadès, I., Lacy, R.C., Converse, S.J., 2016. Adaptive management for improving species conservation across the captive-wild spectrum. Biol. Conserv. 199, 123–131.

Chavez-Ramirez, F., Wehtje, W., 2012. Potential impact of climate change scenarios on whooping crane life history. Wetlands 32 (1), 11–20.

Condon, E., Brooks, W.B., Langenberg, J., Lopez, D., 2018. whooping crane shootings since 1967 (Chapter 23). In: J.B., Jr., French, Converse, S.J., Austin, J.E. (Eds.), Whooping Cranes: Biology and Conservation. Biodiversity of the World: Conservation from Genes to Landscapes. Academic Press, San Diego, CA.

Converse, S.J., Servanty, S., Moore, C.T., Runge, M.C., 2018. Population dynamics of reintroduced whooping cranes (Chapter 7). In: French, Jr., J.B., Converse, S.J., Austin, J.E. (Eds.), Whooping Cranes: Biology and Conservation. Biodiversity of the World: Conservation from Genes to Landscapes. Academic Press, San Diego, CA.

Converse, S.J., Strobel, B.N., Barzen, J.A., 2018. Reproductive failure in the Eastern Migratory Population: interaction of research and management (Chapter 8). In: French, Jr., J.B., Converse, S.J., Austin, J.E. (Eds.), Whooping Cranes: Biology and Conservation. Biodiversity of the World: Conservation from Genes to Landscapes. Academic Press, San Diego, CA.

Dahl, T.E., 2011. Status and Trends of Wetlands in the Conterminous United States 2004 to 2009. U.S. Department of the Interior, Fish and Wildlife Service, Washington, DC, p. 108.

Dellinger, T.A., 2018. Florida's nonmigratory whooping cranes (Chapter 9). In: French, Jr., J.B., Converse, S.J., Austin, J.E. (Eds.), Whooping Cranes: Biology and Conservation. Biodiversity of the World: Conservation from Genes to Landscapes. Academic Press, San Diego, CA.

Doherty, K.E., Ryba, A.J., Stemler, C.L., Niemuth, N.D., Meeks, W.A., 2013. Conservation planning in an era of change: state of the U.S. Prairie Pothole Region. Wildl. Soc. Bull. 37 (3), 546–563.

Duff, J., 2018. The operation of an aircraft-led migration: goals, successes, challenges 2001 to 2015 (Chapter 21). In: French, Jr., J.B., Converse, S.J., Austin, J.E. (Eds.), Whooping Cranes: Biology and Conservation. Biodiversity of the World: Conservation from Genes to Landscapes. Academic Press, San Diego, CA.

Fitzpatrick, M.J., Mathewson, P.D., Porter, W.P., 2018. Ecological energetics of whooping cranes in the Eastern Migratory Population (Chapter 12). In: French, Jr., J.B., Converse, S.J., Austin, J.E. (Eds.), Whooping Cranes: Biology and Conservation. Biodiversity of the World: Conservation from Genes to Landscapes. Academic Press, San Diego, CA.

French, Jr., J.B., Converse, S.J., Austin, J.E., 2018. Whooping cranes past and present (Chapter 1). In: French, Jr., J.B., Converse, S.J., Austin, J.E. (Eds.), Whooping Cranes: Biology and Conservation. Biodiversity of the World: Conservation from Genes to Landscapes. Academic Press, San Diego, CA.

Harris, J., Mirande, C., 2013. A global overview of cranes: status, threats and conservation priorities. Chin. Birds 4 (3), 189–209.

Hartup, B.K., 2018. Health of whooping cranes in the Central Flyway (Chapter 18). In: French, Jr., J.B., Converse, S.J., Austin, J.E. (Eds.), Whooping Cranes: Biology and Conservation. Biodiversity of the World: Conservation from Genes to Landscapes. Academic Press, San Diego, CA.

Hartup, B.K., 2018. Rearing and release methods for reintroduction of captive-reared whooping cranes. In: French, Jr., J.B., Converse, S.J., Austin, J.E. (Eds.), Whooping Cranes: Biology and Conservation. Biodiversity of the World: Conservation from Genes to Landscapes. Academic Press, San Diego, CA.

Johnson, L.A., Haukos, D.A., Smith, L.M., McMurry, S.T., 2012. Physical loss and modification of southern Great Plains playas. J. Environ. Manage. 112, 275–283.

King, S.L., Selman, W., Vasseur, P., Zimorski, S.E., 2018. Louisiana nonmigratory whooping crane reintroduction (Chapter 22). In: French, Jr., J.B., Converse, S.J., Austin, J.E. (Eds.), Whooping Cranes: Biology and Conservation. Biodiversity of the World: Conservation from Genes to Landscapes. Academic Press, San Diego, CA.

Krajewski, C., 2018. Phylogenetic taxonomy of cranes and the evolutionary origin of the whooping crane (Chapter 2). In: French, Jr., J.B., Converse, S.J., Austin, J.E. (Eds.), Whooping Cranes: Biology and Conservation. Biodiversity of the World: Conservation from Genes to Landscapes. Academic Press, San Diego, CA.

Longley, W.L., Solis, R.S., Brock, D.A., Malstaff, G., 1994. Coastal hydrology and the relationships among inflow, salinity, nutrients, and sediments. In: Longley, W.L. (Ed.), Freshwater Inflows to Texas Bays and Estuaries: Ecological Relationships and Methods for Determination of Needs. Texas Water Development Board and Texas Parks and Wildlife Department, Austin, pp. 23–72.

Loomis, J.B., White, D.S., 1996. Economic benefits of rare and endangered species: summary and meta-analysis. Ecol. Econ. 18 (3), 197–206.

Loss, S.R., 2016. Avian interactions with energy infrastructure in the context of other anthropogenic threats. Condor 118 (2), 424–432.

Miller, P.S., Traylor-Holzer, K., Bidwell, M., Harrell, W. (Eds.), 2017. Recovery Planning for the Whooping Crane – Workshop 2: Species Conservation Planning. IUCN SSC Conservation Breeding Specialist Group, Apple Valley, MN.

Mueller, T., Teitelbaum, C.S., Fagan, W.F., Converse, S.J., 2018. Movement ecology of reintroduced migratory whooping cranes (Chapter 11). In: French, Jr., J.B., Converse, S.J., Austin, J.E. (Eds.), Whooping Cranes: Biology and Conservation. Biodiversity of the World: Conservation from Genes to Landscapes. Academic Press, San Diego, CA.

Olsen, G.H., Hartup, B.K., Black, S., 2018. Health and disease treatment in captive and reintroduced whooping cranes (Chapter 19). In: French, Jr., J.B., Converse, S.J., Austin, J.E. (Eds.), Whooping Cranes: Biology and Conservation. Biodiversity of the World: Conservation from Genes to Landscapes. Academic Press, San Diego, CA.

Pearse, A.T., Brandt, D.A., Hartup, B.K., Bidwell, M., 2018. Mortality in Aransas-Wood Buffalo Whooping Cranes: timing, location, and causes (Chapter 6). In: French, Jr., J.B., Converse, S.J., Austin, J.E. (Eds.), Whooping Cranes: Biology and Conservation. Biodiversity of the World: Conservation from Genes to Landscapes. Academic Press, San Diego, CA.

Pearse, A.T., Brandt, D.A., Krapu, G.L., 2016. Wintering sandhill crane exposure to wind development in the central and southern Great Plains, USA. Condor 118 (2), 391–401.

Smith, E.H., Chávez-Ramirez, F., Lumb, L., 2018. Wintering habitat ecology, use, and availability for the Aransas-Wood Buffalo Population of whooping cranes (Chapter 13). In: French, Jr., J.B., Converse, S.J., Austin, J.E. (Eds.), Whooping Cranes: Biology and Conservation. Biodiversity of the World: Conservation from Genes to Landscapes. Academic Press, San Diego, CA.

Songsasen, N., Converse, S.J., Brown, M., 2018. Reproduction and reproductive strategies relevant to management of whooping cranes ex situ (Chapter 17). In: French, Jr., J.B., Converse, S.J., Austin, J.E. (Eds.), Whooping Cranes: Biology and Conservation. Biodiversity of the World: Conservation from Genes to Landscapes. Academic Press, San Diego, CA.

Stoll, J., Johnson, L., 1984. Concepts of value, nonmarket valuation, and the case of the whooping crane. Transactions of the North American Wildlife and Natural Resources Conference 49, 382–393.

Strobel, B.N., Butler, M.J., 2018. Monitoring recruitment and abundance of the Aransas-Wood Buffalo population of whooping cranes: 1950–2015 (Chapter 5). In: French, Jr., J.B., Converse, S.J., Austin, J.E. (Eds.), Whooping Cranes: Biology and Conservation. Biodiversity of the World: Conservation from Genes to Landscapes. Academic Press, San Diego, CA.

Timoney, K., 1999. The habitat of nesting whooping cranes. Biol. Conserv. 89 (2), 189–197.

Urbanek, R.P., Szyszkoski, E.K., Zimorski, S.E., Fondow, L.E.A., 2018. Pairing dynamics of reintroduced migratory whooping cranes (Chapter 10). In: French, Jr., J.B., Converse, S.J., Austin, J.E. (Eds.), Whooping Cranes: Biology and Conservation. Biodiversity of the World: Conservation from Genes to Landscapes. Academic Press, San Diego, CA.

Wilson, S., Bidwell, M., 2018. Population and breeding range dynamics in the Aransas-Wood Buffalo Whooping Crane Population (Chapter 4). In: French, Jr., J.B., Converse, S.J., Austin, J.E. (Eds.), Whooping Cranes: Biology and Conservation. Biodiversity of the World: Conservation from Genes to Landscapes. Academic Press, San Diego, CA.

Wilson, S., Gil-Weir, K.C., Clark, R.G., Robertson, G.J., Bidwell, M.T., 2016. Integrated population modeling to assess demographic variation and contributions to population growth for endangered whooping cranes. Biol. Conserv. 197, 1–7.

Index